Plant and Human Health, Volume 1

Munir Ozturk • Khalid Rehman Hakeem
Editors

Plant and Human Health, Volume 1

Ethnobotany and Physiology

Springer

Editors
Munir Ozturk
Vice President of the Islamic World
Academy of Sciences
Amann, Jordan

Department of Botany and Centre for
Environmental Studies
Ege University
Izmir, Izmir, Turkey

Khalid Rehman Hakeem
Department of Biological Sciences
Faculty of Science
King Abdulaziz University
Jeddah, Saudi Arabia

ISBN 978-3-030-06768-7 ISBN 978-3-319-93997-1 (eBook)
https://doi.org/10.1007/978-3-319-93997-1

This Springer imprint is published by the registered company Springer Nature Switzerland AG
The registered company address is: Gewerbestrasse 11, 6330 Cham, Switzerland

This volume is dedicated to

Abu Rayḥan Muḥammad ibn Aḥmad Al-Biruni

He was born in Hive-Turkmenistan in the year 973 and died on December 13, 1051, in Ghazni in Afghanistan. Al-Biruni was one of the famous scholars, who contributed much to the world of science. A book on the Medicinal Curricula "Kitab al-saydala fi al-Tıb" published by him covered details on 200 herbal drugs. Pic source: Google.com

Foreword

Her Excellency

If biodiversity underpins life on earth, then medicinal plants and traditional knowledge have underpinned the development of modern medicine. At the dawn of a new millennium, one of the most pressing challenges of our time is the continuing, and at times irreversible, loss of biodiversity and its associated precious knowledge on our planet.

Global efforts to reduce biodiversity loss had begun with the establishment of the Convention on Biological Diversity (CBD) in 1992 at the Rio Earth Summit and today the CBD has 193 parties (or governments) as members. This seminal moment has represented a dramatic step forward in the conservation of biological diversity, the sustainable use of its components, and the fair and equitable sharing of benefits arising from the use of genetic resources. Subsequent to this major effort, the Nagoya Convention further enshrines the need to relook at Access and Benefit Sharing.

To keep reminding the world of the urgency of preserving our biodiversity, United Nations has decreed the period 2011–2020 as the Decade on Biodiversity. However, nowhere is the need for conservation and sustainable utilization of biodiversity greater than in sub-Saharan Africa, whose biodiversity wealth is uniquely important from a global conservation viewpoint.

The African continent is home to around 60,000 plant species, of which at least 35,000 are found nowhere else. Africa's biodiversity wealth is unfortunately not uniformly distributed. Countries including Madagascar and South Africa have been classified as "megadiverse" countries and the world's 17 most biologically diverse countries that together account for nearly 70% of global species diversity.

Despite its enormous natural wealth, sub-Saharan Africa faces daunting conservation challenges. Its flora and fauna are under threats and climate change is not helping. Biodiversity loss, in Africa and in other parts of the world, has a significant impact on economic growth and social development. For the rural citizens, it has the effect of removing key sources of food, fuel, and medicines, as well as adversely affecting tourism and pharmaceuticals—from a reduction in the availability of medicinal plants. New knowledge, about conservation and whole plant utilization, is needed, not just to strengthen the conservation effort but to harness this unique patrimony of natural resources to foster economic development, reduce poverty, and protect the environment.

Seen in this context, this volume, *Plant and Human Health, Volume 1: Ethnobotany and Physiology* edited by Munir Ozturk and Khalid Rehman Hakeem, adds new contribution for celebrating traditional knowledge and how the latter has underpinned our well-being providing us with food and medicine. It will help advance our understanding of the increasingly crucial role that plants play in the economic, cultural, medical, and social spheres of our lives. This volume brings on board contributions from several continents, and it is a welcome addition in terms of safeguarding this previous knowledge for humanity as a whole.

The volume also highlights the contributions from a diverse range of scholars who offer fresh, new insights on novel sources of materials and leads including bryophytes and lichens as well as a wide range of important, related topics, including new ethnobotanical explorations that would add to the lore of knowledge for humanity now and into an uncertain future.

Sustainable utilization and management of plant genetic resources is a topic of contemporary significance. By marshaling the latest evidence and cutting-edge as well as age-old knowledge, this volume should find broad appeal among academics, scientists, farmers, policy-makers, and all those who are committed to reducing biodiversity loss on our planet and promoting new leads for the development of new drugs and other products that will sustain our well-being.

Quatre Bornes, Mauritius Ameenah Gurib-Fakim

Preface

The association of living beings on our planet originated with the beginning of life. The use of herbal products for their healing powers can be traced to earliest of myths, traditions, and writings. The plant-based medicine systems evolved primarily within local areas and produced the well-known traditional herbal treatment systems. The history of herbal use now dates back to 60,000 years, because 8 species of flowering plants have been found at the old burial site in a cave in northern Iraq—Shanidar. It is followed by the history of the use of cannabis going back to 12,000 years and olive native to Asia Minor, first domesticated in the Eastern Mediterranean between 8000 and 6000 years ago. It spread from Iran, Syria, and Palestine to the rest of the Mediterranean basin. One of the important herbs used since early times for medicinal purposes is garlic, which has been used for over 7000 years and was found in Egyptian pyramids as well as ancient Greek temples. We come across notations on garlic in medical texts from Greece, Egypt, Rome, China, and India.

The written evidence of herbal use for the treatment of diseases dates back to over 5000 years, to the Sumerians, who created lists of plants. Very sophisticated ships loaded with earthenware amphorae were built solely for the olive oil trade. In fact, olive oil trade may have been the source of wealth for this advanced Minoan civilization. The earliest reference to opium growth and use is in 3400 BC when the opium poppy was cultivated in lower Mesopotamia (Southwest Asia). The Sumerians referred to it as Hul Gil, the "joy plant." The Chinese book on roots and grasses "Pen T'Sao," written by the Emperor Shen Nung circa 2500 BC, treats 365 drugs (dried parts of medicinal plants), many of which are used even today. The Ebers Papyrus, written circa 1550 BC, represents a collection of 800 proscriptions referring to 700 plant species and drugs used for therapy such as pomegranate etc. Pharaonic Egypt used cumin as a medicine around 1550 BC. Saffron was used as medicine on the Aegean island of Thera.

In Homer's epics the *Iliad* and the *Odyssey*, created circa 800 BC, 63 plant species from the Minoan, Mycenaean, and Egyptian Assyrian pharmacotherapy were referred to. Some of them were given the names after mythological characters from these epics; for instance, Elecampane (Inula helenium L. Asteraceae) was named in honor of Elena, who was the center of the Trojan War. As regards the plants from the genus Artemisia, which were believed to restore strength and protect health, their name was derived from the Greek word artemis, meaning "healthy." As a digestive aid, Confucius wrote as far back as 500 BC of never being without ginger when he ate. It was around 500 BC that turmeric emerged as an important part of Ayurvedic medicine.

The works of Hippocrates (460–370 BC) contain 300 medicinal plants classified by physiological action. Wormwood and common centaury were applied against fever; garlic against intestine parasites; opium, henbane, deadly nightshade, and mandrake were used as narcotics; fragrant hellebore and haselwort as emetics; sea onion, celery, parsley, asparagus, and garlic as diuretics; and oak and pomegranate as astringents. Theophrast (371–287 BC) founded botanical science with his books *De Causis Plantarium*. In the books, he generated a classification of more than 500 medicinal plants known at the time. Pliny the Elder (23–79 AD) wrote about approximately 1000 medicinal plants in his book *Historia naturalis*. Pliny's works incorporated all knowledge of medicinal plants at the time.

In 65 AD, Dioscorides wrote his *Materia Medica*, a practical text dealing with the medicinal use of more than 600 plants. Charles the Great (742–814 AD), the founder of the reputed medical school in Salerno, in his "Capitularies" ordered which medicinal plants were to be grown on the state-owned lands. Around 100 different plants were quoted, which have been used to date such as sage, sea onion, iris, mint, common centaury, poppy, and marshmallow. Al-Kindi (800–870) contributed to the history of medicine. This scholar was heavily influenced by the work of Galen and also made unique contributions of his own to the field. In his Aqrabadhin (Medical Formulary), he described many preparations drawn from plant, animal, and mineral sources. "Not only is every sensation attended by a corresponding change localized in the sense-organ, which demands a certain time, but also, between the stimulation of the organ and consciousness of the perception an interval of time must elapse, corresponding to the transmission of stimulus for some distance along the nerves."

Ibn Sina, also known as Avicenna (980–1037 AD), combined the herbal traditions of Dioscorides and Galen with the ancient practices of his own people. His book spread through Europe during the eleventh and twelfth centuries. *Canon Medicinae* and *Liber Magnae Collectionis Simplicum Alimentorum Et Medicamentorum* by Ibn Baitar (1197–1248) included descriptions on 1000 medicinal plants. Vasco da Gama's journeys to India (1498) resulted in many medicinal plants being brought into Europe. Botanical gardens emerged all over Europe, and attempts were made for cultivation of domestic medicinal plants and of the ones imported from the old and the new world. Paracelsus (1493–1541) was one of the proponents of chemically prepared drugs out of raw plants and mineral substances.

The great pharmacologist of the Ming dynasty, Li Shizhen (b1518–d1593 AD), spent 30 years consulting some 800 texts and personally harvesting herbs for use in treatment to write the great classic, *Materia Medica*, containing 52 articles. Withering gives clinical details of how to prescribe extract of foxglove, or digitalis, in the treatment of dropsy and hints that it may be of use for heart disease.

In 1858, Louis Pasteur wrote that garlic killed bacteria. As he maintained, it was effective even against some bacteria resistant to other factors. He also noted that garlic killed *Helicobacter pylori*. Cocaine was first isolated (extracted from coca leaves) in 1859 by the German chemist Albert Nieman. In 1886, the popularity of the drug got a further boost when John Pemberton included coca leaves as an ingredient in his new soft drink, Coca-Cola. From 1966 to 1976, traditional doctors were purged from the schools, hospitals, and clinics, and many of the old practitioners were jailed or killed.

According to WHO, nearly 80% of the population rely on plants for their primary health care globally. Approximately 30,000–70,000 plant taxa are used as medicaments. This means that nearly 14–28% of the 250,000 identified plant taxa in the world and 35–70% of all species are used on our earth; more than 50 major drugs have originated from tropical plants.

The great surge of public interest in the use of plants for medical purposes has been based on the assumption that these resources will be available on a continuing basis. Among the medicinal plants there are many pharmacopoeial ones. From about 250,000 species of higher plants around the world, only 17% have been scholarly investigated for medical potential. The chemical and biological diversity of plants represents a potentially limitless renewable source for the use in the development of new pharmaceuticals. Traditional Chinese medicine used 5000 of them, whereas the Native Americans have used only 2564 herbs as medicine. The botanical wisdom accumulated by indigenous people has led to the establishment of the traditional systems of medicine. The pharmacologist Farnsworth Norman says that 89 plant-derived drugs currently prescribed in the industrial world have been found with the help of ethnobotanical approach.

Definite signs of plant cultivation first appeared in early Neolithic villages in the Near East around 7500–7000 BC. The initiation of food production in what could be called the "nuclear area" was based on the domestication of about 8 species of local grain plants. Olives were probably first brought into cultivation in the Levant. Many condiment and dye plants have been cultivated here for thousands of years, including coriander, cumin, saffron, and safflower. The core of first domestication of the above-mentioned plants and several others is mainly represented by what is called the "Fertile Crescent," considered to be the cradle of civilization and covering the valleys of Tigris and Euphrates Rivers, the southern slopes of the Taurus Mountains, and the eastern shores of the Mediterranean Sea. Herodotus described the amazing fertility of the irrigated plains around this area.

The knowledge of medicinal plants spread widely in this region; even its conservation was achieved by the scholars from different faiths. Numerous treatises, in various languages, were written on the use of medicinal plants. The monumental and celebrated *Materia Medica* about herbal medicine and related medicinal substances was widely read for more than 1500 years. It was written in 78 AD by Pedanius Dioscorides, a physician, pharmacologist, and a botanist, who was born in Anazarba, today's Tarsus in Turkey. Many plants out of the 950 drugs given by him grow wild in the area. The origin, morphological and pharmaceutical features of these plants were given, in addition to the illustrations. Many herbalists wrote numerous treatises. Abul-Abbas Ahmed, lbn ara Rumiya (d. 1239 AD) journeyed in North Africa, Syria, and Mesopotamia and described many plants in his book *The Botanical Journey*. One of the most original botanists of the thirteenth century was Rashid ad-Din ibn as-Suri, who lived in Syria between 1177 and 1243 AD, traveled in the Near East accompanied by a painter, described many unknown plants, and had them painted as fresh plants and drugs. Ibn al Baitar (1248 AD; Damascus) wrote a monumental book *Al Garni* (Collection on Remedies), which is a very valuable book about medicinal plants. One of the most well-known treatises on medicinal plants is *Dhakhirat Uli al-Albab* (Memorandum of Intelligent People) written by Dawud Al-Antaki who was born in Antakya. It contains an alphabetical annotated list of herbal drugs and medical terms. In Iraq, many herbalists wrote about the medicinal plants and their uses and many treatises appeared.

The effectiveness of foxglove from traditional herbal medicine in the eighteenth century has been helpful in the treatment of dropsy. More than 30 cardiac glycosides have been isolated from dried foxglove leaves including digitoxin and digoxin. All these are useful because they increase the force of heart contractions. Nearly 1500 kg of digoxin and 200 kg of digitoxin are prescribed to heart patients globally. The snakeroot plant was traditionally used for the treatment of insomnia in India. In 1949, German chemists extracted alkaloid reserpine from its roots, which is used today for the treatment of high blood pressure. Similarly artemisinin is the biologically active compound used today to fight malaria. It is a sesquiterpene lactone from wormwood and was first isolated in 1972 by Chinese chemists. Quinine is another example coming from *Chinchona* species used to treat malaria. Madagascar periwinkle is used today in the chemotherapy of childhood leukemia and for the treatment of Hodgkin's disease. The compound taxol with anticancer action comes from the bark of Pacific yew tree. The extract from opium poppy has been used since the time of Pharoes as pain killer. It contains morphine, codeine, and heroin alkaloids. The anesthetic drug cocaine too has been widely used as a local pain killer.

Understanding the relationship among medicinal plants used in traditional medicine systems can help identify plant materials with potential constituents applicable to modern medicine. Licorice has been used for the treatment of bronchial asthma in traditional medicine. Illiterate traditional healers living in the forests around the globe have used the herbs correctly for medicinal purposes. They learnt all about these through trial and error; there was some spiritual learning by ritual use of medicinal plants in religious ceremonies, like "invoking hidden power of the plants" and meditation; they followed by observing how apes and other animals use the plants; and finally they strived hard to preserve the oral tradition by passing their knowledge from generation to generation.

Investigation of plants used in traditional medicine to determine biological activities is a complicated process. It requires obtaining reliable ethnobotanical data on use in the traditional system, collecting specimens from the correct genera and species, investigating the activity of crude extracts and active principles, and analyzing the chemical structure, synthesis, and structural modification.

Nearly seven billion people and the plants live together on this planet. We need not forget that plants lived there for millions of years before us. A major difference is that plants can live without people, but people cannot live without plants. For a successful research on potential new sources of medicines from plants, each medicinal plant has hundreds of biologically active chemical compounds that work synergistically together as a result of natural selection. Each herb affects humans directly and indirectly. The former is based on the pharmacological action of its biologically active compounds, whereas the latter is related to interaction with other plants or drugs taken. Search for medicinal plants to cure epidemic diseases should include the plants from the geographical place, where these diseases originated and spread around. If a plant from a genus has significant medical value, it is not necessary that other plants from the same genus may have the same medical value—only difference is potency; other plants from the genus may have more or less potency. The geographical position, habitat, and correct identification of medicinal plants are very important. An identification of chemical compounds and genetic markers alone is not enough; we have to learn about their chemotaxonomy, molecular biology complemented by classical botanical methods. If a particular ethnic group has used plants for several generations continuously from one traditional herbal medicine system, those plants remain the first choice for treatment.

As many people globally depend on medicinal plants for health, their sustainability and conservation must be our first priority. We should do everything possible to preserve the plants for our future generations.

Izmir, Turkey; Amann, Jordan
Jeddah, Saudi Arabia

Munir Ozturk
Khalid Rehman Hakeem

Acknowledgments

We owe a deepest sense of gratitude to our contributors who waited patiently for our long-lasting submission procedure. Our special thanks are due to our families for extending their helping hand to us at every and each step. We would like to convey our most sincere thanks to the staff of Springer Nature, in particular Rahul Sharma, Project Coordinator (Books), and Ms. Mohanarangan Gomathi, Project Manager (Content Solutions) at SPi Global-India, for their whole hearted collaboration.

Contents

About the Editors

Munir Ozturk (Ph.D.) has served at the Ege University Izmir, Turkey, for 50 years in different positions. He has been elected as the Vice President of the Islamic World Academy of Sciences; has received fellowships from Alexander von Humboldt Foundation, Japanese Society for the Promotion of Science, and National Science Foundation of the USA. Dr. Ozturk has served as Chairman of Botany Department and Founding Director of the Centre for Environmental Studies, Ege University, Izmir, Turkey; as Consultant Fellow, Faculty of Forestry, Universiti Putra Malaysia, Malaysia; and as Distinguished Visiting Scientist, ICCBS, Karachi University, Pakistan. His fields of scientific interest are plant eco-physiology; conservation of plant diversity; biosaline agriculture and crops; pollution; biomonitoring; and medicinal and aromatic plants. He has published 40 books, 50 book chapters, and 175 papers in journals with significant impact factor.

Khalid Rehman Hakeem (Ph.D.) is Associate Professor at King Abdulaziz University, Jeddah, Saudi Arabia. He has completed his Ph.D. (Botany) from Jamia Hamdard, New Delhi, India, in 2011. Dr. Hakeem has worked as Postdoctoral Fellow in 2012 and Fellow Researcher (Associate Prof.) from 2013 to 2016 at Universiti Putra Malaysia, Selangor, Malaysia. His speciality is plant eco-physiology, biotechnology and molecular biology, plant-microbe-soil interactions, and environmental sciences and so far he has edited and authored more than 25 books with Springer International, Academic Press (Elsevier), etc. He has also to his credit more than 110 research publications in peer-reviewed international journals, including 40 book chapters in edited volumes with international publishers.

Molecular Biodiversity Convergence with Biogeography and Ethnobotany of Rare and Endangered Medicinal Plants from Northern Vietnam

H. V. Huy, H. V. Hung, R. T. Buckney, and L. F. De Filippis

Introduction

Biogeography, Ethnobotany and Phylogeny

Plants fulfil the basic needs of humans with materials for existence, which can be medicinal, economic, food and fodder values. From any region plants can be lost, and the knowledge enclosed within them is also destroyed; sometimes it can disappear forever. In more recent times, plant conservation can be an emotive and a 'hot' issue; however, it is known that primarily due to overharvesting medicinal and some important food, plant populations have become severely reduced. Overexploitation of plant resources combined with improper harvesting and postharvesting techniques and lack of oversight and protection are increasing pressure on plant biodiversity (Ellegren 2008; Garnatje et al. 2017).

'Let food be thy medicine and medicine be thy food'; a statement attributed to Hippocrates (460–about 370 BC), which reflects the approach of the Greek physician to medicine and food, emphasizing for the first time the importance of diet and living habits in preventing illness and disease. In the past, a large group of plant species were used for the preparation of medicines and were also consumed as foods. This concept was well-established among people who traditionally gathered wild food plants, and the people were also aware of their health-beneficial properties. Nowadays, wild food plants are generally known to have high nutritional values, higher fibre and polyphenol contents and greater antioxidant capacity than the corresponding cultivated species. Moreover, many wild green plants have been demonstrated to be effective in preventing chronic diseases, such as cardiovascular diseases and diabetes. Much of this knowledge has been orally passed from

H. V. Huy · H. V. Hung · R. T. Buckney · L. F. De Filippis (✉)
Faculty of Science, School of Life Sciences, University of Technology Sydney (UTS), Broadway/Sydney, NSW, Australia
e-mail: lou.defilippis@uts.edu.au

M. Ozturk, K. R. Hakeem (eds.), *Plant and Human Health, Volume 1*,
https://doi.org/10.1007/978-3-319-93997-1_1

generation to generation which has led to the development of the traditional health-care system, practised in various countries of the world (Savo et al. 2015; Qureshi et al. 2016).

There are over 300,000 species of seed plants around the globe, of which about 60% of plants have found medicinal and food use in post-Neolithic human history. Nowadays, people collect plants for medicine and food not only from the wild but also through artificial cultivation, which are an indispensable part of human civilization. Medicinal plants are also essential raw materials of many chemical drugs, e.g. the blockbuster drugs for antimalaria and anticancer therapy, and, currently, more than one-third of clinical drugs are derived from botanical extracts and/or their ingredients (Henry 2012; Shaheen et al. 2017). Different vegetation associations classified through biogeography and ethnobotany of local ecological knowledge are called ecotopes (Table 1). The term may also be used to describe how people view, use and occupy their land. Forests around the world in a large part have been transformed into 'cultural ecotypes', since many forests are influenced by natural disturbances, as well as by human disturbances. The vegetation patterns, which result from disturbances, reflect complex interactions between biotic and abiotic characteristics (Kellogg et al. 2016), as well as cultural characteristics (Chivenge et al. 2015; Hao and Xiao 2015).

One selection criterion for plant characteristics used by biogeographers and ethnobotanists is based on phylogeny (Table 1). A non-random distribution of used medicinal plant species across families has been observed in several parts of the world (e.g. Medeiros et al. 2013). Plants within the same family, with close evolution ties, are more likely share similar secondary compounds which may have similar or equal medicinal properties (e.g. Yessoufou et al. 2015), and this has been intuitively discovered by many traditional communities. Furthermore, plants that are evolutionarily closely related have generally more total uses than those that are evolutionarily isolated. Promising predictions of medicinal plant uses have been developed based on the conjunction of ethnobotanical, phytochemical and molecular phylogenetic data (Massana 2015). The use of the same (or closely related) species in the same way in different cultures indicates that different and often noninteracting human groups have independently acquired this knowledge. This results from the fact that some plants have similar morphological characteristics because they have a close phylogenetic placement (Leonti 2011).

Convergence of Ethnobotany and Molecular Biology

Ethnobotanical studies discover plant resources that can be used for targeting novel compounds leading to the development of new medicaments for treating complicated and minor diseases. Today, ethnobotany and ethnopharmacology (Table 1) are being used for targeting new compounds. Tropical regions are rich in plant diversity and may play key roles in providing germplasm with new bioactive compounds (Hedrick 2004; Garrick et al. 2015). Plants and humans are engaged in a dynamic

Table 1 Definitions, terms and features commonly used in this review related to biogeography, ethnobotany, phylogeny, population statistics and population genetics

Term or feature	Definition
ANCOVA	Analysis of covariance
ANOSIM	Analysis of similarity
ANOVA	Analysis of variance
Biodiversity	The number, variety and genetic variation of different organisms found within a specified geographic region. A term that describes the number of different species that live within a particular ecosystem
Biogeography	The study of the geographical distribution of living things. A biogeographic region is a large, generally continuous division of the Earth's surface having a distinctive biotic community
Bioprospecting	The search for plant species from which medicinal drugs and other commercially valuable biocompounds can be obtained. The process of discovery and commercialization of new products based on biological resources, only recently begun to incorporate indigenous knowledge
Cluster analysis	Clustering is the task of grouping a set of objects in such a way that objects in the same group (called a cluster) are more similar (in some sense) to each other than to those in other groups (clusters) used in many fields
Ecotypes	A group of organisms within a species that is adapted to particular environmental conditions and therefore exhibits behavioural, structural or physiological differences from other members of the species
Genomics	An interdisciplinary field of science focusing on the structure, function, evolution, mapping and editing of genomes or an organism's complete set of DNA, including all of its genes, DNA sequencing and analysis
MDS	Multidimensional scaling
PCA	Principle component analysis
Pharmacology	The branch of biology concerned with the study of drug action, where a drug can be broadly defined as any man-made, natural or endogenous (from within body) molecule which exerts a biochemical or physiological effect on the cell, tissue, organ or organism. More specifically, it is the study of the interactions that occur between a living organism and chemicals that affect normal or abnormal biochemical function
Phylogeny	The sequence of events involved in the evolution of a species and genus. The evolutionary development and history of a species or higher taxonomic grouping of organisms, through evaluation of heritable traits, such as DNA sequences or morphology under a model of evolution of these traits
Phytochemical	The study of chemicals derived from plants and strives to describe the structures of the large number of secondary metabolic compounds found in plants, the functions of these compounds in human biology, their biosynthesis and in many cases the health benefits of these compounds
Pleiotropy	When one gene influences two or more seemingly unrelated phenotypic traits and the single gene is capable of controlling or influencing multiple (and possibly unrelated) phenotypic traits
Population genetics	A subfield of genetics that deals with genetic differences within and between populations, due to adaptation, speciation, inheritance and population structure; a part of evolutionary biology usually using statistical analysis

Definitions extracted and modified from the authors

relationship, where plants evolve creating biodiversity and humans develop strategies and solutions to use them. In this relationship, plants evolve secondary metabolites to protect themselves from being used excessively, and people find ways to use these metabolites to their advantage.

Thus, we propose to use the term 'convergence', to label similar uses for plants included in any node of a phylogeny relationship. Determining the phylogenetic and genetic relationships among plant species could be an appropriate tool for discovering new drugs based on recorded plant medicinal uses and analysis of ethnobotanical data. New perspectives have emerged with the development of new molecular tools, especially for DNA sequencing; and these enable phylogenetic reconstruction and clustering of potentially useful plants (Robertson and Richards 2015; Maestri 2017). For example, extracts from *Pterocarpus* Jacq. spp. (a phylogenetic cluster) have the same medicinal uses in geographically distant areas, namely, the neotropics, tropical Africa and Indomalaya. This example of ethnobotanical convergence illustrates that different cultures have discovered related plants that are used to treat similar disorders. Another example of ethnobotanical convergence is provided by the spices used as condiments for two products in different geographical and cultural areas. Pizza in Western cultures is seasoned with *Origanum vulgare* L., and near Eastern similar food (manousheh) is prepared using another species of the same genus as a condiment, *O. syriacum* L. Both taxa are phylogenetically very close, implying a similar chemical composition and thus a similar use. In addition to the phylogenetic approach, the large data sets obtained using 'molecular biology' techniques (e.g. genomics, transcriptomics, proteomics and metabolomics; Table 1) and their analyses using bioinformatic tools are more often used for identifying plants with popular ethnobotanical uses and the most promising taxa (or genes within those taxa) for medicinal and culinary use (Keatinge et al. 2011; Kahane et al. 2013). These molecular methods and the resulting data sets also provide a better understanding of the evolutionary history of medicinal and food plants and are further developments in ethnobotanical convergence with molecular biology (He et al. 2017).

Genetic Diversity and Population Genetics

Biodiversity is the material foundation of human survival and development and also is an important symbol to measure the environmental quality status and degree of ecological state in a region or a country. Biodiversity refers to the sum total of different animals, plants and organisms living on Earth and may include species and genetic diversity, as well as the variety of habitats and ecosystems where they live (Table 1). Biodiversity functions to provide direct and beneficial products to humans, regulation of climate and the environment, formation of unique cultures and other important functions (Holliday et al. 2017).

Tropical and sub-tropical forests cover only about 7% of the Earth's land yet contain up to 50% of all plant species. These regions are important areas of biodiversity, containing many endemic vascular plants, yet the nature and integrity of these important ecological zones are being impacted on at a greater rate than ever. Therefore much of the biodiversity in these areas is unlikely to survive without effective protection. This high diversity is in part due to steep ecological gradients, including microclimatic conditions, sharply defined ecotones and a lower amount of anthropogenic disturbance compared to temperate and dry forests. The distribution of plant species within tropical and sub-tropical national parks at present has been subjected to less human impacts and is likely to be less fragmented (Ford-Lloyd et al. 2011).

A good example is Vietnam where it is stated that 58% (19 million hectares) of total land is legally classified as forest, but ecologically speaking only part of this area actually possesses forest vegetation. Total forest cover has declined steadily throughout the twentieth century, and this decline has accelerated in recent decades. Only three million hectares are considered to possess well-stacked healthy forests; and old-growth forest is estimated to be only two million hectares. In the north and north-west regions of Vietnam, forest cover has been reduced from 95% in 1943 to between 14 and 24% in 1995 (Dang 2015). Conservation of these remaining forests is essential; however, priorities for conservation must use as estimators a number of economic and evolutionary criteria to be effective, and rare and endangered plants must form a solid basis for conservation strategies. Hence, it is important to establish sound criteria and a set of guidelines for the conservation of rare species and at the same time collect genetic diversity data to help formulate a sound management plan for endangered species (Comadran et al. 2012; MacDicken et al. 2016).

Natural Foods and Biopharmaceuticals

It is estimated that 80% of the world's population lives in developing countries and that over 80% of the world's population rely on plant-derived foods and medicines for their primary health care. Based on experience, people in the past knew therapeutic potential of medicinal plants without rationale of their efficacy. Because of advancement, we have a better understanding of the healing powers of plants due to the presence of multifunctional chemical entities for treating complicated health conditions. The plant kingdom is an implicit gold mine of new chemical compounds which are still waiting to be explored. It is estimated that there are approximately 500,000 to 750,000 species of higher plants existing on Earth and less than 10% of them have been examined for their biochemical constituents. The importance of ethnobotany must therefore be as an interdisciplinary science.

Traditional medicinal practice (TMP) encompasses a holistic worldview, which reflects that of the World Health Organizations definition of health, that is, one of 'physical, mental and social wellbeing and not merely the absence of disease or infirmity' (Leonti 2011; Savo et al. 2015). This worldview recognizes good health as a complex system involving interconnection with the land, recognition of spirit and ancestry and social, mental, physical and emotional wellbeing both of the individual and the community. Indigenous people view ill health as the result of one of three causes—a natural physical cause, a spirit causing harm and/or a sickness due to sorcery. Traditional healers (THs) are found in most societies and are often part of a local community, culture and tradition, and they continue to have high social standing in many places, exerting influence on local health practices (Hedrick et al. 2013).

Traditional healing is the oldest form of structured medicine and was originally an integral part of seminomadic and agricultural tribal societies. Archaeological evidence for its existence dates back to only around 6000 BC; but its origins probably date back to well before the end of the last Ice Age (Smith and Eyzaguirre 2007). There were and still are differences between the principles and philosophy of TMP, although there are also many fundamental similarities that arise from the profound knowledge of natural laws, and the understanding of how these influence living things, which are shared by all traditional healers. Major factors which affect medicinal plant diversity loss include razing for pastures, forest encroachment, soil erosion, over-collection, agro-system use, poverty, forest fires and invasive species intensifying in that environment. Climate variations and extremes may be additional influences on dispersal and richness of plant varieties. Deforestation, illegal trade, habitat loss, growing demand for natural products, industrial pollution and lack of adequate knowledge and training are other threats (Leonti 2011; Hao and Xiao 2015).

Aims and Scope

In this chapter we provide an overview of plant biogeography and ethnobotany and the multidisciplinary approach and convergence with molecular biology. We cover traditional phylogenetic approaches to drug discovery and move onto new methods (i.e. gel, fragment, size and sequence based) appropriate for use in molecular biology. We provide web-based resources available for use in plant molecular research, and we describe gel-based methods for use as molecular markers and use in population genetic diversity studies for rare, endangered and drug plants (Datta et al. 2010; Huang et al. 2012; Salgotra et al. 2014; Williams et al. 2014; Unamba et al. 2015). We also describe the current status of resources and technologies for transcriptomics, proteomics and metabolomics; however, some of these fields are more comprehensively described in other literature (Ekblom and Galindo 2011; Alvarez et al. 2012; Egan et al. 2012; Varshney et al. 2014; Guttikonda et al. 2016). Species of plants used in the research described in this review have been divided into sections based on rare and endangered plants and plants used for drugs and foods. Resources and techniques for use in next-generation sequencing (NGS) research will be discussed, and the integration of computer programmes and bioinformatics across plants in comparative genomics is outlined (Llaca 2012; Ray and Satya 2014; Barabaschi et al. 2016; Thottathil et al. 2016; De Filippis 2018). Currently, traditional and deep sequencing users are faced with an abundance of marker and sequencing data and analysis tools, both publicly and commercially available. We intend to point out various aspects to be considered when choosing an analytical tool and emphasize the relevant challenges and possible limitations so as to assist the user in picking the most suitable platforms and methods.

The four plant species under detailed investigation in this study have their highest frequency in the north of Vietnam adjacent to Ba Be and Cat Ba National Parks (Fig. 1), and all four are represented by small fragmented populations with a high risk of local extinction. *Sinocalamus mucclure* (string bamboo) and *Markhamia stipulata* are currently restricted to the volcanic limestone ridges of Ba Be National Park, and only string bamboo has apparently been recorded elsewhere, in a localized region of Southern China. *Cycas fugax* has only been described in Phu Tho Province, 200 m above sea level in very low numbers and in few locations. *Celastrus hindsii* is more widely distributed up to an altitude of 2500 m in regions of Northern Vietnam, China, India and Myanmar. Unfortunately for all four species, their preferred habitat has also been favoured by local tribal people for collection in traditional medicine (Ban 2003; Hung et al. 2011). Land clearance and overuse of these four species have led directly to the endangerment of populations, through direct removal of individuals, and the division of previously continuous populations into smaller and smaller fragments separated by inhospitable terrain (Dinh and Bui 2010; Dang 2015; Shaheen et al. 2017).

Traditional Ethnobotanical Methods

Natural and Social Sciences: Crucial Resources for Humans

Biogeography seeks to understand the underlying biotic and abiotic processes responsible for the spatial and temporal distributions of organisms (Tables 1 and 2). Evolutionary biogeography uses phylogenetic data to integrate concepts from phylogenetic ecology and evolutionary biology (Weckerle et al. 2011) with ecological and historical biogeography (e.g. environmental filters, dispersion, variance statistics); the goal is to elucidate biogeographic patterns and processes in a historical and evolutionary context. Biogeographical studies are traditionally focused on species diversity and distribution. However, in recent decades, the study of traits across spatial and temporal scales has proved useful for explaining and describing the diversity of forms on a biogeographical scale, thus creating the discipline of functional biogeography and 'the analysis of the patterns, causes, and consequences of the geographical distribution of the diversity of form and function' has lead to a convergence of natural and social sciences (Tables 1 and 2).

Ethnobotany is also located at a critical interface of natural and social sciences and has numerous applications to plants, especially related to human health and wellbeing (Table 1). Plants with exceptional traits and market or end-user potential must be identified; genetics and breeding can then help to resolve some of the current issues which are preventing the more extensive use of these species, but it must

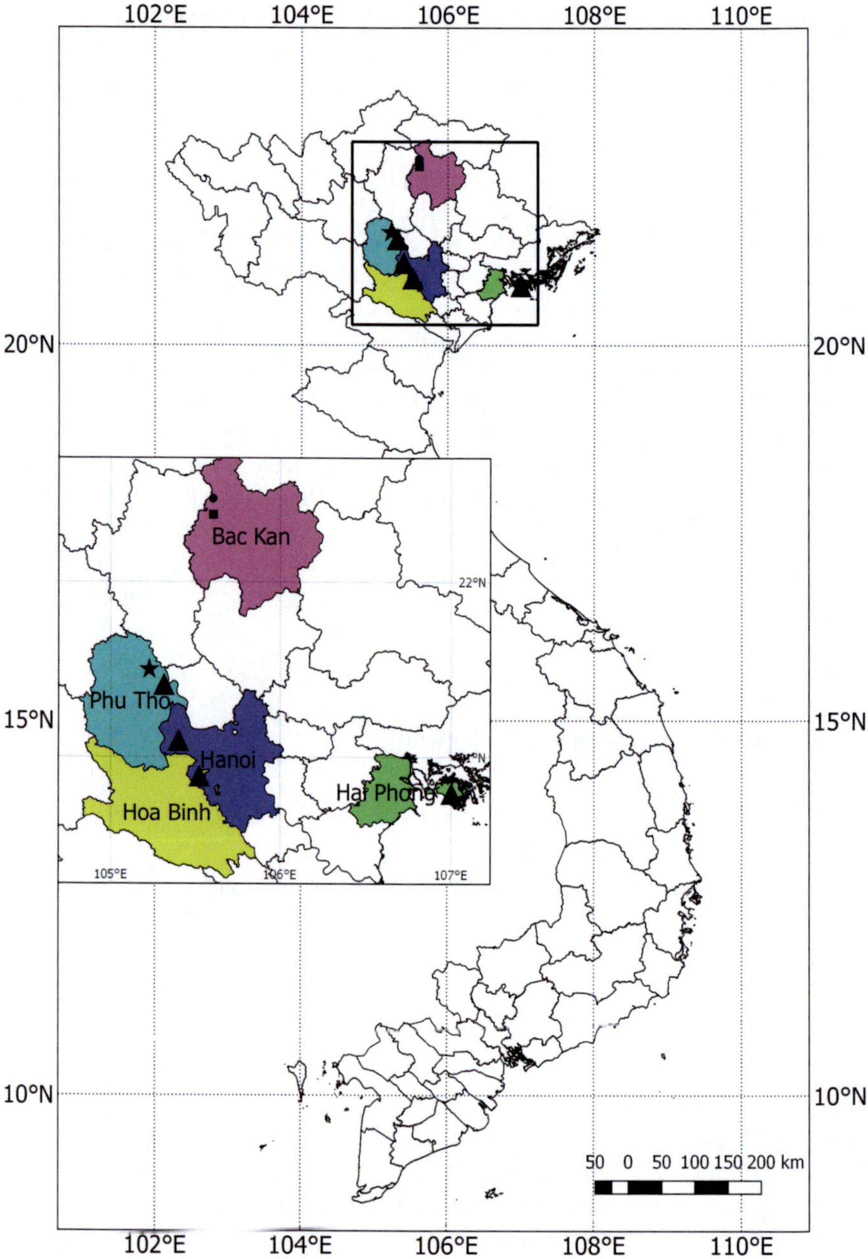

Fig. 1 Map of Vietnam including the locations of populations and sampling sites for the two rare and endangered plants and the two drug and food plants used in this study

Table 2 Definitions, terms and features commonly used in this review related to molecular biology, bioinformatics and next-generation sequencing (NGS)

Term or feature	Definition
AFLP	Amplified fragment length polymorphism
Bioinformatics	An interdisciplinary field that develops methods and software tools for understanding biological data. An interdisciplinary field of science, combining computer science, biology and mathematics and referencing to specific analysis 'pipelines' that are repeatedly used in the field of genomics
DNA barcodes	A taxonomic method that uses a short genetic marker in an organism's DNA to identify it as belonging to a particular species. It differs from molecular phylogeny in that the main goal is not to determine patterns of relationship but to identify an unknown sample in terms of a preexisting classification. The most commonly used barcode region is the internal transcribed spacer between rRNA genes, and in plants multiple regions are now advocated.
EST	Expressed sequence tags
Ethnobotany	The study of a region's plants and their practical uses through traditional knowledge of a local culture and people. The practical uses of local flora for many aspects of life, such as plants as medicines, foods and clothing
Genetic diversity	The total number of genetic characteristics in the genetic make-up of a species and serves as a way for populations to adapt to changing environments. It is distinguished from genetic variability, which describes the tendency of genetic characteristics to vary
ESI-MS/MS	Electrospray ionization tandem mass spectrometry (MS)
NGS	Next-generation sequencing
PCR	Polymerase chain reaction
Proteomics	The large-scale study of proteins, a vital parts of living organisms, with many functions. Proteomics is an interdisciplinary domain, and the proteome is the entire set of proteins that are produced or modified by an organism or system
RADseq	Restriction site-associated DNA sequencing
RAMP	Random amplification microsatellite polymorphism
RAPD	Random amplification polymorphic DNA
RNAseq	RNA sequencing
SDS-PAGE	Sodium dodecyl sulphate-polyacrylamide gel electrophoresis
SNP	Single-nucleotide polymorphism
SSR	Simple sequence repeats
TE	Transposable elements

Definitions extracted and modified from the authors

be a focused effort and is likely to be enabling, rather than transformative. For instance, with regard to geographical distribution, a species that might be underutilized in some regions may not be in other areas (Foley et al. 2011). Metacommunity analysis suggests that the influence of environmental factors on mean trait variation relies heavily on spatial biogeographical clade sorting. This implies that biogeographical lineage distribution should be taken into account in analyses seeking to correlate environmental variables with mean trait variations.

Ethnobotany Information: Bioprospecting and Traditional Uses

During evolution, plants develop tactics of chemical defences, leading to the evolution of specialized metabolites with diverse potencies. A correlation between phylogeny and biosynthetic pathways could offer a predictive approach, enabling more efficient selection of alternative and/or complementary plants for guaranteeing clinical use and novel food discovery. This relationship has been rigorously tested and the potential predictive power subsequently validated. A phylogenetic hypothesis was put forwards for medicinal plants in the subfamily Amaryllidoideae (Amaryllidaceae) based on parsimony and tested whether alkaloid diversity and activity in bioassays related to the central nervous system were significantly correlated with molecular phylogeny. Evidence for a significant phylogenetic signal in these traits has been found, but the effect was not that strong (Fierst 2011; Archmiller et al. 2015).

Bioprospecting for new drugs with a botanical origin and for new food crops has traditionally been based on ethnobotanical information. Ethnobotanically directed bioprospecting has become more powerful than random assays for finding and identifying bioactive compounds from plants. Aspirin (from *Filipendula ulmaria* L. Maxim), codeine and papaverine (from *Papaver somniferum* L.), colchicine (from *Colchicum autumnale* L.), digoxin and digitoxin (from *Digitalis purpurea* L.), tetrahydrocannabinol and cannabidiol (from *Cannabis sativa* L.) and vinblastine and vincristine (from *Catharanthus roseus* L. Don) are among the most important classical drugs developed from ethnobotanical leads (Leonti 2011; Ahmad et al. 2012). The first evidence for the anticancer properties of paclitaxel, from *Taxus* L. spp., came from its toxic effects on murine leukaemia cells, in agreement with the well-known general toxicity of these genera of plants. The success of *Taxus*-related anticancer products highlights the promising role of plant products in drug development. More recently, during the avian flu epidemic, oseltamivir was developed from *Illicium verum* Hook based on ethnobotanical data from Chinese traditional medicine. Ethnobotanical records have also led to the isolation and development of artemisinin (from *Artemisia annua* L.) as a powerful antimalarial drug (Yessoufou et al. 2015), whose relevance was recognized with the 2015 Nobel Prize in Physiology or Medicine.

Examples from a social point of view of high nutritional foods are leafy vegetables, a group of several species used by millions of people in South America and sub-Saharan Africa. Yet poor marketing makes them largely underutilized in economic terms, and any underutilized plant species can make an important contribution to a better diet for local communities (Weinberger 2007; Yang and Keding 2009). Oca (*Oxalis tuberosa*), ulluco (*Ullucus tuberosus*) and mashua (*Tropaeolum tuberosum*), three traditional Andean tuber plants, are richer in vitamin A and vitamin C than the well-known potato. Quinoa (*Chenopodium quinoa*), cañahua (*Chenopodium pallidicaule*) and amaranth (*Amaranthus caudatus*), grains from the Andean region, contain far higher amounts of certain essential amino acids than wheat. The leaves of black nightshade (*Solanum nigrum*) can provide significant amounts of calcium, iron, phosphorous, vitamin A, vitamin C, proteins and the amino acid methionine, scarce in commonly marketed vegetables (Uusiku et al. 2010; Kahane et al. 2013).

Phylogeny and Chemical Methods

Medicinal plants synthesize an arsenal of protective (even toxic) molecules, most of which are secondary metabolites, which can be ingested by animals and humans. Plants evolving in the same lineage have more medicinal uses than evolutionarily isolated species, and the diversity of medicinal uses is correlated with the evolutionary history of the species. Species-rich clades are more likely than species-poor clades to contain taxa with more uses, while ancient taxa are less abundant in the flora and, therefore, are less used in traditional medicine (Savolainen et al. 2013; Rai et al. 2017). Given that chemical properties are evolutionarily conserved (Weckerle et al. 2011), bioscreening could be targeted to the lineages identified as 'hot nodes' for medicinal properties. Current nature-derived drugs come mostly from drug-productive families that tend to be clustered rather than scattered in phylogenetic trees. Only 62 of the 457 families of angiosperms and gymnosperms are used as sources for medicinal drugs (Xu et al. 2011; Memon 2012). As a result of evolution, species that have a wide geographical distribution may be more capable of synthesizing metabolites that enable them to adapt to such a wide distribution compared with species with a restricted distribution and with a local evolutionary history.

Nevertheless, the relationship between one specific bioactive compound and medicinal activity is not always clear, complicating the phylogenetic prediction of plant use. Phylogenomics can be integrated into the flowchart of drug discovery and development and extends the field of pharmacophylogeny at the molecular level. Phyloproteomics can also be used in a proteome-based phylogeny study and may be used to examine the evolutionary relationship at the epigenomic level, and phylometagenomics is also applicable in the exploration of medicinal plant-associated microbiota (Albert 2013; Hao and Xiao 2015).

Metabolomics and Proteomics: Evolutionary-Conserved Traits

The rapid development of the main techniques used in the analyses of metabolites (e.g. gas chromatography, high-performance liquid chromatography and nuclear magnetic resonance) is increasing the application of metabolomics in many aspects of natural drug (and food) discoveries (Zhang et al. 2010; Saxena and Cramer 2013; Jensen et al. 2016). Metabolomics, which is designed to provide general qualitative and quantitative profiles of metabolites in organisms exposed to different conditions, enables us to monitor the spatial and temporal distribution of target phytochemicals. In fact, assigning bioactive compounds from complex mixtures is a central challenge of natural product research. The combination of bioassay-guided fractionation with untargeted metabolite profiling improves the identification of active components (Wong et al. 2014). Metabolomics is also enabling a better understanding of medicinal plants and the identification of important metabolic quantitative trait loci for enhanced breeding. The integration of the metabolomics approach with genome-based functional characterizations of gene products for ethnobotanically important plants is helping to accelerate the discovery of novel biosynthetic pathways of specialized bioactive metabolites.

Biochemical integration has strongly enhanced the potential discovery and production of pharmaceutical and culinary products. For example, the production of the antimalarial drug artemisinin is being enhanced via traditional breeding, with new high-yielding hybrids to convert *A. annua* into a robust cropping system, and by the reconstitution of the biosynthetic artemisinin pathway in re-engineered microbial hosts (Ahmad et al. 2012). Genomics, proteomics and metabolomics are high-throughput technologies that may help speed up the determination of the mode of action of phytomedicines and allow investigation of herbal extracts without prominent active principles. Although metabolomics and proteomics techniques have generally proven valuable, they still face substantial challenges, including large-scale metabolite identification. However, further development of the metabolomics field in general could provide better tools for the discovery of the next generation of natural products inspired by popular knowledge gathered in ethnobotanical studies and enhanced by recent phylogenetic approaches.

Meta-analysis was most succinctly defined as 'the analysis of analyses'. In other words, the authors of a meta-analysis compile and quantitatively synthesize the results of available and pertinent studies using a meaningful common statistic to address a specific research question (Frankham 2015). Such meta-analytical techniques are recommended to synthesize the available literature because they often have higher statistical power than an individual primary study, due to the increased precision of the summary effects (Connor et al. 2011; Madden and Paul 2011). However, the power of meta-analysis relies on very specific methodological and statistical treatment of the individual studied. Meta-analysis can be applied to traditional biogeography studies but is better suited to studies using molecular methods and NGS approaches.

Molecular Ethnobotanical Methods

Molecular DNA Methods

For about 25–30 years, DNA markers have been the most widely used molecular markers in plants, owing to their abundance and polymorphism. Most of these markers can be selectively neutral because they are usually located in non-coding and non-regulatory regions of DNA (Hoang et al. 2009; Allendorf 2017). The first plant DNA markers were based on difficult restriction fragment length polymorphisms (RFLPs) and Southern blot-based methods. Eventually these were replaced by safer, less complex and more cost-effective PCR-based markers including random amplification of polymorphic DNA (RAPD), random amplification of microsatellite DNA (RAMP), amplified fragment length polymorphism (AFLP), microsatellite or simple sequence repeats (SSR), sequence-characterized amplified regions and cleaved amplified polymorphic sequences (see reviews by Carrière et al. 2013; De Filippis 2013; Teixeira da Silva et al. 2016). Recent sequence-based markers (mostly from NGS), especially single-nucleotide polymorphism (SNP) (Gupta et al. 2014; Jeffries et al. 2016), are now the most important and can be applied to a large number of non-model plant species. Direct array technology (DArT) also commonly uses SNP as a base (Sansaloni et al. 2011; Table 2). When some of the non-NGS markers are used for marker-assisted selection in plant breeding, they have limitations owing to some markers being dominant, genetic recombination may give rise to false positives, and some produce inconsistent results. High-throughput sequencing techniques and technical developments in NGS have led to an increase in identification and consistency of NGS molecular markers (Ray and Satya 2014; Garrick et al. 2015).

RAPD-PCR (Table 2) is a simple technique that can be applied where no knowledge of the target DNA is available and no information about the region amplified is possible with RAPD-PCR, but it is likely that middle or highly repetitive DNA is primarily targeted (Hung et al. 2011; Yang et al. 2015). The results from this study suggest that RAPD-PCR was able to identify genetic differences between individuals and populations, which have been isolated/disturbed over a period of time and distance, and this has also been reported for other plants (Heider et al. 2007; Tripathi and Goswami 2011; Fu et al. 2013). These results must be considered and evaluated with some caution since the nature of the RAPD-PCR method means that normally it over estimates genetic differences. This overestimation appears consistent with this study where RAPD-PCR was able to detect a large number of polymorphic band differences between population sites (Treangen and Salzberg 2011; Long et al. 2015).

SSR (RAMP-PCR) (Table 2) analysis also requires no prior knowledge of the target DNA, nor are specific microsatellite primers needed for each species or population tested. Microsatellite makers are codominant and very informative; however, it is uncertain if SSR loci are codominant (Zhao et al. 2016). Comparison of microsatellite primers with six and eight two base repeats demonstrated that each had different abilities to target sites on DNA. It is likely that the ability of an SSR primer to amplify a sequence in target DNA depends on how abundant the sequence repeat is in the genome (Baruah et al. 2017). Differences in the frequency of repetition of dinucleotide microsatellites have been reported in other species of plants (Hung et al. 2011). In this study with a number of different ISSR primers, the bases in the repeat sequences, number of repeats and G + C content were also important, especially the GC content which is best at values above 50%. In general, dinucleotide repeats amplify slightly more DNA bands than trinucleotide repeats, so the longer the repeat sequence is, the fewer bands are amplified. There is some evidence that SSR, and other multilocus microsatellite fingerprinting methods, captures only some of the polymorphisms associated with microsatellites and SSR repeats, and this must be considered in interpretation of data (Treangen and Salzberg 2011).

Microarray and Protein Methods

Microarray flow cytometry has been used to determine the genome size of four *Panax* species with *Oryza sativa* as the internal standard. *P. notoginseng* (San Qi in traditional Chinese medicine) has the largest genome (2454 Mb), followed by *P. pseudoginseng* (2433 Mb), *P. vietnamensis* (2018 Mb) and *P. stipuleanatus* (1947 Mb), but their genomes are smaller than the *P. ginseng* genome (~3.2 Gb) (Angres 2005). A more reliable and sound approach for species identification without a reference genome is a genome survey with non-deep sequencing (30 times coverage), followed by bioinformatics analysis; this can be highly valuable in assessing the genome size, heterozygosity, repeat sequence and GC content (Okou et al. 2007; Hurd and Nelson 2009; Karakach et al. 2010; Farsani and Mahdavi 2011).

Proteomics with 2D SDS-PAGE and electrospray ionization tandem mass spectrometry (ESI-MS/MS) was used to identify the expression of proteins and enzymes present in *Robinia pseudoacacia* L. based on peptide (amino acid) fingerprint sequencing. Proteins strongly expressed in sapwood were responsible for carbohydrate metabolism and flavonoid turnover. Proteins strongly expressed in the transition zone were mainly responsible for flavonoid biosynthesis. Lectins were found in both sapwood and transition zone, and heat stress proteins were detected only in the transition zone. The results were further proof that flavonoids were synthesized directly at the transition zone between sapwood and heartwood (De Filippis and Magel 2012). A pattern based on a small number of peptides usually cannot gain enough confidence in representing the data and also cannot sufficiently reflect divergence during evolution. As more peptides and proteins are sequenced, pattern search procedures applied to all peptides in database will be updated accordingly (He et al. 2012; Marmiroli and Maestri 2014).

Epigenetic mechanisms are important in ecology and evolution, and are beginning to be studied using methylation-sensitive AFLP (MS-AFLP), but this approach provides relatively few anonymous and dominant markers per individual. A more promising approach to address this problem is to apply a reduced representation bisulphite sequencing (RRBS) approach, which may also be based on next-generation sequencing (NGS) methods in an ecological context (Pu et al. 2009; Albers et al. 2010; Hufford et al. 2012; Robertson and Richards 2015). RNA interference mediated by short RNA molecules is another epigenetic mechanism yet to be fully investigated for transmission and effects on genetic diversity in plants (Taylor et al. 2007; Gupta et al. 2014).

Transposable elements (TEs) have the potential to elucidate challenging questions in evolutionary biology, such as the nature of adaptive genetic variation and reproductive isolation. A growing body of work highlights that the merging of divergent genomes goes beyond the reshuffling of existing variation from related species and promotes mutations (Chiu et al. 2010; Abbott et al. 2016). However, to what extent such genome instability generates evolutionary significant variation remains largely elusive (Gupta et al. 2014; Dennenmoser et al. 2017). Transposable elements (TEs) in established hybrids represent an inspiring opportunity to further address the possible association between genome dynamics and 'rapid evolution of hybrid species' (Schoch et al. 2012; Choudhury and Parisod 2017).

Bioinformatics, Computers and Statistics

Molecular population genetics is being invigorated by the ever-growing amount of markers and nucleotide sequence data available. As a result, during the last two decades, considerable efforts have been devoted to designing and applying analytical methods for detecting the 'footprint of natural selection' at the molecular level. Finding genomic regions under selection is one of the first steps required to bridge the gap between the genotype and phenotype of adaptive traits and is thus crucial for understanding the process of adaptation. Multilocus DNA sequence data also provide opportunities to gain detailed insight into population history and structure using explicit models that incorporate demographic features of populations. This represents an important challenge because both selection and population history have important influences on the amount and patterns of genetic variation (Buerkle and Gompert 2013; Jacobs et al. 2013).

Methods developed in the last few decades to test for selection and molecular variation mostly stem from the neutral theory of molecular evolution (Kimura 1993). In a nutshell, the neutral theory posits that: the fate of segregating polymorphism is effectively determined by genetic drift, as most variation is neutral with regard to natural selection. Fixed differences in alleles between species (divergence) are mostly neutral, with a negligible contribution from adaptive substitutions, and neutral loci are not affected by the effects of linked selection. Although this theory has stimulated much debate since its inception (Anderson et al. 2013), it soon became widely used as a 'null hypothesis' in molecular population genetics against which to test for selection. However, several crucial assumptions of the standard neutral model, namely, no population structure, a constant population size and random mating, make it a composite hypothesis (Nielsen et al. 2011; Linquist et al. 2015). Thus, the mere rejection of neutrality does not point unambiguously to an effect of selection but could also result from the violation of one (or several) of the assumptions.

Because even a single high-throughput experiment produces large amount of data, powerful statistical techniques of multivariate analysis are well suited to analyse and interpret data sets. Many different multivariate techniques are available, and often it is not clear which method should be applied to particular data (see reviews by Zhang et al. 2011; Valle and Berdanier 2012; Belcaid and Toonen 2015; Paliy and Shankar 2016; Table 2).

Principal Components Analysis (PCA)

PCA can be used as a simple visualization tool to summarize data set variance and show the dominant gradients in low-dimensional space. PCA results are usually displayed as a two- or three-dimensional scatter plot, where each axis corresponds to a chosen principal component and each object is plotted based on its corresponding PC values.

Multidimensional Scaling (MDS)

MDS is a unique ordination technique in that a small number of ordination axes are explicitly chosen prior to the analysis and the data are then fitted to those dimensions. Thus, if only two or three axes are chosen, there will be no nondisplayed axes of variation at the end of the analysis. Similar to PCA, a matrix of object dissimilarities is first calculated using a chosen distance metric.

Cluster Analysis

The goal of cluster analysis is to separate variables into groups based on the similarity of the variable scores among objects, so that variables within each group (cluster) are more similar to one another than to variables in other groups. The algorithms used usually minimize the within-group distances and maximize between-group distances.

Mantel Test

Mantel (r) test typically compares two distance matrices that were calculated for the same set of objects but are based on two independent sets of variables (e.g. a species dissimilarity matrix and population site distance matrix) (Mantel 1967). The test calculates the correlation between values in the corresponding positions for two matrices. The significance of the linear relationship between matrices is assessed through basic permutation statistics.

ANOSIM and ANOVA

ANOSIM and ANOVA are statistical tests of significance; ANOSIM tests for significant difference between two or more classes of objects based on any (dis)similarity measure (Clark and Gorley 2001, 2006). It compares the ranks of distances between objects of different classes with ranks of object distances within classes. The basis of this approach is similar to the MDS ordination technique described above. As ANOSIM is based on ranks, it has fewer assumptions compared to regression techniques such as analysis of variance (ANOVA).

The power of meta-analysis relies on very specific molecular methodological and statistical treatment and bioinformatics of the individual studies. In practical meta-analysis, F_{ST} and analogues such as genetic differentiation (G_{ST}), Nei genetic diversity (H) and Jost differentiation statistic (D) (Nei 1973, 1987; Jost 2008, 2009) are somewhat constrained by expected within-population heterozygosity, which may hinder cross-study comparisons. Despite this, we recommend the use of standardized analogues G_{ST}, H and D, as meta-analysis may be affected by size;

however, these indices are the best way to alleviate the confounding effects of different maximum possible values across studies (Jost 2008, 2009; Heller and Siegismund 2009). Authors must also consult the literature for advantages and disadvantages, as well as the proper application of these statistics in order to choose the appropriate metric for the research question (Whitlock et al. 2000; Whitlock 2011; Dufresne et al. 2014). Finally, we conclude that $G_{ST,}$ H and D meet the criteria and consistency for including effective size differences and avoid the difficulty of calculating errors for these values (Archmiller et al. 2015).

Advantages and Disadvantages

There has been an enormous increase in the amount of data on DNA sequence polymorphism available for many organisms in the last decade. New sequencing technologies provide great potential for investigating natural selection in plants using population genomic approaches. However, plant populations frequently show significant departures from the assumptions of standard models used to detect selection, and many forms of directional selection do not fit with classical population genetics theory. Evolutionary analysis of genomic data is still in its infancy, and many formidable challenges face the field of evolutionary bioinformatics (for a thorough review, see Pool et al. 2010; Valle and Berdanier 2012).

An important problem involves the sheer amount of data that must be dealt with, which imposes a strong constraint on bioinformatic automation and computational demand. The comparison of observed patterns of variation at thousands of loci makes it all the more difficult to avoid false positives, and inclusion of sequencing errors (e.g. appearing as rare SNPs) can skew diversity estimates and perhaps lead to spurious inferences. One possible solution is removing rare variants (Turner et al. 2010; Griffin et al. 2011), but for many analyses, low-frequency variations are of direct interest when testing for the action of selection. It thus appears that for the first time in population genetics history, the limiting factor is the availability of methods and models and not the data on which to address evolutionary questions. However, improved methods are beginning to appear and more will surely follow. Even if the challenges are daunting, there are grounds for optimism. The parallel improvement of NGS techniques and computational and analytical tools should allow large-scale interspecific comparisons of the historical and contemporary context in which selection operates at the molecular level (Dohm et al. 2008; Gilad et al. 2009).

Multivariate analyses also exhibit some limitations. The outputs of these algorithms are more difficult to interpret compared to those generated by univariate analyses (although it is much less time-consuming to interpret single multivariate analysis than hundreds of individual univariate outputs). The assumptions of a particular method are sometimes not easy to assess or meet. It is also important to remember that revealed associations among variables and patterns of object distribution do not inherently imply causality and that synthetic ordination axes or cluster groups might not necessarily match any biological effects or gradients. Finally, many multivariate techniques are very computationally demanding and require significant computing resources when applied to very large data sets (Gompert and Buerkle 2011; Belcaid and Toonen 2015; Hoban et al. 2016). A common expression of the data is to build a phylogenetic tree, i.e. a picture of species relationships based on differences found within their underlying sequences. Multiple sequence alignments are central to phylogenetics and are the best way to construct a phylogenetic tree (cluster analysis, Table 2). A group of taxa may include an ancestral taxon but not all descendants, and a molecular clock may be calculated by using the hypothesis that nucleotide or amino acid substitutions occur at more or less fixed rates over a long time. The sequences may or may not contain phylogenetic signals for the relationships of interest; these might be too conserved or too variable, because some DNA or proteins might be highly conserved, while others are not (Schmidt-Lebuhna et al. 2017).

Computer science is becoming an essential part in training the next generation of data-enabled biologists, not only as a tool during the inevitable integration of computer science in biology but also to foster productive interactions in the new era of multidisciplinary and large-scale genetics. Undergraduate and graduate programmes are beginning to include bioinformatics; however, precious few students seem to understand the principal computational concepts underlying the tools they use on a regular basis in their research (Felsenstein 2015; Paliy and Shankar 2016; Schmidt-Lebuhna et al. 2017).

Next-Generation Sequencing and Ethnobotany

Molecular Approaches

Phylogeographers have been working to collect multilocus data ever since a series of theoretical papers pertinent to the discipline demonstrated that estimates of key demographic parameters improve as the number of loci increases (e.g. Gupta 2008; Krauss et al. 2013; Garrick et al. 2015). Recent improvements in DNA sequencing technology have led to platforms with greater speed, resolution and/or output. Given that phylogeographic studies often focus on organisms for which few or no genomic resources exist, new technical developments were required to enable detailed investigations of non-model species and expand the complexity and scope of questions that can be addressed. Before NGS, variation within and between plants (genetic diversity) at the genome level was restricted to those working on well-studied model organisms (or close relatives) possessing a wide scope of genomic resources ranging from sequenced EST libraries (Table 2) to whole-genome sequences (Imelfort et al. 2009; Jackson et al. 2011; De Filippis 2017). Ecologists and evolutionary biologists need data from large numbers of individuals, and, until recently, those working on non-model plants were limited to slow and costly gene-by-gene approaches. It is true that even low-coverage data, which does not aim to assemble complete nuclear sequences, provides genomic sequences of DNA sufficiently good to provide high-quality assemblies (Allendorf 2017; De Filippis 2017).

Next-generation sequencing (NGS) technologies have produced a substantial decrease in the cost and the complexity of generating sequence data and are allowing researchers to tackle questions that were not previously possible. Along with this remarkable progress in data acquisition, parallel advances in computational sciences, such as in machine learning and high-performance computing, are allowing researchers to answer complex biological problems using creative computational and quantitative techniques. The big advantage of high-throughput sequencing (HTS) tools in molecular diversity surveys is the huge number of reads obtained at once, obviating the time-consuming cloning step needed in Sanger sequencing. On the other hand, the main challenge is bioinformatic analysis to extract relevant and useful information from these large data sets. Next-generation plant breeding aims to develop more efficient technologies and programmes for low-cost, high-throughput genotyping and screening of large populations in a shorter time (Gepts 2004; Sansaloni et al. 2011; Liu et al. 2012; Nicolai et al. 2012).

All biological disciplines that depend on DNA sequence data have been fundamentally changed in the last few years due to the development and emergence of NGS; and our knowledge of biology, particularly molecular genomics, has grown. NGS creates huge amounts of data, presenting many problems to computational biologists, bioinformaticians and end-users (especially ecologists and taxonomists) endeavouring to assemble and analyse NGS data. A comprehensive discussion of these challenges is outside the scope of this review, but several papers in these disciplines address some of the key issues and possible strategies in dealing with them (e.g. Ellegren 2008; Connor et al. 2011; Tsai et al. 2011; Grover et al. 2012; Kvam et al. 2012). NGS data is very cost-effective, and molecular ecologists are now starting to take advantage of sequencing information and embracing the discipline of 'ecological genomics' (Gilad et al. 2009; Krauss et al. 2013). By shifting genomics from laboratory-based studies of model plant species towards studies of natural populations of non-model plants, NGS has been applied to questions about organisms' history, with the promise of revolutionizing the field. However, no systematic assessment of how phylogeographic data sets have changed over time with respect to overall size and greater information contents is available (Garrick et al. 2015).

In the last 30 years, a number of DNA fingerprinting methods such as RFLP, RAPD, RAMP, AFLP, SSR and DArT, primarily used in marker development for molecular plant breeding, have found their role in ecology, genetic diversity and species and population studies. However, it remains a daunting task to identify highly polymorphic and closely linked molecular markers for targeted traits in many plants (Ochatt and Jain 2007; Ritchie and Krauss 2012; Yoder et al. 2014). NGS technology is far more powerful than any existing genetic DNA fingerprinting methods mentioned above in generating DNA markers and continues to present problems and challenges in plant molecular biology. The widespread implementation of NGS analyses of the whole genome (the entire DNA content) and the whole transcriptome (the genes being expressed) is becoming commonplace. NGS enables the analysis of a vast amount of previously unattainable genetic information. Despite this potential, NGS has yet to be widely implemented in genetic studies of the majority of plants. This is perhaps not surprising as projects involving NGS are still costly in terms of equipment, biochemicals and experienced researchers (Shokralla et al. 2012, 2014; De Filippis 2017, 2018).

Methods to Consider

Restriction Site-Associated DNA Sequencing (RADseq)

RADseq provides researchers with the ability to record genetic polymorphism across thousands of loci for non-model plants, potentially revolutionizing the field of molecular ecology. However, as with other genotyping methods, RADseq is prone to a number of sources of error that may have consequential effects for population genetic inferences, and these have received only limited attention in terms of the estimation and reporting of genotyping error rates. Several sources of bias specific to RADseq are not explicitly addressed by current genotyping tools, namely, restriction fragment bias, restriction site heterozygosity and PCR GC content bias (Buggs et al. 2012; Arnold et al. 2013; Davey et al. 2013; Mastretta-Yanes et al. 2015).

RNA Sequencing (RNAseq)

RNAseq adds large-scale transcriptomics to the toolkit of ecological and evolutionary biologists, enabling differential gene expression (DE) studies in non-model species without the need for prior genomic knowledge. Sequencing costs have plummeted, yet RNAseq studies still underutilize biological replication, as finite research budgets force a trade-off between sequencing effort and replication in RNAseq experimental design; and the state of play of RNAseq in ecology and evolution is difficult to assess. Biological replication is often low, regardless of whether pooled DNA libraries were used, and much of the current RNAseq literature appears to be under-replicated. Pooled study designs were poorly designed. These included the number of true biological replicates. No information available on whether or not library preparation involved pooled samples or not, if so, how many samples were pooled per library not clear (De Wit et al. 2012; Zhou et al. 2015; Jeffries et al. 2016; Todd et al. 2016).

DNA Barcoding

DNA barcoding has the potential to tell all of the world's species apart, and this involves sample sizes of many millions. Given this scale of the task, the crux issue is allocating minimal sequencing effort per sample to achieve the DNA barcoding impacting on biodiversity. The elegant simplicity of establishing massive-scale databases for a few barcode loci is continuing to change our understanding of species diversity patterns and continues to enhance the ability to distinguish among species. In plants, the choice of the standardized barcode(s) has been more complex. The low substitution rates of plant mitochondrial DNA (Schoch et al. 2012; Ruhsam et al. 2015) precluded the use of the 'CO' locus. As a consequence, alternative barcoding regions were investigated, leading to selection of two plastid DNA (ITS) regions, the *rbcL* gene the *matK* gene, without clear and consistent standardisation. The same ITS regions have also been suggested as the core barcode region for fungi (Barley and Thomson 2016; Coissac et al. 2016).

Single-Nucleotide Polymorphisms (SNP)

SNP from multiple individuals per species are increasingly used to study population structure, species delimitation and shallow phylogenetics. Properly chosen data summary approaches to inferring species trees from SNP data may represent a potential alternative to currently available individual-level coalescent analyses especially for quick data exploration. With the rapid increase in production of genetic data from new sequencing technologies, a myriad of new ways to study genomic patterns in non-model organisms are currently possible. Because genome assembly still remains a complicated procedure, and because the functional role of much of the genome is unclear, focusing on SNP genotyping from expressed sequences provides a cost-effective way to reduce complexity while still retaining some functionally relevant information (Bianco et al. 2014; De Wit et al. 2015; Schmidt-Lebuhna et al. 2017).

Epigenetic Modifications

Epigenetic modifications are expected to occur at a much faster rate than genetic mutations, potentially causing isolated populations to stochastically drift apart or, if they are subjected to different selective regimes, to directionally diverge. A high level of genome-wide epigenetic divergence between individuals occupying distinct habitats is often predicted. Bisulphite-converted restriction site-associated DNA sequencing (bsRADseq), an approach to quantify the level of DNA methylation differentiation across multiple individuals, is useful (Abbott et al. 2016; Trucchi et al. 2016). This reduced representation method is flexible in the extent of DNA sequences interrogated. Although DNA methylation frequency between different positions of a genome varies widely, there is a surprisingly high consistency in the methylation profile between individuals thriving in divergent ecological conditions (Gupta et al. 2014; Dennenmoser et al. 2017).

Informatics and Computers

A comprehensive discussion of the genetic and statistical analysis employed in population genetics is beyond the scope of this review, but I refer you to the following books and reviews (Clark and Gorley 2001, 2006; Pu et al. 2009; De Filippis 2013; see also section "Molecular Ethnobotanical Methods"). Population gene family data sets are usually produced by computational procedures, including a step that conducts an all-against-all sequence similarity analysis and then a step for building clusters of inter- and intrapopulation analysis parameters, by methods such as Markov clustering (MCL), multidimensional scaling (MDS) and principal component analysis (PCA) in programmes like PRIMER and Arlequin. Discriminatory methods are an extension of the interpretive multivariate techniques and are usually called discrimination analyses (DA). The goal of DA is to define discriminant functions (synthetic variables) or hyperspace planes that will maximize the separation of objects among different classes. Stand-alone software packages such as CANOCO (R Core Team 2017), PRIMER v6 (Clark and Gorley 2006) and PAST (Price and Casler 2012) are available as alternatives to MATLAB and R statistics. Advanced software statistics (e.g. PopGen, Tree Viewer, IBD genetics, GenAlEx) can yield indices and information from databases that are useful for further phylogenetic studies using analysis of covariance (ANCOVA), analysis of similarity and analysis of variance (ANOVA) (De Filippis 2017, 2018). These methods and indices could contribute significantly to an understanding of population structure but have not been used much in ecology and ethnobotany.

NGS contain short read lengths and high data volumes that have made challenges more difficult from a computational perspective. Repeats create ambiguities in alignment and assembly, which in turn can produce biases and errors when interpreting results. Simply ignoring repeats is not an option as this creates problems of its own and may mean that important biological phenomena are missed. Although some repeats appear non-functional, others have played a part in evolution, at times creating novel functions but also acting as independent 'selfish sequence elements'. Repeats arise from a variety of mechanisms; they come in all shapes and sizes and can be widely interspersed in the plant genome and can constitute over 80% of the genome in maize and over 65% in potato (Treangen and Salzberg 2011).

Advantages and Disadvantages

Kane et al. (2012) sequenced plastid and ribosomal DNA with the goal of producing 'barcodes' (taxon-specific molecular profiles) below the species level. Using Illumina sequencing, they examined whole plastid genomes and nearly 6000 bases of nuclear ribosomal DNA sequences. The large amount of data vastly exceeded that of traditional barcoding, which uses short sequences from defined regions of the genome. They term their approach 'ultrabarcoding' and used it to examine different geographic locations, nine genotypes of three varieties of *Theobroma cacao* L. and an individual of a related species *T. grandiflorum* (Sprengel) Schumann. They obtained 4.2–11 times coverage of the nuclear genome and had enough coverage for plastid and nuclear ribosomal DNA. The data clearly separated *T. cacao* from *T. grandiflorum* and showed two strongly supported clades in *T. cacao*, corresponding to two of the three varieties of *T. cacao* (i.e. Forastero and Criollo), and the maternal lineages of the third variety (Trinitario) came from both Forastero and Criollo. The authors suggest that ultrabarcoding is very useful as a supplement to traditional barcoding methods and showed that taxon-specific profiling can be successful below the species level.

Wet laboratory procedures, parallel sequencing and species specific genome data also contribute to error in several ways, leading to variance in (a) the total number of reads per individual, (b) the number of loci represented in each individual, (c) the read count per locus and (d) the read counts of alternative alleles at polymorphic loci (Hohenlohe et al. 2012). Amplification success during the PCR step may lead to variation in the depth of coverage among loci and individuals, potentially causing locus or allelic dropout. The consequences of sequencing errors and statistical methods to account for it have been widely discussed for other molecular makers, from AFLPs and microsatellites (Price and Casler 2012) to whole-genome sequence data (Gompert and Buerkle 2011; Nielsen et al. 2011; Pool et al. 2010). Errors may lead to incorrect biological conclusions, such as an artificial excess of homozygotes (Jacobsen et al. 2015; Wayne and Shaffer 2016), false departure from Hardy-Weinberg equilibrium, overestimation of inbreeding, unreliable inferences about population structure and incorrectly inferring demographic expansion from the confounding influence of low-frequency error-derived SNPs (Jeffries et al. 2016; Pool et al. 2010; Schmidt-Lebuhna et al. 2017).

Principles of standardization in barcodes must include 'agreed' regions of DNA (so that joint efforts build a shared global resource), quality control (to ensure the library of DNA sequences is reliable) and minimalism (using one or a few regions of DNA to ensure scalability). These so-called extended barcodes or ultrabarcodes should have the following features:

(a) Additional data must increase the phylogenetic signal in the data set, enabling a single data set to work effectively for species discrimination and assessing true phylogenetic relationships.

(b) Generate whole plastid genomes and ribosomal sequences, circumventing preferences for different loci for some meta-barcoding based studies, as all relevant loci are routinely recovered.
(c) Additional sequence data from sequenced plastid genomes and ribosomal repeats should lead to an increase in levels of species discrimination (Ruhsam et al. 2015; Coissac et al. 2016).

At current market rates, the consumables cost for sequencing one gigabase of data is about $80, but before the sequencing step, specific adaptors must be ligated onto each side of the fragmented genomic DNA (i.e. library preparation). The cost of building the library is still relatively high. At the bioinformatic level, a large and complex database and an automated workflow must be designed to process and manage this amount of data (Mardis 2011; Hayden 2014). Counteracting this is that genomic studies of medicinal plants lag behind those of most other plants and important food crop plants. The genome sequences encompass essential information of plant origin, evolution, development, physiology, inheritable traits, epigenomic regulation, etc., which are the premise and foundation of deciphering genome diversity and chemodiversity (especially various secondary metabolites with potential bioactivities) at the molecular level. High-throughput sequencing of medicinal plants could not only shed light on the biosynthetic pathways of medicinal compounds, especially secondary metabolites, but also their regulation mechanisms; this information is vital and plays a major role in the molecular breeding of high-yielding medicinal cultivars and molecular farming of transgenic medicinal strains.

Molecular Biology of Rare and Endangered Plants

Geography and Occurrence

Cycas fugax

Cycads represent a very primitive group of vascular plants that have been in existence for more than 200 million years (Hedrick 2004). Thus, all cycad species have been listed in the Convention on International Trade in Endangered Species of Wild Fauna and Flora. Generally, Vietnam has the greatest diversity of cycads at the regional level and the highest number of species of any country except Australia (Hill and Weir 2011). The genus *Cycas* in Vietnam includes 27 species, and 9 of them have been described as new, 1 species is listed critically endangered (*Cycas fugax*) and 3 are considered endangered (*C. aculeate*, *C. hoabinhensis* and *C. multipinnata*); the others are either vulnerable or near threatened (Osborne et al. 2007). *Cycas fugax* is one of the rarest cycad species assessed as critically endangered (CR) (possibly extinct) in the 2010 IUCN Red List of Vietnam (Osborne et al. 2010). The species is known only from Phu Tho Province in the north of Vietnam, and it is estimated that more than 80% of *Cycas fugax* populations have been reduced due to natural habitat destruction by agriculture encroachment and urbanization (Osborne et al. 2010).

Sinocalamus mucclure

The rare 'string' bamboo (*Ampelocalamus* sp./*Sinocalamus* sp. McClure) is only found in Ba Be National Park, Vietnam, on the limestone forest areas with high humidity around lakes and rivers. This species has been widely known and utilized by local people but has so far not been documented in the better-known bamboo flora of Vietnam (Dinh and Bui 2010; Hung et al. 2011). Bamboo has emerged as one of the important non-timber forest products (NTFP) in the world. However, not much scientific research has been done to prove these claims. *Sinocalamus* sp. have been used for making twine and are believed to possess bioactive compounds and used as a medicinal herb by the local people.

Morphology and Taxonomy

Cycas fugax

Cycas fugax has a subterranean flask-shaped stem, about 18 cm long, 15 cm in diameter at the base and 10 cm at the apex, which may be larger and branched in older specimens. The stem holds one to three leaves at the crown. Leaves are bright to deep green; moderately glossy, with an orange tomentum which is lost as the leaf expands; and from 280 to 380 cm long and flat (not keeled) in section. There are 80–100 leaflets and the rachis ends consistently with a pair of leaflets; the basal leaflets are 12–30 cm long and are abruptly replaced by petiolar spines. The median leaflets are simple, strongly discoloured, angled forwards, decurrent at the base and flat in section, with margins flat or wavy, with apex softly pointed and with midrib raise. The petiole is 150–220 cm long (50–60% of total leaf length) and glabrous and spinescent for 90–100% of its length (Osborne et al. 2007). Male cones are solitary, spindle-shaped, cream in colour, 10–12 cm long and 2.5–4 cm in diameter. Female cones are 20–25 cm long, brown-tomentose and has 2–4 glabrous ovules. Seed is ovoid, 25–27 mm long and 18–21 mm wide (Osborne et al. 2007).

Sinocalamus mucclure

Sinocalamus sp. McClure is a monocotyledon species with healthy root systems, copious branching and strong rhizomes. The stem is long (500–600 cm), separated into merostomial parts of 40–50 cm and diameter from 0.4 to 0.6 cm. The bamboo shoot is tender and immature, and young stalks emerge from the nodes of the pseudo rhizome. It is the most important product in ethnomedicinal use, as an excellent source of vitamins, carbohydrates, proteins and minerals. Bamboo shoots have been regarded as a customary medicinal constituent and proclaimed to be beneficial to human health, by promoting motion and peristalsis of the intestine, helping digestion and preventing and curing cardiovascular diseases and cancers (Staya et al. 2012). Furthermore, bamboo shoots are a rich source of antioxidant compounds like flavonoids, phenols and phenolic acids which are key ingredients in dietary fibre and possess anticancer, antibacterial and antiviral activities.

Examples and Results

The G-statistic for differentiation (G_{ST}), Nei genetic diversity (H) and Jost genetic differentiation (D) demonstrated that both *C. fugax* and *M. stipulata* contained moderate genetic diversity but consistently low G_{ST} values (Table 3). Values for RAPD-PCR analysis showed slightly higher values than using RAMP-PCR. These values suggested that similarities detected between individuals at any one site in this study were only moderate (not low) and suggested that significant genetic differences were present in individuals. Even the similarity between what might have been unrelated and distant sites was moderate (Xiao et al. 2004; Hung et al. 2011). In summary, results from the molecular data showed that a considerable amount of genetic variation between populations was present at the fragmented sites sampled. Similar results were found in a study of the legume *Flemingia macrophylla* and

Table 3 List of population genetics parameters for the two rare and endangered plants used in this review

Population genetics	*C. fugax*		*M. stipulata*	
Parameters	RAPD	RAMP	RAPD	RAMP
Number of populations	4	4	4	4
Number of polymorphic bands	46	48	90	73
Percent of polymorphism (%)	74.2	72.7	64.7	70.2
Nei genetic diversity (H)	0.24	0.25	0.18	0.21
Shannon information index (I)	0.36	0.37	0.31	0.32
Estimation of gene flow (N_{m})	0.80	0.66	0.65	0.59
Genetic differentiation (G_{ST})	0.38	0.40	0.24	0.28
Jost genetic differentiation (D)	0.14	0.15	0.17	0.20
Mantel r test (probability)	−0.14 (*P* = 0.48)	0.22 (*P* = 0.12)	0.33 (*P* = 0.02)	0.68 (*P* = 0.002)

Summary of some data extracted and modified from Hung et al. (2011) and other data provided from the authors other cycads in North Vietnam and China, in which the authors discovered that the genetic composition of lowland accessions differed significantly from those in upland regions (Xiao and Gong 2006; Heider et al. 2007).

It is often the case that variability in smaller populations of plants can appear to contain nearly the same amount of variation as is contained in larger populations, and therefore our data was in agreement with these findings. However care should be taken in drawing conclusions from the study since it is not known how many of those seedlings sampled would have eventually reached maturity (Frankham et al. 2010; Allendorf 2017). The findings provide evidence for a slower than expected loss of genetic diversity and consequently a higher than expected long-term evolutionary potential in small fragmented populations. Recommendation that Nm (estimation of gene flow) should be larger than 0.5–0.6 to avoid inbreeding depression is present and consistent with our data (Table 3; Schou et al. 2017). Genetic diversity at the species level (and at the population level) was consistently different and suggested that much more genetic polymorphisms existed among populations than within populations.

In cycads, low genetic variation within populations and relatively high genetic differentiation between populations are characteristics of the biology and evolution of the genera. Multivariate statistical methods show that all populations of *C. fugax* and *M. stipulata* have low genetic variation and effective methods of sexual reproduction (pollination) should be implemented. PCA and MDS are consistent among RAPD and RAMP data and clearly show that in *C. fugax* populations 1 and 3, 4 are further apart genetically; in *M. stipulata* populations 1 and 4 are most apart (Figs. 2 and 3); however the MDS analysis for *C. fugax* is not as clear as is PCA analysis. Therefore, both species would benefit from the introduction of genetic diversity by breeding and hybridization particularly the natural forest population identified by population 4 (NF). Cluster analysis (dendrograms, Fig. 6) tends to support the structure and genetic diversity between the populations. Therefore, we believe that further genetic rescue efforts on one or a few of the populations would be considered inadequate for either species, and all four populations should be preserved, especially as all four sites in the one species are within a very short geographic zone.

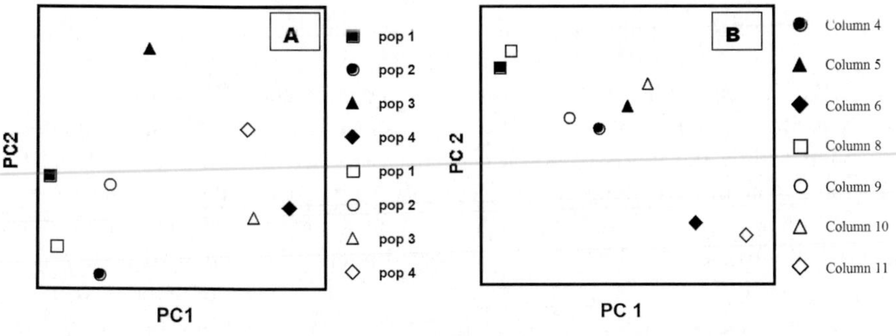

Fig. 2 Principal component analysis (PCA) of the two rare and endangered plants. (**a**) *C. fugax* using RAPD-PCR data (closed symbols) and RAMP-PCR data (open symbols); (**b**) *M. stipulata* using RAPD-PCR data (closed symbols) and RAMP-PCR data (open symbols)

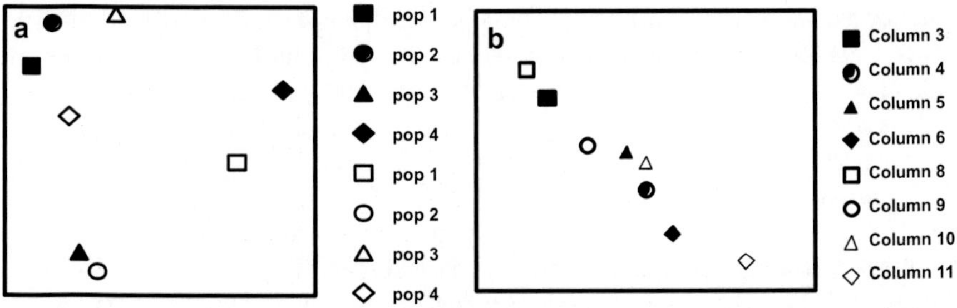

Fig. 3 Multidimensional scaling (MDS) of the two rare and endangered plants. (**a**) *C. fugax* using RAPD-PCR data (closed symbols) and RAMP-PCR data (open symbols); (**b**) *M. stipulata* using RAPD-PCR data (closed symbols) and RAMP-PCR data (open symbols)

Conservation and Protection

Many species have fragmented distribution with small isolated populations suffering some degree of inbreeding depression and/or reduced ability to evolve. Without gene flow from another population within the species (genetic rescue), these populations are likely to be near exhausted. However, there have been only a few published cases of such outcrossing for conservation purposes and probably a very low proportion of populations that would potentially benefit. One impediment to genetic rescues is the lack of an overview of the magnitude and consistency of genetic rescue effects in wild species. Fitness benefits also increase significantly with maternal inheritance (reduction in inbreeding coefficient due to gene flow) and for naturally outbreeding versus inbreeding species. However, benefits did not differ significantly among invertebrates, vertebrates and plants. Evolutionary potential for fitness characters in inbred populations also may benefit from gene flow. There are no scientific impediments to the widespread use of outcrossing to genetically rescue inbred populations of naturally outbreeding species, provided that potential crosses have a low risk of outbreeding depression. Revised guidelines for the management of genetic rescue have been outlined (Primmer 2009; Frankham 2015).

A key theory of evolutionary and conservation biology is that loss of genetic diversity can be predicted from population size, but is this on very shaky ground? In the face of increasing human-induced species depletion and habitat fragmentation, this question and the study of genetic diversity in small populations are paramount to understanding the limits of species responses to environmental change and to providing remedies to endangered species conservation. The authors reveal a slower rate of loss of genetic diversity in small populations under varying conditions than theoretically expected and hence an unexpected retention of genetic diversity. It is entirely plausible that there is associative overdominance, wherein homozygosity of deleterious recessive alleles are especially disfavoured in genomic regions of low recombination. The results in this study contribute to a budding literature on the varying mechanisms underlying genetic diversity in small populations and encourage further such research towards the effective management and conservation of fragmented or endangered populations (Anderson et al. 2013; Fraser 2017).

The risk of extinction now faced by the remaining populations of *C. fugax* and *S. mucclure* is compounded by the species' own biology, i.e. they require a specialized habitat, they have poor seed dispersal mechanisms, and they are slow growing after harvest. These factors make the species particularly susceptible to inbreeding, with subsequent potential loss of genetic variation, accumulation of deleterious alleles and inbreeding depression (Wayne and Shaffer 2016). Previous studies have found that a loss of genetic diversity decreases the ability of wild populations to survive climatic extremes, pollutants, pests, diseases and exploitation (Frankham et al. 2010). According to Frankham (2015), the maintenance of genetic diversity and heterozygosity in natural populations may provide the best general strategy for ensuring the survival of most organisms. Smaller fragmented populations are more susceptible to deterministic effects that include inbreeding and loss of genetic variation. Schou et al. (2017) have claimed that there is no evidence that populations in the wild suffer from inbreeding depression and that catastrophes, over-exploitation and demographic or environmental factors are more important causes of extinction than inbreeding. Waller (2015) however makes a valid point, in that extinctions may be incorrectly attributed to non-genetic factors alone, when it is the interaction between genetic and non-genetic factors that is important.

Molecular Biology of Drugs and Food

Geography and Occurrence

Celastrus hindsii

C. hindsii Benth belong to the genus *Celastrus* which include species of aromatic herbaceous perennials, almost exclusively native to Asia. *Celastrus* belongs to the family Celastraceae and is comprised of about 35 species, distributed in the subtropics and tropics, with a few representatives in temperate areas around the world; the highest densities are in East Asia, Oceania and both North America and Madagascar with the present centre of distribution lying in Southeast Asia (Liang et al. 2016). In Vietnam, the genus *Celastrus* was known to have eight species: *C. hindsii*, *C. gemmatus*, *C. hookeri*, *C. monospermus*, *C. orbiculatus*, *C. annamensis*, *C. stylosus* and *C. paniculatus* (Ban 2003). *Celastrus* have been used as natural insecticides and also as important folk medicine to treat fever, chill, joint pain, oedema, rheumatoid arthritis, muscle pain and paralysis, leprosy and bacterial infection in Asia for a long time. They are well known to produce a large spectrum of structurally and biogenetically diverse secondary metabolites, such as sesquiterpenes, alkaloids, triterpenes, diterpenes and flavonoids.

Markhamia stipulata

M. stipulata is commonly found in evergreen forest on limestone mountains, mixed with *Aglaia gigantea*, *Cinnamomum obtusifolium* and *Dracontomelon dao*. In Vietnam, *M. stipulata* has been recorded with three varieties: *M. stipulata var. kerrii*, *M. stipulata var. pierrei* and *M. stipulata var. canaense* (Pham 1999; Dang 2015). Its distribution is from North to South Vietnam but concentrated in two northern province of Tuyen Quang and Lang Son, while *M. stipulata var. canaense* is known only from a very restricted area of about 20 hectares in Ca Na of Thuan Nam District, Ninh Thuan Province (South Vietnam), at an elevation of about 19 20 m (Dang 2015). The roots, barks, stems and leaves of *Markhamia* species have been used by traditional healers for the treatment of miscellaneous disease conditions such as microbial and parasitic diseases, anaemia, diarrhoea, backache, sore eyes, intercostal pain, pulmonary troubles, gout, scrotal elephantiasis, rheumatoid arthritis and external skin diseases (Kanchanapoom et al. 2002). The plants have also been used in the treatment of diarrhoea, dysentery, pain and inflammation in veterinary patients (Ibrahim et al. 2016).

Morphology and Taxonomy

Celastrus hindsii

The genus *Celastrus* is a deciduous or evergreen woody vine, featured by alternate simple leaves, axillary or terminal cymes with mostly five-merous pale-green bisexual or unisexual flowers, three- to six-seeded capsules and seeds ($\leq$5 mm in length) covered by a distinct red or orange fleshy aril (Mu et al. 2012). The genus also can be distinguished by the typically scandent shrubs, a distinct articulation on the stalk, consistent three-valved fruits and complete cup-shaped aril. Due to overlapping morphologies of inter- and intraspecies, mainly variable leaves and inconspicuous flowers, several species have been defined in the literature. Leaves are elliptic to oblong or broadly ovate to orbicular (Liang et al. 2016). Recent molecular studies investigating the phylogenetic relationships within the genus *Celastrus* have been implemented (Simmons et al. 2008, 2012; Zhang et al. 2010); however, the summary data is still rudimentary.

Markhamia stipulata

Markhamia is a genus of flowering plants in the family Bignoniaceae with about 100 genera and 800 species, and at least 10 species are widely distributed from tropical Africa, China (Zhang and Santisuk 1998), Thailand and Vietnam (Mabberley 2008). The plants of this genus are trees or shrubs with opposite, compound pinnate leaves and yellow-green flowers and are grown mostly for social, agri-horticultural and medicinal purposes. Trees are small (3–6 m) and young branches sparsely puberulous to glabrescent. Leaves are compound, opposite and 14–20 cm long; leaflets 7–9 are opposite, oblong, ovate-oblong or elliptic-oblong; base is broadly acute to rounded; and lateral veins 5–7 on each side of midrib, along with main veins, convex on both sides. The petiole is 2–2.5 cm long, and inflorescences are terminal, 8–14-flowered raceme, and the flowering period is from April to June, with a fruiting period from July to September (Pham 1999; Dang 2015).

Examples and Results

Genetic diversity indices G_{ST}, H and D demonstrated that both *C. hindsii* and *S. mucclure* contained moderate genetic diversity, but low G_{ST} values (Table 4). Values for RAPD-PCR analysis also consistently showed higher values than using RAMP-PCR. These values suggest that similarities detected between individuals at any one site in this study were only moderate (not low) and suggested that significant genetic differences were present in individuals. Even the similarity between what might have been unrelated and distant sites were moderate (Xiao et al. 2004; Xiao and Gong 2006; Hung et al. 2011). In summary, results from the molecular data showed that a considerable amount of genetic variation between populations was present at the fragmented sites sampled. Similar results were found in our study of *C. fugax* and *M. stipulata*, in which we suggest that the genetic composition of the two rare and endangered plants was similar.

Table 4 List of population genetics parameters for the two medicinal drug and food plants used in this review

Population genetics	*C. hindsii*		*S. mucclure*	
Parameters	RAPD	RAMP	RAPD	RAMP
Number of populations	4	4	8	8
Number of polymorphic bands	98	46	102	81
Percent of polymorphism (%)	81.7	66.7	70.8	71.1
Nei genetic diversity (H)	0.28	0.20	0.22	0.25
Shannon information index (I)	0.42	0.31	0.35	0.38
Estimation of gene flow (N_m)	0.95	0.89	0.77	0.79
Genetic differentiation (G_{ST})	0.32	0.36	0.24	0.28
Jost genetic differentiation (D)	0.15	0.10	0.16	0.19
Mantel r test (probability)	0.55 ($P = 0.03$)	0.51 ($P = 0.05$)	0.34 ($P = 0.22$)	0.49 ($P = 0.09$)

Summary of some data extracted and modified from Hung et al. (2011) and other data provided from the authors

Often it is the case that variability in smaller populations of plants can appear to contain nearly as much variation as is contained in larger populations; however, care should be taken in drawing conclusions from such studies in general (Frankham et al. 2010; Allendorf 2017). The findings provide evidence for a slower than expected loss of genetic diversity and consequently a higher than expected long-term evolutionary potential in small fragmented populations. Nm (estimation of gene flow) should be larger than 0.5–0.6 to avoid inbreeding depression, and this is true with our data for *C. hindsii* and *S. mucclure* (Table 4; Schou et al. 2017). Genetic diversity at the species level (and at the population level) was consistently different and suggested that much more genetic polymorphisms existed among populations than within populations.

Natural forest populations for both *C. hindsii* and *S. mucclure* contained higher percentage of polymorphic bands, and both appeared to be in a habitat that was conducive to maintaining diversity in the wild. Perhaps in their natural forests, the populations are subjected to minimal impacts from biotic and abiotic stress. Multivariate statistical methods show that all populations of *C. hindsii* and *S. mucclure* have low genetic variation and effective methods of sexual reproduction (pollination) should be implemented. PCA and MDS are consistent among RAPD and RAMP data and clearly show that in *C. hindsii* populations 1,2 and 3,4 are further apart genetically; in *S. mucclure* populations 1, 2, 3 and 5, 7, 8 are most apart (Figs. 4 and 5). In the situation of *S. mucclure* with eight populations, separating groups was difficult as expected, but it was clear that it was not possible for populations 4 and 6 to be consistently assigned. However both species would benefit from breeding and hybridization. Cluster analysis (dendrograms, Fig. 6) tends to support the structure and genetic diversity between the populations in both species but especially genetic diversity between populations 1, 3 and 7, 8 for *S. mucclure*. We also conclude that further genetic rescue efforts on one or a few

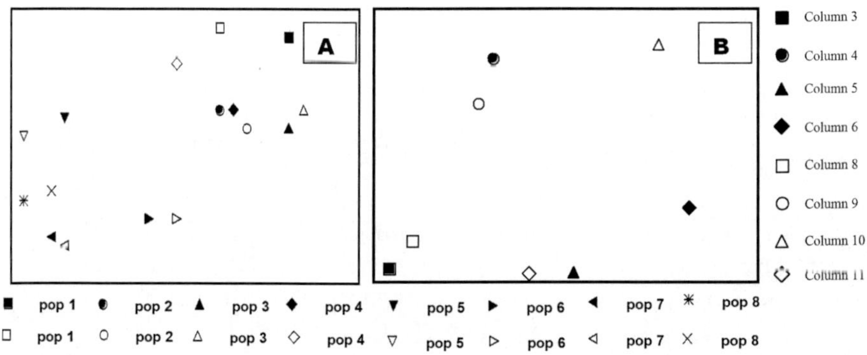

Fig. 4 Principal component analysis (PCA) of the two drug and food plants. (**a**) *C. hindsii* using RAPD-PCR data (closed symbols) and RAMP-PCR data (open symbols); (**b**) *M. stipulata* using RAPD-PCR data (closed symbols) and RAMP-PCR data (open symbols)

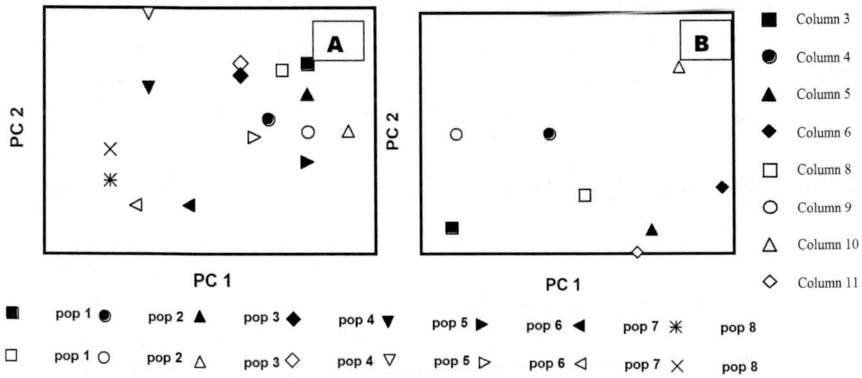

Fig. 5 Multidimensional scaling (MDS) of the two drug and food plants. (**a**) *C. hindsii* using RAPD-PCR data (closed symbols) and RAMP-PCR data (open symbols); (**b**) *M. stipulata* using RAPD-PCR data (closed symbols) and RAMP-PCR data (open symbols)

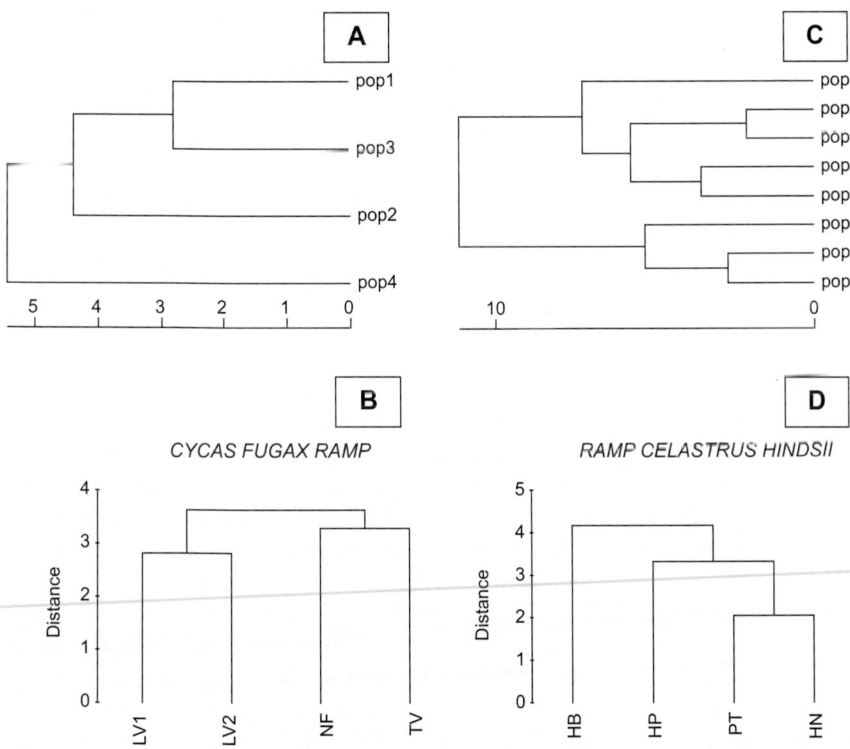

Fig. 6 Dendrograms (cluster analysis using Nei genetic diversity in RAMP-PCR) of the two rare and endangered plants and the two drug and food plants. (**a**) *M. stipulata*, (**b**) *C. fugax*, (**c**) *S. mucclure* and (**d**) *C. hindsii*

of the populations would be considered inadequate for either species, and all four populations should be preserved, especially as all four sites in the one species are within a very short geographic zone.

Ethomedicinal Use and Markets

Forest trees are an unparalleled group of organisms in their combined ecological, economic and societal importance. Forest trees have widespread distribution, predominantly random mating systems and large population sizes; most tree species harbour extensive genetic variation both within and among populations. Yet it has been from forest resources that most medicinally important plants have originated. At the same time, demographic processes associated with Pleistocene climate oscillations and land-use changes have affected contemporary range-wide diversity and may impinge on the potential for future adaptation of forest trees. Forests cover approximately 30% of the Earth's terrestrial surface from boreal to tropical latitudes (Araki and Ishii 2015; MacDicken et al. 2016), and many tree species play foundation or keystone roles in their respective ecosystems while at the same time representing a renewable resource that is a principal driver of economic activity in many areas. Forests also contain approximately three-quarters of the Earth's terrestrial biomass, tightly linking them with atmospheric carbon budgets (Jensen et al. 2016; Holliday et al. 2017). With a growing awareness of the role that different geographic origins (provenances) play in adaptability and productivity, forest genetics emerged as a research discipline in the early part of the twentieth century. A key goal of modern forest genomics is to couple provenance studies of phenotypic variation with genome wide data to characterize the genomic underpinnings of adaptive traits. In addition, in situ sampling of native forests provides insights into past evolutionary history and population dynamics.

For health-care providers, it is important that they have an appreciation and understanding of indigenous belief systems in relation to health care and work to incorporate this understanding into their plant-derived medicinal selection and service delivery. One way to do this would be to adopt a family-centred, integrative approach that works with the individual, in concert with their family and cultural support. Applying this type of approach not only respects indigenous people choice to utilize bush medicine as a part of community treatments and their overall search for health and wellbeing but also recognizes and begins to work with a holistic epidemiology. It is also of value for practitioners to know that their patients may be taking bush medicine because there can be potential risks involved in using both. Plants, leaves and trees used in making bush medicines may be bioactive and can have physiological, emotional and psychological effects. A well-known example of a herbal drug that interacts with biomedicines is St. John's Wort (*Hypericum perforatum*), which is a traditional European herbal drug used to treat mild depression, but may negatively interact with a wide number of biomedicines, including antiretroviral drugs, oral contraceptives and warfarin (Mayes et al. 2012; Tabata and Sawa 2014).

Plant Breeding Cultivation and Management

Plant Breeding and Selection for Sustainability

Small and isolated plant populations face threats from genetic drift and inbreeding. To rescue populations from these threats, conservation biologists can augment gene flow into small populations to increase variation and reduce inbreeding depression. Spectacular success stories have mostly addressed endangered animals (Madsen et al. 1999; Johnson et al. 2010; Hohenlohe et al. 2012). In plants, the only attempts at rescuing inbred populations primarily have described chestnut reintroduction efforts (Wheeler and Sederoff 2009; Miller et al. 2014) and maybe willow (Mosner et al. 2012), although the latter has been poorly documented. However, we also know that performing such crosses risks introducing genes that may be poorly adapted to local conditions or genetic backgrounds. A number of examples of such 'outbreeding depression' have led Frankham (2015) to advocate using assisted gene flow more widely; however, the studies analysed so far in this meta-analysis may not adequately test for latent outcrossing depression (Waller 2015).

Plant breeding is a continuous attempt to alter genetic architecture of plants for efficient utilization as food, fodder, fibre, fuel or other end use. In North America alone, an estimated 3000–5000 species of wild plants were once used as food, but most have been abandoned (Meyer et al. 2012; Keatinge et al. 2011). Although the scientific concepts in plant breeding originated well over 100 years ago, domestication and selection by humans of desirable traits have contributed a great deal to ensure increased production (Ochatt and jain 2007). The review by Jain and Gupta (2013) considered information on breeding achievements in 203 major and minor crop plants compiled across 36 categories, including centres of domestication, changes in phenotype and progression from the wild to the cultivated forms and conservation status. Conclusions from these studies include:

(a) Marginality of species, by which crops are domesticated after removal from their native range, is not common, and many of these cases are recent domesticates produced by intensive agricultural research and not by gradual directed breeding methods.
(b) There are major gaps in the literature for many crops, particularly with regard to ancestors, region of origin, domestication dates, expansions of major civilizations and increased trade effects, giving biogeographers here a key role to play.

Unfortunately, most medicinal plants have not been domesticated, and currently there is no toolkit to improve their medicinal attributes for better clinical efficacy. Immoderate harvesting has led to supply crisis of phytomedicine, exemplified by the taxane-producing *Taxus* sp. plants. On the other hand, successful domestication and improvement are not realistic without deeper insights into the evolutionary pattern of medicinal plant genomes. Artificial selection can be regarded as an accelerated and targeted natural selection. Studies of medicinal plant genome evolution are crucial not only for the understanding of the ubiquitous mechanisms of plant evolution and phylogeny but also for plant-based drug discovery and development, as well as the sustainable utilization of plant pharmaceutical resources (Kroc et al. 2014).

The effects of outcrossing on fitness in rare and endangered plants can be consistently beneficial, resulting in higher composite fitness in stressful conditions and even greater benefits for outbreeding species. There are also highly consistent and substantial benefits of outcrossing on evolutionary adaptation for fitness. The financial costs of augmenting gene flow may also impede their use. However, improvements in fitness are worthwhile in comparison with many other procedures. There are legitimate concerns about spreading diseases, pest and parasites if these differ significantly between the populations to be crossed. However, populations that have experienced gene flow in the recent past are likely to already share these organisms (or will in the future). Movement of individuals across political jurisdictions (countries, states and provinces) may be required for genetic rescue. However, botanic gardens regularly move taxa across state and country boundaries, and several genetic rescues have involved such movements. None of the non-genetic issues impeding genetic rescue attempts is insuperable, especially given the large and consistent benefits typically revealed by genetic rescue attempts. Large improvements in fitness and evolutionary potential can be made by augmenting gene flow into small inbred populations (Frankham 2015).

NGS applications in ecology will greatly increase the contribution that molecular tools can make to ecological restoration and enable progress in restoration genetics and rescue beyond what is possible with current methods. The benefits, addressed in small detail by NGS, are largely due to an enhanced capacity to develop molecular markers, a significantly enhanced sampling of genomes, a greater ability to assess adaptive variation and a new capacity to characterize species composition and/or functioning via meta-barcoding or metagenomics as mass-throughput biomonitoring techniques (Stapley et al. 2010; De Filippis 2018).

Cultivation and Sustainable Active Product Production

Restoration and propagation genetics have their roots in genealogical research stemming back more than 200 years. Common garden and reciprocal transplant provenance trials clearly demonstrated the important role of environmental selection in shaping genetic variation within species (Turesson 1922). In a restoration, propagation and cultivation context, this has been identified as a 'home ground (site) advantage' (Montalvo and Ellstrand 2001). Many of the gel-based molecular methods provide a powerful resource to infer genetic provenance guidelines from life history properties for species with no population genetic knowledge. With the development of PCR, AFLP, microsatellites and recently SNPs have become the tools of choice for population genetics, enabling the assessment of genotypic variability across hundreds, even thousands, of markers and/or highly polymorphic loci.

Restoration ecology through active propagation and cultivation is a young scientific discipline underpinning improvements in the rapid global expansion of ecological information. The application of molecular tools over the past 25 years has made an important contribution to understanding genetic factors influencing ecological success. Novel applications include a dramatically enhanced capacity to measure adaptive variation for optimal seed sourcing, high-throughput assessment and monitoring of natural and restored biological communities above ground and below ground and gene expression analysis as a measure of genetic resilience in restored populations (McKay et al. 2005; Fraser 2017).

Nursery propagation of elite genotypes, cultivation in selected locations (small plots of land) and small-scale breeding can, and has often provided valuable plant resources, however such ventures on a relatively large, professional scale are few. Tissue culture, somatic embryogenesis may be applicable in selecting ‘elite plants’ for field nurseries, and these type of facilities must be developed (Jain and Gupta 2013; De Filippis 2014). Synthetic seed production is another useful approach where seeds can be stored and are viable for a long period of time; however, infrastructure and money must be invested. Long-term monitoring and effects of seed stability and the genetic stability of elite plants may also be required. In this period of genetic advancements, both traditional markers and NGS markers are indispensable; however, funding for further research is urgently required. Efforts must be made to involve the community, business and governments at all levels with sound conservation policies and planning for such ventures.

After decades of discussion, there is little consensus on the extent to which ‘hybrids’ between rare and endangered medicinal species should be protected by law. As increasingly larger genome-scale data sets are developed, we can identify individuals and populations with even trace levels of genetic mixture, making the ‘hybrid problem’ all the more difficult to analyse. A decision tree framework for evaluating hybrid protection must include the processes that produced hybrids (human mediated or natural), propagation and cultivation and the ecological impact of hybrids on natural ecosystems. This decision tree approach for case studies is drawn from the available literature, and management outcomes that this approach provides for cultivation and hybridization are discussed by Wayne and Shaffer (2016).

Management of Resources

Ex situ conservation and cultivation are not always practised. The most effective way to achieve this is to educate and practise sustainable management of resources and provide opportunities to modify cultivation practices. In this way the misuse of native flora might be reduced. The results from this study suggest that small population fragments of *C. fugax*, *S. mucclure*, *M. stipulata* and possibly also *C. hidsii* in Vietnam are well worth conserving as part of an overall strategy to maintain genetic diversity. Management to ensure better conservation of all four species however must be more proactive than at present. A first step would be to present this information to the management of these resources and make sure that they understand the significance of the findings. The second step is to promote ex situ conservation in buffer zones and cultivation steps and in situ monitor the remaining fragmented populations. A third step would be to preserve as much as possible the shallow soils of the area. A fourth step would be to educate the indigenous population not to collect material from the wild and develop a plan to invest in methods of propagation that will allow growing of these valuable species in domestic situations. Domestic plants could be harvested with ease and without the need for collection from wild populations, eventually also providing a wider genetic basis for the species. Finally, the conservation of these endangered plants would benefit considerably by interconnecting the present fragmented populations via corridors to help guarantee continuing genetic exchange (Siol et al. 2010).

All of these aspects cannot be given the coverage that they deserve in a single article like this. Therefore the focus has been on genetic diversity and, specifically, on the potential of new breeding and genetic research to overcome some of the concerns and constraints when dealing with rare, endangered and medicinal plants. Time is another factor that must be taken into consideration. Plant species may suddenly become popular and have the focus of research in one area (or country), while in another area they continue to be poorly studied, marketed and managed, causing confusion that may actually create even more problems and misunderstanding and lead to poor management.

Markets and Benefits

A number of other conservation problems may be present in developing countries like poverty, lack of education or awareness, lack of proper marketing, heavy human and animal pressure and lack of adequate government regulation and providing inadequate courses and training. A significant correlation was observed in some species between populations and geographic distance as determined by the Mantel test, and distance appeared to be a major contributing factor to changes in genetic diversity. This relationship was apparent for both *M. stipulata* and *C. hindsii*, although for *C. fugax* and *S. mucclure*, this relationship was not significant (Tables 3 and 4). Plants from the botanic garden site for *M. stipulata* and forest area for *C. hindsii* (Tables 3 and 4) had high genetic variation but may represent a method of temporal genetic transfer among closely related populations (Schou et al. 2017). Similar results were found in a study of the annual plant *Lesquerella fendleri* in which the author discovered that the genetic composition of plants differed according to geographic distance and how far these populations had been transported away from their point of origin (Staya et al. 2012).

Resequencing methods in forest trees may shed some more light on neutral processes shaping patterns of intra- and interspecific diversity and local adaptation. Functional and epigenomics of adaptation and natural hybrid zones are useful tools to study adaptation. The longer breeding cycles of forest trees mean that practical applications of genomic prediction methods are still in early stages, because a wide variety of marker genotyping platforms are in use. A multilayered approach to incorporate genomic information into an ongoing breeding programme for maritime pine (*Pinus pinaster*) using a 9 k SNP array has been described by Holliday et al. (2017). Divergent natural selection promotes local adaptation and can lead to reproductive isolation of populations in contrasting environments; however, the genetic basis of local adaptation remains largely unresolved in natural populations. Local adaptation might result from antagonistic pleiotropy where alternate alleles are favoured in distinct habitats, and polymorphism is maintained by selection. Alternatively, under conditional neutrality some alleles may be favoured in one environment but neutral at other locations, as in common garden experiments of polygenic local adaptation (Anderson et al. 2013; Waller 2015). Conditional neutrality can result in local adaptation at the organismal level, especially when limited gene flow prevents the assembly of recombinant genotypes carrying conditionally beneficial alleles at different loci (Fournier-Level et al. 2011; Wetterstrand 2014).

Perspective and Future Direction

Biogeography, Ethnobotany and Conservation

The Living Planet Index, which reflects changes in the health of the Earth's ecosystems, declined between 1992 and 2012 by 12% at the global level and by 30% in the tropics (Cao et al. 2014; Ebert 2014). The dramatic decline in the tropics is indicative of a severe degradation of biodiversity due to high deforestation rates of primary forests and transformation into agricultural land and pastures (Jacobsen et al. 2013; Ebert 2014). Juma (2011) lists six measures for sustainable intensification of agriculture: (a) use the same or less land and water, (b) efficient and prudent use of inputs, (c) minimize greenhouse gas emissions, (d) increase natural capital, (e) strengthen resilience and (f) reduce environmental impact. However, understanding the reasons that drive people to select and exploit plants in a certain area is still rudimentary. Our study shows that selection criteria for plants (including medicinal plants) could not be limited to phylogeny, and it is likely that plants are selected for multiple different reasons.

Domestic plants could be harvested with ease and without the need for wild collections, eventually also providing a wider genetic basis for both species. Finally, the conservation of these endangered plants would benefit considerably by interconnecting the present fragmented populations via corridors to help guarantee continuing genetic exchange. A promising recent development in molecular biology involves viewing the genome as a mini-ecosystem, where genetic elements are compared to organisms and the surrounding cellular and genomic structures are regarded as the local environment. Ecological restoration and propagation genetics is a discipline underpinning these issues and is aided by the knowledge that almost all species show spatial genetic structure across their range, with varying levels of genetic variation within populations. Within a propagation context, genetic markers have been used to characterize genetic variation within populations and the differentiation between them, with a primary focus on the issue of 'how local is local'; and importance is placed on the use of local provenance material so as to minimize negative impacts on the environment (Weckerle et al. 2011).

New Molecular Developments

Future improvement in NGS approaches will be the use of tagged samples, wherein multiple individuals/populations can be genotyped simultaneously. These approaches are predicted to change the way we investigate plant breeding, phylogeography, demography and conservation genetics, by massively increasing the number of loci studied (Grover et al. 2012). This improvement will require the development of new software tools to make the analyses feasible from a computational point of view. Realistically, genomics will be applied to plant breeding, biogeography, ethnobotany and ecology when it becomes easy and cost-effective to do so.

As the amount and quality of sequence information generated per run keeps increasing, which allows even higher level analysis and lower costs per samples, it can be anticipated that the high density of markers from NGS will be extensively applied to all plant methods, old and new. NGS provides enormous new potential for delineating locally adapted source populations for restoration and propagation through investigating natural selection in plants using population genomic approaches. However, a major challenge for models used to infer selection comes from confounding demographic variation (e.g. changes in population size and subdivision), which is common in plant populations. Although there is great potential in NGS technology, ultimately it is the science that needs to direct the research, not the technology. What is the issue and what data are needed to answer it are the questions that need to be answered. For many objectives in ecological genetics, current tools such as AFLP and microsatellites will continue to make significant cost-effective contributions, as will traditional field, nursery and cultivation trials.

Cost of genotyping is another determining factor for adopting appropriate NGS technologies in plant ecology and breeding (Hayden 2009; Mardis 2010). Since plant ecology and breeding handle large population sizes, it is an expensive process (Mardis 2011; Hayden 2014). However, a word of caution is that additional associated cost for target-enriched library preparation and bioinformatic analysis that precedes and succeeds the sequencing steps may not decrease as rapidly as the cost of sequencing itself. The cost of data mining and efficiency to extract and use information may be more crucial than genotyping costs themselves. NGS lengths are considerably shorter (50–150 bp) than Sanger sequences (800–1000 bp) which makes assembly more difficult, but the depth of coverage and cost of NGS are considerably better. In a pessimistic sense, the scans may simply produce long lists of candidate genes (where the list appears to be largely dependent on the statistical method chosen; Biswas and Akey 2006; Fournier-Level et al. 2011) for which any follow-up functional study would be extremely difficult and rarely undertaken.

Common Mistakes in Molecular Biology

The most common mistake in molecular biology is giving more attention to genotyping than to sampling. The following specific situations are well worth remembering:

(a) High-throughput sequencing projects have for the time being generally had lower sample sizes than classical genotyping studies, making it more important to know where the included individuals are present. Many tools are available for doing population genetic simulations (Hoban et al. 2016), some of which can simulate realistic geographical scenarios (Meirmans 2015).
(b) Failing to perform or report experimental randomization in the laboratory. PCR may cause differences in band intensity and therefore lead to differences in genotypes when gels are scored. Most types of errors may be gel or plate specific, including stutter bands, ghost bands, mistakes in the estimation of band length and other artefacts. With NGS studies, the coverage, and hence the degree of missing data, may vary greatly.
(c) Equating geopolitical borders with biological borders. In practice, the groups are often made by simply dividing the sampled area into, for example, an 'eastern' and a 'western' cluster. Such groupings can even be based on purely anthropogenic factors, such as geopolitical borders. It is obvious that there is very little insight to be gained from such a 'random split' analysis.
(d) Testing significance of clustering output as nonindependence as the grouping genotyping data is essentially multivariate. Avoid this problem by not reporting the P-values of an ANOVA that is based on clustering, but note that it is perfectly fine to report the G, H and D statistics.
(e) Misinterpreting Mantel's (r) statistic that individuals that are close together tend to exchange more genes than those that are far apart. The main problem is that Mantel's r (or the r^2 from a linear regression) does not provide an accurate decomposition of all the genetic variation.
(f) Forgetting that only a small portion of the genome will be covered. Andrew et al. (2012) tested for climatic adaptation in plants using a set of SNPs. After applying a set of criteria to avoid false positives, they presented a list of only a few SNPs that were actually statistically associated with fitness. The use of multiple methods concurrently may help to select the loci that show the most robust pattern. But be careful, however, that different methods may be sensitive to similar biases, and it is nearly impossible to completely avoid false positives.
(g) Interpreting results should be more focused on biological relevance than on statistical significance. That does not mean that significance is unimportant and

results that have a straightforward interpretation insignificance. On the other hand, one should not be blinded by results that are strongly significant. In the genomics era, with thousands upon thousands of loci, strong significance is easily obtained even for biologically marginal processes.

Management and Sustainability

It is apparent that knowledge of traditional natural remedies for healing human diseases is quickly disappearing. The few people who still retain this knowledge are mostly elderly, and they worry that the chain will soon be broken and modern pharmaceuticals will replaced many natural remedies; and there will someday remain no real traditional medicinal knowledge in the area. For many agricultural people in mountainous regions, swidden agriculture is an integral part of natural resources management and genetic resources conservation but also of ethnic identity and biocultural heritage (Qureshi et al. 2016; Fraser 2017). In areas rich with natural biodiversity, the persistence of swidden cultivation is mostly seen as a 'problem' that obstructs the achievement of conservation objectives, and policies originating from a forest conservation perspective often seek to eradicate this form of agriculture. However, traditional rotational 'shifting cultivation' contributes to the maintenance of diversity of plant genetic resources, and these play important roles in maintaining cultural identity. Because the location of our study sites are neighbouring on China, Laos and Myanmar, only some plant varieties are likely to be unique to the region. Genetic diversity, variety diversity and species diversity of forest resources in the area have declined, and some plant varieties have even disappeared. For any ethnic group engaged in agriculture, better agricultural production is a core part of ethnic culture, and traditional agriculture is an important part of the culture of ethnic minorities. Traditional ecological knowledge, religion, and the concept of the universe are highly interrelated, and many younger villagers have lost interest in traditional culture and are unwilling to return to traditional practices. In situ conservation and sustainability, including techniques developed for propagation, cultivation, on-farm and off-farm management, new variety breeding and scientific studies, are also important for agrobiodiversity to be maintained (Schou et al. 2017; Wayne and Shaffer 2016).

In many mountain/forest areas, local food and medicinal plant security is mostly dependent on conservation of agricultural biological and plant resources, so in situ conservation measures by local farmers are very important. We can recommend the following conservation strategies:

(a) Promote consensus among different stakeholders on the value of agrobiodiversity conservation: A way to alleviate the contradiction between traditional variety conservation and new variety extension. It may be necessary to implement strong promotion of certain plants.
(b) Encourage households to conserve traditional varieties in permanent plots: The advantages of traditional variety conservation through individual household

cultivation may be realized by ensuring local villagers can take personal ownership of germplasm (seed) resources.

(c) Convene seed exchange facilities among farmers: Crop diversity can be enhanced by facilitating seed exchange among farmers. Farmers may be retaining traditional older crop seeds and expect that if, in the future, they want to plant these crops, they will be able to obtain seeds.
(d) Make a visual documentary of the indigenous knowledge related to cultivation: Considering the cultural significance and abundance of traditional knowledge of various resources, it is necessary to record the status, distribution, use and cultural management for policymakers.
(e) Provide traditional agricultural products to tourists: The feasibility of developing processed products using traditional crop varieties for sale to tourists can be explored as a means to promote agrobiodiversity conservation. Local governments could help villagers in tourism centres.
(f) Ex situ conservation of agrobiodiversity: Rare and endangered plant and medicinal species may be particularly appropriate for certain plant varieties. In addition to preserving endangered species, ex situ conservation also provides good sources of plant material for research if needed.

To conclude, Northern Vietnam is a multi-ethnic region considered to be a global biodiversity and cultural 'hotspot'. In the study area, there are abundant farming crops and livestock resources, and that agrobiodiversity is central to local livelihood and traditional culture. However, due to rapid economic development and land-use changes, local agrobiodiversity and related traditional rare and endangered plants have suffered losses and faced tremendous challenges. Some traditional plant and crop resources have declined and some have even disappeared. Conservation measures such as preserving seeds and replanting stocks of rare and endangered species are very important methods for conserving local biodiversity but need to be enhanced (Evans et al. 2014). Thus in future conservation of biodiversity, sustainable protection measures based primarily in the local communities should be considered and most probably adopted.

References

Abbott RJ, Barton NH, Good JM (2016) Genomics of hybridization and its evolutionary consequences. Mol Ecol 25:2325–2332

Ahmad P, Ashraf M, Younis M, Hu X, Kumar A, Akram NA, Al-Quariny F (2012) Role of transgenic plants in agriculture and biopharming. Biotechnol Adv 30:524–540

Albers CA, Lunter G, Mccarthur DG, McVean G, Ouwehand WH, Durbin R (2010) Dindel: accurate indel calls from short-read data. Genome Res 21(6):961–973. https://doi.org/10.1101/gr.112326.110

Albert M (2013) Peptides as triggers of plant defence. J Exp Bot 64:5269–5279

Allendorf FW (2017) Genetics and the conservation of natural populations: allozymes to genomes. Mol Ecol 2:420–430

Alvarez M, Schrey AW, Richards CJ (2012) Ten years of transcriptomics in wild populations: what have we learned about their ecology and evolution? Mol Ecol 24:710–725

Anderson JT, Lee CR, Rushworth CA, Colautti RI, Mitchell-Olds T (2013) Genetic trade-offs and conditional neutrality contribute to local adaptation. Mol Ecol 22:699–708

Andrew RL et al (2012) Adaptation with gene flow across the landscape in a dune sunflower. Mol Ecol 21:2078–2091

Angres B (2005) Cell microarrays. Expert Rev Mol Diagn 5:769–779

Araki M, Ishii T (2015) Towards social acceptance of plant breeding by genome editing. Trends Plant Sci 20:145–149

Archmiller AA, Bauer EF, Koch RE, Wijayawardena BK, Ammu A, Kottwitz JJ, Munsterman AS, Wilson AE (2015) Formalizing the definition of meta-analysis in molecular ecology. Mol Ecol 24:4042–4051

Arnold B, Corbett-Detig RB, Hartl D, Bomblies K (2013) RADseq underestimates diversity and introduces genealogical biases due to nonrandom haplotype sampling. Mol Ecol 22:3179–3190

Ban NT (2003) Catalogue of Vietnam's plant species, vol II. Agriculture Publishing House, Hanoi

Barabaschi D, Tondellia A, Desiderioa F, Volanteb A, Vaccinoc P, Valèb G, Cattivelli L (2016) Next generation breeding. Plant Sci 242:3–13

Barley AJ, Thomson RC (2016) Assessing the performance of DNA barcoding using posterior predictive simulations. Mol Ecol 25:1944–1957

Baruah J, Gogoi B, Das K, Ahmed NM, Sarmah DK, Lal M, Bhau BS (2017) Genetic diversity study amongst *Cymbopogon* species from NE-India using RAPD and ISSR markers. Ind Crop Prod 95:235–243

Belcaid M, Toonen RJ (2015) Demystifying computer science for molecular ecologists. Mol Ecol 24:2619–2640

Bianco A, Cestaro DJ, Sargent E, Banchi S, Derdak M, Di Guardo S, Salvi J, Jansen R, Viola I, Gut F, Laurens D, Chagné R, Velasco E et al (2014) Development and validation of a 20K single nucleotide polymorphism (SNP) whole genome genotyping array for apple (*Malus* × domestica Borkh). PLoS One 9:e110377

Biswas S, Akey JM (2006) Genomic insights into positive selection. Trends Genet 22:437–446

Buerkle CA, Gompert Z (2013) Population genomics based on low coverage sequencing: how low should we go? Mol Ecol 22:3028–3035

Buggs RJA, Renny-Byfield M, Chester IE, Jordon-Thaden LF, Viccini S, Chamala AR, Leitch R et al (2012) Next generation sequencing and genome evolution in allopolyploids. Am J Bot 99:372–382

Cao K, Zheng Z, Wang L, Liu X, Zhu G (2014) Comparative population genomics reveals the domestication history of the peach, *Prunus persica*, and human influences on perennial fruit crops. Genome Biol 15:415

Carrière SM, Rodary E, Méral P, Serpantié G, Boisvert V, Kuli CA, Lestrelin G, Lhoutellier L, Moizo B, Smektala G et al (2013) Rio+20, biodiversity marginalized. Conserv Lett 6:6–11

Chiu RWK, Sun H, Akolekar R, Clouser C, Lee C, McKernan K, Zhou D et al (2010) Maternal plasma DNA analysis with massively parallel sequencing by ligation for non-invasive prenatal diagnosis of trisomy 21. Clin Chem 56:459–463

Chivenge P, Mabhaudhi T, Modi AT, Mafongoya P (2015) The potential role of neglected and underutilised crop species as future crops under water scarce conditions in Sub-Saharan Africa. Int J Environ Res Public Health 12:5685–5711

Choudhury RR, Parisod C (2017) Jumping genes: genomic ballast or powerhouse of biological diversification. Mol Ecol 26:4587–4590

Clark KR, Gorley RN (2001) Primer version 5.2.7 user manual/tutorial. Plymouth Marine Laboratory, PRIMER-E Ltd, Plymouth

Clark KR, Gorley RN (2006) Getting started with Primer version 6. Plymouth Marine Laboratory, PRIMER-E Ltd, Plymouth

Coissac E, Hollingsworth PM, Lavergne S, Taberlet P, Leca F (2016) From barcodes to genomes: extending the concept of DNA barcoding. Mol Ecol 25:1423–1428

Comadran J, Kilian B, Russell J, Ramsay L, Stein N, Ganal M, Shaw P, Bayer M, Thomas W, Marshall D, Hedley P, Tondelli A, Pecchioni N, Francia E, Korzun V, Walther A, Waugh R

(2012) Natural variation in a homolog of *Antirrhinum centroradialis* contributed to spring growth habit and environmental adaptation in cultivated barley. Nat Genet 44:1388–1392
Connor DJ, Loomis RS, Cassman KG (2011) Crop ecology-productivity and management in agricultural systems, 2nd edn. Cambridge University Press, Cambridge, p 562
Dang VS (2015) A new variety of *Markhamia stipulata* (Bignoniaceae) from Southern Vietnam. Taiwania 60:129–132
Datta S, Datta S, Kim S, Chakraborty S, Gill RS (2010) Statistical analyses of next generation sequence data: a partial overview. J Proteomics Bioinform 3:183–190
Davey JW, Cezard TE, Fuentes-Utrilla P, Eland C, Gharbi K, Blaxter ML (2013) Special features of RAD sequencing data: implications for genotyping. Mol Ecol 22:3151–3164
De Filippis LF (2013) Bioinformatic tools in crop improvement. In: Hakeem KR, Ahmad PA, Ozturk MA (eds) Crop improvement – new approaches and modern techniques. Springer Business, Dordrecht, pp 49–122
De Filippis LF (2014) Crop improvement by tissue culture. In: Hakeem KR, Ahmad PA, Ozturk MA (eds) Crop improvement – new approaches and modern techniques. Springer Business, Dordrecht, pp 49–122
De Filippis LF (2017) Plant bioinformatics: next generation sequencing approaches. Hakeem KR, Malik A, Ozturk MA, Sukan FV Plant bioinformatics: decoding the phyta, 1-40, Springer Business Dordrecht
De Filippis LF (2018) Underutilised and neglected crops: next generation sequencing approaches for crop improvement and better food security. In: Hakeem KR, Malik A, Ozturk MA, Sukan FV (eds) Plant bioinformatics: decoding the phyta. Springer Business, Dordrecht, pp 1–44
De Filippis LF, Magel E (2012) Identification of biochemical differences between the sapwood and transition zone in *Robinia pseudoacacia* L. by differential display of proteins. Z Holzforschung 66:543–549
De Wit P, Pespeni MH, Ladner JT, Barshis DJ, Seneca F et al (2012) The simple fool's guide to population genomics via RNA-Seq: an introduction to high-throughput sequencing data analysis. Mol Ecol Resour 12:1058–1067
De Wit P, Pespeni MH, Palumbi SR (2015) SNP genotyping and population genomics from expressed sequences – current advances and future possibilities. Mol Ecol 2:2310–2323
Dennenmoser S, Sedlazeck FJ, Iwaszkiewicz E, Li X-Y, Altmuller J, Nolte AW (2017) Copy number increases of transposable elements and protein coding genes in an invasive fish of hybrid origin. Mol Ecol 26:4712–4724
Dinh NL, Bui DH (2010) Conservation and management for climbing bamboo- Truc Day (*Ampelocalamus* sp) in Ba Be National Park. Vietnam J Sci Technol., Thai Nguyen University 14:138–142
Dohm JC, Lottaz C, Borodina T, Himmelbaucr II (2008) Substantial biases in ultrashort read data sets from high-throughput DNA sequencing. Nucleic Acids Res 36:e105
Dufresne F, Stift M, Vergilino R, Mable BK (2014) Recent progress and challenges in population genetics of polyploidy organisms: an overview of current state-of-the-art molecular and statistical tools. Mol Ecol 23:40–69
Ebert AW (2014) Potential of underutilised traditional vegetables and legume crops to contribute to food and nutritional security, income and more sustainable production systems. Sustainability 6:319–335
Egan AN, Schlueter J, Spooner DM (2012) Applications of next-generation sequencing in plant biology. Am J Bot 99:175–185
Ekblom R, Galindo J (2011) Applications of next generation sequencing in molecular ecology of non-model organisms. Heredity 107:1–15
Ellegren H (2008) Sequencing goes 454 and takes large-scale genomics in the wild. Mol Ecol 17:1629–1631
Evans LM, Slavov GT, Rodgers-Melnick E, Martin J, Ranjan P et al (2014) Population genomics of *Populus trichocarpa* identifies signatures of selection and adaptive trait associations. Nat Genet 46:1089–1096

Farsani SF, Mahdavi MA (2011) Quantification of gene expression based on microarray In: Mahdavi MA (ed), Bioinformatic-trends and methodologies. InTech Publications. http://www.intechopen.com/books/bioinformatics

Felsenstein J (2015) Theoretical evolutionary genetics. Seattle, Washington. http://evolution.gs.washington.edu/pgbook/pgbook.html

Fierst JL (2011) A history of phenotypic plasticity accelerates adaptation to a new environment. J Evol Biol 24:1992–2001

Foley JA, Ramankutty N, Brauman KA (2011) Solutions for a cultivated planet. Nature 478:337–342

Ford-Lloyd BV, Schmidt M, Armstrong SJ, Barazani O, Engels J, Hadas R, Hammer K, Kell SP, Kang D, Khoshbakht K, Li Y, Long C, Lu B-R, Ma K, Nguyen VT, Qiu L et al (2011) Crop wild relatives-undervalued, underutilized and under threat? Bioscience 61:559–565

Fournier-Level A, Korte A, Cooper MD et al (2011) A map of local adaptation in *Arabidopsis thaliana*. Science 334:86–89

Frankham R (2015) Genetic rescue of small inbred populations: metaanalysis reveals large and consistent benefits of gene flow. Mol Ecol 24:2610–2618

Frankham R, Ballou JD, Briscoe DA (2010) Introduction to conservation genetics, 2nd edn. Cambridge University Press, Cambridge

Fraser DJ (2017) Genetic diversity of small populations: not always "doom and gloom"? Mol Ecol 26:6499–6650

Fu J, Yang L, Khan MA, Mei Z (2013) Genetic characterization and authentication *of Lonicera japonica* Thunb. by using improved RAPD analysis. Mol Biol Rep 40:5993–5999

Garnatje T, Pnuelas J, Valles J (2017) Ethnobotany, phylogeny, and 'omics' for human health and food security. Trends Plant Sci 22:187–191

Garrick RC, Bonatelli IAS, Hyseni C, Morales A, Pelletier TA, Perez MF, Rice E, Satler JD, Symula RE, Thome MTC, Carstens BC (2015) The evolution of phylogeographic data sets. Mol Ecol 24:1164–1171

Gepts P (2004) Crop domestication as a long-term selection experiment. Plant Breed Rev 24:1–44

Gilad Y, Pritchard JK, Thornton K (2009) Characterizing natural variation using next-generation sequencing technologies. Trends Genet 25:463–471

Gompert Z, Buerkle CA (2011) A hierarchical Bayesian model for next-generation population genomics. Genetics 187:903–917

Griffin PC, Robin C, Hoffmann AA (2011) A next-generation sequencing methods of overcoming the multiple gene copy problem in polyploid phylogenetics, applied to *Poa* grasses. BMC Biol 9:19

Grover CE, Salmon A, Wendel JE (2012) Targeted sequence capture as a powerful tool for evolutionary analysis. Am J Bot 9:312–319

Gupta PK (2008) Single-molecule DNA sequencing technologies for future genomics research. Trends Biotechnol 26:602–611

Gupta K, Sengupta A, Saha J, Gupta B (2014) The attributes of RNA interference in relation to plant abiotic stress tolerance. Gene Technol 3:1 ISSN:2329–6682 GNT

Guttikonda SK, Marri P, Mammadov J, Ye L, Soe K, Richey K, Cruse J, Zuang M, Gao Z, Evans C, Rounsley S, Kumpatia SP (2016) Molecular characterisation of transgenic events using next generation sequencing approach. PLoS One 11:e0149515. https://doi.org/10.1371/journal.pone.0149515

Hao D-C, Xiao P-G (2015) Genomics and evolution in traditional medicinal plants: road to a healthier life. Evol Bioinforma 11:197–212. https://doi.org/10.4137/EBO.S31326

Hayden EC (2009) Genome sequencing: the third generation. Nature 457:768–769

Hayden EC (2014) The $1000 genome. Nature 507:294–295

He R, Kim MJ, Nelson W, Balbuena TS, Kim R et al (2012) Next-generation sequencing-based transcriptomic and proteomic analysis of the common reed *Phragmites australis* (Poaceae) reveals genes involved in invasiveness and rhizome specificity. Am J Bot 99:232–247

He Q, Prado JR, Knowles LL (2017) Inferring the geographic origin of a range expansion: latitudinal and longitudinal coordinates inferred from genomic data in an ABC framework with the program X-ORIGIN. Mol Ecol 26:6908–6920

Hedrick PW (2004) Recent developments in conservation genetics. Forest Ecol Manag 197:3–19

Hedrick PW, Allendorf FW, Waples RS (2013) Let nature control adaptation. Nature 503:303

Heider B, Anderson MS, Schulze-Kraft R (2007) RAPD variation amongst Vietnamese *Flemingia macrophylla* (Willd.) Kuntze ex Merr. accessions. Biodivers Conserv 16:1617–1631

Heller R, Siegismund HR (2009) Relationship between three measures of genetic differentiation G_{ST}, DE_{ST}, and G_{ST}: how wrong have we been? Mol Ecol 18:2080–2083

Henry RJ (2012) Next-generation sequencing for understanding and accelerating crop domestication. Brief Funct Genomics 11:51–56

Hill WG, Weir BS (2011) Variation in actual relationship as a consequence of Mendelian sampling and linkage. Genet Res 93:47–64

Hoang TML, De Filippis LF, Le XT (2009) Salt tolerance and screening for genetic changes in rice mutants after gamma irradiation using RAPD and microsatellite (RAMP) markers. Open Hort J 2:62–69

Hoban S, Kelley JL, Lotterhos KE et al (2016) Finding the genomic basis of local adaptation: pitfalls, practical solutions, and future directions. Am Nat 188:379–397

Hohenlohe PA, Bassham S, Currey M, Cresko WA (2012) Extensive linkage disequilibrium and parallel adaptive divergence across three spine stickleback genomes. Philos Trans R Soc Lond B Biol Sci 36:395–408

Holliday JA, Aitken SN, Cooke JEK, Fady B, Gonzalez SC, Martinez ME, Correa J, Lexer C, Staton M, Whetten RW, Plomion C (2017) Advances in ecological genomics in forest trees and applications to genetic resources conservation and breeding. Mol Ecol 26:706–717

Huang YL, Fang XT, Lu L, Yan YB, Chen SF et al (2012) Transcriptome analysis of an invasive weed *Mikania micrantha*. Biol Plantarum 56:111–116

Hufford KM et al (2012) Inbreeding and outbreeding depression in *Stylidium hispidum*: implications for mixing seed sources for ecological restoration. Ecol Evol 2:2262–2273

Hung HV, De Filippis LF, Buckney RT (2011) Population structure and genetic diversity of the rare and endangered *Sinicalamus mcclure* and *Markhamia stipulata* in Ba Be National park, Vietnam. Asian J Plant Sci 10:312–322

Hurd PJ, Nelson CJ (2009) Advantages of next-generation sequencing versus the microarray in epigenetic research. Brief Funct Genomic Proteomic 8:174–183

Ibrahim MB, Kaushik N, Sowemimo AA, Odukoya OA (2016) Review of the phytochemical and pharmacological studies of the genus *Markhamia*. Pharmacogn Rev 10:50–59

Imelfort M, Duran C, Batley J, Edwards D (2009) Discovering genetic polymorphisms in next-generation sequencing data. Plant Biotechnol J 7:312–317

Jackson SA, Iwata A, Lee S-H, Schmutz J, Shoemaker R (2011) Sequencing crop genomes: approaches and applications. New Pytol 191:915–926

Jacobs DF et al (2013) A conceptual framework for restoration of threatened plants: the effective model of American chestnut (*Castanea dentata*) reintroduction. New Phytol 197:378–393

Jacobsen S-E, Sørensen M, Pedersen SM, Weiner J (2013) Feeding the world: genetically modified crops versus agrobiodiversity. Agron Sustain Dev 33:651–662

Jacobsen S-E, Sorensen M, Pedersen SM, Weiner J (2015) Using our agrobiodiversity: plant-based solutions to feed the world. Agron Sustain Dev 35:1217–1235

Jain SM, Gupta SD (eds) (2013) Biotechnology of neglected and underutilized crops. Springer, Berlin

Jeffries DL, Copp GH, Handley LL, Akan Olsen KH, Sayer CD, Hanfling B (2016) Comparing RADseq and microsatellites to infer complex phylogeographic patterns, an empirical perspective in the Crucian carp, *Carassius carassius*, L. Mol Ecol 25:2997–3018

Jensen JD, Foll M, Bernatchez L (2016) The past, present and future of genomic scans for selection. Mol Ecol 25:1–4

Johnson WE, Onorato DP, Roelke ME et al (2010) Genetic restoration of the Florida panther. Science 329:1641–1645

Jost L (2008) GST and its relatives do not measure differentiation. Mol Ecol 17:4015–4026

Jost L (2009) D vs G_{ST}: response to Heller and Siegismund (2009) and Ryman and Leimar (2009). Mol Ecol 18:2088–2091

Juma C (2011) The new harvest – agricultural innovations in Africa. Oxford University Press, Inc., Oxford, p 296

Kahane R, Hodgkin T, Jaenicke H, Hoogendoorn C, Hermann M, Keatinge JDH, Hughes JDA, Padulosi S, Looney N (2013) Agrobiodiversity for food security, health and income. Agron Sustain Dev 33:671–693

Kanchanapoom T, Kasai R, Yamasaki K (2002) Phenolic glycosides from *Markhamia stipulata*. Phytochemistry 59:557–563

Kane N, Sveinsson S, Dempewolf H, Yang JY, Zhang DZ, Engels MM, Cronk Q (2012) Ultra-barcoding in cacao (*Theobroma* spp; Malvaceae) using whole chloroplast genomes and nuclear ribosomal DNA. Am J Bot 99:320–329

Karakach TK, Flight RM, Douglas S (2010) An introduction to DNA microarrays for gene expression analysis. Chemom Intell Lab Syst 104:28–52

Keatinge JDH, Yang R-Y, Hughes JDA, Easdown WJ, Holmer R (2011) The importance of vegetables in ensuring both food and nutritional security in attainment of the millennium development goals. Food Sci 3:491–501

Kellogg JJ et al (2016) Biochemometrics for natural products search: comparison of data analysis approaches and application to identification of bioactive compounds. J Nat Prod 79:376–386

Kimura M (1993) The neutral theory of molecular evolution. Cambridge University Press, Cambridge

Krauss SL et al (2013) An ecological genetic delineation of local seed-source provenance for ecological restoration. Ecol Evol 3:2138–2149

Kroc M, Koczyk G, Swiecicki W, Kilian A, Nelson MN (2014) New evidence of ancestral polyploidy in the Genistoid legume *Lupinus angustifolius* L (narrow-leafed lupin). Theor Appl Genet 127:1237–1249

Kvam VM, Liu P, Si Y (2012) A comparison of statistical methods for detecting differentially expressed genes from RNA-seq data. Am J Bot 99:248–256

Leonti M (2011) The future is written: impact of scripts on the cognition, selection, knowledge and transmission of medicinal plant use and its implications for ethnobotany and ethnopharmacology. J Ethnopharmacol 134:542–555

Liang XQ, Ferguson DK, Jacques FMB, Su T, Wang L, Zhou ZK (2016) A new Celastrus species from the middle Miocene of Yunnan, China and its palaeoclimatic and palaeobiogeographic implications. Rev Palaeobot Palynol 225:43–52

Linquist S, Cottenie K, Elliott T, Saylor B, Cremer SC, Gregory TR (2015) Applying ecological models to communities of genetic elements: the case of neutral theory. Mol Ecol 24:3232–3242

Liu L, Li Y, Li S, Hu N, He Y, Pong R, Lin D, Lu L, Law M (2012) Comparison of next-generation sequencing systems. J Biomed Biotechnol 2012:251364. https://doi.org/10.1155/2012/251364

Llaca V (2012) Sequencing technologies and their use in plant biotechnology and breeding. In: Munshi A. DNA sequencing – methods and applications. InTech. ISBN: 978-953-51-0564-0

Long Y, Cheng J, Mei Z, Zhao L, Wei C, Fu S, Khan MA, Fu J (2015) Genetic analysis of litchi (*Litchi chinensis* Sonn.) in southern China by improved random amplified polymorphic DNA (RAPD) and inter-simple sequence repeat (ISSR). Mol Biol Rep 42:159–166

Mabberley DJ (2008) Mabberley's plant-book: a portable dictionary of plants, their classification and uses, 3rd edn. Cambridge University Press, New York, p 1,021

MacDicken K, Jonsson O, Pina L et al (2016) Global forest resources assessment 2015. Food and Agriculture Organizations of the United Nations (FAO), Rome

Madden LV, Paul PA (2011) Meta-analysis for evidence synthesis in plant pathology: an overview. Phytopathology 101:16–30

Madsen T, Shine R, Olsson M, Wittzel H (1999) Restoration of an inbred adder population. Nature 402:34–35

Maestri R (2017) Using phylogenetic clade composition to understand biogeo-graphical variation in functional traits. Front Biogeogr 9(3):e34435

Mantel N (1967) The detection of disease clustering and a generalized regression approach. Cancer Res 27:209–220

Mardis ER (2010) The $1,000 genome, the $100,000 analysis? Genome Med 2:84
Mardis ER (2011) A decade's perspective on DNA sequencing technology. Nature 470:198–203
Marmiroli N, Maestri E (2014) Plant peptides in defense and signaling. Peptides 56:30–44
Massana R (2015) Getting specific: making taxonomic and ecological sense of large sequencing data sets. Mol Ecol 24:2904–2906
Mastretta-Yanes A, Arrigo N, Alvarez N, Jorgensen TH, Pinero D, Emerson BC (2015) Restriction site-associated DNA sequencing, genotyping error estimation and de novo assembly optimization for population genetic inference. Mol Ecol Resour 15:28–41
Mayes S, Massawe FJ, Alderson PG, Roberts JA, Azam-Ali SN, Hermann M (2012) The potential for underutilized crops to improve security of food production. J Exp Bot 63:1075–1079
McKay JK et al (2005) 'How local is local?' – a review of practical and conceptual issues in the genetics of restoration. Restor Ecol 13:432–440
Medeiros PMD, Ladio AH, Santos AMM, Albuquerque UPD (2013) Does the selection of medicinal plants by Brazilian local populations suffer taxonomic influence? J Ethnopharmacol 146:842–852
Meirmans PG (2015) Seven common mistakes in population genetics and how to avoid them. Mol Ecol 24:3223–3231
Memon AR (2012) Transcriptomics and proteomics analysis of root nodules of modern legume plants. In: Asharaf M, Ahmad MSA, Ozturk M, Aksoy A (eds) Crop production for agricultural improvement. Springer Science – Business Media, Berlin
Meyer RS, DuVal AE, Jensen HR (2012) Patterns and processes in crop domestication: an historical review and quantitative analysis of 203 global food crops. New Phytol 196:29–48
Miller AC, Woeste KE, Anagnostakis SL, Jacobs DF (2014) Exploration of a rare population of Chinese chestnut in North America: stand dynamics, health and genetic relationships. AoB Plants 6:plu065
Montalvo AM, Ellstrand NC (2001) Transplantation of the subshrub *Lotus scoparius*: testing the home-site advantage hypothesis. Conserv Biol 14:1034–1045
Mosner E et al (2012) Floodplain willows in fragmented river landscapes: understanding spatio-temporal genetic patterns as a basis for restoration plantings. Biol Conserv 153:211–218
Mu XY, Zhao LC, Zhang ZX (2012) Phylogeny of *Celastrus* L. (Celastraceae) inferred from two nuclear and three plastid markers. J Plant Res 125:619–630
Nei M (1973) Analysis of gene diversity in subdivided populations. Proc Natl Acad Sci U S A 70:3321–3323
Nei M (1987) Molecular evolutionary genetics. Columbia University Press, New York
Nicolai M, Pisani C, Bouchet J-P et al (2012) Discovery of a large set of SNP and SSR genetic markers by high-throughput sequencing of pepper (*Capsicum annuum*). Genet Mol Res 11:2295–2300
Nielsen R, Paul JS, Albrechtsen A, Song YS (2011) Genotype and SNP calling from next-generation sequencing data. Nat Rev Genet 12:443–451
Ochatt S, Jain SM (2007) Breeding of neglected and under-utilized crops, spices and herbs. Science Publishers Inc, Enfield, NH
Okou DT, Steinberg KM, Middle C, Cutler DJ, Albert TJ, Zwick ME (2007) Microarray-based genomic selection for high throughput resequencing. Nat Methods 4:907–909
Osborne R, Hill KD, Nguyen HT, Ke LP (2007) *Cycads* of Vietnam. Osborne and Roy, Singapore
Osborne R, Hill KD, Nguyen HT, Phan KL (2010) *Cycas fugax*, The IUCN red list of threatened species: e.T42039A10635348. https://doi.org/10.2305/IUCN.UK.2010-3.RLTS.T42039A10635348.en
Paliy O, Shankar V (2016) Application of multivariate statistical techniques in microbial ecology. Mol Ecol 25:1032–1057
Pham HH (1999) Illustrated flora of Vietnam. Youth Publishing House, Hanoi (in Vitnamese)
Pool JE, Hellmann I, Jensen JD, Nielsen R (2010) Population genetic inference from genomic sequence variation. Genome Res 20:291–300
Price DL, Casler MD (2012) Simple regression models as a threshold for selecting AFLP loci with reduced error rates. BMC Bioinformatics 13:268

Primmer CR (2009) From conservation genetics to conservation genomics. Ann N Y Acad Sci 1162:357–368

Pu P, Zhang Z, Kang C, Jiang R, Jia Z, Wang G, Jiang H (2009) Downregulation of Wnt2 and beta-catenin by siRNA suppresses malignant glioma cell growth. Cancer Gene Ther 16:351–361

Qureshi R, Ghazanfar SA, Obied H, Vasileva V, Tariq MA (2016) Ethnobotany: a living science for alleviating human suffering. Evid Based Complement Alternat Med 2016:9641692 3 pages

R Core Team (2017) R: a language and environment for statistical computing. R Foundation for Statistical Computing, Vienna

Rai A, Saito K, Yamazaki M (2017) Integrated omics analysis of specialized metabolism in medicinal plants. Plant J 90:764–787

Ray S, Satya PN (2014) Next generation sequencing technologies for next generation plant breeding. Front Plant Sci 5:367

Ritchie AL, Krauss SL (2012) A genetic assessment of ecological restoration success in *Banksia attenuata*. Restor Ecol 20:441–449

Robertson M, Richards C (2015) Opportunities and challenges of nextgeneration sequencing applications in ecological epigenetics. Mol Ecol 24:3799–3801

Ruhsam M, Rai HS, Mathews S et al (2015) Does complete plastid genome sequencing improve species discrimination and phylogenetic resolution in araucaria? Mol Ecol Resour 15:1067–1078

Salgotra RK, Gupta BB, Stewart CN Jr (2014) From genomics to functional markers in the era of next-generation sequencing. Biotechnol Lett 36:417–426

Sansaloni C, et al. (2011) Diversity arrays technology (DArT) and next-generation sequencing combined: genome-wide, high throughput, highly informative genotyping for molecular breeding of Eucalyptus. Proc BMC P54, BioMed Central Ltd

Savo V, Caneva JRG, McClatchey WC (2015) Plant selection for ethnobotanical uses on the Amalfi coast (southern Italy). J Ethnobiol Ethnomed 11:58

Savolainen O, Lascoux M, Merila J (2013) Ecological genomics of local adaptation. Nat Rev Genet 14:807–820

Saxena A, Cramer CS (2013) Metabolomics: a potential tool for breeding nutraceutical vegetables. Adv Crop Sci Technol 1:106. https://doi.org/10.4172/2329.8863.1000106

Schmidt-Lebuhna AN, Aitken NC, Chuah A (2017) Species trees from consensus single nucleotide polymorphism (SNP) data: testing phylogenetic approaches with simulated and empirical data. Mol Phylogenet Evol 116:192–201

Schoch CL, Seifert KA, Huhndorf S et al (2012) Nuclear ribosomal internal transcribed spacer (ITS) region as a universal DNA barcode marker for fungi. Proc Natl Acad Sci U S A 109:6241–6246

Schou MF, Loeschcke F, Bechsgaard J, Schlotterer JC, Kristensen TN (2017) Unexpected high genetic diversity in small populations suggests maintenance by associative overdominance. Mol Ecol 26:6510–6523

Shaheen S, Bibi Y, Hussain M, Iqbal M, Saira H, Safdar I, Mehboob H, Tul Ain Q, Naseem K, Laraib S (2017) A review on *Geranium wallichianum* D-don ex-sweet: an endangered medicinal herb from Himalaya region. Med Aromat Plants (Los Angles) 6:2. https://doi.org/10.4172/2167-0412.1000288

Shokralla S, Spall JL, Gibson JF, Hajibabaei M (2012) Next-generation sequencing technologies for environmental DNA research. Mol Ecol 21:1794–1805

Shokralla S, Gibson JF, Nikbakht H, Janzen DH, Hallwachs W et al (2014) Next-generation DNA barcoding: using next-generation sequencing to enhance and accelerate DNA barcode capture from single specimens. Mol Ecol Res 14:892–901

Simmons MP, Cappa JJ, Archer RH, Ford AJ, Eichstedt D, Clevinger CC (2008) Molecular phy logenetics and evolution phylogeny of the *Celastreae* (Celastraceae) and the relationships of *Catha edulis* inferred from morphological characters and nuclear and plastid genes. Mol Phylogenet Evol 48:745–757

Simmons MP, Mckenna MJ, Bacon CD, Yakobson K, Cappa JJ, Archer RH, Ford AJ (2012) Molecular phylogenetics and evolution phylogeny of *Celastraceae* tribe Euonymeae inferred from morphological characters and nuclear and plastid genes. Mol Phylogenet Evol 62:9–20

Siol M, Wright SI, Spencer C, Barrett H (2010) The population genomics of plant adaptation. New Phytol 188:313–332. https://doi.org/10.1111/j.1469-8137.2010.03401.x
Smith FI, Eyzaguirre P (2007) African leafy vegetables: their role in the World Health Organization's global fruit and vegetables initiative. Afr J Food Agric Nutr Dev 7:1–17
Stapley J, Reger J, PGD F, Smadja C, Galindo J, Ekblom R et al (2010) Adaptation genomics: the next generation. Trends Ecol Evol 25:705–712
Staya S, Singhal P, Bal LM, Sudhakar P (2012) Bamboo shoot: a potential source of food security. Med J Nutrition Metab 5:1–10
Tabata R, Sawa S (2014) Maturation processes and structures of small secreted peptides in plants. Front Plant Sci 5:311
Taylor KH, Kramer RS, Davis JW, Guo J, Duff DJ, Xu D et al (2007) Ultra-deep bisulfite sequencing analysis of DNA methylation patterns in multiple gene promoters by 454 sequencing. Cancer Res 67:8511–8518
Teixeira da Silva JA, Jin X, Dobránszki J, Lu J, Huizhong Wang H, Zotz G, Cardoso JC, Songjun Zeng S (2016) Advances in *Dendrobium* molecular research: applications in genetic variation, identification and breeding. Mol Phylogenet Evol 95:196–216
Thottathil GP, Jayasekaran K, Othman AS (2016) Sequencing crop genomes: a gateway to improved agriculture. Trop Life Sci Res 27:93–114
Todd EV, Black MA, Gemmell NJ (2016) The power and promise of RNA-seq in ecology and evolution. Mol Ecol 25:1224–1241
Treangen TJ, Salzberg SL (2011) Repetitive DNA and next-generation sequencing: computational challenges and solutions. Nature 13:36–46
Tripathi V, Goswami S (2011) Genetic relationship amongst Cassia L., Senna Mill. and Chanaecrita Moench using RAPD markers. Int J Biodivers Conserv 3:92–100
Trucchi E, Mazzarella AB, Gilfillan GD, Lorenzo MT, Schonswetters P, Paun O (2016) BsRADseq: screening DNA methylation in natural populations of non-model species. Mol Ecol 25:1697–1713
Tsai H, Howell T, Nitcher R, Missirian V, Watson B, Ngo KJ, Lieberman M, Fass J, Uauy C, Tran RK, Khan AA, Filkov V, Tai TH, Dubcovsky J, Comai L (2011) Discovery of rare mutations in populations: TILLING by sequencing. Plant Physiol 156:1257–1268
Turesson G (1922) The genotypical response of plant species to their habitat. Hereditas 3:211–227
Turner TL, Bourne EC, Von Wettberg EJ, Hu TT, Nuzhdin SV (2010) Population resequencing reveals local adaptation of *Arabidopsis lyrata* to serpentine soils. Nat Genet 42:260–263
Unamba CIN, Nag A, Sharma RK (2015) Next generation sequencing technologies: the doorway to the unexplored genomics of non-model plants. Front Plant Sci 6:art1074. https://doi.org/10.3389/fpls.2015.01074
Uusiku NP, Oelofse A, Duodu KG, Bester MJ, Faber M (2010) Nutritional value of leafy vegetables of sub-Saharan Africa and their potential contribution to human health: a review. J Food Composit Anal 23:499–509
Valle D, Berdanier A (2012) Computer programming skills for environmental sciences. Bull Ecol Soc Am 93:373–389
Varshney RK, Terauchi R, McCouch SR (2014) Harvesting the promising fruits of genomics: applying genome sequencing technologies to crop breeding. PLoS Biol 12(6):e1001883. https://doi.org/10.1371/journal.pbio.1001883
Waller DM (2015) Genetic rescue: a safe or risky bet? Mol Ecol 24:2595–2597
Wayne RK, Shaffer HB (2016) Hybridization and endangered species protection in the molecular era. Mol Ecol 25:2680–2689
Weckerle CS, Cabras S, Castellanos ME, Leonti M (2011) Quantitative methods in ethnobotany and ethnopharmacology: considering the overall flora—hypothesis testing for over- and underused plant families with the Bayesian approach. J Ethnopharmacol 137:837–843
Weinberger K (2007) Are indigenous vegetables underutilised crops? Some evidence from Eastern Africa and Southeast Asia. Acta Hortic 752:29–34

Wetterstrand KA (2014) DNA sequencing costs: data from the NHGRI genome sequencing program. http://www.genome.gov/sequencingcosts
Wheeler N, Sederoff R (2009) Role of genomics in the potential restoration of the American chestnut. Tree Genet Genomes 5:181–187
Whitlock MC (2011) G_{ST} and D do not replace F_{ST}. Mol Ecol 20:1083–1091
Whitlock M, Ingvarsson PK, Hatfield T (2000) Local drift load and the heterosis of interconnected populations. Heredity 84:452–457
Williams AV, Nevill PG, Krauss SL (2014) Next generation restoration genetics: applications. Trends Plant Sci 19:529–537
Wong KL, Wong RNS, Zhang L, Liu WK, Ng TB, Shaw PC et al (2014) Bioactive proteins and peptides isolated from Chinese medicines with pharmaceutical potential. Chin Med 9:19
Xiao LQ, Gong X (2006) Genetic differentiation and relationships of populations in the *Cycas balansae* complex (Cycadaceae) and its conservation implications. Ann Bot 97:807–812
Xiao L, Ge X, Gong X, Hao G, Zheng S (2004) ISSR variation in the endemic and endangered plant Cycas *guizhouensis* (Cycadaceae). Ann Bot 94:133–138
Xu J, Ge X, Dolan MC (2011) Towards high-yield production of pharmaceutical proteins with plant cell suspension cultures. Biotechnol Adv 29:278–299
Yang R-Y, Keding GB (2009) Nutritional contributions of important African indigenous vegetables. In: Shackleton CM, Pasquini M, Drescher AW (eds) African indigenous vegetables in urban agriculture. Earthscan, London, pp 105–143
Yang H, Li C, Lam H-M, Clements J, Yan G, Zhao S (2015) Sequencing consolidates molecular markers with plant breeding practice. Theor Appl Genet 128:779–795
Yessoufou K, Daru BH, Muasya AM (2015) Phylogenetic exploration of commonly used medicinal plants in South Africa. Mol Ecol Resour 15:405–413
Yoder JB, Stanton-Geddes J, Zhou P, Briskine R, Young ND et al (2014) Genomic signature of adaptation to climate in *Medicago truncatula*. Genetics 196:1263–1275
Zhang ZY, Santisuk T (1998) Flora of China. In: Wu ZY, Raven PH (eds) , vol 18. Science Press, Beijing & Missouri Botanical Garden Press, St. Louis, pp 213–225
Zhang ZH, Lee HK, Mihalek I (2010) Reduced representation of protein structure: implications on efficiency and scope of detection of structural similarity. BMC Bioinformatics 11:155
Zhang Z, Bajic VB, Yu J, Cheung K-H, Townsend JP (2011) Data integration in bioinformatics: current efforts and challenges. In: Mahdavi MA (ed), Bioinformatic – trends and methodologies. InTech Publications. http://www.intechopen.com/books/bioinformatics
Zhao H, Wang Y, Yang D, Zhao X, Li N, Zhou Y (2016) An analysis of genetic diversity in *Marphysa sanguinea* from different geographic populations using ISSR polymorphisms. Biochem System Ecol 64:65–69
Zhou L et al (2015) Identification of domestication-related loci associated with flowering time and seed size in soybean with the RAD-seq genotyping method. Sci Rep 5:9350. https://doi.org/10.1038/srep09350

Health and Illness as a State of Being Human

Nevin Turgay, Özlem Yılmaz, and Fehmi Akçiçek

Introduction

Humans have been contemplating about life and death ever since their existence on the planet earth. They are regarded as the only living beings aware of being mortal, and knowing this has led us to think about the concept of health and illness. This concept and the struggle to cope with ailments are a common characteristic of global human communities. Nevertheless, it is clear that health-disease concept is not a stable one and that there is a continuous transformation related to the age and culture.

Concept of Health from Past to Present

The earliest definition of health was a very simple one: a person *with no illness is a healthy one*. Today, when it comes to the concepts of health and illness, the first that comes to mind is the field of medicine and the relevant concepts and practices. However, to be able to understand all aspects of health and illness concepts, they cannot be limited only to their biological aspects, but social and cultural aspects need to be included. While the illness is a universal phenomenon, its perception and treatment methods are different. Health and illness occur within a social structure and are shaped by it, what makes these concepts meaningful are the way that the

N. Turgay
Faculty of Medicine, Department of Parasitology, Ege University, Izmir, Turkey

Ö. Yılmaz
Faculty of Medicine, Department of Physiology, Ege University, Izmir, Turkey

F. Akçiçek (✉)
Faculty of Medicine, Department of Geriatrics, Ege University, Izmir, Turkey

M. Ozturk, K. R. Hakeem (eds.), *Plant and Human Health, Volume 1*,
https://doi.org/10.1007/978-3-319-93997-1_2

society perceives them (Mechanic 1978; Kleinman 1981). For this reason, the concept of health-illness cannot be considered different from the era, geography, and society where human beings live. Capra (1982) notes that health depends on a person's perspective of a living organism and the interactions with its environment, and as this perspective changes in the course of time and in relation to the cultures, our understanding of health also changes. Therefore, the consideration of health and illness varies, based on historical periods.

In the historical process, the concepts of health-illness have changed considerably and acquired their present meaning after passing through different stages. The first stage was the period which lasted for thousands of years until the creation of alphabet and initiation of settled life. Findings from archaeological excavations indicate that diseases occurred at the same time as humans came into existence. In the prehistoric period, it was believed that the body was robust and healthy as long as there was no effective external factor (Bayat 2016). During the hunter-gatherer period, people were losing their lives due to changes in the climatic conditions and natural disasters. The most important factor determining health/illness status then was the natural conditions. At that time, a very large number of babies died at birth, while the average life span for those who did survive was between 18 and 20 years (Cirhinlioğlu 2001). Under these circumstances, feeling powerless and helpless against nature, man believed that diseases emerged from supernatural powers or as a punishment from the Gods due to their sinful behaviors. This period, when polytheistic religions ruled, is the one when superstitious beliefs and magic were dominant. Therefore, the clergies and shamans were consulted for the treatment of diseases, and there was no systematic method for treating diseases; the primary goal being to protect life (Rivers and Köksaldı 2004).

Nearly 12000 years ago, a significant change occurred in the diet due to the onset of sedentism and farming, and people started to feed on the products they grew and to benefit from the animals domesticated by them. In this period, food-borne diseases and infectious and parasitic diseases resulting from these involvements went up as more people started to live together and with their animals, diseases became widespread, and epidemics became the greatest cause of death. Approximately 6000 years ago, a different lifestyle emerged in the form of urban life, which was a transition from the rural living conditions, and different illnesses began to appear (Fitzpatrick 1991).

In Mesopotamia, which is regarded as the first place where settled life began, tablets have been discovered showing treatments such as prayers and magic used to soothe the anger of Gods and to expel the evil spirits capturing the body, and specific medicines for illnesses have been recorded in here. The Sumerian tablets (2000 AD), where hundreds of prescriptions tested for centuries were inscribed, are the world's oldest documents describing how to prepare medicines for different diseases (*Materia Medica*) and show that the pharmacy was highly developed at that time (Bayat 2016).

The mythical stories that have survived until today also tell the pursuit of humans to protect their health and to be immortal. In this regard, the epic of Gilgamesh is the oldest known text. The epic of Gilgamesh, estimated to be written between 2500 and 3000 BC, is the first written epos in history that was found in Mesopotamia. In the epic of Gilgamesh mentioned in Akkad and Sumer mythologies, a king pursuing immortality is narrated. This saga reflects a sense of eternal life and fear of death, as well as other human emotions such as loneliness, love, grief, hate, revenge, and addiction to fame and power. Therefore, it continued to exist as a popular epic in Mesopotamia and neighboring lands for thousands of years and in different time periods. The people found a piece of themselves in this epos that reflects transition from a hunter-gatherer lifestyle to a settled life. A serpent swallows the immortality herb that the king tries to reach and rejuvenates by changing skin (Çığ 2002). The serpent, which has an important place in Eastern mythology, was later accepted as a symbol of eternal life and also adopted as a symbol of eternal life in Egyptian, in Indian, and later in ancient Greek mythologies. Asclepius, who was considered as the God of Health and Healing in ancient Greece during the polytheistic periods, is shown with a serpent-entwined staff symbolizing the tree of life in paintings and sculptures. Asclepius continued to exist in the western civilization influenced by the ancient Greek. In classical Greek civilization, all physicians were known as *Asclepiad* (Asclepius disciples). Even after the adoption of Christianity, Asclepius cult continued. Asclepius is also regarded as a talented physician who achieved God's inspiration in the Islamic civilization. Throughout the ages, the serpent figure has continued to symbolize health-related issues. At present, health professions and institutions, in particular the schools of medicine, are symbolized by the image of a snake entwining itself around various objects such as a tree, a rod, a goblet, etc. (Bayat 2016).

In Asia Chinese and Indian civilizations, which emerged at the same time as the Mesopotamian and Egyptian civilizations, approaches to health/illness were rather different. Recent studies have shown that the influence of Chinese and Indian civilizations influenced Greek and later Islamic civilization through Mesopotamia. Medical information about the Chinese civilization dates back to 3000 BC. According to Taoism, the religious and philosophical system of the Chinese, the universe, the living being, and the matter are the work of two great powers, called Yin and Yang, which complement each other but, at the same time, which are quite opposite of each other. Diseases occur when the balance of Yin and Yang or the harmony between the internal energy of the body (Chi) and the external energies is disrupted (Bayat 2016). In the therapy book by the Chinese Emperor Huang-Di (2700 BC) in the East, named *Yellow Emperor's Canon of Internal Medicine*, he emphasizes the need to take measures to prevent health problems before they occur (Unschuld 2003).

In Indian civilization, it was believed that the seasonal changes, the environment, our life style and eating habits, and our mental health determine our overall health. This approach was different in the Ayurvedic period that started in 3300 BC and continued until 800 BC and that focused on mainly religious information and in Brahmanic period between 800 BC and 1000 AD when physicians were prominent (Tez 2010; Bayat 2016).

In the ancient Greek period, religion was part of everyday life, and it was believed that every aspect of life was under the protection of Gods. It was believed that the

Gods had divine powers such as granting of life and death, healing a patient, or extending life. During these periods, healing temples called *Asclepion* were built in the name of Asclepius, who was regarded as the God of Health and Healing, starting from the sixth century BC. The clergies and physicians provided health services here. At the same time, it is known that there are around 200 Asclepions which also served as educational places. The most famous ones were located in Epidaurus, Knidos, Kos, and Pergamon – all around the Aegean Sea. Religious treatments in these temples were carried out along with practices such as diet, music, mud bath, and herbal medicine. On the entrance gate of the temple of Asclepion in Bergama (Pergamon), built in the name of Asclepius the health god, there is an inscription mentioning that "The place where death cannot enter." This temple is close to the State of Izmir in Turkey and still standing today; the patients with incurable deadly diseases were not allowed to enter this temple (Tez 2010). Asclepius was a philosopher physicians who lived around sixth and fifth centuries BC and tried to understand the world and the humans through free and systematic thought. He wanted to explain supernatural events with cause-effect relations and to understand medical cases with observations and experiments.

The first one to explain the occurrence of diseases from a scientific point of view was Hippocrates (430–370 BC), from the island of Kos (Tez 2010). He argued that the emergence of diseases is not because of a supernatural and mysterious power, but there is a physical and rational cause. Since he came from the Asclepion tradition, he believed that the physicians benefit from both the power of God and the healing power of nature. He first established the rules of medical ethics and practices, and, in this regard, he was the founder of the first medical school teaching medical ethics and practices (Bayat 2016). Hippocratic oath, which has been regarded as the moral law of medical profession throughout the ages and has preserved its essence until today, is an oath that should be taken by medical graduates who will start their professional lives as a physician. Hippocrates's understanding of rational medicine from the fifth century BC eventually evolved into today's modern medicine.

Dioscorides, a citizen of Anazarba, nowadays Tarsus in Turkey, who wrote *De Materia Medica*, was a Roman physician who lived in the first century. He is regarded as one of the most important herbal doctors who wrote the medicine guide books of ancient times and has been a source for books published in Europe later on.

Galen of Pergamon (130–200 AD), who was regarded as the greatest physician after Hippocrates and as the father of pharmacy in ancient times, systematized the Hippocrates' principles. Many medicines and methods developed for the treatment of diseases by Galen have been used for centuries (Bayat 2016). Physicians such as Avicenna (Ibn Sina), Al-Razi, Al-Zahravi, and Al-Biruni, who were educated during the Abbasid period, also wrote about a large number of herbal medicines against diseases in their books. The works *El Kanun* (*The Canon of Medicine*) and *Kitab-ül Şifa* (*Book of Healing*), in which Ibni Sina compiled medical information, have remained among the classical books for medical education for centuries. According to the Great Medical theorists, *The Canon of Medicine* is the greatest book of all times, including twentieth century. Until the eighteenth century, it was used as a main text book in medical schools around the world (Nasser et al. 2009).

During the Medieval times, when the Roman Empire got disintegrated in the eighth to fourteenth centuries, the Islamic civilization was ruling over a vast geography from Spain to India. Islam has served as a binding link for civilizations of Egypt-Mesopotamia and Greek-Roman cultures and today's western civilization. The most common diseases leading to death during the Medieval times were epidemics such as tuberculosis, cholera, and plague. At that time, the strict dogmatic attitudes of the churches in Europe led to the oblivion of the old developments, and the mysticism/magical thought gained dominance; however, there were significant developments in the description and treatment of diseases in Islamic medicine (Bayat 2016).

Since human body is considered sacred in many belief systems, no study has been done on human body. As the religion has lost its social power over time, there has been an increasing interest for medicine in the society. After the seventeenth century, when positive sciences started to develop, the religious institutions had to abandon health-related functions to medicine. The roles of the physicians and the clergymen got separated, and the churches were replaced by clinics (Cirhinlioğlu 2001). An unprecedented era began with the discovery of microscope and germs and, thus, with better hygienic measures. In the eighteenth and nineteenth centuries, with the introduction of industrialization, the concept of health has changed a lot.

The eighteenth century can be regarded as the initiation of an era when health-illness concepts were discussed from theoretical and methodological perspectives. The Cartesian thought (dualism) argues that the body and the spirit are independent from each other. This has played an important role in the emergence of the concept of illness (Locker 1983). The body began to be analyzed as a whole, which is functioning within itself and has its own rules. The sociopolitical and socioeconomic incidents such as the French Revolution (1789–1799) and the Industrial Revolution (1760–1830) were influential in understanding that health-illness interact with the social environment as well. Physiologist Claude Bernard (1813–1878), who introduced the concept of homeostasis, suggests that the disruption of the internal physiological balance, the inability of the individual to cope with environmental factors, and the loss of ability of adaptation to the external environment lead to illnesses (Noble 2008). In the nineteenth century, when the era of modern medicine began, the discovery of germs and the vaccines, developed against different diseases, has resulted in the significant improvements in human health (Hilleman 2000). Pasteur and Koch, who emphasized the necessity of an external cause for the emergence of any infectious disease, made an important contribution to the concept of public health.

After the nineteenth century, during the intensive period of industrialization and mechanization, the studies on human body were legalized; however, the view on health started to be limited only to the human body (Baloğlu 2006). Physicians started to make experiments with reproducible results and publish them in scientific journals. Significant progress was made in the treatment of diseases. Nowadays, the fact that the average life span is 70 years and over, especially in the developed countries, is a result of this progress. Most of the infectious diseases have disappeared; however, more chronic diseases have come to the forefront with chronic aging, changes in diets, and lifestyle. A multidimensional approach to health-illness concept is needed to explain chronic diseases. For this reason, the idea that diseases occur with the interaction of many biological, social, and psychological factors has begun to be adopted.

Parsons (2001) has defined health-illness concepts from a sociological point of view. Every society has a pattern for patient's role in how the individual will behave when sick. The patient's role also includes the behaviors that the patient must display and certain privileges. According to him, health is the ability of an individual to function and fulfill his role and responsibilities in society (Parsons 2001). On the other hand, health has also been regarded as a matter of adaptation, an ability to adapt to changes in the environment and ability to grow, to age, to heal, to suffer, and to be able to expect death in a peaceful way (Illich 1995). Medical institutions take control of the individual's control over their own lives. Some also argue that health-illness concepts do not emerge as a natural phenomenon. The *Birth of the Clinic* according to Foucault (2012) has been dominating power in the society (power), tending to control individuals through their bodies in order to impose a desired mode of life. As such, a new understanding of health has emerged in this direction in the modern age (Foucault 2012).

Therefore, it is clear that there are no precise health-disease concepts and that these concepts transform continuously with time and culture.

Food consumption and staying healthy or getting recovery from an illness has always been integrated closely to each other. In essence food was ever at the heart of being healthy. From the time of Hippocrates, physicians had counseled their patients on what, how, and when to eat. What they considered "healthy to eat," and why. This changed over time. Beyond everything, people at least living in the developed countries seem to be wary of the relationship between food, diet, and health as never before. Rather, recent obsession we have with "nutritionism" evolves from the media supported medicalization of food intake and diet. So we are anxious more on the nutrition particular foodstuffs provided, instead of a joyful moment they might give or the social grace related to their consumption.

An Overview of Health at Present

While "being healthy" was defined as "having no sign of any disease" until the 1940s in medical terms, it is seen that the World Health Organization (WHO) started to define "health" as "being physically, mentally and socially healthy/a complete well-being" considering the spiritual destructions following the Second World War in 1948 over the masses. As a biological condition, being healthy is defined as the ability of the individual to function in harmony with the environment physically, emotionally, mentally, and socially and to have well-being in the body and soul (Callahan 1973).

It is thought that the main factors affecting the quality of life of the individuals are in fact associated with the state of health. Among the determinants in this area include the occupation of an individual and daily routines (career well-being), being economically competent and sufficient (financial well-being), having a powerful relationship/love in his/her life (social/emotional well-being), and having good physical conditions and sufficient energy and the ability to perform daily activities (physical/mental well-being), the ability to adapt to the community (societal/community well-being), and the desire and effort to read and learn new and various things constantly through a high level of awareness of the world around him/her trying to increase the level of education (intellectual well-being). In a study conducted by the Gallup research company in more than 150 countries from Zimbabwe to Afghanistan, individuals were asked how good their life is in terms of health, wealth, and social relationships and *how they feel about these items*. A big percentage, 66% of those included in the survey, defined themselves as good in only one of these headlines, while only 7% rated themselves as good in all (Rath and Harter 2010).

What Is a Healthy Diet Actually?

In modern times, the concept of a "healthy diet" is directly associated to being healthy. However, when it comes to diet, the question is "should we eat to lower certain numerical values (cholesterol, blood sugar, etc.) or to get pleasure?" Centuries ago, as a Hellenistic philosopher, Epicurus (341–270 BC) said, "before you eat or drink anything, consider carefully who you eat or drink with rather than what you are to eat or drink: for feeding without a friend is the life of a lion or a wolf." This is an answer to the question. When we eat, the emotions which are hard to explain, happiness, and the feelings such as the repair of the losses and the expansion in one's life span are realized to appear. Therefore, when eating, the fact that one should ask him/herself is "what is good for me" and be aware of these feelings will heal him/her.

Different Dietary Habits Between Societies and Possible Consequences

It is known that the definition of a healthy diet varies based on eating habits. The "French paradox," one of the most important examples of this, is known as the low rate of coronary heart diseases (CHD) in France against diets rich in saturated fat. In the early 1990s, Serge Renaud from the University of Bordeaux used the definition of "French paradox" for the first time and saw in a crowded study consisting of a group of middle-aged French men that they took their 15% of calorie from saturated fats with daily foods containing butter, cheese, eggs, and cream. Nevertheless, it is noteworthy that the proportion of CHD in the group followed was only about 40% of that in the United States. It has been concluded that the consumption of 2–3 glasses of red wine per day reduces the mortality rates due to all diseases by 30%, but some side effects of increasing this amount to four glasses can occur. It was also concluded that alcohol and antioxidants in red wine generally reduced the incidence of CHD (Renaud and de Lorgeril 1992; Simini 2000). In particular, regular wine consumption is considered a noteworthy; however, high vegetable and fruit consumption rates and regular physical activity are also considered to support this protection. In addition to the presence of resveratrol in the red wine, it has been shown that the polyphenol compound is also present in apples, blackberries, blueberries, red plums, and peanuts. In recent years, in addition to the cardiovascular effects, it has been shown to have a positive effect in anti-inflammatory, neurodegenerative, metabolic, and age-related conditions (Bonnefont-Rousselot 2016). However, resveratrol has a very short-time bioavailability. It should be also kept in mind that excessive use of alcohol will cause various malignancies, primarily breast cancer, especially in women (Weiskirchen and Weiskirchen 2016), and mortality rate in traffic accidents due to the use of alcoholic beverages in France is very high.

The first large-scale study examining the effects of nutrition on health has been carried out by the Rockefeller Foundation in 1948. After the Second World War, Crete Island was taken as a model for analyzing the dietary problems encountered in Europe and their effects on health. However, on the contrary to what was expected, researchers determined that the Cretan people could not have a better diet in terms of health. At that time, when the food consumed by the local people was recorded, it was observed that there were little amount of wine, milk, and dairy products in the coastal regions as well as vegetable proteins, fresh vegetables, and plants/herbs collected from the nature during the right season and little amount of goat meat and game meat. It has been found that the need for dessert is usually covered from seasonal fruits. In addition, when the amount of fat consumed per day is examined, it is noteworthy that approximately 120 mL olive oil per person is consumed. At that time, drinking olive oil from small glasses on an empty stomach in the morning was observed as a surprising habit. Over the years, the diet of the islanders has changed, but the Crete cuisine, still one of the most successful examples of Mediterranean-style food, is at the head of the healthiest cuisines ever known (Coultier and Adamson 2009).

While the average life span of the world population was 50 years in the 1940s and 65 in the 1970s, it is expected to reach 75 in 2025. On the other hand, in Okinawa, the largest island of the Ryukyu Islands in the south of Japan, where 1.5 million people live, the average age is reported to be 80 for men and 88 for women today. Okinawa Island attracts attention as the place where most people over the age of 100 live in the world. Okinawa eating habits include sweet potatoes, green vegetables, and soya products. All these traditionally seem to reduce the risks of cardiovascular diseases. Eating unprocessed carbohydrates, vegetable proteins, and fish-based diets just as in the Mediterranean diet, the DASH diet, and the Portfolio diet helps to have a better survival. It is worth noting that in the island every meal is consumed consciously in the way of being healthy like a healing tool. The rhythm of life is slower, and the elderly are active in all stages of life and in all activities. However, it is also noteworthy that the rate of illness has increased due to the fact that this philosophy of life has not continued among the young people who left the island and settled in the United States or started consuming fast food (Willcox et al. 2014).

Tips for a Life Full of Wellness

Genetic inheritance is important for a lifetime wellness. However, it is seen in the studies that it is effective at only 30% level (Driscoll 1990). In addition, when the main factors affecting health are examined, it is observed that social conditions count to 10%, medical conditions to 5%, and climatic conditions to 5%, whereas lifestyle and pollution brought about by modern life count up to 50% and later have gained much more importance in our present-day living conditions. When it comes to modern pollution, the first thing that comes to mind is the electromagnetic fields and environmental pollution (air, water, and food pollution) that we are exposed to in our homes and in outdoor environment in everyday life (WHO 2009).

It is known that it is crucial to focus on peace and serenity, sending messages of love to his/her own body, mind, and soul for individual wellness. The emotional stability of an individual is of great importance in the way of being healthy. One of the most important components of this is having strong family/friendship relationships. The basic approach in relationships is to avoid being hurt and hurting someone's feelings. It is noteworthy that individuals with much stronger sense of belonging have a much healthier/stronger physical conditions in situations where they act with feelings of love, tolerance, forgiveness, and compassion. While it seems that all kinds of stress in daily life trigger serious risks for health, it has been seen that strong individuals, especially those with spiritual feelings, are living their lives more healthy.

One of the things that can be taken care of personally is the personal investment to protect the memory. The desire to learn new things, the efforts to increase intellectual capacity, and the various hobbies are known to be very important in the long run for mental wellness. It may be beneficial in the long run to do regular reading, to follow new and intellectual publications, to do some memory exercises (such as puzzles), and to take folic acid, vitamin C, and vitamin E supplements. It should not be forgotten that antacids, aluminum containers, and foil papers, which are heavily used in daily life, may cause toxicity to the brain (Lo 2017). It is observed that the individuals with habits of smoking and intensive alcohol use and drug addiction have more health problems in the general sense and the risk for malignancies is higher for them. Given the eating habits, eating less is of great importance. It is necessary to be careful about the amount of food in the people over 50 years old. It is seen that Eskimos eat the small fish with bones, thus taking more than 2000 mg calcium per day, which results in the over-hardening of their bones and an increase in hip joint fractures after falling down. It should also be kept in mind that daily eating habits might lead to unexpected risks. At certain ages there are risks caused by some eating habits (Mazess and Mather 1974).

From the perspective of eating habits, strong muscles and bones over 60 years of age are of great importance. It is especially important to have a physically active life, and attention should be paid to bone health. In an active and dynamic daily life, it is very valuable to do exercises regularly for many years that can strengthen muscles (Ferrucci et al. 2014). It should be known that bone fracture risk may increase with long-term usage of medications regulating stomach acidity, diuretics, and corticosteroids, excess smoking or alcohol usage, and various thyroid, liver, and adrenal gland diseases in women especially after menopause (Duncan et al. 2015). The fact that the fractures in the hip, shoulder, and wrists possibly affect the quality of life adversely should be considered, and necessary precautions should be taken. It is also beneficial to use calcium supplements if necessary.

The hazards of excessive food intake over 70 years old should not be forgotten. Eating less should be emphasized, and it should be kept in mind that excess food may cause serious health problems. When making all choices, the goal should be getting older in a quality manner.

In order to strengthen the immune system, importance should be given to vaccination when necessary. Vaccination programs are important for protection from infections such as hepatitis, HPV, influenza, pneumonia, and for the maintenance of health. It is also known that *Echinacea*, vitamin C, zinc, beta-glucan, and vitamin B supplements serve as natural supports to strengthen the immune system when necessary (Sheerin et al. 2017).

Successfully managing the stress in daily life and avoiding all kinds of unnecessary drug intakes (especially antibiotics) are also very important.

It is known that adequate and quality sleep is beneficial to health. The resting period supported by quality sleep has a vital importance for the organisms to recover themselves, as well as for spiritual and physical wellness. Psychological or physical problems that cause sleep to be interrupted should be defined properly. Etiological causes and treatment options of sleep apnea and snore complaints should be investigated (Jike et al. 2017). *It is also beneficial to try various herbal alternatives (such as essential herbal oils) in the first place to support sleep. It would be beneficial to seek medications if the herbal solutions remain inadequate.*

For vascular health, it is also important to follow blood lipids regularly. It is beneficial to regularly measure the levels of blood lipids such as LDL, HDL, total cholesterol, triglycerides, and lipoproteins. In addition, regular arterial blood pressure measurements are useful to prevent problems due to long-term high-pressure exposure of the arterial wall structure. It is important to limit the intake of salt, take care of keeping fit, and make the necessary adjustments between stress and sleep, so that the blood pressure can be followed and maintained at normal limits. Programs that are supported by regular exercises to take weight under control will ensure for long years of vascular health (Qian et al. 2016).

It is also advisable to have thyroid hormone levels tested every year to monitor the function of the thyroid gland, which acts like the orchestra conductor for all the systems of the body.

It is beneficial not to delay the routine checkups, considering the mouth and dental health. To maintain a sense of enjoyment from what you eat is becoming more important as you get older. Care for oral and dental hygiene, gingival health, and regular dental checkups allow for the maintenance of a healthy diet.

It should not be forgotten that the personal efforts of an individual are also very valuable in the long run for health. It is important to consciously see life from a positive point of view and to make efforts to be more cheerful. It is best for a person to live in a circle of love with friends, family, and loved ones, to go out on short or long journeys, and to have enjoyable moments. Supporting all of these with a strong spirituality will allow us to make great steps for a life full of health and wellness.

Conclusion

Being under the control of a health center and a physician on a regular basis with access of health information, guidance, proper follow-up, and treatment options are important as a guarantee of a healthy life. Just as even though the slow-changing process of seasons is not felt for some months, it is important to be prepared at the beginning of an early winter, and it is also important to be aware of what expects us as when we get older. When a long, energetic lifestyle that focuses on a pleasant, healthy, and balanced soul/body is embraced, it will be easier to get in physical form, to achieve success in the workplace and in social life. It should be kept in mind that life means making a fresh start on an every new day.

References

Baloğlu B (2006) Ekonomik ve sosyolojik bakış açısıyla sağlık ve hastalık. Der Yayınevi, İstanbul

Bayat AH (2016) Tıp Tarihi. Merkezefendi Geleneksel Tıp Derneği, İstanbul

Bonnefont-Rousselot D (2016) Resveratrol and cardiovascular diseases. Nutrients 8(5):E250

Callahan D (1973) The WHO definition of 'health'. Stud Hastings Cent., The Concept of Health 1(3):77–87

Capra F (1982) The turning point: a new vision of reality. Futurist 16(6):19–24

Çığ Mİ (2002) Gilgameş Tarihte İlk Kral Kahraman. Kaynak Yayınları, İstanbul

Cirhinlioğlu Z (2001) Sağlık Sosyolojisi. Nobel Kitapevi, Ankara

Cloutier M, Adamson E (2009) The Mediterranean diet. Harper Collins Publisher, New York

Driscoll CE (1990) Chapter 215: The family history. In: Walker HK, Hall WD, Hurst JW (eds) Clinical methods: the history, physical, and laboratory examinations, 3rd edn. Butterworths, Boston

Duncan R, Francis RM, Jagger C, Kingston A, McCloskey E, Collerton J, Robinson L, Kirkwood TB, Birrell F (2015) Magnitude of fragility fracture risk in the very old--are we meeting their needs? The Newcastle 85+ study. Osteoporos Int 26(1):123–130

Ferrucci L, Baroni M, Ranchelli A, Lauretani F, Maggio M, Mecocci P, Ruggiero C (2014) Interaction between bone and muscle in older persons with mobility limitations. Curr Pharm Des 20(19):3178–3197

Fitzpatrick R (1991) Surveys of patients satisfaction: I-Important general considerations. BMJ 302(6781):887

Foucault M (2012) The birth of the clinic. Routledge, London and Newyork

Hilleman MR (2000) Vaccines in historic evolution and perspective: a narrative of vaccine discoveries. Vaccine 18(15):1436–1447
Illich I (1995) Sağlığın gaspı: medical nemesis. Süha Sertabiboğlu (Çev) Ayrıntı, İstanbul
Jike M, Itani O, Watanabe N, Buysse DJ, Kaneita Y (2018) Long sleep duration and health outcomes: a systematic review, meta-analysis and meta-regression. Sleep Med Rev 39:25–36 S1087-0792(17)30027-8. https://doi.org/10.1016/j.smrv.2017.06.011
Kleinman A (1981) Patients and healers in the context of culture-an exploration of the borderland between anthropology, medicine and psychiatry. University of California Press, London
Lo RY (2017) The borderland between normal aging and dementia. TCMJ 29(2):65–71
Locker D (1983) Disability and disadvantage: the consequences of chronic illness. Tavistock, London
Mazess RB, Mather W (1974) Bone mineral content of north Alaskan Eskimos. Am J Clin Nutr 27(9):916–925
Mechanic D (1978) Medical sociology. The Free Press, New York
Nasser M, Tibi A, Savage-Smith E (2009) Ibn Sina's canon of medicine: 11th century rules for assessing the effects of drugs. J R Soc Med 102(2):78–80
Noble D (2008) Claude Bernard, the first systems biologist, and the future of physiology. Exp Physiol 93(1):16–26
Parsons T (2001) Sağlık ve Hastalık: Sosyolojik Bir Eylem Perspektifi. Tülin Kurtarıcı (Çev) Toplumbilim Dergisi (13)99–108
Qian F, Korat AA, Malik V, Hu FB (2016) Metabolic effects of monounsaturated fatty acid-enriched diets compared with carbohydrate or polyunsaturated fatty acid-enriched diets in patients with type 2 diabetes: a systematic review and meta-analysis of randomized controlled trials. Diabetes Care 39:1448–1457
Rath T, Harter J (2010) Wellbeing: the five essential elements. Gallup Press, New York
Renaud S, de Lorgeril M (1992) Wine, alcohol, platelets and the French paradox for coronary heart disease. Lancet 339(8808):1523–1526
Rivers WHR, Köksaldı İE (2004) Tıp, Büyü ve Din. Epsilon, İstanbul
Sheerin D, Openshaw PJ, Pollard AJ (2017) Issues in vaccinology: present challenges and future directions. Eur J Immunol 47:2017. https://doi.org/10.1002/eji.201746942
Simini B (2000) Serge Renaud: from French paradox to Cretan miracle. Lancet 355(9197):48
Tez Z (2010) Tıbbın Gizemli Tarihi. Hayy Kitap, İstanbul
Unschuld PU (2003) Huang Di Nei Jing Su Wen: Nature, Knowledge, Imagery in an Ancient Chinese Medical Text: With an appendix: The Doctrine of the Five Periods and Six Qi in the Huang Di Nei Jing Su Wen. University of California Press, Berkeley, CA
Weiskirchen S, Weiskirchen R (2016) Resveratrol: how much wine do you have to drink to stay healthy? Adv Nutr 7(4):706–718
WHO (2009) Global health risks. Mortality and burden of disease attributable to selected major risks. WHO Press, Geneva, http://www.who.int/healthinfo/global_burden_disease/GlobalHealthRisks_report_full.pdf. Accessed 19 Sept 2017
Willcox DC, Scapagninid G, Willcox BJ (2014) Healthy aging diets other than the Mediterranean: a focus on the Okinawan diet. Mech Ageing Dev 136–137:148–162

Ethnobotanical Explorations in Telangana, the Youngest State in Union of India: A Synoptic Account

Sateesh Suthari, S. Raju Vatsavaya, and Narasimha Vara Prasad Majeti

Introduction

On the planet Earth, all the heterotrophic organisms, including human beings, directly or indirectly depend on the autotrophic plants for their survival. Over millennia, the humans have learnt the art of using and domesticating the other life for the provisions such as food, fodder, milk, medicine, fuel, textiles, fibre and scores of others. The human success, both in cultural and economic, depended on the ability to utilize and exploit the natural resources.

Plants are the primary source for the prevention and treatment of different diseases or disorders of human beings and animals. With the advent of human civilization, many therapeutical systems have been developed majorly based on plants. Ayurveda, Unani, Sidda and Homeopathy are our traditional systems of medicines. More than three-quarters of population of the world rely on the plant-based traditional medical systems for the primary health care. According to the report of the World Health Organization (WHO), ca. 80% of the world population mainly depend on traditional system of medicine for the primary health care (Akerele 1992).

Indigenous herbal treatment is an integral part of the tradition, and it is a dominant therapeutical method in most of the developing countries. These traditional phytoremedies, with a considerable extent of effectiveness, are socially accepted and economically viable and mostly are the only available means. Even today, one-third of the modern pharmaceutical preparations have been originated from

S. Suthari · N. V. P. Majeti (✉)
Department of Plant Sciences, School of Life Sciences, University of Hyderabad, Hyderabad, Telangana, India
e-mail: mnvsl@uohyd.ac.in

S. R. Vatsavaya
Department of Botany, Kakatiya University, Warangal, Telangana, India

M. Ozturk, K. R. Hakeem (eds.), *Plant and Human Health, Volume 1*,
https://doi.org/10.1007/978-3-319-93997-1_3

plants only. Medicinal plant trade at international level is increasing enormously due to the intensified adoption of crude drug extracts from the plants for self-consumption by the local people in developed countries. In India huge number of medicinal plants were used for the treatment of common ailments since ancient times. It was officially recognized that 2500 plant species have medicinal value, while over 6000 plants are estimated to be explored in traditional, folk and herbal medicine (Huxley 1984).

Ethnobotany deals with the aboriginal people and plants. Ethnobotanical research can provide a wealth of information regarding both past and present relationships between plants and the traditional people. Investigations into traditional use and management of local plant wealth have demonstrated the existence of extensive traditional knowledge of many plant species not only about the physical and chemical properties but also the phenological and ecological features in the case of domesticated species. In addition to its traditional roles in economic botany and exploration of human cognition, ethnobotanical research has also been applied to current areas of study such as biodiversity prospecting and vegetation management. It is hoped that, in the future, ethnobotany may play an important role in biodiversity conservation and sustainable development (Rajasekaran and Warren 1994). In other words, it is all about the 'plants and people'. Ethnobotanical research can provide the value of plants in the day-to-day life and wealth of information from the past through present, with an insight into the future relationships between plants and the primitive societies. Ethnomedicine is the study of useful plants used by the ethnic people to cure different ailments and shows the relationship between plants and ethnic people.

Traditional Botanical Knowledge (TBK), broadly speaking, includes all types of knowledge pertaining to identification, processing and management of plants used in subsistence, material culture and medicine. It includes its original, spiritual and sociological context as well. It is the overall botanical knowledge held by any non-industrial community. It incorporates the utilitarian, ecological and cognitive aspects of plant use as well as the vegetation management (Cotton 1996). Parallel to the advancement of medicine and biochemistry, there were sincere attempts made to utilize the plant world to benefit human kind and their pet animals. Rapid developments in phytochemical isolation methods, screening techniques for bioactive compounds and reliable biochemical and pharmacological tools on one hand and the synthetic compounds as 'drugs' for disease control, on the other, have their profound impact on *ethnomedicine*. When the modern drug use evinced harmful side-effects on the patients, lack of effective medicine for threatening diseases like AIDs, cancer and the troublesome ailments like rheumatism made the scientific community to look for effective alternatives. Furthermore, WHO and the Rio Earth Summit have directed the nations towards the use of traditional knowledge of the people for health-care management. The ever-increasing prices of drugs along with the spurious ones are taxing the marginalized sections of the society. The gestation period and the costs of discovering a new drug with good efficacy are the hurdles to the present-day allopathic practice.

Ethnobotanical investigations have led to the documentation of a large number of wild plants used by the indigenous people to meet their multifarious needs (Anonymous 1990). In India, the Botanical Survey of India (Howrah) initiated an official programme in the Economic Botany Section, and with this, the ethnobotanical exploration studies were carried out by Janaki Ammal for the first time in 1954. Consequently, Jain started intensive field exploration studies among tribal areas of central India (Jain 1963a, b, c, 1964a, b, c, Jain 1965a, b). All these publications have triggered the young minds of botanists, anthropologists and Ayurvedic medical practitioners in ethnobotanical activities in the early 1960s. The All India Coordinated Research Project (AICRP) on Ethnobiology came into operation from 1982 at NBRI, Lucknow, and four centres/circles, namely, Shillong, Howrah, Coimbatore and Port Blair of Botanical Survey of India. Mudgal (1987) documented a synoptic account of ethnobotanical studies and Binu et al. (1992) reported the ethnobotanical work carried out in India. Later, Lalramnghinglova and Jha (1999) reviewed the work on ethnobotany of the world with special reference to India. An important prerequisite for proper utilization of raw materials of the country is the survey of its natural resources and the preparation of an inventory. It is necessary to have a full knowledge about the occurrence, abundance, distribution and phenology of various plants for their sustainable utilization. The forests of Telangana have great potential in terms of economic benefits to ethnic people. The Telangana State is one of the timber and non-timber-rich forests in the country.

Probably for the first time in Telangana, the ethnobotanical uses of few medicinal plants were recorded in the *Forest Flora of Hyderabad* (Khan 1953). Kapoor and Kapoor (1980) listed the wealth of medicinal plants from Karimnagar district. Ramarao (1988) visited Rangapur and Thupakulagudem hamlets and interacted with Koyas and Lambadis of Warangal district for his doctoral work. Ravishankar (1990) worked on the ethnobotany of Adilabad and Karimnagar districts for his doctoral degree. Ravishankar and Henry (1992) published a brief account of ethnobotany of Gonds of Adilabad district. Later on, notes on ethnomedicinal uses of some important plant taxa of Mahabubnagar district (Pullaiah and Kumar 1996; Kumar and Pullaiah 1998) and folk treatment of bone fracture in Ranga Reddy district (Padmarao and Reddy 1999) were published. Reddy et al. (1998) recorded the ethnoveterinary plants from Warangal district, and of these, 49 plant taxa were included in *Dictionary of Ethnoveterinary Plants of India* by Jain (1999) as an appendix. Upadhyay and Chauhan (2000) provided an account of the ethnobotany of Gundala mandal in Khammam district. The floristic studies of Karimnagar (Naqvi 2001) and Warangal (Reddy 2001) districts and the chemotaxonomic study of *Crotalaria* (Samata 2007) included ethnobotanical information. Reddy and Rao (2002) documented the folklore and ethnomedicinal drugs from Ranga Reddy district. Reddy (2003) documented ethnobotanical plants for Khammam district. Reddy and Raju (2002) published the ethnobotanical uses by Konda Reddis of Mothugudem, and Raju and Reddy (2005) enlisted the ethnobotanical-medicinal plants for diarrhoea and dysentery. A brief report on the ethnoveterinary practices by Koyas in Pakhal wildlife sanctuary by Murthy et al. (2007). Sudharani et al. (2007) made a survey of ethnobotanical plants whereas Reddy (2008) recorded the medicines from bio-fencing plants form Nalgonda district. Sreeramulu (2008) recorded 313 ethnomedicinal plants from Nalgonda and Warangal districts for his doctoral degree. Reddy et al. (2010) reported 82 medicinal plant species of ethnic use in Medak district. The utilization pattern and diversity of NTFPs from Adilabad district was documented by Omkar et al. (2012). Later, Sreeramulu et al. (2013) reviewed the ethnobotanical medicine for common human ailments in northern (Warangal) and southern (Nalgonda) Telangana districts. Suthari et al. (2014a) investigated the traditional knowledge of medicinal plants used by Koya community inhabiting in and around *Eturnagaram wildlife sanctuary* area, and they recorded 237 species of 75 families of Magnoliophyta and four ferns of Pteridophyta as the resources of medicine. Suthari et al. (2014b) enumerated 204 climbing plants of northern Telangana and their ethnomedicinal uses. Saidulu et al. (2015) published the ethnobotany of *Pocharam wildlife sanctuary* in the former Medak and Nizamabad districts. Recently, Suthari and Raju (2016) reported the traditional knowledge on 124 flowering plant taxa for poisonous snake bites used by Koyas of Warangal district, whereas Suthari et al. (2016) documented 470 species of 318 genera and 95 angiosperm families used by Koyas of Warangal North Forest Division of northern Telangana. Mohan et al. (2017a) enlisted 198 ethnomedicinal plants of 165 genera belonging to 72 families, whereas Mohan et al. (2017b) recorded 22 antirheumatic plant taxa of 22 genera and 17 families from Kawal wildlife sanctuary of Mancherial district (Table 1).

Multidisciplinary Subjects of Ethnobotany

Ethnobotany is majorly based on botany and anthropology, of which can provide a valid information about useful plants which were practiced by the indigenous people. Ethnobotany is the interdisciplinary science between botany and ethnology. It includes interdisciplinary subjects like medicine, pharmacology, narcotics, ecology, etc. and the subdisciplines of ethnobotany have in different subjects and parts of botany like ethnophycology, ethnomycology, ethnolichenology, ethnobryology and ethnopteridology which is the knowledge and use of that particular group of plants by the different ethnic groups. It is carved out of several interrelated subjects like agriculture (*ethnoagriculture*), ecology (*ethnoecology*), forestry (*ethnoforestry*), taxonomy (*ethnotaxonomy*), phytochemicals (*ethnophytochemistry*), medicine (*ethnomedicine*), pharmacology (*ethnopharmacology*), gynic issues (*ethnogynaecology*), narcotics (*ethnonarcotics*), health care of children (*ethnopediatrics*), cosmetics (*ethnocosmetics*), knowledge of soils (*ethnopedology*), archaeological sites (*archaeoethnobotany*), toxicology (*ethnotoxicology*), music (*ethnomusicology*), etc.

Table 1 Ethnobotanical work carried out in Telangana State, India, during 1953–2017

Work carried out	District/area covered	Year	Author/s
Ethnic uses of some medicinal plants	Hyderabad State	1953	Khan
Gonds of Adilabad	Adilabad	1979	Heimondorf
Medicinal plant wealth	Karimnagar	1980	Kapoor and Kapoor
Ethnobotanical knowledge from Rangapur and Thupakulagudem	Warangal	1988	Ramarao
Medicinal value of plants	Karimnagar and Warangal	1990	Hemadri
Ethnobotanical studies	Adilabad and Karimnagar	1990	Ravishankar
Ethnobotany of Gonds	Adilabad	1992	Ravishankar and Henry
Utility of herbal plants from Mannanur forest	Mahabubnagar	1996	Pullaiaha and Kumar
Ethnomedicinal use of plants	Mahabubnagar	1998	Kumar and Pullaiah
Ethnoveterinary practices	Warangal	1998	Reddy et al.
Folk treatment of bone fractures	Ranga Reddi	1999	Padmarao and Reddy
Folklore biomedicine for veterinary diseases	Nalgonda	2000	Reddy and Raju
Ethnobotany of Gundala mandal	Khammam	2000	Upadyay and Chauhan
Note on rare and endemic medicinal plant	Khammam	2001	Reddy et al.
Ethnomedicinal use of plant taxa	Karimnagar	2001	Naqvi
A review on ethnoveterinary medicine	Andhra Pradesh	2001	Raju
Ethnic uses of important medicinal plants	Warangal	2001	Reddy
Ethnobotanical observations on Konda Reddis of Mothugudem	Khammam	2002	Reddy and Raju
Ethnomedicine and folklore drugs from Ranga Reddy	Ranga Reddy	2002	Reddy and Rao
Ethnomedicinal plants from Khammam district	Khammam	2003	Reddy
Ethnobotanical medicine for dysentery and diarrhoea	Khammam	2005	Raju and Reddy
Ethnobotanical information of some *Crotalaria*	Warangal	2007	Samata
Plants used in ethnoveterinary practices by Koyas	Warangal	2007	Murthy et al.
Traditional botanical knowledge on wild food plants	Andhra Pradesh	2007	Reddy et al.
Survey on ethnomedicinal plants from Nalgonda	Nalgonda	2007	Sudharani et al.
Traditional botanical knowledge of local people of Nalgonda and Warangal districts	Nalgonda and Warangal	2008	Sreeramulu
Ethnomedicinal uses of bio-fencing plants	Nalgonda	2008	Reddy

(continued)

Table 1 (continued)

Work carried out	District/area covered	Year	Author/s
Documentation of the traditional medicine knowledge of Gond community	Warangal	2008	Reddy et al.
Ethnomedicinal observations among the Konda Reddis of Khammam	Khammam	2008	Reddy et al.
Ethnobotanical uses of orchids	Andhra Pradesh	2008	Raju et al.
Ethnobotanical uses of sedges	Andhra Pradesh	2008	Raju et al.
Ethnobotanical knowledge from Adilabad district	Adilabad	2009	Swamy
Medicinal plant taxa of ethnic use	Medak	2010	Reddy et al.
Utility pattern and diversity of major NTFPs	Adilabad	2012	Omkar et al.
Ethnomedicinal plants used by Gonds of Adilabad	Adilabad	2012	Murthy
Useful plants for bone fractures used by indigenous people	Nizamabad	2012	Dinesh and Sharma
A review on ethnobotanical medicine for common human ailments	Warangal and Nalgonda	2013	Sreeramulu et al.
Documentation of ethnomedicinal plants used by tribal traditional healers of Adilabad	Adilabad	2013	Lingaiah and Rao
Ethnobotanical uses of some plant barks used by Gondu tribes	Adilabad	2013	Kumar et al.
Study on ethnomedicine for rheumatism by the tribal people	Khammam	2013	Manjula et al.
Traditional uses of Euphorbiaceae	Khammam	2013	Padal and Sathyavathi
Traditional botanical knowledge of plant medicines of Koya tribe	Warangal	2014a	Suthari et al.
Ethnomedicinal and economic uses of 204 climbing plants	Northern Telangana	2014b	Suthari et al.
Utilization of ethnomedicinal plants by ethnic tribes of Adilabad	Adilabad	2014	Krishna et al.
Medicinal plants used by ethnic people of Adilabad	Adilabad	2014	Ramakrishna and Saidulu
Ethnobotany of Pocharam wildlife sanctuary	Medak and Nizamabad	2015	Saidulu et al.
Documentation of traditional knowledge of medicinal plants used by tribes	Mahabubnagar	2015	Singh and Singh
Studies on ethnomedicinal plants of Leguminaceae family	Adilabad	2015	Ramakrishna et al.
Documentation of ethnomedicinal plants of Parnasala sacred grove	Khammam	2015	Rao et al.
Study on ethnomedicinal plants used by the tribals of Achampet forest division	Mahabubnagar	2015	Reddy

(continued)

Table 1 (continued)

Work carried out	District/area covered	Year	Author/s
Ethnic knowledge on 124 angiospermous plant taxa for poisonous snake bites used by Koyas	Warangal	2016	Suthari and Raju
Four hundred and seventy medicinal plants documented with ethnic uses	Warangal	2016	Suthari et al.
Ethnomedicinal uses of 198 plant taxa from Kawal wildlife sanctuary	Adilabad	2017a	Mohan et al.
Ethnic use of 22 antirheumatic plants from Kawal wildlife sanctuary	Adilabad	2017b	Mohan et al.

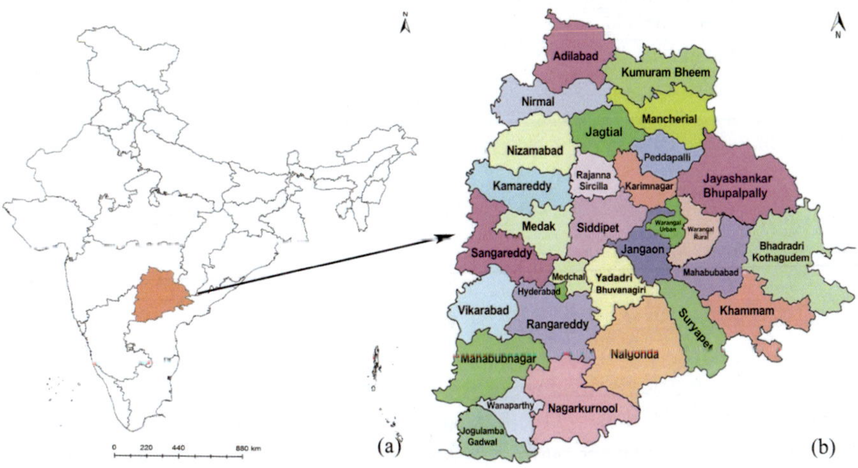

Fig. 1 Map showing the study area: (**a**) India; (**b**) Telangana State with 31 districts

Study Area

Telangana is the 29th state of India and was formed on 2 June 2014 with the capital of Hyderabad district, and it is the youngest and 12th largest state. It lies in between latitudes 15°48′32″ to 19°55′46″N and longitudes 77°09′02″ to 81°18′ 51″E. It is bounded on the north and north-west by Maharashtra, west by Karnataka, north-east by Chhattisgarh, east by Odisha and south by Andhra Pradesh. The government has reorganized the then 10 districts into 31 districts in 2016 (Fig. 1). Geographically, it spreads over an area of 1,12,077 km^2 (Table 2) and predominantly drained by two major rivers, namely, Godavari and Krishna. The state is very warm and dry during the summer (March–May), and the temperature rises up to 50 °C in coal mining regions. It is the hottest in Godavarikhani, Ramagundam, Kothagudem, Manuguru and the coal belt Bhupalpally. The rainfall is high along the river Godavari. In December and January, the temperature drops during nights and is very cold. Some parts of the state drops up to 4 °C (Suthari 2013).

Table 2 Demographical statistics of Telangana State, India

	Demography	Details
1	Geographical area (km^2)	112,077
2	Number of districts	31
3	Capital city	Hyderabad
4	Revenue divisions	68
5	Population (in lakhs)	351.94
6	Male	177.04
7	Female	174.90
8	Sex ratio (M:F)	1000:988
9	Rural population (in lakhs)	215.85
10	Urban population	136.09
11	Tribal population	31.78
12	Literacy rate (%)	66.46
13	State tree	Jammi (Khejri: *Prosopis cineraria*)
14	State flower	Tangedu (Tanner's Cassia: *Senna auriculata*)
15	State bird	Pala pitta (Indian roller: *Coracias benghalensis*)
16	State animal	Jinka (deer: *Antilope cervicapra*)

Source: Anonymous (2017)

Ethnic People of Telangana

There are about 20 scheduled tribes, inhabiting both the hilly and plain regions of Telangana State. Of these, 11 ethnic tribal groups of central India are found in Telangana, viz. Koyas, Konda Reddis, Gonds, Kolams, Naikpods, Pardhans (Pradhans), Thotis, Andhs, Mannewars, Bhils and Gowaris. The Yerukulas and Lambadis are largely found in the plains. There are Chenchus of Nallamalais settled in Nallamalai forest region of Mahabubnagar and some pockets of Warangal district (Regonda mandal), Vikarabad forest region (Ranga Reddy district) and Nalgonda district on the bank of Krishna river. The total schedule tribe population is 31.78 lakh and accounts for 9.03% of total population of the state (Table 3). Most of the ethnic tribes (ca. 53%) inhabit in the erstwhile districts of Khammam (20.68%), Warangal (16.7%) and Adilabad (15.6%). The predominant tribes are Lambadis (20.46%), Koyas (4.86%), Gonds (2.98%) and Yerukalas (1.44%). Nakkala and Dhulia communities were recognized as tribes in 2002–2003, and they distributed sporadically in the state. The newly formed government has announced that the communities such as Boyas and Mathura Lambadis can also be considered under scheduled tribes in 2017. Most of the tribal people were settled in nine districts, namely, Adilabad, Komaram Bheem Asifabad, Mancherial, Jayashankar Bhupalpally, Warangal Rural, Mahabubabad, Bhadradri Kothagudem, Khammam and Nagarkurnool (Kannabiran et al. 2017).

Table 3 District-wise geographical, demographical and forest cover information of Telangana State

	District	Geo. area (km^2)	Population (2011 census)	Sex ratio (F/1000 M)	Literacy rate	Tribal population	Forest area (km^2)	Forest cover (%)
1	Adilabad	4153	708,972	989	63.46	224,622	1706.89	41.1
2	Kumaram Bheem Asifabad	4878	515,812	998	56.72	133,627	2420.17	49.61
3	Nirmal	3845	709,418	949	57.77	80,576	1213.08	31.55
4	Mancherial	4016	807,037	911	64.35	56,969	1761.17	43.85
5	Jagtial	2419	985,417	1036	60.26	23,351	554.91	22.94
6	Nizamabad	4288	1,571,022	1044	64.25	107,035	853.21	19.9
7	Peddapalli	2236	795,332	992	65.52	14,945	304.91	13.64
8	Kamareddy	3652	972,625	1033	56.51	81,656	868.99	23.79
9	Rajanna Sircilla	2019	552,037	1014	62.71	22,990	379.14	18.78
10	Karimnagar	2128	1,005,711	993	69.16	12,779	3.47	0.16
11	Jayashankar Bhupalpally	6175	711,434	1009	60.33	123,544	4505.05	72.96
12	Sangareddy	4403	1,527,628	965	64.08	86,710	203.68	4.63
13	Medak	2786	767,428	1027	56.12	72,900	563.41	20.22
14	Siddipet	3632	1,012,065	1008	61.61	25,010	252.18	6.94
15	Jangaon	2188	566,376	997	61.44	62,662	30.75	1.41
16	Warangal Urban	1309	1,080,858	997	76.17	33,306	28.76	2.2
17	Warangal Rural	2175	718,537	994	61.26	105,300	163.59	7.52
18	Mahabubabad	2877	774,549	996	57.13	292,778	1490.7	51.81
19	Bhadradri Kothagudem	7483	1,069,261	1008	66.4	392,034	4286.98	57.29
20	Vikarabad	3386	927,140	1001	57.91	94,623	432.97	12.79
21	Medchal	1084	2,440,073	957	82.49	55,244	89.19	8.23
22	Hyderabad	217	3,943,323	954	83.25	48,937	1.43	0.66
23	Yadadri Bhuvanagiri	3092	739,448	973	65.53	43,318	116.97	3.78
24	Suryapet	3607	1,066,560	996	64.11	141,271	117.78	3.27
25	Khammam	4361	1,401,639	1005	65.95	199,342	621.89	14.26
26	Ranga Reddy	5031	2,446,265	950	71.95	138,710	292.73	5.82
27	Nalgonda	7122	1,618,416	978	63.75	209,252	644.85	9.05
28	Mahabubnagar	5285	1,486,777	995	56.78	132,131	370.0.84	7.02
29	Nagarkurnool	6924	861,766	968	54.38	106,880	2498.49	36.08
30	Wanaparthy	2152	577,758	960	55.67	46,062	118.74	5.52
31	Jogulamba Gadwal	2928	606,660	972	49.87	9376	6.79	0.23

Source: Anonymous (2017)

Fig. 2 The Koya tribal women and one of the others with the Koya tribe

Koyas

The land of Koyas is obscure. Largely, they are inhabitants of Godavari valley or Koya-Konda Reddi region, along the Godavari valley. They are Gonds and believed to have migrated to southeast from Bastar in Chhattisgarh. The word 'koya' comes from the word 'koi' or 'koitor' (Ramarao and Henry 1996). They started speaking Telugu and became Koya. They are nomadic by nature but now settled agriculturists. They still indulge in *podu* cultivation though occasionally. As Koyas live proximate to forests, their lifestyle revolves around the local flora and fauna. Majority of the Koyas practice agriculture, though some have turned out or pushed to be labourers in their own lands (Haimendorf 1979). They also gather non-timber forest products (NTFPs), largely tendu leaf, gum karaya, broom grass, mahua flowers, cleaning nuts, etc. The process of deculturation and modernization has affected this group tremendously. The Koyas ruled parts of Jayashankar Bhupalpally district, as tribal kings. The war waged against Kakatiya Kings by Sammakka and Saralamma is well-known, and the episode is now almost mythology. The place of this historic war is near Medaram, in Tadvai mandal of Jayashankar Bhupalpally district.

The Koyas are divided into two groups, namely, those who speak in Gondi dialect and those who speak Telugu. They usually choose the foot hills and plains in the forest areas as their places of abode (Fig. 2). The Koyas are more progressive than Konda Reddi and other tribal groups though they nowhere stand near Lambadis, the exotic counterparts, in their political power. The settlements of Koyas consist of groups of hamlets scattered over the village land. Although Koyas raise the same food crops as the Konda Reddis, both exchange certain household articles and agricultural implements on *barter basis*. For instance, the Koyas provide iron implements to Konda Reddis, while the Konda Reddis weave baskets and supply them to the Koyas (Reddy 2003).

The Koyas are divided into a number of groups after their occupations or the tracts of inhabitation. The Koyas have a Kula Panchayat which exercises control over the community. The marriages in Koyas are either by proposal or by force which include elopement and capture. In the case of a marriage by proposal, the bridegroom gives a wedding feast and pays the bride's price. The most striking feature of marriage is the dance with bison horn. The Telugu-speaking Koyas refer to their community as 'Dorala Sattam' and do not intermarry the other groups.

Konda Reddis

Konda Reddis are one of the most primitive tribes of northern Telangana. They call themselves Pandava Reddis. Konda Reddis are largely in Khammam district. The small settlements of Konda Reddis are usually perched on hillsides or neatly tucked in the jungle clearings. The settlements on hills or jungle clearings are exclusively inhabited by Konda Reddis. In foothills, they live in association with Koyas. Even if they live in the same village along with Koyas, their houses are distanced from Koyas as they treat them inferior in social status. Their thatched huts are square to rectangular in shape. Bamboo wattle with mud plastering or mud walls provide the outer walls and inner partitions of the house. The roof is usually constructed with locally available bamboo while timber stakes providing the support beams and poles. The roof is covered with thatch or palm leaves.

Physically, Konda Reddis are of small height with sturdy stature. Legs are short in comparison with the length of the body. The skin colour varies from a very light copper brown to dark chocolate. The hair is usually wavy but almost straight as well as curly. Konda Reddi women dress generally with a sari, a small loin-cloth and in some areas a bodice. In the hills, the women wear short, narrow pieces of saris, which they wrap round the hips. Men wear a loin-cloth called 'gochi'. Dhoti and shirt have now become common with some of the Konda Reddi youth. Their primitive technology is of preagricultural times and characterized by digging stick and bow and arrow (Fig. 3). Podu (*shifting*) cultivation is still practiced by Konda Reddi.

Authority is patriarchal, i.e. father or elderly male member is the highest authority in the family. Divorce and widow remarriage are socially approved. Polyandry is strictly prohibited. Bride price or 'oli' is invariably paid to the parents of the bride. Marriage negotiations are initiated by bridegroom's parents. Marriage is also performed in the house of the bridegroom. 'Pedda Kapu' is the chief of the Konda Reddi village. His decisions are governed by the majority opinion of the family heads in the village. The highest institution of social control is 'Kula Pedda'. Disputes between two Konda Reddi are referred to 'Kula Pedda' whose decision is final.

The main food of Konda Reddis is jowar. It is often substituted with bajra, ragi and other millets. Rice is also increasingly consumed, especially by wage earners like forest labourers. Red gram is the most popular of the pulses which is also raised on *podu* fields, for family consumption. Green leaves, immature fruits, roots and tubers from the forest around are cooked and eaten as curries. Chilly chutney and tamarind pulp boiled in water with salt are the regular items of their food. In periods of scarcity, wild tuber and roots like 'Adda dumpa' and 'Niluvu dumpa' form their diet. Both indigenously brewed and liquors locally purchased are consumed

Fig. 3 The Konda Reddi tribe with his bow and arrow in the forest for hunting

frequently. Toddy is tapped by Konda Reddis from *Caryota urens*, and *Borassus flabellifer* (palms) fills the 'cup of joy' for them. It is a must drink during the rituals.

To ensure the protection and success of agricultural operations, several deities are propitiated. They may be categorized as benevolent and malevolent. The former are revered and propitiated to secure their sustained blessings, while the later are feared and offered bloody sacrifices to escape their wrath.

Gonds

The Gonds are an important ethnic tribe in central India. They gave the name the 'Gondwanaland', the southern land mass of Pangea, the super continent. In northern Telangana, Gonds are called 'Raj Gonds' and 'Durve Gonds'. In this region, Naikpod is equal to the Raj Gonds (as per the approved list of Scheduled Tribes). But, in tribal areas of erstwhile Adilabad district, Naikpod is a separate tribe. The Gonds not only constitute the principal rural population in greater parts of central India but also once formed a ruling race equal in power and material status

Fig. 4 Collection of data from the Gond tribes of Kumaram Bheem Asifabad district

comparable to that of many a contemporary Hindu Prince in the neighbourhood (Haimendorf 1979). The Gonds are known for their geographic, ethnic, cultural and linguistic diversity and the extent of dependence on the local forests (Fig. 4).

Gonds are a dominant tribal community in northern Telangana in the land north of Godavari and west of Pranahita rivers. The Gonds also reside further down in parts of Warangal and Khammam districts of Godavari valley. They speak Gondi, Marathi and Telugu as per their stay with other local communities in the region. Moreover, the population of Gonds in Jayashankar Bhupalpally district is steadily brought under the cultural influence of other tribes (the descendant Koyas) or non-tribes, which form the local commanding community.

Kolams

The Kolam is a primitive tribal group in northern Telangana. These people call themselves as Kolavar. Kola, in their dialect, is bamboo or stick. As Kolams prepare baskets, wattles and winnowing fans with bamboo, they might have been calling themselves as Kolavar. The suffix 'war' or 'var' means people of that family or sect. Kolams use herbal medicines, diagnose the diseases and forecast events by measuring a charmed magic stick of bamboo called 'Kolam'. The Raj Gonds call Kolams as 'Pujaris' (priests). Both enjoy equal social status. The Kolams and Gonds celebrate a colourful 'Ghusadi' festival after Dasara and perform 'Dandari' dance during the festival, which resembles the association of Adivasis with the nature.

Naikpods

The Naikpods live in forest areas of northern Telangana, like Nayaks. The generic word Nayak means 'leader' but the suffix pod is corruption of wad which means that person. They prefer podu cultivation. Naikpod community is largely found inside scheduled areas and sparsely outside them; they are divided into a number of exogamous groups on the basis of surnames or sects. The surname only regulates the matrimonial relations. These Naikpods are not to be confused with Lambadis, who keep Naik as suffix to their names to identify their community.

Pardhans (Pradhans)

The Pardhans are traditional bards to Gonds and recite mythologies, folk tales and songs of their deities at various festivals, ceremonies and fairs for which they are paid. Each Gond family has a Pardhan family, as traditional bard. Gonds call them 'Patadi', meaning singer. Pardhans claim that they are called Pardhans since they served as ministers and advisors to past Gond kings in the region.

Thotis

The Thotis are inhabitants of erstwhile Adilabad and Karimnagar districts and listed as a Scheduled Tribe. They are numerically the smallest tribes in the district. The tribe is divided into four exogamous phratries, just as saga of Gonds. The mother tongue of Thotis is Gondi. Gonds call the Thotis as Birdal (receiver), while Thotis call the Gonds as Dhani (donor).

Andhs

The tribe Andh is one of the branches of Gonds. Probably, the word 'Andh' is originated from the Sanskrit 'Andhra', a connection given by the ancient Aryans to an aboriginal tribe dwelling in the Andhra Desh. Andhs appear in dark complexion and thick lips and resemble the Gonds. They worship the Hindu gods and employ Brahmins for religious and ceremonial purposes. They eat all kinds of items of foods available in the regions and drink spirituous and fermented liquors. They spread sporadically in Adilabad and Mancherial districts. They are usually considered born hunters, and many of them are sustaining as labourers. They collect non-timber forest produce like honey, firewood, wild fruits, etc. for their day-to-day livelihood (Ul-Hassan 1920).

Mannewars

The tribe Mannewar is an ancient tribal group of India and belonging to Dravidian family. They live in and around forest areas of Adilabad and Mancherial districts. They are the smallest group of Adivasis in the region and isolated from Koya tribe. They usually speak Telugu themselves, but with others, they speak either Hindi or Marathi. The word 'Mannewar' is derived from Telugu word, meaning 'a jungle' or 'forest'. The Telugu-speaking people call them as 'Mannewarlu' meaning people living in the forest areas. They worship usually lord Vishnu or Shiva. They normally eat chapathi of jawar, bajra and wheat, rarely take rice. Both the genders drink countrymade liquor during festivals. They collect forest produce for their livelihood and firewood for cooking.

Bhils

Bhil tribe is a primitive and backward group. They are yet assimilated enough to be hardly distinguished. They are inhabited in Adilabad district only. They live along with other ethnic communities like Kolams and Andhs. Bhils are the recognized tribal group in Telangana, but the population is negligible and is only 604 as per Census 2011 (Satyanarayana 2014).

Gowari

The word 'Gowari' is derived from 'gao' or 'gai', which means 'cow' and Gowari means 'cowherd'. They are basically cattle herders. They live in and around forests and graze their cattle in the forests. They found in Adilabad about 8–10 families and work mainly as labourers. The tribe is the subgroup of the 'Gond' in the Government of India list for Scheduled Tribe. They are distinct community. They worship god in nature but today most of them are Hindus. They are very poor in facilities and lacking in education and opportunities like other communities in STs.

Yerukalas

These people are found in the plains, in the outskirts of towns or employed in government service. They are chiefly engaged in making baskets and brooms and rearing pigs (Fig. 5). They have small houses constructed with bamboo poles and thatched grass. The women of this tribe are noted for their soothsaying (*Sodhi* in Telugu, which is now a fading art). They are mostly wanderers.

Fig. 5 A Yerukala woman is weaving a basket

Lambadis

Lambadis are emerging as one of the major tribes in Telangana districts and predominated tribe with 20.46%. These are an exotic tribe, living in *thandas*, exclusively of their communities. Each *thanda* has a chief, known as Naik (Headman). The word Lambadi might have been derived from the Sanskrit word, *lavana*, meaning salt. The forefathers of Lambadis were traders of salt across Rajasthan desert. In northern India, they are Banjaras. The word 'Banjara' must have evolved from Prakrit, Hindi and Rajasthani words 'bana/ban' or 'van/van' meaning forest or moorland and 'chara' meaning 'movers' or dwellers.

The Lambadis have a good stature with fair complexion. Men dress simply with a shirt of handloom cloth, dhoti and a turban on their head. The women wear patchwork, petticoats and light fitting bodies of some material with several rows of bead necklaces, while their arms are covered with bracelets up to the elbow though this ornamentation is now hardly seen in this part of the country (Fig. 6). They are almost Hindus and worship Sevalal Maharaj. Since a decade, the religious conversion has taken place to Christianity in the community by offering attractive emoluments to the poor families.

The Lambadis eat all kinds of meat including fouls and pork. *Gudumba* is their favourite drink which is distilled from jaggery though they drink toddy from wild date palm (*Phoenix sylvestris*; eatha), Palmyra palm (*Borassus flabellifer*; thaati) and fishtail palm (*Caryota urens*) 'giraka thaati' in Warangal and 'jiluga' in Khammam. The Lambadis better utilize all opportunities for their socio-economic development. Some of them have taken to pastoralism, agriculture and various types of labour, while their women collect forest produce (NTFPs), sell fire wood, etc. The literates invariably migrate to cities and are better employed.

Fig. 6 The Lambadi women on their way to collect tendu leaves (*Diospyros melanoxylon*) in traditional dresses

Chenchus

These are aboriginal tribes and native of Nallamalais, in Mahabubnagar district of southern Telangana. They are migrated and settled in a few pockets of Vikarabad forest region of Ranga Reddi and Regonda mandal of Jayashankar Bhupalpally district. The Chenchus are hunter gatherers. They have good skills in collecting the honey and other forest produce like tendu leaves, *Litsea* bark, bamboo culms, amla fruits, etc.

Boyas

The Boya is one of the primitive tribes in India. Boyas are the descendants of Maharshi Valmiki, who wrote 'Ramayana', an ancient Indian epic. These are primitive, economically poor, have no specific profession and inhabitants of agencies and platues. This tribe was under Scheduled Tribe category before 1976; later they were included in BC-A. The present Telangana government is planning to include this tribe into Scheduled Tribe category again.

Mathura Lambadis

The tribe Mathura Lambadis is confined to only a few villages of erstwhile Adilabad and Nizamabad districts. These are nomadic people, their settlements are known as 'Thandas' and hereditary leader is 'Naik'. These people are almost Hindus and worship Lord Krishna. They are very fare looking and cleaner in habits than other communities and also wear sacred thread. They cook their own food and do not allow any other community to offer or cook their food. They are very specific in that culture. They do not eat flesh.

Nakkala

In 2003, the Nakkala community was included in the Scheduled Tribe. The tribe is a wandering community, nomadic hunters and traders. They hunt small animals and birds for livelihood. These people are also known as 'dove' catchers (guvvalavallu), 'bird' catchers (pittalollu), 'oosikora' (needle sellers) and shikaris (hunters). The women play a significant role in all the economic activities of the family. Their dialect is 'Nakkala basha', with no script.

Dhulia

In Telangana, the Dhulia community is a Scheduled Tribe since 2003. The word 'Dhulia' was derived from 'Dhuli', which means 'dust'. Dhulias found sporadically in Mancherial and Kumaram Bheem Asifabad districts. They are either farmers or agricultural labourers. They also collect forest produce from the surrounding forest areas for their day-to- day survival. Their deity is 'Nisani Devanta', which is a small boulder installed under a huge tree outside the village.

Gothi Koya

Gothi Koyas are migrants from Chhattisgarh. They have started occupying the forest lands and made their permanent settlement by clearing forest lands majorly in the northern part of Telangana. Primarily, they are agricultural labourers, hunters and NTFP gatherers. Gothi Koyas are settling in the forest habitations of the region along the southern bank of river Godavari. Their presence is perceived as a threat to the forest vegetation and wild life. They are indulging the podu cultivation (Fig. 7).

Fig. 7 Gothi Koya community at a hamlet and podu cultivation by Gothi Koya tribe in nearby forest areas

Fig. 8 The huts of Raj Koya tribe at Dudekulapalli of Jayashankar Bhupalpally district

Raj Koya

Raj Koyas are migrants from the neighbouring Chhattisgarh state and are rehabilitated in some hamlets of Jayashankar Bhupalpally and Bhadradri Kothagudem districts. They are hunters and collect non-timber forest products for their survival; now they are sustaining agricultural coolies without any facilities. These people have come predominantly for the land and labour work. They live in forest fringes and very far from other villages (Fig. 8). They walk for kilometres to get water for their domestic use, and there are no electricity facilities in their villages. The hamlets are known as *Raj Koya gudems*.

Forests

Telangana State has a forest cover of 20,419 km^2, which is 18.22% of its geographical area (ISFR, 2017). The erstwhile districts in Telangana with good forest cover are Adilabad (35.32%; 5688 sq km), Khammam (33.42%; 4433 sq km) and Warangal (22.72%; 2918 sq km). These occupy 64% of the total forest cover of the state (ISFR 2017). The forests in northern Telangana are largely of tropical dry deciduous type, with teak dominating and forming pure stands in Adilabad district which extend to south and southeastern part where it forms *Tectona-Terminalia* transition zone and then *Terminalia-Hardwickia* association, further to *Madhuca-Terminalia-Cleistanthus* zone along the river Godavari towards the east (Raju et al. 2014).

Prior Informed Content (PIC)

The study of traditional knowledge usually relies on interactions with a few members of the community to represent the knowledge held by the entire community. Many ethnobotanists adopt the cultural consensus model developed by Romney et al. (1986). Although this model has some limitations, it devalues differences in knowledge and focuses on consensus. Prior informed content (PIC) is an important concern underlying the philosophical and political environment of bioprospecting research involving indigenous societies (Rosenthal 2006). In compliance with the convention on biological diversity (CBD), the PIC should establish for relevant discussions with the tribal communities regarding the intended use of their knowledge in any form. The mutual understanding was that any benefit derived from research pursued after their medicinal plant knowledge shall belong to them. In India, there is a general and urgent need to pattern the process of PIC after educating the local ethnic groups in different regions of IPR, as per a standard scale. The scientific approach towards the collection of ethnomedicinal data from the forest inhabitants was initiated through University Grants Commission. Furthermore, the majority of the local people in India are of the opinion that the knowledge could be displayed in full text for non-commercial and academic purposes (Singh 2008; Suthari et al. 2014a, b).

Wildlife Sanctuaries

There are a good number of wildlife sanctuaries in the study area created by the then Nizam Government and then the erstwhile Government of Andhra Pradesh, for on-site conservation of wildlife (Table 4). To mention the major ones, district-wise:

Table 4 Information on wildlife sanctuaries in Telangana State, India

District/s covered	Sanctuary name	Area (km^2)
Mancherial	Kawal	893
Mancherial	Pranahita	136
Mancherial	Sivaram	38.66
Jayashankar Bhupalpally	Eturnagaram	803
Mahabubabad	Pakhal	860
Bhadradri Kothagudem	Kinnerasani	635.41
Medak and Nizamabad	Pocharam	129.85
Sangareddy	Manjeera (Crocodile)	20
Khammam, East and West Godavari	Papikondalu	591
Mahabubnagar, Nalgonda, Guntur, Kurnool and Prakasam	NSTR[a]	3568

[a]*NSTR* Nagarjunasagar-Srisailam Tiger Reserve

Ethnic Tribes: Health Care and Status

Literacy rate is an indicator of development achieved by a community in a region, which indicates the health condition, social status and hygiene. According to the Census 2011, Jogulamba Gadwal district bears the lowest literacy rate, i.e. 49.87%, followed by Nagarkurnool (54.38%) and Wanaparthy (55.67%) where the highest rate was shown by Hyderabad district with 83.25, immediately followed by Medchal (82.49%) and Warangal Urban (76.17%). Health status plays an important role in the development of a society or community. When compared to the towns, the tribal people have resistance to many ailments. The ethnic or aboriginal people have their thoughts generated from nature and way of living. They have nothing to deal with teaching or textbook procedures. The cognitive ability of the ethnic groups is different. It is connected to thinking or mental process. Even within the same tribe (intra-cultural), how the cognizance related to utilization of plant resource varies with the local, forest around and level of economic development was demonstrated for the first time in India by Suthari et al. (2014a), with Koyas of Eturunagarm wildlife sanctuary as the case study. The primitive tribes of Telangana usually inhabited in forest fringes, and these people are safe from some health problems like high blood pressure and diabetes due to their food habits compared to urban people. These tribes do not have proper transport facilities to check up their health condition even today. They are in very pathetic condition in health status due to unavailability of minimum facilities and lack of awareness.

During rainy season, in agency areas the floods do not allow them to travel to nearby towns even in emergency time also. Primarily, most of the pregnant ladies and the persons who are suffering from fever or some other diseases have lead them to lost their lives only because of lacking hospital facilities (Fig. 9). Even today, the government medical services are unable to meet the minimum health needs of the tribal people. Poorly equipped facilities, appearance of old buildings, poor performance and non-availability of qualified doctors in remote areas during day and night led to reduce the quality of treatment.

Fig. 9 (**a**) The tribal people carrying ill person on a cot during rainy season due to unavailability of transport in remote areas. (**b**) The dead fatal baby, a common phenomenon in the region due to lack of proper medication

For typical diseases or in serious conditions, the tribal people used to approach nearby government or private hospitals; otherwise, they depend on traditional practitioners and local healers for common ailments. In Telangana, the ethnic practitioners and local healers greatly depend on plant resources to cure diseases from nearby forest areas. They used to collect plant parts such as roots/rhizomes of *Rauvolfia serpentina* and *Gloriosa superba*; leaves of *Gymnema sylvestre*, *Moringa concanensis*, etc.; fruits of *Aegle marmelos*, *Balanites roxburghii*, *Terminalia bellirica*, *Terminalia chebula* and *Phyllanthus emblica*; seeds of *Strychnos nux-vomica*, *Strychnos potatorum* and *Abrus precatorius*; whole plants of *Andrographis paniculata*, *Anodendron paniculatum* and *Centella asiatica*; and bark of *Litsea glutinosa*, *Bombax ceiba*, *Lannea coromandelica*, etc. (Fig. 10). The opportunities for gathering medicinal plants for self-consumption and sale for the resident ethnic groups are immense in the erstwhile Adilabad, Karimnagar, Warangal and Khammam districts.

Ethnic Tribes: Agriculture

The main occupation of the tribes in the study area is agriculture and gathering forest produce. They are also professional cattle breeders. The tribes indulged in *podu* cultivation. Now, they are largely agricultural labourers and receive supplementary income through gathering non-timber forest produce (NTFP). On the other, the modern agriculture has its profound impact on the use of inputs and price rise leading to farmer suicides, while the traditional agriculture was remunerative, cost-effective and with no adverse effects on the fertility of the soil and the health of the consumers. They cultivate majorly cotton, maize, green gram, chickpea, ground nut, jowar and cow pea due to insufficient irrigation facilities. Most of the ethnic people are illiterate, landless and marginal agricultural coolies. The local people are not well equipped to do agriculture because of their poorly developed skills and restricted access to the resources required.

Fig. 10 Ethnomedicinal plants used by the tribes with high importance. (**a**) *Abrus precatorius* for insect bite; (**b**) *Andrographis paniculata* leaves for ephemeral fever; (**c**) *Centella asiatica* for memory boosting; (**d**) *Anodendron paniculatum* for bone fracture; (**e**) *Paederia foetida* leaves for dysentery; and (**f**) *Gloriosa superba* tubers for snake bite

Ethnic Tribes: Beliefs

Basically the tribes worship god in nature. They offer either vegetables, fruits or animals, which depend on the specific deity type. The tribes celebrate many occasions connected manly with the agricultural seasons and birth-death ceremonies. There are about 20 tribal communities that celebrate the occasions of their own which is purely dependent on their community and beliefs. Each community has its own identity and celebrations. For example, the Koya tribe could not cut mahuwa (*Madhuca longifolia* var. *latifolia*); they treat that tree as a sacred one, which provides fresh corolla, fruits, etc. for their sustenance. Ancestor worship is prevalent among the community members. The dead are buried (Ramarao and Henry 1996). Sickness, illness and other situations of travail and tribulations are attributed to the work of angered malevolent deities. To identify the trouble-making deity, the services of an established spirit medium are commissioned. Diagnosis and treatment of sickness involve a series of magico-religious practices besides administering herbal medicine. While spirit medium is called *Manthragadu*, the local doctor is known as *Vejju*.

Ethnic Tribe: Sacred Groves

The cultural, ancestral and magico-religious beliefs have great influence on the availability of a species, as well as the vegetation of a particular area. A sacred grove is a protected area where the natural vegetation patch is present and is protected by the local communities considering it as the abode of their ancestral deities or spirits. The IUCN considers the sacred groves as sacred natural sites (SNSs), classified as 'the natural areas of special spiritual significance to people and communities which include the natural areas recognized as sacred by the indigenous and traditional people as well as natural areas recognized by institutionalized religious faiths as the places of worship and remembrance'. The WWF-Telangana has recorded 65 sacred groves in the state as of erstwhile district-wise, namely, Adilabad 2, Nizamabad 7, Karimnagar 4, Warangal 3, Khammam 4, Medak 4, Hyderabad 13, Nalgonda 9, Ranga Reddy 10 and Mahabubnagar 9 (Suthari et al. 2016), but there are hundreds of sacred groves which are available and protected in the state where natural vegetations with rare and threatened species are highly protected by the local people in the name of beliefs in their deities and spirits (Fig. 11). Very important plant taxa such as *Anodendron paniculatum*, *Paederia foetida*, *Flagellaria indica* and *Dysolobium pilosum* are seen only inside the Mallur sacred grove where there is no single record of availability of these taxa in the study area. Some of the species like *Madhuca longifolia* var. *latifolia*, *Phyllanthus emblica*, *Terminalia chebula* and *Ficus benghalensis* are common within the sacred grove, but we can rarely observe outside.

Fig. 11 Sacred groves in the forests: places for the worship by the ethnic tribes

Ethnic Tribes: Politics

The constitution of India made special provision for political participation of STs, their social and economic welfare. The tribes have 6% reservations in Telangana State. The current state government is planning to hike reservations from 6 to 12% as per the population of STs through TS Reservation Act, 2017. The primitive tribes of Telangana are illiterate, very poor and innocent compared to exotic tribes. The provided reservations were fully occupied by the migrated tribes. The Lambadis are predominantly utilizing all the opportunities in many aspects.

Ethnic Tribes: Associated Festivals and Fairs

Like other religious fests, the tribal groups have also their own identical and ancestral fairs and festivals per year. Each community has their own festivals and fairs. They celebrate according to their rituals, deities and beliefs (Fig. 12). There are many fairs and festivals celebrated by the tribal people in Telangana, but there are two major tribal festivals have been celebrating in Telangana, i.e. Sri Sammakka-Sarakka Jatara and Nagoba Jatara in Tadvai and Indravelli mandals, respectively (Fig. 13).

Fig. 12 A bride groom of the Gond tribe on his marriage ceremony

Fig. 13 The historic massive tribal fairs Sri Sammakka-Sarakka (**a**, **b**) and Nagoba (**c**) Jataras in Telangana

Sri Sammakka-Sarakka Jatara is the biennial tribal congregation which about 1.2 crore people attend only in 4 days of main jatara. It is the event of Koya tribe. The festival attracts tribes and nontribal people across the country and the states which include mainly Telangana, Andhra Pradesh, Chhattisgarh, Orissa, Karnataka, Maharashtra and Madhya Pradesh. It is the biggest tribal congregation in Asia and popularly known as 'South Kumbhmela'.

Nagoba Jatara is the annual gathering of Gond tribe at Keslapur of Adilabad district. The Gonds used to worship an anthril under a tree. This was initiated in 1941 and is organized by Mesram clan of Gond tribes. For the convenient of tribal people, Christopher von Furer-Haimendorf started the 'darbar' (to address the grievances of the Gonds) on the third day of Jatara in 1944.

Ethnic Groups: Economic Source from Major NTFPs

The ethnic people, who live in and around the forests, collect, use and sell non-timber forest products (NTFPs). Chandrasekharan (1995) has broadly categorized the NTFPs into food and food additives, fodder, fibril and flosses, fertilizer, herbal portions, phytochemical and aroma chemicals, oils, latex, resins and exudates, organic construction materials, decorative articles, different kinds of animal products and services such as grazing and recreation. Around 70 million ethnic people, spread all over the globe, largely depend on forests for their livelihood, of which 50 million people obtain their livelihood from forests. The New Forest Policy 1988 has clearly envisaged the need and means for involving forest-dependent communities (mainly ethnic groups) as partners in the management of the forest resources. Although 3000 NTFP species are known to be in use in India, only 126 have marketability (Maithani 1994). It is because the NTFPs are associated with traditional uses and are not widely known and often linked to subsistence. The NTFPs have been identified as one of key income sources for rural households, and the examples indicate an income share greater than that from cash crops or informal cash incomes (Dovie 2003). The diversity of NTFPs in Telangana is a rich forest resource to the ethnic people to exploit for livelihood by self-employment. The gums (karaya, tiruman and konda gogu), seeds (visha mushti, chilla, Pongamia, chinta), roots (sarpagandha) and leaves (tumiki) provide the bulk of the income, and a considerable proportion of the poor households use NTFPs for self-consumption than their wealthier counterparts. The NTFPs alone have such potential to provide good income to the local communities, provided they are harvested on a sustainable basis (Omkar et al. 2012). The NTFPs are a good economic source for Telangana, Andhra Pradesh, Chhattisgarh, Kerala and several other north-east states in India. Combined state of Andhra Pradesh earned about 10 crore rupees on yearly average from 2005–2006 to 2015–2016 (GCC 2016).

The tribes and scores of other forest-dependent people are actively involved in the collection of NTFP species like tendu leaves (*Diospyros melanoxylon*), bamboo culms (*Dendrocalamus strictus*, *Bambusa arundinacea*), gum karaya (*Firmiana simplex*), axle wood gum (*Anogeissus latiofolia*), nux-vomica seeds (*Strychnos nux-vomica*), cleaning nut seeds (*Strychnos potatorum*), mahua corolla/seeds (*Madhuca longifolia* var. *latifolia*), bastard myrobalan seeds (*Terminalia bellirica*), myrobalan seeds (*Terminalia chebula*), amla fruits (*Phyllanthus emblica*), tallow laurel bark (*Litsea glutinosa*), silk cotton gum (*Cochlospermum religiosum*), soap nut fruits (*Sapindus emarginatus*), marking nut seeds (*Semecarpus anacardium*), etc. (Table 5). Women folk were found excellent with the collection of NTFPs for the family (Omkar et al. 2008). The Girijan Cooperative Corporation Limited (GCC) in Eturnagaram Division (Telangana) alone procured NTFPs from the local people for worth rupees more than one crore from April 2014 to February 2015 (GCC 2015). Conversely, NTFPs not only have the potential for significant contribution to improve the rural livelihoods at household level but also the economy of states at macro level (Suthari et al. 2018).

Table 5 The NTFPs availability in Telangana for which Girijan Cooperative Corporation has a price index

	Scientific name	Trade name	NTFP	Grade	Price (per kg)
1	*Firmiana simplex*	Gum karaya	Gum	I	270
				II	220
				III	125
2	*Anogeissus latifolia*	Axle wood	Gum	I	80
				II	60
				III	50
3	*Cochlospermum religiosum*	Karita gum	Gum	I	200
				II	150
				III	120
4	*Rauvolfia serpentina*	Indian snakeroot	Root		150
5	*Litsea glutinosa*	Indian laurel	Stem bark		32
6	*Strychnos nux-vomica*	Poison nut	Seed		30
7	*Strychnos potatorum*	Cleaning nut	Seed		18
8	*Pongamia pinnata*	Kanuga	Seed		10
9	*Tamarindus indica*	Tamarind	Fruit	With seed	18
				Seedless	32
10	*Sapindus emarginatus*	Soap nut	Fruit		6
11	*Madhuca longifolia* var. *latifolia*	Mahuwa	Corolla		10
			Seed		14
13	*Thysanolaena maxima*	Broom grass	Shoot		18
14	*Terminalia chebula*	Karakkaya	Fruit		6
15	*Phyllanthus emblica*	Amla	Fruit		15
16	*Gardenia resinifera*	Dikamali	Resin		12
17	*Semecarpus anacardium*	Marking nut	Seed		12
18	Honey	Honey	Honey		130
19	Honey bees wax	Wax	Wax		120

Bioprospecting and Ethnobotany

Bioprospecting and ethnobotany indicate the screening of bioactive compounds or active principles from ethnomedicinally important plant for a particular purpose. For any new drug discovery or biochemical resources, formulations, generally the scientists rely on the guidance of ethnic or local people, who have practical knowledge traditionally for a particular disease or desired activity.

Traditional and ancient knowledge about natural resources still exists in many locations of the study area. The knowledge among individuals could pass by oral from generation to generation, or acquired by selected persons and requires systematic documentation. The knowledge of tribes about plants around them and their usage can be studied by personal observation and close association with the primitive tribes. The help of reliable old people and medicine men of the particular area is taken in collection of information about parts of the plant of economic value, mode of collection, processing, preparation and administration of drug, dosages, duration, etc.

Conclusion

Telangana region is known for its primitive tribes as well as the forest richness in southern India. The present study served the purpose of exploring Telangana State for its ethnobotany. The present paper on the review of ethnobotanical account from Telangana State is reporting 403 vascular plant taxa of 93 families under Pteridophyta and Magnoliophyta. The identified taxa with promising medicinal properties are arranged alphabetically. Each taxon is provided with the currently accepted name with vernacular name, family, distribution pattern in the study area, ethnic utility and useful part (Table 6). The study resulted in proving that the ethnic people in the state have different cognitive abilities and perspectives of the natural resources around and how to use them need-based. Of the ethnic tribes, the Koyas and Konda Reddis are found to possess more traditional knowledge about the plant kinds, distribution and their uses, followed by Gonds and the related sects. The study also brought to light the similar use of different medicinal plants for the treatment of various ailments used by the ethnic people (Table 7) and their dependence on forests and the need for protecting the local ecosystems for their services to the needy. Furthermore, the area under forests is steadily decreasing on one hand, and the plant invasions and exotic tree plantations on the other are eating away the economic opportunities of the local people. There is a need to document the pattern of plant resource utilization in traditional ways for subsistence and economic development of the poor and largely not so literate local people.

Acknowledgements Sateesh Suthari is grateful to the Science and Engineering Research Board (SERB-DST) for financial support through Start-Up Research Grant (Young Scientists) (No. SB/YS/LS-70/2014 dated March 11, 2015). The authors are obliged to the Heads of Department of Plant Sciences, University of Hyderabad, Hyderabad, and Department of Botany, Kakatiya University, Warangal, for providing facilities.

Table 6 List of some promising and interesting medicinal plants used by the ethnic tribal groups from Telangana State along with their vernacular name, family, growth form, distribution pattern, utility pattern and useful parts to cure various ailments

	Botanical name	Local name	Family	Growth form	Distribution	Ethnic use (part/s useful)
1	*Abelmoschus moschatus*	Kasturi benda	Malvaceae	Herb/I	Wild	Abdominal pain (F)
2	*Abrus precatorius*	Gurivinda	Fabaceae	Climber/I	Wild	Anthrax, insect bite, snake bite (L, Wh)
3	*Abutilon indicum*	Tutturu benda	Malvaceae	Herb/I	Wild	Dysentery, helminthiasis, insect bite (L)
4	*Acacia chundra*	Sandra	Fabaceae	Tree/I	Wild	Asthma, fever, trypanosomiasis, wound healing (Sb)
5	*Acacia farnesiana*	Muriki tumma	Fabaceae	Shrub/I	Wild	Dog bite (F)
6	*Acacia leucophloea*	Tella tumma	Fabaceae	Tree/I	Wild	Boils, ephemeral fever, wound healing (Sb)
7	*Acacia nilotica*	Nalla tumma	Fabaceae	Tree/Intr	Wild	Burns (Sb)
8	*Acacia pennata*	Korinda	Fabaceae	Shrub/I	Wild	Fits (Sb)
9	*Acacia torta*	Konda korinda	Fabaceae	Shrub/I	Wild	Labour pain (Sb)
10	*Acalypha indica*	Muripinda	Euphorbiaceae	Herb/I	Wild	Maggot-infected sores, skin diseases (L)
11	*Acalypha lanceolata*	Penta puti	Euphorbiaceae	Herb/I	Wild	Laxative (L)
12	*Achyranthes aspera*	Uttareni	Amaranthaceae	Herb/I	Wild	Boils, insect bite, snake bite, wound healing (L, R, Wh)
13	*Acmella paniculata*	Chinna akkala karra	Asteraceae	Herb/E	Natur	Toothache (Infl)
14	*Acorus calamus*	Vasa	Acoraceae	Herb/I	Cult	Stomachache (Rh)
15	*Adenostemma lavenia*	Adavi jilakara chettu	Asteraceae	Herb/I	Wild	Wound healing (Wh)
16	*Adiantum lunulatum*	Hamsapadi	Pteridaceae	Herb/I	Wild	Dysentery, epilepsy (Wh)
17	*Aegle marmelos*	Maredu	Rutaceae	Tree/I	Wild	Corneal opacity, dysentery, impaction (F, L)
18	*Aerva lanata*	Pindi kura	Amaranthaceae	Herb/I	Wild	Earache (L)
19	*Aerva scandens*	Konda pindi	Amaranthaceae	Herb/I	Wild	Dysentery (L)
20	*Agave americana*	Saga nara	Asparagaceae	Herb/E	Planted	Ephemeral fever (L)
21	*Ageratum conyzoides*	Gabbu chettu	Asteraceae	Herb/E	Natur	Rheumatism (L)

22	*Ailanthus excelsa*	Peddamanu	Simaroubaceae	Tree/E	Planted/R_ wild	Piles, tympany (Sb)
23	*Alangium salviifolium*	Oodugu	Cornaceae	Tree/I	Wild	Bone fracture, oedema, purgative, snake bite (Sb, R)
24	*Albizia procera*	Tella chinduga	Fabaceae	Tree/I	Wild	Rheumatism (Sb)
25	*Aloe vera*	Kala banda	Xanthorrhoeaceae	Herb/I	Planted	Insect bite, piles (L)
26	*Amaranthus spinosus*	Mulla thotakura	Amaranthaceae	Herb/E	R_wild	Appetite stimulant (L)
27	*Ammannia baccifera*	Agni vednapaku	Amaranthaceae	Herb/I	Wild	Skin diseases (L)
28	*Ampelocissus latifolia*	Dobba teega	Vitaceae	Climber/I	Wild	Anorexia (S)
29	*Ampelocissus tomentosa*	Atukula baddu	Vitaceae	Shrub/I	Wild	Anthrax, bone fracture, insect bite (L, R)
30	*Andrographis paniculata*	Nela vemu	Acanthaceae	Herb/I	Wild	Dyspepsia/indigestion, ephemeral fever, snake bite (L, R)
31	*Anisomeles indica*	Dayyam marri	Lamiaceae	Herb/I	Wild	Ephemeral fever, snake bite (L)
32	*Anisomeles malabarica*	Dayyam chettu	Lamiaceae	Herb/I	Wild	Fever (Wh)
33	*Annona squamosa*	Sheetaphalam	Annonaceae	Shrub/E	Planted/R_ wild	Ephemeral fever, snake bite (L)
34	*Anodendron paniculatum*	Atukudu teega	Apocynaceae	Climber/I	Wild	Bone fracture (Wh)
35	*Anogeissus latifolia*	Thiruman	Combretaceae	Tree/I	Wild	Anthrax, asthma, insect bite (Sb)
36	*Argemone mexicana*	Brahma dandi	Papaveraceae	Tree/E	Wild	Skin disease (Lx)
37	*Argyreia nervosa*	Chandra podi	Convolvulaceae	Climber/I	Wild	Skin disease, tympany (L)
38	*Aristolochia bracteolata*	Gadida garapa	Aristolochiaceae	Herb/I	Wild	Anthrax, insect bite, purgative, snake bite, wound healing (R, L)
39	*Aristolochia indica*	Nalla usiri	Aristolochiaceae	Climber/I	Wild	Earache, hemiplegia (partial paralysis), snake bite (R)
40	*Artabotrys hexapetalus*	Muddasampenga	Annonaceae	Climber/I	Cult	Fertility (L)

(continued)

Table 6 (continued)

	Botanical name	Local name	Family	Growth form	Distribution	Ethnic use (part/s useful)
41	*Asparagus gonoclados*	Guddelugu bochu	Asparagaceae	Climber/I	Wild	Aphrodisiac, skin disease, snake bite (T)
42	*Asparagus racemosus*	Ellamma gaddalu	Asparagaceae	Climber/I	Wild/cult	Anthrax, aphrodisiac, dysentery, galactogogue, insect bite (T, Sh)
43	*Asystasia gangetica*	Lavana valli	Acanthaceae	Herb/I	Wild	Rheumatism (Wh)
44	*Atalantia monophylla*	Adavi nimma	Rutaceae	Tree/I	Wild	Rheumatism (Sb)
45	*Azadirachta indica*	Vepa	Meliaceae	Tree/E	Planted/R_ wild	Ephemeral fever (Sb, Sh)
46	*Azima tetracantha*	Tella uppi	Salvadoraceae	Shrub/I	Wild	Infant diseases, rheumatism (Sb)
47	*Baccharoides schimperi*	Davijilakara	Asteraceae	Herb/I	Wild	Wound healing (L)
48	*Bacopa monnieri*	Brahmi	Plantaginaceae	Herb/I	Wild	Tonic (Wh)
49	*Balanites roxburghii*	Gare	Zygophyllaceae	Tree/I	Wild	Corneal opacity, ephemeral fever, Pertusis (Fl, F)
50	*Baliospermum solanifolium*	Danthi	Euphorbiaceae	Shrub/I	Wild	Purgative (R)
51	*Barleria montana*	Adavi dishambaralu	Acanthaceae	Herb/I	Wild	Skin disease (L)
52	*Barleria prionitis*	Mulla gorinta	Acanthaceae	Herb/I	Wild	Scorpion sting (R)
53	*Barringtonia acutangula*	Nir kanki	Lecythidaceae	Tree/I	Wild	Rheumatism (Sb)
54	*Bauhinia malabarica*	Puli are	Fabaceae	Tree/I	Wild	Ephemeral fever (Sb)
55	*Bauhinia purpurea*	Deva kanchanam	Fabaceae	Tree/I	Wild	Diarrhoea (Sb)
56	*Bauhinia racemosa*	Aare/aari	Fabaceae	Tree/I	Wild	Dysentery (Sb, Fl)
57	*Bauhinia semla*	Goddeti are	Fabaceae	Tree/I	Wild	Neck pain, tonic (Sb)
58	*Bauhinia vahlii*	Addaku	Fabaceae	Tree/I	Wild	Dyspepsia/indigestion (Sd)
59	*Biophytum sensitivum*	Atti patti	Oxalidaceae	Herb/I	Wild	Blisters, boils, cuts (L)
60	*Blepharis maderaspatensis*	Nela pariki	Acanthaceae	Climber/I	Wild	Purgative (L)
61	*Blepharis repens*	Nela pariki	Acanthaceae	Herb/I	Wild	Heat and summer stroke (R)
62	*Blumea axillaris*	Kukka pogaku	Asteraceae	Herb/E	Wild	Cough, cuts (L)
63	*Blumea bifoliata*	Kukka pogaku	Asteraceae	Herb/E	Wild	Cough (L)
64	*Blumea lacera*		Asteraceae	Herb/E	Wild	Blisters (L)

65	*Blumea obliqua*		Asteraceae	Herb/E	Wild	Wound healing (L)
66	*Blumea virens*	Adavi pogaku	Asteraceae	Herb/E	Wild	Skin disease (L)
67	*Boerhavia diffusa*	Atikamamidi	Nyctaginaceae	Herb/E	Wild	Trypanosomiasis (L)
68	*Boerhavia erecta*	Punarnava	Nyctaginaceae	Herb/I	Wild	Tonic (Wh)
69	*Bombax ceiba*	Buruga	Malvaceae	Tree/I	Wild	Anthrax, dysentery, fertility (Sb, Sd)
70	*Boswellia serrata*	Anduga	Burseraceae	Tree/I	Wild	Dog bite, rheumatism, scorpion sting (Sb)
71	*Breynia retusa*	Tella pulicheru	Euphorbiaceae	Shrub/I	Wild	Aphrodisiac, cough (L)
72	*Bridelia montana*	Panchotkam	Euphorbiaceae	Tree/I	Wild	Blisters, boils, cuts (L, Sb)
73	*Bryophyllum pinnatum*	Ranapala	Crassulaceae	Herb/E	Planted	Snake bite (R)
74	*Buchanania axillaris*	Pedda morri	Anacardiaceae	Tree/I	Wild	Wound healing (Fl)
75	*Buchanania cochinchinensis*	Chinna morli	Anacardiaceae	Tree/I	Wild	Chest pain (Fl, F)
76	*Butea superba*	Teega moduga	Fabaceae	Climber/I	Wild	Labour pains (Fl)
77	*Byttneria herbacea*	Erra teega	Malvaceae	Herb/I	Wild	Anthrax, dysentery, impaction (L)
78	*Cadaba fruticosa*	Vutharasi chettu	Capparaceae	Shrub/I	Wild	Blisters, boils, cuts (Sb)
79	*Caesalpinia bonduc*	Gatchikai	Fabaceae	Climber/I	Wild	Ephemeral fever, hydrocele, rheumatism (Sd, L)
80	*Cajanus cajan*	Kandulu	Fabaceae	Shrub/I	Cult	Contraceptive (Sd)
81	*Calotropis gigantea*	Tella jilledu	Apocynaceae	Shrub/E	Natur	Constipation, earache, fever, rheumatism, snake bite (Fl, Lx)
82	*Canavalia cathartica*	Adavi chemma	Fabaceae	Climber/I	Wild	Sexually transmitted diseases, scorpion sting (R, S)
83	*Canavalia gladiata*	Advi chemma	Fabaceae	Climber/I	Wild	Rheumatism (L)
84	*Canthium coromandelicum*	Balusu	Rubiaceae	Shrub/I	Wild	Insect bite (Sb, F)
85	*Capparis divaricata*	Nalla uppi	Capparaceae	Climber/I	Wild	Trypanosomiasis (Sb, F)
86	*Capparis sepiaria*	Uppi	Capparaceae	Climber/I	Wild	Anthrax, contraceptive, rheumatism, skin disease (Sb, F)
87	*Capparis zeylanica*	Adonda	Capparaceae	Shrub/I	Wild	Diabetes, dyspepsia/indigestion, impaction (Sb, F)

(continued)

Table 6 (continued)

	Botanical name	Local name	Family	Growth form	Distribution	Ethnic use (part/s useful)
88	*Caralluma umbellata*	Kundeti kommu	Apocynaceae	Herb/I	Wild	Burns (Sh)
89	*Cardiospermum halicacabum*	Budda kakara	Sapindaceae	Climber/I	Wild	Ephemeral fever, hydrocele (L, R)
90	*Careya arborea*	Budda darmi	Lecythidaceae	Tree/I	Wild	Labour pains, snake bite (Fl)
91	*Casearia elliptica*	Kanusiri	Salicaceae	Shrub/I	Wild	Corneal opacity, ephemeral fever (Sb, L, F)
92	*Cassia fistula*	Rela	Fabaceae	Tree/I	Wild	Tympany (L)
93	*Cassytha filiformis*	Paachi teega	Lauraceae	Climber/I	Wild	Bone fracture (Wh)
94	*Cayratia pedata*	Edakula mandulamari	Vitaceae	Climber/I	Wild	Sexually transmitted diseases (L)
95	*Celastrus paniculatus*	Malleru teega	Celastraceae	Climber/I	Wild	Abortion, snake bite (Sb)
96	*Celosia argentea*	Gunugu	Amaranthaceae	Herb/E	Natur	Galactogogue (L)
97	*Centella asiatica*	Saraswathi aku	Apiaceae	Herb/I	Wild	Memory booster (L)
98	*Ceratopteris thalictroides*	Medha	Pteridaceae	Herb/I	Wild	Memory (Wh)
99	*Ceriscoides turgida*	Tella elaka	Rubiaceae	Tree/I	Wild	Diarrhoea, leucorrhoea, piles, tympany (Sb)
100	*Ceropegia hirsuta*	Dusari manda	Apocynaceae	Climber/I	Wild	Diarrhoea (T)
101	*Ceropegia juncea*	Bella gadda	Apocynaceae	Herb/I	Wild	Galactogogue (T)
102	*Chamaesyce hirta*	Reddivari nanubalu	Euphorbiaceae	Herb/I	Wild	Blisters, boils, cuts, skin disease (Wh)
103	*Chamaesyce thymifolia*	Reddivari nanubalu	Euphorbiaceae	Herb/E	Planted	Laxative (L)
104	*Cheilocostus speciosus*	Kepu kanda	Costaceae	Shrub/I	Wild	Asthma (Rh)
105	*Chlorophytum tuberosum*	Sarala pagada	Asparagaceae	Herb/I	Wild	Aphrodisiac (T)
106	*Chloroxylon swietenia*	Billudu	Rutaceae	Tree/I	Wild	Neck pain, shivering (Sb)
107	*Chromolaena odorata*	Juttu puvvu	Asteraceae	Shrub/E	Wild	Wound healing (L)
108	*Cinnamomum zeylanicum*	Dalchina chekka	Lauraceae	Tree/I	Wild	Red leucorrhoea (Sb)
109	*Cissampelos pareira*	Boddi	Menispermaceae	Climber/I	Wild	Snake bite, stomachache, tonic (R)
110	*Cissus adnata*	Kakkitayaralu	Vitaceae	Climber/I	Wild	Wound healing (T)
111	*Cissus quadrangularis*	Nalleda	Vitaceae	Climber/I	Wild	Anorexia, asthma, bone fracture, helminthiasis (Sb, L)

112	*Cissus vitiginea*	Adavi draksha	Vitaceae	Climber/I	Wild	Stomachache (L)
113	*Cleistanthus collinus*	Nalla kodisha	Euphorbiaceae	Tree/I	Wild	Blisters, boils, wound healing (Sb, L)
114	*Cleome gynandra*	Vaminta	Cleomaceae	Herb/I	Wild	Head-ache (L)
115	*Cleome monophylla*	Gaddi vaminta	Cleomaceae	Herb/I	Wild	Skin disease (L)
116	*Cleome viscosa*	Kukka vaminta	Cleomaceae	Herb/I	Wild	Blisters, boils, skin disease, wound healing (L)
117	*Clerodendrum phlomidis*	Takkali	Lamiaceae	Shrub/I	Wild	Rheumatism (L)
118	*Clitoria ternatea*	Gantena	Fabaceae	Climber/I	Planted/R_ wild	Dysentery (L)
119	*Coccinia grandis*	Kaki donda	Cucurbitaceae	Climber/I	Wild/cult	Blisters, boils, cuts, dysentery, tympany (L, F)
120	*Cocculus hirsutus*	Dusara teega	Menispermaceae	Climber/I	Wild	Epistaxis, red leucorrhoea, urinary problems (R)
121	*Cochlospermum religiosum*	Konda gogu	Bixaceae	Tree/I	Wild	Piles (L)
122	*Coldenia procumbens*	Cheppu thattaku	Boraginaceae	Herb/I	Wild	Wound healing (Wh)
123	*Combretum latifolium*	Yada teega	Combretaceae	Climber/I	Wild	Diarrhoea (L)
124	*Commelina benghalensis*	Yennadari	Commelinaceae	Herb/I	Wild	Helminthiasis (Sb)
125	*Corallocarpus epigaeus*	Naga donda	Cucurbitaceae	Climber/I	Wild	Diabetes, snake bite (T)
126	*Crateva magna*	Uskia tammidi	Capparaceae	Tree/I	Wild/Cult	Tympany (Fl)
127	*Crinum asiaticum*	Penjari gadda	Amaryllidaceae	Herb/I	Wild	Snake bite, wound healing (Sb, T)
128	*Crotalaria albida*	Adavi janumu	Fabaceae	Herb/I	Wild	Labour pains (R)
129	*Crotalaria medicaginea*	Konda janumu	Fabaceae	Herb/I	Wild	Rheumatism (R)
130	*Crotalaria verrucosa*	Tella usiri	Fabaceae	Herb/I	Wild	Ephemeral fever, fits, insect bite, snake bite (L, R)
131	*Cryptolepis dubia*	Adavi pala teega	Apocynaceae	Climber/I	Wild	Galactogogue, skin diseases (L, Lx)
132	*Ctenolepis garcinii*	Gargoo	Cucurbitaceae	Climber/I	Wild	Throat disorders (Wh)
133	*Curculigo orchioides*	Nela thadi	Hypoxidaceae	Herb/I	Wild	Aphrodisiac, ephemeral fever, Galactogogue, trypanosomiasis (T)
134	*Curcuma longa*	Pasupu	Zingiberaceae	Herb/I	Cult	Antiseptic, blisters, bone fracture, cuts (T)
135	*Curcuma pseudomontana*	Adavi pasupu	Zingiberaceae	Herb/I	Wild	Anthrax, wound healing (T)

(continued)

Table 6 (continued)

	Botanical name	Local name	Family	Growth form	Distribution	Ethnic use (part/s useful)
136	*Cuscuta chinensis*	Pashi teega	Convolvulaceae	Climber/E	Natur	Galactogogue (Wh)
137	*Cuscuta reflexa*	Lanja savaram	Convolvulaceae	Climber/E	Natur	Purgative (Wh)
138	*Cyanthillium cinereum*	Sahadevi	Asteraceae	Herb/E	Natur	Fever (R)
139	*Cynodon dactylon*	Garika	Poaceae	Herb/I	Wild	Trypanosomiasis (Wh)
140	*Cyperus rotundus*	Tunga	Cyperaceae	Herb/I	Wild	Ephemeral fever (T)
141	*Cyphostemma setosum*	Barre bachali	Vitaceae	Climber/I	Wild	Dysentery, rheumatism (Wh)
142	*Dalbergia lanceolaria* ssp. *paniculata*	Patchari	Fabaceae	Tree/I	Wild	Baldness (Sb)
143	*Dalbergia latifolia*	Jitregi	Fabaceae	Tree/I	Wild	Stomachache (Sb)
144	*Dalbergia volubilis*	Teega patchari	Fabaceae	Climber/I	Wild	Skin disease (Sb)
145	*Datura metel*	Ummetta	Solanaceae	Herb/I	Wild	Snake bite (R)
146	*Dendrocalamus strictus*	Veduru	Poaceae	Tree/I	Wild	Oedema (T)
147	*Dendrophthoe falcata*	Vepa vajinika	Loranthaceae	Shrub/I	Wild	Fever, tuberculosis (L, Sb)
148	*Derris scandens*	Nalla teega	Fabaceae	Climber/I	Wild	Anthrax, impaction, trypanosomiasis (L, Sb)
149	*Desmodium gangeticum*	Konda saru	Fabaceae	Herb/I	Wild	Scorpion sting (R)
150	*Desmodium heterocarpon*	Cheppu thatta	Fabaceae	Herb/I	Wild	Cough (Wh)
151	*Desmodium velutinum*	Teega veluga	Fabaceae	Climber/I	Wild	Scorpion sting (R)
152	*Dichrostachys cinerea*	Velturu	Fabaceae	Shrub/I	Wild	Blisters, boils, cuts, rheumatism (L)
153	*Dicliptera paniculata*	Chebura	Acanthaceae	Herb/I	Wild	Skin disease (L)
154	*Dillenia pentagyna*	Revadi	Dilleniaceae	Tree/I	Wild	Anthrax, bone fracture, dysentery, rheumatism (L, F, Fl)
155	*Dioscorea alata*	Bellam gadda	Dioscoreaceae	Climber/I	Wild	Aphrodisiac (T)
156	*Dioscorea bulbifera*	Chenna gadda	Dioscoreaceae	Climber/I	Wild	Bone fracture, dysentery, dyspepsia/indigestion (T)
157	*Dioscorea pentaphylla*	Adavi ginsu teega	Dioscoreaceae	Climber/I	Wild	Cough, rheumatism (T)
158	*Diospyros chloroxylon*	Illinda	Ebenaceae	Tree/I	Wild	Anthrax, snake bite (Fl, R)

159	*Diospyros montana*	Muchi tuniki	Ebenaceae	Tree/I	Wild	Anorexia (Sb, F)
160	*Diplocyclos palmatus*	Putaka kaya	Cucurbitaceae	Climber/I	Wild	Diarrhoea, fever, snake bite (L)
161	*Dodonaea viscosa*	Puli vailu	Sapindaceae	Shrub/I	Wild	Bone fracture (Fl, Sb)
162	*Dolichandrone falcata*	Oddi	Bignoniaceae	Tree/I	Wild	Anthrax, corneal opacity (Sb, F)
163	*Dregea volubilis*	Bandi gurija	Apocynaceae	Climber/I	Wild	Neck pain, paralysis, rheumatism, snake bite, tonic (R, Wh)
164	*Drimia indica*	Adavi ulli	Asparagaceae	Herb/I	Wild	Aphrodisiac, ephemeral fever (L)
165	*Drypetes roxburghii*	Putran jivika	Euphorbiaceae	Tree/I	Wild	Cough (Sb)
166	*Ehretia laevis*	Pala danthi	Boraginaceae	Tree/I	Wild	Dyspepsia/indigestion (F)
167	*Ehretia microphylla*	Boore	Boraginaceae	Tree/I	Wild	Snake bite (R)
168	*Elephantopus scaber*	Enugu adugu	Asteraceae	Herb/I	Wild	Stomachache, wound healing (R)
169	*Eleusine coracana*	Taidalu	Poaceae	Herb/I	Cult	Dysentery (L)
170	*Elytraria acaulis*	Eddu adugu padam	Acanthaceae	Herb/I	Wild	Tonic (R)
171	*Enicostemma axillare*	Resca	Gentianaceae	Herb/I	Wild	Blisters, boils, diabetes, wound healing (Wh)
172	*Erythrina suberosa*	Tella moduga	Fabaceae	Tree/I	Wild	Leucorrhoea (Sd)
173	*Erythrina variegata*	Tella vajram	Fabaceae	Tree/I	Planted	Heat and summer stroke, impaction, trypanosomiasis (L)
174	*Erythroxylum monogynum*	Devadaru	Erythroxylaceae	Tree/I	Wild	Bone fracture (Sb, F)
175	*Euphorbia antiquorum*	Burre jemudu	Euphorbiaceae	Tree/I	Wild	Bone fracture, gout (lx, L)
176	*Euphorbia meenae*	Pala gaddalu	Euphorbiaceae	Herb/I	Wild	Blisters, boils, cuts (L)
177	*Euphorbia tirucalli*	Jemudu	Euphorbiaceae	Tree/I	Wild	Skin disease (L, Lx)
178	*Evolvulus alsinoides*	Vishnu krantha	Convolvulaceae	Herb/I	Wild	Blisters, boils, ephemeral fever, wound healing (Wh)
179	*Ficus altissima*		Moraceae	Tree/I	Wild	Wound healing (Lx)
180	*Ficus benghalensis*	Marri	Moraceae	Tree/I	Wild	Rheumatism (Lx)
181	*Ficus infectoria*		Moraceae	Tree/I	Wild	Snake bite (R)

(continued)

Table 6 (continued)

	Botanical name	Local name	Family	Growth form	Distribution	Ethnic use (part/s useful)
182	*Ficus mollis*	Banda juvvi	Moraceae	Tree/I	Wild	Wound healing (Sb)
183	*Ficus racemosa*	Medi	Moraceae	Tree/I	Wild	Bone fracture, infant disease (F)
184	*Ficus talboti*	Juvvi	Moraceae	Tree/I	Wild	Diarrhoea (Sb)
185	*Ficus virens*	Banda juvvi	Moraceae	Tree/I	Wild	Rheumatism, skin disease (Sb, L)
186	*Firmiana simplex*	Tapsi	Malvaceae	Tree/I	Wild	Menstruation pains (L)
187	*Flacourtia indica*	Kan regu	Salicaceae	Tree/I	Wild	Blisters, boils, cuts (L)
188	*Flagellaria indica*	Kokkem teega	Flagellariaceae	Climber/I	Wild	Contraceptive (Rh)
189	*Flemingia macrophylla*	Erri puvvu	Fabaceae	Shrub/I	Wild	Skin disease (R)
190	*Flemingia strobilifera*	Adavi chappa	Fabaceae	Shrub/I	Wild	Skin disease (R)
191	*Gardenia latifolia*	Pedda karinga	Rubiaceae	Tree/I	Wild	Blisters, boils, piles, wound healing (L)
192	*Garuga pinnata*	Garugu	Burseraceae	Tree/I	Wild	Leucorrhoea (Sb, F)
193	*Geodorum densiflorum*	Donthula dumpa	Orchidaceae	Herb/I	Wild	Ephemeral fever (T)
194	*Glinus lotoides*	Chatarasi kura	Molluginaceae	Herb/I	Wild	Abdominal pain (Wh)
195	*Glinus oppositifolius*	Santrasi	Molluginaceae	Herb/I	Wild	Stomachache (L)
196	*Globba marantina*	Konda pasupu	Zingiberaceae	Herb/I	Wild	Asthma, earache (T)
197	*Gloriosa superba*	Potti dumpa	Colchicaceae	Climber/I	Wild	Abortion, anthrax, insect bite, snake bite (L, T)
198	*Gmelina asiatica*	Kavva gummudu	Lamiaceae	Shrub/I	Wild	Epistaxis (L)
199	*Gomphrena celosioides*	Tella pagada banthi	Amaranthaceae	Herb/E	Planted	Cough (R)
200	*Grewia damine*	Adavi jana	Malvaceae	Shrub/I	Wild	Laxative (F)
201	*Grewia hirsuta*	Jibilika	Malvaceae	Shrub/I	Wild	Laxative (F)
202	*Grewia rothii*	Chinna jana	Malvaceae	Shrub/I	Wild	Laxative (F)
203	*Gymnema sylvestre*	Poda patri	Apocynaceae	Climber/I	Wild	Anthrax, diabetes, ephemeral fever, Galactogogue, snake bite (L, Wh)
204	*Gymnosporia emarginata*	Danthi	Celastraceae	Shrub/I	Wild	Mouth ulcers (L)
205	*Habenaria marginata*	Pasupu suddulu gadda	Orchidaceae	Herb/I	Wild	Tonic (T)

206	*Habenaria roxburghii*	Malle sudulu	Orchidaceae	Herb/I	Wild	Tonic (L)
207	*Haldina cordifolia*	Bandaru	Rubiaceae	Tree/I	Wild	Stomachache (L)
208	*Helicteres isora*	Nultada	Malvaceae	Shrub/I	Wild	Insect bite, trypanosomiasis, tympany (L, Sb)
209	*Heliotropium indicum*	Naga danthi	Boraginaceae	Herb/I	Wild	Dog bite (L)
210	*Heliotropium supinum*		Boraginaceae	Herb/I	Wild	Stomachache (S)
211	*Hemidesmus indicus*	Sugandi pala	Apocynaceae	Climber/I	Wild	Blood purifier, impaction, rheumatism (Wh, L)
212	*Hemidesmus indicus* var. *pubescens*	Sugandhi	Apocynaceae	Climber/I	Wild	Diabetes, galactogogue (Wh)
213	*Hibiscus panduriformis*	Adavi benda	Malvaceae	Shrub/I	Wild	Skin disease (R)
214	*Holarrhena pubescens*	Istari pala/ palakodisa	Apocynaceae	Tree/I	Wild	Cough, dysentery, headache (R, Sb)
215	*Holoptelea integrifolia*	Nemali nara	Ulmaceae	Tree/I	Wild	Rheumatism, skin disease (R)
216	*Holostemma ada-kodien*	Pala gurugu	Apocynaceae	Climber/I	Wild	Mouth ulcers (R)
217	*Hybanthus enneaspermus*	Nela kobbari	Violaceae	Herb/I	Wild	Aphrodisiac, urinary problems (Wh)
218	*Hydrolea zeylanica*	Langali	Hydroleaceae	Herb/I	Wild	Antiseptic (L)
219	*Hygrophila auriculata*	Neeli gorimidi	Acanthaceae	Herb/I	Wild	Aphrodisiac, oedema (L)
220	*Hymenodictyon orixense*	Chedippa	Rubiaceae	Tree/I	Wild	Bone fracture, menstrual pains (Sb)
221	*Ichnocarpus frutescens*	Nalla teega	Apocynaceae	Climber/I	Wild	Diabetes (R)
222	*Indigofera cassioides*	Karu kandi	Fabaceae	Herb/I	Wild	Scorpion sting (R)
223	*Indigofera trifoliata*	Baragadamu	Fabaceae	Herb/I	Wild	Aphrodisiac (S)
224	*Indigofera trita*	Jidi vempali	Fabaceae	Herb/I	Wild	Impaction (L)
225	*Iphigenia indica*	Kaki katuka	Colchicaceae	Herb/I	Wild	Gout (T)
226	*Ipomoea carnea*	Tuti kada	Convolvulaceae	Climber/I	Wild	Neck pain, tonic (Wh)
227	*Ipomoea eriocarpa*	Elika chevi	Convolvulaceae	Climber/I	Wild	Skin disease, stomachache (L)
228	*Ipomoea hederifolia*	Kashi ratnam	Convolvulaceae	Climber/I	Wild	Tonic (L)
229	*Ipomoea obscura*	Golla jiddaku	Convolvulaceae	Climber/I	Wild	Stomachache (L)
230	*Ipomoea turbinata*	Katla kaya	Convolvulaceae	Climber/I	Wild	Constipation (L)

(continued)

Table 6 (continued)

	Botanical name	Local name	Family	Growth form	Distribution	Ethnic use (part/s useful)
231	*Ixora arborea*	Korivi	Rubiaceae	Shrub/I	Wild	Wound healing (R)
232	*Jasminum auriculatum*	Adavi malli	Oleaceae	Climber/I	Wild	Anthrax, blisters, boils, cough, tympany, wound healing (L, Sb)
233	*Jatropha curcas*	Nepalam	Euphorbiaceae	Shrub/E	Planted	Maggot-infected sores (L)
234	*Justicia adhatoda*	Addasaram	Acanthaceae	Shrub/I	Wild	Anthrax, cough, epistaxis (L)
235	*Justicia procumbens*		Acanthaceae	Herb/I	Wild	Laxative (L)
236	*Kydia calycina*	Konda patti	Malvaceae	Tree/I	Wild	Dyspepsia/indigestion (Sb)
237	*Kyllinga odorata*	Thunga	Cyperaceae	Herb/I	Wild	Diuretic (Rh)
238	*Lagerstroemia parviflora*	Chennangi	Lythraceae	Tree/I	Wild	Blisters, boils, cuts (L)
239	*Lannea coromandelica*	Dumpidi	Anacardiaceae	Tree/I	Wild	Bone fracture, cracked heals, rheumatism (L)
240	*Lawsonia inermis*	Gorinta	Lythraceae	Shrub/I	Wild/planted	Jaundice (L)
241	*Ledebouria hyacinthoides*	Adavi tella gadda	Asparagaceae	Herb/I	Wild	Aphrodisiac, urinary problems (B)
242	*Leea asiatica*	Neerteega	Vitaceae	Shrub/I	Wild	Insect bite, skin disease, snake bite (R)
243	*Leea indica*	Velama sandi	Vitaceae	Shrub/I	Wild	Skin disease (R)
244	*Leonotis nepetifolia*	Ranabheri	Lamiaceae	Herb/E	Planted	Rheumatism (L)
245	*Leptadenia reticulata*	Mukku tummudu teega	Apocynaceae	Climber/I	Wild	Aphrodisiac (Wh)
246	*Leucas aspera*	Thummi	Lamiaceae	Herb/I	Wild	Wound healing (L)
247	*Leucas decemdentata*	Konda thummi	Lamiaceae	Herb/I	Wild	Skin disease (L)
248	*Limnophila indica*	Ambuja	Plantaginaceae	Herb/I	Wild	Antiseptic (L)
249	*Limonia acidissima*	Velaga	Rutaceae	Tree/I	Wild	Dyspepsia/indigestion (Sb)
250	*Lindernia ciliata*	Nela rampi	Linderniaceae	Herb/I	Wild	Skin disease (L)
251	*Litsea glutinosa*	Narra mamidi	Lauraceae	Tree/I	Wild	Bone fracture, labour pain, red leucorrhoea (F, Fl)
252	*Ludwigia adscendens*	Neeti bachali	Onagraceae	Herb/I	Wild	Dysentery (L)
253	*Ludwigia hyssopifolia*	Neeti lalangam	Onagraceae	Herb/I	Wild	Diarrhoea (L)

254	*Luffa aegyptiaca*	Nethi beera	Cucurbitaceae	Climber/I	Wild	Jaundice (F)
255	*Lygodium flexuosum*	Meka sannu	Lygodiaceae	Herb/I	Wild	Fertility, skin disease (L)
256	*Madhuca longifolia* var. *latifolia*	Ippa	Sapotaceae	Tree/I	Wild	Galactogogue (Sb, F)
257	*Maerua oblongifolia*	Bhu chakram	Capparaceae	Climber/I	Wild	Diabetes, fertility, snake bite (T)
258	*Mallotus philippensis*	Kunkuma chettu	Euphorbiaceae	Tree/I	Wild	Shivering (Sb, F)
259	*Mangifera indica*	Konda mamidi	Anacardiaceae	Tree/I	Wild	Blisters, boils, wound healing (Wh)
260	*Marsdenia tenacissima*	Adavi juttuku	Apocynaceae	Climber/I	Wild	Purgative, snake bite, wound healing (R)
261	*Martynia annua*	Telukondikaya chettu	Martyniaceae	Herb/E	Wild	Blisters, boils, wound healing (Fl)
262	*Melastoma malabarthricum*	Nekkarika	Melastomataceae	Shrub/I	Wild	Skin disease (L, Sb)
263	*Melia azedarach*	Turkavepa	Meliaceae	Tree/E	Planted	Rheumatism (Sb)
264	*Memecylon umbellatum*	Alli	Melastomataceae	Shrub/I	Wild	Leucorrhoea (L)
265	*Mentha spicata*	Pudina	Lamiaceae	Herb/E	Cult	Trypanosomiasis (L)
266	*Merremia aegyptiaca*	Eluka chevvu aku	Convolvulaceae	Climber/I	Wild	Jaundice (L)
267	*Merremia emarginata*	Chevi aaku	Convolvulaceae	Climber/I	Wild	Blisters, boils, cuts (Wh)
268	*Merremia hederacea*	Thalantu teega	Convolvulaceae	Climber/I	Wild	Hair tonic/shampoo (Wh)
269	*Merremia tridentata*	Sitha savaram	Convolvulaceae	Climber/I	Wild	Rheumatism (Wh)
270	*Merremia tridentata* ssp. *hastata*	Soorapu teega	Convolvulaceae	Climber/I	Wild	Toothache (Wh)
271	*Merremia turpethum*	Leenaku	Convolvulaceae	Climber/I	Wild	Earache (R)
272	*Microstachys chamaelea*	Bapana buraku	Euphorbiaceae	Herb/I	Wild	Skin disease (R)
273	*Mimosa hamata*	Magadadi	Fabaceae	Shrub/I	Wild	Aphrodisiac (Sb)
274	*Mimosa pudica*	Atti patti	Fabaceae	Herb/E	Natur	Diarrhoea, dysentery, maggot-infected sores (L)
275	*Mollugo nudicaulis*	Parpatakam	Molluginaceae	Herb/I	Wild	Boils (L)
276	*Mollugo pentaphylla*	Verri chatarasi	Molluginaceae	Herb/I	Wild	Stomachache (L)
277	*Moringa concanensis*	Yerri munaga	Moringaceae	Tree/I	Wild	Abortion, Anthrax, cough (L)

(continued)

Table 6 (continued)

	Botanical name	Local name	Family	Growth form	Distribution	Ethnic use (part/s useful)
278	*Moringa pterygosperma*	Mulaga	Moringaceae	Tree/I	Wild	Aphrodisiac, labour pains (Sb)
279	*Mucuna pruriens* var. *pruriens*	Dulagunda	Fabaceae	Climber/I	Wild	Abortion, blisters, boils, wound healing (R)
280	*Mukia maderaspatana*	Noogu dosa	Cucurbitaceae	Climber/I	Wild	Toothache (R)
281	*Naringi crenulata*	Torri elka	Rutaceae	Tree/I	Wild	Piles, snake bite (R)
282	*Nyctanthes arbor-tristis*	Parijat	Oleaceae	Tree/I	Wild/planted	Bone fracture, fits (L)
283	*Ochna obtusata*	Sonnari	Ochnaceae	Tree/I	Wild	Bone fracture, snake bite (Sb)
284	*Ocimum americanum*	Kukka tulasi	Lamiaceae	Herb/E	Natur	Wound healing (L)
285	*Ocimum basilicum*	Bhu tulsi	Lamiaceae	Herb/E	Natur	Corneal opacity, heat and summer stroke, tympany (L)
286	*Ocimum tenuiflorum*	Tulasi	Lamiaceae	Herb/E	Planted/R_ wild	Fever (L)
287	*Olax scandens*	Turaka toppe	Olacaceae	Shrub/I	Wild	Diarrhoea, stomachache (L, Fl)
288	*Oldenlandia umbellata*	Chiruveru	Rubiaceae	Herb/I	Wild	Asthma (Wh)
289	*Operculina turpethum*	Tagada	Convolvulaceae	Climber/I	Wild	Purgative (Wh)
290	*Oroxylum indicum*	Dundilam	Bignoniaceae	Tree/I	Wild	Purgative, rheumatism (Sb)
291	*Osbeckia stellata*	Adavi gulabi	Melastomataceae	Herb/I	Wild	Abortion (R)
292	*Oxalis corniculata*	Puli chinta	Oxalidaceae	Herb/E	Natur	Cracked heals (Wh)
293	*Oxystelma esculentum*	Dudipala	Apocynaceae	Climber/I	Wild	Galactogogue (Wh)
294	*Paederia foetida*	Surya bhakta	Rubiaceae	Climber/I	Wild	Diarrhoea, dysentery, rheumatism (L)
295	*Paracalyx scariosus*	Adavi tella kandi	Fabaceae	Climber/I	Wild	Skin disease (R)
296	*Passiflora foetida*	Pasi jumiki	Passifloraceae	Climber/E	Natur	Headache (L)
297	*Pavetta indica*	Papidi	Rubiaceae	Shrub/I	Wild	Anthrax, maggot-infected sores, ophthalmic diseases (Sb, L)
298	*Pavonia odorata*	Chitti benda	Malvaceae	Herb/I	Wild	Stomachache (R)
299	*Pentanema indicum*	Adavi chamanti	Asteraceae	Herb/I	Wild	Abortion (R)

300	*Pergularia daemia*	Juttupu	Apocynaceae	Climber/I	Wild	Blisters, boils, corneal opacity, gout, snake bite, stomachache, wound healing (L)
301	*Phyla nodiflora*	Bokkena	Verbenaceae	Herb/I	Wild	Stomachache (Wh)
302	*Phyllanthus amarus*	Nela usiri	Phyllanthaceae	Herb/E	Wild	Ephemeral fever, jaundice (Sb)
303	*Phyllanthus emblica*	Usiri	Phyllanthaceae	Tree/I	Wild/cult	Anorexia, Anthrax, impaction, rheumatism (L)
304	*Phyllanthus hookeri*	Erra usiri	Phyllanthaceae	Tree/I	Wild	Jaundice (Wh)
305	*Phyllanthus nozeranii*	Erra usirikee	Phyllanthaceae	Tree/I	Wild	Appetite stimulant (S)
306	*Phyllanthus reticulatus*	Nalla pulicheru	Phyllanthaceae	Shrub/I	Wild	Bone fracture, dysentery, insect bite (L)
307	*Phyllodium pulchellum*	Sarivi	Fabaceae	Shrub/I	Wild	Scorpion sting (R)
308	*Plectranthus mollis*	Nela marri	Lamiaceae	Herb/I	Wild	Wound healing (Wh)
309	*Plumbago zeylanica*	Chitra mulam	Plumbaginaceae	Shrub/I	Wild	Abortion, fits, rheumatism, skin disease, tympany (L, R)
310	*Polygala arvensis*	Chitti janumu	Polygalaceae	Herb/I	Wild	Snake bite (R)
311	*Pongamia pinnata*	Kanuga	Fabaceae	Tree/I	Wild/planted	Skin disease (Sh, L)
312	*Premna mollissima*	Nelli	Lamiaceae	Tree/I	Wild	Oedema (L)
313	*Pseudarthria viscida*	Nayaku ponna	Fabaceae	Herb/I	Wild	Rheumatism (R)
314	*Pteris multifida*		Pteridaceae	Herb/I	Wild	Dysentery (Rh)
315	*Pteris vittata*		Pteridaceae	Herb/I	Wild	Wound healing (Wh)
316	*Pterocarpus marsupium*	Peddegi	Fabaceae	Tree/I	Wild	Diabetes, fertility, labour pain (F, L)
317	*Pueraria tuberosa*	Nela gummadi	Fabaceae	Climber/I	Wild	Ephemeral fever, rheumatism (R)
318	*Pupalia lappacea*	Yerra uttareni	Amaranthaceae	Herb/I	Wild	Wound healing (L)
319	*Rauvolfia serpentina*	Sarpagandha	Apocynaceae	Herb/I	Wild	Snake bite (R)
320	*Rhynchosia minima*	Adavi kandi	Fabaceae	Climber/I	Wild	Abortion (R)
321	*Rhynchosia suaveolens*	Adavi kandi	Fabaceae	Climber/I	Wild	Dysentery (S)
322	*Ricinus communis*	Amudam	Euphorbiaceae	Shrub/E	Cult/R_wild	Gout (Sh)
323	*Rivea hypocrateriformis*	Boddi kura	Convolvulaceae	Climber/I	Wild	Purgative, snake bite (R)
324	*Rotheca serrata*	Gantu barangi	Lamiaceae	Herb/I	Wild	Skin disease (Wh)

(continued)

Table 6 (continued)

	Botanical name	Local name	Family	Growth form	Distribution	Ethnic use (part/s useful)
325	*Rotula aquatica*	Jana pamba	Boraginaceae	Shrub/I	Wild	Shivering (Sb)
326	*Sapindus emarginatus*	Kunkudu	Sapindaceae	Tree/I	Wild	Hair tonic/shampoo (F)
327	*Saraca indica*	Asoka	Fabaceae	Tree/I	Planted	Urinary problems (Sb)
328	*Sarcostemma acidum*	Konda pala	Apocynaceae	Climber/I	Wild	Bone fracture, Galactogogue, snake bite (Sb)
329	*Sarcostemma intermedium*	Soma	Apocynaceae	Climber/I	Wild	Emetic (Lx)
330	*Sarcostemma secamone*	Pala teega	Apocynaceae	Climber/I	Wild	Galactogogue (Lx)
331	*Schleichera oleosa*	Pusugu	Sapindaceae	Tree/I	Wild	Chest pain (T)
332	*Scleria corymbosa*	Ashta medha	Cyperaceae	Herb/I	Wild	Blood purifier (L)
333	*Scoparia dulcis*	Goda tulasi	Plantaginaceae	Herb/E	Natur	Menstrual pains (Wh)
334	*Secamone emetica*	Konda pala	Apocynaceae	Climber/I	Wild	Emetic (F)
335	*Selaginella bryopteris*	Pitta kalu	Selaginellaceae	Herb/I	Wild	Infant disease, leucorrhoea, memory (Wh)
336	*Selaginella involvens*	Pitta kalu	Selaginellaceae	Herb/I	Wild	Memory (Wh)
337	*Semecarpus anacardium*	Nalla jeedi	Anacardiaceae	Tree/I	Wild	Dog bite, fits (Sb)
338	*Senna absus*	Chanupala vittulu	Fabaceae	Herb/I	Wild	Skin disease (S)
339	*Senna alata*	Tamara chettu	Fabaceae	Shrub/E	Natur	Skin disease (L)
340	*Senna angustifolia*	Nela tangedu	Fabaceae	Herb/E	Natur	Constipation (L, F)
341	*Senna italica*	Nela tangedu	Fabaceae	Herb/E	Natur	Constipation (F)
342	*Senna obtusifolia*	Tagarisa	Fabaceae	Herb/E	Natur	Insect bite (L)
343	*Senna occidentalis*	Advi chennangi	Fabaceae	Herb/E	Natur	Rheumatism (L)
344	*Senna tora*	Tagirisa	Fabaceae	Herb/E	Natur	Insect bite (Sb)
345	*Sesamum alatum*	Adavi nuvvulu	Pedaliaceae	Herb/E	Natur	Blisters, boils, cuts (Wh)
346	*Sida cordata*	Gayapaku	Malvaceae	Herb/I	Wild	Tonic (R)
347	*Sida cordifolia*	Chiru benda	Malvaceae	Herb/I	Wild	Appetite stimulant (R)
348	*Sida spinosa*	Naga bala	Malvaceae	Herb/I	Wild	Fever (R)
349	*Smilax perfoliata*	Nageti dumpa	Smilacaceae	Climber/I	Wild	Abortion, aphrodisiac (T)
350	*Smilax zeylanica*	Firangi	Smilacaceae	Climber/I	Wild	Aphrodisiac, sexually transmitted diseases (R)

351	*Solanum surattense*	Tella mulaka	Solanaceae	Climber/E	Natur	Fertility (S)
352	*Solanum virginianum*	Nela mulaka	Solanaceae	Herb/E	Natur	Laxative, asthma (F)
353	*Solena amplexicaulis*	Adavi donda	Cucurbitaceae	Climber/I	Wild	Spermatorrhoea (R)
354	*Sophora velutina*	Adavi kanuga	Fabaceae	Shrub/I	Wild	Appetite stimulant (Sb)
355	*Soymida febrifuga*	Somi	Meliaceae	Tree/I	Wild	Corneal opacity, gout, shivering, snake bite, tonic (S, L)
356	*Sphaerostephanos unitus*	Hamsapadi	Thelypteridaceae	Climber/I	Wild	Scorpion sting (Wh)
357	*Stachytarpheta jamaicensis*	Eduru uttareni	Verbenaceae	Herb/E	Planted	Tonic (L)
358	*Stemona tuberosa*	Kanepu teega	Stemonaceae	Climber/I	Wild	Fever (T)
359	*Striga angustifolia*	Jonna malli	Orobanchaceae	Herb/I	Wild	Appetite stimulant (Wh)
360	*Strychnos nux-vomica*	Mushti	Loganiaceae	Tree/I	Wild	Dysentery, insect bite (Sd)
361	*Symphorema involucratum*	Konda thakkali	Lamiaceae	Shrub/I	Wild	Tonic (L)
362	*Synedrella nodiflora*	Mudi puvvu chettu	Asteraceae	Herb/E	Natur	Epilepsy (L)
363	*Syzygium cumini*	Neredu	Myrtaceae	Tree/I	Wild	Anthrax, diabetes, epistaxis (R, Sb)
364	*Tacca leontopetaloides*	Adavi kanda	Dioscoreaceae	Herb/I	Wild	Piles (T)
365	*Tamarindus indica*	Chinta	Fabaceae	Tree/E	Planted	Piles, scorpion sting (Sb, F)
366	*Tarenna asiatica*	Papidi	Rubiaceae	Shrub/I	Wild	Fits (R)
367	*Tectona grandis*	Teak	Lamiaceae	Tree/I	Wild	Elephantiasis, pregnancy (Sb, F)
368	*Telosma pallida*	Konda malle	Apocynaceae	Climber/I	Wild	Galactogogue (L)
369	*Tephrosia purpurea*	Vempalli	Fabaceae	Herb/I	Wild	Anthrax, cough, scorpion sting (Sd)
370	*Terminalia alata*	Nalla maddi	Combretaceae	Tree/I	Wild	Wound healing (Sb)
371	*Terminalia arjuna*	Yeru maddi	Combretaceae	Tree/I	Wild	Blisters, boils, ephemeral fever, wound healing (Sb, Sh)
372	*Terminalia bellirica*	Tani	Combretaceae	Tree/I	Wild	Cough, gout (L)
373	*Terminalia chebula*	Kara kaya	Combretaceae	Tree/I	Wild	Constipation, cough (L, F)
374	*Theriophonum minutum*	Adavi champa	Araceae	Herb/I	Wild	Aphrodisiac (T)
375	*Thunbergia alata*	Nallakandla pula teega	Acanthaceae	Climber/I	Cult/Natur	Bone fracture, head-ache (L)

(continued)

Table 6 (continued)

	Botanical name	Local name	Family	Growth form	Distribution	Ethnic use (part/s useful)
376	*Tinospora cordifolia*	Tippa teega	Menispermaceae	Climber/I	Wild	Aphrodisiac, bone fracture, impaction, insect bite, snake bite (S)
377	*Toddalia asiatica*	Konda kasinda	Rutaceae	Climber/I	Wild	Blisters, boils, cuts, dog bite (F)
378	*Tragia involucrata*	Dula gondi	Euphorbiaceae	Climber/I	Wild	Stomachache (R)
379	*Trema orientalis*	Kaka mushti	Cannabaceae	Tree/I	Wild	Diarrhoea (R)
380	*Tribulus terrestris*	Palleru	Zygophyllaceae	Climber/I	Wild	Sexually transmitted diseases, urinary problems (L)
381	*Trichodesma indicum*	Nela nakshatralu	Boraginaceae	Herb/I	Wild	Wound healing (L)
382	*Trichosanthes cucumerina*	Adavi potla	Cucurbitaceae	Climber/I	Wild	Dysentery, skin disease (L)
383	*Trichuriella monsoniae*	Yerra pindi	Amaranthaceae	Herb/I	Wild	Purgative (Wh)
384	*Tridax procumbens*	Nalla alam	Asteraceae	Herb/E	Natur	Antiseptic, blisters, boils, cuts (L)
385	*Triumfetta rhomboidea*	Marla benda	Malvaceae	Herb/I	Wild	Galactogogue (R)
386	*Tylophora indica*	Meka meyani aaku	Apocynaceae	Climber/I	Wild	Asthma (S)
387	*Urena lobata*	Pedda benda	Malvaceae	Herb/I	Wild	Skin disease (R)
388	*Vanda tessellata*	Kodikalla chettu	Orchidaceae	Herb/I	Wild	Ephemeral fever, scorpion sting, snake bite (Wh)
389	*Ventilago maderaspatana*	Surala teega	Rhamnaceae	Climber/I	Wild	Aphrodisiac, stomachache (R)
390	*Vernonia arborescens*	Sahdevi	Asteraceae	Herb/E	Natur	Fever (Wh)
391	*Vigna trilobata*	Pilli pesalu	Fabaceae	Climber/I	Wild	Skin disease (L)
392	*Vitex negundo*	Vavili	Lamiaceae	Tree/I	Wild	Anthrax, ephemeral fever (Wh, L)
393	*Waltheria indica*	Dubba dulicheru	Malvaceae	Herb/E	Natur	Insect bite (L)
394	*Withania somnifera*	Dommadolu gadda	Solanaceae	Herb/I	Cult/R_wild	Anthrax, blisters, boils, fertility, wound healing (R)
395	*Woodfordia fruticosa*	Jaaji	Lythraceae	Shrub/I	Wild	Blood purifier, diarrhoea (L)
396	*Wrightia arborea*	Putta pala	Apocynaceae	Tree/I	Wild	Menstrual pains (L)
397	*Wrightia tinctoria*	Pala kodisha	Apocynaceae	Tree/I	Wild	Anthrax, blisters, boils, ephemeral fever, wound healing (L)

398	*Xanthium strumarium*	Matangi	Asteraceae	Herb/I	Wild	Galactogogue (L)
399	*Xylia xylocarpa*	Bojja	Fabaceae	Tree/I	Wild	Diarrhoea (Sb)
400	*Zehneria scabra*	Adavi dosa	Cucurbitaceae	Climber/I	Wild	Diarrhoea, toothache (L)
401	*Ziziphus oenopolia*	Pariki	Rhamnaceae	Climber/ shrub/I	Wild	Dysentery (L)
402	*Ziziphus rugosa*	Enuga pariki	Rhamnaceae	Tree/I	Wild	Bone fracture (Sb)
403	*Ziziphus xylopyrus*	Gotti	Rhamnaceae	Tree/I	Wild	Ephemeral fever, snake bite, wound healing (Sd, Sb)

After Sreeramulu et al. (2013), Suthari et al. (2014a, b), (2016), and Suthari and Raju (2016)

Note: *Cult* cultivated, *R_wild* running wild, *I* indigenous, *E* exotic, *Intr* introduced, *Natur* naturalized, *R* root, *S* stem, *Sb* stem bark, *L* leaf, *Sh* shoot, *Lx* latex, *Fl* flower, *F* fruit, *Infl* inflorescence, *Sd* seed, *Wh* whole plant, *B* bulb, *Rh* rhizome, *T* tuber

Table 7 Similar use of different medicinal plants for the treatment of various diseases by different ethnic tribes from Telangana State

	Category	Plant species	Genus	Species
1	Abdominal pain	*Abelmoschus moschatus* and *Glinus lotoides*	2	2
2	Abortion	*Celastrus paniculatus*, *Gloriosa superba*, *Moringa concanensis*, *Mucuna pruriens* var. *pruriens*, *Osbeckia stellata*, *Pentanema indicum*, *Plumbago zeylanica*, *Rhynchosia minima* and *Smilax perfoliata*	9	9
3	Anorexia	*Ailanthus excelsa*, *Ampelocissus latifolia*, *Cissus quadrangularis*, *Diospyros montana* and *Phyllanthus emblica*	5	5
4	Anthrax	*Abrus precatorius*, *Ampelocissus tomentosa*, *Anogeissus latifolia*, *Aristolochia bracteolata*, *Asparagus racemosus*, *Bombax ceiba*, *Byttneria herbacea*, *Capparis sepiaria*, *Curcuma pseudomontana*, *Derris scandens*, *Dillenia pentagyna*, *Diospyros chloroxylon*, *Dolichandrone falcata*, *Gloriosa superba*, *Gymnema sylvestre*, *Jasminum auriculatum*, *Justicia adhatoda*, *Moringa concanensis*, *Pavetta indica*, *Phyllanthus emblica*, *Syzygium cumini*, *Tephrosia purpurea*, *Vitex negundo*, *Withania somnifera* and *Wrightia tinctoria*	25	25
5	Antiseptic	*Curcuma longa*, *Hydrolea zeylanica*, *Limnophila indica* and *Tridax procumbens*	4	4
6	Aphrodisiac	*Asparagus gonoclados*, *A. racemosus*, *Breynia retusa*, *Chlorophytum tuberosum*, *Curculigo orchioides*, *Dioscorea alata*, *Drimia indica*, *Hybanthus enneaspermus*, *Hygrophila auriculata*, *Ledebouria hyacinthoides*, *Indigofera trifoliata*, *Leptadenia reticulata*, *Moringa pterygosperma*, *Tinospora cordifolia*, *Smilax perfoliata*, *S. zeylanica*, *Theriophonum minutum* and *Ventilago maderaspatana*	16	18
7	Appetite stimulant	*Amaranthus spinosus*, *Phyllanthus nozeranii*, *Sida spinosa*, *Sophora velutina* and *Striga angustifolia*	5	5
8	Asthma	*Acacia chundra*, *Anogeissus latifolia*, *Cheilocostus speciosus*, *Cissus quadrangularis*, *Globba marantina*, *Oldenlandia umbellata*, *Solanum virginianum* and *Tylophora indica*	8	8
9	Baldness	*Dalbergia latifolia* subsp. *paniculata*	1	1

(continued)

Table 7 (continued)

	Category	Plant species	Genus	Species
10	Blisters	*Biophytum sensitivum*, *Blumea lacera*, *Bridelia montana*, *Cadaba fruticosa*, *Chamaesyce hirta*, *Cleistanthus collinus*, *Cleome viscosa*, *Coccinia grandis*, *Curcuma longa*, *Dichrostachys cinerea*, *Enicostemma axillare*, *Euphorbia meenae*, *Evolvulus alsinoides*, *Flacourtia indica*, *Gardenia latifolia*, *Jasminum auriculatum*, *Lagerstroemia parviflora*, *Mangifera indica*, *Martynia annua*, *Merremia emarginata*, *Mucuna pruriens* var. *pruriens*, *Pergularia daemia*, *Sesamum alatum*, *Terminalia arjuna*, *Toddalia asiatica*, *Tridax procumbens*, *Withania somnifera* and *Wrightia tinctoria*	28	28
11	Blood purifier	*Hemidesmus indicus*, *Scleria corymbosa* and *Woodfordia fruticosa*	3	3
12	Boils	*Acacia leucophloea*, *Achyranthes aspera*, *Biophytum sensitivum*, *Bridelia montana*, *Cadaba fruticosa*, *Chamaesyce hirta*, *Cleistanthus collinus*, *Cleome viscosa*, *Coccinia grandis*, *Curcuma longa*, *Dichrostachys cinerea*, *Enicostemma axillare*, *Euphorbia meenae*, *Evolvulus alsinoides*, *Flacourtia indica*, *Gardenia latifolia*, *Jasminum auriculatum*, *Lagerstroemia parviflora*, *Mangifera indica*, *Martynia annua*, *Merremia emarginata*, *Mollugo nudicaulis*, *Mucuna pruriens* var. *pruriens*, *Pergularia daemia*, *Sesamum alatum*, *Terminalia arjuna*, *Toddalia asiatica*, *Tridax procumbens*, *Withania somnifera* and *Wrightia tinctoria.*	30	30
13	Bone fracture	*Alangium salviifolium*, *Ampelocissus tomentosa*, *Anodendron paniculatum*, *Cassytha filiformis*, *Cissus quadrangularis*, *Curcuma longa*, *Dillenia pentagyna*, *Dioscorea bulbifera*, *Dodoanea viscosa*, *Erythroxylum monogynum*, *Euphorbia antiquorum*, *Ficus racemosa*, *Hymenodictyon orixense*, *Lannea coromandelica*, *Litsea glutinosa*, *Nyctanthes arbor-tristis*, *Ochna obtusata*, *Phyllanthus reticulatus*, *Sarcostemma acidum*, *Tinospora cordifolia*, *Ziziphus rugosa* and *Thunbergia alata*	22	22
14	Burns	*Acacia nilotica* and *Caralluma umbellata*	2	2
15	Chest pain	*Buchanania cochinchinensis* and *Schleichera oleosa*	2	2
16	Constipation	*Calotropis gigantea*, *Ipomoea turbinata*, *Senna angustifolia*, *S. italica* and *Terminalia chebula*	4	5
17	Contraceptives	*Cajanus cajan*, *Flagellaria indica* and *Capparis sepiaria*	3	3
18	Corneal opacity	*Aegle marmelos*, *Balanites roxburghii*, *Casearia elliptica*, *Dolichandrone falcata*, *Ocimum basilicum*, *Pergularia daemia* and *Soymida febrifuga*	7	7

(continued)

Table 7 (continued)

	Category	Plant species	Genus	Species
19	Cough	*Blumea bifoliata, Breynia retusa, Desmodium heterocarpon, Dioscorea pentaphylla, Drypetes roxburghii, Gomphrena celosioides, Holarrhena pubescens, Jasminum auriculatum, Justicia adhatoda, Moringa concanensis, Tephrosia purpurea, Terminalia bellirica* and *T. chebula*	12	13
20	Cracked heals	*Lannea coromandelica* and *Oxalis corniculata*	2	2
21	Cuts	*Blumea axillaris, Biophytum sensitivum, Bridelia montana, Cadaba fruticosa, Chamaesyce hirta, Coccinia grandis, Curcuma longa, Dichrostachys cinerea, Euphorbia meenae, Flacourtia indica, Lagerstroemia parviflora, Merremia emarginata, Sesamum alatum, Toddalia asiatica* and *Tridax procumbens*	15	15
22	Diabetes	*Capparis zeylanica, Corallocarpus epigaeus, Enicostemma axillare, Gymnema sylvestre, Hemidesmus indicus* var. *pubescens, Ichnocarpus frutescens, Maerua oblongifolia, Pterocarpus marsupium* and *Syzygium cumini*	9	9
23	Diarrhoea	*Bauhinia purpurea, Ceropegia hirsuta, Ceriscoides turgida, Combretum latifolium, Diplocyclos palmatus, Ficus talboti, Ludwigia hyssopifolia, Mimosa pudica, Olax scandens, Paederia foetida, Trema orientalis, Woodfordia floribunda, Xylia xylocarpa* and *Zehneria mysorensis*	14	14
24	Diuretic	*Kyllinga odorata*	1	1
25	Dog bite	*Acacia farnesiana, Boswellia serrata, Heliotropium indicum, Semecarpus anacardium* and *Toddalia asiatica*	5	5
26	Dysentery	*Abutilon indicum, Adiantum lunulatum, Aegle marmelos, Aerva scandens, Asparagus racemosus, Bauhinia racemosa, Bombax ceiba, Byttneria herbacea, Clitoria ternatea, Coccinia grandis, Cyphostemma setosum, Dillenia pentagyna, Dioscorea bulbifera, Eleusine coracana, Holarrhena pubescens, Ludwigia adscendens, Mimosa pudica, Paederia foetida, Phyllanthus reticulatus, Pteris multifida, Rhynchosia suaveolens, Strychnos nux-vomica, Trichosanthes cucumerina* and *Ziziphus oenopolia*	24	24
27	Dyspepsia/ indigestion	*Andrographis paniculata, Bauhinia vahlii, Capparis zeylanica, Dioscorea bulbifera, Ehretia laevis, Kydia calycina* and *Limonia acidissima*	7	7
28	Earache	*Aerva lanata, Aristolochia indica, Calotropis gigantea, Globba marantina* and *Merremia turpethum*	5	5
29	Elephantiasis	*Tectona grandis*	1	1
30	Emetic	*Sarcostemma intermedium* and *Secamone emetica*	2	2

(continued)

Table 7 (continued)

	Category	Plant species	Genus	Species
31	Ephemeral fever	*Acacia leucophloea*, *Agave americana*, *Andrographis paniculata*, *Anisomeles indica*, *Annona squamosa*, *Azadirachta indica*, *Balanites roxburghii*, *Bauhinia malabarica*, *Caesalpinia bonduc*, *Cardiospermum halicacabum*, *Casearia elliptica*, *Crotalaria verrucosa*, *Curculigo orchioides*, *Cyperus rotundus*, *Drimia indica*, *Evolvulus alsinoides*, *Geodorum densiflorum*, *Gymnema sylvestre*, *Phyllanthus amarus*, *Pueraria tuberosa*, *Terminalia arjuna*, *Vanda tessellata*, *Vitex negundo*, *Wrightia tinctoria* and *Ziziphus xylopyrus*	25	25
32	Epilepsy	*Adiantum lunulatum* and *Synedrella nodiflora*	2	2
33	Epistaxis	*Cocculus hirsutus*, *Gmelina asiatica*, *Justicia adhatoda* and *Syzygium cumini*	4	4
34	Fertility	*Anisomeles malabarica*, *Artabotrys hexapetalus*, *Bombax ceiba*, *Lygodium flexuosum*, *Maerua oblongifolia*, *Mimosa hamata*, *Pterocarpus marsupium*, *Solanum surattense* and *Withania somnifera*	9	9
35	Fever	*Acacia chundra*, *Calotropis gigantea*, *Dendrophthoe falcata*, *Diplocyclos palmatus*, *Ocimum tenuiflorum*, *Sida spinosa*, *Stemona tuberosa*, *Vernonia arborescens*, and *V. cinerea*	8	9
36	Fits	*Acacia pennata*, *Chomelia asiatica*, *Crotalaria verrucosa*, *Nyctanthes arbor-tristis*, *Plumbago zeylanica* and *Semecarpus anacardium*	6	6
37	Galactogogue	*Asparagus racemosus*, *Celosia argentea*, *Ceropegia juncea*, *Cryptolepis dubia*, *Curculigo orchioides*, *Cuscuta chinensis*, *Gymnema sylvestre*, *Hemidesmus indicus* var. *pubescens*, *Madhuca longifolia* var. *latifolia*, *Oxystelma esculentum*, *Sarcostemma acidum*, *Sarcostemma secamone*, *Telosma pallida*, *Triumfetta rhomboidea* and *Xanthium strumarium*	15	15
38	Gout	*Euphorbia antiquorum*, *Iphigenia indica*, *Pergularia daemia*, *Ricinus communis*, *Soymida febrifuga* and *Terminalia bellerica*	6	6
39	Hair tonic/ shampoo	*Merremia hederacea* and *Sapindus emarginatus*	2	2
40	Headache	*Cleome gynandra*, *Holarrhena pubescens*, *Passiflora foetida* and *Thunbergia alata*	4	4
41	Heat and summer stroke	*Blepharis repens*, *Erythrina variegata* and *Ocimum basilicum*	3	3
42	Helminthiasis	*Abutilon indicum*, *Cissus quadrangularis* and *Commelina benghalensis*	3	3
43	Hydrocele	*Caesalpinia bonduc* and *Cardiospermum halicacabum*		

(continued)

Table 7 (continued)

	Category	Plant species	Genus	Species
44	Impaction	*Aegle marmelos*, *Byttneria herbacea*, *Capparis zeylanica*, *Derris scandens*, *Erythrina variegata*, *Hemidesmus indicus*, *Indigofera trita*, *Phyllanthus emblica* and *Tinospora cordifolia*	9	9
45	Infant	*Azima tetracantha*, *Ficus racemosa* and *Selaginella bryopteris*	3	3
46	Insect bite	*Abrus precatorius*, *Abutilon indicum*, *Achyranthes aspera*, *Aloe vera*, *Ampelocissus tomentosa*, *Anogeissus latifolia*, *Aristolochia bracteolata*, *Asparagus racemosus*, *Canthium parviflorum*, *Crotalaria verrucosa*, *Gloriosa superba*, *Helicteres isora*, *Leea asiatica*, *Phyllanthus reticulatus*, *Senna obtusifolia*, *S. tora*, *Strychnos nux-vomica*, *Tinospora cordifolia* and *Waltheria indica*	18	19
46	Jaundice	*Lawsonia inermis*, *Luffa aegyptiaca*, *Merremia aegyptiaca*, *Phyllanthus amarus* and *P. hookeri*	4	5
47	Labour pains	*Acacia torta*, *Butea superba*, *Careya arborea*, *Crotalaria albida*, *Litsea glutinosa*, *Moringa pterygosperma* and *Pterocarpus marsupium*	7	7
48	Laxative	*Acalypha lanceolata*, *Chamaesyce thymifolia*, *Grewia damine*, *G. hirsuta*, *G. rothii*, *Justicia procumbens* and *Solanum virginianum*	5	7
49	Leucorrhoea	*Cerascoides turgida*, *Erythrina suberosa*, *Garuga pinnata*, *Memecylon umbellatum* and *Selaginella bryopteris*	5	5
50	Maggot-infected sores	*Acalypha indica*, *Jatropha curcas*, *Mimosa pudica* and *Pavetta indica*	4	4
51	Memory	*Centella asiatica*, *Ceratopteris thalictroides*, *Selaginella bryopteris* and *S. involvens*	3	4
52	Menstrual pains	*Firmiana simplex*, *Hymenodictyon orixense*, *Scoparia dulcis* and *Wrightia arborea*	4	4
53	Mouth ulcers	*Holostemma ada-kodien* and *Maytenus emarginata*	2	2
54	Neck pain	*Bauhinia semla*, *Chloroxylon swietenia*, *Dregea volubilis* and *Ipomoea carnea*	4	4
55	Oedema	*Alangium salviifolium*, *Dendrocalamus strictus*, *Hygrophila auriculata* and *Premna latifolia*	4	4
56	Ophthalmic diseases	*Pavetta indica*	1	1
57	Paralysis	*Dregea volubilis*	1	1
58	Hemiplegia (partial paralysis)	*Aristolochia indica*	1	1
59	Pertussis	*Balanites roxburghii*	1	1

(continued)

Table 7 (continued)

	Category	Plant species	Genus	Species
60	Piles	*Ailanthus excelsa*, *Aloe vera*, *Ceriscoides turgida*, *Cochlospermum religiosum*, *Gardenia latifolia*, *Naringi crenulata*, *Tacca leontopetaloides* and *Tamarindus indica*	8	8
61	Pregnancy	*Tectona grandis*	1	1
62	Purgative	*Alangium salviifolium*, *Aristolochia bracteolata*, *Baliospermum solanifolium*, *Blepharis maderaspatensis*, *Cuscuta reflexa*, *Marsdenia tenacissima*, *Operculina turpethum*, *Oroxylum indicum*, *Rivea hypocrateriformis* and *Trichuriella monsoniae*	10	10
63	Red leucorrhoea	*Cinnamomum zeylanicum*, *Cocculus hirsutus* and *Litsea glutinosa*	3	3
64	Rheumatism	*Albizia procera*, *Ageratum conyzoides*, *Asystasia gangetica*, *Atalantia monophylla*, *Azima tetracantha*, *Barringtonia acutangula*, *Boswellia serrata*, *Caesalpinia bonduc*, *Calotropis gigantea*, *Canavalia gladiata*, *Capparis sepiaria*, *Clerodendrum phlomidis*, *Crotalaria medicaginea*, *Cyphostemma setosum*, *Dichrostachys cinerea*, *Dillenia pentagyna*, *Dioscorea pentaphylla*, *Dregea volubilis*, *Ficus benghalensis*, *F. virens*, *Hemidesmus indicus*, *Holoptelea integrifolia*, *Lannea coromandelica*, *Leonotis nepetifolia*, *Melia azedarach*, *Merremia tridentata*, *Oroxylum indicum*, *Paederia foetida*, *Phyllanthus emblica*, *Plumbago zeylanica*, *Pseudarthria viscida*, *Pueraria tuberosa* and *Senna occidentalis*	32	33
65	Scorpion sting	*Barleria prionitis*, *Boswellia serrata*, *Canavalia cathartica*, *Cyclosorus unitus*, *Desmodium gangeticum*, *D. velutinum*, *Phyllodium pulchellum*, *Tamarindus indica*, *Tephrosia purpurea* and *Vanda tessellata*	9	9
66	Sexually transmitted diseases	*Canavalia virosa*, *Cayratia pedata*, *Smilax zeylanica* and *Tribulus terrestris*	4	4
67	Shivering	*Chloroxylon swietenia*, *Mallotus philippensis*, *Rotula aquatica* and *Soymida febrifuga*	4	4

(continued)

Table 7 (continued)

	Category	Plant species	Genus	Species
68	Skin disease	*Acalypha indica, Ammannia baccifera, Argemone mexicana, Argyreia nervosa, Asparagus gonoclados, Barleria montana, Blumea virens, Capparis sepiaria, Chamaesyce hirta, Cleome monophylla, C. viscosa, Cryptolepis dubia, Dalbergia volubilis, Dicliptera paniculata, Dolichos trilobatus, Euphorbia tirucalli, Ficus virens, Flemingia macrophylla, F. strobilifera, Hibiscus panduriformis, Holoptelea integrifolia, Ipomoea eriocarpa, Leea indica, Leucas decemdentata, Lindernia ciliata, Lygodium flexuosum, Melastoma malabarthricum, Microstachys chamaelea, Paracalyx scariosus, Plumbago zeylanica, Pongamia pinnata, Rotheca serrata, Senna absus, S. alata, Trichosanthes cucumerina* and *Urena lobata*	34	37
69	Snake bite	*Abrus precatorius, Achyranthes aspera, Alangium salviifolium, Andrographis paniculata, Anisomeles indica, Annona squamosa, Aristolochia bracteolata, A. indica, Asparagus racemosus, Bryophyllum pinnatum, Calotropis gigantea, Careya arborea, Celastrus paniculatus, Cissampelos pareira, Corallocarpus epigaeus, Crinum asiaticum, Crotalaria verrucosa, Datura metel, Diospyros chloroxylon, Diplocyclos palmatus, Dregea volubilis, Ehretia microphylla, Ficus infectoria, Gloriosa superba, Gymnema sylvestre, Leea asiatica, Maerua oblongifolia, Marsdenia tenacissima, Naringi crenulata, Ochna obtusata, Pergularia daemia, Polygala arvensis, Rauvolfia serpentina, Rivea hypocrateriformis, Sarcostemma acidum, Soymida febrifuga, Tinospora cordifolia, Vanda tessellata* and *Ziziphus xylopyrus*	38	39
70	Spermatorrhoea	*Solena amplexicaulis*	1	1
71	Stomachache	*Acorus calamus, Cissampelos pareira, Cissus vitiginea, Dalbergia latifolia, Elephantopus scaber, Glinus oppositifolius, Haldina cordifolia, Heliotropium supinum, Ipomoea eriocarpa, I. obscura, Mollugo pentaphylla, Olax scandens, Pavonia odorata, Pergularia daemia, Phyla nodiflora, Tragia involucrata* and *Ventilago maderaspatana*	16	17
72	Throat disorders	*Ctenolepis garcinii*	1	1
73	Tonic	*Bacopa monnieri, Boerhavia erecta, Cissampelos pareira, Elytraria acaulis, Habenaria marginata, H. roxburghii, Ipomoea hederifolia, Sida cordata, Soymida febrifuga, Stachytarpheta jamaicensis* and *Symphorema involucratum*	10	11
74	Tonsils	*Bauhinia semla, Dregea volubilis* and *Ipomoea carnea*	3	3

(continued)

Table 7 (continued)

	Category	Plant species	Genus	Species
75	Toothache	*Acmella paniculata*, *Melothria maderaspatana*, *Merremia tridentata* ssp. *hastata*, *Mukia maderaspatana* and *Zehneria scabra*	5	5
76	Trypanosomiasis	*Acacia chundra*, *Boerhavia diffusa*, *Capparis divaricata*, *Curculigo orchioides*, *Cynodon dactylon*, *Derris scandens*, *Erythrina variegata*, *Helicteres isora* and *Mentha spicata*	9	9
77	Tuberculosis	*Dendrophthoe falcata*	1	21
78	Tympany	*Ailanthus excelsa*, *Argyreia nervosa*, *Cassia fistula*, *Cerascoides turgida*, *Coccinia grandis*, *Crateva magna*, *Helicteres isora*, *Jasminum auriculatum*, *Ocimum basilicum* and *Plumbago zeylanica*	10	10
79	Urinary problems	*Cocculus hirsutus*, *Hybanthus enneaspermus*, *Ledebouria revoluta*, *Saraca indica* and *Tribulus terrestris*	5	5
80	Wound healing	*Acacia chundra*, *A. leucophloea*, *Achyranthes aspera*, *Adenostemma lavenia*, *Aristolochia bracteolata*, *Baccharoides schimperi*, *Blumea obliqua*, *Buchanania axillaris*, *Chromolaena odorata*, *Cissus adnata*, *Cleistanthus collinus*, *Cleome viscosa*, *Coldenia procumbens*, *Crinum asiaticum*, *Curcuma pseudomontana*, *Elephantopus scaber*, *Enicostemma axillare*, *Evolvulus alsinoides*, *Ficus altissima*, *F. mollis*, *Gardenia latifolia*, *Indigofera caerulea*, *Ixora arborea*, *Jasminum auriculatum*, *Leucas aspera*, *Mangifera indica*, *Marsdenia tenacissima*, *Martynia annua*, *Mucuna pruriens* var. *pruriens*, *Ocimum americanum*, *Pergularia daemia*, *Plectranthus mollis*, *Pteris vittata*, *Pupalia lappacea*, *Terminalia alata*, *T. arjuna*, *Trichodesma indicum*, *Withania somnifera*, *Wrightia tinctoria* and *Ziziphus xylopyrus*	37	40

References

Akerele O (1992) WHO guideline for assessment of herbal medicines. Fitoterapia 63:99–118

Anonymous (1990) Ethnobiology in India: a status report. Ministry of Environment and Forests, New Delhi

Anonymous (2017) Telangana State Information.pdf. Website: http://www.telangana.gov.in/about/districts

Binu S, Nayar TS, Pushpangadan P (1992) An outline of ethnobotanical research in India. J Econ Tax Bot 10:405–428

Chandrasekharan C (1995) Non-wood forest products – a global view of potentials and challenges. Minor Forest Produces News 5(1 & 2):23–31

Cotton CM (1996) Ethnobotany: principles and applications. John Willy & Sons, Chichester, NY, p 424

Dinesh V, Sharma PP (2012) Plants used for bone fracture by indigenous folklore of Nizamabad district, Andhra Pradesh. Int Multidiscip Res J 2(12):14–16

Dovie DBK (2003) Rural economy and livelihoods from the non-timber forest products trade. Compromising sustainability in southern Africa? Int J Sust Dev World 10:247–262. https://doi.org/10.1080/13504500309469803

GCC (2015) Brief note on G.C.C. activities in Warangal & Karimnagar Districts. Girijan Cooperative Corporation, Eturnagaram Division, Warangal, pp 1–3

GCC (2016) Commodity-wise procurement of MFP/AP from 2005–2016. Girijan Co-op Corporation Limited, Visakhapatnam, Andhra Pradesh

Haimendorf CVF (1979) The Gonds of Andhra Pradesh. Vikas, New Delhi

Hemadri K (1990) Contribution to the medicinal flora of Karimnagar and Warangal districts (A.P.). Indian Med 2:16–28

Huxley A (1984) Green inheritance: the World Wildlife Fund book of India. Collins/Harvel, London

ISFR (2017) India State of Forest Report 2017. Forest Survey of India, Ministry of Environment, Forestry & Climate Change, Dehra Dun

Jain SK (1963a) Studies in Indian ethnobotany-less known uses of fifty common plants from the tribal areas of Madhya Pradesh. Bull Bot Surv India 5:223–226

Jain SK (1963b) Observations of ethnobotany of the tribals of Madhya Pradesh. Vanyajati 11:177–183

Jain SK (1963c) Studies in Indian ethnobotany-plants used in medicine by the tribals of Madhya Pradesh. Bull Reg Res Lab Jammu 1:126–128

Jain SK (1964a) The role of botanist in folklore research. Folklore 5:145–150

Jain SK (1964b) Wild plant foods of the tribals of Bastar. Gramodyog 10:557–561

Jain SK (1964c) Native plant remedies for snake-bite among Adivasis of Central India. Indian Med J 57:307–369

Jain SK (1965a) Wooden musical instruments of the Gonds of Central India. Ethnomusicology 9:39–42

Jain SK (1965b) Medicinal plant-lore of the tribals of Bastar. Economic Bot 19:236–250

Jain SK (1999) Dictionary of ethnoveterinary plants of India. Deep Publications, New Delhi

Kannabiran K, Jeyaranjan J, Swaminathan P (2017) Telangana social development report. Council for Social Development, Hyderabad

Kapoor L, Kapoor LD (1980) Medicinal plant wealth of the Karimnagar district of Andhra Pradesh. Bull Med Ethnobotanical Res 1:120–144

Khan MS (1953) Forest Flora of Hyderabad State. Government Press, Hyderabad

Krishna NR, Varma YNR, Saidulu C (2014) Ethnobotanical studies of Adilabad district, Andhra Pradesh. J Pharmacogn Phytochem 3(1):18–36

Kumar TD, Pullaiah T (1998) Ethnomedicinal uses of some plants of Mahaboobnagar district, Andhra Pradesh, India. J Econ Taxon Bot 23:341–345

Kumar RS, Venkateshwar C, Samuel G, Rao SG (2013) Ethnobotanical uses of some plant barks used by Gondu tribes of Seethagondi grampanchayath, Adilabad district, Andhra Pradesh, India. J Nat Prod Plant Resour 3(5):13–17

Lalramnghinglova H, Jha LK (1999) Ethnobotany: a review. J Econ Taxon Bot 23(1):1–27

Lingaiah M, Rao N (2013) An ethnobotanical survey of medicinalplants used by traditional healers of Adilabad district, Andhra Pradesh, India. Biolife 1(1):17–23

Maithani GP (1994) Management perspectives of minor forest produce. MFP News, October–December, 1994

Manjula RR, Rao KJ, Reddi TVVS (2013) Ethnomedicine for rheumatism by the tribals of Khammam district, Andhra Pradesh. J Nat Rem 13(2):138–141

Mohan AC, Suthari S, Kandagatla R, Harikrishna L, Ragan A (2017b) Antirheumatic plants used by indigenous people of Kawal wildlife sanctuary, Telangana. Int J Adv Res Sci Technol 6(1):666–669

Mohan AC, Suthari S, Ragan A (2017a) Ethnobotanical plants of Kawal wildlife sanctuary, Telangana, India. Ann Plant Sci 6(2):1537–1542

Mudgal V (1987) Recent ethnobotanical works on different States/Tribes of India – A synoptic treatment. In: Jain SK (ed) A manual of ethnobotany. Deep Publications, New Delhi, pp 58–68
Murthy EN (2012) Ethnomedicinal plants used by gonds of Adilabad district, Andhra Pradesh, India. Int J Pharm Life Sci 3(10):2034–2043
Murthy EN, Reddy CS, Reddy KN, Raju VS (2007) Plants used in ethnoveterinary practices by Koyas of Pakhal wildlife sanctuary, Andhra Pradesh, India. Ethnobot Leaflets 11:1–5
Naqvi AH (2001) Flora of Karimnagar District, Andhra Pradesh, India. Ph.D. Thesis, Kakatiya University, Warangal
Omkar K, Ragan A, Raju VS (2008) Economic empowerment of tribal women of Adilabad district of Andhra Pradesh, utilizing non-timber forest products. In: Proceedings of the National Seminar on 'Globalization-Agricultural Development of Tribes, Issues and Challenges'. Kakatiya University, Warangal, Telangana, pp 96–98
Omkar K, Suthari S, Alluri S, Ragan A, Raju VS (2012) Diversity of NTFPs and their utilization in Adilabad district of Andhra Pradesh, India. J Plant Stud 1(1):33–46
Padal SB, Sathyavathi K (2013) Traditional uses of Euphorbiaceae family of Khammam district, Andhra Pradesh, India. Int J Pharm Biol Sci 3(2):585–591
Padmarao P, Reddy PR (1999) A note on folk treatment of bone fractures in Ranga Reddy district, Andhra Pradesh. Ethnobotany 11:107–108
Pullaiah T, Kumar DCT (1996) Herbal plants in Mannanur forest of Mahabubnagar district, Andhra Pradesh. J Econ Taxon Bot 12:218–220
Rajasekaran B, Warren DM (1994) Indigenous knowledge for socio-economic development and biodiversity conservation: the Kolli hills. Indigenous Knowl Dev Monit 2:13–17
Raju VS (2001) Ethnoveterinary medicine in Andhra Pradesh. National Symposium '21st Century Perspectives in Plant Sciences', July 29–31, Andhra University, Waltair, India
Raju VS, Krishna PG, Suthari S (2014) Environmental assessment of climate of a habitat through floristic life-form spectra, a case study of Warangal north forest division, Telangana, India. J Nat Sci 2(1):77–93
Raju, V.S., Ragan, A., Omkar, K. and Geetha, S. 2008. Sedges of Andhra Pradesh: Biodiversity at species and ecosystem levels, along with economic and ethnobotanical value. *In*: B. Bahadur (ed.), Proceedings of Andhra Pradesh Akademi of Sciences: Special Issue on 'Plant Wealth of Andhra Pradesh'. 12(1&2): 214–226. Osmania University Campus, Hyderabad
Raju VS, Reddy KN (2005) Ethnobotanic medicine for dysentery and diarrhea from Khammam district of Andhra Pradesh, India. Indian J Tradit Knowl 4(4):443–447
Raju, V.S., Reddy, C.S., Reddy, K.N., Seshagirirao, K. and Bahadur, B. 2008. Orchid wealth of Andhra Pradesh, India. *In*: B. Bahadur (ed.), Proceedings of Andhra Pradesh Akademi of Sciences: special issue on 'Plant Wealth of Andhra Pradesh'. 12(1&2): 180–192. Osmania University Campus, Hyderabad
Ramakrishna N, Saidulu C (2014) Medicinal plants used by ethnic people of Adilabad district, Andhra Pradesh, India. Int J Pharm Res Allied Sci 2:51–59
Ramakrishna N, Sunitha EM, Saidulu C, Rajani A (2015) Studies on some medicinal plants of Leguminaceae family in Adilabad district, Telangana State, India. Int J Pharm Sci Rev Res 30(1):315–318
Ramarao N (1988) The ethnobotany of Eastern Ghats in Andhra Pradesh, India. Ph. D. Thesis. Bharathiar University, Coimbatore
Ramarao N, Henry AN (1996) The ethnobotany of Eastern Ghats in Andhra Pradesh, India. Botanical Survey of India, Calcutta
Rao DS, Rao VS, Murthy PP, Rao GMN, Rao YV (2015) Some ethno medicinal plants of Parnasala sacred grove area Eastern Ghats of Khammam district, Telangana, India. J Pharam Sci Res 7(4):210–216
Ravishankar T (1990) Ethnobotanical studies in Adilabad and Karimnagar districts of Andhra Pradesh, India. Ph.D. Thesis, Bharathiar University, Coimbatore
Ravishankar T, Henry AN (1992) Ethnobotany of Adilabad district Andhra Pradesh, India. Ethnobotany 4:45–52

Reddy CS (2001) Floristic Studies in Warangal District, Andhra Pradesh, India. Ph.D. Thesis, Kakatiya University, Warangal
Reddy KN (2003) Ethnobotany in Khammam District of Andhra Pradesh, India. Ph.D. Thesis, Kakatiya University, Warangal
Reddy AVB (2008) Use of various bio-fencing plants in the control of human disease by Lambada tribe inhabiting Nalgonda district, Andhra Pradesh, India. Ethnobot Leaflets 12:520–523
Reddy DS (2015) Ethnomedicinal plants used by the tribals of Achampet forest division in Nallamalais, Telangana, India. Int J Plant Anim Environ Sci 5(2):65–73
Reddy KN, Bhanja MR, Raju VS (1998) Plants used in ethnoveterinary practices in Warangal district, Andhra Pradesh, India. Ethnobotany 10:75–84
Reddy CS, Gopalkrishna P, Raju VS (2008) Phytotherapy at rural communities: a case study from the Gonds of Warangal district, Andhra Pradesh, India. Res J Bot 3(2):97–102
Reddy KN, Pattanaik C, Reddy CS, Raju VS (2007) Traditional knowledge on wild food plants in Andhra Pradesh. Indian J Tradit Knowl 6(1):223–229
Reddy CS, Raju VS (2000) Folklore biomedicine for common veterinary diseases in Nalgonda district, Andhra Pradesh, India. Ethnobotany 12:113–117
Reddy KN, Raju VS (2002) Ethnobotanical observations on Konda Reddis of Mothugudem in Khammam district, Andhra Pradesh. Abstract published in National seminar on Conservation of Eastern Ghats 24–26th March, Tirupati
Reddy PR, Rao PP (2002) A survey of plant crude drugs in folklore from Ranga Reddy district, Andhra Pradesh, India. Indian J Tradit Knowl 1(1):20–25
Reddy KN, Reddy CS, Jadhav SN (2001) *Heterostemma deccanense* (Talb.) Swarup & Mangaly (Asclepiadaceae): An endangered and endemic taxon in Andhra Pradesh. Indian For 127:1403–1404
Reddy KN, Reddy CS, Raju VS (2008) Ethnomedicinal observations among the Kondareddis of Khammam district, Andhra Pradesh, India. Ethnobot Leaflets 12:916–926
Reddy KN, Trimurthulu G, Reddy CS (2010) Medicinal plants used by ethnic people of Medak district, Andhra Pradesh. Indian J Tradit Knowl 9(1):184–190
Romney AK, Weller SC, Batchelder WH (1986) Culture as consensus: a theory of culture and informant accuracy. Am Anthropol 88(2):313–338
Rosenthal JP (2006) Politics, culture, and governance in the development of prior informed consent in indigenous communities. Curr Anthropol 47(1):119–142
Saidulu P, Suthari S, Ramesh K, Ragan A, Raju VS (2015) Ethnobotanical knowledge studied in Pocharam Wildlife Sanctuary, Telangana, India. Notulae Scientia Biologicae 7(2):164–170
Samata A (2007) Chemosystematics of the genus Crotalaria (Papilionaceae) in Andhra Pradesh, India. Ph.D. Thesis, Kakatiya University, Warangal
Satyanarayana D (2014) Tribes of Telangana: society, religion and culture. Negadu, Telangana Resource Centre, Hyderabad, Telangana
Singh RK (2008) Implications of prior informed consent for the conservators of indigenous biological diversity of Northeast India. Indian J Tradit Knowl 7(4):655–665
Singh PS, Singh DSRR (2015) Documentation of tribal's traditional knowledge of medicinal plants from Renlagadda Thanda, Kodangal Mandal, Mahabubnagar Dist, Telangana, India. Int Lett Nat Sci 8:13–24
Sreeramulu N (2008) Traditional botanical knowledge of local people in Nalgonda and Warangal Districts of Telangana, Andhra Pradesh, India. Ph.D. Thesis, Kakatiya University, Warangal
Sreeramulu N, Suthari S, Omkar K, Raju VS (2013) Ethno-botanico-medicine for common human ailments in Nalgonda and Warangal districts of Andhra Pradesh, India. Ann Plant Sci 2(7):220–229
Sudharani T, Umadevi M, Rajani B, Padmavathi V, Maiti RK (2007) Ethnobotanical survey of Nalgonda district, Andhra Pradesh, India. Res Crops 8(3):700–715
Suthari S (2013) Biodiversity characterization and aboveground vegetation carbon pool assessment in Northern Telangana at landscape level using geospatial technique. Ph.D. Thesis, Kakatiya University, Warangal

Kandagatla R, Ragan A, Raju VS (2016) Plant wealth of a sacred grove: Mallur Gutta, Telangana State, India. Int J Gen Med 9:369–381

Suthari S, Prasad MNV, Ragan A, Raju VS (2018) Are non-timber forest products boon to people and bane to forests? pp: 116–125. In: Pindi PK, Rathod V (eds.), *Recent Trends in Life Sciences: Current Archives in Life Sciences*. Lambert Academic Publishing, Mauritius (ISBN: 978-613-9-84380-0)

Suthari S, Raju VS (2016) Antidote botanicals for snake-bite from Koyas of Warangal district, Telangana, India. J Herbs Spices Med Plants 22:57–68

Suthari S, Sreeramulu N, Omkar K, Raju VS (2014a) The climbing plants of northern Telangana in India and their ethnomedicinal and economic uses. Indian J Plant Sci 3(1):86–100

Suthari S, Sreeramulu N, Omkar K, Reddy CS, Raju VS (2014b) Intracultural cognizance of medicinal plants of Warangal North Forest Division, Northern Telangana, India. Ethnobot Res Appl 12:211–235

Swamy NSN (2009) Ethnobotanical knowledge from Adilabad District, Andhra Pradesh, India. Ph.D. Thesis, Andhra University, Visakhapatnam

Ul-Hassan SS (1920) The castes and tribes of H.E.H. The Nizam's Dominions, vol I. The Times Press, Bombay, p 651

Upadhyay R, Chauhan SVS (2000) Ethnobotanical observations on Koya tribe of Gundala Mandal of Khammam district, Andhra Pradesh. Ethnobotany 12:93–99

Ethnobotany and Pharmacological Uses of *Elaeocarpus floribundus* Blume (Elaeocarpaceae)

Mohamad Fawzi Mahomoodally and Veedooshee Sookhy

Introduction

Plants have always been a reservoir of therapeutic molecules since the dawn of humanity. The use of plants to cure minor and major ailments has been passed on through old scriptures, through cultures, and orally. While plants were used on a trial and error basis for its therapeutic roles, these traditional approaches have metamorphosed into evidence-based approach. Presently, there has been a drive to unveil new plants and its proposed therapeutic aspects. Researchers are ambitious to find safer, cheaper, and easily accessible alternate that plant-based drugs offer. Today, ethnobotanical and pharmacological studies have been used as a medium to spot new plants and compounds (Gurib-Fakim 2006).

According to WHO, medicinal plants, food plants, and herbs are among the prevailing forms of medication, with 85% of drugs used in primary health care derived from plants (Abbasi et al. 2010). Moreover, the acceptability of plant-based drugs will not pose an issue since men have been using them without knowledge of its efficacy. The plant kingdom is an implicit gold mine of new chemical compounds, which are still waiting to be explored. It is estimated that there are approximately 500,000 to 750,000 species of higher plants existing on earth and less than 10% of them are examined for their biochemical constituents. In the past few years, ethnobotany has contributed to the discovery and the isolation of numerous active compounds like morphine from opium, cocaine, codeine, digitoxin, and quinine (Butler 2004). Keeping this thought in mind, researchers are on the hunt for new plants for their pharmacological properties. One family of species that has gained much interest is the Elaeocarpaceae owing to their resemblance with the Cucurbitaceae, which are known for its multiple pharmacological activities. The Elaeocarpaceae family

M. F. Mahomoodally (✉) · V. Sookhy
Faculty of Science, Department of Health Sciences, University of Mauritius, Reduit, Mauritius
e-mail: f.mahomoodally@uom.ac.mu; veedooshee.sookhy@umail.uom.ac.mu

M. Ozturk, K. R. Hakeem (eds.), *Plant and Human Health, Volume 1*,
https://doi.org/10.1007/978-3-319-93997-1_4

has attracted much interest since they harbour both indolizidine alkaloids, which have exhibited ability to inhibit enzymatic activity of glucosidases and cucurbitacin with chemotherapeutic potential (Wiart 2006). Belonging to the Elaeocarpaceae family, *Elaeocarpus floribundus* Blume, also locally known as "olive," has been used as phytomedicine in African and Asian folkloric medicine to treat and/or manage ailments such as dysentery, diabetes, and inflamed gums. *Elaeocarpus floribundus* Blume belongs from genus *Elaeocarpus* consisting of 350 species (Das 2014). The plant is widely distributed across Madagascar, India, Southeast Asia, Malaysia, China, and Japan as well as Australia, Fiji, and Hawaii (Sukari et al. 2013). It is known by different names in several countries, namely, medang teja (Malaysia), Jolpai or Indian olive (Bangladesh), rugged oil fruit in England, and "Olivier" in Mauritius (Sukari et al. 2013). The members of this genus usually grow in hill slopes and ridges with sandy to clay soil. Nevertheless, they also grow well in sunny to moderately shaded areas only requiring fertile, moist, but well-drained soil (Das 2014). Few pharmacological studies on this plant have expressed promising results.

In this present chapter, we attempt to review the traditional, pharmacological, and ethnobotanical studies amassed till date on *Elaeocarpus floribundus* Blume.

The Elaeocarpaceae Family

The species of Elaeocarpaceae are mostly tropical and subtropical with most species evergreen. They are commonly distributed in Madagascar, Southeast Asia, Malaysia, Eastern Australia, New Zealand, West Indies, and Chile (Zaman 2016). The plants are hermaphrodite or dioecious and bear flowers clustered in inflorescences (Wiart 2006). These species are distinguished by three main botanical features: the blade, the petiole, and the inflorescences (Wiart 2006). The blades are dark green, elliptical, and glossy with petioles that are straight, woody, and kneeled at the apex. The inflorescences are the most recognizable feature with terminal bell-shaped flowers that matures into olive-like fleshy drupes. The plant's fruiting stage is usually all round the year, while its flowering stage happens during the September to December period (Fig. 1). The fruits of some species of Elaeocarpaceae are edible with sour, sweet taste, and the stones are often used as beads for rosaries,

Fig. 1 Common botanical features of Elaeocarpaceae: (**a**) bell-shaped flowers, (**b**) olive-like drupes, (**c**) straight woody petioles

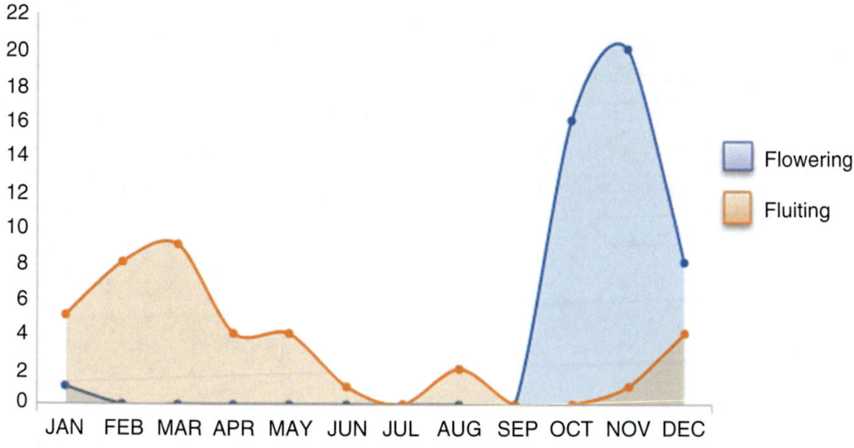

Fig. 2 Flowering and fruiting stage of Elaeocarpaceae species

necklaces, bracelets, and other ornamental objects. The leaves and the barks are also used as local medicines (Burkill et al. 1966) (Fig. 2).

As mentioned previously, one attractive attribute that the Elaeocarpaceae family has to offer is their ability to elaborate an array of steroids or cucurbitacins which so far has been expressed by the Cucurbitaceae and Begoniaceae families (Wiart 2006). This flourishing source of cucurbitacins in this family indicates enthralling potential as cytotoxic agents (Fang et al. 1984; Rodriguez et al. 2003; Ito 2002). One chemotherapeutic evaluation carried by Ito and colleagues on the bark of *Elaeocarpus mastersii* resulted in the isolation of cucurbitacin D and cucurbitacin F as cytotoxic principles and two new ellagic acid derivatives, 4′-*O*-methylellagic acid 3-(2″,3″-di-*O*-acetyl)-*α*-L-rhamnoside and 4,4′-*O*-methylellagic acid 3-(2″,3″-di-*O*-acetyl)-*α*-L-rhamnoside (Sukari et al. 2013). Other prominent indolizidine alkaloids, such as elaeocarpenine, isoelaeocarpine, isoelaeocarpicine, and elaeocarpine, were successfully isolated from leaves of *Elaeocarpus fuscoides* (Ito 2002; Katavic et al. 2006; Piao et al. 2009). Furthermore, another study isolated three flavonoids from leaves of *Elaeocarpus lanceofolius* and were identified as 4′-methylmyricetin, myricetin, and 3-*O*-rhamnoside (Ray et al. 1976).

Another line of interest in Elaeocarpaceae is the presence of abundant indolizidine alkaloids which has fascinated researchers due to their enzyme inhibitory effect on glucosidases. Their ability to inhibit the enzyme activity of glucosidases is credited to similar structural differences to glucose. Hence, Elaeocarpaceae has promising potential to be unveiled and be delved into the treatment/management of diabetes, cancer, and HIV (Wiart 2006). However, only a small number of plants from the Elaecarpaceae species have been studied. Thus, it is of absolute must to bring to light the other members of the family and probe for valuable pharmacological activities (Fig. 3).

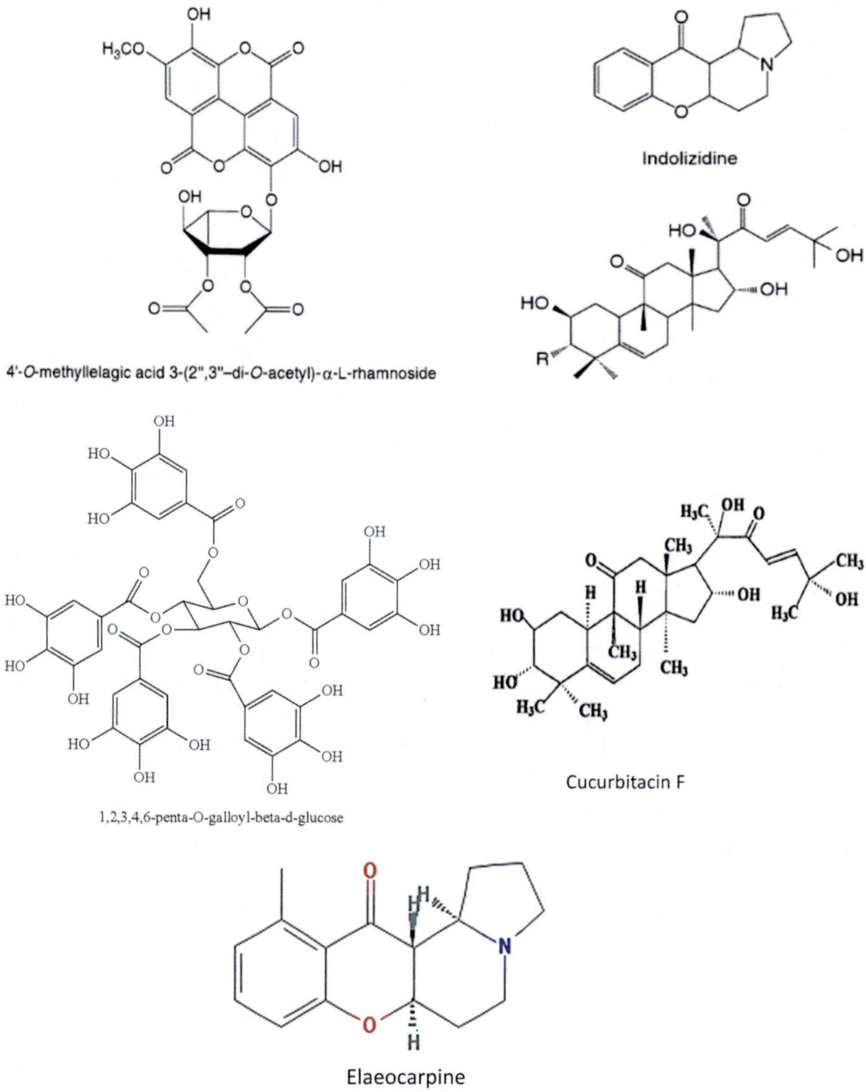

Fig. 3 Examples of bioactive compounds from the Elaeocarpaceae family

Elaeocarpus Genus

The name *Elaeocarpus* is derived from a Greek word form by a combination of the words Elaei and Carpus which mean "wild olive" tree and fruit, respectively; thus, the seed from the fruit is known as *Elaeocarpus*. This genus constitutes more than 200 species which are widely distributed in warm regions from Madagascar, Socotra, and Mauritius westward to Hawaii eastward, including the whole

Southeast Asia regions, Himalaya to southern China, Formosa and southern Japan, Celebes, Moluccas, New Guinea, the eastern part of Australia, New Caledonia, New Zealand, and other islands of the Pacific regions. The center of the distribution is New Guinea with over 100 species (Coode 1984). Many species of *Elaeocarpus*, especially in Southeast Asia regions, are rather small trees not reaching to timber size. However, some from New Guinea and Australia regions are large enough, attaining 30–40 m tall and 100 cm in diameter. Woods of light colors are suitable for drawer sides, boxes, match box and splints, interior finish, moldings, lower class furniture, and veneer. *Elaeocarpus* trees are planted for ornamental purposes as bunches of blue fruits all over the tree which give a very scenic look (Shah et al. 2011).

This genus has been used traditionally to treat ulcers, rheumatism, piles, pneumonia, and leprosy (Nandy et al. 2015a; b). Plants of this genus have been reported to be used in traditional medicine particularly in India. Species from this genus that is much publicized is *Elaeocarpus ganitrus* commonly known as Rudraksha which holds a very important place in the Hindu religion. It is used in folk medicine in treatment of stress, anxiety, depression, palpitation, nerve pain, epilepsy, migraine, lack of concentration, asthma, hypertension, arthritis, and liver diseases (Nandy et al. 2015a, b). Studies on the *Elaeocarpus ganitrus* have revealed sedative, anti-epileptic hypnotic, tranquillizing, and antihypertensive activities (Sakat et al. 2009). Other species of this genus has expressed worthy pharmacological properties like *Elaeocarpus grandis* indolizidine alkaloids grandisine A and isoelaeocarpiline compounds bind to opoid receptor and have analgesic effect (Carroll et al. 2005). Another study showed that *Elaeocarpus sylvestris* which has 1,2,3,4,6-penta-*O*-galloyl-b-D-glucose as a major chemical constituent possesses several biological effects, such as antitumor, vasodilatory, anti-inflammatory, neuroprotective, hepatoprotective, and anti-angiogenic activities (Choi et al. 2002; Kang et al. 2005; Lee et al. 2003; Oh et al. 2001; Pae et al. 2006). Numerous alkaloids and phytochemicals have been isolated over the past years on different species of *Elaeocarpus*. Fig 4 shows the various compounds isolated from this genus. A chloroform-soluble extract of the bark of *Elaeocarpus mastersii* was found to exhibit significant cytotoxic activity when investigated against a series of human cancer cell lines (Kinghorn et al. 1999). In addition, various compounds have been isolated from this genus whixh tend to justify the need for further studies to probe into its therapeutic properties.

Elaeocarpus floribundus Blume

Elaeocarpus floribundus Blume, also locally known as "olive," has been used as phytomedicine in African and Asian folkloric medicine to treat and/or manage ailments such as dysentery, diabetes, and inflamed gums. *Elaeocarpus floribundus* Blume belongs to the Elaeocarpaceae family with its genus *Elaeocarpus* consisting of 350 species (Das 2014). The plant is widely distributed across

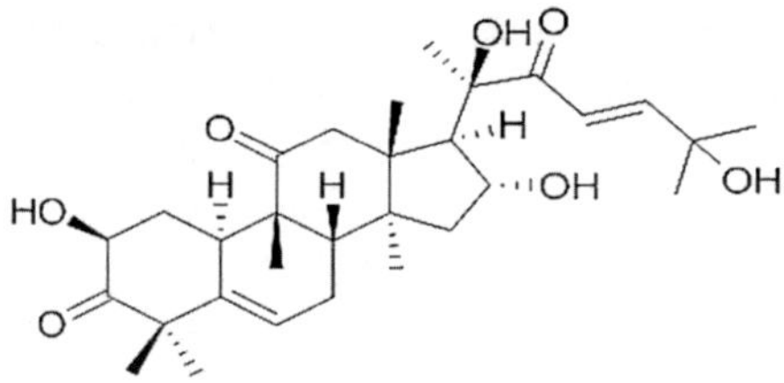

Cucurbitacin D

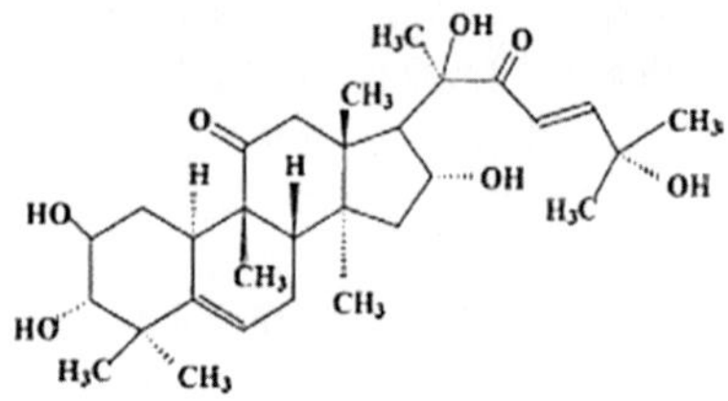

Cucurbitacin F

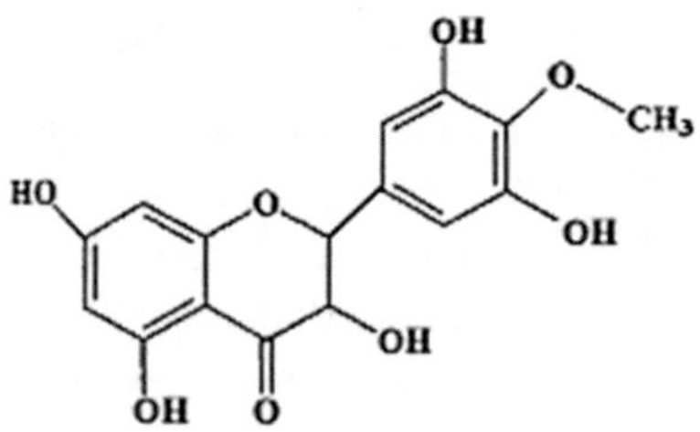

Mearnsetin

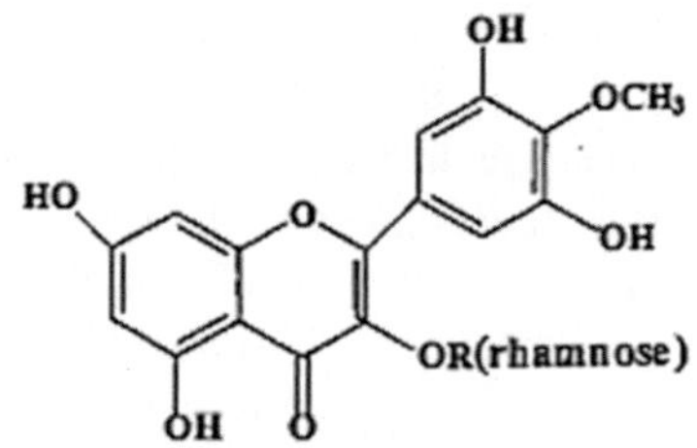

Mearnsitrin

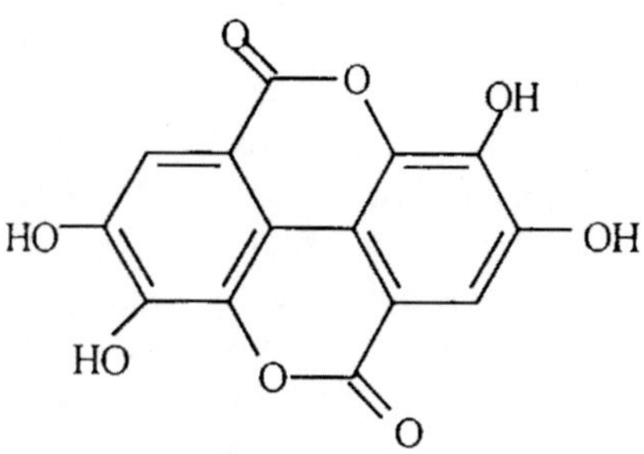

Ellagic acid

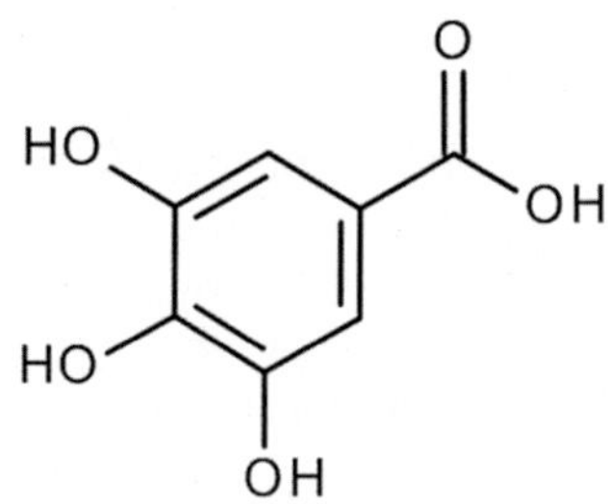

Gallic acid

Fig. 4 Various isolates of *Elaeocarpus* genus

Madagascar, Mauritius, India, Southeast Asia, Malaysia, China, and Japan as well as Australia, Fiji, and Hawaii (Sukari et al. 2013). This species has been introduced in some warm countries like Madagascar and Mauritius. It is known by different names in several countries, namely, medang teja (Malaysia), Jolpai or Indian olive (Bangladesh), rugged oil fruit in England, and "Olivier" in Mauritius (Sukari et al. 2013). The members of this genus usually grow in hill slopes and ridges with sandy to clay soil; nevertheless they also grow well in sunny to moderately shaded areas only requiring fertile, moist but well-drained soil (Das 2014). It is commonly grown in lowland hills and mountains up to an altitude of 1500 m in India, Burma, Thailand, Vietnam, Malaysia, Indonesia, and the Philippines (Wiart 2006).

Fig. 5 *Elaeocarpus floribundus* Blume: (**a**) tree, (**b**) flower, (**c**) leaf, (**d**) seed, (**e**) fruits

Botanical Description

The name *Elaeocarpus* has been attributed to the plant in reference to the shape of its fruit which has the appearance of the olive fruit (Crayn 2006). It is an evergreen medium-sized tree growing around 49 m tall with spreading crown and clean bole of around 12–16 m in length (Zaman 2016).The fruits are green, fleshy drupes about 2.5–3.5 cm long, narrow at the base and with pointed apex (Zaman 2016). Additionally, theses acidic drupes have smooth surfaces with a sour taste. The single seed embedded inside the fruit consists of a stone 3-celled having a spindle shape (Das 2014). The leaves are elliptical, ovate shaped with pointed tip and toothed margin with a length varying from 6.5 to 19 cm and width 3 to 9 cm (Zaman 2016). Some of the leaves are also red or orange in color. The plant usually has white and creamy white flowers which always occur in cluster form with finely fringed petals (Zaman 2016). The barks of the trees are brown, rough, and lenticelled with the inner bark pale yellow in color. The leaves, buds, flowers, and fruits are often dotted with scabby or pimply pustules (Wiart 2006). Fruit development shows distinct phases. During 4–9 weeks after flowering, fruits grow fast; during 9–17 weeks, they grow rather slow; and from 17 weeks onward, growth is again fast until maturity is reached about 26 weeks after flowering. *Elaeocarpus floribundus* can be propagated by stones, which should be sown in the shade and have about 15% germination in 4–8 months (Figs. 5 and 6).

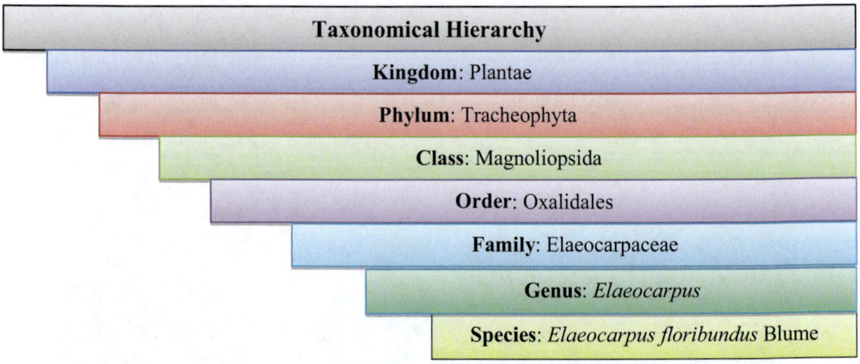

Fig. 6 Taxonomical classification of *Elaeocarpus floribundus* Blume

Traditional Uses of Elaeocarpus floribundus *Blume*

In Mauritius, the fruit is famously known for its pickle in vinegar or brine or as chutney and also enjoyed by many as "fruit cristallisés" also known as candied fruit (Das 2014). The wood of the plant, although under exploited, is used by some for light interior construction and as plywood (Das 2014).

Medicinally, various parts of the plant are used to treat several diseases. In Mauritius, a leaf decoction is used as a treatment for diabetes and hypertension (Gurib-Fakim et al. 1996), while in other parts of the world, such as Sumatra, infusion of leaves and barks is utilized as mouthwash for treating inflamed gums (Wiart 2006). In Malaysia, the leaves are used as poultice to treat ulcers and its extract is consumed as tonic (Corner 1988). Additionally, the leaves are also used in rheumatism, and its fruit has been recommended in dysentery and diarrhea (Zaman 2016). Regardless, little literature has been documented on the traditional used of *Elaeocarpus floribundus* Blume even though it is highly used in some countries for treatment of diseases such as India.

Chemical Composition of Elaeocarpus floribundus *Blume*

According to the limited literature on its chemical composition, the leaves of the plant have been reported to compose of vitamin C, myricetin, myricitrin, mearnsetin, and ellagic acid (Zaman 2016). The fruit of elaeocarpus consists mainly of tannin and an extensive amount of plant acid including citric acid which is responsible for the acidic taste to the fruit such as acidic portions of water-soluble dietary fiber of the fruits comprised of rhamnose (11%), arabinose (26%), galactose (35%), and uronic acid (27%). The polymeric portion is composed of α-n-galactopyranosyl uronic acid chains with side chains of rhamnose, galactose, and arabinose residues (Zaman 2016). Figure 7 shows some of the chemical constituent in the leaves and fruit of this plant.

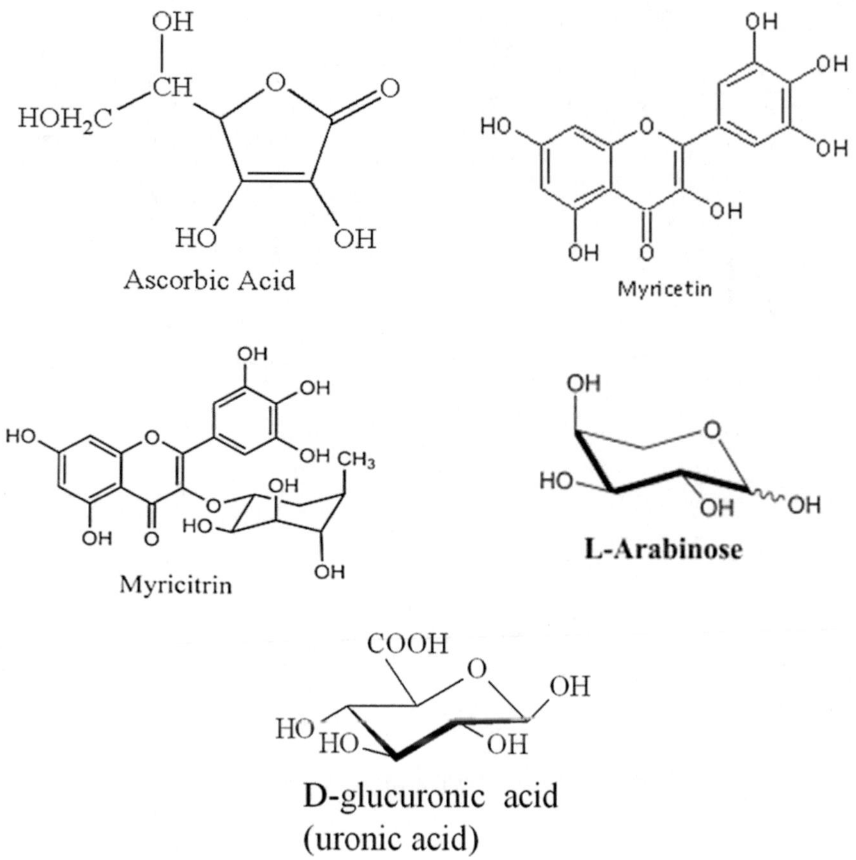

Fig. 7 Some chemical constituents of *Elaeocarpus floribundus* Blume fruits and leaves

Pharmacological Activity of *Elaeocarpus floribundus* Blume

Till now, only a few studies have been conducted on the plant. According to Zaman (2016), skin care products consisting of *Elaeocarpus floribundus* Blume extracts expressed excellent active oxygen scavenging action which is postulated to have notable antiaging and skin whitening activities. The biological activities of *Elaeocarpus floribundus* Blume have been reported to be attributed to the array of phytochemicals. Various phytochemical studies have expressed high level of phenolic compounds such as tannin, flavonoids, phenolic acid, and anthocyanins (Zaman 2016). In confer with phytochemical evaluation, Sukari et al. (2013) reported that the methanolic extracts of leaves possess more phenolic substances than the bark of the plant. In the same study, Sukari and colleagues were the first researchers to isolate three triterpenoid, namely, friedelin, epifriedelanol, and β-sitosterol (Fig. 8). When friedelin and epifriedelanol were tested, they have

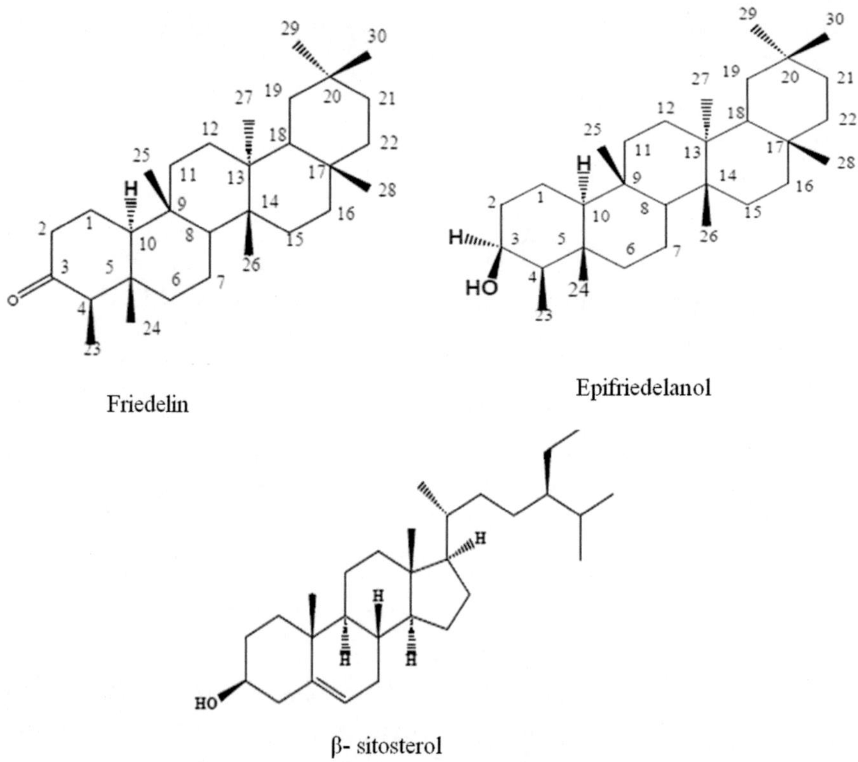

Fig. 8 First three isolation compounds of *Elaeocarpus floribundus* Blume

shown to have potential cytotoxic activities against cancer cells. Among these two isolates, friedelin showed the most potent inhibitory effect against HeLa cancer cell with IC_{50} value of 3.54 ± 0.30 μg/mL (Sukari et al. 2013). Moreover, friedelin advertises strong cytotoxic activities on the proliferation of four human cancer cells, namely, A375, L292, HeLa, and THP-1 (Lu et al. 2010). These results converge to promising avenue in the field of cancer and antitumor effects from the plant. *Elaeocarpus floribundus* has also been reported to exhibit potential antioxidant activity (Zaman 2016). However, in this scenario, the methanolic extract of the stem bark was found to have better free radical scavenging ability against DPPH compared to the leaf extract (Sukari et al. 2013).

Traditionally the plant has been used as an antiseptic and to treat certain infection (Pullaiah 2006). Antibacterial properties of the leaf extract have been validated against Gram-positive and Gram-negative human pathogenic bacteria, having zone diameter of inhibition (ZDI) of 18–22 mm (Zaman 2016). The extracts were also found to be natural sources of both reducing and stabilizing agents for the synthesis of nanoparticles with antibacterial activity against some spoilage and pathogenic bacteria *like Staphylococcus aureus*, *Escherichia coli*, and *Pseudomonas aeruginosa* (Khalil et al. 2013). Moreover, the fruits of the plant have disclosed to have growth inhibitory activity against food-borne bacteria (Mandal et al. 2017). From well diffusion technique, Mandal and co-workers found that leaves extracts were effective against a number of *S. aureus*, for which the ZDIs ranged 10–22 mm. Effective susceptibility was also found against *E. coli* (ZDI, 16 mm*) and P. aeruginosa* (ZDI, 18 mm). Thus, *Elaeocarpus floribundus* could provide advancement as non-antibiotic phytotherapeutics against bacterial infection to humans.

Elaeocarpus floribundus has also been utilized in treatment of infections in folk medicines. However there is no scientific validation of this practice. One study contributed to the validation of the use of the plant against infection. An in vitro study on fruits of *Elaeocarpus floribundus* by Mandal et al. (2017) in combating MRSA infection by the agar well diffusion method showed that mesocarp-epicarp extract expressed potential bacterial growth inhibitory activity against the tested MRSA isolates (Mandal 2017). The seed, however, exhibited no inhibitory activities. Antibacterial inhibition against MRSA tend to confirm the evidence that the fruit and leaves of the plant will unfold an outlet for its use in opposition to bacterial infections.

Based on the theraputic potential of plants from the Elaecarpaceae to inihibit key enzymes of clinical relevance, a recent study was conducted to evaluate the enzyme inhibitory potential of the leaves of the plant against four key enzymes: α-amylase, α-glucosidases, acetyl-cholinesterase, and elastase. As established by the Elaeocarpaceae family, the methanolic extract of the leaves expressed excellent α-glucosidase inhibition with an IC_{50} value lower than the positive control acarbose which is known as an antidiabetic drug. Followed by excellent α-glucosidase inhibition, the extracts were effective against α-amylase and elastase. Phytochemical and antioxidant analyses also divulge the presence of very high content of phenolic compound and good free radical scavenging power which could be responsible for these biological activities. These preliminary data warrant further pharmacological validation of its use as antidiabetic, antiaging components and also its emergence as a functional food. To sum, *Elaeocarpus floribundus* should be further exploited for its therapeutic and pharmacological activities since little scientific elucidation has been carried out on its use in treatment of diseases.

Conclusion

Enclosing all the present evidences till date on *Elaeocarpus floribundus* Blume, this species could potentially be used as a multi-targeted intervention with its multiple therapeutic effects in diabetes, cancer, infection, and aging. More research needs to be designed to discover new chemical entities and pharmacological activities that this plant has to offer. More ethnobotanical and ethnopharmacological studies should be executed to provide more drive and weight to support and validate its uses.

References

Abbasi A, Khan M, Ahmad M, Zafar M, Jahan S, Sultana S (2010) Ethnopharmacological application of medicinal plants to cure skin diseases and in folk cosmetics among the tribal communities of North-West Frontier Province, Pakistan. J Ethnopharmacol 128(2):322–335

Burkill IH, Birtwistle W, Foxworthy FW, Scrivenor JB, Watson JG (1966) A dictionary of the economic products of the Malay peninsula. Kuala Lumpur, Malaysia, Governments of Malaysia and Singapore, Ministry of Agriculture and Cooperative

Butler M (2004) The role of natural product chemistry in drug discovery. J Nat Prod 67(12):2141–2153

Choi BM, Kim HJ, Oh GS, Pae HO, Oh H, Jeong S (2002) 1,2,3,4,6- Penta-O-galloyl-beta-D-glucose protects rat neuronal cells (Neuro 2A) from hydrogen peroxide-mediated cell death via the induction of heme oxygenase-1. Neurosci Lett 328:185–189

Coode MJE (1984) *Elaeocarpus* in Australia and New Zealand. Kew Bull 39:509–586

Crayn D, Rossetto M, Maynard D. 2006. "Molecular phylogeny and dating reveals an Oligo-Miocene radiation of dry-adapted shrubs (former Tremandraceae) from rainforest tree progenitors (Elaeocarpaceae) inAustralia". American Journal of Botany 93(9):1328–1342

Dadhich A, Rishi A, Sharma G, Chandra S (2011) Phytochemicals of *Elaeocarpus* with their therapeutic value: a review. Int J Pharm Bio Sci 4(3):591–598

Das N (2014) The effect of seed sources variation and presowing treatments on the seed germination of *Acacia catechu* and *Elaeocarpus floribundus* species in Bangladesh. Int J For Res 2014:1–2

Fang X, Phoebe CH, Pezzuto JM, Fong HHS, Farnsworth NR, Yellin B, Hecht SM (1984) Plant anticancer agents, XXXIV. Cucurbitacins from *Elaeocarpus dolichostylus*. J Nat Prod 47(6):988–993

Gurib-Fakim A (2006) Medicinal plants: traditions of yesterday and drugs of tomorrow. Mol Asp Med 27(1):1–93

Gurib-Fakim A, Guého J, Bissoondoyal MD (1996) Plantes médicinales de Maurice, tome 2. Editions de l'Océan Indien, Rose-Hill, Mauritius, p 532

Ito A (2002) Ellagic acid derivatives and cytotoxic cucurbitacins from *Elaeocarpus mastersii*. Phytochemistry 61(2):171–174

Kang DG, Moon MK, Choi DH, Lee JK, Kwon TO, Lee HS (2005) Vasodilatory and anti-inflammatory effects of the 1,2,3,4,6-penta-O-galloylbeta- D-glucose (PGG) via a nitric oxide-cGMP pathway. Eur J Pharmacol 524:111–119

Katavic P, Venables D, Forster P, Guyme G, Carroll A (2006) Grandisines C–G, indolizidine alkaloids from the Australian rainforest tree *Elaeocarpus grandis*. J Nat Prod 69(9):1295–1299

Khalil MMH, Ismail EH, Baghdady KZ, Mohamed D (2013) Green synthesis of silver nanoparticles using olive leaf extract and its antibacterial activity. Arab J Chem:1–23

Kinghorn AD, Farnsworth NR, Soejarto DD, Cordell GA, Pezzuto JM, Udcani GO, Wani MC, Wall ME, Navarro HA, Kramer RA, Menendez AT, Fairchild CR, Lane KE, Forenza S, Vyas DM, Lam KS, Shu YZ (1999) Novel strategies or the discovery of plant-derived anticancer agents. Pure Appl Chem 71:1611–1618

Lee S, Lee I, Mar W (2003) Inhibition of inducible nitric oxide synthase and cyclooxygenase-2 activity by 1,2,3,4,6-Penta-O-galloyl-β-D-glucose in murine macrophage cells. Arch Pharm Res 26(10):832–839

Nandy B, Hardainiyan S, Kumar K (2015a) *Elaeocarpus Ganitrus* (Rudraksha): a reservoir plant with their pharmacological effects. Int J Pharm Sci Rev Res 34(1):55–64

Nandy B, Hardainiyan S, Saxena R (2015b) Phytochemical investigation of fruit extract of *Elaeocarpus ganitrus*. Int J Pharm Pharm Sci 7(6):415–418

Oh GS, Pae HO, Oh H, Hong SG, Kim IK, Chai KY (2001) In vitro anti-proliferative effect of 1,2,3,4,6-penta-O-galloyl-beta-D-glucose on human hepatocellular carcinoma cell line, SK-HEP-1 cells. Cancer Lett 174:17–24

Pae HO, Oh GS, Jeong SO, Jeong GS, Lee BS, Choi BM (2006) 1,2,3,4,6-penta-O-galloyl-beta-D-glucose up-regulates heme oxygenase-1expression by stimulating Nrf2 nuclear translocation in an extracellular signal-regulated kinase-dependent manner in HepG2 cells. World J Gastroenterol 12:214–221

Piao M, Kang K, Zhang R, Ko D, Wang Z, Lee K, Chang W, Chae S, Jee Y, Shin T, Park J, Lee N, Hyun J (2009) Antioxidant properties of 1,2,3,4,6-penta-O-galloyl-β-d-glucose from *Elaeocarpus sylvestris var. ellipticus*. Food Chem 115(2):412–418

Pullaiah T (2006) Encyclopedia of world medicinal plants. Regency Publication, New Delhi, pp 852–853

Qureshi R, Ghazanfar S, Obied H, Vasileva V, Tariq M (2016) Ethnobotany: a living science for alleviating human suffering. Evidence-Based Complementary and Alternative Medicine 2016:9641692 3 pages

Ray AB, Dutta SC, Dasgupta S (1976) Flavanoids of *Elaeocarpus lanceofolius*. Phytochemistry 15:1797–1798

Rodriguez N, Vasquez Y, Hussein A, Coley P, Solis P, Gupta M (2003) Cytotoxic Cucurbitacin constituents from *Sloanea zuliaensis*. J Nat Prod 66(11):1515–1516

Sakat S, Bodhankar S, Juvekar M, Wankhede S, Mali V, Juvekar A (2009) Study of antihypertensive activity of *Elaeocarpus ganitrus* water extract in renal artery-occluded hypertensive rats. Planta Med 74:9

Shah G, Singh P, Mann A, Shri R (2011) Scientific basis for the chemical constituent and therapeutic use of *Elaeocarpus* species: a review. Int J Institutional Pharm Life Sci 1(1):267–278

Sircar B, Mandal S (2017) Screening of *Elaeocarpus floribundus* fruit extracts for bioactive phytocomponents and antibacterial activity against food-borne bacteria. Int J Res Med Sci 5(8):3665

Sircar B, Mandal M, Mandal S (2017a) Indian olive, *Elaeocarpus floribundus* fruits in combating MRSA infection. J Coastal Life Med 5:501–503

Sircar B, Mandal M, Mondal M, Mandal S (2017b) High performance liquid chromatography analysis and anti-methicillin resistant *Staphylococcus aureus* activity of olive fruit ethanolic extract. Int Res J Pharm 8(7):126–130

Sukari M, Utami R, Khalid N, Rahmani M, Abdul A, Dachriyanus (2013) Phenolic contents, antioxidant and cytotoxic activities of *Elaeocarpus floribundus* Blume. Pak J Pharm Sci 26(2):245

Wiart C (2006) Medicinal plants of Asia and the Pacific. CRC/Taylor & Francis, Boca Raton

Zaman S (2016) Exploring the antibacterial and antioxidant activities of *Elaeocarpus floribundus* leaves. Indo Am J Pharm Sci 3(2):92–97

Medicinal Plants Against Cancer

Sameen Ruqia Imadi, Isra Mahmood, and Alvina Gul

Introduction

"*Medicinal plants are the valuable flora which possess active chemical constituents in any of their part(s) and when used in the treatment of various ailments; create a physiological response*." Since time immemorial the practice of using medicinal plants is prevalent in countries like China, India, Japan, Nepal, Pakistan, Sri Lanka, and Thailand (Hamayun et al. 2006a; Ahmad 2007).

The earliest records regarding the use of medicinal plants are obtained from Mesopotamian civilizations and are as old as 2600 BC (Gurib-Fakim 2006). History shows that plants with medicinal properties had been in use in Assyrians, Babylonians, Chinese, Greeks, and Hebrews civilizations (Hamayun 2007). Furthermore, people in rural areas have always used native plants and herbs as medicines (Shinwari and Khan 2000) partly due to the wide gap of educational, research, and health facilities between the rural and urban areas and partly because of socioeconomic issues (Hamayun et al. 2006b).

The mid of twentieth century witnessed an advancement in various fields of life and simultaneously the use of synthetic chemical compounds for therapeutic purposes became widespread (Hamayun et al. 2006a). However, in a short time, the worth of medicinal plants came to spotlight again when the researchers failed to produce comparatively affordable and safe medicines and thus herbal medicine experienced a revival especially in Western society. It is still in practice throughout the world along with corresponding therapies such as traditional Chinese medicine (TCM), homeopathy, and osteopathy (Hamayun et al. 2006b).

According to an estimate, 20,000 species of plants are used throughout the world for medicinal purposes (Hamayun 2007) and 50% of all the clinically used drugs are obtained from natural products and their derivatives (Gurib-Fakim 2006;

S. R. Imadi · I. Mahmood · A. Gul (✉)
Atta-ur-Rahman School of Applied Biosciences, National University of Sciences and Technology, Islamabad, Pakistan

M. Ozturk, K. R. Hakeem (eds.), *Plant and Human Health, Volume 1*,
https://doi.org/10.1007/978-3-319-93997-1_5

Rosangkima and Prasad 2004). World Health Organization (WHO) reckons that 80% of the world's population depends on plants for health services (Shinwari and Gilani 2003; Saumya et al. n.d.; Madhuri and Pandey 2008) and in developed countries 25% of all medicines prescribed contain ingredients derived from medicinal plants (Qureshi et al. 2009).

Cancer is the second leading cause of death after cardiovascular diseases worldwide (Jackson 2000). According to American Cancer Society, 3500 million people are killed every year from cancer (American Cancer Society n.d.). Cancer is defined as "a group of diseases that are characterized by the presence of malignant cells which proliferate in an unrestricted manner and ultimately invade the tissues and disseminate the organs (metastasis) to form secondary areas of growth known as secondary tumors" (Moura et al. 2001).

According to National Cancer Institute (NCI), 35,000 species of plants out of 250,000–350,000 are used for anti-cancerous purposes worldwide whereas, in Pakistan, 400–600 out of 6000 species of wild plants are considered to be of medicinal worth (Hamayun 2007). Saumya reported that 60% of anti-cancerous drugs used clinically are obtained from natural sources (Saumya et al. n.d.).

After many years of extensive research on medicinal plants possessing chemotherapeutic properties (*Aphanamixis polystachya, Alhagi pseudalhagi, Annona squamosa, Calamus rotang, Cirsium rhinoceros, Terminalia arjuna, Euphorbia jolkini, Polygonum cuspidatum, Myrica rubra Sieb et Zucc, Centella asiatica, Bupleurum kaoi, Ochrosia elliptica Labill, Stephania tetrandra, Ophiorrhiza mungos, Ornithogalum umbellatum, Taxus brevifolia, Tabernaemontana divaricata, Paederia scandens, Elephantopus scaber, Impatiens balsamina, Coix lachryma, Rhei Rhizoma, Taxus wallichiana, Moringa oleifera, Vitex negundo*, and many others) (Kuo et al. 2005a), scientists were successful in the discovery of anethol, allicin, catechins, curcumin, capsaicin, diallyl sulfide, ellagic acid, eugenol, genistein, lycopene, resveratrol, silymarin, S-allyl cysteine, 6-gingerol, ursolic acid, and other such compounds (Saumya et al. n.d.) with anti-cancerous properties.

The discovery of anti-cancerous compounds met success when North American medicinal plants successfully completed clinical trials and were marketed for use. These plants were licensed for use in ovarian cancer (*Taxus brevifolia*), leukemia, lymphoma, and lung and testicular cancer (*Podophyllum peltatum*) (Gurib-Fakim 2006).

Medicinal plants and their derivatives are effective, economical, and safe with negligible side effects (Hamayun et al. 2006b; Feng et al. 2005; Rahmawati et al. 2006). Furthermore, they are more accessible (Feng et al. 2005), have better compatibility (Kamboj 2000), and are highly acceptable (Feng et al. 2005; Ghimire et al. 2005) with respect to cultural, ethical, and historical issues.

A thorough review of literature shows that countless medicinal plants have been exploited for their antitumor and anticancer potential. This chapter highlights some of the renowned anti-cancerous plants and the mechanism of action of their active constituents as expounded by research. The parts found to be effective for different ailments are summarized in Table 1.

Table 1 Effective parts of medicinal plants

Medicinal plant	Effective part(s)	References
Achyranthes aspera Linn.	Whole plant	Ahmad (2009)
Agrimonia pilosa Ledeb.	Roots	Miyamoto (1988), Wang and Jin (2011)
Allium sativum Linn.	Bulb	Wilson (1997)
Aloe vera Linn. (Burm. f)	Pulp of leaves	Liao (2004)
Annona squamosa Linn.	Fruit, seeds, and leaves	Shankar (2006)
Aphanamixis polystachya	Bark	Arguello (1998)
Aronia melanocarpa L. (Pers.)	Berries, leaves, and juice	Missouri Botanical Garde (n.d.)
Bacopa monnieri L. (Pennell)	Whole plant	Warrier (1993)
Brassica campestris Linn.	Seed and seed oil	Bilov (2005)
Brassica oleracea Linn.	Leaves	Gray (1982)
Camellia sinensis Linn. (Kuntze)	Leaves, flowers, and leaf stems	Top Tropicals (n.d.)
Centella asiatica (L.) Urban	Aerial parts	Physicians Desk Reference for Herbal Drugs (2000)
Crocus sativus Linn.	Leaves, stigma, petals, and stem	Srivastava (2010)
Curcuma longa Linn.	Rhizome	Singh and Khar (2006)
Echinacea purpurea L. (Moench)	Roots	Hu (2000)
Elephantopus Scaber Linn.	Whole-plant extract	Poli (1992)
Nigella sativa Linn.	Seeds and seed oil	
Petiveria alliacea	Roots	Soladoye (2010)
Phaleria macrocarpa (Scheff.) Boerl	Leaves, fruit, and bark	Hutapea (n.d.), Gotama (1999), Yoshida (2000), Harmanto (2001), Winarto (2003), Tsuda (2004)
Tabernaemontana divaricata	Root, leaves, and bulb	
Thespesia populnea L. sol ex Correa	Bark	Friday (2006)
Thuja occidentalis Linn.	Whole pant	Johnston (n.d.)

Achyranthes aspera Linn.

Achyranthes aspera L., also known as devil's horsewhip or prickly chaff flower (Eng), is an annual or perennial (Arshad and Ahmad 2004) herb that belongs to family Amaranthaceae (Table 2) (Arshad and Ahmad 2004; Ayyanar and Ignacimuthu 2008). Although the whole plant (Ahmad et al. 2009) (fruit, stem, leaves, and roots) (Hamayun et al. 2006b; Arshad and Ahmad 2004; Aziz et al. 2005; Subbarayan et al. 2012) is used for medicinal purposes roots of the plant are considered to be most effective due to the presence of triterpenoid saponins.

Table 2 Families of medicinal plants

Plant	Family
Achyranthes aspera Linn.	Amaranthaceae
Agrimonia pilosa Ledeb.	Rosaceae
Allium sativum Linn.	Amaryllidaceae
Aloe vera Linn. (Burm. f)	Xanthorrhoeaceae
Annona squamosa Linn.	Annonaceae
Aphanamixis polystachya	Meliaceae
Aronia melanocarpa L. (Pers.)	Rosacea
Bacopa monnieri L. (Pennell)	Scrophulariaceae
Brassica campestris Linn.	Brassicaceae
Brassica oleracea Linn.	Brassicaceae
Camellia sinensis Linn. (Kuntze)	Theaceae
Centella asiatica (L.) Urban	Apiaceae (Umbelliferae)
Crocus sativus Linn.	Iridaceae
Curcuma longa Linn.	Zingiberaceae
Echinacea purpurea L. (Moench)	Asteraceae
Elephantopus scaber Linn.	Asteraceae
Nigella sativa Linn.	Ranunculaceae
Petiveria alliacea	Phytolaccaceae
Phaleria macrocarpa (Scheff.) Boerl	Thymelaeaceae
Tabernaemontana divaricata	Apocynaceae
Terminalia arjuna (Roxb. ex DC.) Wight & Arn.	Combretaceae
Thespesia populnea L. sol ex Correa	Malvaceae
Thuja occidentalis Linn.	Cupressaceae

Since historic times, *Achyranthes aspera L.* has been in use as a medicinal plant in Indian traditional medical system known as Ayurveda (Bagavan et al. 2008). The plant is known to posses antitumor (Chakrabortya et al. 2002), antibacterial (Aziz et al. 2005), antifertility (Prakash 1996), and anti-inflammatory (Neogy et al. 1969; Vetrichelvan and Jegadeesan 2003) properties.

Achyranthes aspera L. has been used for the treatment of fever (Chakrabortya et al. 2002; Bhom and Liersch 1992; Girach and Khan 1992; Tang and Eisenbrand 1992), asthma (Ahmad et al. 2009; Bhom and Liersch 1992; Girach and Khan 1992; Tang and Eisenbrand 1992; Singh 1995), dysentery (Ahmad et al. 2009; Chakrabortya et al. 2002; Bhom and Liersch 1992; Girach and Khan 1992; Tang and Eisenbrand 1992), hypertension (Chakrabortya et al. 2002; Bhom and Liersch 1992; Girach and Khan 1992; Tang and Eisenbrand 1992), diabetes (Bhom and Liersch 1992; Girach and Khan 1992; Tang and Eisenbrand 1992), pneumonia, colic (Misra et al. 1991), diarrhea, cold (Borthakur and Goswami 1995), constipation (Hamayun et al. 2006a), renal dropsy, leprosy (Jain 1991), skin diseases, and abdominal pains (Ahmad et al. 2009) and for removing kidney stones (Hamayun et al. 2006a). Ash of plant mixed with honey is given to treat seasonal cough (Ahmad et al. 2009; Jain 1991). A decoction of plant is used for diuretic purposes (Hamayun et al. 2006a; Misra et al. 1991) and for treating toothache (Ahmad et al. 2009).

The plant has also been exploited for its anti-cancerous potential and has been used for the treatment of cancer for centuries (Goyal et al. 2007). Either used alone or in combination, *Achyranthes aspera L.* can treat solid tumors and pancreatic cancers (Subbarayana et al. 2010).

Chakrabortya et al. (Chakrabortya et al. 2002) showed that *Achyranthes aspera L.* exerts inhibitory effects on a tumor promoter (12-O-tetradecanoylphorbol-13-acetate) in Raji cells in the process of carcinogenesis. He and colleagues studied the antitumor effects of the methanolic extract (MeOH) consisting of alkaloid, non-alkaloid, and saponin fractions on Raji cells and found that methanolic extract (MeOH) inhibits tumor formation by 76%. The non-alkaloid fraction (containing mainly ecdysterone and dihydroxyketone) was found to be the most effective among the aforementioned fractions and possess 96.9% (at 100 mg/mL) anticarcinogenic activity (Chakrabortya et al. 2002). Several mechanisms have been proposed by which methanolic extract (MeOH) and its non-alkaloid fraction are thought to inhibit tumor formation. According to Chakrabortya et al. (2002), tumor inhibition is the consequence of the delay in tumor formation which results in the formation of fewer papillomas (Chakrabortya et al. 2002). Tahiliani and Kar (2000) proposed that plant extract works by free radical scavenging mechanism (Tahiliani and Kar 2000).

Leaf extract (LE) of *Achyranthes aspera L.* has antiproliferative activity on pancreas, prostrate, lung, and colon cancer cell lines in a time- and dose-dependent manner as documented by Subbarayana et al. (Subbarayana et al. 2010). Pancreatic cancer cells have found to be the most sensitive to leaf extract (LE) as compared to other cancer cell lines mentioned before. Leaf extract (LE) inhibits angiogenic and metastatic genes, two of the vital genes required for the cancer cells to successfully survive, proliferate, and spread to other body parts. Therefore, pancreatic cancer cells have found to perish through the prevention of tumor enlargement and metastasis (Subbarayana et al. 2010).

In a second study published 2 years later, Subbarayan et al. (2012) proposed another mechanism by which *Achyranthes aspera L.* inhibits pancreatic cancer in athymic mice. They suggested that leaf extract (LE) of *Achyranthes aspera* induces the transcription of caspase cascade especially caspase-3 and it suppresses or blocks the expression of Akt-1 kinase. Caspase-3 transduces the apoptotic signal and aids the cell in the execution of apoptosis (Salvesen and Dixit 1997). However according to Subbarayan et al. (2012), activation of caspase-3 and suppression of Akt-1 kinase hampers the proliferation of cancer cells and mediates their death through apoptosis (Subbarayan et al. 2012).

Agrimonia pilosa LEDEB

Agrimonia pilosa LEDEB. is a perennial plant (Anonymous n.d.-a, n.d.-b) that belongs to family Rosaceae (Table 2). According to Miyamoto et al. (1988), roots of the plant are used for the treatment of cancer due to the presence of tannins (Miyamoto et al. 1988; Wang and Jin 2011).

Agrimonia pilosa Ledeb. is known to posses antiviral (Khanina et al. 2010), antioxidant (Khanina et al. 2010; Zhu et al. 2009), nitric oxide (NO)-scavenging (Taira et al. 2009; Wang et al. 1984), antihemorrhagic (Wang et al. 1984), anticarcinogenic (Khanina et al. 2010), anti-inflammatory (Khanina et al. 2010), antiplatelet (Wang et al. 1985), acetylcholinesterase inhibitory (Jung and Park 2007), vasodilating, and antibacterial activities (Khanina et al. 2010).

In their pursuit to explore the anti-cancerous properties, Koshiura et al. (1985) conducted research on the methanol extract obtained from the roots of *Agrimonia pilosa Ledeb.* to study its effects on murine syngeneic and allogenic tumors in vivo. They deduced that the extract works by stimulating the macrophages which in turn activate cytotoxic lymphocytes. Their results clearly indicated that premedication of tumors with methanolic extract was more effective than post-medication with it (Koshiura et al. 1985).

According to Sugi (1977), *Agrimonia pilosa Ledeb.* is famous for its antitumor properties and is even used today in China (Sugi 1977). Agrimoniin, a tannin compound, is considered to be the main antitumor component of this plant. Miyamoto et al. (1987) studied the effects of agrimoniin on solid and ascites-type tumors of rodents. They found that when administered intraperitoneally to mice at doses of 10 mg/kg of the body weight, agrimoniin strongly dejected the growth of tumors. In addition, this compound also increased the number of peripheral white blood cells (WBCs) and monocytes. Spleen size was also increased due to the intraperitoneal injection of agrimoniin. From their research, Miyamoto et al. (1987) concluded that agrimoniin, antitumor tannin obtained from *Agrimonia pilosa Ledeb.*, inhibits tumor formation by boosting the immune response through the action on tumor cells (Miyamoto et al. 1987).

Apart from boosting immune response, agrimoniin can reduce tumor formation by other means too as elucidated by Wang and Jin (2011). Agrimoniin can also induce apoptosis in the tumor cells through the generation of reactive oxygen species (ROS), by increasing Ca^{2+} in the cells and by decreasing mitochondrial transmembrane potential. By decreasing transmembrane potential, it opens transition pore channels in mitochondria, thus causing the release of Ca^{2+} from them. When Ca^{2+} concentration in the cell increases, it starts the mechanism of apoptosis (Wang and Jin 2011).

According to a research, water extract of *Agrimonia pilosa Ledeb.* is effective against hepatocellular cancer line SMMC-7721. The growth of SMMC-7721 cells is inhibited by water extract and apoptosis is induced in them by downregulation of Bcl-2 protein and upregulation of p53 protein (Cancer Research 2012). p53 protein, also known as "tumor-suppressor protein," once activated activates a series of other proteins which combine together to strengthen the apoptotic signal through the activation of pro-apoptotic proteins and inhibition of anti-apoptotic proteins.

Allium sativum Linn.

Allium sativa is among the oldest cultivated plants (Thomson and Ali 2003) and belongs to family Amaryllidaceae (Chase et al. 2009) (Table 2). Onion, leek, and chive are its close relatives (Block 2010). It is commonly known as serpent garlic and its bulb is used for medicinal purposes (Wilson et al. 1997).

Medicinal properties of *Allium sativum* rely mostly on the presence of organosulfur compounds in it (Iciek et al. 2009). There are at least 33 different organosulfur compounds in garlic (Bottone et al. 2002). Many studies have shown the chemopreventive effects of garlic extracts and organosulfur compounds derived from garlic. It has antioxidant properties and tumor growth inhibition (Thomson and Ali 2003). The organosulfur compounds in garlic inhibit carcinogen activation by increasing phase II detoxifying processes, helping in cell cycle arrest, fueling the mitochondrial apoptotic pathway, and boosting the acetylation of histones (Iciek et al. 2009).

Active allyl sulfur compounds of garlic have a major contribution in anticancer properties (Song and Milner 2001). Garlic and its components help to retard chemically induced cancer at multiple sites of human and murine cell lines, by inhibiting its proliferation and inducing apoptosis (Milner 2010). *Allium sativa* has the ability to inhibit adenosine deaminase activity in cancerous human bladder tissues (Durak et al. 2007).

Chemically induced tumors can be efficiently blocked by water- and lipid-soluble allyl sulfur compounds of garlic. Its components can block the initiation and promotion phases of carcinogenicity of many compounds including polycyclic hydrocarbons. Some of the antitumorigenic properties are due to changes in cellular thiol and phosphorylation stains (Milner 2001).

Allyl sulfides present in garlic inhibit the initiation as well as propagation of tumorigenesis of many types of cancers in carcinogenesis models. They suppress cell growth and induce apoptosis in cancer cells (Ariga and Seki 2006). Diallyl trisulfide, applied on human colon cancer cell line HCT-15, shows a reduction in cell proliferation rate (Seki et al. 2008). Rats with aflatoxin B-induced hepatocellular carcinoma were fed with garlic powder containing different levels of alliin in them. A strong reduction in cancer was observed in rats fed with garlic containing highest level of alliin (Bergès et al. 2004).

Organosulfur compounds increase the activity of metabolizing enzymes that activate cytochrome P450s and detoxify the carcinogens. They also inhibit the formation of DNA adducts in target tissues, which causes apoptosis and alters the cell cycle, which proves it to be a potent chemopreventive agent (Bianchini and Vainio 2001). Garlic is rich in flavanols, especially kaempferol. This compound contributes to the detoxification of carcinogens (Bilyk and Sapers 1985; Hertog et al. 1992).

In rodents, diallyl disulfide (DADS) inhibits chemically induced colon carcinogenesis. It has anti-promoting effect on tumor cell proliferation (Druesne et al. 2004a). Intake of garlic powder or garlic extract reduces chemically induced skin and mammary gland carcinogenesis (Rao et al. 1990; Schaffer et al. 1996; Sadhana et al. 1988). DADS are also involved in reducing genotoxicity of various carcinogenic compounds (Le Bon et al. 1997; Guyonnet et al. 2000; Sheen et al. 2001).

Diallyl trisulfide (DATS) induces apoptosis in human lung cancer cell lines (A549) by mitochondrium-dependent caspase cascade through decrease in anti-apoptotic Bcl-2, which results in upregulation of ratio of Bax/Bcl-2 and activity of caspase-3, -8, and -9. DATS could be an ideal anticancer drug (Li et al. 2012). Allyl sulfur compounds exhibit protective effects on colonic carcinogenesis by mechanisms including inhibition of carcinogen-induced DNA adduct formation (Amagase and Milner 1996), blockage of cell growth (Matsuura et al. 2006), blockage of cell proliferation (Knowles and Milner 2003; Druesne et al. 2004b), blockage of angiogenesis (Sundaram and Milner 1996), induction of apoptosis, enhancement of carcinogen detoxifying enzymes (Shirin et al. 2001), suppression of carcinogen-activating enzymes (Chen et al. 1998; Chung et al. 2004), inhibition of cyclooxygenase-2 expression (Sengupta et al. 2004), scavenging carcinogen-induced free radicals (Wu et al. 2005), and inhibition of lipid peroxidation (Sengupta and Das 2003).

Garlic extract when administered to colon cancer colo 205 cells reduces percentage of viable cells, induces apoptosis, and increases the levels of Bax, cytochrome c, and caspase-3, but decreases the level of Bcl-2. Raw extract of garlic decreases mitochondrial membrane potential and increases caspase-3 activity and gene expression (Su et al. 2006). Fresh extracts of garlic arrest the growth and alter the morphology of MCF7 breast cancer cells. Intake of this extract results in downregulation of cyclin D1, reduced phosphorylation of ERK1, and increased phosphorylation of elF2-α, which results in morphological changes in cells. Growth is reduced due to reduced expression of hsp27 and sam68, and elevated levels of Rb and p21 (Modem et al. 2012).

Aged garlic extract (AGE) consists of compounds such as S-allylcysteine and S-allylmercaptocysteine. These compounds have potent anticarcinogenic effects. AGE suppresses the sizes and concentration of colorectal adenomas (Tanaka et al. 2006). AGE suppresses the proliferation of three different colorectal cell lines, HT29, SW480, and SW620. AGE enhances adhesion of endothelial cells to collagen and fibronectin and suppresses cell motility and invasion. It also inhibits proliferation and tube formation of endothelial cells. This proves that AGE can be a potent chemopreventive and therapeutic agent for colorectal cancer (Matsuura et al. 2006)

DATS suppresses growth of human colon cancer cell lines HCT-15 and DLD-1. It reduces the amount of cells residing at G1 and S phases. It also results in increased activity of caspase-3. DATS inhibited tubulin polymerization in in vitro cell-free system (Hosono et al. 2005).

Aloe vera Linn. (Burm. f)

Aloe vera is a plant belonging to the family Xanthorrhoeaceae (Akinyele and Odiyi 2007) (Table 2). Pulp of leaves is used for medicinal purposes (Liao et al. 2004). Glycoproteins (lectins) and polysaccharides from *Aloe vera* are claimed to have anti-cancerous effects (Reynolds and Dweck 1999). Isolated polysaccharides, especially Acemannan, have been investigated for antitumor activity in *in vitro* and animal models. Studies have shown the antitumor activity as it reduces tumor burden, shrinks tumor, and increases survival rates by necrosis of tumor cells (Boudreau and Beland 2006). It is also proposed that the anticancer property of *Aloe vera* gel is by stimulating the immune system (Steenkamp and Stewart 2007).

Fresh leaf pulp of *Aloe vera* when given to mice resulted in increase in the level of acid-soluble sulfhydryl content, NADPH cytochrome P450 reductase, NADH cytochrome b5 reductase, glutathione S-transferase, DT-diaphorase, superoxide dismutase, catalase, glutathione peroxidase, and glutathione reductase in liver. It reduces the level of cytochrome P450 and b5. Besides liver, other organs including lung, kidney, and forestomach are positively influenced by *Aloe vera* for detoxifying chemical carcinogens. Hence *Aloe vera* can be used as a potent chemopreventive drug (Singh et al. 2000).

Aloe vera administration affects pleural tumor in rats. This was proved when growth of Yoshida AH-130 ascites hepatoma cells injected into pleura of male inbred fisher rats was evaluated at different times. Aloe proved to be a therapeutic method for cancer treatment in vivo (Corsi et al. 1998). *Aloe vera* pulp was administered to Ehrlich ascites tumors in mice. Aloe extracts were injected at an amount of 55 mg protein/kg of mice, two times in a week for a period of 21 days. Best effects are shown when Aloe is administered prophylactically, but reduction in tumor sizes is also observed. Aloe could be used as a prophylactic medicine for cancer prevention (Akev et al. 2007a). Lectin present in *Aloe vera* leaf pulp also shows a prophylactic effect for tumors but this effect is low as compared to that produced by Aloe extracts. Prophylactic administration of lectin results in decrease in the level of tumor necrosis factor α and serum sialic acid (Akev et al. 2007b).

Aloe emodin (AE), a compound isolated from Aloe species, inhibits the proliferation of adherent Merkel cell carcinoma (MCC) cells. Hence it can be used as an adjunct, in combination with chemotherapeutic agents to increase the anticancer effect (Fenig et al. 2004). AE, when administered in T24 human bladder cancer cells, inhibits cell viability, and induces G2/M arrest and apoptosis of T24 cells. It increases the level of Wee1 and cdc25c which leads to inhibition of cyclin-dependent kinase 1 and cyclin B1 and causes G2/arrest. It also induces p53 and p21 expression and activates caspase-3, which is associated with apoptosis of cells. It also causes increase in Fas/APO1 receptor and Bax expression, but it inhibits Bcl-2 expression (Lin et al. 2005). AE is nontoxic to normal cells but it possesses specific toxicity for neuroectodermal cancer cells (Lina et al. 2002). AE induces cell death is in a dose- and time-dependent manner in human gastric carcinoma cell lines, AGS, and NCI-N87. It causes the release of apoptosis-inducing factors and cytochrome c, which activates caspase-3 and leads to nuclear shrinkage and apoptosis (Chen et al. 2007). AE also shows effects against melanoma cell proliferation. In light of all experiments done on AE, it can be said that AE is a newborn differentiation therapy of cancer (Tabolacci et al. 2010).

Succus aloes treatment in tumorous mice and rats results in reduction of tumor mass, metastatic foci, and metastatic frequency at different stages of tumor development without major tumor growth. *Succus aloe* can be used in combination with 5-fluorouracil and cyclophosphamide as chemotherapy (Gribel and Pashinki 1986).

Effects of *Aloe vera* were measured for phase I and phase II carcinogen-metabolizing enzymes. The results showed that *Aloe vera* induces phase II enzymes as it increases levels of acid-soluble sulfhydryl content, NADPH cytochrome P450 reductase, NADH cytochrome b5 reductase, glutathione S transferase, DT diaphorase, superoxide dismutase, catalase, glutathione peroxidase, and glutathione reductase in liver. Thus it can detoxify chemical carcinogens and drugs (Singh et al. 2000). Di (2-ethylhexyl)phthalate (DEHP) is isolated from *Aloe vera*. It exerts growth inhibitory effects against three leukemic cell lines, K562, HL60, and U937, at concentration less than 100 μg/mL. This compound is considered as an active antileukemic ingredient of *Aloe vera* (Lee et al. 2010). *Aloe vera* shows anticancer effects against Walker tumor. This effect was shown in Wistar rats implanted by Walker 256 carcinoma. It modulates tumor growth by reducing cell proliferation and increasing apoptosis (Tomasin and Gomes-Marcondes 2011).

Annona squamosa Linn.

Annona squamosa belongs to the family Annonaceae (Agroforestry Tree Database 2008) (Table 2). It is commonly known as custard apple (Andrade et al. 2001). Fruit, seeds, and leaves of the plant are used for medicinal purposes (Shanker et al. 2007).

The solvent extract of *Annona squamosa* fruit pericarp has shown cytotoxicity in in vitro cultures for Dalton's lymphoma cells as well as Hela cells (Joy and Remani 2008). Squadiolins A and B isolated from *Annona squamosa* are cytotoxic to Hep G2 hepatoma cells and MDA-MB-231 breast cancer cells. Squafosacin B also had effects against HepG2 and 3B hepatoma cells and MCF-7 breast cancer cells (Liaw et al. 2008). Effects of aqueous and organic seed extracts of *Annona squamosa* are studied on histiocytic tumor cell line, AK-5. It causes apoptotic tumor cell death due to enhanced caspase-3 activity, downregulation of anti-apoptotic genes Bcl-2 and Bcl XL, enhanced generation of intracellular ROS, and decreased level of GSH. Aqueous extracts have more potent activity in killing tumor cells (Pradhasaradhi et al. 2004).

Ethanolic extracts of leaves and stem of *Annona squamosa* have anti-cancerous effects (Bhakuni et al. 1969). Leave extracts of *Annona squamosa* are potent chemotherapeutic agents as they reduce chemically induced oral squamous cell carcinomas in Syrian golden hamsters, at an oral dose of 500 mg/kg of body weight. Leaf extract of *Annona squamosa* also modifies the abnormality in cell surface glycoconjugates in neoplastic formation (Suresh et al. 2007). MCF-7 and K-562 cell lines treated with leaf extracts of *Annona squamosa* showed condensation of nucleus, fragmentation of DNA, downregulation of Bcl-2, generation of reactive oxygen species (ROS), and reduced intracellular glutathione level which results in apoptosis of tumor cell lines (Pradhasaradhi et al. 2005).

Annonaceous acetogenins are extracted from the plant of *Annona squamosa*. They are potent chemopreventive agent, as they deplete ATP level by inhibition of complex 1 of mitochondria and inhibition of NADH oxidase of tumor cell plasma membrane (Alali et al. 1992). Aqueous and ethanolic extracts of *Annona squamosa* reduce the number of tumor, tumor burden, tumor volume, and tumor incidence in 7,12-dimethylbenzanthracene-induced hamster buccal pouch carcinogenesis (Suresh et al. 2010). Antitumor activity of extract containing Annonaceous acetogenins, showed potent antitumor activity against human tumor cell lines especially MCF-7 at IC50 of 0.25 μg/mL and Hep G2 at IC50 0.36 μg/mL. It also inhibited the growth of H22 tumor cells in mice at a rate of 69.55% by oral administration (Chen et al. 2012a).

Squamocin, an acetogenin isolated from *Annona squamosa*, inhibits growth of human leukemia cell line HL-60. Its IC50 value is 0.17 μg/mL. It does not affect bcl-2, bax, and p21WAF1 in cells, but it inactivates MAPK, which results in apoptosis (Ning et al. 2012).

Ethanolic herbal extracts of *Annona squamosa* increase the life span of mice with Dalton lymphoma ascites (DLA) tumor cells. Annoglacin-B is the active ingredient in *Annona squamosa* (Ganesan and Muthuchelian 2011). Annonaceous acetogenins suppress tumor growth in a dose-dependent manner in HepS and S180 (Chen et al. 2012b). Six Annonaceous acetogenins (ACGs) were isolated from seed extracts of *Annona squamosa*. These include annosquacins A–D, annosquatin A, and annosquatin B. All the six show potent antitumor activity against five human tumor cell lines but annosquatin A is more sensitive to MCF-7 and annosquatin B is more sensitive to A549 cell lines (Chen et al. 2012c).

Eleven compounds were isolated from 95% ethanolic extract of *Annona squamosa*. Among these compounds annosquamocin C, 15,16-epoxy-17-hydroxy-ent-kau-ran-19-oic acid, 16,17-dihydroxy-ent-kau-ran-19-oic acid, ent-kaur-16-en-19-oic acid, and annosquamosin B showed inhibitory activity against 95D lung cancer cells at IC50 of 7.78 μmol/L. They also suppressed the growth of A2780 ovarian cancer cells (Sun et al. 2012). *Annona squamosa* bark extract yields an acetogenins known as squamotacin. This acetogenin shows cytotoxicity against human prostate tumor cell line PC-3 (Hopp et al. 1996).

Aphanamixis polystachya

Aphanamixis polystachya belongs to the family Meliaceae (Table 2). It is commonly known as rohituka (World Conservation Monitoring Centre 2006). Its bark is used for medicinal purposes (Arguello et al. 1998).

Rohitukine (Flavopiridol), a compound isolated from the bark of *Aphanamixis polystachya*, has apoptotic effects in xenografts of human hematopoietic tumors HL-60, SUDHL-4, and Nalm/6 when administered intravenously (Arguello et al. 1998). Extract of the plant bark protects the mice bone marrow cells from chromosomal aberrations and mutations caused by gamma radiations, which may result in decreased rate of tumor and cancer genesis (Jagetia and Venkatesha 2006). A triterpene acid known as Amooranin, which is isolated from *Aphanamixis polystachya*, results in inhibition of proliferation of breast and cervical cancers by arresting cell cycle at the stage of G2/M and inducing apoptosis (Govind 2011a).

Flavopiridol inhibits tumor necrosis factor-induced nuclear factor KB activation. It also inhibits the activation of activator protein-1, which is induced by carcinogens. It causes the inhibition of TNF-induced activation of MAPK, including p38MAPK and p44/42MAPK. It also suppresses activation of Akt, which is a cell survival kinase. Flavopiridol inhibits tumorigenesis mediators, adhesion molecule-1, c-Myc, and c-Fos. Hence it is known to cause apoptosis in cancerous cells (Takada et al. 2008).

Ethanolic extracts of *Aphanamixis polystachya* when tested in Swiss albino mice transplanted with Ehrlich ascites carcinoma (EAC) increased the effect of radiation therapy and thus help in treating cancer (Jagetia and Venkatesha 2005). Bark extracts of *Aphanamixis polystachya* cause reduced cancer growth and apoptosis of cancer cells (Dhanamani et al. 2011).

Aronia melanocarpa L. (Pers.)

Aronia melanocarpa belongs to renowned flower family of plants, the Rosacea (Table 2). It is commonly known as chokeberry. Its berries, leaves, and juice are used for medicinal purposes (Missouri Botanical Garden n.d.).

Aronia melanocarpa shows potent anticancer effects. Administration of polyphenol-rich extract of *Aronia melanocarpa* inhibits the growth of breast cancer cells in vivo and in vitro (Kedzierska et al. 2009). Anthocyanins isolated from *Aronia melanocarpa* inhibit the growth of breast cancer cells and human HT-29 colon cancer cell lines (Valcheva-Kuzmanova and Belcheva 2006). Extract of *Aronia melanocarpa* at the concentration of 50 μg/mL for 5 min at 37° has invasive effect on breast cancer cell lines and patients with benign breast cancer cells in vitro. This effect is due to presence of different thiols in the extract of *Aronia melanocarpa* (Olas et al. 2010).

Acetone extract of *Aronia melanocarpa* shows inhibitory effects on L1210 leukemia cells at a concentration of 50 μg/mL. It inhibits topoisomerase I, which leads to apoptosis of leukemia cells (Sueiro et al. 2006). Extract of *Aronia melanocarpa* berries significantly reduces the oxidative stress of platelets in breast cancer cell lines in vitro, which results in reduced tumor growth and tumor volume (Kedzierska et al. 2010, 2012). Extract of *Aronia melanocarpa* is rich in anthocyanins. These anthocyanins induce apoptosis in glioblastoma cell line U373, when treated for a period of 48 h at an IC50 value of 200 μg/mL. Apoptosis of cancer cells is associated with downregulation of MMP-2, -14, -16, and -17. These results show that chokeberry has potent anticancer effects (Thani et al. 2012).

Aronia melanocarpa juice when administered in lymphoblastic leukemia jurkat cell lines inhibits cell proliferation, associated with cell cycle arrest and induction of apoptosis. The results are associated with upregulated expression of tumor-suppressor p73 and active caspase-3, and downregulated expression of cyclin B1 and epigenetic integrator UHRF1. It also increases the release of reactive oxygen species and cytochrome c in cytoplasm. Juice of *Aronia melanocarpa* is efficient in treating human lymphoblastic leukemia HSB-2, Molt-4, and CCRF-CEM cell lines. Chemotherapeutic activity of *Aronia melanocarpa* is due to its selective targeting on lymphoblast-derived tumor cells (Sharif et al. 2012). Semipurified anthocyanin-rich extract of *Aronia melanocarpa* fruit when administered in human HT-29 colon carcinoma cell lines at a concentration of 50 μg/mL results in 60% growth inhibition of cells. This inhibition is associated with G1/G0- and G2/M-phase cell cycle arrest (Malik et al. 2003).

Bacopa monnieri L. (Pennell)

Bacopa monnieri is a common Ayurvedic plant which belongs to Scrophulariacea family (Table 2) of plants (Kumar et al. 1998). It is known as Brahmi in common language. Its whole plant is used as a medicine for various purposes (Warrier et al. 1993).

Bacopa monnieri is shown to have potent antitumor and anticancer activities (Warrier et al. 1993). Whole-plant ethanolic extract of *Bacopa monnieri* when tested in sarcoma-180 culture cells showed inhibition of cell growth with the increase in concentration of extract. The results proved that *Bacopa monnieri* shows anticancer effects in a dose-dependent manner (Elangovan et al. 1995). Ethanolic extract of whole plant of *Bacopa monnieri* shows cytotoxicity for Dalton's lymphoma cells at a concentration of 150 μg/mL. Oral administration of this extract reduces the risk of solid tumor formation (Kumar et al. 1998). Ethanolic extract of *Bacopa monnieri* when administered in mouse S-180 cells results in cytotoxicity in a dose- and time-dependent manner. The peak cytotoxic effect is shown at 48 h of administration of 550 μg/mL. Apoptosis was associated with decreased glutathione level in treated cells (Rohini and Devi 2008).

Stigmasterol, a phytosterol which is isolated from *Bacopa monnieri*, decreases the tumor volume, packed cell volume, and viable cell count and increases the life span of Ehrlich ascites carcinoma transplanted in Swiss albino mice. Antitumor activity of stigmasterol is probably due to activation of protein phosphatase 2A by ceramide causing apoptosis (Ghosh et al. 2011). Extract of *Bacopa monnieri* inhibits tumor cell proliferation and malignant ascites fluid accumulation. Its solvent extract shows apoptosis in Ehrlich ascites tumor cell lines, associated with DNA fragmentation, increased expression of Bax, and decreased expression of anti-apoptotic protein Bcl-2. Possible mechanism of apoptosis is Bax-related caspase-3 activation (Kalyani et al. 2013).

Brassica campestris Linn.

Brassica campestris, commonly known as mustard, belongs to the family Brassicaceae (Table 2). It is commonly known as mustard (Madhuri and Govind 2008). Its seeds and seed oil are used as medicines (Bilov 2005).

Brassica campestris, also known as Brassica rapa, possesses anticancer activity against some malignant and benign tumors (Madhuri and Govind 2008). Seed oil of *Brassica campestris* contains dithiolthiones and isothiocyanates, which are effective in treating different types of cancers and tumors (Govind 2011b). Brassinin is a phytoalexin isolated from *Brassica campestris*. This compound proves its chemopreventive activity in preclinical models mediated by inhibition of indoleamine 2,3-dioxygenase (IDO). Brassinin can suppress the growth of highly aggressive melanoma tumors and can inhibit autochthonous mammary gland tumor in MMTV Neu mice (Banerjee et al. 2008).

Isothiocyanates present in *Brassica* can inhibit cancer cell proliferation (Lim et al. 2009). Active component known as dithiolthiones, present in mustard plant are antitumorous (Craig 2006). Cyclobrassinin is a biologically derived product of Brassinin. It inhibits the preneoplastic mammary lesions in cultures. It can be used as a chemopreventive agent during initiation and promotion of carcinogenesis (Mehta et al. 1995). Rapeseed peptide obtained from this plant has anticancer activity. This peptide causes inhibition in growth of Hela cells by cell cycle arrest and induction of apoptosis at the concentration of 320 mg/L administered for 4 days (Xue et al. 2010).

Bee pollen of *Brassica campestris* is used to increase resistance of body for cancer. Steroid fraction of chloroform extract from bee pollen of *Brassica campestris* shows potent cytotoxicity against prostate cancer cell line PC-3 by enhancing caspase-3 activity, decreasing expression of anti-apoptotic protein Bcl-2, and leading to apoptosis (Wu and Lou 2007). *Brassica campestris* seed extract shows cytotoxicity for cancer cells in vivo and in vitro at a concentration of 30 g (Pezzuto et al. 2007). 70% Ethanolic extract of *Brassica campestris* seed is used for treating cancer due to its antioxidant potential (Ryu et al. 1994). Glucosinolates present in *Brassica campestris* are precursors of isothiocyanates. Glucosinolates and their broken product isothiocyanates are potent anticarcinogen. They inhibit the growth of human-derived hepatoma cell line (HepG2) at IC50 value of 24.5 µM (Hong and Kim 2008). Isothiocyanates are cytotoxic for NMBA-induced esophageal tumorigenesis by inhibition of cytochrome P450 enzymes (Stoner and Morse 1997). Isothiocyanates reduce tumor size and growth in F344 rats transplanted with tobacco-specific nitrosamine 4-(methylnitrosamino)-1-(3-pyridyl)-1-butanone (NNK)-induced lung cancer (Chung et al. 1996). Isothiocyanates induce apoptosis in human colon adenocarcinoma LS-174 and Caro-2 cells (Govind and Madhuri 2006). Glucobrassicin isolated from *Brassica campestris* reduces the effect of tobacco-specific lung carcinogen 4-(methylnitrosamino)-1-(3-pyridyl)-1-butanone (Hecht et al. 2004). Isothiocyanates enhance phase I and phase II biotransformation enzyme activities, and thereby metabolize many chemical carcinogens (Verhoeven et al. 1997). Male CF-1 mice transplanted with colon tumors are protected from tumor formation by *Brassica campestris* seed extracts. Thus *Brassica campestris* seeds can be used as a potent chemopreventive agent (Barrett et al. 1998). Alcoholic extracts of *Brassica rapa* when administered in mice transplanted with Ehrlich ascites carcinoma showed reduction in tumor volume and tumor cell count and increase in survival time and life span. Increase in RBC count and hemoglobin and decrease in WBC count were significant. Treated mice were shown to have decreased levels of malondialdehyde dehydrogenase, increased levels of catalase, and reduced glutathione (Manoj 2010). *Brassica campestris* is reported to protect from prostate, lung, breast, and colon cancers (Sarikamis 2009).

Brassica oleracea Linn.

Brassica oleracea is a cruciferous vegetable. It belongs to the renowned family of Kingdom Plantae, named as Brassicaceae (Brassica oleracea L. n.d.) (Table 2). The plant usually lives at the sea cliffs and coasts (Brassica oleracea-Wild Cabbage (Brassicacea) n.d.). *Brassica oleracea* is a species which contains multiple cultivars (Gray 1982). The most common edible varieties of this plant are broccoli, Brussels sprouts, tronchuda cabbage, cabbage, cauliflower, kale, and kohlrabi (Brassica oleracea n.d.). Leaves of the plant are used for medicinal purposes (Gray 1982).

It shows anticancer properties against many malignant and benign tumors (Madhuri and Govind 2008). The leaves have antioxidant activity, which is dependent on the given concentration of leaves in vitro (Vrchovska et al. 2006). Juice obtained from leaves of cabbage reduces the cell proliferation in human breast cancer cell lines associated with reduction in CDK6 expression and increase in level of p27. Decrease in retinoblastoma protein phosphorylation is also seen. When this juice was given in higher concentration, cell death occurs (Brandi et al. 2005). Broccoli is involved in reduced risk of different cancers. Selenium found in broccoli decreases the incidence of abnormality in rats with chemically induced colon cancers. Selenium from high selenium broccoli also reduces the formation of mammary tumors in rats which were treated with 7,12-dimethylbenz (a) anthracene (Finley 2003).

Sprouts and leaves of *Brassica oleracea* contain ascorbigens, dithiolthiones, and isothiocyanates, which are involved in prevention of bladder, lung, stomach, colon, rectum, and breast cancer (Govind 2011b). Mechanisms by which *Brassica oleracea* is supposed to prevent breast cancer are induction of apoptosis (Ge et al. 1996) and interference with cell cycle-regulatory proteins (Cover et al. 1998). Glucosinolate present in *Brassica oleracea* is a precursor for indole-3-carbinol, indole-3-acetonitrile, and 3,3'-diindolylmethane. All these indoles are potent anticarcinogens (Bradfield and Bjeldanes 1987).

Female Swiss mice transplanted with 1,2-dimethylhydrazine-induced colon tumor when administered with selenium and cabbage showed a reduction in tumor growth and size. Reduction in size of adenomas is also seen (Temple and Basu 1987). Broccoli induces phase II detoxifying enzymes including quinone reductase, NADPH (quinone-acceptor) oxidoreductase, and glutathione-S-transferase leading to apoptosis of cancerous cells (Zhang et al. 1992). Sulforane, an isothiocyanate isolated from *Brassica oleracea*, decreases cell population in human breast cancer cell lines and reduces the size and number of mammospheres formed during breast cancer. Sulforane is also efficient in removing breast cancer stem cells in vivo in mice (Li et al. 2010).

Brassica oleracea shows protective effects from early stages of prostate carcinogenesis (Giovannucci et al. 2003). Combination of tomato and *Brassica oleracea* when administered into Dunning R-3327H prostate adenocarcinoma model showed reduction in tumor growth to up to 52%. It was associated with reduced tumor proliferation and increased apoptosis of tumor cells (Cannene-Adams et al. 2007). Brassica vegetables are xenobiotic metabolizing enzyme modulators which do not let the DNA damage, and prevent prostate cancer (Kristal and Lampe 2002).

Sulforaphane, a compound present in *Brassica oleracea*, inhibits the proliferation of prostate cancer cell line PC-3 by induction of apoptosis. The effect it causes is characterized by appearance of cells with sub-G0/G1 DNA, formation of histone-associated DNA fragments, and cleavage of poly (ADP-ribose) polymerase. Apoptosis is induced due to upregulated Bax, downregulated Bcl2, and activated caspase-3, -9, and -8 (Singh et al. 2004). High intake of *Brassica oleracea* is inversely proportional to the incidence of colorectal cancer and colon cancers in human beings (Wu et al. 2012).

Hydrolysis products of glucosinolates show anti-cancerous properties for lung, stomach, colon, and rectal cancer but have low effects on prostatic, endometrial, and ovarian cancer (Verhoeven et al. 1996). Phenyl isothiocyanates (PEITC) found in *Brassica oleracea*, when administered into transgenic mice with adenocarcinoma of prostate cancer, decreased the incidence of poorly differentiated tumors. Inhibition of carcinogenesis is associated with induction of autophagy and overexpression of E-cadherin (Powolny et al. 2011). *Brassica oleracea*, when consumed in raw form, showed inverse effects for bladder cancer growth. The mechanism it follows is not understood, but when it is thought that these effects are due to presence of isothiocyanates (Tang et al. 2008).

3-Day-old sprouts of *Brassica oleracea* contain a high amount of glucoraphanin. Extract of 3-day-old sprouts when administered to rats transplanted with mammary tumors reduced the incidence, multiplicity, and rate of cancer development (Fahey et al. 1997).

Camellia sinensis Linn. (Kuntze)

Camellia sinensis belongs to a family of evergreen plant, known as Theaceae (Khalaf et al. 2008) (Table 2). This plant is used all around the world as most common beverage known as tea. Leaves, flowers, and leaf stems are used for their medicinal properties (Top Tropicals n.d.).

Tea drinking may prevent diseases which are caused as a result of oxidative stress like cancer. Green tea polyphenols inhibit formation and development of many types of cancers in animal models, including skin, lung, oral cavity, esophagus, stomach, intestine, colon, liver, pancreas, bladder, mammary gland, and prostate cancer (Yang and Wang 2010). Catechins present in *Camellia sinensis* inhibit the production of reactive oxygen species and thus reduce the risk of cancers (Aucamp et al. 1997). Green tea extracts contain catechins which are antioxidants and hence prevent various forms of cancer (Cooper et al. 2005). *Camellia sinensis* inhibits an enzyme named topoisomerase II. This inhibition results in chemopreventive activity of the plant. It is cytotoxic to hepatoma cell line HepG2 (Ramirez-Mares et al. 2004).

Epicatechin isolated from *Camellia sinensis* reduces poorly differentiated epithelial ovarian cancer cells. Epigallocatechin 3-gallate (EGCG), another polyphenol isolated from green tea, suppresses the proliferation of benign and metastatic prostate cancer, and moderately and poorly differentiated epithelial ovarian cancer cells (Ravindranath et al. 2009). Green tea is a promising agent for chemoprevention of prostate cancer, but further studies are needed to reveal its chemopreventive effects (Johnson et al. 2010).

EGCG is the most abundant and most active polyphenol in *Camellia sinensis* (Yang and Wang 2011). It suppresses cell invasion, angiogenesis, and metastasis of cancer (Yang and Wang 2010). It shows chemotherapeutic effects in mice transplanted with Ehrlich ascites carcinoma. It shrinks the tumor by reducing levels of leukocytosis, malondialdehyde, and C-reactive protein (Ei-Mowafy et al. 2010). Nude mice subcutaneously injected by heterotopic tumor SGC-7901 cells were positively affected by administration of EGCG, which inhibits the growth of gastric cancer by 60.4%. Inhibition of gastric cancer is associated with decrease in level of vascular endothelial growth factor protein, endothelial cell proliferation, migration, and tube formation (Zhu et al. 2007). EGCG reduces cell proliferation, induces apoptosis, and decreases androgen receptor, insulin-like growth factor-1, IGF-1 receptor, phosphor-extracellular signal-regulated kinases 1 and 2, cyclo-oxygenase-2, and inducible nitric oxide synthase in ventral prostate cancer cells (Harper et al. 2007).

PANC-1 pancreatic cancer cells injected into nude Balb/c mice were gavaged with EGCG after the formation of tumor. Reduced tumor growth and apoptosis were induced by upregulation of Bim and activation of caspase-3 (Shankar et al. 2013). EGCG inhibits 2D and 3D migration of bronchial cancer cells, so is active to inhibit invasion and migration of tumor cells (Hazgui et al. 2008). EGCG inhibits the growth of cancerous cells but do not affect normal cells. It induces cell damage, DNA damage, cell cycle arrest, and apoptosis in cancer cells. EGCG when administered in colon carcinoma cell lines inhibits topoisomerase I, which is essential for cell survival and has a critical role in DNA metabolism and structure. Inhibition of topoisomerase I may be one of the possible causes of cancer cell destruction by EGCG (Berger et al. 2001). EGCG affects many intracellular signaling pathways which are involved in development of prostate carcinoma. Thus it prevents the risk of prostate cancer (Henning et al. 2011; Zheng et al. 2011).

Intake of green tea reduces the risk for upper gastrointestinal tract cancer, lung cancer, and hepatocellular carcinogenesis. It inhibits the progression of prostate premalignant lesions, in phase II clinical trials. It has shown positive effects against mammary cell carcinogenesis and recurrence of breast cancer (Yuan et al. 2011). EGCG inhibits melanoma-metastasized cell lines SE-0154, NS-1176, GE-0208, and LF-0023 (Ravindranath et al. 2009). Catechins found in *Camellia sinensis* inhibit pancreatic ductal adenocarcinoma (Kurbitz et al. 2011). Regular intake of green tea inhibits carcinogens present in tobacco and thus reduces the risks of lung cancer in smokers (Liang et al. 2007).

Centella asiatica (L.) Urban

Centella asiatica (L.) Urban, also known as Asiatic pennywort (Eng), is a small tropical herbaceous plant that belongs to the family Apiaceae (Umbelliferae) (Jayashree et al. 2003) (Table 2). The plant is rich in saponin-containing triterpene acids; the most worthy of them is Asiatic acid (Singh and Rastogi 1969; Kartnig and Hoffmann-Bohm 1992; The Wealth of India 1992; Inamdar et al. 1996; Brinkhaus et al. 2007; Cheng and Koo 2000; Mauri and Pietta 2000; Physician Desk Reference for Herbal Drugs 2000; Coldren et al. 2003; Yoshida et al. 2005; Bonfill et al. 2006; Pan et al. 2007; Jia and Lu 2008). Aerial parts of the plant (stem and leaves) are generally used for medicinal purposes (Physician Desk Reference for Herbal Drugs 2000).

Centella asiatica (L.) Urban has been used for hundreds of years for treating different types of diseases because of its known pharmacological properties. Chinese herbalist used to call this plant as the "fountain of youth" as it is known for its potential to prolong the life span. Since older times, the plant has been used for wound healing, mental disorders and fatigue, memory improvement (Goh et al. 1995), atherosclerosis, bronchitis, asthma, dysentery, kidney trouble, leucorrhoea, urethritis (Jaganath and Ng 1999), inflammations, tuberculosis, and various skin lesions like leprosy, keloid, psoriasis, and lupus (Utami et al. 2011). The plant is also known to possess fungicidal, antibacterial, antioxidant, anti-allergic (Utami et al. 2011; Kan 1986), anti-ulcer (Cheng and Koo 2000), anti-herpes simplex virus (Yoosook et al. 2000), anti-hepatoma (Lin et al. 2002), and anticancer properties (Kan 1986; Utami et al. 2011).

Anticancer properties of this plant are of particular interest to herbalist and researchers around the world because of increasing rates of cancers and lack of proper and side effect-free treatment for this group of diseases. Asiatic acid, one of the main triterpenes in the plant extract, is known to possess anticancer effect against skin cancer (Park et al. 2005), human breast cancer, gastric cancer, and uterine cancer cells (Yoshida et al. 2005).

Babu et al. (1995) investigated that the crude methanolic extract (CE) and acetone fraction (AF) of *Centella asiatica (L.) Urban* hinder the growth of Dalton's lymphoma ascites tumor cells (DLA) and Ehrlich ascites tumor cells (EAC). Acetone fraction (AF) was found to be more effective than crude methanolic extract (CE) in inhibiting the proliferation of transformed cell lines in a dose-dependent manner. But both the fractions (AF and CE) reduced the ascites and solid tumors by affecting DNA synthesis (Babu et al. 1995).

Inhibition of DNA synthesis and cell cycle arrest impairs the capability of cells to proliferate and survive which results in their death. Hsu et al. (2004) examined the cell cycle arrest of breast cancer cell lines MCF-7 and MDA-MB-231 at S-G2/M phase by *Centella asiatica (L.) Urban*. They found that defect in cell cycle causes the cell cycle arrest and induction of apoptosis (Hsu et al. 2004).

Macrophages are an important defense system of our body and are involved in defense against invading pathogens through the production of nitric oxide (NO). According to Puntureea et al. (2004) *Centella asiatica (L.) Urban* leads to tumor inhibition through the production of nitric oxide (NO). The water extract of *Centella asiatica (L.) Urban* induces the expression of TNF-α gene either in the presence or absence of lipopolysaccharide (LPS) which leads to an increase in the production of nitric oxide (NO) (Puntureea et al. 2004). Water extract of this plant can prevent the tumor development through the modulation of TNF-α gene and NO production by macrophages.

In their research, Bunpo et al. (2004) showed that *Centella asiatica (L.) Urban* extract inhibits the formation of azoxymethane (AOM)-induced aberrant crypt foci (ACF) and intestinal tumorigenesis in male F344 rats. Their results indicated that *Centella asiatica* extract inhibits the formation of AOM-induced aberrant crypt foci (ACF) through the modification of cellular proliferation and induction of apoptosis in colonic crypts (Bunpo et al. 2004).

Bunpo et al. (2005) found that the crude water extract of *Centella asiatica (L.) Urban* reduced the proliferation rate of human colon adenocarcinoma-derived Caco-2 cells. The crude water extract exhibits antitumor activity and is known to arrest the cell cycle in S and G2-M phases through the accumulation of cyclin B1 protein. Caco-2 cells face a significant reduction in their proliferation in a concentration- and time-dependent manner (Bunpo et al. 2005).

Park et al. (2007) proved that Asiatic acid inhibits 12-O-tetradecanoylphorbol 13-acetate (TPA)-induced skin tumorigenesis. 12-O-tetradecanoylphorbol

13-acetate (TPA) induces the production of nitric oxide (NO), and increases the expression of inducible NO synthase (iNOS) and cyclooxygenase-2 (COX-2) (Park et al. 2007). Nosho et al. (2005) and Yagihashi et al. (2000) showed that the overexpression of NO synthase (iNOS) and cyclooxygenase-2 (COX-2) is correlated with colorectal cancers (Yagihashi et al. 2000; Nosho et al. 2005). Moreover COX-2, which is a product of phorbol ester-responsive gene, is known to play a significant role in tumor development. This protein has been found to be upregulated in human epithelial cells (Chen et al. 2004; Chang et al. 2005) as well as in a variety of human cancers. Asiatic acid inhibits tumorigenesis by blocking NO-COX-2 pathway and may also hinder the expression of iNOS and COX-2 independently (Park et al. 2005). These inhibitions collectively suppress tumor development process mediated by 12-O-tetradecanoylphorbol 13-acetate (TPA).

Crocus sativus Linn.

Crocus sativus belongs to the family Iridaceae (Table 2). It is commonly known as saffron. It has been used for medicinal purposes since centuries. Its leaves, stigma, petals, stem, etc. are used as medicines (Srivastava et al. 2010).

Crocus sativus when studied in animal models demonstrated its antitumor and cancer-preventive activities (Abdullaev 2002). Saffron is a dried, dark red stigma of *Crocus sativus*. It possesses potent anticancer and antitumor properties (Abdullaev 2002). It retards tumorigenesis in large number of in vivo animal models (Salomi et al. 1990). Extract of saffron when administered topically results in inhibition of DMBA-induced skin carcinogenesis at a concentration of 100 mg/kg of body weight (Salomi et al. 1990). Oral administration of saffron extract inhibits methylcoanthracene-induced soft-tissue sarcoma in mice (Salomi et al. 1991). Saffron extract is shown to increase the life span of mice transplanted with cisplatin-induced carcinogenesis. This effect is associated with prevention of decrease in body weight, hemoglobin levels, and leukocyte count (Nair et al. 1991a).

Oral administration of saffron extract at a concentration of 200 mg/kg of body weight inhibits the growth of S-180, Ehrlich ascites carcinoma, and Dalton's lymphoma ascites by 111%, 83.5%, and 112.5%, respectively, in mice, and results in increase of life span of tumor-bearing mice (Nair et al. 1991b). Crocin, crocetin, picrocrocin, and safranal are cytotoxic compounds isolated from ethanolic extract of *Crocus sativus* dried stigmas. These isolates show cell growth inhibition in HeLa cells at LD50 of 2.3 mg/mL. Crocin is the most promising saffron compound to be chemopreventive and chemotherapeutic (Escribano et al. 1996). *Crocus sativus* extract is found to be cytotoxic to P38B, S-180, EAC, and DLA tumor cell lines in vitro (Nair et al. 1991b).

Saffron exhibits cytotoxicity to hepatocellular carcinoma cell HepG2 and human cervix carcinoma HeLa cell lines at IC50 value of 800 and 950 μg/mL after 48 h of treatment, respectively. This cytotoxicity is independent of generation of reactive oxygen species (Tavakkol-Ashfari et al. 2008). Crocin, when administered into tumor-transplanted rats, showed increase in life span of rats and decrease in rate of tumor growth. Crocin has potent side effects on animal adenocarcinoma cell lines HT-29 (Garc-Olmo et al. 1999). Crocin, when injected into different colorectal cancer cell lines, showed that proliferation of cells was restricted most efficiently in HCT-116 cells at a concentration of 1.0 mg/mL. It also reduces the proliferation of SW-480 cells and HT-29 cells. Mechanism which Crocin follows has to be investigated (Aung et al. 2007).

Active compounds of saffron prevent cancer in human malignant cell lines and animal models by inhibiting cell growth. They also reduce the growth of tumor cells in vivo (Abdullaev n.d.). *Crocus sativus* styles show cytotoxicity to breast cancer cell line MCF-7 and MDA-MB-231 when incubated for 48 h (Chryssanthi et al. 2007). Saffron extract at a concentration of 200–2000 μg/mL decreases the viability of cells of MCF-7 breast cancer cell lines in a dose- and time-dependent manner. Possible cause of decrease in cell viability is apoptosis which is associated with upregulated expression of Bax protein (Mousavi et al. 2009). *Crocus sativus* extract is cytotoxic for a broad range of murine tumors, sarcomas, and human leukemia cell lines in a dose-dependent manner. Cytotoxicity is due to inhibition in synthesis of nucleic acids and topoisomerase II which is necessary for DNA synthesis of cells (Nair et al. 2009).

Saffron administration in animal bearing Dalton's lymphoma shows increase in life span to up to 37% and reduction in risk of formation of solid tumor by 95% (Bakshi et al. 2009). *Crocus sativus* extract, when injected in lung cancer cell lines L929, shows significant decrease in cell viability at a concentration of 1500 μg/mL for 24 h and 565 μg/mL for 48 h (Samarghandian et al. 2010). *Crocus sativus* is a potent antioxidant. This antioxidant effect of saffron can be used positively for the treatment of hepatocellular carcinoma cell in mice. In HepG2 cells it inhibits nuclear factor kappa B activation, and increases the cleavage of caspase-3, DNA damage, and cell cycle arrest, which leads to apoptosis of cancer cells (Amin et al. 2011).

Carotenoids in saffron are chemopreventive in nature. They are cytotoxic to HeLa cells, and they are nontoxic, nonmutagenic, non-anti-mutagenic, and non-co-mutagenic (Figueroa-Hernandez et al. 2005). Petal and stigma extract of *Crocus sativus* show antitumor activity at IC50 of 10.8 and 5.3 mg/mL, respectively (Hosseinzadeh et al. 2013). Saffron ingestion inhibits formation of skin papillomas in animals and reduces the size of preformed papillomas. Saffron also inhibits DMBA-induced skin carcinogenesis in mice (Das et al. 2010). Saffron induces apoptosis in carcinomic human alveolar basal epithelial cells A549 at IC50 of 650 μg/mL for 48 h in a dose-dependent manner (Samarghandian et al. 2011). Crocin decreases human tongue squamous cell carcinoma cell line Tca8113 cell viability and cell growth, and induces apoptosis (Pandey 2011).

Curcuma longa Linn.

Curcuma longa L., commonly known as turmeric (Eng), is a perennial herb that belongs to the family Zingiberaceae (Roth et al. 1997; Araújo and Leon 2001; Beneficial Ayurvedic Herbs 2010) (health mango) (ginger family) (Beneficial Ayurvedic Herbs 2010; Aggarwal et al. 2003) (Table 2). In Pakistan it is known by the name Haldi and its rhizome and tubers are used against cancer (Anonymous 2012a). The major component present in *Curcuma longa L.* which gives it a characteristic yellow color differentiable from other plants is curcumin. Curcumin by nature is a polyphenol and is scientifically known as diferuloyl-methane (Anonymous 2012b; Huang et al. 1998). Curcumin is further divided into curcumin I, curcumin II, and curcumin III; curcumin I (diferuloyl-methane) is considered to be most abundantly present in *Curcuma longa* L. (Rubya et al. 1995).

Medicinal properties of *Curcuma longa L.* have been known since olden times. These properties have been attributed to *Curcuma longa L.* specifically because of the presence of curcumin, a hydrophobic compound (Oetari et al. 1996) that is extracted from the rhizome of the plant (Singh and Khar 2006).

Curcumin acts as a strong antioxidant, antibacterial, anti-nematocidal (Araújo and Leon 2001), antiviral, antifungal (Nita 2003), and anti-inflammatory agent by reducing the levels of histamine in our body and increasing the production of cortisone by adrenal glands (Anonymous 2012c). It also possesses strong angioinhibitory effects (inhibition of proliferation of blood vessels) because of its ability to downregulate proangiogenic genes (genes that promote angiogenesis) such as "VEGF" and "angiopoietin" and reduce the migration and incursion of endothelial cells (Singh and Khar 2006).

Anti-cancerous properties of curcumin have also been highlighted by many researchers. Curcumin works in many different ways to inhibit tumor formation and cancer progression. Several mechanisms of inhibition of tumor and cancer development have been proposed after extensive research on animal models especially rats and mouse. A thorough review of the literature brings into limelight many of the possible mechanisms due to which *Curcuma longa L.* is thought effective against cancer.

Suppression of tumor cells, downregulation of enzymes, growth factors, cell surface adhesion molecules, and transcription factors (Aggarwal et al. 2003) are just the few of the many possible ways by which tumor progression is hampered by *Curcuma longa* L.

Kuttan et al. (1985) evaluated the anticancer activity of rhizomes of *Curcuma longa L.* (turmeric) by using Dalton's lymphoma cells in ascites form. They found that curcumin—the active constituent in turmeric extract—was cytotoxic to lymphocytes and Dalton's lymphoma cells in vitro (Kuttan et al. 1985). Experimentally curcumin was also found to inhibit the development of animal tumors and it is currently being developed as an anticancer agent by National Cancer Institute (NCI) (Kelloff 1996).

In another research conducted on 62 patients, Kuttan et al. (1985) applied an ethanol extract of *Curcuma longa L.* and an ointment of curcumin on the skin of the patients with external cancerous lesions. Both of the applied substances were found to reduce lesion size (in 10% of cases), itching (in almost all patients), and pain (in 10% of patients) (Kuttan et al. 1987).

12-O-tetradecanoylphor-bol-13-acetate (TPA) is a compound the application of which stimulates the formation of reactive oxygen species (ROS) (Copeland 1983; Cerutti 1985); according to Copeland (1983), and Cerutti (1985) ROS play an important role in the formation of tumors. The tumor formation mediated by TPA can be inhibited by curcumin as proposed by Huang et al. (1998). Curcumin application on skin inhibits epidermal DNA synthesis, TPA-induced epidermal ornithine decarboxylase activity, B[a] P-mediated formation of B[a] P-DNA adducts, and promotion of skin tumors in mice (Cerutti 1985; Huang et al. 1992).

Research of Azuinea and Bhidea (1992) has revealed that turmeric inhibits benzo[a]pyrene-(BP)-induced forestomach neoplasia in a time- and dose-dependent manner. Curcumin inhibits B[a] P-induced forestomach tumorigenesis, ENNG-induced duodenal tumorigenesis, and AOM-induced colon tumorigenesis (Huang et al. 1994).

Inhibitory effects of curcumin on human colon cancer cells have been reported by many of the researchers. Curcumin inhibits the proliferation of HT-29 and HCT-15 human colon cancer cell lines as suggested by Hanif et al. (Hanif et al. 1997). Cyclooxygenase-2 (COX-2) plays an important role in tumor development (Singh et al. 1996). The activity of cyclooxygenase-2 (COX-2) is decreased in the colon tumors by curcumin which leads to the inhibition of colon carcinogenesis (Reddy et al. 1993; Rao et al. 1995a, b; Boolbol et al. 1996). Kawamori et al. (1999) suggested that administration of curcumin before, during, or after colon carcinogenesis retards the growth of neoplastic colon lesions by inducing apoptosis in them (Kawamori et al. 1999).

It is thought that curcumin inhibits the activation of carcinogens by modulating phase I and phase II metabolic enzymes (Huang et al. 1992; Azuinea and Bhidea 1992; Mukundan et al. 1993); because of this activity, DNA adducts and tumor formation are prevented (Singh and Khar 2006). It exerts strong inhibitory effects on the proliferation of cancer cells (Huang et al. 1992) and hampers the synthesis of DNA in cancer cell lines (Nagabhushan et al. 1996; Huang et al. 1997a).

Surendra et al. (1997) tested the inhibitory effects of curcumin (turmeric) and genistein (soya bean) on the proliferation of estrogen-positive human breast MCF-7 cells induced either by 17-β-estradiol or by a mixture of pesticides (chlordane, DDT, and endosulfan). They founded that both of the compounds inhibited the proliferation of estrogen positive human breast MCF-7 cells while curcumin was more effective in its inhibitory activity than genistein. Possibly both of them block the binding sites of estrogen receptor or meddle with the process that is induced after the binding of estrogenic pesticides to the receptors. This research clearly indicates that the extensive proliferation of estrogen-positive cells induced by pesticides or other carcinogenic compounds like 17-β-estradiol can be controlled by anti-cancerous compounds like curcumin and genistein (Surendra et al. 1997).

Busquets et al. (2001) discovered that administration of curcumin to rats bearing the highly cachectic Yoshida AH-130 ascites hepatoma inhibits the tumor growth (Busquets et al. 2001). Menon et al. (1995) found that oral administration of curcumin decreases the number of lung tumor nodules and increases the life span. It was also discovered that proliferation of androgen-dependent and androgen-independent prostate cancer cells is inhibited by curcumin. Apoptosis-suppressor proteins are also modulated by curcumin which results in apoptosis (Menon et al. 1995).

Singh and Khar (2006) found that curcumin has a potential to induce apoptosis (Villegas et al. 2008) in otherwise apoptosis- and radiation-resistant cell lines. NFkB is a gene that induces proliferation of cells and suppresses apoptosis. Curcumin works to downregulate NFkB and thereby induces apoptosis and inhibits proliferation of cancer cells. It also inhibits IKB kinase which ultimately hinders the phosphorylation process critical for the provision of energy and activation and deactivation of many different proteins in the cell. Curcumin is an effective chemopreventive compound that suppresses colonic aberrant crypt focus formation in the cancers of skin, stomach, and colon (Singh and Khar 2006). Curcumin works best when it is present in its demethylated form because methylation puts an end to its antioxidant and antitumor activities.

Curcumin arrests the cell cycle and induces apoptosis (Samaha et al. 1997) in cancer cells both in vivo and in vitro as researched by Shankar and Srivastava (2007a, b). Bax and Bak are the members of Bcl-2 (pro-apoptotic) family. According to Shankar and Srivastava (2007a, b) curcumin-based cancer treatment induces the expression of Bax and Bak proteins, which in turn causes the release of apoptogenic molecules (molecules that induce apoptosis) from mitochondria to cytosol. Once released in the cytosol, Smac/DIABLO and cytochrome c (apoptogenic molecules) cause the activation of caspases which ultimately leads to apoptosis of cancer cells (Shankar and Srivastava 2007a).

Bcl xL is an anti-apoptotic gene that hinders the apoptosis of cancerous cells. Any mechanism that inhibits Bcl xL is considered effective in the inhibition of cancer progression. Curcumin is known to downregulate Bcl xL gene by increasing the expression of c-myc protein because c-myc suppresses Bcl-xL gene (Duyao et al. 1990; Merino et al. 1995; Susin et al. 1999; Sun et al. 1999). Shankar and Srivastava (2007a, b) found much similar results in another research conducted on prostate cancer LNCaP cells. They concluded from this research that curcumin downregulates Bcl-2 and Bcl-XL proteins, upregulates Bax and Bak, and inhibits PI3-K–Akt pathway in prostate cancer LNCaP cells (Shankar and Srivastava 2007b).

In addition to curcumin, another compound, ar-turmerone, isolated from *Curcuma longa L.* has shown anti-cancerous properties. Effects of ar-turmerone on DNA of human leukemia cell lines (Molt 4B, HL-60) and stomach cancer (KATO III) cells were studied by Aratanechemuge et al. (2002). They found that ar-turmerone causes fragmentation of DNA very effectively in human leukemia cell lines as compared to stomach cancer cells. As DNA fragmentation is a characteristic sign of apoptosis it was proposed that ar-turmerone inhibits cellular proliferation by inducing apoptosis in them (Aratanechemuge et al. 2002).

All in all, studies by different scientists have shown that curcumin inhibits skin (Conney et al. 1991; Lu et al. 1994; Limtrakul et al. 1997; Huang et al. 1997b), forestomach (Huang et al. 1994; Piper et al. 1998), liver (Aggarwal et al. 2003; Chuang et al. 2000), and mammary carcinogensis (Singletary et al. 1996; Chan et al. 1998; Inano et al. 1999); B- and T-cell leukemia (Aggarwal et al. 2003; Kuo et al. 1996; Abe et al. 1999; Han et al. 1999; Piwocka et al. 1999); epidermoid carcinoma (Korutla and Kumar 1994); breast carcinoma (Aggarwal et al. 2003;

Mehta et al. 1997; Simon et al. 1998; Ramachandran and You 1999); multiple myeloma (Bharti et al. 2003); and cervical, pancreatic, prostrate, and gastric cancers (Aggarwal et al. 2003).

Echinacea purpurea L. (Moench)

Echinacea purpurea is a herbaceous plant which belongs to the daisy family of plants known as Asteraceae (O'Hara et al. 1998) (Table 2). The habitats of plants are usually open woods and cultivated beds (Zimmerman n.d.). It is commonly known as purple coneflower (O'Hara et al. 1998), Missouri snakeroot, and broad-leaf cone flower (Miller n.d.). Root of the plant is used for medicinal purposes (Hu and Kitts 2000).

Echinacea is one of the most widely used medicinal plants in the world. Patients with malignant tumors take Echinacea as medicine but the proper route which it follows in curing cancer is not known (Chicca et al. 2007). The root extracts of the plant have in vitro cytotoxic effects on human pancreatic cancer and colon cancer cell lines (Chicca et al. 2007). Root extracts of *Echinacea purpurea* increase the number of white blood cells circulating in blood (Bauer et al. 1988) and hence this results in increased phagocytosis (Roesler et al. 1991). It stimulates the cytokine production and activates the alternate complement pathway (Luettig et al. 1989). *Echinacea purpurea* also increases the production of certain cytokines which include interferon, tumor necrosis factor (Murray 1995), interleukin-1, and interleukin-6 (Braunig et al. 1992). Due to all these effects on the body's immune system *Echinacea* is considered as an anticancer and antineoplastic agent (Lersch et al. 1992).

Echinacea preparation causes decrease in effects caused by human lung and kidney cancer cells (Rogala et al. 2008). 50% aqueous ethanolic extract of *Echinacea purpurea* flowers and cichoric acid (a compound present in *Echinacea purpurea*), when administered in colon cancer cell line, resulted in apoptosis, but the exact mechanism it follows is not known (Tsai et al. 2012). Glycerol extracts of roots of *Echinacea purpurea* show cytotoxic activity to camptothecin-induced cancer (Miller 2012).

Elephantopus scaber Linn.

Elephantopus scaber is a perennial plant from renowned sunflower family, the Asteraceae (Table 2), formerly known as Compositae family (Ho et al. 2011). Natural habitat of plant is subtropical or tropical moist forests (Poli et al. 1992). The plant is commonly known as Prickly-leaved elephant's foot (Panda 2004). Whole-plant extracts are used for medicinal purposes (Poli et al. 1992).

The ethanolic extract of *Elephantopus scaber* plant shows cytotoxicity against human breast cancer cell lines MCF-7. IC50 value of *Elephantopus scaber* is 15 μg/mL. It upregulates the expression of tumor-suppressor gene p53 and hence induces p53-dependent apoptosis (Ho et al. 2011). Deoxyelephantopin, a natural compound found in *E. scaber* inhibits nasopharyngeal cancer cell line proliferation. It induces apoptosis by dysfunction mitochondria. It also arrests cell cycle in S and G2 phases. Due to all these evidences, *Elephantopus scaber* can be used as a chemotherapeutic agent for nasopharyngeal cancer (Su et al. 2011).

Chloroform extract of *Elephantopus scaber* contains two medicinally active compounds, which are deoxyelephantopin (DET) and isodeoxyelephantopin. DET is a potent chemotherapeutic agent. These compounds cause a dose-dependent response and reduce the viability of L-929 tumor cells in 72-h culture at IC50 of 2.7 μg/mL. DET causes maximum apoptosis at the amount of 3 μg/mL. It also shows in vivo antitumor activity against DLA tumor cells (Geetha et al. 2012). Deoxyelephantopin shows cytotoxic effects against mammary adenocarcinoma. It inhibits cell proliferation, arrests cell cycle at G2 phase, and induces apoptosis. Several mechanisms which it follows show that the compound also has chemopreventive activity for breast cancer cells (Huang et al. 2010).

DET which is a Sesquiterpene lactone, induces reactive oxygen species which results in formation of centrosomal ubiquitinated proteins which leads to the restriction of cancer cell growth in mammary adenocarcinoma cells transplanted in mice (Lee and Shyur 2012). Scabertopin, isoscabertopin, deoxyelephantopin, and isodeoxyelephantopin are four sesquiterpenes isolated from *Elephantopus scaber*. Scabertopin, deoxyelephantopin, and isodeoxyelephantopin show antitumor activity in a dose-dependent manner. Hela cells when exposed to deoxyelephantopin change morphologically and lead towards apoptosis (Xu et al. 2006).

Nigella sativa Linn.

Nigella sativa belongs to family of flowering plants known as Ranunculaceae (Table 2). It is commonly known as black cumin or black seeds (Ali and Blunden 2003). Its seeds and seed oil are used for medicinal purposes.

Seed extract of *Nigella sativa* is cytotoxic to P815 cell lines and vero cell lines. Administration of seed extract in vivo in DBA2/P815 mouse model showed that it inhibits the solid tumor formation and development (Mbarek et al. 2007). Active ingredient of *Nigella sativa* showed 50% cytotoxicity against Ehrlich ascites carcinoma cells, Dalton's lymphoma ascites, and sarcoma 180 cells at IC50 of 1.5 μg, 3 μg, and 1.5 μg, respectively (Salomi et al. 1992). Effects of extract of *Nigella sativa* in hepatoma HepG2 cells induced in mice showed that after 24 h of exposure to *Nigella sativa* extract, apoptosis gets initiated in HepG2 cells (Thabrew et al. 2005). Methanolic extract of *Nigella sativa* shows cytotoxicity against HeLa cells at IC50 of 2.28 μg/mL, its hexane extracts showed cytotoxicity at IC50 of

2.20 μg/mL, and its chloroform extracts showed similar results at IC50 of 0.41 ng/mL (Shafi et al. 2009).

Thymoquinone and α-hederin are two of the most active ingredients in *Nigella sativa*. These two are potent chemotherapeutic agents as they show cytotoxicity for lung carcinoma cell line (A549), larynx epidermoid carcinoma (Hep-2), colon adenocarcinoma (HT-29), and pancreas carcinoma (MIA PaCa-2) cells in vitro (Rooney and Ryan 2005). Thymoquinone (TQ) also showed apoptosis of prostate cancer cells, when exposed for 24–48 h at a concentration of 25–150 μmol/mL. It is shown to inhibit the growth of androgen receptor-independent prostate cancer cells (C4-2B) and PC-3 cells. After 1 h of delivery of TQ the cells showed increased level of reactive oxygen species and decreased glutathione levels. Cell death is supposed to occur due to these two mechanisms (Koka et al. 2010).

Aqueous extracts of *Nigella sativa* enhances the cytotoxic activity of natural killer cells against YAC-1 tumor cells, which proves the antitumor effects of *Nigella sativa* (Majdalawieh et al. 2010). *Nigella sativa* shows cytotoxicity for mammary carcinoma cells and breast cancer cells. It also reduces the carcinogenicity of certain carcinogens which shows its chemopreventive effects (Abd El-Aziz et al. 2005). *Nigella sativa* shows chemoprevention for ferric nitrilotriacetate-induced renal oxidative stress and renal carcinogenesis in Wistar rats. It increases renal lipid peroxidation, xanthane oxidase, and hydrogen peroxide (Khan and Sultana 2005). *Nigella sativa* oil shows antitumor activity for fibrosarcoma cell line HT1080 in vitro. Apoptosis of cells is associated with tissue-type plasminogen activator, urokinase-type plasminogen activator, and plasminogen activator inhibitor type 1 (Awad 2005).

Thymoquinone and thymohydroquinone show cytotoxic activity against L929 cancer cell lines and tumor cell lines in vitro in a dose-dependent manner (Ivankovic et al. 2006). TQ is cytotoxic to cervical squamous carcinoma cells SiHa at IC50 of 10.7 and 9.3 μg/mL. TQ is less toxic to normal cells and specifically kills cancerous cells. It results in increased level of p53, and downregulation of anti-apoptotic Bcl-2 protein without any change in the level of Bax protein (Ng et al. 2011). Topical application of *Nigella sativa* extract results in inhibition of formation and proliferation of skin carcinogenesis in mice. Administration of *Nigella sativa* intraperitoneally resulted in restriction of tumor growth to 33.3% (Find me a Cure n.d.). *Nigella sativa* can be used as a natural and potent chemopreventive and chemotherapeutic agent.

Petiveria alliacea

Elephantopus scaber is a common perennial shrub which belongs to the family Phytolaccaceae (Find me a Cure n.d.) (Table 2). It is commonly known as guinea-hen weed and Anamu (Mendes n.d). Roots are used for treating cancer (Soladoye et al. 2010).

Elephantopus scaber has a promising anticancer activity (Lowe et al. 2012). It is found to cure different types of cancers including neuroblastoma, primary bladder carcinoma, mammary carcinoma, sarcoma, and melanoma. Its anticancer properties are due to dibenzyl trisulfide (DTS), a lipophilic compound present in it (Jamaica n.d.). DTS is a signal transduction molecule (Williams et al. 2007). It downregulates the cytokines produced from Th1 pathway and upregulates the cytokines of Th2 pathway; by this mechanism it exhibits cytotoxic activity against cancer cells (Williams et al. 2007). Fractions of root of plant can induce apoptosis in mitochondrium-dependent pathway and downregulate HSP70 expression in vitro. This activity can result in its antitumoral properties (Cifuentes et al. 2009). The mechanism which it follows for cytotoxicity is not well understood, but it is hypothesized that *Petiveria alliacea* induces G2 cell cycle arrest and apoptosis in mitochondrium-independent pathway (Uruena et al. 2008).

Phaleria macrocarpa (Scheff.) Boerl

Phaleria macrocarpa (Scheff.) Boerl (*Phaleria papuana Warb var. Wichnannii (Val) Back*) commonly known as mahkota dewa or made in Indonesia (Faried et al. 2007) belongs to the family Thymelaeaceae (Faried et al. 2007; Backer and van den Brink 1965; Anonymous n.d.-c; Ali et al. 2012; Katrin and Selvie 2011) (Table 2). The plant is notably known as the "Crown of God" (Backer and van den Brink 1965; Ali et al. 2012).

Leaves, fruit, and bark of mahkota dewa contained alkaloids, terpenoids, flavonoids, polyphenols, saponins, resins, lignin, and benzophenones (Hutapea n.d.; Gotama et al. 1999; Harmanto 2001; Winarto and Dewa 2003). All of these compounds are known anti-cancerous agents (Yoshida et al. 2000; Tsuda et al. 2004).

Phaleria macrocarpa (Scheff.) Boerl has been in use for a long time to treat diabetes mellitus, high blood pressure and other blood diseases, hypertension, dysentery, allergies, liver disease, lung disease, heart disease, impotency, kidney disorders, hemorrhoid, stroke, migraine, acne, diuretic conditions, and rheumatism (Aditama 2001; Harmanto 2003, 2005; Ao et al. 2008; Kurnia et al. 2008; Diantini et al. 2012). According to Harmanto (2003) the seeds of this plant are toxic and hence used for treating skin diseases while its fruit is used for flu treatment (Harmanto 2003) and stems are effective against bone cancer (Yekti 2010).

Phaleria macrocarpa has verified medicinal effects against tumor and cancer (Fariza et al. 2012). Harmanto (2005) elucidates that the plant has antihistamine, antioxidant, and anti-cancerous properties (Harmanto 2005) whereas according to Tjandrawinata et al. (2011) an extract of *Phaleria macrocarpa* possesses anti-inflammatory, antineoplastic, and anti-angiogenic activities (Tjandrawinata et al. 2011).

Researchers have not lagged behind in the investigation of chemical constituents of *Phaleria macrocarpa* having valuable anti-cancerous potential. A literature review on this plant shows that it can act as a potent source of anticancer drug.

For the sake of bearing minimal side effects and to increase their effectiveness, researchers have combined chemotherapeutic drugs with the active constituents of plant extracts. One such approach was carried out by Miki et al. (2001) in their study on C57 black mice transplanted with lung cancer cells (LL-2). They studied the in vitro effects of an anticancer drug (cisplatin) alone and in combination with gallic acid (GA) on the aforementioned cancer cells. Their findings indicated the increase in the effectiveness of cisplatin when combined with gallic acid as observed from elevated rate of apoptosis in tumor cells (Miki et al. 2001).

The reduction in the weight, size, and volumes of tumors induced after transplantation was analyzed by Rahmawati et al. (2006). Rahmawati and associates administered 70% ethanol extract of mahkota dewa fruit pulp [*Phaleria macrocarpa (Scheff.) Boerl.*] orally to C_3H mouse after tumor transplantation. Rahmawati et al. (2006) concluded from their research that the extract was not able to inhibit tumor formation after transplantation but it significantly increased the rate of apoptosis in them (Rahmawati et al. 2006). Apoptosis alone is worthy for the death of cancerous cells as there are certain plant products that specifically induce apoptosis in cancerous cells but not in normal cells (Chiao et al. 1995; Hirano et al. 1995).

Faried et al. (2007) investigated the anti-cancerous potential of gallic acid (GA); gallic acid (GA-3, 4, 5-trihydroxybenzoic acid) is an antioxidant naturally present in the fruits of *Phaleria macrocarpa* and has been known to induce apoptosis in leukemia, lung cancer, and colon adenocarcinoma cell lines (Inoue et al. 1994; Kawada et al. 2001; Salucci et al. 2002; Sohi et al. 2003). GA is also known to prevent the development of process of carcinogenesis (Taraphdar et al. 2001). Faried et al. (2007) reported that gallic acid (GA) inhibits the process of carcinogenesis by activating multiple pathways and acts on cancer cells in a time- and dose-dependent manner. They investigated that a synergy between the extrinsic and intrinsic (mitochondrial) pathways is required to make the esophageal cancer cells (TE-2) susceptible to gallic acid. The extrinsic pathway downregulates Akt/mTOR survival pathway and creates an imbalance in the anti-apoptotic Bcl-2 family (Bax and Bcl-2), ultimately which are responsible for the induction of apoptosis via the activation of caspase-cascade and pro-apoptosis protein (Bax). The intrinsic pathway involves mitochondria and generates reactive oxygen species (ROS) and the influx of Ca+2 ions which causes the activation of calmodulin (Inoue et al. 1994). GA also cleaves PARP which executes cancer cell death (Faried et al. 2007). The events involved in intrinsic pathway also lead to apoptosis just like extrinsic pathway. A combination of extrinsic and intrinsic pathways activated by gallic acid (GA) caused death of esophageal cancer cells (TE-2) in a time- and dose-dependent manner.

Winarno and Katrin (2009) isolated a benzophenone glucoside compound named as 6,4'-dihydroxy-4-methoxybenzophenone-2-O-β-D-glucopyranoside from the ethyl acetate extract of the bark of *Phaleria macrocarpa*. The inhibitory activity of this compound was tested against leukemia cell line (L1210) and was found to inhibit the growth of these cells at 50% inhibitory concentration (IC_{50}) of 5.1 μg/mL

(Winarno and Katrin 2009). Thus 6,4'-dihydroxy-4-methoxybenzophenone-2-O-β-D-glucopyranoside has a potential to treat leukemic patients.

Tandrasasmita et al. (2010) reported a much similar mechanism of inhibition of proliferation of the cancerous cells as elucidated by Faried et al. (2007). However, Tandrasasmita et al. (2010) discovered the inhibitory effects of phalerin instead of gallic acid. They studied the inhibitory effects of an extract of *Phaleria macrocarpa (Scheff.) Boerl* on breast cancer cell line (MDA-MB-231). Detailed analysis of this extract of *Phaleria macrocarpa (Scheff.) Boerl* revealed that it contains phalerin (4, 5-dihydroxy, 4'-methoxybenzophenone-3-O-β-D-glucoside)—an important chemical constituent of *Phaleria macrocarpa*. DLBS1425 works against MDA-MB-231 breast cancer cells via the downregulation of phosphoinositide-3 (PI3)-kinase/protein kinase B (AKT) signaling pathway (PI3-K/AKT signaling pathway) and by the induction of apoptosis (Tandrasasmita et al. 2010). PI3-K/AKT signaling pathway is required by the cell for its proliferation and survival. This pathway also protects the cells from apoptosis (Izhou et al. 2001; Shaw and Cantley 2006; Wee et al. 2008; Jia et al. 2009). But if overexpressed AKT causes the activation of PI3-K/AKT signaling pathway in many cells as seen in breast, ovarian, pancreatic, and thyroid cancers (Nicholson and Anderson 2002; Fujiwara et al. 2006; Engelman 2009; Liu et al. 2009). This is responsible for the massive proliferation and survival of these cancerous cells in addition to assisting them in evading apoptosis. Phosphatase and tensin homolog (PTEN) is a tumor-suppressor protein and has been found to be mutated in a number of breast cancer cases (Cantley and Neel 1999; Hennessy et al. 2005; Carnero et al. 2008). If activated, PTEN leads to the inactivation of PI3-K/AKT signaling pathway and vice versa (Cantley and Neel 1999; Hennessy et al. 2005; Uddin et al. 2004; Barber and Welch 2006; Daikoku and Dey 2008; Cain and Ridley 2009; Kang et al. 2010). DLBS1425 downregulates PI3-K transcript levels and hence reduces phosphorylation of AKT through the upregulation of PTEN. AKT is also known to phosphorylate Bad, which is a pro-apoptotic member of Bcl2 family and mediates cell death by forming a heterodimer with anti-apoptotic protein-Bcl-xL (Nicholson and Anderson 2002; Kim 2005). Due to the reduction of AKT phosphorylation, pro-apoptotic genes (BAX, BAD, and PUMA) are activated and anti-apoptotic gene (Bcl-xL) is suppressed. Apoptosis is induced as a result of both of these events and consequently breast cancer cell death occurs via the activation of caspase-9, PARP cleavage, and DNA fragmentation (Tandrasasmita et al. 2010). DLBS1425 not only acts against breast cancer cell lines but has also been found effective against certain types of other cancer cell lines such as HepG2 liver cancer cells, PC3 prostate cancer cells, HCT116 colon cancer cells, and AGS gastric cancer cells in a dose-dependent manner (Tandrasasmita et al. 2010).

In continuation with their previous research, Tjandrawinata et al. (2010) discovered a compound DLBS1425 from the fruit extract of *Phaleria macrocarpa* with known potential to be used as an anti-cancerous agent. Breast cancer cell line (MCF-7) employed in research was found to be vulnerable to DLBS1425 as indicated by reduction in the growth and proliferation rate. DLBS1425 adopts eicosanoid pathway to hinder the proliferation of MCF-7 breast cancer cell line. The extract regulates the expression of Bcl and Bax genes at mRNA level in addition to the activation of caspase 9 and DNA fragmentation. All of these events lead to apoptosis. Bcl and Bax genes are the members of Bcl-2 (pro-apoptotic) family and work as antagonists. While Bcl downregulates apoptosis, Bax is known to induce it. DLBS1425 maintains equilibrium between the expressions of these two genes and hence prevents the progression of breast cancer. Cyclooxygenase-2 (COX-2) plays a significant role in tumor development and its expression is related to the tumor growth; hence it is a biological marker of breast cancer. High expression of COX-2 is present in highly invasive breast cancer cells with estrogen-independent characteristics (such as MDA-MB-231 cells), while less invasive breast cancer cells and estrogen-dependent characteristics (such as MCF-7 cells) did not express COX-2. Prostaglandins and thromboxanes are the end products of COX-2 which mediate the steps in cancer cell progression (Tjandrawinata et al. 2011). DLBS1425 downregulates the expression of this protein (Tjandrawinata et al. 2010) along with the suppression of HER-2/neu and cytoplasmic phospholipase A2 (cPLA2) through the downregulation of NFkB. Thence, fruit extract-DLBS1425 of *Phaleria macrocarpa* has strong antiproliferative potential against breast cancer.

Aripin et al. (2011) isolated two compounds (DLBS1425E2.2 and DLBS1425F1) from the fruits of *Phaleria macrocarpa* and investigated their anti-cancerous activity against breast cancer cell line (MDA-MB-231). Both the fractions exert antiproliferative activity against MDA-MB-231 but DLBS1425E2.2 was found to be more potent as it also induced apoptosis. These compounds can be used against anal intraepithelial neoplasia, cervical intraepithelial neoplasia, and leukemia (Aripin et al. 2011).

Hendra et al. (2011) studied the cytotoxic activity of methanol extract of pericarp, mesocarp, and seeds of *Phaleria macrocarpa (Boerl.) Scheff* on Chang, HeLa, HT-29, and MCF-7 cancer cell lines. The methanol extract of pericarp, mesocarp, and seeds was found to be cytotoxic against two of the four cell lines used (MCF-7 and HeLa). But the HT-29 cell line was only vulnerable to cytotoxicity by seed extract (Hendra et al. 2011).

Fevicordin A is a bioactive compound present in seeds of *Phaleria macrocarpa (Boerl.) Scheff.* and was successfully isolated by Kurnia and his colleagues (Kurnia et al. 2008). Diantini et al. (2012) investigated the cytotoxicity of fevicordin A on leukemia (P 338), cervical cancer (HeLa and CasKi), and esophageal cancer (TE-2 and TE-8) cell lines. Fevicordin A strongly inhibited the proliferation of leukemia cells (P 388) and cervical cancer cells (HeLa) but was found to have weak antiproliferative activity against cervical cancer (CasKi) and esophageal cancer cells (TE-8). From this study, it can be concluded that fevicordin A is a successful candidate for cervical cancer and leukemia (Diantini et al. 2012).

Tabernaemontana divaricata

Tabernaemontana divaricata belongs to the family Apocynaceae (Table 2). It is commonly known as crape jasmine. The root, leaves, and bulb are used as medicines; all of these have a bitter taste and sharp smell (ITIS Report n.d.).

Ethyl acetate extract of *Tabernaemontana divaricata* is effective against 502713 colon cancer cell line whereas chloroform extract of *Tabernaemontana divaricata* shows cytotoxicity against colon carcinoma cells HCT-15, HT-29, and 502713 in vitro. It is observed that chloroform extract of plant inhibits the unwinding of DNA and inhibition of topoisomerases I and II (Thind et al. 2008). Hydroalcoholic extract of *Tabernaemontana divaricata* shows anticancer activities against HeLa cell lines at IC50 more than 100 μg/mL (Dantu et al. 2012).

Ethanolic extract of leaves of *Tabernaemontana divaricata* shows cytotoxicity against tumor cells and can be further analyzed as a candidate for chemoprevention (Khan and Islam 2012). Indole alkaloids obtained from ethanolic extracts of *Tabernaemontana divaricata* show inhibition of Sarcoma-180 cell lines in vitro at a concentration of 10 μg/kg/day. It also shows selective cytotoxicity against Chinese hamster V79 cells (Bharat et al. 2011).

Terminalia arjuna (Roxb. ex DC.) Wight and Arn.

Terminalia arjuna (Roxb. ex DC.) Wight and Arn is a large deciduous tree (Verma and Vinayak 2009) which belongs to the family Combretaceae (Pettit et al. 1996; Pasquini et al. 2002) (Table 2). The plant is a source of tannins, triterpenoid saponins (arjunic acid, arjunolic acid, arjungenin, arjunglycosides), flavonoids (arjunone, arjunolone, luteolin), gallic acid, ellagic acid, oligomeric proanthocyanidins (OPCs), phytosterols, calcium, magnesium, zinc, and copper (Verma and Vinayak 2009; Pasquini et al. 2002; Kapoor 1990; Bone 1996).

Traditionally the plant was used for the treatment of certain ailments like anemia, aphrodisiac, asthma, biliousness, blood diseases, cardiac failure, diuresis, dysentery, excessive perspiration, fractures, heart diseases, helminthes, hypertension, intoxications, leukoderma, tumors, ulcers, and wounds (Verma and Vinayak 2009; Oudhia n.d.).

According to Hartwell (1982) and Jain et al. (1992) several species of *Terminalia arjuna* have been used for the treatment of cancer (Hartwell 1982; Jain et al. 1992) while several anti-cancerous constituents have been documented from the leaves, stem, and bark (Pettit et al. 1996). Bark extract of *Terminalia arjuna* has been found to have antimutagenic (Kaur et al. 1997), antibacterial (Samy et al. 1998), and anticancer (Kandil and Nassar 1998) properties.

Research has been conducted on chemical compounds from *Terminalia arjuna* for their potential to treat cancer and associated diseases. Three important compounds worth mentioning are the gallic acid (GA), ethyl gallate, and flavone luteolin isolated from the bark, stem, and leaves of *Terminalia arjuna*. Among them, luteolin is known to exhibit antitumor (Asaka et al. 1992; Middleton et al. 1987) and antimutagenic activities (Hertog et al. 1992; Huang et al. 1983). Furthermore, luteolin is reported to inhibit a number of ascites (NK/LY) (Molnar et al. 1981), solid tumors (renal A-549, HCT15, gastric HGC-27, ovary SK-OV-3, melanoma SK-MEL-2, and XF-498) (Matsukawa et al. 1993; Ryu et al. 1994), and leukemia (CEM-C1, CEM-C7, and P388) (Post and Varma 1992; Chen et al. 1992) cell lines.

Mice treated with 12-O-tetradecanoylphorbol-13-acetae (TPA) were subjected to hydrolysable tannins, ellagic acid, and gallic acid (GA) from *Terminalia arjuna*. It was found that tannins were most efficient at inhibiting the tumor formation in TPA-treated mice than ellagic acid and gallic acid (GA) (Hu et al. 1992).

Acetone and methanol extracts from the bark of *Terminalia arjuna* were tested by Nagpal and his colleagues (2000) against human normal fibroblasts (WI-38), osteosarcoma (U2OS), and glioblastoma (U251) cells in vitro. These extracts worked in a similar fashion to casuarinin as reported by Kuo et al. (2005a, b) in their study. Tumor-suppressor protein, p53, was found to be activated in osteosarcoma (U2OS) cells but not in human normal fibroblasts (WI-38) and glioblastoma (U251) cells as the former has wild-type and latter has mutated p53. As a result, p53 is responsible for the activation of $p21/^{WAFI}$ protein which arrests the cell cycle. Glioblastoma (U251) cells were found to have moderate levels of p53 although they have mutated gene for this protein. This shows that the activation of $p21/^{WAFI}$ is required for the cancer cell growth inhibition by either p53-dependent or -independent pathways (Nagpal et al. 2000). Hence the acetone and methanol extracts inhibit the proliferation of human normal fibroblasts (WI-38), osteosarcoma (U2OS), and glioblastoma (U251) cells in vitro in a dose-dependent manner.

Sivalokanathan et al. (2005) studied the effects of ethanolic extract of *Terminalia arjuna* on the levels of carbohydrate-metabolizing enzymes in N-nitrosodiethylamine-induced hepatocellular carcinoma (HCC). In their research conducted on Wistar albino rats with HCC, Sivalokanathan and his colleagues found that the levels of glycolytic enzymes (hexokinase, phosphoglucoisomerase, and aldolase) in the liver and plasma were elevated. But the levels of glyconeogenic enzyme (glucose-6-phosphatase) were found to be reduced in liver and plasma of cancerous rats. Administration of *Terminalia arjuna* ethanolic extract to these rats for 28 days significantly modulated the enzyme levels and led to cancer growth inhibition (Sivalokanathan et al. 2005).

The anti-cancerous effects of casuarinin on human non-small cell lung cancer (A549) (Kuo et al. 2005b) and on human breast adenocarcinoma (MCF-7) cells (Kuo et al. 2005c) were reported by Kuo et al. (2005a, b). Casuarinin is a hydrolyzable tannin which is isolated from the bark of *Terminalia arjuna L.* Casuarinin exerts its antiproliferative activity on A549 and MCF-7 cells by two mechanisms: by induction of apoptosis and by blockage of cell cycle at G0/G1 phase. Apoptosis is induced due to the augmentation of two Fas ligand (FasL), namely membrane-bound Fas ligand (mFasL) and soluble Fas ligand (sFasL), and of Fas/APO-1. An error in the cell cycle leads to the activation of p53 which stimulates the transcription of a protein called p21. This protein binds to G1/S-Cdk (cyclin-dependent kinase) and S-Cdk and inhibits their activities that prevent the cell cycle progression (Alberts et al. 2002). Casuarinin blocks the cell cycle in G0/G1 phase via the p53-dependent induction of cyclin-dependent kinase inhibitor-p21/^{WAF1} (Kuo et al. 2005b, c). This shows that blockages in cell cycle progression change the fate of proliferating cancerous cells.

A year later, in 2006, Sivalokanathan and colleagues conducted another research on human hepatoma cell line (HepG2), to investigate its propensity to an ethanolic extract of *Terminalia arjuna* (Sivalokanathan et al. 2006). Like many of the researches conducted in the earlier years, this one also indicated that apoptosis is the mechanism responsible for induced cytotoxicity in HepG2 cells. The mechanism of apoptosis induction is no more different from ones reported by Nagpal et al. (2000), Kuo et al. (2005a, b). The genome guardian p53 also commonly known as tumor-suppressor gene controls the cell cycle, DNA repair, and apoptosis (Bharat et al. 2011; Hartwell 1982; Sivalokanathan et al. 2006; Levine et al. 1993) and is found to be mutated in 50% of human cancers (Pettit et al. 1996). HepG2 cells treated with *Terminalia arjuna* extract regulated the expression of p53 protein which resulted in accumulation of p53, DNA fragmentation, and proteolytic cleavage of procaspase-3 protein. Proteolytic cleavage of procaspase-3 cascade activates it to caspase-3 which transduces and executes apoptotic signal (Salvesen and Dixit 1997). When treated with *Terminalia arjuna* extract, HepG2 cells also face a significant reduction in GSH levels. GSH is considered to be a main antioxidant system of the cells and in the cells where its concentration drops reactive oxygen species (ROS) start to accumulate (Miyajima et al. 1997). The ultimate result is the death of cells through the induction of apoptosis because of depleted stores of GSH (Sivalokanathan et al. 2006; Fernandes and Cotter 1994; Beaver and Waring 1995).

Saxena et al. (2007) isolated arjunoglucoside I, arjunic acid, arjunetin, and arjungenin from the bark of *Terminalia arjuna* and evaluated them for their cytotoxic activity against four cancer cell lines. Among four of them, arjunic acid was found to be considerably active against ovarian (PA 1), human oral (KB), and liver (HepG-2 and WRL 68) cancer cell lines (Saxena et al. 2007). Arjunic acid hence has a potential to be used against cancer cells.

Human malignant tumors (HCT-15 and AGS) when cultured with tannic acid experience a significant suppression in their growth. According to Kamei et al. (2009), tannic acid totally blocks S phase of the cell cycle (Kamei et al. 2009). The cells replicate their DNA during S phase; hence blockage of S phase not only creates a hindrance in DNA replication but also leads to impairment in completion of cell cycle.

Classical apoptotic pathway is not the solitary one that can incline the cancerous cells to death as proved by Chowdhury et al. (2009). A cervical cancer cell line (HeLa) was subjected to the leaf extract of *Terminalia arjuna* along with an extract obtained from *Azadirachta indica*. *Terminalia arjuna* extract was found to induce death in a dose-dependent manner in 95% of the cells exposed to it. However, the cells evaded the normal apoptotic pathway as DNA fragmentation was not observed (Chowdhury et al. 2009). Cell death that occurs without DNA fragmentation is apparently considered to be necrosis and this type of death requires a protein phosphorylation-linked signal (Akhand et al. 1998).

Moulisha and associates (2010) isolated a pentacyclic triterpenoid named as ursolic acid from the methanolic extract of *Terminalia arjuna*. Ursolic acid is an anti-cancerous and anti-leishmanial compound but the mechanism by which it exerts its anti-cancerous properties is not understood yet. Moulisha et al. (2010) opted the cancer cell line K562 and proposed that the compound may work in the similar manner as that of vincristine, vinblastine (anti-cancerous compounds obtained from *Vinca roseus*), and calcitonin (Moulisha et al. 2010). Vincristine and vinblastine block the metaphase of the cell cycle while calcitonin hinders the cellular proliferation of K562 cell line (Erba et al. 1996).

Thespesia populnea L. sol ex Correa.

Thespesia populnea belongs to mallow family of plants known as Malvaceae (Table 2). It is commonly known as Portia tree. Its bark is used for medicinal purposes (Friday and Okano 2006).

Sequesterpenoids isolated from *Thespesia populnea* can be used as a potent cytotoxic agent for many types of cancer cell lines. Most active sequesterpenoid of *Thespesia populnea* are mansonone and gossypol. These are shown to induce apoptosis in MCF-7 breast cancer cell lines, HeLa cell lines, HT-29, and KB carcinoma cell lines in vitro (Boonsri et al. 2008). Methanol extract of *Thespesia populnea* inhibits the growth of solid tumor in vivo in mice. Inhibition of tumor cell growth is associated with decreased level of glutathione, serum gamma glutamyl transpeptidase, and nitric oxide. This can be used as a natural chemopreventive agent (Mika and Guruvayoorappan 2013). Ethanolic extract of bark of *Thespesia populnea* shows activity against DMBA-induced hamster buccal pouch carcinogenesis in Syrian golden hamsters. Oral administration of this extract at a dose of 300 mg/kg of body weight for 14 weeks prevents tumor incidence, tumor volume, and tumor burden. Further studies are needed to isolate specific active ingredient against cancer from the bark of *Thespesia populnea* (Dhanarasu et al. 2010).

Thuja occidentalis Linn.

Thuja occidentalis, a plant commonly known as red or white cedar, belongs to famous cypress family, the Cupressaceae (Sunila et al. 2011) (Table 2). The plant as a whole is used as medicinal herb for centuries and is now shown to have potent chemopreventive and chemotherapeutic effects (Johnston n.d.).

Administration of ethanolic extracts of leaves of *Thuja occidentalis* decreases the viability of A549 lung carcinoma cells, after 24 h of exposure. It inhibits cell proliferation in a dose-dependent manner. Inhibition of growth of cells is associated with upregulation of Bax and downregulation of Bcl-2 (Mukherjee et al. 2012).

Silver nanoparticles biosynthesized by ethanolic extract of *Thuja occidentalis* show cytotoxicity for A375 carcinoma cell lines. Cytotoxicity of cells is due to arrest of cell cycle at G2/M phase (Das et al. 2013). Crude extracts of *Thuja occidentalis* contain an active compound, thujone, which is used as homeopathic medicine. It induces apoptosis and is cytotoxic to A375 carcinoma cell lines in vitro. Decrease in cell viability, increase in DNA fragmentation, increase in ROS generation, release of cytochrome c, and activation of caspase-3 are the major events which lead to thujone-induced apoptosis in A375 cells (Biswas et al. 2011).

Administration of *Thuja occidentalis* and its polysaccharides in B16F-10 metastatic melanoma cells transplanted in mice results in increase in activity of natural killer cells, antibody-dependent cell-mediated cytotoxicity, and antibody-dependent complement-mediated cytotoxicity. A decrease in level of cytokines IL-1beta, Il-6, GM-CSF, and tumor necrosis factor alpha is seen in these mice. Moreover increase in levels of antitumor cytokine IL-2 is observed (Sunila et al. 2011). Ethanolic extract of *Thuja occidentalis* shows reduction in tumor weight, tumor volume, and glutathione levels in DMBA-induced breast cancer cells (Ojeswi et al. 2010).

Extract of *Thuja occidentalis* inhibits lung metastasis induced by B16F-10 melanoma cells in C57BL/6 mice. Reduction in tumor nodule formation by 74% is observed after the administration of extract. This extract decreases the level of elevated lung collagen hydroxyproline and uronic acid. Treated mice showed increased life span (Sunila and Kuttan 2006). Administration of thujone in C57BL/6 mice transplanted with lung metastatic B16F-10 melanoma cells resulted in inhibition of tumor nodule formation and increase in survival rate of animals associated with reduction in tumor cell proliferation, adhesion and invasion, and regulation in expression of MMPs, VEGF, ERK-1, ERK-2, TIMPs, nm23, and pro-inflammatory cytokines (Siveen and Kuttan 2011).

References

Abd El Aziz MA, Hassan HA, Mohamed MH, Meki AMA, Abdel-Ghaffar SKH, Hussein MR (2005) The biochemical and morphological alterations following administration of melatonin, retinoic acid and Nigella sativa in mammary carcinoma: an animal model. Int J Exp Pathol 86(6):383–396

Abdullaev FI (2002) Cancer chemopreventive and tumoricidal properties of Saffron (Crocus sativus). Exp Biol Med 227(1):20–25

Abdullaev FI (n.d.) Antitumor effect of saffron (Crocus sativus L.): overview and perspectives. International symposium on saffron biology and biotechnology. ISHS Acta Horticulture

Abe Y, Hashimoto S, Horie T (1999) Curcumin inhibition of inflammatory cytokine production by human peripheral blood monocytes and alveolar macrophages. Pharmacol Res 39(01):41–47

Aditama TY (2001) Kanker. Med J Kedokteran 02:1–5

Aggarwal BB, Kumar A, Bharti AA (2003) Anticancer potential of curcumin, preclinical and clinical studies. Anticancer Res 23:363–398

Agroforestry Tree Database (2008) Annona squamosa. International Centre for research in Agroforestry. http://www.worldagroforestry.org/Sea/Products/AFDbases/AF/asp/SpeciesInfo.asp?SpID=214.

Ahmad SS (2007) Medicinal wild plants from Lahore-Islamabad Motorway (M-2). Pak J Bot 39(02):355–375

Ahmad SS, Mahmood F, Dogar Z, Khan ZI, Ahmad K, Sher M, Mustafa I, Valeem EE (2009) Prioritization of medicinal plants of Margala Hills National Park, Islamabad on the basis of available information. Pak J Bot 41(05):2105–2114

Akev N, Turkay G, Can A, Gurel A, Yildiz F, Yardibi H, Ekiz EE, Uzun H (2007a) Effect of Aloe vera leaf pulp extract on Ehrlich ascites tumor in mice. Eur J Canc Prevent 16(2):151–157

Akev N, Turkay G, Can A, Gurel A, Yildiz F, Yardibi H, Ekiz EE, Uzun H (2007b) Tumor preventive effect of Aloe vera leaf pulp lectin (Aloctin I) on Ehrlich acscites tumours in mice. Phytother Res 21(11):1070–1075

Akhand AA, Kato M, Suzuki H, Miyata T, Nakashima I (1998) Level of HgCl2-mediated phosphorylation of intracellular proteins determines death of thymic T-lymphocytes with or without DNA fragmentation. J Cell Biochem 71:243–253

Akinyele BO, Odiyi AC (2007) Comparative study of the vegetative morphology and the existing taxonomic status of Aloe vera L. J Plant Sci 2(5):558–563

Alali FQ, Liu XX, McLaughlin JL (1992) Annonaceous acetogenins: recent progress. J Nat Prod 62(3):504–540

Alberts B, Johnson A, Lewis J, Raff M, Roberts K, Walter P (2002) Molecular biology of the cell, 4th edn. Garland Science, New York, pp 2558–2559

Ali BH, Blunden G (2003) Pharmacological and toxicological properties of Nigella sativa. Phytother Res 17(4):299–305

Ali RB, Atangwho IJ, Kuar N, Mohamed EAH, Mohamed AJ, Asmawi MZ, Mahmud R (2012) Hypoglycemic and anti-hyperglycemic study of Phaleria macrocarpa fruits pericarp. J Med Plant Res 6(10):1982–1990

Amagase HSE, Milner JA (1996) Dietary components modify the ability of garlic to suppress 7,12-dimethylbenz(a)anthacene-induced mammary DNA adducts. J Nutr 126:817–824

American Cancer Society (n.d.) A biotechnology company dedicated to cancer treatment. Am Cancer Soc. www.cancervax.com/info/index.htm

Amin A, Hamza AA, Bajbouj K, Ashraf SS, Daoud S (2011) Saffron: a potential candidate for a novel anticancer drug against hepatocellular carcinoma. Hepatology 54(3):857–867

Andrade EHA, Zoghbi MDCB, Maia JGS, Fabricius H, Marx F (2001) Chemical characterization of the fruit of Annona squamosa L. occurring in Amzaon. J Food Compos Anal 14(2):227–232

Anonymous (2012) Anti-tumor effects of the Agrimonia pilosa ledeb. on liver cancer Smmc-7721 cells in vitro. Cancer Res Cent. 2012; http://www.rescancer.com/stomach-cancer/34400.html

Anonymous (2012a) Herb Information. Holistic Online.Com. http://www.holistic-online.com/herbal-med/_herbs/h27.htm. Accessed 15 Oct 2012

Anonymous (2012b) What is curcumin, or turmeric? Cancer active. http://www.canceractive.com/cancer-active-page-link.aspx?n=1571. Accessed 14 Oct 2012

Anonymous (2012c) Turmeric—benefits and side effects. The Herbal resource. http://www.herbal-supplement-resource.com/turmeric-benefits.html. Accessed 15 Oct 2012

Anonymous (n.d.-a) Agrimonia pilosa herb benefit for flu and viral infections. http://www.acetyll-carnitine.info/agrimoniapilosa.html. Accessed 09 Oct 2012

Anonymous (n.d.-b) Agrimonia pilosa - Ledeb. Plants for a Future. Weblog [Online]. http://www.pfaf.org/user/Plant.aspx?LatinName=Agrimonia+pilosa. Accessed 09 Oct 2012

Anonymous (n.d.-c) Tropical Plant Catalog. Top tropicals. http://toptropicals.com/catalog/uid/phaleria_macrocarpa.htm. Accessed 11 Oct 2012

Ao C, Li A, Elzaawely AA, Xuan TD, Tawata S (2008) Evaluation of antioxidant and antibacterial activities of Ficus microcarpa L. fil. extract. Food Control 19:940–948

Aratanechemuge Y, Komiya T, Moteki M, Katsuzaki H, Imai K, Hibasami H (2002) Selective induction of apoptosis by ar-turmerone isolated from turmeric (Curcuma longa L) in two human leukemia cell lines, but not in human stomach cancer cell line. Int J Mol Med 9(5):481–484

Araújo CAC, Leon LL (2001) Biological Activities of Curcuma longa L. Mem Inst Oswaldo Cruz 96(05):723–728

Arguello F, Alexander M, Sterry JA, Tudor G, Smith EM, Kalavar NT, Greene JF Jr, Koss W, Morgan CD, Stinson SF, Siford TJ, Alvord WG, Klabansky RL, Sausville EA (1998) Flavopiridol induces apoptosis of normal lymphoid cells, causes immunosuppression and has potent antitumor activity in vivo against human leukemia and lymphoma xenografts. Blood 91(7):2482–2490

Ariga T, Seki T (2006) Antithrombotic and anticancer effects of garlic-derived sulfur compounds: a review. Biofactors 26:93–103

Aripin A, Aritin PF, Tjandrawinata RR (2011) Isolated compounds from Phaleria Macrocarpa as anti-cancer agents. Publication No: US 2011/0257112 A1

Arshad M, Ahmad M (2004) Ethnobotanical study of Galliyat for botanical demography and bio-ecological diversification. Department of Botany, University of Arid Agriculture Rawalpindi, Pakistan

Asaka Y, Ohsaki A, Kubota T, Matsukawa Y, Satomi Y, Nishino H (1992) 5, 7, 3', 4'-Tetrahydroxy-3-methoxyflavone, a potent antitumor promoter isolated from Gnaphalium indicum. Kyoto-furitsu Ika Daigaku Zasshi 101:353–359

Aucamp J, Gaspar A, Hara Y, Apostolides Z (1997) Inhibition of xanthine oxidase by Catechins from tea (Camellia sinensis). Anticancer Res 17(6):4381–4385

Aung HH, Wang CZ, Ni M, Fishbein A, Mehendale SR, Xie JT, Shoyama AY, Yuan CS (2007) Crocin from Crocus sativus possesses significant anti-proliferation effects on human colorectal cancer cells. Exp Oncol 29(3):175–180

Awad EM (2005) In vitro decreases of the fibrinolytic potential of cultured human fibrosarcoma cell line, HT1080, by Nigella sativa oil. Phytomedicine 12(1–2):100–107

Ayyanar M, Ignacimuthu S (2008) Medicinal uses and pharmacological actions of five commonly used indian medicinal plants: a mini-review. Iranian J Pharmacol Ther 07(01):107–114

Aziz A, Rahman M, Mondal AK, Muslim T, Rahman A, Quader A (2005) 3-Acetoxy-6-benzoyloxyapagamide from Achyranthes aspersa. Pharm J 04:1820

Azuinea MA, Bhidea SV (1992) Chemopreventive effect of turmeric against stomach and skin tumors induced by chemical carcinogens in swiss mice. Nutr Cancer 17(01):77–83

Babu TD, Kuttan G, Padikkala J (1995) Cytotoxic and anti-tumour properties of certain taxa of umbelliferae with special reference to Centella Asiatica (L.) Urban. J Ethnopharmacol 48:53–57

Backer C, van den Brink R (1965) Flora of Java (Spermatophytes only), vol 02. Groningen, Noordhoff, The Netherlands

Bagavan A, Rahuman AA, Kamaraj C, Geetha K (2008) Larvicidal activity of saponin from Achyranthes aspera against Aedes aegypti and Culex quinquefasciatus (Diptera:Culicidae). Parasitol Res 103:223–229

Bakshi HA, Sam S, Feroz A, Ravesh J, Shah GA, Sharma M (2009) Crocin from kashmiri saffron (Crocus sativus) induces in vitro and in vivo xenograft growth inhibition of Dalton's lymphoma (DLA) in mice. Asian Pac J Cancer Prev 10:887–890

Banerjee T, Duhadaway JB, Gaspari P, Sutanto-Ward E, Munn DH, Mellor AL, Malachowski WP, Prendergast GC, Muller AJ (2008) A key in vivo antitumor mechanism of action of natural product based brassinins in inhibition of indoleamine 2,3-dioxygenase. Oncogene 27:2851–2857

Barber MA, Welch HC (2006) PI3K and RAC signalling in leukocyte and cancer cell migration. Bull Cancer 93:44–52

Barrett JE, Klopfenstein CF, Leipold HW (1998) Protective effects of cruciferous seed meals and hulls against colon cancer in mice. Cancer Lett 127(1):83–88

Bauer V, Jurcic K, Puhlmann J, Wagner V (1988) Immunologic invivo and invitro examination of Echinacea extracts. Arzeim Forsch 38:276–281

Beaver JP, Waring P (1995) A decrease in intracellular glutathione concentration precedes the onset of apoptosis in murine thymocytes. Eur J Cell Biol 68:47–54

Beneficial Ayurvedic Herbs (2010) Health Mango; http://www.healthmango.com/herbal/beneficial-ayurvedic-herbs/. Accessed 15 Oct 2012

Berger SJ, Gupta S, Belfi CA, Gosky DM, Mukhtar H (2001) Greentea constituent (--)-epigallo-catechin-3-gallate inhibits topoisomerase I activity in human colon carcinoma cells. Biochem Biophys Res Commun 288(1):101–105

Bergès R, Siess MH, Arnault I, Auger J, Kahane R, Pinnert MF, Vernevaut MF, Bon AM (2004) Comparison of the chemopreventive efficacies of garlic powders with different alliin contents against aflatoxin B1, carcinogenicity in rats. Carcinogenesis 25(10):1953–1959

Bhakuni DS, Dhar ML, Dhar MM, Dhawan BN, Mehrotra BB (1969) Screening of Indian plants for biological activity: Part-II. Indian J Exp Biol 7:250–262

Bharat S, Ram AS, Govind KV (2011) Antimicrobial, antineoplastic and cytotoxic activities of indole alkaloids from Tabernaemontana divaricata (L.) R.Br. Curr Pharm Anal 7(2):125–132

Bharti AC, Donato N, Singh S, Aggarwal BB (2003) Curcumin (diferuloylmethane) down-regulates the constitutive activation of nuclear factor-kappa B and IkappaBalpha kinase in human multiple myeloma cells, leading to suppression of proliferation and induction of apoptosis. Blood 101(03):1053–1062

Bhom KH, Liersch R (1992) Achyranthes. Hagers Handbuch der Pharmazeutischen Praxis V. Springer-Verlag, Berlin, pp 54–59

Bianchini F, Vainio H (2001) Allium vegetables and organosulphur compounds: do they help prevent cancer. Environ Health Perspect 109(9):893–902

Bilov M (2005) Image of Brassica campestris. Santiago, Metropolitania

Bilyk A, Sapers GM (1985) Distribution of quercetin and kaempferol in lettuce, kale, chive, garlic, leek, horseradish, red radish and red cabbage tissues. J Agric Food Chem 33:226–228

Biswas R, Mandal SK, Dutta S, Bhattacharyya SS, Boujedaini N, Khuda-Bukhsh AR (2011) Thujone-rich fraction of Thuja occidentalis demonstrates major anticancer potentials: evidences from in vitro studies on A375 cells. Evid Based Complem Alternat Med

Block E (2010) Garlic and other Alliums. Royal Soc Chem, p 1

Bone K (1996) Clinical applications of ayurvedic and Chinese herbs. Phytotherapy Press, Warwick, Queensland, pp 131–133

Bonfill M, Mangas S, Cusidó RM, Osuna L, Piñol MT, Palazón J (2006) Identification of triterpenoid compounds of Centella asiatica by thin-layer chromatography and mass spectrometry. Biomed Chromatogr 20:151–153

Boolbol SK, Dannenberg AJ, Chadburn A, Martucci C, Guo X, Ramonettti JT, Abreu-Goris M, Newmark HL, Lipkin ML, DeCosse JJ, Bertagnolli MM (1996) Cyclooxygenase-2 overexpression and tumor formation are blocked by sulindac in a murine model of familial adenomatous polyposis. Cancer Res 56:2556–2560

Boonsri S, Karalai C, Ponglimanont C, Chantrapromma S, Kanjana-Opas A (2008) Cytotoxic and antibacterial sesquiterpenes from Thespesia populnea. J Nat Prod 71(7):1173–1177

Borthakur SK, Goswami N (1995) Herbal remedies from Dimoria of Kamrup district of Assam in northeastern India. Fitoterapia 66:333–340

Bottone FG, Baek SJ, Nixon JB, Eling TE (2002) Diallyl disulfide (DADS) induces the antitumorigenic NSAID-activated gene (NAG-1) by a p53-dependent mechanism in human colorectal HCT 116 cells. J Nutr 132:773–778

Boudreau MD, Beland FA (2006) An evaluation of the biological and toxicological properties of Aloe Barbadensis (Miller), Aloe vera. J Environ Sci Health C 24:103–154

Bradfield CA, Bjeldanes LF (1987) High performance liquid chromatography analysis of anticarcinogenic indoles in Brassica oleracea. J Agric Food Chem 35(1):46–49

Brandi G, Schiavano GF, Zaffaroni N, De Marco C, Paiardini M, Cervasi B, Magnani M (2005) Mechanism of action of and antiproliferative properties of Brassica oleracea juice in Human breast cancer cell lines. Am Soc Nutr Sci 135(6):1503–1509

Brassica oleracea (n.d.) Encyclopedia Britannica. http://www.britannica.com/EBchecked/topic/77916/Brassica-oleracea

Brassica oleracea L. (n.d.) PFAF. http://www.pfaf.org/user/plant.aspx?latinname=Brassica+olera cea.

Brassica oleracea-Wild Cabbage (Brassicacea) (n.d.) First nature. http://www.first-nature.com/flowers/brassica_oleracea.php

Braunig B, Dorn M, Limburg E, Knick E (1992) Enhancement of resistance in common cold by Echinacea purpurea. Z Phytother 13:7–13

Brinkhaus B, Lindner M, Schuppan D, Hahn EG (2007) Chemical, pharmacological and clinical profile of the East Asian medicinal plant Centella asiatica. Phytomedicine 07:427–448

Bunpo P, Kataoka K, Arimochi H, Nakayama H, Kuwahara T, Bando Y, Izumi K, Vinitketkumnuen U, Ohnishi Y (2004) Inhibitory effects of Centella asiatica on azoxymethane-induced aberrant crypt focus formation and carcinogenesis in the intestines of F344 rats. Food Chem Toxicol 42:1987–1997

Bunpo P, Kataoka K, Arimochi H, Nakayama H, Kuwahara T, Ohnishi Y, Vinitketkumnuen U (2005) Centella asiatica extract induces cell cycle arrest In Caco-2 human colon cancer cells. Chiang Mai Med Bull 44(01):21–28

Busquets S, Carbo N, Almendro V, Quiles MT, Lopez-Soriano FJ, Argiles JM (2001) Curcumin, a natural product present in turmeric, decreases tumor growth but does not behave as an anticachectic compound in a rat model. Cancer Lett 167(01):33–38

Cain RJ, Ridley AJ (2009) Phosphoinositide 3-kinases in cell migration. Biol Cell 101:13–29

Cannene-Adams K, Lindshield BL, Wang S, Jeffery EH, Clinton SK, Erdman JW Jr (2007) Combinations of tomato and broccoli enhance antitumor activity in dunning R3327-H prostate adenocarcinomas. Cancer Res 67:836

Cantley LC, Neel BG (1999) New insights into tumor suppression: PTEN suppresses tumor formation by restraining the phosphoinositide 3-kinase/AKT pathway. Proc Natl Acad Sci 96:4240–4245

Carnero A, Blanco-Aparicio C, Renner O, Link W, Leal JF (2008) The PTEN/PI3K/AKT signalling pathway in cancer, therapeutic implications. Curr Cancer Drug Targets 08:187–198

Cerutti PA (1985) Prooxidant states and tumor promotion. Science 227:375–381

Chakrabortya A, Brantnera A, Mukainakab T, Nobukunib Y, Kuchideb M, Konoshimac T, Tokudab H, Nishinob H (2002) Cancer chemopreventive activity of Achyranthes aspera leaves on Epstein–Barr virus activation and two-stage mouse skin carcinogenesis. Cancer Lett 177:1–5

Chan MM, Huang HI, Fenton MR, Fong D (1998) In vivo inhibition of nitric oxide synthase gene expression by curcumin, a cancer preventive natural product with anti-inflammatory properties. Biochem Pharmacol 55(12):1955–1962

Chang MS, Chen BC, Yu MT, Sheu JR, Chen TF, Lin CH (2005) Phorbol 12-myristate 13-acetate upregulates cyclooxygenase-2 expression in human pulmonary epithelial cells via Ras, Raf-1, ERK, and NF-kappaB, but not p38 MAPK. Pathways Cell Signal 17:299–310

Chase MW, Reveal JL, Fay MF (2009) A subfamilial classification for the expanded asparagalean families Amaryllidaceae, Asparagaceae and Xanthorrhoeceae. Botan J Linnean Soc 161(2):132–136

Chen J, Huang S, Li F, Fang S, Chen Y (1992) Chemical constituents in Inula salsoloides (Turcz). Ostenf Zhiwu Xuebao 34:62–65

Chen GW, Chung JG, Hsieh CL, Lin JG (1998) Effects of the garlic components diallyl sulfide and diallyl disulfide on arylamine N-acetyltransferase activity in human colon tumour cells. Food Chem Toxicol 36:761–770

Chen BC, Yu CC, Lei HC, Chang MS, Hsu MJ, Huang CL, Chen MC, Sheu JR, Chen TF, Chen TL, Inoue H, Lin CH (2004) Bradykinin B2 receptor mediates NF-kappa B activation and cyclooxygenase-2 expression via the Ras/Raf-1/ERK pathway in human airway epithelial cells. J Immunol 173:5219–5228

Chen SH, Lin KY, Chang CC, Fang CL, Lin CP (2007) Aloe-emodin induced apoptosis in human gastric carcinoma cells. Food Chem Toxicol 45(11):2296–2303

Chen Y, Chen JW, Wang Y, Xu SS, Li X (2012a) Six cytotoxic Annonaceous acetogenins from Annona squamosa seeds. Food Chem 135(3):960–966

Chen Y, Xu SS, Chen JW, Wang Y, Xu HQ, Fan NB, Li X (2012b) Antitumor activity of Annona squamosa seeds extract containing Annonaceous acetogenin compounds. J Ethnopharmacol 142(2):462–466
Chen Y, Chen JW, Xu SS, Wang Y, Li X, Cai BC, Fan NB (2012c) Antitumor activity of Annonaceous acetogenins in HepS and S180 xenografts bearing mice. Bioorg Med Chem Lett 22(8):2717–2719
Cheng CL, Koo MWL (2000) Effects of Centella asiatica on ethanol induced gastric mucosal lesions in rats. Life Sci 67:2647–2653
Chiao C, Carothers AM, Grunberger D, Solomon G, Preston GA, Barrett JC (1995) Apoptosis and altered redox state induced by caffeic acid phenethyl ester (CAPE) in transformed rat fibroblast cells. Cancer Res 55:3576–3583
Chicca A, Adinolfi B, Martinotti E, Fogli S, Breschi MC, Pellati F, Benvenuti S, Nieri P (2007) Cytotoxic effects of Echinacea root hexanic extracts on human cancer cells lines. J Ethnopharmacol 110(1):148–153
Chowdhury MK, Ahsan N, Akhand AA (2009) Leaf extracts of Azadirachta indica and Terminalia arjuna induce death of HeLa cells without DNA degradation. Dhaka Univ J Pharm Sci 08(01):75–79
Chryssanthi DG, Lamari FN, Iatrou G, Pylara A, Karamanos NK, Cordopatis P (2007) Inhibition of breast cancer cell proliferation by style constituents of different Crocus species. Anticancer Res 27(1A):357–362
Chuang SE, Cheng AL, Lin JK, Kuo ML (2000) Inhibition by curcumin of diethylnitrosamin-einduced hepatic hyperplasia, inflammation, cellular gene products and cell-cycle-related proteins in rats. Food Chem Toxicol 38(11):991–995
Chung FL, Kelloff G, Steele V, Pittman B, Zang E, Jiao D, Rigotty J, Choi CI, Rivenson A (1996) Chemopreventive efficacy of arylalkyl isothiocyanates and N-Acetylcysteine for lung tumorigenesis in Fischer rats. J Canc Res 56:772
Chung JG, Lu HF, Yeh CC, Cheng KC, Lin SS, Lee JH (2004) Inhibition of N-acetyltransferase activity and gene expression in human colon cancer cell lines by diallyl sulfide. Food Chem Toxicol 42:195–202
Cifuentes MC, Castaneda DM, Uruena CP, Fiorentino S (2009) A fraction from Petiveria alliacea induces apoptosis via mitochondria dependent pathway and regulates HSP70 expression. Univ Sci 14(2):125–134
Coldren CD, Hashim P, Ali JM, Oh SK, Sinskey AJ, Rha C (2003) Gene expression changes in the human fibroblast induced by Centella asiatica triterpenoids. Planta Med 69:725–732
Conney AH, Lysz T, Ferraro T, Abidi TF, Manchand PS, Laskin JD, Huang MT (1991) Inhibitory effect of curcumin and some related dietary compounds on tumor promotion and arachidonic acid metabolism in mouse skin. Adv Enzyme Regul 31:385–396
Cooper R, Morre DJ, Morre DM (2005) Medicinal benefits of Green Tea: Part II. Review of anticancer properties. J Altern Complement Med 11(4):639–652
Copeland ES (1983) National Institutes of Health Workshop Report. Free radicals in promotion â€" a chemical pathology study section workshop. Cancer Res 43:5631–5637
Corsi MM, Bertelli AA, Gaja G, Fulgenzi A, Ferrero ME (1998) The therapeutic potential of Aloe vera in tumor bearing rats. Int J Tissue React 20(4):115–118
Cover CM, Hsieh SJ, Tran SH, Hallen G, Kim GS, Bjeldanes LF, Firestone GL (1998) Indole-3-carbinol inhibits the expression of cyclin dependent kinase-6 and induces a G1 cell cycle arrest of human breast cancer cells independent of estrogen receptor signaling. J Biol Chem 273:3838–3847
Craig, W.J. (2006) Phytochemicals: guardians of our health website
Crazy Jamaica (n.d.). Guinea henweed (Petiveria alliacea) (Anamu) This Jamaican herb is said to be a powerful cure for cancer. Zooming in on Jamaica. http://crazyjamaica.com/?page_id=331
Daikoku T, Dey SK (2008) Two faces of PTEN. Nat Med 14:192–193
Dantu AS, Shankarguru P, Ramya DD, Vedha HBN (2012) Evaluation of in vitro anticancer activity of hydro-alcoholic extract of Tabernaemontana divaricata. Asia J Pharma Clin Res 5(4):59–61

Das I, Das S, Saha T (2010) Saffron suppresses oxidative stress in DMBA-induced skin carcinoma: a histopathological study. Acta Histochem 112(4):317–327
Das S, Das J, Samadder A, Bhattacharyya SS, Das D, Khuda-Bukhsh AR (2013) Biosynthesized silver nanoparticles by ethanolic extracts of Phytolacca decandra, Gelsemium sempervirens, Hydrastis canadensis and Thuja occidentalis induce differential cytotoxicity through G2/M arrest in A375 cells. Colloids Surf B Biointerfaces 101:325–336
Dhanamani M, Devi SL, Kannan S (2011) Ethnomedicinal plants for cancer therapy: a review. J Drug Med 3(1):1–10
Dhanarasu S, Al-hazimi AM, Sethuraman P, Selvam M (2010) Chemopreventive and anti lipid peroxidative potential of Thespesia populnea (L.) on experimental buccal pouch carcinogenesis. Ibnosina J Med Biomed Sci 2(6):269–277
Diantini A, Subarnas A, Levita JS, Abdulah R, Achmad TH, Faried A, Faried LS, Julaeha E, Kurnia D, Wardhani SR, Koyama H (2012) Cytotoxicity of Fevicordin-A from Phaleria Macrocarpa (Scheff.) Boerl on P 388, Hela, Caski, Te-2, Te-8 and Prepuce's Fibroblast Cells. E3. J Med Res 01:001–005
Druesne N, Pagniez A, Mayeur C, Thomas M, Cherbuy C, Duee PH, Martel P, Chaumontet C (2004a) Diallyl disulfide (DADS) increases histone acetylation and p21waf1/cip1 expression in human colon tumor cell lines. Carcinogenesis 25:1227–1236
Druesne N, Pagniez A, Mayeur C, Thomas M, Cherbuy C, Duée PH, Martel P, Chaumontet C (2004b) Repetitive treatments of colon HT-29 cells with diallyl disulfide induce a prolonged hyperacetylation of histone H3 K14. Ann N Y Acad Sci 1030:612–621
Durak I, Biri H, Erguder IB, Devrim E, Senocak C, Avci A (2007) Effects of garlic and black grape extracts on the activity of adenosine deaminase from cancerous and non cancerous human urinary bladder tissues. Med Chem Res 16(6):259–265
Duyao MP, Kessler DJ, Spicer DB, Sonenshein GE (1990) Binding of NF-KB-like factors to regulatory sequences of the c-myc gene. Curr Top Microbiol Immunol 166:211–220
Ei-Mowafy AM, Al-Gayyar MM, Salem HA, ME EI-M, Darweish MM (2010) Novel chemotherapeutic and renal protective effects for the green tea (EGCG): a role of oxidative stress and inflammatory cytokine signaling. Int J Phytother Phytopharmacol 17(14):1067–1075
Elangovan V, Govindasamy S, Ramamoorthy N, Balasubramanian K (1995) Invitro studies on anticancer activity of Bacopa monnieri. Fitoterapia 66(3):211–215
Engelman JA (2009) Targeting PI3K signalling in cancer: opportunities, challenges and limitations. Nat Rev Cancer 09:550–562
Erba E, Sen S, Negri M (1996) Synchronization of cancer cell lines with methotrexate in vitro. Methods Cell Sci 18:149–163
Escribano J, Alonso G, Coca-Prados M, Fernandez J (1996) Crocin, saffranal and picrocrocin from saffron (Crocus sativus L.) inhibit the growth of human cancer cells in vitro. Cancer Lett 100(1–2):23–30
Fahey JW, Zhang Y, Talalay P (1997) Broccoli sprouts: an exceptionally rich source of inducers of enzymes that protect against chemical carcinogens. PNAS 94(19):10367–10372
Faried A, Kurnia D, Faried LS, Usman N, Miyazaki T, Kato H, Kuwano H (2007) Anticancer effects of gallic acid isolated from Indonesian herbal medicine, Phaleria macrocarpa (Scheff.) Boerl, on human cancer cell lines. Int J Oncol 30(03):605–613
Fariza IN, Fadzureena J, Zunoliza A, Chuah AL, Pin KY, Adawiah I (2012) Anti-inflammatory activity of the major compound from the methanol extract of phaleria macrocarpa leaves. J Appl Sci 12(11):1195–1198
Feng XU, Zhang S, Shao R, Zhen Y (2005) Anticancer activity of sodium caffeate and its mechanism. Acta Pharmacol Sin 26(10):1248–1252
Fenig E, Nordenberg J, Beery E, Sulkes J, Wasserman L (2004) Combined effect effect of aloe-emodin and chemotherapeutic agents on the proliferation of an adherent variant cell line of merkel cell carcinoma. Oncol Rep 11(1):213–217
Fernandes RS, Cotter TG (1994) Apoptosis or necrosis: intracellular levels of glutathione influence mode of cell death. Biochem Pharmacol 48:675–681

Figueroa-Hernandez JL, Gonzalez GS, Ascecio VJ, Figueroa-Espitia JL, Saavedra GF (2005) Plant products with anticancer properties employed in the treatment of bowel cancer: literature review 1985–2004. Proc West Pharmacol Soc 48:77–83

Find me a Cure (n.d.) Petiveria alliacea. Alternative Medicine. http://findmeacure.com/2012/06/30/petiveria-alliacea/

Finley JW (2003) Reduction of cancer risk by consumption of Selenium enriched plants: enrichment of broccoli with Selenium increases the anticarcinogenic properties of Broccoli. J Med Food 6(1):19–26

Friday JB, Okano D (2006) Thespesia populnea (milo). Species profile for Pacific Island Agroforestry. http://www.agroforestry.net/tti/Thespesia-milo.pdf

Fujiwara Y, Kawada K, Takano D, Tanimura S, Ozaki K, Kohno M (2006) Inhibition of the PI3 kinase/Akt pathway enhances doxorubicin-induced apoptotic cell death in tumor cells in a p53-dependent manner. Biochem Biophys Res Commun 340:560–566

Ganesan RM, Muthuchelian K (2011) Antitumor potential of an acetogenin isolated from the seed extracts of Annona squamosa Linn. J Canc Res Exp Oncol 3(8):95–104

Garc-Olmo DC, Riese HH, Escribano J, Onta J, Fernandez JA, Ateinzar M, Garci-Olmo D (1999) Effects of long term treatment of colon adenocarcinoma with crocin, a carotenoid from saffron (Crocus sativus L.): an experimental study in the rat. Nutr Cancer 35(2):120–126

Ge X, Yannai S, Rennet G, Gruener N, Fares FA (1996) 3,3'-diindolylmethane induces apoptosis in human cancer cells. Biochem Biophys Res Commun 228:153–158

Geetha BS, Nair MS, Latha PG, Remani P (2012) Sesquiterpene lactones isolated from Elephantopus scaber L. inhibits human lymphocyte proliferation and the growth of tumour cell lines and induces apoptosis in vitro. J Biomed Biotechnol:1–8

Ghimire SK, McKey D, Aumeeruddy-Thomas Y (2005) Conservation of Himalayan medicinal plants: harvesting patterns and ecology of two threatened species Nardostachys grandiflora and Neopicrorhiza scrophulariiflora. Biol Conserv 124:463–475

Ghosh T, Maity TP, Singh J (2011) Evaluation of anti tumor activity of stigmasterol a constituent isolated from Bacopa monnieri Linn aerial parts against Ehrlich ascites carcinoma in mice. Orient Pharm Exp Med 11:41–49

Giovannucci E, Rimm EB, Liu Y, Stampfer MJ, Willett WC (2003) A prospective study of cruciferous vegetables and prostate cancer. Cancer Epidemiol Biomarkers Prev 12(12):1403

Girach RD, Khan ASA (1992) Ethnomedicinal uses of Achyranthes aspera leaves in Orissa (India). Int J Pharm 30:113–115

Goh SH, Chuah MJSL, Soepadmo E (1995) Malaysian medicinal plants for the treatment of cardiovascular diseases. Pelanduk Publication Sdn, Bhd, Petaling Jaya

Gotama IBI, Sugiarto S, Nurhadi M, Widiyastuti Y, Wahyono S, Prapti IJ (1999) Inventory of Indonesian medicinal plants. Research units and development, vol 05. Ministry of Health, Jakarta, pp 147–148

Govind P (2011a) Some important anticancer herbs: a review. Int Res J Pharm 2(7):45–52

Govind P (2011b) Antioxidant vegetables act against cancer and other diseases. Int J Pharm Stud Res 2(1):32–38

Govind P, Madhuri S (2006) Medicinal plants: better remedy for neoplasm. Indian Drugs 43:869–874

Goyal BR, Goyal RK, Mehta AA (2007) Phyto-pharmcology of Achyranthes aspera: a review. Pharmacogn Rev 01:143–150

Gray AR (1982) Taxonomy and evolution of broccoli (Brassica oleracea var. italica). Econ Bot 36(4):397–410

Gribel NV, Pashinki VG (1986) Antimetastatic properties of Aloe juice. Vopr Onkologii 32(12):38–40

Gurib-Fakim A (2006) Medicinal plants: traditions of yesterday and drugs of tomorrow. Mol Aspects Med 27:1–93

Guyonnet D, Belloir C, Suschetet M, Siess MH, Le Bon AM (2000) Liver subcellular fractions for rats treated by organosulphur compounds from Allium modulate mutagen activation. Mutat Res 466(1):17–26

Hamayun M (2007) Traditional uses of some medicinal plants of swat valley, Pakistan. Ind J Trad Know 06(04):636–641

Hamayun M, Khan MA, Chudhary MF, Ahmad H (2006a) Studies on traditional knowledge of medicinal herbs of Swat Kohistan, District Swat, Pakistan. J Herbs Spices Med Plants 12(04):11–28

Hamayun M, Khan SA, Sohn EY, Lee IJ (2006b) Folk medicinal knowledge and conservation status of some economically valued medicinal plants of District Swat, Pakistan. J Ecol Appl 11(02):101–113

Han SS, Chung ST, Robertson DA, Ranjan D, Bondada S (1999) Curcumin causes the growth arrest and apoptosis of B cell lymphoma by downregulation of egr-1, c-myc, bcl-XL, NF-kappa B, and p53. Clin Immunol 93(02):152–161

Hanif R, Qiao L, Shiff SJ, Rigas B (1997) Curcumin, a natural plant phenolic food additive, inhibits cell proliferation and induces cell cycle changes in colon adenocarcinoma cell lines by a prostaglandin-independent pathways. J Lab Clin Med 130:576–584

Harmanto N (2001) Mahkota Dewa Obat Pusaka Para Dewa, Cetakan Kedua, Agromedia Pustaka, Jakarta 01

Harmanto N (2003) Conquering disease in Unison with Mahkota Dewa, Phaleria macrocarpa, 1st edn. PT Mahkotadewa Jakarta, Indonesia, p 14

Harmanto N (2005) Mahkota Dewa Obat Pusaka Para Dewa. In: A medicine the legacy of the Gods, 6th edn. Agro Media Pustaka, Jakarta

Harper CE, Patel BB, Wang J, Eltoum IA, Lamartiniere CA (2007) Epigallocatechin-3-gallate suppresses early stage, but not late stage prostate cancer in TRAMP mice: mechanism of action. Prostate 67(14):1576–1589

Hartwell JL (1982) Plants used against cancer. Quarterman Publications Inc., Lawrence, MA

Hazgui S, Bonnomet A, Nawrocki-Raby B, Milliot M, Terryn C, Cutrona J, Polette M, Birembaut P, Zahm JM (2008) Epigallocatechin-3-gallate (EGCG) inhibits the migratory behavior of tumor bronchial epithelial cells. Respir Res 9:33

Hecht SS, Carmella SG, Kenney PMJ, Low SH, Arakawa K, Yu MC (2004) Effects of cruciferous vegetable consumption on urinary metabolites of tobacco specific lung carcinogen 4-(Methylnitrosamino)-1-(3-Pyridyl)-1-Butanone in Singapore Chinese. Cancer Epidemiol Biomarkers Prev 13:997

Hendra R, Ahmadd S, Oskoueian E, Sukari A, Antioxidant SMY (2011) Antiinflammatory and cytotoxicity of phaleria macrocarpa (Boerl.) Scheff Fruit. BMC ComplemAltern Med 11(110)

Hennessy BT, Smith DL, Ram PT, Lu Y, Mills GB (2005) Exploiting the PI3K/AKT pathway for cancer drug discovery. Nat Rev Drug Discov 04:988–1004

Henning SM, Wang P, Heber D (2011) Chemopreventive effects of tea in prostate cancer: green tea versus black tea. Mol Nutr Food Res 55(6):905–920

Hertog MGL, Hollman CH, Katan MB (1992) Content of potentially anticarcinogenic flavanoids of 28 vegetables and 9 fruits commonly consumed in the Netherlands. J Agric Food Chem 40:2379–2383

Hirano T, Abe K, Gotoh M, Oka K (1995) Citrus flavone tangeretin inhibits leukaemic HL-60 cell growth partially through induction of apoptosis with less cytotoxicity on normal lymphocytes. Br J Cancer 72:1380–1388

Ho WY, Yeap SK, Ho CL, Raha AR, Suraini AA, Alitheen NB (2011) Elephantopus scaber induces cytotoxicity in MCF-7 Human breast cancer cells via p53-induced apoptosis. J Med Plant Res 5(24):5741–5749

Hong E, Kim GH (2008) Anticancer and antimicrobial acivities of β-Phenylethyl Isothiocyanate in Brassica rapa L. Food Sci Tech Res 14(4):377

Hopp DC, Zeng L, Gu Z, Mclaughlin JL (1996) Squamotacin: an Annonaceous acetogenin with cytotoxic selectivity for the human prostate tumor cell line (PC-3). J Nat Prod 59(2):97–99

Hosono T, Fukao T, Ogihara J, Ito Y, Shiba H, Seki T, Ariga T (2005) Diallyl trisulphide suppresses the proliferation and induces apoptosis of human colon cancer cells through oxidative modification of β-tubulin. J Biol Chem 280:41487–41493

Hosseinzadeh H, Behravan J, Mohammad R, Ajgan KH (2013) Antitumor and cytotoxic evaluation of Crocus sativus L. stigma and petal extracts using brine shrimp and potato disc assays. Sci Info Database

Hsu YL, Kuo P, Lin LT, Lin CC (2004) Asiatic acid, a Triterpene, induces apoptosis and cell cycle arrest through activation of extracellular signal-regulated kinase and p38 mitogen-activated protein kinase pathways in human breast cancer cells. J Pharmacol Exp Ther 313(01):333–344

Hu C, Kitts DD (2000) Studies on the antioxidant activity of Echinacea root extract. J Agric Food Chem 48(5):1466–1472

Hu G, Percehellet EM, Klish DS, Johnson JM, Perchellet JP (1992) Hydrolyzable tannins: potent inhibitors of hydroperoxide production and tumor promotion in mouse skin treated with 12-O-tetradecanoylphorbol-13-acetate in vivo. Int J Cancer 51(3):425–432

Huang MT, Wood AW, Newmark HL, Sayer JM, Yagi H, Jerina DM, Conney AH (1983) Inhibition of mutagenicity of Bay region diol epoxides of polycyclic aromatic hydrocarbons by phenolic plant flavonoids. Carcinogenesis 04:1631–1638

Huang MT, Wang ZY, Georgiadis CA, Laskin JD, Conney AH (1992) Inhibitory effects of curcumin on tumor initiation by benzo[a]pyrene and 7,12-dimethylbenz[a]anthracene. Carcinogenesis 13:2183–2186

Huang MT, Lou YR, Ma W, Newmark HL, Reuhl KR, Conney AH (1994) Inhibitory effects of dietary curcumin on forestomach, duodenal, and colon carcinogenesis in mice. Cancer Res 54(22):5841–5847

Huang MT, Ma W, Yen P, Xie J, Han J, Frenkel K, Grunberger D, Conney AH (1997a) Inhibitory effects of topical application of low doses of curcumin on 12-Otetradecanoylphorbol-13-acetate-induced tumor promotion and oxidized DNA bases in mouse epidermis. Carcinogenesis 18:83–88

Huang M, Newmark HL, Frenkel K (1997b) Inhibitory effects of curcumin on tumorigenesis in mice. J Cell Biochem Suppl 27:26–34

Huang M, Smart RC, Wong CQ (1998) Inhibitory effect of curcumin, chlorogenic acid, caffeic acid, and ferulic acid on tumor promotion in mouse skin by 12- O-tetradecanoylphorbol-13-acetate. Cancer Res 48:5941–5946

Huang CC, Lo CP, Chiu CY, Shyur LF (2010) Deoxyelephantopin, a novel multifunctional agent, suppresses mammary tumor growth and lung metastasis and doubles survival time in mice. Br J Pharmacol 159(4):856–871

Hutapea JR (n.d.) Inventaris Tanaman Obat Indonesia, Jilid V, Badan Penelitian dan Pengembangan Kesehatan—Departemen Kesehatan Republik Indonesia, Jakarta 1999; 03

Iciek M, Kwiecień I, Włodek L (2009) Biological properties of garlic and garlic derived organosulphur compounds. Environ Mol Mutagen 50(3):247–265

Inamdar PK, Yeole RD, Ghogare AB, de Souza NJ (1996) Determination of biologically active constituents in Centella asiatica. J Chromatogr A 742:127–130

Inano H, Onoda M, Inafuku N, Kubota M, Kamada Y, Osawa T, Kobayashi H, Wakabayashi K (1999) Chemoprevention by curcumin during the promotion stage of tumorigenesis of mammary gland in rats irradiated with gamma-rays. Carcinogenesis 20(06):1011–1018

Inoue M, Suzuki R, Koide T, Sakaguchi N, Ogihara Y, Yabu Y (1994) Antioxidant, gallic acid, induces apoptosis in HL-60RG cells. Biochem Biophys Res Commun 204:898–904

ITIS Report (n.d.). Tabernaemontana divaricata (L.). And Schult. 2013; http://www.itis.gov/servlet/SingleRpt/SingleRpt?search_topic=TSN&search_value=505424

Ivankovic S, Stojkovic R, Jukic M, Milos M, Milos M, Jurin M (2006) The antitumor activity of thymoquinone and thymohydroquinone in vitro and in vivo. Exp Oncol 28(3):220–224

Izhou X, Jiang G, Zhao A, Bondeva T, Hirszel P, Balla T (2001) Inhibition of Na,K-ATPase activates PI3 kinase and inhibits apoptosis in LLC-PK1 cells. Biochem Biophys Res Commun 285:46–51

Jackson BG (2000) Mechanism based target identification and drug discovery in cancer research. Science 287:1969
Jaganath IB, Ng LT (1999) Herbs: the green pharmacy of Malaysia, Kuala Lumpur. Vinpress Sdn. Bhd
Jagetia GC, Venkatesha VAK (2005) Enhancement of radiation effect by Aphanamixis polystachya in mice transplanted with Ehrlich ascites carcinoma. Boil Pharma Bullet 28(1):69–77
Jagetia JC, Venkatesha VA (2006) Treatment of mice with stem bark extract of Aphanamixis polystachya reduces radiation induced chromosome damage. Int J Rad Boil 82(3):197–209
Jain SK (1991) Dictionary of Indian folk medicine and ethnobotany, vol 609. Deep Publications, New Delhi, p 610
Jain V, Poonia A, Agarwal RP, Panwar RB, Kochar DK, Misra SN (1992) Effect of Terminalia arjuna in patients of angina pectoris. Ind Med Gaz 36:56–59
Jayashree G, Muraleedhara GP, Sudarslal S, Jacob VB (2003) Anti-oxidant activity of Centella asiatica on lymphoma-bearing mice. Fitoterapia 74:431–434
Jia GT, Lu XY (2008) Enrichment and purification of madecassoside and asiaticoside from Centella asiatica extracts with macroporous resins. J Chromatogr A 1193:136–141
Jia S, Roberts TM, Zhao JJ (2009) Should individual PI3 kinase isoforms be targeted in cancer? Curr Opin Cell Biol 21:199–208
Johnson JJ, Bailey HH, Mukhtar H (2010) Green tea polyphenols for prostate cancer chemoprevention: a translational perspective. Int J Phytother Phytopharmacol 17(1):3–13
Johnston WF (n.d.) Northern White Cedar. Thuja occidentalis L. http://www.na.fs.fed.us/Spfo/pubs/silvics_manual/Volume_1/thuja/occidentalis.htm
Joy B, Remani P (2008) Antitumor constituents from Annona squamosa fruit pericarp. Med Chem Res 2(7):345–355
Jung M, Park M (2007) Acetylcholinesterase inhibition by flavonoids from Agrimonia pilosa. Molecules 12:2130–2139
Kalyani MI, Lingaraju SM, Salimath BP (2013) A pro-apoptotic 15-kDa protein from Bacopa monnieri activated caspase-3 and downregulates Bcl-2 gene expression in mouse mammary carcinoma cells. J Nat Med 67(1):123–136
Kamboj VP (2000) Herbal medicine. Curr Sci 78(01):35–39
Kamei H, Koide T, Hashimoto Y, Kojima T, Hasegawa M (2009) Tumor cell growth suppression by tannic acid. Cancer Biother Radiopharm 14(02):135–138
Kan WS (1986) Pharmaceutical botany. National Research Institute of Chinese Medicine, p 416
Kandil FE, Nassar ML (1998) A tannin anti-cancer promoter from Terminalia arjuna. Phytochemistry 47:1567–1568
Kang MH, Kim JS, Seo JE, Oh SC, Yoo YA (2010) BMP2 accelerates the motility and invasiveness of gastric cancer cells via activation of the phosphatidylinositol 3-kinase (PI3K)/Akt pathway. Exp Cell Res 316:24–37
Kapoor LD (1990) Handbook of ayurvedic medicinal plants. CRC Press, Boca Raton, FL, pp 319–320
Kartnig T, Hoffmann-Bohm K (1992) Centella. In: Hager's Handbuchder Pharmazeutischen Praxis. Band 4. Springer Verlag, Berlin
Katrin E, Selvie WH (2011) Chromatogram profiles and cytotoxic activity of irradiated Mahkota Dewa (Phaleria Macrocarpa (Scheff.) Boerl) Leaves. Atom Indonesia 37(1):17–23
Kaur S, Grover IS, Knmar S (1997) Antimutagenic potential of ellagic acid isolated from Terminalia arjuna. Indian J Exp Biol 35:478–482
Kawada M, Ohno Y, Ri Y, Ikoma T, Yuugetu H, Asai T, Watanabe M, Yasuda N, Akao S, Takemura G, Minatoguchi S, Gotoh K, Fujiwara H, Fukuda K (2001) Anti-tumor effect of gallic acid on LL-2 lung cancer cells transplanted in mice. Anticancer Drugs 12:847–852
Kawamori T, Lubet R, Steele VE, Kelloff GJ, Kaskey RB, Rao CV, Reddy BS (1999) Chemopreventive effect of curcumin, a naturally occurring anti-inflammatory agent, during the promotion/progression stages of colon cancer. Cancer Res 59:597–601

Kedzierska M, Olas B, Wachowicz B, Stochmal A, Oleszek W, Jeziorski A, Piekarski J, Glowacki R (2009) An extract from berries of Aronia melanocarpa modulates the generation of superoxide anion radicals in blood platelets from breast cancer patients. Planta Med 75(13):1405–1409

Kedzierska M, Olas B, Wachoxwicz B, Stochmal A, Oleszek W, Jeziorski A, Piekarski J (2010) The nitrative and oxidative stress in blood platelets isolated from breast cancer patients: the protectory action of Aronia melanocarpa extract. Platelets 21(7):541–548

Kedzierska M, Olas B, Wachowicz B, Glowacki R, Bald E, Czernek U, Szydlowska-Pazera K, Potemski P, Piekarski J, Jeziorski A (2012) Effects of commercial extract of Aronia on oxidative stress in blood platelets isolated from breast cancer patients after the surgery and various phases of chemotherapy. Fitoterapia 83(2):310–317

Kelloff GJ (1996) J Cell. Biochem:72–85

Khalaf NA, Shakya AK, Al-Othman A, El-Agbar Z, Farah H (2008) Antioxidant activity of some common plants. Turk J Biol 32:51–55

Khan MAA, Islam MT (2012) Analgesic and cytotoxic activity of Acorus calamus L., Kigelia pinnata L., Mangifera indica L. and Tabernaemontana divaricata. J Pharm Bioallied Sci 4(2):149–154

Khan N, Sultana S (2005) Inhibition of two stage renal carcinogenesis, oxidative damage and hyperproliferative response by Nigella sativa. Eur J Cancer Prev 14(2):159–168

Khanina MG, Khanina MA, Rodin AP (2010) Element structure of Agrimonia pilosa LEDEB. Chem Plant Raw Mater 02:99–104

Kim R (2005) Recent advances in understanding the cell death pathways activated by anticancer therapy. Cancer 103:1551–1560

Knowles LM, Milner JA (2003) Diallyl disulfide induces ERK phosphorylation and alters gene expression profiles in human colon tumor cells. J Nutr 133:2901–2906

Koka PS, Mondal D, Schultz M, Abdel-Mageed AB, Agrawal KC (2010) Studies on molecular mechanisms of growth inhibitory effects of thymoquinone against prostate cancer cells: role of reactive oxygen species. Exp Biol Med 235(6):751–760

Korutla L, Kumar R (1994) Inhibitory effect of curcumin on epidermal growth factor receptor kinaseactivity in A431 cells. Biochim Biophys Acta 1224(03):597–600

Koshiura R, Miyamoto K, Ikeya Y, Taguchi H (1985) Antitumor activity of methanol extract from roots of Agrimonia pilosa LEDEB. Japan J Pharmacol 38:9–16

Kristal AR, Lampe JW (2002) Brassica vegetables and prostate cancer risk: a review of the epidemiological evidence. Nutr Cancer 42(1):1–9

Kumar EP, Elshurufa AA, Elango K, Subburaju T, Suresh B (1998) Cytotoxic and anti-tumor properties of ethanolic extract of Bacopa monnieri (L) Penn. Anc Sci Life 17(3):228–234

Kuo ML, Huang TS, Lin JK (1996) Curcumin, an antioxidant and anti-tumor promoter, induces apoptosis in human leukemia cells. Biochim Biophys Acta 1317(02):95–100

Kuo PL, Hsu YL, Lin TC, Lin LT, Chang JK, Lin CC (2005a) Casuarinin from the bark of terminalia arjuna induces apoptosis and cell cycle arrest in human breast adenocarcinoma MCF-7 Cells. Planta Med Biochem Mol Biol 71(03):237–243

Kuo PL, Hsu YL, Lin CC (2005b) The chemopreventive effects of natural products against human cancer cells. Int J App Sci Eng:203–214

Kuo PL, Hsu YL, Lin TC, Chang JK, Lin CC (2005c) Induction of cell cycle arrest and apoptosis in human non-small cell lung cancer A549 cells by casuarinin from the bark of Terminalia arjuna Linn. Anticancer Drugs 16(04):409–415

Kurbitz C, Heise D, Redmer T, Goumas F, Arlt A, Lemke J, Rimbach G, Kalthoff H, Trauzold A (2011) Epicatechin gallate and catechin gallate are superior to epigallocatechin gallate in growth suppression and anti-inflammatory activities in pancreatic tumor cells. Cancer Sci 102(4):728–734

Kurnia D, Akiyama K, Hayashi H (2008) 29-norcucurbitacin derivatives isolated from the Indonesian medicinal plant, Phaleria macrocarpa (Scheff.) Boerl. Biosci Biotechnol Biochem 72(02):618–620

Kuttan R, Bhanumathy P, Nirmala K, George MC (1985) Potential anticancer activity of turmeric (Curcuma longa). Cancer Lett 29(02):197–202
Kuttan R, Sudheeran PC, Tumori JCD (1987) Turmeric and curcumin as topical agents in cancer therapy. UK Pubmed Central 73(01):29–31
Le Bon AM, Roy C, Dupont C, Suschetet M (1997) Invivo antigenotoxic effects of dietary allyl sulphides in rats. Cancer Lett 114(1-2):131–134
Lee WL, Shyur LF (2012) Deoxyelephantopin impedes mammary adenocarcinoma cell motility by inhibiting calpain-mediated adhesion dynamics and inducing reactive oxygen species and aggresome formation. Free Radic Biol Med 52(8):1423–1436
Lee KH, Kim JH, Lim DS, Kim CH (2010) Antileukaemic and antimutagenic effects of Di(2-ethylhexyl)phthalate isolated from Aloe vera Linne. J Pharm Pharmacol 52(5):593–598
Lersch C, Seuner M, Bauer A, Siemens M, Hart R, Drescher M (1992) Non specific immunostimulation with low doses of cyclophosphamide (LDCY) thymostimulin, and Echinacea purpureae extracts (echinacin) in patients with far advanced colorectal preliminary results. Cancer Invest 10:343–348
Levine ES, Friedman HS, Griffith OW, Colvin OM, Raynor JH, Lieberman M (1993) Cardiac cell toxicity induced by 4-hydroperoxycyclophosphamide is modulated by glutathione. Cardiovasc Res 27:1248–1253
Li Y, Zhang T, Korkaya H, Liu S, Lee HF, Newman B, Yu Y, Clouthier SG, Schwartz SJ, Wicha MS, Sun D (2010) Sulfophorane, a dietary component of broccoli/ broccoli sprouts, inhibits breast cancer stem cells. Clin Cancer Res 16:2580
Li W, Tian H, Li L, Li S, Yue W, Chen Z, Qi L, Hu W, Zhu Y, Hao B, Gao C, Si L, Gao F (2012) Diallyl trisulphide induces apoptosis and inhibits proliferation of A549 cells in vitro and in vivo. Acta Biochim Biophys Sin 44(7):577–583
Liang W, Binns CW, Jian L, Lee AH (2007) Does the consumption of green tea reduce the risk of lung cancer among smokers? ECAM 4(1):17–22
Liao Z, Chen M, Tan F, Sun X, Tang K (2004) Micropropagation of endangered Chinese aloe. Plant Cell Tissue Organ Cult 76(1):83–86
Liaw CC, Yang YL, Chen M, Chang FR, Chen SL, Wu SH, Wu YC (2008) Mono-tetrahydrofuran Annonaceous acetogenins from Annona squamosa as cytotoxic agents and calcium ion chelators. J Nat Prod 71(5):764–771
Lim S, Lee J, Kim JK (2009) Analysis of isothiocyanates in newly generated vegetables, Baemuchae (Brassicoraphanus) as affected by growth. Int J Food Sci Tech 44(7):1401–1407
Limtrakul P, Lipigorngoson S, Namwong O, Apisariyakul A, Dunn FW (1997) Inhibitory effect of dietary curcumin on skin carcinogenesis in mice. Cancer Lett 116(02):197–203
Lin LT, Liu LT, Chiang LC, Lin CC (2002) In vitro antihepatoma activity of fifteen natural medicines from Canada. Phytother Res 16:440–444
Lin JG, Chen GW, Li TM, Chouh ST, Tan TW, Chung JG (2005) Aloe-emodin induces apoptosis in T24 human bladder cancer cells through the p53 dependent apoptotic pathway. J Urol 175(1):343–347
Lina W, Smadar A, Einat B, Jardena N, Eyal F (2002) Effect of Aloe emodin on the proliferation of a new merkel carcinoma cell line. Am J Dermatopath 24(1):17–22
Liu P, Cheng H, Roberts TM, Zhao JJ (2009) Targeting the phosphoinositide 3-kinase pathway in cancer. Nat Rev Drug Discov 08:627–644
Lowe HIC, Watson CT, Badal S, Ateh EN, Toyang NJ, Bryant J (2012) Anti-angiogenic properties of the Jamaican ball moss, (Tillandsia recurvata L.). Int Res J Biol Sci 1(4):73–76
Lu YP, Chang RL, Lou YR, Huang MT, Newmark HL, Reuhl KR, Conney AH (1994) Effect of curcumin on 12-O-tetradecanoylphorbol-13-acetate- and ultraviolet B light-induced expression of c-Jun and c-Fos in JB6 cells and in mouse epidermis. Carcinogenesis 15(10):2363–2370
Luettig B, Steinmuller C, Gifford G, Wagner-Mathes M (1989) Nacrophage activation by the polysaccharide arabinogalactan isolated from plant cell cultures of Echinace purpurea. J Natl Cancer Inst 81:669–675
Madhuri S, Govind P (2008) Some dietary agricultural plants with anticancer

Madhuri S, Pandey G (2008) Some dietary agricultural plants with anticancer properties. Plant Arch 8(1):13–16

Majdalawieh AF, Hmaidan R, Carr RI (2010) Nigella sativa modulates spleenocyte proliferation, Th1/Th2 cytokine profile, macrophage function and NK anti-tumor activity. J Ethnopharmacol 131(2):268–275

Malik M, Zhao C, Schoene N, Guisti MM, Moyer MP, Magnuson BA (2003) Anthocyanin rich extract from Aronia melanocarpa E. induces a cell cycle block in colon cancer but not normal colonic cells. Nutr Cancer 46(2):186–196

Manoj VR (2010) Evaluation of the antitumor activity of the ethanol extract of Brassica rapa on ehrlich ascites carcinoma in mice. Rajiv Gandhi Uni Health Sci, Karnataka

Matsukawa Y, Marui N, Sakai T, Satomi Y, Yoshida M, Matsumoto K, Nishino K, Aoike A (1993) Genistein arrests cell cycle progression at G-2-M. Cancer Res 53:1328–1331

Matsuura N, Miyamae Y, Yamane K, Nagao Y, Hamada Y, Kawaguchi N, Katsuki T, Hirata K, Sumi SI, Ishikawa H (2006) Aged garlic extract inhibits angiogenesis and proliferation of colorectal carcinoma cells. J Nutr 136:S842–S846

Mauri P, Pietta P (2000) Electrospray characterization of selected medicinal plant extracts. J Pharm Biomed Anal 23:61–68

Mbarek LA, Mouse HA, Elabbadi N, Bensalah M, Gamouh A, Aboufatima R, Benharref A, Chait A, Kamal M, Dalal A, Zyad A (2007) Anti-tumor properties of black seed (Nigella sativa L.) extracts. Braz J Med Biol Res 40(6):839–847

Mehta RG, Liu J, Constantinou A, Thomas CF, Hawthorne M, You M, Gerhuser C, Pezzuto JM, Moon RC, Moriarty RM (1995) Cancer chemopreventive activity of brassinin, a phytoalexin from cabbage. Carcinogenesis 16(2):399–404

Mehta K, Pantazis P, McQueen T, Aggarwal BB (1997) Antiproliferative effect of curcumin (diferuloylmethane) against human breast tumor cell lines. Anticancer Drugs 08(05):470–481

Mendes J. Cote ce Cote La. Trinidad and Tobago dictionary. Arima, trindiad. p. 95

Menon LG, Kuttan R, Kuttan G (1995) Inhibition of lung metastasis in mice induced by B16F10 melanoma cells by polyphenolic compounds. Cancer Lett 95(1–2):221–225

Merino R, Grillot DA, Simonian PL, Kumar SM, Fanslow WC, Bondada S, Nunez G (1995) Modulation of anti-IgM-induced B cell apoptosis by Bcl-xL and CD40 in WEHI-231 cells. Dissociation from cell cycle arrest and dependence on the avidity of the antibody –IgM receptor interaction. J Immunol 155:3830–3838

Middleton E Jr, Fujiki H, Savliwala M, Drzewiecki G (1987) Tumor-promoter-induced basophil histamine release: Effect of selected flavonoids. Biochem Pharmacol 36:2048–2052

Mika D, Guruvayoorappan C (2013) Experimental study on antitumor and anti inflammatory effect of Thespesia populnea phytochemical extract in mice models. Immunopharmacol Immunotoxicol 35(1):157–163

Miki K, Yasushi O, Yunmo R, Tetsuro I, Hideyuki Y, Toshihiro A, Motohiro W, Norio Y, Seigou A, Genzou T, Shinya M, Kohshi G, Hisayoshi F, Kazunori (2001) Anti-tumor effect of gallic acid on LL-2 lung cancer cells transplanted in mice. Anticancer Drugs 12(10):847–852

Miller SC (2012) Can herbs be useful in cancer therapy? A review of studies on the influence of Echinacea on cells of the immune system and on tumor amelioration. Biomed Res 23 SI:9–16

Miller RA (n.d.): Echinacea 2000: technical crop report. Richters herbs. Series 2. pp.2–5

Milner JA (2001) A historical perspective on garlic and cancer. Am Soc Nutr Sci 131(3):1027–1031

Milner JA (2010) Garlic and cancer prevention. In: Milner JA, Romagnolo DF (eds) Nutrition and health: bioactive compounds and cancer. Springer, New York, Dordrecht, Heidelberg, London, pp 567–588

Misra TN, Singh RS, Pandey HS, Prasad C (1991) An aliphatic dihydroxyketone from Achyranthes aspera. Phytochemistry 30:2076–2078

Missouri Botanical Garden (n.d.) Aronia melanocarpa. http://www.missouribotanicalgarden.org/gardens-gardening/your-garden/plant-finder/plant-details/kc/j420/aronia-melanocarpa.aspx

Miyajima A, Nakashima J, Yoshioka K, Tachibana M, Tazaki H, Murai M (1997) Role of reactive oxygen species in cis-dichlorodiammineplatinum- induced cytotoxicity on bladder cancer cells. Br J Cancer 76:206–210

Miyamoto K, Kishi N, Koshiura R (1987) Antitumor effect of agrimoniin, a tannin of Agrimonia pilosa Ledeb., on transplantable rodent tumors. Jap J Pharmacol 43(2):187–195

Miyamoto K, Kishi N, Murayama T, Furukawa T, Koshiura R (1988) Induction of cytotoxicity of peritoneal exudates cells by agrimoniin, a novel immunomodulatory tannin of Agrimonia pilosa Ledeb. Cancer Immunol Immunother 27:59–62

Modem S, DiCarlo SE, Reddy TR (2012) Fresh garlic extract induces growth and arrest and morphological differentiation of MCF7 breast cancer cells. Genes Cancer 3(2):177–186

Molnar J, Beladi I, Domonkos K, Foldeak S, Boda K, Veckenstedt A (1981) Antitumor activity of flavonoids on NK/LY ascites tumor cells. Neoplasma 28:11–18

Moulisha B, Kumar GA, Kanti HP (2010) Antileishmanial and anticancer activities of pentacyclic triterpenoid isolated from the leaves of Terminalia arjuna Combretacea. Trop J Pharm Res 9(2):135–140

Moura MD, Torres AR, Oliveira RAG, Diniz MFFM, Barbosa-Filho JM (2001) Natural products inhibitors of models of mammary neoplasia. Publicado Brit J Phytother 05(03):124–145

Mousavi SH, Tavakkol-Afshari J, Brook A, Jafari-Anarkooli I (2009) Role of caspases and Bax protein in saffron-induced apoptosis in MCF-7 cells. Food Chem Toxicol 47(8):1909–1913

Mukherjee A, Sikdar S, Bishayee K, Paul A, Ghosh S, Boujedaini N, Khuda-Bukhsh AR (2012) Ethanolic extract of Thuja occidentalis blocks proliferation of A549 cells and induces apoptosis in vitro. Zhong Xi Yi Jie He Xue Bao 10(12):1451–1459

Mukundan MA, Chacko MC, Annapurna VV, Krishnaswamy K (1993) Effect of turmeric and curcumin in BP-DNA adducts. Carcinogenesis 14:493–496

Murray M (1995) The healing power of the herbs, 2nd edn. Prima Publishing, Rocklin, CA

Nagabhushan M, Rangnekar VN, Ranjan D (1996) Curcumin is cytotoxic to cancer cells. Proc Am Assoc Cancer Res 37(409)

Nagpal A, Meena LS, Kaur S, Grover IS, Wadhwa R, Kaul SC (2000) Growth suppression of human transformed cells by treatment with bark extracts from a medicinal plant, Terminalia arjuna. In Vitro Cell Dev Biol Anim 36:544–547

Nair SC, Salomi MJ, Pannikar B, Pannikar KR (1991a) Modulatory effects of the extracts of saffron and Nigela sativa against cisplatinum induced toxicity in mice. J Ethnopharmacol 31:75–83

Nair SC, Pannikar B, Pannikar KR (1991b) Antitumor activity of saffron (Crocus sativus). Cancer Lett 57(2):109–114

Nair SC, Kurumboor SK, Hasegawa JH (2009) Saffron chemoprevention in biology and medicine: a review. Cancer Biother Radiopharmaceut 10(4):257–264

Neogy N, Rathore RS, Shreshth AD, Bannerji BK (1969) Studies on the anti-inflammatory and anti-arthritic activity of achyranthine. Indian J Pharm 01(47)

Ng WK, Yazan LS, Ismail M (2011) Thymoquinone from Nigella sativa was more potent than cisplastin in eliminating SiHa cells via apoptosis with down-regulation of Bcl-2 protein. Toxicol In Vitro 25(7):1392–1398

Nicholson KM, Anderson NG (2002) The protein kinase B/ Akt signalling pathway in human malignancy. Cell Signal 14:381–395

Ning ZN, Feng ZX, Fen XB, Chao LZ, Can FG, Qil JW, Hui XH, Jun WS, Zhou YR, Yi WX (2012) Squamocin induces apoptosis in leukemia cell line HL-60. Chin J Cancer

Nita CW (2003) Safety and anti-inflammatory activity of curcumin: a component of tumeric (Curcuma Longa). J Altern Complem Med 9(01):161–168

Nosho K, Yoshida M, Yamamoto H, Taniguchi H, Adachi Y, Mikami M, Hinoda Y, Imai K (2005) Association of Ets-related transcriptional factor E1AF expression with overexpression of matrix metalloproteinases, COX-2 and iNOS in the early stage of colorectal carcinogenesis. Carcinogenesis 26:892–899

O'Hara M, Kiefer D, Farell K, Kemper K (1998) A review of 12 commonly used Medicinal herbs. Arch Fam Med 7:523–536

Oetari S, Sudibyo JN, Commandeur R, Samhoedi NP, Vermeulen NP (1996) Effects of curcumin on cytochrome P450 and glutathione S-transferase activities in rat liver. Biochem Pharmacol 51:39–45

Ojeswi BK, Khoobchandani M, Hazra DK, Srivastava MM (2010) Protective effect of Thuja occidentalis against DMBA induced breast cancer with reference to oxidative stress. Hum Exp Toxicol 29(5):369–375

Olas B, Kedzierska M, Wachowicz B, Stochmal A, Oleszek W, Jeziorski A, Piekarski J, Glowacki R (2010) Effect of Aronia on thiol leaves in plasma of breast cancer patients. Cent Eur J Biol 5(1):38–46

Oudhia P (n.d.) Arjun or Koha [Terminalia arjuna (Roxb.) W. & A.]. Society for Parthenium Management. (SOPAM). http://www.hort.purdue.edu/newcrop/CropFactSheets/terminalia.html. Accessed 25 Oct 2012

Pan J, Kai GQ, Yuan CX, Jin RS (2007) Seperation and determination of the structural isomers of madecassoside by HPLC using beta-cyclodextrin as mobile phase additive. Chromatographia 66(1-2):121–123

Panda H. Elephantopus scaber. Handbook on medicinal herbs with uses. Asia Specific Bussiness Press, Delhi 2004; pp.491-493.

Pandey G (2011) Antioxidant vegetables act against cancer and other diseases. Int J Pharm Stu Res 2(1):32–38

Park BC, Bosire KO, Lee ES, Lee YS, Kim JA (2005) Asiatic acid induces apoptosis in SK-MEL-2 human melanoma cells. Cancer Lett 218(01):81–90

Park BC, Paek SH, Lee YS, Kim SJ, Lee ES, Choi HG, Yong CS, Kim JA (2007) Inhibitory effects of asiatic acid on 7,12-dimethylbenz[a]anthracene and 12-O-tetradecanoylphorbol 13-acetate-induced tumor promotion in mice. Biol Pharm Bull 30(01):176–179

Pasquini R, Scassellati-Sforzolini G, Villarini M, Moretti M, Marcarelli M, Fatigoni C, Kaur S, Kumar S, Grover IS (2002) In vitro protective effects of Terminalia arjuna bark extract against the 4-nitroquinoline-N-oxide genotoxicity. J Environ Pathol Toxicol Oncol 21:33–44

Pettit GR, Hoard MS, Doubek DL, Schmidt JM, Pettit RK, Tackett LP, Chapuis JC (1996) Antineoplastic agents 338. The cancer cell growth inhibitory. Constituents of Terminalia Arjuna (Combretaceae). J Ethnopharmacol 53:57–63

Pezzuto JM, Moon RC, CKH C, Chang CJ (2007) New approach for evaluating anti breast cancer activity of traditional Chinese medicine. Altern Treatment Cancer 3:325

Physician Desk Reference for Herbal Drugs (2000) Medical Economics Company Montvale, New Jersy, pp 359–361

Piper JT, Singhal SS, Salameh MS, Torman RT, Awasthi YC, Awasthi S (1998) Mechanisms of anticarcinogenic properties of curcumin: the effect of curcumin on glutathione linked detoxification enzymes in rat liver. Int J Biochem Cell Biol 30(04):445–456

Piwocka K, Zablocki K, Wieckowski MR, Skierski J, Feiga I, Szopa J, Drela N, Wojtczak L, Sikora E (1999) A novel apoptosis-like pathway, independent of mitochondria and caspases, induced by curcumin in human lymphoblastoid T (Jurkat) cells. Exp Cell Res 249(02):299–307

Poli A, Nicolau M, Simoes CMO, Nicolau RMR, Zanin M (1992) Preliminary pharmacologic evaluation of crude whole extract of Elephantopus scaber. Part I: in vivo studies. J Ethnopharmacol 37(1):71–76

Post JFM, Varma RS (1992) Growth inhibitory effects of bioflavonoids and related compounds on human leukemic CEM-C1 and CEM-C7 cells. Cancer Lett 67:207–213

Powolny AA, Bommareddy A, Hahm ER, Normolle DP, Beumer JH, Nelson JB, Singh SV (2011) Chemopreventative potential of the cruciferous vegetable constituent phenethyl isothiocyanate in mouse model of prostate cancer. J Natl Cancer Inst 103(7):571–584

Pradhasaradhi BVV, Madhurima R, Mubarak AA, Leela KA, Ashok K (2004) Antitumor activity of Annona squamosa seed extracts is through the generation of free radicals and induction of apoptosis. Indian J Biochem Biophys 41(4):167–172

Pradhasaradhi BVV, Madhurima R, Mubarak AA, Leela KA, Ashok K (2005) Differential cytotoxic effects of Annona squamosa seed extracts on human tumor cell lines: role of reactive oxygen species and glutathione. J Biosci 30(2):237–244
Prakash AO (1996) Potentialities of some indigenous plants for anti-fertility activity. Int J Crude Drug Res 24(1):19–24
Puntureea K, Wildb CP, Vinitketkumneuna U (2004) Thai medicinal plants modulate nitric oxide and tumor necrosis factor-α in J774.2 mouse macrophages. J Ethnopharmacol 95:183–189
Qureshi R, Waheed A, Arshad M, Umbreen T (2009) Medico-ethnobotanical inventory of tehsil Chakwal, Pakistan. Pak J Bot 41(2):529–538
Rahmawati E, Dewoto HR, Wuyung PE (2006) Anticancer activity study of ethanol extract of mahkota dewa fruit pulp (phaleria macrocarpa (scheff.) boerl.) in c3h mouse mammary tumor induced by transplantation. Med J Indonesia 15(4):217–222
Ramachandran C, You W (1999) Differential sensitivity of human mammary epithelial and breast carcinoma cell lines to curcumin. Breast Cancer Res Treat 54(03):269–278
Ramirez-Mares MV, Chandra S, De Mejia EG (2004) In vitro chemopreventive activity of Camellia sinensis, Ilex paraguariensis and Ardisia compressa tea extracts and selected polyphenols. Mut Res/Fundamental Mol Mech Mutagen 554(1–2):53–65
Rao AR, Sadhana AS, Goel HC (1990) Inhibition of skin tumors in DMBA-induced complete carcinogenesis system in mice by garlic (Allium sativum). Indian J Exp Biol 28:405–408
Rao CV, Rivenson A, Simi B, Zang E, Kelloff G, Steele V, Reddy BS (1995a) Chemoprevention of colon carcinogenesis by sulindac, a nonsteroidal anti-inflammatory agent. Cancer Res 55:1464–1472
Rao CV, Rivenson A, Simi B, Reddy BS (1995b) Chemoprevention of colon carcinogenesis by dietary curcumin, a naturally occurring plant phenolic compound. Cancer Res 55:259–266
Ravindranath MH, Ramasamay V, Moon S, Ruiz C, Muthugounder S (2009) Differential growth suppression of human melanoma cells by tea (Camellia sinensis) epicatechins (ECG, EGC and EGCG). ECAM 6(4):523–530
Reddy BS, Rao CV, Rivenson A, Kelloff G (1993) Inhibitory effect of aspirin on azoxymethane-induced colon carcinogenesis in F344 rats. Carcinogenesis (Lond.) 14:1493–1497
Reynolds T, Dweck AC (1999) Aloe vera leaf gel: a review update. J Ethnopharmacol 68:3–37
Roesler J, Steinmuller C, Kiderlen A, Emmendorfer A, Wagner H, Lohmann-Mathes M (1991) Application of purified polysaccharides fromcell cultures of the plant Echinacea purpurea to test subjects mediated activation of the phagocyte system. Int J Immunopharmacol 13:931–941
Rogala E, Skopinska-Rozewska E, Wasiutinski A, Siwicki AK, Pastewka K (2008) Echinacea purpurea diminishes neovascular reaction induced in mice skin by human cancer cells and stimulate non-specific cellular immunity in humans. Cent Eur J Immunol 33(3):127–130
Rohini G, Devi CSS (2008) Bacopa monnieri extract induces in murine sarcoma cells (S-180). Phytother Res 22(12):1595–1598
Rooney S, Ryan MF (2005) Effects of alpha-hederin and thymoquinone constituents of Nigella sativa, on human cancer cell lines. Int J Cancer Res Treatment 25(3B):2199–2204
Rosangkima G, Prasad SB (2004) Antitumour activity of some plants from Meghalaya and Mizoram against murine ascites Dolton's lymphoma. Indian J Exp Biol 42:981–988
Roth GN, Chandra A, Nair MG (1997) Novel bioactivities of Curcuma longa constituents. J Nat Prod 61(04):542–545
Rubya AJ, Kuttana G, Babub KD, Rajasekharanb KN, Kutta R (1995) Anti-tumour and antioxidant activity of natural curcuminoids. Cancer Lett 94:79–83
Ryu SY, Choi SU, Lee CO, Lee SH, Ahn JW, Zee OP (1994) Antitumor activity of some phenolic components in plants. Arch Pharm Res 17:42–44
Sadhana AS, Rao AR, Kucheria K, Bijani V (1988) Inhibitory action of garlic oil on the initiation of benzo(a)pyrene-induced skin carcinogenesis in mice. Cancer Lett 40:193–197
Salomi MJ, Nair SC, Panikkar PR (1990) Inhibitory effects of Nigella sativa and saffron (Crocus sativus) on chemical carcinogenesis in mice and its non mutagenic activity. Proc Ker Sci Congr 3:125–126

Salomi MJ, Nair SC, Panikkar KR (1991) Inhibitory effects of Nigella sativa and saffron (Crocus sativus) on chemical carcinogenesis in mice. Nutr Cancer 16(1):67–72
Salomi NJ, Nair SC, Jayawardhanan KK, Varghese CD, Panikkar KR (1992) Antitumor principles from Nigella sativa seeds. Cancer Lett 63(1):41–46
Salucci M, Stivala LA, Maiani G, Bugianesi R, Vannini V (2002) Flavoids uptake and their effect on cell cycle of human colon adenocarcinoma cells (Caco2). Br J Cancer 86:1645–1651
Salvesen GS, Dixit VM (1997) Caspases: Intracellular signaling by proteolysis. Cell 91:443–446
Samaha HS, Hamid R, El-Bayoumy K, Rao CV, Reddy BS (1997) The role of apoptosis in the modulation of colon carcinogenesis by dietary fat and by the organoselenium compound 1,4-phenylenebis(methylene)selenocyanate. Cancer Epidemiol Biomarkers Prev 06:699–704
Samarghandian S, Boskabady MH, Dovoodi S (2010) Use of in vitro assays to assess the potential anti-proliferative and cytotoxic effects of saffron (Crocus sativus L.) in human lung cancer cell line. Cancer Biother Radiopharmaceut 6(24):309–314
Samarghandian S, Afshai JT, Davoodi S (2011) Suppression of pulmonary tumor promotion and induction of apoptosis by Crocus sativus L. extraction. Appl Biochem Biotechnol 164(2):238–247
Samy PR, Ignacimuthu S, Sen A (1998) Screening of 34 Indian medicinal plants for antibacterial properties. J Ethnopharmacol 62:173–182
Sarikamis G (2009) Glucosinolates in crucifers and their potential effects against cancer: review. Can J Plant Sci 89(5):953–959
Saumya D, Priya B, Kumar DM, Sanjita D. (n.d.) Plants as rich source of antimalignant agents. Pharma tutor. Pharmacy infopedia. http://www.pharmatutor.org/articles/plants-used-as-source-of-antimalignant-agent
Saxena M, Faridi U, Mishra R, Gupta MM, Darokar MP, Srivastava SK, Singh D, Luqman S, Khanuja SPS (2007) Cytotoxic agents from Terminalia arjuna. Planta Med 73:1486–1490
Schaffer EM, Liu JZ, Green J, Dangler CA, Milner JA (1996) Garlic and associated allyl sulfur components inhibit N-methyl-N-nitrosourea induced rat mammary carcinogenesis. Cancer Lett 102:199–204
Seki T, Hosono T, Hosono-Fukao T, Inada K, Tanaka R, Oqihara J, Ariqa T (2008) Anticancer effects of diallyl trisulphide derived from garlic. Asia Pac J Clin Nutr 17(1):249–252
Sengupta A, Das S (2003) Tomato and garlic can modulate azoxymethane-induced colon carcinogenesis in rats. Eur J Cancer Prev 12:195–200
Sengupta A, Ghosh S, Das S (2004) Modulatory influence of garlic and tomato on cyclooxygenase-2 activity, cell proliferation and apoptosis during azoxymethane induced colon carcinogenesis in rat. Cancer Lett 208:127–136
Shafi G, Munshi A, Hasan TN, Alshatwi AA, Jyothy A, Lei DK (2009) Induction of apoptosis in HeLa cells by chloroform fraction of seed extracts of Nigella sativa. Cancer Cell Int 9:29
Shankar S, Srivastava RK (2007a) Bax and Bak genes are essential for maximum apoptotic response by curcumin, a polyphenolic compound and cancer chemopreventive agent derived from turmeric, Curcuma longa. Carcinogenesis 28(06):1277–1286
Shankar S, Srivastava RK (2007b) Involvement of Bcl-2 family members, phosphatidylinositol 3#-kinase/AKT and mitochondrial p53 in curcumin (diferulolylmethane)- induced apoptosis in prostate cancer. Int J Oncol 30(4):905–918
Shankar S, Marsh L, Srivastava RK (2013) EGCG inhibits growth of human pancreatic tumors orthotopically implanted in Balb C nude mice through modulation of FKHRL1/FOXO3a and neuropilin. Mol Cell Biochem 372(1-2):83–94
Shanker KS, Kanjilal S, Rao BVSK, Kishore KH, Prasad RBN (2007) Isolation and antimicrobial evaluation of isomeric hydroxy ketones in leaf cuticular waxes of Annona squamosa. Phytochem Anal 18(1):7–12
Sharif T, Alhosin M, Auger C, Minker C, Kim JH, Etienne-Selloum N, Bories P, Gronemeyer H, Lobstein A, Bronner C, Fuhrmann G, Schini-Kerth VB (2012) Aronia melanocarpa juice induces a redox sensitive p73 related caspase-3 dependent apoptosis in human leukemia cells. Plos One 7(3):e32526

Shaw RJ, Cantley LC (2006) Ras.PI(3)K and mTOR signaling controls tumour cell growth. Nature 441:424–430

Sheen LY, Wu CC, Lii CK, Tsai SJ (2001) Effect of diallyl sulphide and diallyl disulphide, the active principle of garlic, on aflatoxin B(1)-induced DNA damage in primary rat hepatocytes. Toxicol Lett 122(1):45–52

Shinwari ZK, Gilani SS (2003) Sustainable harvest of medicinal plants at Bulashbar Nullah, Astore, Northern Pakistan. J Ethnopharmacol 84:289–298

Shinwari MI, Khan MA (2000) Folk use of medicinal herbs of Margalla hills National Park, Islamabad. J Ethnopharmacol 69:45–56

Shirin H, Pinto JT, Kawabata Y, Soh JW, Delohery T, Moss SF, Murty V, Rivlin RS, Holt PR, Weinstein IB (2001) Antiproliferative effects of S-allylmercaptocysteine on colon cancer cells when tested alone or in combination with sulindac sulfide. Cancer Res 61:725–731

Simon A, Allais DP, Duroux JL, Basly JP, Durand-Fontanier S, Delage C (1998) Inhibitory effect of curcuminoids on MCF-7 cell proliferation and structure-activity relationships. Cancer Lett 129(01):111–116

Singh V (1995) Traditional remedies to treat asthma in the North West and Trans-Himalayan region in J. & K. state. Fitoterapia 66:507–509

Singh S, Khar A (2006) Biological effects of curcumin and its role in cancer chemoprevention and therapy. Anticancer Agents Med Chem 06(03):259–270

Singh B, Rastogi RP (1969) A reinvestigation of triterpenes of Centella asiatica. Phytochemistry 08:917

Singh AK, Sidhu GS, Deepa T, Maheshwari RK (1996) Curcumin inhibits the proliferation and cell cycle progression of human umbilical vein endothelial cell. Cancer Lett 107:109–115

Singh RP, Dhanalakshmi S, Rao AR (2000) Chemomdoulatory action of Aloe vera on the profiles of enzymes associated with carcinogen metabolism and antioxidant status regulation in mice. Phytomedicine 7(3):209–219

Singh AV, Xiao D, Lew KL, Dhir R, Singh SV (2004) Sulphorane induces caspase-mediated apoptosis in cultured PC-3human prostate cancer cells and retards growth of PC-3 xenografts in vivo. Carcinogenesis 25(1):83–90

Singletary K, MacDonald C, Wallig M, Fisher C (1996) Inhibition of 7,12-dimethylbenz[a] anthracene (DMBA)-induced mammary tumorigenesis and DMBA-DNA adduct formation by curcumin. Cancer Lett 103(02):137–141

Sivalokanathan S, Ilayaraja M, Balasubramanian MP (2005) Efficacy of Terminalia arjuna (Roxb.) on N-Nitrosodiethylamine induced hepatocellular carcinoma in rats. Indian J Exp Biol 43(03):264–267

Sivalokanathan S, Vijayababu MR, Balasubramanian MP (2006) Effects of Terminalia arjuna bark extract on apoptosis of human hepatoma cell line HepG2. World J Gastroenterol 12(07):1018–1024

Siveen KS, Kuttan G (2011) Thujone inhibits lung metastasis induced B16F-10. Melanoma cells in C57BL/6 mice. Can J Physiol Pharmacol 89(10):691–703

Sohi KK, Mittal N, Hundal MK, Khanduja KL (2003) Gallic acid, an antioxidant, exhibits anti apoptotic potential in normal human lymphocytes: a Bcl-2 independent mechanism. J Nutr Sci Vitaminol 49:221–227

Soladoye MO, Amasu NA, Raji-Esan SO, Chukwuma EC, Taiwo AA (2010) Ethnobotanical survey of anti-cancer plants in Ogun state, Nigeria. Ann Biol Res 1(4):261–273

Song K, Milner JA (2001) The influence of heating on the anticancer properties of garlic. Am Soc Nutr Sci 131(3):1054S–1057S

Srivastava R, Ahmed H, Dixit RK, Dharamveer SSA (2010) Crocus sativus L.: a comprehensive review. Pharmacogn Rev 4(8):200–208

Steenkamp V, Stewart MJ (2007) Medicinal applications and toxicological activities of Aloe products. Pharm Biol 45:411–420

Stoner GD, Morse MA (1997) Isothiocyanates and plant polyphenols as inhibitors of lung and esophageal cancer. Cancer Lett 114(1–2):113–119

Su CC, Chen GW, Tan TW, Lin JG, Chung JG (2006) Crude extract of garlic induced caspase-3 gene expression leading to apoptosis in human colon cancer cells. In Vivo 20:85–90
Su M, Chung HY, Li Y (2011) Deoxyelephantopin from Elephantopus scaber L. induces cell cycle arrest and apoptosis in the human nasopharyngeal cancer CNE cells. Biochem Biophys Res Commun 411(2):342–347
Subbarayan PR, Sarkar M, Rao SN, Philip S, Kumar P, Altman N, Reis I, Ahmed M, Ardalan B, Lokeshwar BL (2012) Achyranthes aspera (Apamarg) leaf extract inhibits human pancreatic tumor growth in athymic mice by apoptosis. J Ethnopharmacol 142:523–530
Subbarayana PR, Sarkarb M, Impellizzeric S, Raymoc F, Lokeshward BL, Kumare P, Agarwala RP, Ardalana B (2010) Anti-proliferative and anti-cancer properties of Achyranthes aspera: Specific inhibitory activity against pancreatic cancer cells. J Ethnopharmacol 131:78–82
Sueiro L, Yousef GG, Seigler D, De Mejia EG, Grace MH, Lila MA (2006) Chemopreventive potential of flavonoid extracts from plantation-bred and wild Aronia melanocarpa (Black chokeberry) fruits. J Food Sci 71(8):C480–C488
Sugi M (1977) Cancer therapy by Chinese crude drugs. In: Kondo K (ed) Cancer therapy in China today, pp 95–96
Sun XM, MacFarlane M, Zhuang J, Wolf BB, Green DR, Cohen GM (1999) Distinct caspase cascades are initiated in receptor-mediated and chemical-induced apoptosis. J Biol Chem 274:5053–5060
Sun L, Zhu H, Gan L, Mo J, Feng F, Zhou C (2012) Constituents from the bark of Annona squamosa and their antitumor activity. Zhongguo Zhong Yao Za Zhi 37(14):2100–2104
Sundaram SG, Milner JA (1996) Diallyl disulfide inhibits the proliferation of human tumor cells in culture. Biochim Biophys Acta 1315:15–20
Sunila ES, Kuttan G (2006) A preliminary study on antimetastatic activity of Thuja occidentalis L. in mice model. Immunopharmacol Immunotoxicol 28(2):269–280
Sunila ES, Hamsa TP, Kuttan G (2011) Effect of Thuja occidentalis and its polysaccharide on cell-mediated immune responses and cytokine levels of metastatic tumor bearing animals. Pharm Biol 49(10):1065–1073
Surendra P, Salamone VE, Goldin B (1997) Curcumin and genistein, plant natural products, show synergistic inhibitory effects on the growth of human breast cancer Mcf-7 cells induced by estrogenic pesticides. Biochem Biophys Res Commun 233(03):692–696
Suresh K, Manoharan S, Panjamurthy K, Senthil N (2007) Modifying effects of Annona squamosa glycoconjugates levels in 7,12-dimethylbenz(a)Anthracene induced hamster buccal pouch carcinogenesis. J Med Sci 7(1):100–105
Suresh K, Manoharan S, Vijayaanand MA, Sugunadevi G (2010) Modifying effects of Annona squamosa linn. On glycoconjugates levels in 7,12-dimethylbenz(a)Anthracene (DMBA) induced hamster buccal pouch carcinogenesis. J App Sci Res (6, 8):973–979
Susin SA, Lorenzo HK, Zamzami N, Marzo I, Snow BE, Brothers GM, Mangion J, Jacotot E, Costantini P, Loeffler M, Larochette N, Goodlett DR, Aebersold R (1999) Mitochondrial release of caspase-2 and -9 during the apoptotic process. J Exp Med 189:381–394
Tabolacci C, Lentini A, Mattioli P, Provenzano B, Oliverio S, Carlomosti F, Beninati S (2010) Antitumor properties of aloe-emodin and induction of transglutaminase 2 activity in B16-F10 melanoma cells. Life Sci 87:316–324
Tahiliani P, Kar A (2000) Achyranthes aspera elevates thyroid hormone level and decreases hepatic lipid peroxidation in male rats. J Ethnopharmacol 71:527–532
Taira J, Nanbu H, Ueda K (2009) Nitric oxide-scavenging compounds in Agrimonia pilosa Ledeb on LPS-induced RAW264.7 macrophages. Food Chem 115:1221–1227
Takada Y, Sethi G, Sung B, Aggarwal BB (2008) Flavopiridol suppresses tumor necrosis factor induced activation of activator protein-1, c-Jun N-terminal kinase, p38 Mitogen activated protein kinase (MAPK), p44/p42 MAPK, and Akt, inhibits expression of antiapoptotic gene products, and enhances apoptosis through cytochrome c release and caspase activation in human myeloid cells. Mol Pharmacol 73(5):1549–1577

Tanaka S, Haruma K, Yoshihara M, Kajiyama G, Kira K, Amagase H, Chayama K (2006) Aged garlic extract has potential suppressive effect on colorectal adenomas in human. J Nutr 136:821S–826S

Tandrasasmita OM, Lee JS, Baek SH, Tjandrawinata RR (2010) Induction of cellular apoptosis in human breast cancer by DLBS1425, a Phaleria macrocarpa compound extract, via downregulation of PI3-kinase/AKT pathway. Cancer Biol Ther 10(08):814–823

Tang W, Eisenbrand G (1992) Achyranthes bidentata BI. In: Chinese drugs of plant origin. Springer-Verlag, Berlin, pp 13–17

Tang L, Zirpoli GR, Guru K, Moysich KB, Zhang Y, Ambrosone CB, McCann SE (2008) Consumption of raw cruciferous vegetables is inversely associated with bladder cancer risk. Cancer Epidemiol Biomarkers Prev 17:938

Taraphdar AK, Roy M, Bhattacharya RK (2001) Natural products as inducers of apoptosis: Implication for cancer therapy and prevention. Curr Sci 80:1387–1396

Tavakkol-Ashfari J, Brook A, Mousavi SH (2008) Study of cytotoxic and apoptogenic properties of saffron extract in human cancer cell lines. Food Chem Toxicol 46(11):3443–3447

Temple NJ, Basu TK (1987) Selenium and cabbage and colon carcinogenesis in mice. J Natl Cancer Inst 79(5):1131–1134

Thabrew MI, Mitry RR, Morsy MA, Hughes RD (2005) Cytotoxic effects of a decoction of Nigella sativa, Hemidesmus indicus and Smilax glabra on human hepatoma HepG2 cells. Life Sci 77(12):1319–1330

Thani ANA, Sallis B, Nuttall R, Schubert FR, Ahsan M, Davies D, Purewal S, Cooper A, Rooprai HK (2012) Induction of apoptosis and reduction of MMP gene expression in U373cell line by polyphenolics in Aronia melanocarpa and by curcumin. Oncol Rep 28(4):1435–1442

The Wealth of India (1992): A dictionary of Indian raw materials and industrial products-raw materials series, vol. 3. Publications and Information Directorate, CSIR, New Delhi. Rev Ser., pp 428–430.

Thind TS, Agrawal SK, Saxena AK, Arora S (2008) Studies on cytotoxic, hydroxyl radical scavenging and topoisomerase inhibitory activities of extracts of Tabernaemontana divaricata (L.) R.Br. ex Rome. and Schult. Food Chem Toxicol 46(8):2922–2927

Thomson M, Ali M (2003) Garlic [Allium sativum]: a review of its potential use as an anticancer agent. Curr Cancer Drug Targets 3(1):67–81

Tjandrawinata RR, Aritin PF, Tandrasasmita OM, Rahmi D, Aripin A (2010) DLBS1425, a Phaleria macrocarpa (Scheff.) Boerl. extract confers anti proliferative and proapoptosis effects via eicosanoid pathway. J Exp Ther Oncol 08(03):187–201

Tjandrawinata RR, Aripin A, Aritin PF, Rahmi D (2011) Extract of Phaleria macrocarpa as an anti-neoplastic, anti-inflammatory and antiangiogenic agent. United States. Patent Application Publication

Tomasin R, Gomes-Marcondes MCC (2011) Oral administration of Aloe vera and honey reduces walker tumor growth by decreasing cell proliferation and increasing apoptosis in tumor tissues. Phytother Res 25(4):619–623

Top Tropicals (n.d.) Camellia sinensis. Available online http://toptropicals.com/downloads/tea.pdf

Tsai YL, Chiu CC, Chen JY, Chan KC, Lin SD (2012) Cytotoxic effects of Echinacea purpurea flower extracts and cichoric acid on human colon cancer cells through induction of apoptosis. J Ethnopharmacol 143(3):914–919

Tsuda H, Ohshima Y, Nomoto H, Fujita K, Matsuda E, Iigo M, Takasuka N, Moore MA (2004) Cancer prevention by natural compounds. Drug Metab Pharmacokinet 19:245–263

Uddin S, Hussain A, Al-Hussein K, Platanias LC, Bhatia KG (2004) Inhibition of phosphatidylinositol 3'-kinase induces preferentially killing of PTEN-null T leukemias through AKT pathway. Biochem Biophys Res Commun 320:932–938

Uruena C, Cifuentes C, Castaneda D, Arango A, Kaur P, Asea A, Fiorentino S (2008) Petiveria alliacea extracts uses multiple mechanisms to inhibit growth of human and mouse tumoral cells. BMC Complement Altern Med 8(60)

Utami CV, Hatane SE, Gorjian M (2011) The application of three herbs; Chrysanthemum indicum, Centella asiatica, and Andrographis paniculata to reduce bacteria in cow milk. Int J Comp Inter Manag 19(02):38–43

Valcheva-Kuzmanova SV, Belcheva A (2006) Current knowledge of Aronia melanocarpa as a medicinal plant. Folia Med (Plovdiv) 48(2):11–17

Verhoeven DT, Goldbohm RA, Van Poppel G, Verhagen H, Van Den Brandt PA (1996) Epidemiological studies on Brassica vegetables and cancer risk. Cancer Epidemiol Biomarkers Prev 5:733

Verhoeven DTH, Verhagen H, Goldbohm RA (1997) Ven den Brandt PA, Van Poppel G. A review of mechanism underlying anticarcinogenicity by Brassica vegetables. Chem Biol Interact 103(2):79–129

Verma N, Vinayak M (2009) Effect of terminalia Arjuna on antioxidant defense system in cancer. Mol Biol Rep 36:159–164

Vetrichelvan T, Jegadeesan M (2003) Effect of alcohol extract of Achyranthes aspersa Linn. on acute and sub-acute inflammation. Phytother Res 17(1):77–79

Villegas I, Sanchez-Fidalgo S, de la Lastra CA (2008) New mechanisms and therapeutic potential of curcumin for colorectal cancer. Mol Nutr Food Res 52:1040–1061

Vrchovska V, Sousa C, Valentao P, Ferreres F, Pereira JA, Seabra RM, Andrade PB (2006) Antioxidative properties of tronchuda cabbage (Brassica oleracea L. var. costata DC) external leaves against DPPH, superoxide radical, hydroxyl radical and hypochlorous acid. Food Chem 98(3):416–425

Wang B, Jin Z (2011) Agrimoniin induced SGC7901 cell apoptosis associated mitochondrial transmembrane potential and intracellular calcium concentration. J Med Plant Res 5(15):3513–3519

Wang JP, Hsu MF, Teng CM (1984) Antihemostatic effect of Hsien-Ho-T0sao (Agrimonia pilosa). Am J Chin Med 12:116–123

Wang JP, Hsu MF, Teng CM (1985) Antiplatelet effect of Hsien-Ho-T0sao (Agrimonia pilosa). Am J Chin Med 13:109–118

Warrier PK, Nambiar VPK, Kutty RC (1993) Indian medicinal plants, vol 1. Orient Longman Ltd., Madras, p 209

Wee S, Wiederschain D, Maira SM, Loo A, Miller C, deBeaumont R et al (2008) PTEN-deficient cancers depend on PIK3CB. Proc Natl Acad Sci U S A 105:13057–13062

Williams LAD, Rosner H, Levy HG, Barton EN (2007) A critical review of the therapeutic potential of dibenzyl trisulphide isolated from Petiveria alliacea L (Guinea henweed, Anamu). West Indian Med J 56(1):17–21

Wilson CL, Solar JM, Ghaouth AEL, Wisnicwski MC (1997) Rapid evaluation of plant extracts and essential oils for antifungal activity against *Botrytis Cinerea*. Plant Dis 81(z):204–210

Winarno H, Katrin E (2009) Benzophenone glucoside isolated from the ethyl acetate extract of the bark of Mahkota dewa [Phaleria macrocarpa (Scheff.) Boerl.] and its inhibitory activity on Leukemia L1210 cell line. Indo J Chem 09(01):142–145

Winarto WP, Mahkota Dewa (2003), Budi Daya & Pemanfaatan Obat, Cetakan Kelima, Penebar Swadaya. Jakarta; 03.

World Conservation Monitoring Centre (2006) In: IUCN 2006. 2006 IUCN Red List of Threatened Species

Wu YD, Lou YJ (2007) A steroid fraction of chloroform extract from bee pollen of Brassica campestris induces apoptosis in human prostate cancer PC-3 cells. Phytother Res 21(11):1087–1091

Wu X, Kassie F, Mersch-Sundermann V (2005) Induction of apoptosis in tumor cells by naturally occurring sulfur-containing compounds. Mutat Res 589:81–102

Wu QJ, Yang Y, Vogtmann A, Wang J, Han LH, Li LH, Xiang YB (2012) Cruciferous vegetables intake and the risk of colorectal cancer: a meta-analysis of observational studies. Ann Oncol http://annonc.oxfordjournals.org/content/early/2012/12/03/annonc.mds601.abstract

Xu G, Liang Q, Gong Z, Yu W, He S, Xi L (2006) Antitumor activities of the four sesquiterpene lactones from Elephantopus scaber L. Exp Oncol 28(2):106–109
Xue Z, Liu Z, Wu M, Zhuang S, Yu W (2010) Effect of rapeseed peptide on DNA damage and apoptosis in HeLa cells. Exo Toxicol Pathol 62(5):519–523
Yagihashi N, Kasajima H, Sugai S, Matsumoto K, Ebina Y, Morita T, Murakami T, Yagihashi S (2000) Increased in situ expression of nitric oxide synthase in human colorectal cancer. Virchows Arch 436:109–114
Yang CS, Wang X (2010) Green tea and cancer prevention. Nutr Cancer 62(7):931–937
Yang CS, Wang H (2011) Mechanistic issues concerning cancer prevention by tea Catechins. Mol Nutr Food Res 55(6):819–831
Yekti Y (2010) Phaleria Macrocarp. Free traditional medicinal tips; http://tmedical.blogspot.com/2012/04/phaleria-macrocarpa.html. Accessed 21 Oct 2012
Yoosook C, Bunyapraphatsara N, Boonyakiat Y, Kantasuk C (2000) Anti-herpes simplex virus activities of crude water extracts of Thai medicinal plants. Phytomedicine 06:411–419
Yoshida T, Hatano T, Ito H (2000) Chemistry and function of vegetable polyphenols with high molecular weights. Biofactors 13:121–125
Yoshida M, Fuchigami M, Nagao T, Okabe H, Matsunaga K, Takata J, Karube Y, Tsuchihashi R, Kinjo J, Mihashi K, Fujioka T (2005) Antiproliferative constituents from Umbelliferae plants VII. Active terpenes and rosmarinic acid from Centella asiatica. Biol Pharm Bull 28(1):173–175
Yuan JM, Sun C, Butler LM (2011) Tea and cancer prevention: epidemiological studies. Off J Ital Pharmacol Soc 64(2):123–135
Zhang Y, Talalay P, Cho CG, Posner GH (1992) A major inducer of anticarcinogenic protective enzymes from broccoli: isolation and elucidation of structure. PNAS 89(6):2399–2403
Zheng J, Yang B, Huang T, Yang J, Li D (2011) Green tea and Black tea consumption and prostate cancer risk: an exploratory meta-analysis of observational studies. Nutr Cancer 63(5):663–672
Zhu BH, Zhan WH, Li ZR, Wang Z, He YL, Peng JS, Cai SR, Ma JP, Zhang CH (2007) (-)-Epigallocatechin-3-gallate inhibits growth of gastric cancer by reducing VEGF production and angiogenesis. World J Gastroenterol 13(8):1162–1169
Zhu L, Tan J, Wang B, He R, Liu Y, Zheng C (2009) Antioxidant activities of aqueous extract from Agrimonia pilosa Ledeb and its fractions. Chem Biodivers 06:1716–1726
Zimmerman B (n.d.) Echinacea. Open line garden show. Articles 2007; http://www.brucezimmerman.com/ARTICLES/ECHINACEA.htm

Herbals in Iğdır (Turkey), Nakhchivan (Azerbaijan), and Tabriz (Iran)

Munir Ozturk, Volkan Altay, Ernaz Altundağ, S. Jamshid Ibadullayeva, Behnaz Aslanipour, and Tuba Mert Gönenç

Introduction

The families, communities, nations, and future generations will have to depend on biodiversity as an essential resource in their daily life (Cocks 2006). Bio and human diversities represent the link between biocultural diversity. The role played by humans in the conservation of biodiversity is great, since the latter represents a source of raw material in which the processes of evolution depend. The lesser the diversity, the greater the chances for living beings to get destroyed due to loss of resilience to environmental changes. Therefore it is important that we maintain our biodiversity, as it provides living beings with different ways of understanding and interacting with each other on our planet and ultimately offers different possibilities for the future of humans (Milton 1996; Cocks 2006; Bedelov et al. 2014).

The biodiversity is perceived and appreciated by different cultures and people in different ways. This is because of their distinct heritage and experience (Posey 1999).

M. Ozturk (✉)
Vice President of the Islamic World Academy of Sciences, Amann, Jordan

Department of Botany, Centre for Environmental Studies, Ege University, Izmir, Turkey

V. Altay
Faculty of Science and Arts, Department of Biology, Hatay Mustafa Kemal University, Hatay, Turkey

E. Altundağ
Faculty of Science and Arts, Department of Biology, Düzce University, Düzce, Turkey

S. Jamshid Ibadullayeva
Botany Institute of the Azerbaijan NAS, Ethnobotany, Bakü, Azerbaijan

B. Aslanipour
Centre for Science and Technology, Ege University, Izmir, Turkey

T. M. Gönenç
Faculty of Pharmacy, Department of Pharmacognosy, Ege University, Izmir, Turkey

M. Ozturk, K. R. Hakeem (eds.), *Plant and Human Health, Volume 1*,
https://doi.org/10.1007/978-3-319-93997-1_6

The intricate connection between biodiversity conservation and cultural diversity is revolving around the topic if cultural diversity can sustain a wide variety of use practices together with conservation of natural resources (Posey 1999; McNeely 2000). We know well that, all through the past centuries, owing to their several beneficial features, the plants have played an important role in the daily life of humans, providing food and medicines, together with plant fibers for clothes, wood for building, etc. (Han and Bulut 2015). Plants are an integral part of all living organisms of the planet Earth and MAPS are distributed worldwide (Singh et al. 2012).

The use of natural products with therapeutic properties is as ancient as human civilization, and, for a long time, mineral, plant, and animal products have remained as the main source of drugs (De Pasquale 1984; Rates 1991). The MAPS used for various illnesses are placed first in order of importance as natural resources. Treatment with traditional folk medicine is still highly accepted, particularly by those lacking an access to the modern healthcare. Inevitably the so-called traditional folk medicine mostly originates from plants, and around 80% of the world population use plants to treat health problems (WHO 1993; Han and Bulut 2015). The MAPS are an important source of modern drugs. Nearly 25% of the drugs prescribed worldwide come from plants (Rates 1991). While estimates vary between 35.000 and 70.000 plant species, these probably have been used for medicaments (Farnsworth and Soejarto 1991; Akerele 1992; Leman 2006). This level of usage suggests from 10 to 25% of the present day 310.129 plant species, thought to occur in the world, could have medicinal active constituents (Chapman 2009; Mamedov and Craker 2012). Identifying medicinal plants and their beneficial medicinal qualities, however, requires active investigations (Mamedov and Craker 2012). Fabricant and Farnsworth (2001) have identified 123 plant-derived drugs currently prescribed in the industrial world. The vast majority of these drugs, many of which are used daily in allopathic practices, were mostly discovered by studying the plants used in the traditional medicines of a number of countries (Farnsworth 1992).

One of the most powerful means for addressing the issue of change and variability of medicinal plant uses and heritage is cross-cultural and cross-border ethnobotany. Further studies in Caucasus and nearby regions and beyond need to be addressed with the trajectory proposed here. This will provide an opportunity to establish if disparate sociocultural, economic, and political situations have induced some differences in the local plant use in communities, basically sharing a historical legacy in the same environment (Sõukand and Pieroni 2016). Ethnobotanical investigations have mainly considered the recording of traditional ecological knowledge in a specific place (De Almeida et al. 2012; Quiroga et al. 2012; Mattalia et al. 2013; Kidane et al. 2014; Zlatković et al. 2014; Menendez-Baceta et al. 2015). But many of the studies have compared the species traditionally used in different regions or countries (Madaleno 2010; Bradacs et al. 2011; Ellena et al. 2012; Ghorbani et al. 2012; Sõukand et al. 2013). Cross-cultural ethnobotanical investigations usually focus on how different cultures select and use plants, some trying to analyze how traditional ecological knowledge changes over space and time and which variables explain such patterns (Leporatti and Ivancheva 2003; Hadjichambis et al. 2008; Leporatti and Ghedira 2009; Łuczaj 2010; Pieroni et al. 2011; Menendez-Baceta et al. 2015).

Study Areas

Iğdır is located in Turkey, in the Eastern Anatolia Region around Erzurum–Kars. The area is 3.539 km^2 (Ozturk et al. 2016). Aras River serves as the border between Turkey and Armenia; it makes up the northern and northeastern borders of this province. Nakhchivan (Azerbaijan) is located in the east, and on the south is the Turkey–Iran border of the province (Fig. 1). It is the only province in Turkey which has borders with three countries. Continental climate prevails here but is totally different from the surrounding areas. It has typical microclimatic features, because the temperatures are higher and rainfall is less, thus consisting a "regional climate area" in the Eastern Anatolia Region. During the year, there is a short, but significant winter season (December to February) and a relatively long summer (May to September). In winter the temperatures do fall from −25 to −30 °C. The maximum temperature in July and August is around 35–40 °C. The highest rainfall is recorded in spring, while winter is notable as the least rainy season (Ozturk et al. 2016).

Nakhchivan is a part of the Azerbaijan Republic, located in the southwestern part of the Lesser Caucasus mountains, with a total border length of 398 km. The region covers 5363 km^2 and borders Armenia (221 km) to the east and north, Iran (179 km)

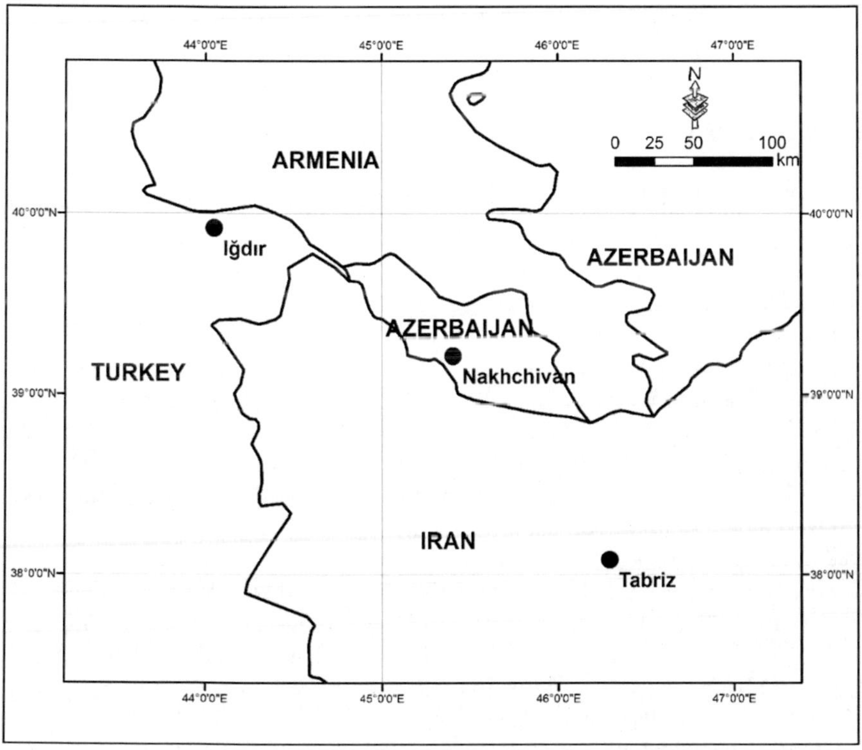

Fig. 1 Map showing the study areas

to the south and west, and Turkey (15 km) to the northwest (Fig. 1). The climate is extreme continental type, with hot summers and severe winters. Average annual temperature is 10–14 °C, but areas above 2300–2400 m altitude show a mean annual air temperature below 4 °C. The maximum air temperature in the lower part of the state is 18 °C in January and 41–43 °C in July to August; the relative humidity varies, being 74–76% in December to February, but 39–40% in July to August, in the city proper. In the middle mountain zone, the values change between 69–78% and 52–55% in December to February and July to August, respectively. This situation is similar to the foothills of the Lesser Caucasus. The main bulk of precipitation falls in spring (March to May) and the minimum in July to August. In the lowlands annual rainfall is 210–310 mm; in the mid-mountainous area, it varies between 365 and 550 mm. In the alpine zone, it is 660 mm. Nakhchivan is regarded as a separate climatic and a physical-geographical region of Azerbaijan (Mirzeyev 1972; Seyidova and Hüseyin 2012).

Tabriz (38°05′, 46°17′) is the largest city in northwest Iran, with a population exceeding 1.5 million inhabitants and an area of 45.481 km^2 (Taghipour and Mosaferi 2009). The city is regarded as the capital of East Azerbaijan placed in northwest of Iran (Fig. 1). Tabriz experiences warm summers and cold winters and has an average annual temperature of 12.2 °C, with an annual precipitation of 311.1 mm, which falls mostly in winter and spring and accounts for almost half of annual potential evapotranspiration in this semiarid region (Amiri et al. 2009).

Data Analysis

This data published by Altundağ (2009), Altundag and Ozturk (2011), and Ozturk et al. (2012) in Iğdır, Turkey; by Mir-Babayev and Waigh (1997), Hasanova et al. (2000), Mehtiyeva and Zeynalova (2008), and Novruzova et al. (2015) in Nakhchivan, Azerbaijan; and by Ghazanfar (2011) as well as Joudi and Bibalani (2010) in Tabriz, Iran, has been evaluated in this chapter in the light of latest findings, together with other ethnobotanical investigations undertaken in the neighboring areas. The MAPS whose status or name has changed, or has become synonyms, or has been included under new combinations have been corrected following the "List of Turkish Vascular Plants" Güner et al. (2012). For each of the MAPS, the scientific name, vernacular name, preparations, part used, and its use have been recorded at length in the Appendix 1. The information recorded has been symbolized from the three states in order to follow the appendix easily as follows:

Local names: *AZ* Nakhchivan-Azerbaijan, *IR* Tabriz-Iran, *TR* Iğdır-Turkey

Part used: *AP* aerial parts, *BD* buds, *BLE* basal leaves, *BR* barks, *BU* bulb, *CA* capitulum, *COS* corn silk, *FL* flowers, *FR* fruit, *GU* gum, *HE* herb, *IF* inflorescences, *LA* latex, *LE* leaves, *PET* petiole, *RH* rhizome, *RO* root, *SE* seed, *SH* shoots, *SP* spores; *ST* stem, *WP* whole plants

Preparations: *BO* boiled, *CH* chewing, *CR* crushed, *DAP* direct application, *DE* decoction, *DI* distillate, *FE* fresh, *GA* gargle, *IH* inhalation, *IN* infusion, *IT* internal,

OI oil, *PA* pasture, *PI* pickled, *PO* poultice, *PU* pounded, *PW* powdered, *SM* smoke, *VA* vaporization

In addition to this, in the areas studied by us, MAPS have been grouped on the basis of diseases as given below. This has allowed us to make a comparison if the MAPS used in these areas have same or similar uses:

- *Digestive system*: Stomach disorders; stomachache; appetizing/orexigenic; abdominal pain/colic; anthelmintic; carminative/antiflatulence; ulcer; enteralgia; digestive; intestinal disorders; emetic; abdominal ailments; dysentery; gastrointestinal disorders; cholagogue/bile secretion enhancer; gastritis; gallbladder disorders/bile diseases; enteritis/intestinal inflammation; indigestion; stomach tonic; stomach cancer; stomach cramps; stomach bleeding; nausea; gallstones; gastrointestinal tract inflammations; inflammation of the stomach; gastrointestinal bloating; disinfection of intestine
- *Dermal system*: Vulnerary; wounds/swollen wounds/inflamed wounds; mycodermatitis/dermatitis; hair care; abscess; antiseptic; scabies; sunstroke; itching; eczema; erythema on skin; balding; skin diseases; burn; skin cancer; antiverrucous/warts; skin care/wrinkles and skin irritations; acne; leprosy; hives; skin infection; improves the skin's ability to protect against ultraviolet rays
- *Respiratory system*: Colds; cough/antitussive; asthma; bronchitis; pulmonic disorders/lung diseases; expectorant; catarrh; flu/influenza; tuberculosis; respiratory system; whooping cough; pectoral; pleuritic; emollient; chest pain; shortness of breath; smoothing chest; sterilization of chest; lung infection; inflammation of trachea; bronchi infection
- *Cardiovascular system*: Hypertension; astringent; cardiotonic; anemia; hypercholesterolemia/high cholesterol/cholesterol; cardialgia; embolism/atherosclerosis; malaria; cardiac/cardiovascular system; tachycardia; exudative; hemafacient; blood purifier; lowering blood pressure; tension; increasing blood flow; heart attack; palpitations
- *Urogenital system*: Diuretic; hemorrhoid; kidney stones; constipation; nephralgia; diarrhea; laxative; kidney diseases; urinary tract antiseptic; urinary system disorders; enuresis/urinary retention; nephritis/kidney inflammations; cystitis/bladder inflammation; Addison disease; anal fissure; kidney infection; urinary infection; bladder discomfort; strengthens the kidneys; gonorrhea; prostate cancer; prostate inflammation; inflammation of the urinary tract system
- *Gynecological system*: Gynecological diseases; emmenagogue; galactagogue/to increase milk in women; women's sterility/infertility; menstrual disorders; abortive; menstrual pain; uterine diseases; birth control; useful for pregnancy; vaginal discharge
- *Ear, nose and throat*: Sore throat; throat diseases; sinusitis; angina; tonsillitis; laryngitis; otitis; pharynx diseases; vertigo; swelling of throat; throat cancer; ear pain; inflammations of the pharynx; nasal inflammations
- *Neurological and psychological*: Headache; sedative; analgesic; epilepsy; neuralgia/nerve pain; insomnia; nervous diseases; somniferous; Parkinson; antispasmodic; antidepressants; migraine; strengthening nerves; sciatica pain; relaxation;

stimulate; melancholy diseases; hypnotic; anti-stress; brain hemorrhage; meningitis; strengthen the body and spirit; dizziness; hysteria; paralysis.

- *Mouth and teeth ailment*: Toothache; stomatitis; mouth diseases; mouth wounds; gingivitis; gum diseases; prevent dental decay; halitosis
- *Skeletal–muscular system*: Rheumatism; edema; backache; calcification (knee); rachitism/rickets; arthritis; arthralgia/joint pain; myalgia/muscle pain; buckling of legs; gout
- *Other ailments*: Diabetes; anti-inflammatory; antipyretic; tonic; cancer; jaundice; internal medicine; goiter; snake, scorpion and bee bites; antiparasitic; vitamin and energy supports; internal bleeding; diathesis; scurvy; measles; pancreatitis; scarlatina; obesity; hepatic/liver diseases; hepatitis; facilitator of the liver function; spasm of the liver; liver pain; spleen disorders; swollen spleen; spasm of the spleen; fever; for veterinary purposes; anxiety treatment; pains, disinfection; hiccups; diaphoretic; antiexcitement; improves immune system; boosting sexual power; antifungal; antiviral; antiallergy; antioxidant; antibacterial; typhoid; treatment of cholera and eye diseases (against itching in the eyes, eye diseases, cataract)

Comparative Evaluation on Country Basis

A total of 341 taxa of MAPS naturally distributed in the evaluated areas have been recorded with uses in traditional medicine. These belong to 65 families and have been recorded here alphabetically with their botanical name and local name (if known), part used, ailment treated, and information on the preparations used in Iğdır (Turkey), Nakhchivan (Azerbaijan), and Tabriz (Iran) (Appendix 1).

The families with the highest number of taxa are Asteraceae (63 taxa), Lamiaceae (39 taxa), Apiaceae (24 taxa), Fabaceae (23 taxa), and Rosaceae (23 taxa). These five families constitute 50.44% of the MAPS distributed in these States (Fig. 2). On the basis of parts mainly used, we found the numbers as follows: leaves used (100 taxa), herbs (98 taxa), fruits (57 taxa), and roots (54 taxa) (Fig. 3). The most common preparations used are decoction (161 taxa), followed by infusion (72 taxa), fresh (54 taxa), and poultice (40 taxa). Other uses and their taxa number are given in Fig. 4.

In general, the MAPS determined in these states show the following distribution: 164 taxa in Iğdır, 78 taxa in Nakhchivan, and 159 taxa in Tabriz. An evaluation of these on the basis of diseases has revealed that a major number of taxa are used for digestive (119 taxa), followed by respiratory (92 taxa), urogenital (80 taxa), dermal (59 taxa), cardiovascular (37 taxa), gynecological (31 taxa), skeletal–muscular (30 taxa), neurological and psychological (21 taxa), and ear, nose, and throat systems disorders (6 taxa), together with mouth and teeth ailment system (4 taxa) and other ailments (101 taxa) **in Iğdır** (Turkey); for digestive (130 taxa), followed by

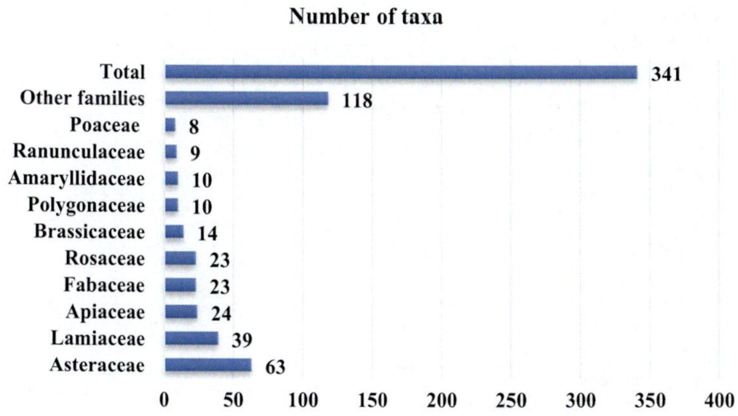

Fig. 2 Families in the studied areas with highest number of taxa

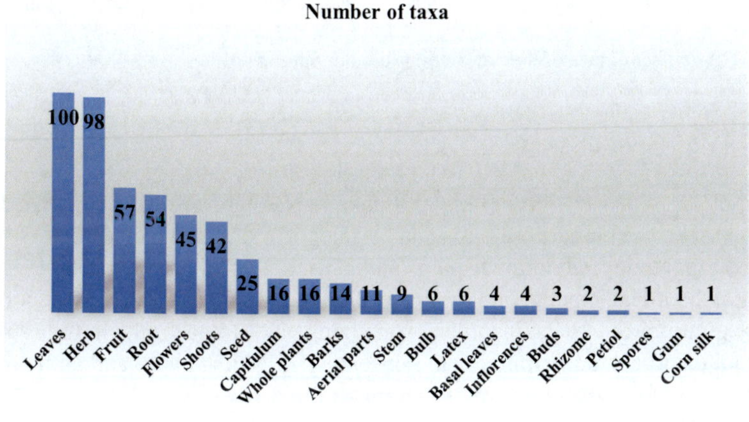

Fig. 3 The parts of MAPS used in three states

respiratory (87 taxa), dermal (59 taxa), urogenital (55 taxa), cardiovascular (44 taxa), skeletal–muscular (32 taxa), neurological and psychological (25 taxa), ear, nose, and throat systems disorders (18 taxa), mouth and teeth ailments (12 taxa), and gynecological system disorders (1 taxa) as well as other ailments (71 taxa) **in Nakhchivan** (Azerbaijan); and for urogenital (102 taxa), followed by respiratory (83 taxa), neurological and psychological (70 taxa), digestive (68 taxa), cardiovascular (51 taxa), dermal (49 taxa), skeletal–muscular (32 taxa), gynecological (12 taxa), ear, nose, and throat systems disorders (12 taxa), as well as mouth and teeth (7 taxa) and other ailments (146 taxa) **in Tabriz** (Iran). An evaluation on the basis of treatment of diseases from the three investigated states is presented in Tables 1–11.

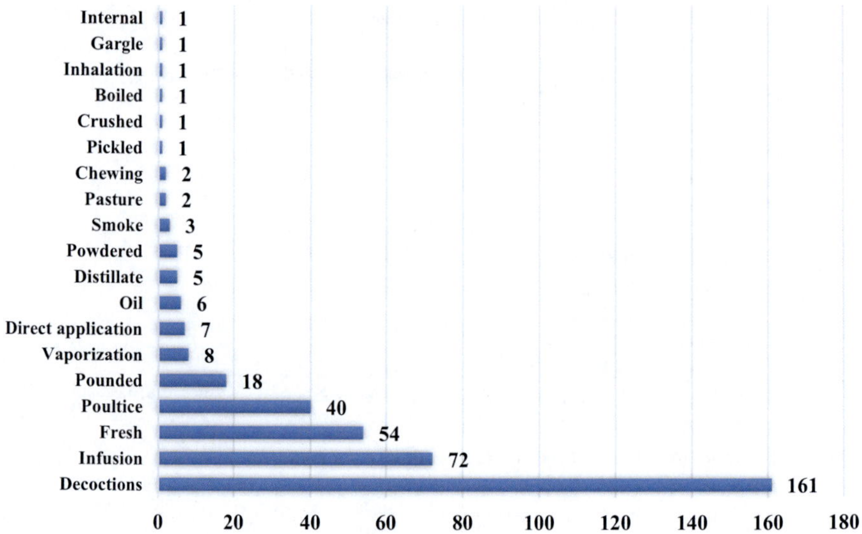

Fig. 4 The number of taxa of the preparations used in three states

Digestive

A total of 119 taxa are used in Iğdır, 130 taxa in Nakhchivan, and 68 taxa in Tabriz (Table 1). The use on purpose basis can be categorized as stomach disorders, stomachache, appetizing, abdominal pain, anthelmintic, carminative, ulcer, digestive, intestinal disorders, emetic, and dysentery (Table 1). On state basis, most common disease groups are stomach disorders, appetizing, and stomachache in Iğdır; gastrointestinal disorders, anthelmintic, and dysentery in Nakhchivan; and anthelmintic, carminative, and stomach disorders in Tabriz (Table 1).

When the similar uses of the taxa are evaluated, *Achillea millefolium*, *Artemisia absinthium*, *Helichrysum plicatum*, *Hypericum perforatum*, *Origanum vulgare* ssp. *gracile*, *Plantago major*, *Scutellaria orientalis*, and *Thymus transcaucasicus* are used both in **Iğdır and Nakhchivan**; *Hypericum scabrum*, *Mentha longifolia*, *Polygonum aviculare*, *Rheum ribes*, *Tragopogon pratensis*, and *Ziziphora clinopodioides* in **Iğdır and Tabriz**; whereas in **Nakhcivan and Tabriz**, *Agrimonia eupatoria*, *Malus sylvestris* ssp. *orientalis*, and *Matricaria chamomilla* are used similarly. The taxa commonly used in **Igdir, Nakhchivan, and Tabriz** are *Glycyrrhiza glabra*, *Rosa canina*, and *Urtica dioica* (Appendix 1).

Dermal

In this disease group, 59 taxa are used in each state (Iğdır and Nakhchivan), but only 49 taxa are used in Tabriz (Table 2). Categorization of the medicinal uses shows that the main use is the treatment of wounds, hair care, and eczema in all three research

Table 1 Number of "MAPS" used in the digestive system disorders

Medicinal use category	Iğdır	Nakhchivan	Tabriz
Stomach disorders	28	2	7
Stomachache	20	8	3
Appetizing/orexigenic	21	10	3
Abdominal pain/colic	9	2	1
Anthelmintic	9	20	8
Carminative/antiflatulence	7	5	8
Ulcer	7	8	2
Enteralgia	5	–	–
Digestive	5	7	5
Intestinal disorders	3	2	1
Emetic	2	1	4
Abdominal ailments	2	–	–
Dysentery	1	13	5
Gastrointestinal disorders	–	24	–
Cholagogue/bile secretion enhancer	–	12	1
Gastritis	–	10	–
Gallbladder disorders/bile diseases	–	2	1
Enteritis/intestinal inflammation	–	2	2
Indigestion	–	2	1
Stomach tonic	–	–	4
Stomach cancer	–	–	2
Stomach cramps	–	–	2
Treatment of stomach bleeding	–	–	2
Nausea	–	–	1
Gallstones	–	–	1
Gastrointestinal tract inflammations	–	–	1
Inflammation of the stomach	–	–	1
Gastrointestinal bloating	–	–	1
Disinfection of intestine	–	–	1
Total (number of taxa)	119	130	68

areas (Table 2). However, the most common disease groups are wounds, vulnerary, and dermatitis in Iğdır; wounds, skin diseases, and eczema in Nakhchivan; and skin diseases, wounds, and skin care in Tabriz (Table 2). An evaluation of same and/or similar uses at the taxon level has revealed that *Glycyrrhiza glabra*, *Hypericum perforatum*, *Origanum vulgare* ssp. *gracile*, and *Plantago major* are used in **Iğdır and Nakhchivan**; *Arctium platylepis*, *Cichorium intybus*, *Fumaria asepala*, and *Ranunculus arvensis* in **Iğdır and Tabriz**. On the other hand, *Matricaria chamomilla* and *Rosa canina* are used for the same purpose in **Nakhchivan and Tabriz** (Appendix 1).

Table 2 Number of "MAPS" used in the dermal system disorders

Medicinal use category	Iğdır	Nakhchivan	Tabriz
Vulnerary	13	1	–
Wounds/swollen wounds/inflamed wounds	15	17	8
Mycodermatitis/dermatitis	6	–	–
Hair care	5	1	5
Abscess	6	3	–
Antiseptic	5	–	4
Scabies	2	5	–
Sunstroke	2	–	–
Itching	2	–	2
Eczema	1	8	1
Erythema on skin	1	–	–
Balding	1	–	–
Skin diseases	–	14	13
Burn	–	8	1
Skin cancer	–	1	–
Anti-verrucous/warts	–	1	2
Skin care/wrinkles and skin irritations	–	–	6
Acne	–	–	2
Leprosy	–	–	2
Hives	–	–	1
Skin infection	–	–	1
Improves the skin's ability to protect against ultraviolet rays	–	–	1
Total (number of taxa)	**59**	**59**	**49**

Respiratory

Nearly 92 taxa are used in Iğdır, 87 taxa in Nakhchivan, and 83 taxa in Tabriz (Table 3). The uses as categorized on treatment basis show colds, cough, asthma, bronchitis, lung diseases, expectorant, and tuberculosis are evaluated in all three states (Table 3). Most common disease groups are colds, cough, and asthma in Iğdır; colds, cough, and tuberculosis in Nakhchivan; and cough, asthma, and bronchitis in Tabriz (Table 3). When the similar uses at the taxon level are evaluated, our findings show that *Artemisia absinthium*, *Berberis vulgaris*, *Origanum vulgare* ssp. *gracile*, and *Urtica dioica* are used in **Iğdır and Nakhchivan**; *Malva neglecta*, *Mentha longifolia*, *Polygonum aviculare*, *Stachys lavandulifolia*, and *Ziziphora clinopodioides* are used in **Iğdır and Tabriz**, whereas *Malva sylvestris*, *Melilotus officinalis*, and *Plantago major* are the generally used taxa in **Nakhchivan and Tabriz**. The taxa commonly used in **Igdir, Nakhchivan, and Tabriz** are *Glycyrrhiza glabra* and *Malus sylvestris* ssp. *orientalis* (Appendix 1).

Table 3 Number of "MAPS" used in the respiratory system disorders

Medicinal use category	Iğdır	Nakhchivan	Tabriz
Colds	32	14	6
Cough/antitussive	24	13	20
Asthma	15	4	10
Bronchitis	6	11	9
Pulmonic disorders/lung diseases	5	7	2
Expectorant	4	3	4
Catarrh	3	–	6
Flu/influenza	1	–	3
Tuberculosis	1	12	1
Respiratory system	1	11	–
Whooping cough	–	6	6
Pectoral	–	3	–
Pleuritic	–	3	–
Emollient	–	–	4
Chest pain	–	–	3
Shortness of breath	–	–	2
Smoothing chest	–	–	2
Sterilization of chest	–	–	2
Lung infection	–	–	1
Inflammation of trachea	–	–	1
Bronchi infection	–	–	1
Total (number of taxa)	92	87	83

Cardiovascular

In all 37 taxa are used in Iğdır, 44 taxa in Nakhchivan, and 51 taxa in Tabriz, but categorizing on the basis of use shows that MAPS are evaluated mainly as astringent, cardiotonic, anemia, and malaria in all three states (Table 4). Most common disease groups are hypertension, astringent, and cardiotonic in Iğdır; anemia, cardiovascular diseases, and malaria in Nakhchivan; and as an astringent, blood purifier and in cardiovascular diseases in Tabriz (Table 4). The same and/or similar uses at the taxon level in the three states reveal that *Thymus transcaucasicus* is used in **Iğdır and Nakhchivan**; *Capsella bursa-pastoris*, *Crataegus azarolus* var. *azarolus*, *Medicago sativa*, *Polygonum aviculare*, *Rosa canina*, and *Urtica dioica* in **Iğdır and Tabriz** (Appendix 1).

Table 4 Number of "MAPS" used in the cardiovascular system disorders

Medicinal use category	Iğdır	Nakhchivan	Tabriz
Hypertension	10	–	1
Astringent	10	2	18
Cardiotonic	5	1	3
Anemia	4	10	2
Hypercholesterolemia/high cholesterol/cholesterol	3	–	2
Cardialgia	2	4	–
Embolism/atherosclerosis	2	–	2
Malaria	1	8	1
Cardiac/cardiovascular system diseases	–	9	6
Tachycardia	–	2	–
Exudative	–	4	–
Hemafacient	–	4	–
Blood purifier	–	–	8
Lowering blood pressure	–	–	4
Tension	–	–	1
Increasing blood flow	–	–	1
Heart attack	–	–	1
Palpitations	–	–	1
Total (number of taxa)	37	44	51

Urogenital

The number of taxa used in Iğdır is 80 taxa, in Nakhchivan 55, and in Tabriz 102 (Table 5). Their use for medicinal purposes can be categorized as diuretic, kidney stones, constipation, diarrhea, and kidney diseases as recorded from three research areas (Table 5). A statewise evaluation shows that most common disease groups are diuretics, hemorrhoid, and kidney stones in Iğdır; diuretics, diarrhea, and kidney diseases in Nakhchivan; and diuretics, diarrhea, and laxative in Tabriz (Table 5). On the basis of same and/or similar uses at the taxon level, we find *Achillea millefolium*, *Glycyrrhiza glabra*, *Hypericum perforatum*, and *Peganum harmala* are used in **Iğdır and Nakhchivan**; *Chenopodium album*, *Hypericum scabrum*, *Malva neglecta*, *Melilotus officinalis*, *Rheum ribes*, and *Urtica dioica* in **Iğdır and Tabriz**; and *Equisetum arvense*, *Malus sylvestris* ssp. *orientalis*, and *Rubia tinctorium* in **Nakhchivan and Tabriz**. The taxon commonly used in **Igdir, Nakhchivan, and Tabriz** is *Rosa canina* (Appendix 1).

Gynecological

Nearly 31 taxa are used in Iğdır, 1 taxon in Nakhchivan, and 12 taxa in Tabriz in this disease group (Table 6). On area basis, most common disease groups are gynecological diseases, emmenagogue, and women's sterility in Iğdır, whereas in Tabriz

Table 5 Number of "MAPS" used in the urogenital system disorders

Medicinal use category	Iğdır	Nakhchivan	Tabriz
Diuretic	20	14	24
Hemorrhoid	16	–	8
Kidney stones	12	1	2
Constipation	9	2	10
Nephralgia	8	2	–
Diarrhea	7	14	15
Laxative	5	–	15
Kidney diseases	2	8	4
Urinary tract antiseptic	1	–	1
Urinary system disorders	–	3	4
Enuresis/urinary retention	–	3	1
Nephritis/kidney inflammations	–	2	5
Cystitis/bladder inflammation	–	2	1
Addison disease	–	1	–
Anal fissure	–	1	–
Kidney infection	–	1	–
Urinary infection	–	1	–
Bladder discomfort	–	–	5
Strengthens the kidneys	–	–	3
Gonorrhea	–	–	1
Prostate cancer	–	–	1
Prostate inflammation	–	–	1
Inflammation of the urinary tract system	–	–	1
Total (number of taxa)	80	55	102

Table 6 Number of "MAPS" used in the gynecological system disorders

Medicinal use category	Iğdır	Nakhchivan	Tabriz
Gynecological diseases	8	–	–
Emmenagogue	7	–	–
Galactagogue/to increase milk in women	5	–	4
Women's sterility/infertility	6	–	–
Menstrual disorders	2	–	2
Abortive	2	–	–
Menstrual pain	1	–	–
Uterine diseases	–	1	1
Venereal diseases	–	–	2
Birth control	–	–	1
Useful for pregnancy	–	–	1
Vaginal discharge	–	–	1
Total (number of taxa)	31	1	12

Table 7 Number of "MAPS" used in the ear, nose, and throat system disorders

Medicinal use category	Iğdır	Nakhchivan	Tabriz
Sore throat	4	–	2
Throat diseases	1	–	–
Sinusitis	1	–	–
Angina	–	11	1
Tonsillitis	–	4	–
Laryngitis	–	1	–
Otitis	–	1	–
Pharynx diseases	–	1	–
Vertigo	–	–	3
Swelling of the throat	–	–	2
Throat cancer	–	–	1
Ear pain	–	–	1
Inflammations of the pharynx	–	–	1
Nasal inflammations	–	–	1
Total (number of taxa)	6	18	12

these are galactagogue, menstrual disorders, and venereal diseases (Table 6). As far as the same and/or similar uses at the taxon level are concerned, *Achillea millefolium*, *Chenopodium album*, and *Hypericum scabrum* are used in **Iğdır and Tabriz** for the same treatments (Appendix 1).

Ear, Nose, and Throat

6 taxa are used in Iğdır, 18 taxa in Nakhchivan, and 12 taxa in Tabriz in their treatments (Table 7). On area basis, most common disease groups are sore throat in Iğdır, angina and tonsillitis in Nakhchivan, and vertigo, sore throat, and swelling of the throat in Tabriz (Table 7). In this disease group, no common taxon with the same or similar use at the taxon level has been recorded (Appendix 1).

Neurological and Psychological System

In this disease group, 21 taxa are used in Iğdır, 25 taxa in Nakhchivan, and 70 taxa in Tabriz (Table 8). Their use for medicinal purposes can be categorized as headache, sedative, analgesic, and epilepsy for all three research areas (Table 8). The most common disease groups are headache, sedative, and analgesic in Iğdır; neuralgia and epilepsy in Nakhchivan; and sedative, insomnia, and nervous diseases in Tabriz (Table 8). The same and/or similar uses at the taxon level on area basis are *Artemisia absinthium* and *Origanum vulgare* ssp. *gracile* in **Iğdır and Nakhchivan**;

Table 8 Number of "MAPS" used in the neurological and psychological system disorders

Medicinal use category	Iğdır	Nakhchivan	Tabriz
Headache	8	2	5
Sedative	7	3	12
Analgesic	5	3	3
Epilepsy	1	4	2
Neuralgia/nerve pain	–	6	3
Insomnia	–	3	6
Nervous diseases	–	1	6
Somniferous	–	1	–
Parkinson	–	1	–
Antispasmodic	–	1	2
Antidepressants	–	–	4
Migraine	–	–	4
Strengthening nerves	–	–	3
Sciatica pain	–	–	3
Relaxation	–	–	3
Stimulate	–	–	2
Melancholy diseases	–	–	2
Hypnotic	–	–	2
Anti-stress	–	–	2
Brain hemorrhage	–	–	1
Meningitis	–	–	1
Strengthen the body and spirit	–	–	1
Dizziness	–	–	1
Hysteria	–	–	1
Paralysis	–	–	1
Total (number of taxa)	21	25	70

Table 9 Number of "MAPS" used in mouth and teeth disorders

Medicinal use category	Iğdır	Nakhchivan	Tabriz
Toothache	3	1	2
Stomatitis	1	7	–
Mouth diseases	–	1	1
Mouth wounds	–	1	–
Gingivitis	–	2	1
Gum diseases	–	–	1
Prevent dental decay	–	–	1
Halitosis	–	–	1
Total (number of taxa)	4	12	7

Table 10 Number of "MAPS" used in the skeletal–muscular system disorders

Medicinal use category	Iğdır	Nakhchivan	Tabriz
Rheumatism	18	24	9
Edema	6	1	1
Backache	5	–	1
Calcification (knee)	1	–	–
Rachitism/rickets	–	3	1
Arthritis	–	1	5
Arthralgia/joint pain	–	1	5
Myalgia/muscle pain	–	1	4
Buckling of legs	–	1	–
Gout	–	–	6
Total (number of taxa)	30	32	32

Glycyrrhiza glabra, *Hypericum scabrum*, *Melilotus officinalis*, and *Mentha longifolia* in **Iğdır and Tabriz**; and in **Nakhcivan and Tabriz**, *Achillea millefolium* is used similarly (Appendix 1).

Mouth and Teeth Ailments

A total of 4 taxa are used in Iğdır, 12 taxa in Nakhchivan, and 7 taxa in Tabriz (Table 9). Their categorization is as follows; toothache is seen in all three areas (Table 9). On regional basis, the most common disease group is toothache in Iğdır and Tabriz and stomatitis in Nakhchivan (Table 9). The same and/or similar use at the taxon level is *Origanum vulgare* ssp. *gracile* which is used similarly in **Iğdır and Nakhchivan** (Appendix 1).

Skeletal–Muscular System

30 taxa are used in Iğdır, 32 taxa in Nakhchivan, and 32 taxa in Tabriz for the treatment of this disease (Table 10). They can be categorized as rheumatism and edema as seen in all three areas (Table 10), but on regional basis, most common disease groups are rheumatism, edema, and backache in Iğdır, rheumatism and rachitism in Nakhchivan, and rheumatism and gout in Tabriz (Table 10). The same and/or similar uses at the taxon level on regional basis are *Peganum harmala* in **Iğdır and Nakhchivan**; *Melilotus officinalis* and *Ranunculus arvensis* in **Iğdır and Tabriz** (Appendix 1).

Table 11 Number of "MAPS" used in the other ailments

Medicinal use category	Iğdır	Nakhchivan	Tabriz
Diabetes	22	9	2
Anti-inflammatory	21	2	3
Antipyretic	19	2	5
Tonic	9	5	7
Cancer	8	1	2
Jaundice	6	6	9
Internal medicine	5	–	–
Goiter	5	–	–
Snake, scorpion and bee bites	4	1	2
Antiparasitic	1	–	4
Eye diseases	1	1	2
Vitamin and energy supports	–	10	8
Internal bleeding	–	6	–
Diathesis	–	5	–
Scurvy	–	4	3
Measles	–	2	1
Pancreatitis	–	1	–
Scarlatina	–	1	–
Obesity	–	1	–
Hepatic disorders/liver diseases	–	11	10
Hepatitis	–	3	2
Facilitator of the liver function	–	–	1
Spasm of the liver	–	–	1
Liver pain	–	–	1
Spleen disorders	–	–	3
Swollen spleen	–	–	2
Spasm of the spleen	–	–	1
Fever	–	–	22
For veterinary purposes	–	–	17
Anxiety treatment	–	–	11
Pains	–	–	9
Disinfection	–	–	2
Hiccups	–	–	3
Diaphoretic	–	–	2
Antiexcitement	–	–	2
Improves immune system	–	–	1
Boosting sexual power	–	–	1
Antifungal	–	–	1
Antiviral	–	–	1
Antiallergy	–	–	1
Antioxidant	–	–	1

(continued)

Table 11 (continued)

Medicinal use category	Iğdır	Nakhchivan	Tabriz
Antibacterial	–	–	1
Typhoid	–	–	1
Treatment of cholera	–	–	1
Total (number of taxa)	101	71	146

Others Ailments

In this disease group, 101 taxa are used in Iğdır, 71 taxa in Nakhchivan, and 146 taxa in Tabriz; these can be categorized as diabetes, anti-inflammatory, antipyretic, tonic, cancer, jaundice, snake, scorpion and bee bites, and eye diseases as seen in all three areas (Table 11). For other ailments a comparison of the MAPS used shows that for diabetes (22 taxa), anti-inflammatory (21 taxa) and antipyretic (19 taxa) are used in Iğdır; in the case of hepatic diseases (11 taxa), vitamin and energy supports (10 taxa), and diabetes (9 taxa) in Nakhchivan and for treatment of fevers (22 taxa), for veterinary purposes (17 taxa), and anxiety treatment (11 taxa) in Tabriz (Table 11). The information on fever, use for veterinary purposes and anxiety treatment is available only for Tabriz, not for other two states (Table 11). Same and/or similar uses at the taxon level evaluation on regional basis have revealed the use as follows: *Berberis vulgaris* (for diabetes), *Malus sylvestris* ssp. *orientalis* (for diabetes), and *Urtica dioica* (for diabetes) in **Iğdır and Nakhchivan** and *Rosa canina* (for a lack of vitamin) in **Nakhchivan and Tabriz** (Appendix 1).

If the data presented above is taken into consideration, similarity ratios can be calculated, using Jaccard similarity index, with disease groups (Table 13), and medicinal flora (Table 12) in three studied areas. This index allows the percentage of similarity to be calculated in each area. The formula to calculate this index is as follows:

Index of Jaccard = $100 \times C/(A + B - C)$

where A is the number of species of the sample A, B is the number of species of the sample, and C is the number of species common to A and B (González-Tejero et al. 2008).

Tables 12 and 13 show the highest degree of similarity between Iğdır and Tabriz. This degree of similarity appears surprising, on the basis of floristic differences, but the fact that Iğdır, Nakhchivan, and Tabriz have borders with each other and have similar cultures with same language in itself explains this similarity. The population in these areas belongs mainly to Azeri ethnicity. However, small similarity ratio differences based on disease groups are because of the folk physicians who have frequently practiced in these areas. These areas have had to travel long geographical distance in cold and cold winters and transportation difficulties have added to that.

Table 12 Jaccard similarity index for the medicinal flora in three areas

	Iğdır	Nakhchivan	Tabriz
Iğdır	–	7.55	12.54
Nakhchivan	7.55	–	6.28
Tabriz	12.54	6.28	–

Table 13 Jaccard similarity index related to disease groups in the study areas

	Iğdır-Nakhchivan	Iğdır-Tabriz	Nakhchivan-Tabriz
Digestive system	4.62	5.06	3.13
Dermal system	3.51	3.85	1.89
Respiratory system	3.47	4.17	3.03
Cardiovascular system	1.25	7.32	0.00
Urogenital system	3.85	4.00	2.61
Gynecological system	0.00	7.50	0.00
Ear, nose and throat system	0.00	0.00	0.00
Neurological and psychological system	4.55	4.60	1.06
Mouth and teeth ailment system	6.67	0.00	0.00
Skeletal–muscular system	1.64	3.33	0.00
Other ailments	1.78	0.00	0.46

Conclusions

MAPS represent the oldest and most widespread form of medication as per the evidence published on diverse lines in this connection. The plant and animal sources have served as a source for many medicines until the last century. In spite of an increase in the use of industrial based synthetic drugs, natural organic healing sources have persisted as the "treatment of choice" for a large number of health problems, particularly in the poor populations all over the globe (Halberstein 2005). Tremendous data is being pooled up, processed, and compiled through double-blind clinical trials, biochemical assays of plant taxa, analyses of interviews from the informants and anecdotal reports, and the observations of traditional herbalists plus their diagnostic and herbal preparation procedures in different cultures. All this data published till now has produced a great pile of information related to different cultures and is expected to serve as cross-cultural evidence for future comparative investigations on a larger scale (Hylands and Stuart 1981; Etkin 1988; Halberstein 1997a, b, 2005).

Historically the herbalists from varying cultures have paid much attention to their target plants, their habitats and microenvironments, geographical locations, edaphic conditions, climatic features, biotic impacts, and pollution impacts. For collection, processing, storing, and preservation of the botanical products particular methods are characteristically used (Halberstein 1997a, b, 2005; Halberstein and Saunders 1978; Wong 1976; Whistler 1985). The origin of ethnobotanical knowledge lies in the complex interaction between humans and their natural resources

(Sõukand and Kalle 2010). The cultural factors generally include local classification systems, mediated through language (Maffi 2005; Ellen 2009; Saslis-Lagoudakis et al. 2014), human cognition and cultural history (Leonti and Casu 2013), beliefs and religion (Pieroni and Quave 2005; Pieroni et al. 2011; Rexhepi et al. 2013), or social networks and access to information (Bandiera and Rasul 2006; van den Broeck and Dercon 2011; Labeyrie et al. 2014).

A connection between linguistic areas and peculiar ethnofloras suggests that linguistic differences have limited and are limiting the knowledge diffusion (Maffi 2005). However, languages are a part and parcel of cultural knowledge, beliefs, and practices developed by human societies (Maffi 2005). As a matter of fact, linguistic areas generally correspond with coherent cultural groups, and differences in accent or lexicon likely reflect the existence of different sociolinguistic communities (Menendez-Baceta et al. 2015). In view of this, each sociolinguistic community has a strong sense of identity based on belonging to the local community, which is shaped by kinship, neighborhood, geographical distance, or local history and constitutes a strong basis for the affinities and identity of their members. As such, absence of exchange between linguistic groups is not necessarily due to language but due to preferences for exchanging with individuals belonging to the same cultural group (Menendez-Baceta et al. 2015). The second pathway through which culture can shape the distribution of knowledge on MAPS is related to the knowledge communities built around social links: knowledge on MAPS is mainly transmitted among closely related people belonging to the same social networks as the complicity required for trying a new remedy is high (Menendez-Baceta et al. 2015). This social link has started gaining much importance at present due to snatching of property rights from the indigenous communities through biopiracy.

The data published reveals the fact that folk medicinal "MAPS" are mainly shaped by two major types of remedies: one being those common in broad areas and the other those only important for closely related communities. An availability of suitable and sustainable ecological conditions and other environmental factors together with pharmacological factors like chemical composition of the plants is self-explanatory why there are widely used plant remedies. The cultural side like language, social networks, and the curative meaning given to plants does influence the regular diffusion of traditional knowledge and explains why some species are only used in particular regions despite their availability throughout the territory. These cultural factors make a remedy available and are determinant in its medicinal effectiveness (Menendez-Baceta et al. 2015). Indeed, folk medicine is a complex cultural domain, with a high cultural and symbolic component, and this fact seems to become especially sensitive to the cultural boundaries (Menendez-Baceta et al. 2015).

At the end, another highlighted factor is the effectiveness of medicinal remedies, at least partially, on its cultural meaning (Moerman and Jonas 2002). We come across many symbolic remedies in the folk medicine, where this meaning response seems to be essential. The cultural meaning at the same time does play an important role in many so-called empirical remedies where the effectiveness apparently relies only on the chemical composition, since the psychological context of its consumption is essential for its effectiveness (Menendez-Baceta et al. 2015).

Appendix 1: MAPS Distributed in the Study Areas (*Cultivated)

	Family–Taxa	Local name(s)	Preparations	Parts used	Treatment		
					Iğdır (TR)	Nakhchivan (AZ)	Tabriz (IR)
	Lycopodiophyta						
	Lycopodiaceae						
1	*Lycopodium clavatum* L.			SP		Diathesis, abscess	
	Pteridophyta						
	Dryopteridaceae						
2	*Dryopteris filix-mas* (L.) Schott	AZ: Ayıdöşəyi		RO		Anthelmintic, antirheumatic, myalgia, vulnerary, ulcer	
	Equisetaceae						
3	*Equisetum arvense* L.	AZ: Qatırquyruğu; IR: Dom-e-asb (per.)	DE	HE, ST		Urinary system disorder, pulmonic disorder, dysentery, antirheumatic, tuberculosis, stomatitis, diarrhea	Diuretic
	Pinophytina						
	Cupressaceae						
4	*Juniperus communis* L.	AZ: Ardıc		FR		Antirheumatic, neuralgia, antitussive, gallbladder disorders, diuretic, digestive	
	Pinaceae						
5	*Pinus sylvestris* L.	AZ: Adi şam		BD, LE		Rachitism, skin disorders, scabies, angina, antitussive, diuretic	

(continued)

	Family–Taxa	Local name(s)	Preparations	Parts used	Treatment		
					Iğdir (TR)	Nakhchivan (AZ)	Tabriz (IR)
	Magnoliophytina						
	Acoraceae						
6	*Acorus calamus* L.			RO		Appetizing, digestive, carminative, cholagogue	
	Adoxaceae						
7	*Sambucus nigra* L.			FR, IF		Cold, cardialgia, angina, respiratory system disorders, avitaminosis, malaria, diabetes, measles, kidney diseases, edema, diuretic	
8	*Viburnum lantana* L.	TR: Germeşc	DE	FR	Diabetes, diarrhea		
9	*Viburnum opulus* L.			BR, FR		Tuberculosis, tonsillitis, angina, whooping cough, gastrointestinal disorders, measles, skin disorders	
	Amaranthaceae						
10	*Beta corolliflora* Zosimovic ex Buttler	TR: Kızılca, Sırk	DE	RO	Antihemorrhoidal		
11	*Chenopodium album* L.	TR: Unluca; IR: Salmak, salmeh tareh (per.) salman tareh (tur.)	DE, PO, VA	AP, FL, LE	Diuretic, women's sterility		Strong laxatives, constipation, fever, menstruation
12	*Chenopodium murale* L.	TR: Salmanca	DE, VA	HE	Women's sterility		
13	*Salsola rigida* Pall.	IR: Alafe shoor	DE	AP			Removing intestinal worms (all the animals)

	Amaryllidaceae						
14	*Allium akaka* S.G. Gmelin	TR: Dana soğanı; IR: Dagh soghani (tur.) Piaz-e-Kouhi (per.)	FE, PO	BU	Analgesic		Anxiety treatment
15	*Allium ampeloprasum* L.	IR: Tare-kuhi	FE	WP			Disinfection, antiparasitic, anti-asthma, lowering blood pressure, atherosclerosis
16	*Allium armenum* Boiss. & Kotschy	TR: Silim soğanı	FE	LE	Orexigenic		
17	*Allium atroviolaceum* Boiss.	TR: Silim soğanı	FE	LE	Orexigenic		
18	*Allium cepa* L.*	IR: Soghan (tur.) Piaz (per.)	PO	BU			Anxiety treatment
19	*Allium hooshidaryae* Mashayekhi, Zarre & R.M. Fritsch	IR: Mandala	FE	AP			Bronchitis, amoebic dysentery, whooping cough, shortness of breath, blood pressure, blood purifier, vertigo, strengthening the nerves, arthritis
20	*Allium longisepalum* Bertol.	IR: Sire kuhi	FE	WP			Atherosclerosis, diarrhea, intestinal, influenza, bronchitis, diseases of amoebic dysentery, whooping cough, shortness of breath, blood pressure, blood purifier, vertigo, strengthening the nerves, arthritis

(continued)

	Family–Taxa	Local name(s)	Preparations	Parts used	Treatment		
					Iğdir (TR)	Nakhchivan (AZ)	Tabriz (IR)
21	*Allium sativum* L.*	IR: Sarmsakh (tur.) Sir (per.)	PO	BU			Anxiety treatment
22	*Allium scorodoprassum* L. ssp. *rotundum* (L.) Stearn	TR: İt soğanı	DE, FE	BU, SE	Orexigenic, hypertension, anthelmintic, diuretic, antiseptic, goiter		
23	*Allium* sp.	IR: Sir-e-vahshi (per.)	PO	BU			Anxiety treatment
	Anacardiaceae						
24	*Rhus coriaria* L.			FR, LE		Malaria, stomachache, diarrhea, dysentery, diabetes, appetizing	
	Apiaceae						
25	*Anethum graveolens* L.*	IR: Shood (tur.) Shevid (per.)	DE	SE, SH			Anxiety treatment
26	*Anthriscus nemorosa* (Bieb.) Spreng.	TR: Gımıgımı	DE	FR	Carminative		
27	*Anthriscus sylvestris* (L.) Hoffm.	TR: Gımıgımı	DE	FR	Carminative		
28	*Carum carvi* L.	AZ: Zire		FR		Gastrointestinal disorders, respiratory system disorders, kidney disorders, stomachache, digestive, appetizing	
29	*Chaerophyllum aureum* L.	IR: Aghjeh bash (tur.) Jafari-farangi khaldar (per.)	DE	SH			Anxiety treatment

30	*Chaerophyllum macrospermum* (Wild. ex Spreng.) Fisch. & C.A.Mey. ex Hohen.	IR: Jafari-farangi	DE, FE	WP			Powerful antioxidant
31	*Coriandrum sativum* L.	IR: Gashnish (tur.) Gheshniz (per.)	DE	SE, SH			Anxiety treatment
32	*Echinophora platyloba* DC.	IR: Tulukh-oti, Tiyakh-oti (tur.) Khosharizeh (per.)	DE	SH			Anxiety treatment
33	*Eryngium* sp.	IR: Zol, Boghnagh (per.)	DE	LE			Anxiety treatment
34	*Falcaria vulgaris* Bernh.	IR: Ghaz ayaghi, Ghaz yaghi (per. and tur.)	DE	SH			Blood purification, vascular diseases and prevent dental decay
35	*Ferula caspica* Bieb.	TR: Gırmızı bolu	DE	HE	Stomachache, gynecologic diseases, diabetes		
36	*Ferula rigidula* Fisch. ex DC.	TR: Çaşır	PO	HE	Diabetes, hypercholesterolemia		
37	*Foeniculum vulgare* Mill.	AZ: Razyana		FR, HE		Cold, respiratory system disorders, gastrointestinal disorders, stomachache, cholagogue	
38	*Grammosciadium platycarpum* Boiss. & Hausskn.	IR: Sourulu (tur.) Shevide kouhi (per.)	DE	SH			Birth control, sedative, arthritis and asthma
39	*Heracleum lasiopetalum* Boiss.	IR: Golpar-e-barfi	DE	AP			Gastrointestinal bloating, to increase milk in women, antiseptic

(continued)

	Family–Taxa	Local name(s)	Preparations	Parts used	Treatment		
					Iğdir (TR)	Nakhchivan (AZ)	Tabriz (IR)
40	*Heracleum persicum* Desf.	IR: Baldirgan (tur.) Golpar (per. and tur.)	DE	FR, LE			Jaundice
41	*Heracleum trachyloma* Fisch. & C.A. Mey.	TR: Baldırgan	DE, FE	ST	Galactagogue, stomachache, cancer		
42	*Laser trilobum* (L.) Borkh.			FR, LE, RO		Malaria, analgesic, tonic, antipyretic	
43	*Oliveria decumbens* Vent.	IR: La'le koohestan	DE	AP			Diarrhea (all animals)
44	*Petroselinum crispum* (Miller) A.W. Hill*	IR: Jafari (per. and tur.)	DE	SH			Jaundice
45	*Peucedanum longifolium* Waldst. & Kit.	TR: Çaşır	PI	HE	Diabetes, hypercholesterolemia		
46	*Prangos ferulacea* (L.) Lindl.	IR: Jashir	DE	AP, LA		For treatment of uterus	For veterinary purposes (ruminants)
47	*Torilis leptophylla* (L.) Rchb. f.	IR: Mastunak-e-nazok barg (per.)	DE	FL, SH			Jaundice
48	*Zosima absinthifolia* (Vent.) Link	TR: Bolu	DE	LE	Diabetes		
	Apocynaceae						
49	*Nerium oleander* L.	IR: Kharzahre	SM	LE			Oestrus ovis larvae (sheep)
	Aristolochiaceae						
50	*Aristolochia clematitis* L.	IR: Ziravand	DE	AP			Cleaning the wounds (all animals)

	Asparagaceae						
51	*Asparagus officinalis* L.			RO		Antirheumatic, diabetes, pulmonic disorders, whooping cough, nephralgia, urinary disorders, hepatic disorders, epilepsy, diuretic	
52	*Ornithogalum kurdicum* Bornm.	IR: Gol agha (tur.) Shir morgh (per.)	DE, FE	LE			Skin care, cataract, improves immune system
	Asteraceae						
53	*Achillea arabica* Kotschy	TR: Ormaderen	DE, PO	CA, HE	Diuretic, asthma, cardiotonic, stomachic, carminative, orexigenic, tonic, colds, nephralgia, gynecologic diseases, women' sterility, emmenagogue, jaundice, abscess, vulnerary, astringent, edema, erythema on skin		
54	*Achillea millefolium* L.	TR: Civanperçemi; AZ: Boymaderen; IR: Boimadaran (tur.), Boomadaran (per.)	DE, PO	CA, FL, HE, LE, SH	Diuretic, carminative, menstrual disorders, stomachic, urinary antiseptic, antitussive, tonic, abdominal pain, colds, vulnerary, astringent	Gastritis, ulcer, diarrhea, internal bleeding, analgesic	Nerve diseases, venereal diseases (in women)
55	*Achillea santolinoides* Lag. subsp. *wilhelmsii* (K.Koch) Greuter	TR: Civanperçemi; IR: Boimadaran (tur.), Boomadaran (per.)	DE, IN	CA, FL, HE, LE, SH	Diuretic, abdominal pain, stomachic, emmenagogues, women's sterility, antihemorrhoidal		

(continued)

	Family–Taxa	Local name(s)	Preparations	Parts used	Treatment		
					Iğdir (TR)	Nakhchivan (AZ)	Tabriz (IR)
56	*Achillea tenuifolia* Lam.	TR: Çoban kirpiği	IN	LE	Hypercholesterolemia, diabetes, asthma, bronchitis, cough		
57	*Anthemis cotula* L.	TR: Hozan çiçeği	DE, IN	CA, HE, LE	Jaundice, dysentery, intestinal disorders, cough, stomachache, hair care	To avoid bee bites (for beekeepers)	
58	*Arctium platylepis* (Boiss. & Bal.) Sosn. ex Grossh.	TR: Devetabanı; IR: Baba adam (per.)	DE, PO, PU	BLE, FL, RO, SH	Edema, inflamed wounds, calcification (knee), back ache, snakebites		Skin diseases
59	*Artemisia absinthium* L.	TR: Yavşan, Pire otu; AZ: Yovşan	DE	HE, IF, LE	Stomachic, sedative, anthelmintic, orexigenic, antipyretic, diuretic, digestive, headache, abdominal pain, asthma, diabetes, tonic	Anthelmintic, gastritis, anemia, cholagogue, hepatic disorders, gastrointestinal disorder, neuralgia, antirheumatic, bronchitis, asthma, eczema	
60	*Artemisia austriaca* Jacq.	TR: Yavşan	DE, IN	HE	Abdominal pain, orexigenic, digestive		
61	*Artemisia chamaemelifolia* Vill.	TR: Yavşan	DE	FL, HE	Orexigenic, diabetes, antipyretic, diuretic		
62	*Artemisia herba-alba* Asso	IR: Dermane	DE	AP			Removing intestinal worms (all animals)
63	*Bellis perennis* L.*	IR: Mina-e-chamani (per.)		FL			Skin diseases

64	*Bidens tripartita* L.	AZ: Yatıqqangal		BLE, HE		Rachitism, diathesis, jaundice, skin disorders, arthritis, hepatic diseases, digestive, appetizing	
65	*Calendula officinalis* L.*	IR: Gol-e-hamisheh bahar (per.)	DE	FL			Skin diseases
66	*Centaurea glastifolia* L.	TR: Kötangoparan	DE, PU	CA, LE	Orexigenic, astringent		
67	*Centaurea iberica* Trev. ex Spreng	TR: Çakırdikeni	PU	LE	Vulnerary		
68	*Centaurea saligna* (K.Koch) Wagenitz	TR: Hol	PU	LE	Astringent		
69	*Chondrilla juncea* L.	TR: Ağ sakız	CH	LA	Stomach disorders		
70	*Cichorium intybus* L.	TR: Cızdankuş; IR: Kasni (per.)	DE, PW	RO, WP	Dermatitis, vulnerary, balding	Dysentery, anemia, jaundice, cholagogue, hepatic disorders, kidney disorders, diabetes, antirheumatic, malaria, cold	Skin diseases
71	*Cirsium arvense* (L.) Scop.	TR: Çakırdikeni; IR: Kangar-e-vahshi (per.)	DE, FE	FL, RO, ST	Orexigenic, tonic, antihemorrhoidal, cough, bronchitis		Skin diseases
72	*Cirsium* sp.	IR: Kangar-e-vahshi (per.)	FE	FL, RO, ST			Skin diseases, disinfectant, tonic, antipyretic, digestive
73	*Cnicus benedictus* L.	IR: Khar moghaddas (per.)	DE	FL, SH			Wrinkles, headache, palpitations, jaundice, diarrhea, appetizer, liver and spleen problems

(continued)

	Family–Taxa	Local name(s)	Preparations	Parts used	Treatment		
					Iğdir (TR)	Nakhchivan (AZ)	Tabriz (IR)
74	*Cota tinctoria* (L.) J.Gay ex Guss var. *pallida* (DC.) U. Özbek & Vural	TR: Papatya	DE, IN	CA	Stomachache, cough, intestinal disorders, hair care		
75	*Cota tinctoria* (L.) J.Gay ex Guss. var. *tinctoria*	TR: Sarı papatya	IN, GA	CA	Jaundice, stomachic, anthelmintic, antipyretic, colds, sore throat		
76	*Cyanus cheiranthifolius* (Willd.) Sojak var. *cheiranthifolius*	TR: Perpatyan, Perpatikanı	PO	LE	Mycodermatitis		
77	*Cyanus depressus* (M.Bieb.) Sojak	TR: Göybaş; IR: Gol-e- gandom (per.)	DE	FL, HE, SH	Cardiotonic, asthma, expectorant, orexigenic		Skin diseases
78	*Echinops cephalotes* DC.	IR: Shekar tighal (per.)	DE, IN	WP			Cough, anti-fever
79	*Gundelia tournefortii* L.	IR: Kangar-e-khoraki (per.)		LE, ST			Stomachache, anti-catarrh
80	*Helichrysum arenarium* (L.) Moench subsp *rubicundum* (C. Koch) Davis & Kupicha	TR: Yayla çiçeği	IN	HE	Diuretic, nephralgia, kidney stones		
81	*Helichrysum pallasii* (Sprengel) Ledeb.	TR: Altunbaş otu	IN	HE	Diuretic, cough, kidney stones, nephralgia		

82	*Helichrysum plicatum* DC.	TR: Yayla çiçeği; AZ: Solmazçiçeyi	DE, IN	CA, HE	Kidney stones, nephralgia, diabetes, cough, diarrhea, diuretic, stomachic	Intestinal disorders, anthelmintic, hemafacient, skin disorders	
83	*Inula helenium* L.	AZ: Andız		RO		Gastrointestinal disorder, bronchitis, cold, antitussive, anthelmintic	
84	*Jurinella moschus* (Habl.) Bobrov subsp. *pinnatisecta* (Boiss.) Danin & Davis	TR: Gazangulpu	DE	HE	Diabetes, diarrhea		
85	*Matricaria chamomilla* L.	AZ: Mollabaşı; IR: Baboone-e-almani, Baboone-e-mamooli (per.)	IN	CA, FL		Colic, carminative, diarrhea, inflamed wounds, swollen wounds, antirheumatic, bronchitis, asthma, eczema, gastritis, enteritis	Sedative, astringent, strengthening the stomach, hair care
86	*Onopordum acanthium* L.	IR: Khar-e-panbeh, Khar-e-pirzan (per.)	DE	FL, SH			Skin care, stomach diseases, liver, fevers, paralysis, leprosy, backache
87	*Picnomon acarna* (L.) Cass.	IR: Zard khar	FE	AP			Appetizer, disinfection of intestine and stomach (goat)
88	*Reichardia dichotoma* (Vahl) Freyn	TR: Gara sakgız	IT	LA	Stomach disorders		

(continued)

	Family–Taxa	Local name(s)	Preparations	Parts used	Treatment		
					Iğdir (TR)	Nakhchivan (AZ)	Tabriz (IR)
89	*Rhaponticum repens* (L.) Hidalgo	IR: Talkheh (per.)	DE	FL, SH			Nerve diseases, venereal diseases (in women)
90	*Scorzonera cana* (C.A. Meyer) Griseb. var. *jacquiniana* (W. Koch) Chamberlain	TR: Yemlik	FE	LE	Orexigenic, galactagogue		
91	*Scorzonera laciniata* L. subsp. laciniata	TR: Yemlik	FE	LE	Orexigenic, galactagogue		
92	*Scorzonera suberosa* K.Koch subsp. *suberosa*	TR: Yemlik	FE	HE, RO	Orexigenic, galactagogue		
93	*Senecio vernalis* Waldst. & Kit.	TR: Acı papatya	DE	HE	Anti-inflammatory		
94	*Senecio vulgaris* L.	IR: Pir-giah (per.)	DE	WP			Regulate the menstrual cycle in women
95	*Silybum marianum* (L.) Gaertn.	IR: Khar Maryam (per.)	DE	FL			Anti-allergy, anticancer, antidepressants, antifungal, anti-edema, anti-inflammation, antiviral, laxative, diaphoretic
96	*Tanacetum aureum* (Lam.) Greuter var. *aureum*	TR: Çeren	DE	CA	Pulmonic disorders, colds, kidney stones, antipyretic		

97	*Tanacetum balsamita* L.	IR: Shasparan (tur.) Shah sparam, Shah sparghan (per.)	PO	LE			Strengthen the stomach, carminative, antiseptic, mouth and digestive system, relieve pain, for halitosis, headache, nervous, cough, inflammation of the urinary tract system
98	*Tanacetum parthenium* (L.) Sch.-Bip.	IR: Baboone-e-gavi, Mokhleseh (per.)	PO	FL, LE			Stomach tonic, appetizer
99	*Tanacetum polycephalum* Sch. Bip. subsp. *argyrophyllum* (K.Koch) Podlech	TR: Çeren	DE, IN	CA, HE	Pulmonic disorders, colds, antipyretic, anti-inflammatory, scabies		
100	*Tanacetum punctatum* (Desr.) Grierson	TR: Sendel	IN	HE	Emmenagogue, anti-inflammatory		
101	*Tanacetum vulgare* L.			CA, LE		Gastrointestinal disorder, diarrhea, exudative, headache, arthralgia, jaundice, hepatic disorders, scabies, skin disorders, anthelmintic	
102	*Taraxacum androssovii* Schischkin	TR: Zeze	IN, PO	LE	Antirheumatic, wounds, stomach disorders, internal medicine, kidney stones, anti-inflammatory		

(continued)

	Family–Taxa	Local name(s)	Preparations	Parts used	Treatment		
					Iğdir (TR)	Nakhchivan (AZ)	Tabriz (IR)
103	*Taraxacum fedtschenkoi* Hand.-Mazz.	TR: Zeze	IN, PO	LE	Antirheumatic, wounds, stomach disorders, internal medicine, kidney stones, anti-inflammatory		
104	*Taraxacum macrolepium* Schischkin	TR: Zeze	IN, PO	LE	Antirheumatic, wounds, stomach disorders, kidney stones, anti-inflammatory		
105	*Taraxacum montanum* (C.A. Mey.) DC.	IR: Khabar chin (tur.), Gol-e-ghased-e-kouhi (per.)	DE				Stomach tonic, bile secretion enhancer, acne, snakebite, itching, liver and kidney diseases
106	*Taraxacum officinale* Wigg.	AZ: Zencirotu		HE, RO		Constipation, eczema, skin disorders, burns, hepatic disorders, kidney disorders, malaria, anemia, antirheumatic, insomnia	
107	*Tragopogon coloratus* C. A. Meyer	TR: At yemliği	FE	HE	Stomachache		
108	*Tragopogon dubius* Scop.	TR: Yemlik	FE	HE	Stomachache		
109	*Tragopogon graminifolius* DC	IR: Yemlik (tur.) Sheng (per.)	FE	LE			Treatment of stomach bleeding, rheumatism
110	*Tragopogon pratensis* L.	TR: At yemliği; IR: Sheng, yelmih (tur.)	FE	AP, HE	Stomachache		Eliminate warts, stomach bleeding, rheumatism

111	*Tragopogon reticulatus* Boiss. & Huet	TR: At yemliği	FE	HE	Stomachache		
112	*Tripleurospermum monticolum* (Boiss. & Huet) Bornm.	TR: Kır papatyası	DE, IN	CA	Hair care, colds, cough, antipyretic, stomachache		
113	*Tripleurospermum parviflorum* (Willd.) Pobed.	TR: Papatya	DE, IN	CA	Hair care, colds, cough, antipyretic, stomachache		
114	*Tussilago farfara* L.	AZ: Devedabanı		CA, LE		Gastrointestinal disorder, respiratory system disorders, nephritis, tuberculosis, inflamed wounds, skin disorders	
115	*Xanthium spinosum* L.	IR: Zardineh-e-khar dar (per.)	DE	FL, FR, SH			Colds and rheumatism, skin problems, itching, malaria, fever, nasal inflammation
	Berberidaceae						
116	*Berberis crataegina* DC.	TR: Zirinç	DE	FR	Colds, diabetes		
117	*Berberis vulgaris* L.	TR: Zirinç; AZ: Zirinc	DE	BR, FR, LE, RO	Colds, diabetes	Rachitism, anemia, malaria, antirheumatic, jaundice, diabetes, gastrointestinal disorder, angina, antitussive, eye diseases, cholagogue, internal bleeding	

(continued)

	Family–Taxa	Local name(s)	Preparations	Parts used	Treatment		
					Iğdir (TR)	Nakhchivan (AZ)	Tabriz (IR)
	Betulaceae						
118	*Betula pendula* Roth	AZ: Tozağacı		BD, IF, LE		Diuretic, cholagogue, avitaminosis, eczema, antirheumatic, scabies, anthelmintic, stomachache	
	Boraginaceae						
119	*Anchusa azurea* Miller	TR: Sığırdili	DE	BLE, HE, RO	Vulnerary, women's sterility		
120	*Anchusa* sp.	IR: Ghavzaban-e-badal (per.)	DE	FL			Blood purifier, sedative, diuretic, strengthens the kidneys, cold, cough, bronchitis
121	*Caccinia macranthera* (Banks & Sol.) Brand	IR: Gavzaban asa (per.)	DE	LE			Sedative, diuretic, strengthens the kidneys, cold, cough, bronchitis, antidepressants
122	*Echium strigosum* Sw.	IR: Gav zabane - kharakdar	PO	FL			Abscess (all animals)
123	*Echium vulgare* L.			RO		Astringent, anticolic	
124	*Nonea macrosperma* Boiss. & Heldr.	TR: Sormuk	PO	RO	Inflamed wounds		
125	*Nonea persica* Boiss.	IR: Cheshm gorbeyi-e-irani (per.)	PO	FL, LE			Emollient, sedative and cardiotonic

126	*Onosma* sp.	IR: Zangooleyi (per.)	DE	RO			Laxative, stimulate, rheumatic, heart disease
	Brassicaceae						
127	*Armoracia rusticana* (Lam.) P. Gaertn			RO		Scurvy, antirheumatic, otitis, angina, expectorant, appetizing, digestive	
128	*Bunias orientalis* L.	TR: Galatürpenk	FE	ST	Orexigenic		
129	*Capsella bursa-pastoris* (L.) Medik.	TR: Çobançantası; IR: Kise keshish (per.)	IN, PO, PU	HE, SH	Kidney stones, antitussive, diuretic, astringent		Anti-inflammation, wounds, epilepsy, astringent
130	*Cardamine uliginosa* M.Bieb.	TR: Tere	DE, PU	HE	Orexigenic, cardialgia, gynecological diseases		
131	*Descurainia sophia* (L.) Webb. ex Prantl	IR: Shuvaran (tur.) Khakshir-e-irani (per.)	BO, DE	SE, SH			Wounds, diarrhea, fever, inflammation of the kidneys and useful for pregnancy
132	*Diplotaxis tenuifolia* (L.) DC.	TR: Türpenk	DE	ST	Anti-inflammatory, orexigenic		
133	*Lepidium draba* L.	IR: Ozmak (per.)		FR			Astringent
134	*Lepidium sativum* L.	TR: Tere; IR: Shahi (tur. and per.)	DE	FR, LE, SE	Goiter		Liver and spleen diseases, asthma, cough, hemorrhoids, lack of vitamin C
135	*Nasturtium officinale* R. Br.	IR: Boolagh oti (tur.) alaf-e-cheshmeh, tartizak-e-abi (per.)	FE	WP			Antipyretic, anthelmintic, diabetes, expectorant

(continued)

	Family–Taxa	Local name(s)	Preparations	Parts used	Treatment		
					Iğdir (TR)	Nakhchivan (AZ)	Tabriz (IR)
136	*Raphanus raphanistrum* L.	IR: Torob-e-vahshi (per.)	PO	LE			Lack of vitamin C, kidney inflammation, kidney stones, chronic rheumatism, gout, asthma, jaundice
137	*Raphanus sativus* L.*	IR: Torobche noghli (per.)	DE, PO	FR, LE, RO			Cough, anti-diarrhea, digestive, fever, for hair growth
138	*Rapistrum rugosum* (L.) All.	TR: Türpek	FE	HE	Orexigenic		
139	*Sisymbrium irio* L.	IR: Khakshir-e-talkh (per.)	DE	SE			Diuretic, antipyretic, kidney inflammation, for measles treatment, hives
140	*Sobolewskia clavata* (Boiss.) Fenzl	TR: Yel otu	PO	HE	Antirheumatic		
	Cannabaceae						
141	*Humulus lupulus* L.			IF		Antispasmodic, analgesic, sedative, somniferous, appetizing, anti-inflammatory, diuretic, anthelmintic	
	Capparaceae						
142	*Capparis sicula* Veill. subsp. *herbacea* (Willd.) Inocencio	TR: Yılan yemişi	PO	FR	Antirheumatic		

143	*Capparis spinosa* L.	IR: Dagh garpouzi (tur.) Kabar, Kavar, alaf-e-mar(per.)	DE	BR, FL, RO			Sedative, the removal and disposal of asthma attacks, cough
	Caprifoliaceae						
144	*Cephalaria procera* Fisch. & Ave-Lall.	TR: Ganteper	DE	CA	Colds, cough, pulmonic disorders, cardiotonic		
	Caryophyllaceae						
145	*Herniaria glabra* L.			HE		Kidney disorders, urinary system disorders, enuresis, nephritis, antirheumatic	
146	*Silene compacta* Fisch. ex Hornem.	TR: Horoz pipiği	PU	LE	Vulnerary		
147	*Silene latifolia* Poir. subsp. *eriocalycinae* (Boiss.) Greuter & Burdet	TR: Garagile	DE	HE	Eczema		
148	*Vaccaria hispanica* (Mill.) Rauschert	IR: Sabunak (per.)	PO	WP			Carminative, stomach and intestine diseases, stimulate, increasing blood flow and cardiac activity
	Convolvulaceae						
149	*Convolvulus arvensis* L.	TR: Dolaşkan; IR: Pichak, pichak-e-sahrayi (per.)	DE, PO	LE, RO	Stomachic		Laxatives, vascular diseases, for liver infection
150	*Convolvulus scammonia* L.	TR: Dolaşkan	PO	LE	Stomachic		
151	*Cressa cretica* L.		CR, DE	LE		Diuretic, wounds	

(continued)

	Family–Taxa	Local name(s)	Preparations	Parts used	Treatment		
					Iğdir (TR)	Nakhchivan (AZ)	Tabriz (IR)
	Cornaceae						
152	*Cornus mas* L.			FR, LE		Diarrhea, gastrointestinal disorders, diabetes, malaria, dysentery, digestive, tonic	
	Cucurbitaceae						
153	*Bryonia alba* L.			RO		Antirheumatic, cardiotonic	
154	*Citrullus colocynthis* (L.) Schrad.	IR: Hendevaneye abujah	DE	FR, LE			Wound healing and disinfection (dog and donkey)
	Elaeagnaceae						
155	*Elaeagnus angustifolia* L.	TR: İğde; IR: Ideh (tur.) Senjed (per.)	DE, FE, IN	BR, FL, FR, LE, SE	Diarrhea, constipation, sunstroke		Flu, bronchi infection, fever, strengthening the stomach, liver diseases
156	*Elaeagnus rhamnoides* (L.) A.Nelson			FR		Avitaminosis, wounds, stomatitis, respiratory system disorders, tuberculosis, cancer, ulcer, burns	
	Ericaceae						
157	*Vaccinium myrtillus* L.			FR, LE		Enuresis, antirheumatic, gastritis, tonsillitis, stomatitis, arthralgia, diarrhea, dysentery, anthelmintic, gastrointestinal disorders	

158	*Vaccinium vitis-idaea* L.			FR, LE		Hepatic disorders, gastritis, scurvy, antirheumatic, enuresis, anthelmintic, diarrhea	
	Euphorbiaceae						
159	*Euphorbia heteradena* Jaub. & Spach.	IR: Farfion, Shir-sag (per.)	DE	RO, SH			Laxative, emetic, nerve pain and arthritis
160	*Euphorbia macroclada* Boiss.	TR: Sütlüyen	DAP	LA	Constipation, inflamed wounds, scorpion and bee bites		
161	*Euphorbia marschalliana* Boiss. subsp. *armena* (Prokh.) Oudejans	TR: Sütlübiyan	DAP	LA	Constipation, inflamed wounds, scorpion and bee bites		
162	*Euphorbia seguieriana* Necker subsp. *seguieriana*	TR: Sütlüyen	DAP	LA	Constipation, inflamed wounds, scorpion and bee bites		
163	*Ricinus communis* L.	IR: Karchak	OI	FR			Treatment of bloat and rumen indigestion (ruminants)
	Fabaceae						
164	*Alhagi maurorum* Medik. subsp. *graecorum* (Boiss.) Awmack & Lock	IR: Goy tikan (tur.) Khar shotor, Toranjabin (per.)	DE	FL, RO			Laxative, anti-fever in contagious disease
165	*Alhagi maurorum* Medik. subsp. *maurorum*	TR: Devedişi	DE	HE	Tonic		
166	*Astragalus aureus* Willd.	IR: Geven	CH, DE	GU, RO	Stomachache, sore throat, jaundice		

(continued)

	Family–Taxa	Local name(s)	Preparations	Parts used	Treatment		
					Iğdir (TR)	Nakhchivan (AZ)	Tabriz (IR)
167	*Astragalus* sp.	IR: Gavan (per. and tur.)	FE, IN, SM	AP, FR			Analgesic, anti-flatulence; for veterinary purposes (*Oestrus ovis* larvae (sheep))
168	*Galega officinalis* L.			HE		Cardiovascular system, diabetes	
169	*Glycyrrhiza glabra* L.	TR: Meyan; AZ: Biyan; IR: Shirin bayan, mak (per.)	DAP, DE	LE, RO	Cough, bronchitis, stomachic, asthma, nephralgia, diuretic, epilepsy, cancer, kidney stones, sunstroke	Ulcer, inflamed wounds, Addison disease, antitussive, tuberculosis, bronchitis, eczema	Sedative, cough, bronchitis, inflammation of trachea, ulcer, to relieve pain
170	*Lathyrus cicera* L.	TR: Gürül	PU	HE	Edema		
171	*Lathyrus rotundifolius* Willd. subsp. *miniatus* (Bieb. Ex Stev.) Davis	TR: Gülçiçeği	DE, PO	HE, LE	Goiter, antirheumatic		
172	*Lotus corniculatus* L. var. *corniculatus*	TR: Gazalboynuzu	DE	HE	Sedative, antihemorrhoidal, abdominal pain, diuretic, stomachache, nephralgia		
173	*Medicago sativa* L.	TR: Karayonca; IR: Yonja (tur.) Yonjeh. (per.)	DE, FE, PU	HE, LE, SH	Wounds, astringent		Vascular diseases, vitamin for infant nutrition, for rickets
174	*Melilotus officinalis* (L.) Desr.	TR: Kokulu yonca; AZ: Heşenbül; IR: Sari yonja (tur.) Yonjeh-e-zard, Aklilolmalek (per.)	FE, IN	FL, HE, LE	Anemia, sedative, constipation, antirheumatic	Respiratory system disorders, expectorant, pectoral, cold, swollen wounds, inflamed wounds	Emollient, insomnia, diuretic, nerve pain, migraines, rheumatic

175	*Onobrychis stenostachya* Freyn subsp. *sosnowskyi* (Grossh.) Hedge	TR: Körülgen	DE	HE	Diuretic		
176	*Onobrychis transcaucasica* Grossh.	TR: Gorunga	DE	HE	Diuretic		
177	*Phaseolus vulgaris* L.	IR: Loobia (tur. and per.)	DE	SE			Kidney stones, emollient, cardiotonic, diuretic, cholesterol, lowering blood pressure, constipation
178	*Securigera orientalis* (Mill.) Lassen subsp. *orientalis*	TR: Yonca	FE	LE	Nephralgia		
179	*Sophora alopecuroides* L.	TR: Acıbiyan; IR: Aji bayan (tur.) talkh bayan (per.)	DE, DI	LE, RO	Scabies		Laxative, hemorrhoids, ulcers, fever, vascular diseases, brain hemorrhage
180	*Trifolium ambiguum* Bieb.	TR: Alma otu	DE, FE	HE	Galactagogue, antihemorrhoidal, tonic, intestinal disorders		
181	*Trifolium pratense* L.	IR: Ouch-yarpakh (tur.) Shabdar-e-ghermez, Shabdar-e-chamanzari (per.)	DE, FE	FL			Laxatives, astringent, sterilization of chest, anti-asthma, stomach cancer
182	*Trifolium repens* L.	IR: Ouch-yarpakh (tur.) Shabdar-e-sefid, shabdar-e-khazandeh (per.)	DE, FE	FL			Diarrhea, astringent, cough, chest sterilization, asthma, stomach and throat cancer

(continued)

	Family–Taxa	Local name(s)	Preparations	Parts used	Treatment		
					Iğdir (TR)	Nakhchivan (AZ)	Tabriz (IR)
183	*Trigonella foenum-graceum* L.	IR: Shanbalileh (tur. and per.)	DI	SH			Tonic, diuretic, swollen spleen, liver, pain, bladder discomfort, hemorrhoids
184	*Trigonella* sp.	IR: Shanbalileh (tur. and per.)	DI	SH			Tonic, diuretic, swollen spleen, liver, pain, bladder discomfort, hemorrhoids
185	*Vicia cracca* L. subsp. *cracca*	TR: Gürül	FE	SE	Stomachache, headache		
186	*Vicia sativa* L.	IR: Mash (tur.) Mashak, Gav-daneh, karsaneh (per.)	FE, IN	SE			Bladder inflammation, tonic, cough, skin diseases
	Fagaceae						
187	*Fagus orientalis* Lipsky			BR, FR		Respiratory system disorder, pulmonic disorder, tuberculosis, skin disorder	
188	*Quercus robur* L.	AZ: Meşe		BR, FR		Stomatitis, gingivitis, tonsillitis, intestinal diseases, stomachache, burn, skin disorder	
	Gentianaceae						
189	*Centaurium erythraea* Rafn	AZ: Qızılçetir		HE		Cholagogue, hepatic diseases, gastritis, indigestion, wounds, pleuritic, pulmonic disorders, anthelmintic, diuretic	

	Geraniaceae						
190	*Erodium cicutarium* (L.) L'Her.	IR: Nok laklaki-e-harz (per.)	DE, PO	SE			Astringent, galactagogue, wash on animal bites, skin infection, diuretic, lack of vitamin K
191	*Geranium tuberosum* L.	IR: Shamdani-e-vahshi, soozan-e-choupan-e-ghodeh dar (per.)	OI	SH			Treatment of cholera, wrinkles and skin irritations, antidepressants
	Grossulariaceae						
192	*Ribes nigrum* L.	AZ: Qarağat		FR, LE		Avitaminosis, gastritis, cold, anemia, antirheumatism, kidney disorders, respiratory system disorders, skin disorders, cardiovascular system disorders	
	Hypericaceae						
193	*Hypericum montbretii* Spach	TR: Çay otu	DE	HE	Kidney stones, stomach disorders, ulcer, antihemorrhoidal		
194	*Hypericum perforatum* L.	TR: Çay otu; AZ: Dazı	DE	HE	Stomachache, ulcer, antiseptic, vulnerary, sedative, kidney disorders, antihemorrhoidal	Gastrointestinal disorders, hepatitis, stomatitis, angina, tonsillitis, cardialgia, antirheumatic, tuberculosis, internal bleeding, swollen wounds, inflamed wounds, abscess, wounds, burns, anthelmintic, diuretic	

(continued)

	Family–Taxa	Local name(s)	Preparations	Parts used	Treatment		
					Iğdir (TR)	Nakhchivan (AZ)	Tabriz (IR)
195	*Hypericum scabrum* L.	TR: Çay otu, Mide otu; IR: Gol-e-raee-e-deyhimi, Alaf-e-chai-e-deyhimi (per.)	DE, IN	FL, HE, SH	Constipation, antihemorrhoidal, kidney disorders, stomach disorders, ulcer, jaundice, menstrual disorders, analgesic, sedative		Energy providers, digestive, sedative, urinary tract antiseptic, anti-catarrh, diuretic, astringent, strengthening the nerves, uterine diseases
	Iridaceae						
196	*Iris x germanica* L.			RH		Tuberculosis, gastrointestinal disorders, respiratory system disorders, kidney and urinary infection, pleuritic, hepatic disease	
197	*Iris reticulata* M. Bieb.	IR: Norous-gouli (tur.) Zanbagh-e-moshabbak (per.)	IN				Diuretic, lung infection, anthelmintic, laxative
	Ixioliriaceae						
198	*Ixiolirion tataricum* (Pall.) Schult. & Schult.f.	IR: Khiarak (per.)	DE	BU			Blood purification, diarrhea, gout disease
	Juglandaceae						
199	*Juglans regia* L.	IR: Girdakan (tur.) Gerdoo (per.)	DE	FR, LE			Astringent
	Lamiaceae						
200	*Ajuga chamaepitys* (L.) Schreb. subsp. *chia* (Schreb.) Arcang.	TR: Mayasılotu	DE	HE	Tonic, antipyretic, emmenagogue, antihemorrhoidal diuretic, vulnerary		

201	*Dracocephalum moldavica* L.	IR: Badrashbi (per.)	PO	SE, SH			Painkiller, for relaxation, insomnia
202	*Lamium album* L.			FL		Internal bleeding, wound healing, gastrointestinal disorder, insomnia, neuralgia, cystitis	
203	*Lavandula vera* DC.*	IR: Ostokhoddus (tur. and per.)	DE, OI	FL, LE			Headaches, asthma, arthritis, gout, stomach tonic, insomnia, sedative, anti-stress, and antidepressants
204	*Leonurus glaucescens* Bunge	TR: Öküzguyruğu	IN, PO	HE, RO	Cardiotonic, inflamed wounds		
205	*Leonurus quinquelobatus* Gilib.	AZ: Şirguyruğu		HE		Sedative, neuralgia, cardiac diseases	
206	*Marrubium astracanicum* Jacq. subsp. *astracanicum*	TR: Dağ çayı	IN	HE	Colds, antipyretic		
207	*Marrubium parviflorum* Fisch. & Mey. subsp. *oligodon* (Boiss.) Seybold	TR: Dağ çayı	IN	HE	Colds, antipyretic		
208	*Mentha longifolia* (L.) L.	TR: Yarpuz; IR: Yarpuz (tur.) Pouneh (per.)	DE, FE, IN, PO, PW	HE, LE, SH	Colds, flu, cough, catarrh, abdominal pain, menstrual pain, stomachic, bronchitis, headache, pulmonic disorders, diarrhea, asthma, antihemorrhoidal		Expectorant, whooping cough, carminative, jaundice, gum diseases, relieve epilepsy, urinary retention
209	*Mentha x piperita* L.	IR: Naana (tur. and per.) Naana felfeli (per.)	DE, FE, PO	SH			Painkiller, energy providers, stomachache, diarrhea, fever, colds, carminative, jaundice

(continued)

				Parts	Treatment		
	Family–Taxa	Local name(s)	Preparations	used	Iğdir (TR)	Nakhchivan (AZ)	Tabriz (IR)
210	*Nepeta haussknechtii* Bornm.	IR: Pooneh say-e-eraghi (per.)	DE	SH			Stomachache, relaxation, anti-fever
211	*Ocimum basilicum* L.	IR: Reyhan, Reyhan banafsh (tur. and per.) Tokhme sharbati (per.)	IN	SE, SH			Boost digestion, diuretic, headaches, nervous, dizziness, nausea, cough, angina, whooping cough, increases milk
212	*Origanum vulgare* L. subsp. *gracile* (K. Koch) Ietsw.	TR: Eşek kekiği; AZ: Qaraqınıq	DE, IN	HE	Vulnerary, stomachache, hypertension, headache, asthma, toothache	Stomatitis, toothache, abscess, skin disorders, antirheumatic, carminative, stomachache, epilepsy, paralysis, antitussive, appetizing	
213	*Phlomis armeniaca* Willd.	TR: Çöl çayı	IN	HE	Antipyretic, colds, asthma, bronchitis		
214	*Rosmarinus officinalis* L.*	IR: Rozmari (tur. and per.) Aklil-e-kouhi (per.)	OI	FL, LE			Strengthening hair and prevent hair loss
215	*Salvia hydrangea* DC. ex Bentham	TR: Koç otu: IR: Maryam goli (tur. and per.) Maryam goli-e-tamashayi (per.)	DE, IN	FL, HE, SH	Colds, diabetes, stomach disorders, antipyretic, emmenagogue		Strengthen the body and spirit, wounds, skin irritations, swelling of throat
216	*Salvia nemorosa* L.	TR: Çöl çayı	DE, IN, PU	HE	Astringent, colds, catarrh		
217	*Salvia palaestina* Benth.	IR: Maryam - goli phelestini	PA	FL, LE			Appetizer (ruminant)

218	*Salvia verticillata* L. subsp. *amasiaca* (Freyn. & Bornm.) Bornm.	TR: Karabaş otu	IN	HE	Laxative, colds		
219	*Salvia verticillata* L. subsp. *verticillata*	TR: Karabaş otu	DE	HE	Catarrh, colds, laxative		
220	*Satureja hortensis* L.	IR: Marzeh (tur. and per.)	DE, IN	SE, SH			Muscle pain, rheumatism
221	*Satureja sahendica* Bornm.	IR: Dagh marzasi (tur.) Marzeh-e-sahandi (per.)	IN	SH			Cough and expectorant, carminative, antiparasitic, stomach cramps
222	*Scutellaria orientalis* L.	TR: Sancı otu	IN	HE, RO	Abdominal pain, nephralgia, carminative	Cardiovascular diseases, neuralgia, headache, insomnia, whooping cough, epilepsy, antirheumatic, bronchitis, pulmonic disorders, anthelmintic, dysentery, tonic	
223	*Stachys annua* (L.) L. subsp. *annua* var. *lycaonica* Bhattacharjee	TR: Dağ çayı	DE	HE	Colds, antipyretic		
224	*Stachys iberica* Bieb. subsp. *georgica* Rech. f.	TR: Dağ çayı	DE	HE	Colds, antipyretic		
225	*Stachys iberica* Bieb. subsp. *stenostachya* (Boiss.) Rech. f.	TR: Dağ çayı	DE	HE	Colds, antipyretic, stomachache		

(continued)

	Family–Taxa	Local name(s)	Preparations	Parts used	Treatment		
					Iğdir (TR)	Nakhchivan (AZ)	Tabriz (IR)
226	*Stachys inflata* Benth.	IR: Sonbole-e-badkonaki, sonbole-e-arghavani (per.)	DE	LE			Nervous disorders and hysteria, antispasmodic, sedative, fever, insomnia, stomach, anthelmintic, antiexcitement, migraine, anxiety, melancholy diseases, sciatica pain, hiccups, chest pain, emetic
227	*Stachys lavandulifolia* Vahl.	TR: Dağ çayı; IR: Tooklije (tur.) chai-e-alafi, chai-e-kouhi, sonbole-e-ziba (per.)	DE, IN	HE, LE, SH	Antipyretic, cough		Stomach, anthelmintic, antiexcitement, migraine, anxiety, melancholy diseases, sciatica pain, hiccups, chest pain, emetic
228	*Teucrium polium* L.	TR: Daşkekiği	DE, FE	HE	Stomach, diarrhea, antihemorrhoidal, internal medicine, diabetes, analgesic, anti-inflammatory, edema, stomachache, digestive, tuberculosis, abdominal pain		
229	*Thymus fallax* Fisch. & C.A.Mey.	TR: Kekik	DE, IN	HE	Backache, hypertension, anti-inflammatory, cancer, enteralgia, anthelmintic, diabetes, colds, abdominal ailments		

230	*Thymus kotschyanus* Boiss. & Hohen. subsp. *kotschyanus*	TR: Kekik	DE, IN	HE	Backache, hypertension, anti-inflammatory, cancer, enteralgia, anthelmintic, diabetes, colds, abdominal ailments		
231	*Thymus migricus* Klokov & Des.-Shost.	TR: Kekik	IN	HE	Backache, hypertension, enteralgia, anti-inflammatory, cancer, anthelmintic		
232	*Thymus praecox* Opiz. subsp. *grossheimii* (Ronniger) Jalas	TR: Kekik	IN	HE	Hypertension, enteralgia, anti-inflammatory, cancer, anthelmintic, colds		
233	*Thymus pubescens* Boiss. & Kotschy ex Celak.	IR: Avishan-e-kork alood (per.)	IN	SH			Cough, bronchitis, joint pain, muscle pain
234	*Thymus* sp.	IR: Kahlik oti (tur.) Avishan (per.)	IN	SH			Disinfection, antiparasitic, strengthness the kidneys, cough, bronchitis, joint pain, muscle pain
235	*Thymus transcaucasicus* Ronniger	TR: Kekik	IN	HE	Backache, hypertension, enteralgia, anti-inflammatory, cancer, anthelmintic	Whooping cough, bronchitis, dysentery, angina, antitussive, cold, carminative, stomachache, tachycardia	
236	*Ziziphora clinopodioides* Lam.	TR: Reyhan; IR: Kakoti kouhi, Naana kouhi (per)	IN	HE, LE	Stomachache, carminative, orexigenic, colds		Fever, smoothing chest, carminative, cough

(continued)

	Family–Taxa	Local name(s)	Preparations	Parts used	Treatment		
					Iğdir (TR)	Nakhchivan (AZ)	Tabriz (IR)
237	*Ziziphora taurica* M.Bieb. subsp. *taurica*	TR: Merze	IN	HE	Carminative, orexigenic, stomachache, hypertension		
238	*Ziziphora tenuior* L.	IR: Kakoti (per.)	IN	LE			Fever, smoothing chest, carminative
	Lythraceae						
239	*Punica granatum* L.	AZ: Nar		BR, FR		Gastrointestinal disorders, wound, anthelmintic, cardialgia, gingivitis, kidney disease, diabetes, antipyretic, scurvy, cold, antitussive, antirheumatic, tonic	
	Malvaceae						
240	*Alcea striata* (DC.) Alef. subsp. *rufescens* (Boiss.) Cullen	TR: Gül çiçek	DE, IN	FL, LE	Anti-inflammatory, anemia, cough		
241	*Althaea officinalis* L.	AZ: Balgamotu		RO		Respiratory system disorders, ulcer, diarrhea, gastritis, stomachache, anthelmintic, cystitis, angina	
242	*Althea* sp.	IR: Khatmi (tur. and per.)	DE	FL, RO			Cough, skin diseases, fever, diuretic, analgesic, antibacterial, anti-inflammatory, antiseptic, anthelmintic

243	*Malva neglecta* Wallr.	TR: Ebemkömeci; IR: Aman-komanji (tur.) Panirak-e-mamouli (per.)	DE, PO, PU, PW	FL, FR, HE, LE, RO	Asthma, abdominal pain, ulcer, colds, stomachic, digestive, sore throat, constipation, emmenagogue, anti-inflammatory, abscess, vulnerary, gynecologic diseases, women's sterility, edema, analgesic, abortive		Cough, diuretic
244	*Malva sylvestris* L.	AZ: Emekömeci; IR: Aman-komanji (tur.) Panirak (per.)	DE	FL, LE		Respiratory system disorders, gastrointestinal disorder, angina, antitussive, cold	Anti-infection, cough
	Moraceae						
245	*Morus alba* L.	AZ: Tut		FR, LE, RO		Cardiovascular system disorders, anemia, scarlatina, ulcer, antirheumatic, anthelmintic, cold, epilepsy, scabies	
	Nitrariaceae						
246	*Peganum harmala* L.	TR: Üzerlik; AZ: Üzerrik; IR: Uzarik (tur.) Espand, Esfand (per.)	DE, SM	HE, RO, SE	Antirheumatic, antihemorrhoidal	Sedative, neuralgia, Parkinson, cold, malaria, antirheumatic, scabies, diuretic, exudative, stomach disorders	Disinfectant, snakebite (all the animals)
	Oleaceae						
247	*Fraxinus excelsior* L.*	IR: Goush-dili (tur.) Zaban gonjeshk (per.)	DE	BR, LE			Antipyretic, expectorant, astringent, diuretic, laxative, diarrhea, rheumatism, gout

(continued)

	Family–Taxa	Local name(s)	Preparations	Parts used	Treatment: Iğdir (TR)	Treatment: Nakhchivan (AZ)	Treatment: Tabriz (IR)
	Papaveraceae						
248	*Chelidonium majus* L.	AZ: Ziyilotu, Dəmrovotu		HE		Anti-verrucous, wounds, ulcer, skin cancer, cholagogue, jaundice	
249	*Fumaria asepala* Boiss.	TR: Şetere; IR: Shah tareh-e-bikasbarg (per.)	DE, PW	HE, SH	Headache, itching, antiseptic		Sore throat, diuretic, diaphoretic, stomach tonic, blood purifier, appetizer, leprosy, skin diseases
250	*Fumaria officinalis* L.	AZ: Şahtere		HE		Avitaminosis, gastrointestinal disorders, carminative, anemia, jaundice, astringent, appetizing	
251	*Fumaria schleicheri* Soy-Will. subsp. *microcarpa* (Hausskn.) Liden	TR: Şetere	PW	HE	Headache, itching, antiseptic		
252	*Papaver lacerum* Popov	TR: Haşhaş	FE	BD	Goiter		
253	*Papaver orientale* L.	TR: Haşhaş	IN	LE	Asthma		
254	*Papaver rhoeas* L.	IR: Ghincha (tur.) Shaghayegh (per.)	DE	FR			Emollient, cough, bronchitis, catarrh, lung diseases, asthma, ear pain, dysentery
255	*Roemeria hybrida* (L.) DC.	IR: Gol-e-arousak-e-banafsh (per.)	IN				Relaxation

	Plantaginaceae						
256	*Plantago atrata* Hoppe	TR: Dartul otu	FE	LE	Astringent, inflamed wounds, vulnerary		
257	*Plantago lanceolata* L.	TR: Sinir otu; IR: Garni-yarikh (tur.) Barhang-e-sarneyzeyi, Kardi (per.)	DE, FE, VA	LE, WP	Abscess, antiparasitic, vulnerary, astringent, anti-inflammatory, gynecologic diseases, stomachic, ulcer		Anti-catarrh, urinary tract diseases, kidney diseases, constipation, bladder discomfort
258	*Plantago major* L.	TR: Bağa yaprağı; AZ: Bağayarpağı; IR: Bozousha (tur.), Barhang-e-kabir (per.)	DE	HE, LE, WP	Abscess, vulnerary, anti-inflammatory, stomachic, ulcer	Ulcer, gastrointestinal disorders, antitussive, wounds, whooping cough, burns	Anti-catarrh, urinary tract, kidney diseases, bladder discomfort
259	*Plantago media* L.	TR: Kılıçotu; IR: Garni-yarikh (tur.) Barhang-e-moattar (per.)	DE, FE	LE, WP	Astringent, anti-inflammatory		Anti-catarrh, urinary tract and kidney diseases, constipation, bladder discomfort
260	*Veronica ceratocarpa* C.A.Mey.	IR: Sizab-e-moshabbak, Sizab-e-miveh shakhi (per.)	FE	FL			Anti-scurvy, diuretic, wound, constipation
261	*Veronica orientalis* Mill.	TR: Gözmuncuğu çiçeği	IN	HE	Kidney stones, pulmonic disorders		
	Poaceae						
262	*Alopecurus myosuroides* Huds.	IR: Dom roubahi moushi (per.)	IN				Sedative
263	*Avena sativa* L.*	IR: Youlaf (tur. and per.)	DE	SE			Heart disease, laxative, anticancer
264	*Bromus japonicus* Thunb.	IR: Jarou alafi-e-japoni (per.)	OI				Accelerate hair growth and stop hair loss

(continued)

	Family–Taxa	Local name(s)	Preparations	Parts used	Treatment		
					Iğdir (TR)	Nakhchivan (AZ)	Tabriz (IR)
265	*Cynodon dactylon* (L.) Pers.	IR: Chayer (tur. and per.)	DI	WP			Vascular diseases
266	*Hordeum spontaneum* K. Koch.	IR: Arpa (tur.) Jo-e-vahshi (per.)	DE				Indigestion, fever, loss of milk production, astringent, headaches, flatulent, swelling of throat, sore throat, constipation
267	*Lolium perenne* L.	IR: Chacham-e-chand saleh (per.)	IN	SE			Hypnotic, joint pain
268	*Phragmites australis* (Cav.) Trin. ex Steud.	IR: Ghamish (tur.) Ney (per.)	IN	RH			Spasms of the liver and spleen, toothache, chest pains
269	*Zea mays* L.*	IR: Makka (tur.) Zorrat (per.)	OI	COS			Tonic, eczema and high blood cholesterol
	Polygonaceae						
270	*Polygonum aviculare* L.	TR: Kuşdili; IR: Alaf-e-haft band (per.)	DE, IN, PO	HE, LE, ST	Cough, antirheumatic, anemia, stomach disorders		Astringent, tonic, wounds, diarrhea, dysentery, intestinal inflammation, vaginal discharge, asthma
271	*Polygonum bistorta* L. subsp. *bistorta*	TR: Çimen eveliği	IN	HE	Sore throat, expectorant		
272	*Polygonum cognatum* Meissn.	TR: Kuş eppeği	DE, IN	HE	Abscess, emetic, cough, antirheumatic, anemia, stomach disorders		

273	*Rheum ribes* L.	TR: Işgın; IR: Rivas	DE, FE, PU	PET, RO, SH	Ulcer, diarrhea, anthelmintic, expectorant, antihemorrhoidal, digestive, stomachic, diabetes, tonic, emetic, constipation, hypertension		Jaundice, typhoid, diarrhea, bile diseases
274	*Rheum rupestre* Litv.			RO		Appetizer, skin disorders, cholagogue	
275	*Rumex confertus* Willd.	AZ: Eveliyi		FR, LE, RO		Gastrointestinal disorders, indigestion, inflammation of intestine, diarrhea, dysentery, avitaminosis, pulmonic disorders, mouth wounds, gastritis, anal fissure, cholagogue	
276	*Rumex crispus* L.	TR: Evelik	DE, PU	FR, LE	Cough, colds, asthma, anti-inflammatory, antihemorrhoidal, gynecologic diseases, antirheumatic, goiter		
277	*Rumex patientia* L.	TR: At eveliği; IR: Avalik (tur.) Torshak-e-shafadahandeh, Torshak-e-bimarkhiz (per.)	DE, IN	LE, SH	Internal medicine		Urinary tract diseases, gingivitis, skin care, acne, colic, hepatitis, chronic constipation
278	*Rumex scutatus* L.	TR: Taş turşusu	FE, IN	LE, RO	Diuretic, antipyretic, orexigenic		
279	*Rumex tuberosus* L. subsp. *horizontalis* (Koch) Reich. f.	TR: Köme turşusu	FE, IN	LE, RO	Diuretic, antipyretic, orexigenic		

(continued)

	Family–Taxa	Local name(s)	Preparations	Parts used	Treatment		
					Iğdir (TR)	Nakhchivan (AZ)	Tabriz (IR)
	Portulacaceae						
280	*Portulaca oleracea* L.	IR: Parpane (tur.) Khorfeh (tur. and per.)	IN	WP			Diuretics, anti-scurvy, fever, blood purifier, pain, gastrointestinal tract inflammations, cough, and insomnia
	Primulaceae						
281	*Primula auriculata* Lam.	TR: Mustafa çiçeği	IN	HE	Stomachic		
	Ranunculaceae						
282	*Adonis aestivalis* L.	IR: Jin-lalasi (tur.) Chesh khorous-e-tabestaneh (per.)	DE, FE	SH			Hair growth, boosting sexual power, old wounds, joint pain
283	*Caltha palustris* L.	TR: Lulpar	PO	PET	Internal medicine		
284	*Clematis orientalis* L.	TR: Pamukotu	PU	HE	Antirheumatic, abscess, mycodermatitis		
285	*Nigella sativa* L.			SE		Pancreatitis, hepatitis, gastritis, anthelmintic, exudative, diuretic, eczema, antitussive	
286	*Pulsatilla violacea* Rupr. subsp. *armena* (Boiss.) Lufenov	TR: Dağ lalesi	IN, VA	LE	Respiratory system disorders, headache, sinusitis, tonic, diuretic, expectorant		
287	*Ranunculus arvensis* L.	TR: Yara otu; IR: Alaleh (per.)	DE, PU	HE, RO, WP	Swollen wounds, antirheumatic		Skin illness, rheumatism, gout, neuralgia, influenza, and meningitis

288	*Ranunculus caucasicus* Bieb. subsp. *subleiocarpus* (Som. & Lev.) Davis	TR: Yara otu	DE, PU	HE, RO	Swollen wounds, antirheumatic		
289	*Ranunculus oreophilus* M.Bieb.	TR: Yara otu	DE, PU	HE, RO	Swollen wounds, antirheumatic		
290	*Thalictrum minus* L. var. *microphyllum* Boiss.	TR: Kahraman kaytaran	DE, PO, VA	HE	Asthma, cardialgia, headache		
	Resedaceae						
291	*Reseda lutea* L.	IR: Afsani (per.)	DE	RO			Painkiller, tonic
	Rhamnaceae						
292	*Paliurus spina-christi* P. Mill.	IR: Konar	PA	FR, LE			Treatment of intestinal infections (sheep and goat)
293	*Ziziphus jujuba* Mill.			FR, LE, RO		Pulmonic disorders, antitussive, tonic	
	Rosaceae						
294	*Agrimonia eupatoria* L.	IR: Ghafes (per.)	DE	FL, HE		Hepatic disorders, antirheumatic, gastrointestinal disorder, mouth and pharynx diseases, buckling of legs, skin disorders	Astringent, diuretic, diarrhea, dysentery, hemorrhoids, heart attack, gallstones
295	*Amygdalus lycioides* Spach	IR: Badam	FE	SE			Sedative, analgesic
296	*Armenica vulgaris* Lam.	AZ: Erik		FR, SE		Pectoral, expectorant, digestive, anthelmintic, cardiovascular system disorder	

(continued)

	Family–Taxa	Local name(s)	Preparations	Parts used	Treatment		
					Iğdir (TR)	Nakhchivan (AZ)	Tabriz (IR)
297	*Cerasus microcarpa* (C.A. Mey.) Boiss.	IR: Mahlab	DE	LE			Disinfection of wounds (all animals)
298	*Cotoneaster integerrimus* L.	TR: Dağ muşmulası	DE	BR	Jaundice, cough		
299	*Crataegus azarolus* L. var. *azarolus*	TR: Gurmut; IR: Agh-yemishan (tur.) Zalzalak (per.)	DE, FE	BR, FL, FR	Embolism, laxative, diabetes		Cardiotonic, digestive, fever, nervous disorders, vertigo
300	*Crataegus meyeri* Pojark.	TR: Gurmut	DE	FR, RO	Embolism, laxative, diabetes		
301	*Crataegus pentagyna* Waldst. et. Kit. ex Willd.	AZ: Yemişan		FL, FR		Tachycardia	
302	*Cydonia oblonga* Mill.*	AZ: Heyva		FR, LE, SE		Gastrointestinal disorder, asthma, chronic bronchitis, tuberculosis, anemia, diathesis, cholagogue, diuretic, cardialgia	
303	*Filipendula ulmaria* (L.) Maxim.			FL, HE, RO		Cold, diarrhea, dysentery, skin disorders, anthelmintic, diuretic, internal bleeding	
304	*Geum urbanum* L.	IR: Alaf-e-mobarak (per.)	DE	RO			Astringent, diarrhea, inflammation of the pharynx, disinfectants, strengthening the nerve, muscle pain, fever

305	*Malus sylvestris* (L.) Mill. subsp. *orientalis* (Uglitzkich) Browicz var. *orientalis*	TR: Alma; AZ: Alma; IR: Alma (tur.) Sib, Sib-e-jangali (per.)	DE, FE	FR, SE	Colds, diabetes	Diarrhea, anemia, antirheumatic, dysentery, diabetes, cardiac diseases, obesity, bronchitis	Constipation, laxatives, antipyretic, anthelmintic, hypnotic, diuretic, kidney inflammation, cold
306	*Potentilla reptans* L.	IR: Panjeh-barg (per.)	IN	WP			Astringent, fever, diarrhea, jaundice, hemorrhoids, treatment of pain of the liver, lungs, and joints and sciatica
307	*Rosa canina* L.	TR: İtburnu; AZ: İt itburnu; IR: Gildik, It-bourni (tur.) Nastaran, sag-gol (per.)	DE, IN	FR, LE, RO	Antihemorrhoidal, cough, stomachic, constipation, malaria, diabetes, tonic, bronchitis diuretic, colds, asthma, kidney stones	Avitaminosis, scurvy, diathesis, hepatitis, kidney and bladder disorders, gastrointestinal disorder, burn	Anti-scurvy, astringent, lack of vitamin C, diarrhea, antispasmodic, anthelmintic, wounds and burns
308	*Rosa damascena* Mill.	IR: Ghizil-goul (tur.) Gol-e-mohammadi (per.)	PO	FL			Laxative, constipation (in children)
309	*Rosa hemisphaerica* J. Herrm.	TR: Yemişen	DE	FR	Cough		
310	*Rosa pulverulenta* M.Bieb.	TR: Gıllıca	DE	FR	Colds, cough		
311	*Rosa spinosissima* L.	TR: Koyungözü	DE	FR, RO	Colds, stomachache, cardiotonic, antihemorrhoidal, antiseptic		

(continued)

	Family–Taxa	Local name(s)	Preparations	Parts used	Treatment		
					Iğdir (TR)	Nakhchivan (AZ)	Tabriz (IR)
312	*Rubus caesius* L.			FR, LE, RO		Diarrhea, dysentery, anemia, avitaminosis, gastrointestinal disorders, cardiac diseases, diabetes, tuberculosis, angina, burn, eczema, cold	
313	*Rubus idaeus* L.			FR, LE		Bronchitis, cold, emetic, appetizing, diuretic, hemafacient	
314	*Sanguisorba minor* L.	IR: Tout-e-roubahi (per.)	IN	LE, RO			Astringent, anti-hemorrhoids
315	*Sanguisorba officinalis* L.			RO		Enteritis, stomachic, diarrhea, anthelmintic, dysentery, burn, stomatitis, tuberculosis	
316	*Sorbus aucuparia* L.	AZ: Quşarmudu		FR		Gastrointestinal disorder, dysentery, diabetes, avitaminosis, diuretic, exudative, hemafacient	
	Rubiaceae						
317	*Galium aparine* L.	IR: Shir-panir (per.)	DE	SH			Diuretic, gout, hepatitis, prostate inflammation
318	*Galium humifusum* M.Bieb.	TR: Koşacı; IR: Bitirakh (tur.) Shir-panir (per.)	DI, IN	HE, SH	Gynecologic diseases		Anti-tension
319	*Galium tricornutum* Dandy	TR: Koşacı	IN	HE	Gynecologic diseases		

320	*Rubia tinctorum* L.	IR: Biakh-bashi (tur.) Ronas (per.)	DE, IN	RO		Kidney stone, cholagogue, nephralgia	Diuretic, astringent
	Salicaceae						
321	*Salix aegyptiaca* L.	IR: Bid-e-meshk (per.)	PO	BR, FL, LE			Fever, laxative
322	*Salix alba* L.	IR: Bid-e-sefid (per.)	PO	BR, LE			Fever, laxative
	Santalaceae						
323	*Viscum album* L.	IR: Darvash		LE, ST			Sedative (ruminants), wound healing (dog and donkey)
	Scrophulariaceae						
324	*Verbascum agrimoniifolium* (K. Koch) Hub.-Mor. subsp. *agrimoniifolium*	TR: Sığır guyruğu	DAP	BLE	Mycodermatitis		
325	*Verbascum densiflorum* Bertol.	AZ: Keçiqulağı, Sığırquyruğu				Whooping cough, eczema, bronchitis, asthma, gastrointestinal disorder, pleuritic, pectoral, anti-inflammatory	
326	*Verbascum oreophilum* K.Koch subsp. *joannis* (Bordz.) Hub.-Mor.	TR: Sığır guyruğu	DAP	LE	Mycodermatitis		
327	*Verbascum speciosum* Schrader	TR: Sığır guyruğu	DAP	LE	Mycodermatitis		

(continued)

	Family–Taxa	Local name(s)	Preparations	Parts used	Treatment		
					Iğdir (TR)	Nakhchivan (AZ)	Tabriz (IR)
328	*Verbascum* sp.	IR: Gol-e-mahoor, Khargooshak (per.)	IN	FL, LE			Anti-stress, anemia, stomach cramps, skin care, vitamin supports, migraine, warts, insomnia
	Solanaceae						
329	*Capsicum frutescens* L.*	IR: Bibar (tur.) Felfel (per.)	FE	FR			Toothache, hemorrhoids, gonorrhea, liver problems, painkiller
330	*Datura stramonium* L.	TR: Deli patpat	DE, VA	HE, LE, SE	Toothache, antipyretic, antirheumatic, sedative		
331	*Hyoscyamus niger* L.	TR: Patpat; IR: Dali-bat bat (tur.) Bazrol-banj, Bang-daneh (per.)	DE, IH, VA	LE, SE	Toothache, against itching in the eyes, stomatitis		Whooping cough, tuberculosis, and bronchitis
332	*Lycium ruthenicum* Murray	IR: Gorg-tigh, div-e-khar-majary (per.)	DE	FR			Diabetes, eye diseases, hypertension, skin diseases, relieve hiccups, emetic, diarrhea
333	*Lycopersicon esculentum* Mill.*	IR: Bamador, Ghirmizi bademjan (tur.) Gojeh-farangi (per.)	FE	FR, SH			Colds, prostate cancer, improves the skin's ability to protect against ultraviolet rays

	Thymelaeceae						
334	*Daphne oleoides* Schreb.	TR: Mundarça	DE	BR	Abortion		
	Ulmaceae						
335	*Ulmus minor* Mill.	TR: Karaağaç	DE, PU	BR, RO	Anti-inflammatory, wounds, cough, asthma		
	Urticaceae						
336	*Urtica dioica* L.	TR: Isırgan otu; AZ: Gicitkan; IR: Dalama (tur.) Gazaneh (tur. and per.)	DE, FE, PO	HE, LE, WP	Cancer, antirheumatic, diabetes, stomachic, cough, colds, throat diseases, analgesic, edema, sedative, laxative, anti-inflammatory, emmenagogue, asthma, hypertension, hair care	Internal bleeding, avitaminosis, antitussive, anthelmintic, diabetes, hair care	Diuretic, astringent, antiparasitic, anthelmintic, kidney inflammations
	Vitaceae						
337	*Vitis sylvestris* C.C. Gmel.	AZ: Üzüm		FR, LE		Tuberculosis, bronchitis, laryngitis, angina, cardiac diseases, diathesis, constipation, nerve system, hepatic disorders, hemafacient	
338	*Vitis vinifera* L.*	IR: Uzum (tur.) Angour, Tak, Mo (per.)	FE	FR, LE, SE			Constipation, inflammation of the stomach, intestinal inflammation, whooping cough, spleen and liver diseases, anemia
	Xanthorrhoeaceae						
339	*Aloe vera* (L.) Burm. f.	IR: Sir zard	DE	LE			Lenient (all animals)

(continued)

	Family–Taxa	Local name(s)	Preparations	Parts used	Treatment		
					Iğdir (TR)	Nakhchivan (AZ)	Tabriz (IR)
	Zygophyllaceae						
340	*Tribulus terrestris* L.	IR: Chagher Tikani (tur.) Khar-Khasak (per.)	FE	FR			Diuretic, facilitator of the liver function, blood purification
341	*Zygophyllum fcbago* L.	TR: İt üzerliği	DE, VA	HE	Infertility, antirheumatic		

References

Akerele O (1992) Importance of medicinal plants: WHO's programme. In: Baba S, Akerele O, Kawaguchi Y (eds) Natural resources and human health. Elsevier Science Publishers B.V, Amsterdam, pp 63–72

Altundağ E (2009) Iğdır İli'nin (Doğu Anadolu Bölgesi) doğal bitkilerinin halk tarafından kullanımı. Ph D. thesis. Istanbul University, Istanbul, Turkey

Altundag E, Ozturk M (2011) Ethnomedicinal studies on the plant resources of east Anatolia, Turkey. Procedia Soc Behav Sci 19:756–777

Amiri R, Weng Q, Alimohammadi A, Alavipanah SK (2009) Spatial-temporal dynamics of land surface temperature in relation to fractional vegetation cover and land use/cover in the Tabriz urban area, Iran. Remote Sens Environ 113(12):2606–2617

Bandiera O, Rasul I (2006) Social networks and technology adoption in northern Mozambique. Econ J 116(514):869–902

Bedelov H, Aliyev A, Zeynalov I (2014) Xınalıqda Yetişen Derman Bitkileri. Gothe University, Frankfurt

Bradacs G, Heilmann J, Weckerle CS (2011) Medicinal plant use in Vanuatu: a comparative ethnobotanical study of three islands. J Ethnopharmacol 137(1):434–448

Chapman AD (2009) Numbers of living species in Australia and the World. Heritage, 2nd (September), p 84

Cocks M (2006) Biocultural diversity: moving beyond the realm of "indigenous" and "local" people. Hum Ecol 34(2):185–200

De Almeida CDFCBR, Ramos MA, Silva RRV, De Melo JG, Medeiros MFT, Arajo TADS et al (2012) Intracultural variation in the knowledge of medicinal plants in an urban-rural community in the Atlantic Forest from northeastern Brazil. Evid Based Complement Alternat Med 2012:1. https://doi.org/10.1155/2012/679373

De Pasquale A (1984) Pharmacognosy: the oldest modern science. J Ethnopharmacol 11:1–16

van den Broeck K, Dercon S (2011) Information flows and social externalities in a Tanzanian banana growing village. J Dev Stud 47(2):231–252

Ellen RF (2009) Classification. In: Barnard A, Spencer J (eds) The Routledge encyclopedia of social and cultural anthropology. Routledge, London and New York, pp 129–133

Ellena R, Quave CL, Pieroni A (2012) Comparative medical ethnobotany of the senegalese community living in Turin (northwestern Italy) and in Adeane (southern Senegal). Evid Based Complement Alternat Med 2012(4):1. https://doi.org/10.1155/2012/604363

Etkin N (1988) Ethnopharmacology: biobehavioral approaches in the anthropological study of indigenous medicines. Ann Rev Anthrop 17:23–42

Fabricant DS, Farnsworth NR (2001) The value of plants used in traditional medicine for drug discovery. Environ Health Perspect 109(Suppl. 1):69–75

Farnsworth NR (1992) Preclinical assessment of medicinal plants. In: Akerele O, Kawaguchi Y, Baba S (eds) Natural resources and human health. Elsevier Science Publishers, B.V., Amsterdam, pp 87–91

Farnsworth NR, Soejarto DD (1991) Global importance of medicinal plants. In: Akerele OV, Heywood V, Synge H (eds) Conservation of medicinal plants. Cambridge University Press, New York, pp 25–51

Ghazanfar S (2011) Medicinal and aromatic plants Arabia and Iran. In: Ethnopharmacology. Encyclopedia of Life Supports Systems (EOLSS), New York

Ghorbani A, Langenberger G, Sauerborn J (2012) A comparison of the wild food plant use knowledge of ethnic minorities in Naban River watershed National Nature Reserve, Yunnan, SW China. J Ethnobiol Ethnomed 8:17

González-Tejero MR, Casares-Porcel M, Sánchez-Rojas CP, Ramiro-Gutiérrez JM, Molero-Mesa J et al (2008) Medicinal plants in the Mediterranean area: synthesis of the results of the project Rubia. J Ethnopharmacol 116(2):341–357

Güner A, Aslan S, Ekim T, Vural M, Babaç MT (2012) Türkiye Bitkileri Listesi (Damarlı Bitkiler). Nezahat Gökyiğit Botanik Bahçesi ve Flora Araştırmaları Derneği Yayını, Istanbul
Hadjichambis AC, Paraskeva-Hadjichambi D, Della A, Elena Giusti M, De Pasquale C et al (2008) Wild and semi-domesticated food plant consumption in seven circum-Mediterranean areas. Int J Food Sci Nutr 59(5):383–414
Halberstein RA (1997a) Traditional botanical remedies on a small Caribbean island: middle (grand) Caicos, West Indies. J Altern Complement Med 3:227–239
Halberstein RA (1997b) Health and disease in the Caribbean: an historical perspective. J Carib Stud 12:1–5
Halberstein RA (2005) Medicinal plants: historical and cross-cultural usage patterns. Ann Epidemiol 15(9):686–699
Halberstein RA, Saunders AB (1978) Traditional medical practices and medicinal plant usage on a Bahamian Island. Cult Med Psychiatry 2(2):177–203
Han MI, Bulut G (2015) The folk-medicinal plants of Kadişehri (Yozgat-Turkey). Acta Soc Bot Pol 84(2):237–248
Hasanova G, Ozturk M, Akçiçek E (2000) Azerbaijan'da geleneksel tedavide kullanılan bitkiler. In: XIII. Bitkisel İlaç Hammaddeleri Toplantısı, pp 221–229, Marmara University, Istanbul, Turkey
Hylands P, Stuart M (1981) The medicinal uses of plants. In: Stuart M (ed) The encyclopedia of herbs and herbalism. Crescent Books, New York, pp 47–65
Joudi L, Bibalani GH (2010) Exploration of medicinal species of Fabaceae, Lamiaceae and Asteraceae families in Ilkhji region, eastern Azerbaijan Province (northwestern Iran). J Med Plants 4(11):1081–1084
Kidane B, Van Der Maesen LJG, Van Andel T, Asfaw Z (2014) Ethnoveterinary medicinal plants used by the Maale and Ari ethnic communities in southern Ethiopia. J Ethnopharmacol 153(1):274–282
Labeyrie V, Rono B, Leclerc C (2014) How social organization shapes crop diversity: an ecological anthropology approach among Tharaka farmers of Mount Kenya. Agric Hum Values 31(1):97–107
Leman DJ (2006) Sustainable wild collection of medicinal and aromatic plants: development of an international standard. In: Bogers RJ, Craker LE, Lange D (eds) Medicinal and aromatic plants: agricultural, commercial, ecological, legal. Pharmacol Soc Aspects Springer, Heidelberg, pp 97–107
Leonti M, Casu L (2013) Traditional medicines and globalization: current and future perspectives in ethnopharmacology. Front Pharmacol 25:4–92
Leporatti M, Ghedira K (2009) Comparative analysis of medicinal plants used in traditional medicine in Italy and Tunisia. J Ethnobiol Ethnomed 5:31
Leporatti ML, Ivancheva S (2003) Preliminary comparative analysis of medicinal plants used in the traditional medicine of Bulgaria and Italy. J Ethnopharmacol 87(2–3):123–142
Łuczaj Ł (2010) Changes in the utilization of wild green vegetables in Poland since the 19th century: a comparison of four ethnobotanical surveys. J Ethnopharmacol 128(2):395–404
Madaleno IM (2010) Local use of front and backyard medicinal species – a comparative study in six Latin American cities. WIT Trans Ecol Environ 129:637–646
Maffi L (2005) Linguistic, cultural and biological diversity. Annu Rev Anthropol 29:599–617
Mamedov NA, Craker LE (2012) Man and medicinal plants: a short review. In: Ghaemghami L, et al. Proc. IS on Medicinal and aromatic plants, IMAPS 2010 and "History of Mayan Ethnopharmacology", IMAPS 2011, Acta Hort. 964, ISHS 2012
Mattalia G, Quave CL, Pieroni A (2013) Traditional uses of wild food and medicinal plants among Brigasc, Kyé, and Provençal communities on the western Italian alps. Genet Resour Crop Evol 60(2):587–603
McNeely JA (2000) Cultural factors in conserving biodiversity. In: Wilkes Y, Tillman A, Salas H, Grinter M, Shaoting T (eds) Links between cultures and biodiversity, Proceedings of the

Cultures and Biodiversity Congress. Yunnan Science and Technology Press, China, Kunming, pp 128–142
Mehtiyeva N, Zeynalova S (2008) Medicinal and aromatic plants of Azerbaijan. In: Ethnopharmacology. Encyclopedia of Life Support Systems (EOLSS), New York
Menendez-Baceta G, Aceituno-Mata L, Reyes-García V, Tardío J, Salpeteur M, Pardo-De-Santayana M (2015) The importance of cultural factors in the distribution of medicinal plant knowledge: a case study in four Basque regions. J Ethnopharmacol 161:116–127
Milton K (1996) Environmentalism and cultural theory. Exploring the role of anthropology in environmental discourse. Routledge, London and New York https://books.google.com.tr/books
Mir-Babayev NF, Waigh RD (1997) Plants of the republic of Azerbaijan with potential medicinal applications. Pharm Biol 31(1):47–54
Mirzeyev PS (1972) Nakhchivan MSSR-in Agroiglim Sedzhijjesi. Azerneshr, Baku
Moerman DE, Jonas WB (2002) Deconstructing the placebo effect and finding the meaning response. Ann Intern Med 136:471–476
Novruzova LA, Maharramov SH, Ibadullayeva SJ (2015) Management of gastroenterostomy illness with herbs in veterinary practice in Nakhchivan Autonomous Republic (Azerbaijan). Int J Vet Sci 4(4):224–226
Ozturk M, Altundağ E, Gücel S (2012) Medicinal and aromatic plants (Turkey). In: Ethnopharmacology. Encyclopedia of Life Support Systems (EOLSS), New York
Ozturk M, Altay V, Altundağ E, Gücel S (2016) Halophytic plant diversity of unique habitats in Turkey: salt mine caves of Çankırı and Iğdır. In: Khan MA et al (eds) Halophytes for food security in dry lands. Academic press is an imprint of Elsevier, Cambridge, MA
Pieroni A, Quave CL (2005) Traditional pharmacopoeias and medicines among Albanians and Italians in southern Italy: a comparison. J Ethnopharmacol 101(1–3):258–270
Pieroni A, Giusti ME, Quave CL (2011) Cross-cultural ethnobiology in the western Balkans: medical ethnobotany and ethnozoology among Albanians and Serbs in the Pešter plateau, Sandžak, south-western Serbia. Hum Ecol 39(3):333–349
Posey DA (1999) Cultural and spiritual values of biodiversity. A complementary contribution to the global biodiversity assessment. In: Posey da (ed) cultural and spiritual values of biodiversity. UNEP and Intermediate Technology Publications, London, pp 1–19
Quiroga R, Meneses L, Bussmann RW (2012) Medicinal ethnobotany in Huacareta. J Ethnobiol Ethnomed 8:29
Rates SMK (1991) Plants as source of new drugs. Toxicon 39(5):603–613
Rexhepi B, Mustafa B, Hajdari A, Rushidi-Rexhepi J, Quave CL, Pieroni A (2013) Traditional medicinal plant knowledge among Albanians, Macedonians and Gorani in the Sharr Mountains (republic of Macedonia). Genet Resour Crop Evol 60(7):2055–2080
Saslis-Lagoudakis CH, Hawkins JA, Greenhill SJ, Pendry CA, Watson MF et al (2014) The evolution of traditional knowledge: environment shapes medicinal plant use in Nepal. Proc R Soc B Biol Sci 281(1780):20132768
Seyidova H, Hüseyin E (2012) Macrofungi of Nakhchivan (Azerbaijan) Autonomous Republic. Turk J Bot 36(6):761–768
Singh RJ, Lebeda A, Tucker AO (2012) Medicinal plants-nature's pharmacy. In: Singh R (ed) Genetic resources, chromosome engineering and crop improvement: medicinal plants. CRC Press, LLC, Taylor & Francis, Boca Raton, FL, pp 13–51
Sõukand R, Kalle R (2010) Herbal landscape: the perception of landscape as a source of medicinal plants. Trames J Humanit Soc Sci 14(3):207–226
Sõukand R, Pieroni A (2016) The importance of a border: medical, veterinary, and wild food ethnobotany of the Hutsuls living on the Romanian and Ukrainian sides of Bukovina. J Ethnopharmacol 185:17–40
Sõukand R, Quave CL, Pieroni A, Pardo-de-Santayana M, Tardío J, Kalle R et al (2013) Plants used for making recreational tea in Europe: a review based on specific research sites. J Ethnobiol Ethnomed 9:58. https://doi.org/10.1186/1746-4269-9-58

Taghipour H, Mosaferi M (2009) Characterization of medical waste from hospitals in Tabriz, Iran. Sci Total Environ 407(5):1527–1535

Whistler WA (1985) Traditional and herbal medicine in the Cook Islands. J Ethnopharmacol 13:239–280

WHO (1993) Guidelines on the conservation of medicinal plants. World Health Organization (WHO), Gland

Wong W (1976) Some folk medicinal plants from Trinidad. Econ Bot 30(2):103–142

Zlatković BK, Bogosavljević SS, Radivojević AR, Pavlović MA (2014) Traditional use of the native medicinal plant resource of Mt. Rtanj (eastern Serbia): ethnobotanical evaluation and comparison. J Ethnopharmacol 151(1):704–713

The Utilization and Conservation of Plants of Medicinal Value by Local Traditional Medicinal Practitioners and the Associated Indigenous Knowledge in Dawuro Zone of Ethiopia: Northeast Africa—An Ethnobotanical Approach

Moin Ahmad Khan, Mathewos Agize, Abraham Shonga, and Asfaw Tora

Introduction

Background of the Study

Ethnobotany is the study of how people of a particular culture and religion make use of indigenous plants. Martin (1995) had defined ethnobotany as "local people's interaction with the natural environment: how they classify, manage and use plants available around them." From the beginning of humanity, indigenous people have developed their own locality-specific knowledge on plant use, management, and conservation (Cotton 1996). This complex knowledge, systems of beliefs, and practices generally termed as indigenous or traditional knowledge develop and change with time and space, with change of resources and culture.

Indigenous knowledge has developed because of human interaction with their environment. In this view, ethnobotanical studies are useful in documenting, analyzing, and communicating knowledge and interaction between biodiversity and human society, and how diversity in nature is used and influenced by human activities (Martin 1995; Cotton 1996; Balick and Cox 1996).

Since ancient times, plants have been indispensable sources of both preventive and curative traditional medicine preparations for human beings and livestock. According to Farnsworth and Soejarto (1991) 70–80% of the world's population uses plants to solve basic medical problems. Furthermore, approximately 99% of veterinary care in developing countries is based on the use of plant extracts (Letchamo and Craker 1996). It was estimated that 25,000–75,000 species of higher plants have been used in traditional medicine worldwide (Farnsworth 1985).

M. A. Khan (✉) · M. Agize · A. Shonga · A. Tora
Department of Biology, Wolaita Sodo University, Wolaita Sodo, Ethiopia

M. Ozturk, K. R. Hakeem (eds.), *Plant and Human Health, Volume 1*,
https://doi.org/10.1007/978-3-319-93997-1_7

Ethiopia is a country with various types of climatic, topographic, soil features and different altitudes. This makes the country to have a rich and diverse source of flora and fauna. With the direct and intimate relationships as well as indispensable dependency of human beings upon plants for their livelihood, plants provide multiple and diverse uses for indigenous societies (Amenu 2007). Various plants and plant products have been closely associated with many social cultures, customs, and mythological rituals such as personal decoration (e.g., cosmetics and tattooing) and entertainment (e.g., musical instruments), arts and crafts, and even magic-religious beliefs (Jain 1986). Plants have significant medicinal value both in developing and developed countries. Over 75% of the world's rural people rely on traditional herbal medicine. About half of the world's medicinal compounds are still extracted from plants (Frankel et al. 1995). Moreover, since modern drugs are too expensive particularly in rural areas, over 95% of the Ethiopian population is not satisfied with the services and benefits of modern medicine. Due to this reason, medicinal plants play a significant role in the health care of local people.

It is reported that nearly 80% of the population in Ethiopia use plant-based traditional medicine as their primary healthcare system (Abebe 2001a, b) and this wide use could be mainly attributed to the fact that it makes use of locally available plant resources (Abebe and Ayehu 1993). The majority of medicinal plants, with exceptions, are harvested from wild habitats, which are currently under great threat (WCMC 1992). There are reports indicating that many potentially useful plants are disappearing throughout the world, and Ethiopia is not exceptional. This is attributed to conversion and destruction of habitats, overexploitation, and use of destructive harvesting techniques (IUCN 1993; Yineger 2005; Giday 2007).

The study of Ethiopian medicinal plants has not been realized as fully as that of India or other traditional communities elsewhere. Some organized ethnobotanical studies in the recent past in different parts of Ethiopia have been carried out by Getahun (1976), Jansen (1981), Taddese (1986), Abebe (1986), Taddese and Demissew (1992), Abebe and Ayehu (1993), Abbink (1995), Giday (1999), Tamene (2000), Hunde (2001), Asfaw (2001), Fullas (2001), Birhanu (2002), Balemie et al. (2004), Lulekal (2005), Gebre (2005), Amenu (2007), Giday (2007), and Teklehaymanot and Giday (2007). However, considering the country's varied flora and sociocultural diversity, these studies are not enough and have covered only a few areas of the country. Since the plant-based medicinal healing systems vary across cultures in different parts of the country, more studies are required on this aspect. Like most other parts of the country, there is no such ethnomedicinal research and documentation carried out in Dawuro zone, in Southern Ethiopia. This study therefore is aimed at documentation of indigenous knowledge on use and conservation of medicinal plants by the people of Dawuro and assessment of the existing threats to medicinal plants.

Statement of the Problem

There is a wide gap in our knowledge about ethnobotanical data and information from various parts of Ethiopia although we have rich and diverse ethnolinguistic groups throughout the country. Complete data collection, identification, and documentation of ethnobotanical works have not yet been made in the study area.

Biodiversity loss due to natural and anthropogenic factors, distribution and undermining of indigenous knowledge, and traditional practices and cultures by the younger generation are becoming evident due to limited integration of traditional practices with modern science in the study area. Traditional home gardening in the conservation and management of medically important plant species by indigenous people of the study area is recognized through such practices but varying from area to area. However, some important medicinal plants have been planted like live fence and observed as fragments in and around the farmlands. What is more is that most of the natural vegetation and forests of the study area are almost lost by human impact and hence monoculture practices are becoming common phenomena.

Objectives

General Objective

The general objective of the study is to assess and record medicinal plants and associated indigenous knowledge of the people in Dawuro zone, Southern Ethiopia.

Specific Objectives

The specific objectives are:

- To gather, record, and document indigenous knowledge of the people on medicinal plants
- To collect, identify, and document traditional medicinal plants used in the study area for the treatment of human and livestock health problems
- To determine the most popular medicinal plants used in the study area
- To analyze factors, if any, contributing to depletion or conservation of medicinal plants and/or associated knowledge in the study area
- To provide recommendations that would contribute to the development of strategies for conservation and sustainable management of medicinal plants in the study area
- To contribute to the ongoing efforts towards building the ethnobotanical database of Ethiopia in order to facilitate further actions in the management and utilization of medicinal plants

Review of Literature

Ethnobotany

The term "ethnobotany" was coined by J. W. Hershberger in 1895 to indicate plants used by the Aboriginals: from "ethno," study of people, and "botany," study of plants. Ethnobotany is considered as a branch of ethnobiology. It deals with the study and evaluation of plant-human relations in all phases and the effect of plant environment on human society (Sharma and Kumar 2011). It is the scientific study of the relationships that exists between people and plants. Since the beginning of civilization, people have used plants as medicine. Perhaps since Stone Age, plants are believed to have healing powers on man. Ancient Vedas dating back between 3500 B.C. and 800 B.C. reveal many references on medicinal plants (Venkataswamy et al. 2010).

The field approach of study of ethnobotany plays a vital role because of the direct contact that can be established with the authentic information on the uses of plants both wild and cultivated. The wild plants in Indian folklore have been and are used to meet the various needs of the tribal and poor people. These plants are used for purposes of food, fodder, medicine, drugs, clothing, agricultural implements, hunting, narcotics, poison, gums, dyes, insecticides, etc. (Pareek and Trivedi 2011).

Medicinal Plants

Ethiopia is characterized by a wide range of ecological, edaphic, and climatic conditions that account for the wide diversity of its biological resources in terms of both flora and fauna (Jansen 1981). It is well known that traditional medicines are widely used especially in the low-income rural parts of the country. It is reported that nearly 80% of the population in the country use plant-based traditional medicines as their healthcare system. The wide utilization of plant-based traditional health care is mainly attributed to the fact that it makes use of locally available plant resources. The majority of medicinal plants, with few exceptions, are harvested from wild habitats, which are currently under great threat.

Medicinal plants play a vital role in providing health care to human beings since the dawn of civilization. It is evident that the Indian people have tremendous passion for medicinal plants and they use them for a wide range of health-related applications. The demand for medicinal plants is increasing in both developing and developed countries and the bulk of their material trade is still from wild harvested plants (Pareek and Trivedi 2011).

The World Health Organization (WHO) has estimated that 80% of the populations of developing countries still rely on traditional medicines, mostly plant drugs, for their primary healthcare needs. Demand for medicinal plant is increasingly felt, in both developing and developed countries due to growing needs of natural products being nontoxic and bereft of side effects, apart from availability at affordable prices. The medicinal plant sector has traditionally occupied a pivotal position in the socio-cultural, spiritual, and medicinal areas of rural and tribal families (WHO 2002).

Indigenous Knowledge

Many aboriginal cultures have retained traditional knowledge concerning the medicinal utility of the native flora. Tribal communities living in biodiversity-rich areas possess a wealth of knowledge on the local utilization and conservation of food and medicinal plants. Today there is a realization to preserve the enormous wisdom, traditional knowledge, and also cultures associated with them. Not only the flora and fauna have been protected but also the knowledge database often treasured in the memories of traditional healers. The knowledge of medicinal plants has been accumulated in the course of many centuries based on different medicinal systems (Venkataswamy et al. 2010).

Folk medicines, mainly based on plants, enjoy a respectable position today, especially in the developing countries, where modern health service is limited. Safe, effective, and inexpensive indigenous remedies are gaining popularity among the people of both urban and rural areas (Pareek and Trivedi 2011). WHO (2003) defined traditional medicine as health practices, approaches, knowledge, and beliefs incorporating plant-, animal-, and mineral-based medicines, spiritual therapies, manual techniques, and exercises applied to treat, diagnose, and prevent illnesses or maintain well-being.

Utilization, Conservation, Management, and Threats to Traditional Medicinal Plants

People are dependent upon their surrounding environment for all of their needs. They use many wild species of plants for food, medicine, clothing and shelter, fuel, fiber, income generation, and fulfilling of cultural and spiritual needs throughout the world (Asfaw 2001). Ethiopia's traditional medicine as elsewhere in Africa is facing the problems of continuity and sustainability (Kelbessa et al. 1992). The primary cause of this problem is loss of taxa of medicinal plants, loss of habitats of medicinal plants, and loss of indigenous knowledge. Some studies have shown that most of the medicinal plants utilized by Ethiopian people are harvested from wild habitats (Giday 1999; Asfaw 1999). And hence, this aggravates the rate of loss of taxa with related indigenous knowledge and loss widely occurring in medicinal plant species.

Manmade and natural causes are the two main sources of treats to medicinal plants. Rapid increase in population, need for fuel, urbanization, timber production, overharvesting, destructive harvesting, invasive species commercialization, degradation, agriculture expansion, and habitat distraction are human-caused threats to medicinal plants. Likewise, natural causes include recurrent drought, bushfire, diseases, and pest outbreaks (Kelbessa et al. 1992). As elsewhere, in Ethiopia, the problem is also manifested in Dawuro zone due to the above-mentioned factors. As is seen around the world, as human populations grow, there is a great impact of human activities on vegetation as well as on the quality of human life.

Conservation is defined as the sustainable use of biological resources. The concept of sustainability is now seen as the guiding principle for economic and social development, particularly with reference to biological resources. According to Asfaw (2001), medicinal plants are considered to be at conservation risk due to overuse and destructive harvesting (roots and bark collection). In a broad sense, conservation is achieved through in situ and ex situ means. In situ conservation is conservation of species in their natural habitat. Some traditional plants have to be conserved in situ due to difficulty for domestication and management (Asfaw 2001). Moreover, some plants fail to produce the desired amount and quantity of the active principles under cultivation out of their natural habitats. Medicinal plants can also be conserved by ensuring and encouraging their growth in special places, as they have been traditionally (Asfaw 2001). This can be possible in places of worship (churches, Mosques, graveyards, etc.), sacred groves, farm margins, river banks, road sides, live fences of gardens, and fields. According to Asfaw (2001), medicinal plants can be conserved using appropriate conservational methods in gene banks and botanical gardens. As the earth's population continues to grow, more resources are in demand. Thus, it is understood that an increase in population is demanding more resources, but with careful management of our natural resources a sustainable balance can be achieved. According to Cunningham (1993), sustainable management of traditional medicinal plant resources is important, not only because of their value as a potential source of new drugs, but also due to reliance on traditional medicine for health.

The diversity of plants in Ethiopia is on the process of erosion due to anthropogenic pressure like habitat destruction and deforestation by commercial timber production and for agriculture which harbor useful medicinal plants over the past several decades (Demisse 2001). Conservation of medicinal plant resource is a critical ecological, cultural, and economic issue (Vanon and Bich 2001). In order to achieve this, Ethiopia has policies and strategies that support the development and utilization of plant resources in a sustainable manner. The policies are reflected under various sectors including environmental protection, development of the natural resources, and diversification of the domestic and export commodities.

The country also has developed policy and a guideline for intellectual property rights protection of traditional medicine (Bekelle 2007). The policies are in line with the convention on biological diversity (CBD) which was adopted at the earth summit in Rio de Janeiro, Brazil, in June 1992, and entered into force on December 1993. CBD provides the legal framework for biodiversity conservation. Moreover, documenting the indigenous knowledge through ethnobotanical studies is important for the conservation and utilization of biological resources particularly traditional medicinal plants (Sirvastava 2000). These plants play an important role in the biodiversity of the region and have great conservation value for global biodiversity. Meanwhile, the planning for natural resource development should continue by establishing close dialogue and communication with indigenous peoples using ethnobotanical approaches. This will also ensure local people participation in future management, and avoid the adverse impact on the local people and the environment that might be caused otherwise (Pei 1991). According to Cunningham (1993), there are some conservation measures that have been undertaken around the world aimed at protecting threatened medicinal plant species from further destruction including in situ and ex situ conservation.

During the last few years, the habitats of medicinal plants all across the world have been under pressure due to urbanization and exploitation of raw materials by pharmaceutical companies. In order to develop strategies for the conservation of medicinal plants, it is necessary to collect traditional knowledge of the plants, their distributions, and any traditional conservation practices.

Study Area

Geographical Location

The research area, Dawuro zone, is located at 6.59°–7.34° N of latitude and 36.68°–37.52° E of longitude and at an altitudinal range between 550 and 2820 m above sea level in Southern Nations, Nationalities, and Peoples Region (SNNPR). It is one of the 14 zones in the region and bounded with Hadiya zone in the North, Kembata & Tembaro zone in the Northeast, Wolayta zone in the East, Gamo Gofa zone in the South, and Konta special woreda in the West within SNNPR and Jimma Zone in Oromia Region. It is also found in between Omo river from North to South and Gojeb river from Northwest to North. Tarcha is the main town about 507 km Southwest of Addis Ababa across Shashemene and Wolayta, 282 km away from Awassa, town of SNNPR, and 140 km from Jimma. It has an area of 466,082 ha. It has a population of about 586,005 people according to the projected CSA final report of 2005 E.C. Out of the five woredas of the zone, the study area covers four Woredas, Essera, Tocha, Loma, and Gena Bosa (except Mareka woreda which is bounded by the four woredas, at the center of zone) and one administrative town—Tarcha (Fig. 1).

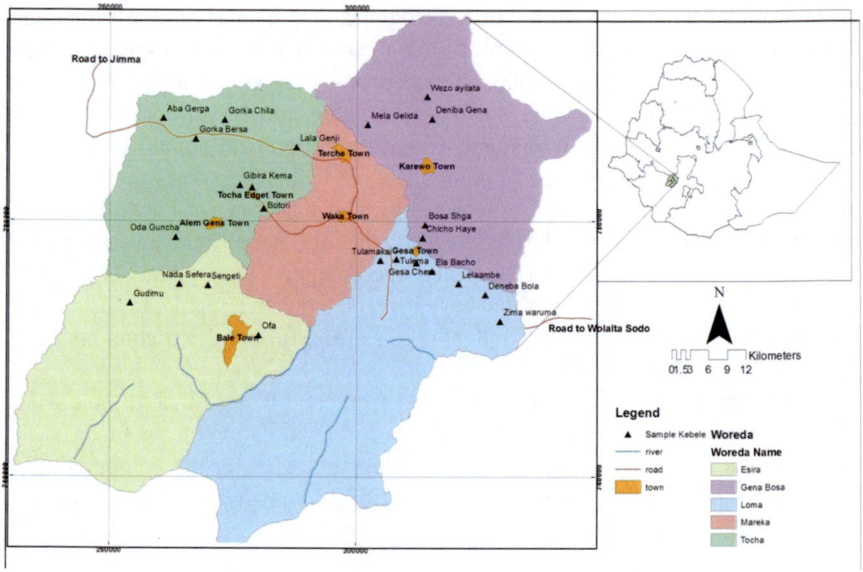

Fig. 1 Location of the study area in Dawuro zone and the sampling sites

Location, Agro Ecology, Soil, Climate, and Vegetation

Based on the 18 recently classified agroecological zones, the area consists of subhumid types of agroecological zones containing deciduous woodland with elevation 550–2820 m having *Boswellia papyrifera*, *Combretum molle, Terminalia brownii, Acacia senegal, Balanites aegyptiaca, Lannea fruticosa*, and others along the Omo and Gojeb river valleys (Agize et al. 2013). The vegetation varies from semidesert types in lowland areas to broad-leaved evergreen forest types in the highlands. Some of the dominant woody species in the lowlands of the study area include *Terminalia brownii, Piliostigma thonningii, Syzygium guineense* var. *guineense*, and *Acacia spp.* In highlands and at midaltitude, bamboo tree and eucalyptus tree are the dominant tree species; in addition, *Ensete ventricosum*, source of the staple food in the area, provides vegetation cover and green scenery.

The area has rugged topography and the terrain of the land is sloppy in nature. The mountains and high plateaus have cool temperature moderated by the altitude and abundant rainfall. In contrast, the lowlands near the Omo and Gojeb rivers are dry and hot. The majority of the population is engaged in subsistence agricultural activities combining both crop production and livestock rearing (mixed farming). In general the population in Dawuro zone shares a cultural identity of what is referred to in Ethiopia as “Enset Culture Complex,” which is commonly characterized by high population density sustained by high yields of enset from small plot of land.

The major economic activity of the people in the project area is mixed agriculture (rearing of animals and growing crops). These woredas have ragged topography comprising lowlands, middle, and highland agroecologies. The project areas are naturally gifted lands with diverse topography, diverse climate, and varied ecology. Hence, it is provided with a wide range of flora and fauna diversity in wildlife and botanical resources. The Chebera-Churchura national park, which includes the Essera and Tocha woredas, is a natural habitat for many wild animals in the area.

The study area in the Dega receives rainfall almost throughout the year, for 9 months, and heavy rain comes between June and September. The mean annual rainfall is 1705.4 mm at Gasa Chere Station while in Tarcha Station the annual mean is 1424.9 mm. The maximum and minimum mean annual temperatures are (22.3 °C and 12.4 °C) and (29.3 °C and 16.8 °C) in Gasa Chere and Tarcha Stations, respectively (Agize et al. 2013).

People in the study area are the Dawuro people and have unique cultural practices and social structure. The word "Dawuro" means impregnable, powerful, and heroic people. Dawuro belongs to the family of the Omotic peoples in the Southern Ethiopia. The language, which is locally called "Dawuro thuwa or Dawuro qaalaa," uses a Latin script and it is classified as a dialect of the central Omotic languages along with Gofa, Gamo, Wolayta, Konta, and others (Anon 2005a, Agize et al. 2013).

Land Use and Management System

The study area is endowed with varied agro-climate. The soil is fertile which has great potential for agriculture. The rugged topography particularly causes difficulties in road, irrigation, and other infrastructure development. The steep slope mountain sides are not suitable for tilling and are vulnerable to soil erosion. The mountain areas are more suitable for livestock rearing. Cattle are indispensable for rural households both as a means of subsistence and source of cash income. In rural areas the number of cattle owned and enset cultivated by a household usually determine the wealth status of the household.

In general, the people are dependent on subsistence agriculture for their livelihoods and the majority of the populations are involved in this sector, whereas the nonagricultural sectors contribute a smaller amount to the livelihoods of the people. The main food crops grown in the area are enset, maize, taro, sweet potato, sorghum, millet, teff, pulses, and yam. Enset is the staple food in Dawuro, particularly in mid- and high-altitude areas, while maize is the most important crop in the lowlands. They have developed traditional medicinal practices cultivating common medicinal plants and use them effectively before they go to clinics or health centers.

The domestic animals include cattle, horses, mules, donkeys, sheep, goats, and poultry. In the *highlands*, herds remain at the holding or in the village throughout the year, tied with rope and cut and carry system of feeding. In the *midlands*, they have permanent settlement and their herds remain in the vicinity, and are sent to hillside. In the *lowlands*, they have permanent place and send their herds tend with herdsman. Herds in the study area live with farmers in the home but in some tin-roofed houses they are left alone in other houses with herdsman or others.

Land forms, rivers, gorges, and mountains have cultural importance for the people of the study area. People living there have long years of experience of interaction with each other and the natural resources of their surroundings.

Preliminary Survey

Preliminary survey of the study area was conducted in April 2012 specifically from April 4 to 9, 2012. During this survey, information about the physical features of the study area was collected.

From five districts, four districts, namely, Essera, Tocha, Gena Bosa, and Loma, were selected by the help of zonal administrative office and agricultural department of Dawuro zone. From these, a total of 22 kebeles were selected purposefully for ethnobotanical data collection based on availability of traditional healers and different agroclimatic zone (*Dega*, *Woina Dega*, and *Kola*) of the region identified with the assistance of woreda and local authorities, elders, and knowledgeable persons to gather diversified information on the management, use, and conservation of medicinal plants.

Methods of Data Collection

Ethnobotanical data were collected using purposive sampling. This sampling technique was preferred because the study focuses on specific issues that were gathered from the most knowledgeable representatives of the society. The full names and residential addresses of traditional healers residing in the 22 kebeles of the four districts selected were exhaustively identified and registered with the help of local administrators, local people, and field assistants.

Individuals who were selected to know and practice at least four/five medicinal plant species were considered as traditional healers in this study. A total of 91 respondents from the entire study sites who were identified as healers were interviewed. Semi-structured interviews were then employed and observations made to collect ethnomedicinal data with the help of local people and field assistants. Data on human and livestock diseases treated, local names of plants used, degree of management (wild/cultivated), status, parts used, methods of preparation, routes of administration, noticeable adverse effects of remedies, indigenous knowledge transfer, other uses of the ethnomedicinal plant species, existing threats to these species, and traditional conservation practices were gathered during the interviews (Table 1). The collected specimens were then pressed, dried, and identified through the flora of Ethiopia and Eritrea in the National Herbarium of Addis Ababa University.

Table 1 Traditional healers of the study area

Serial No.	Name of informant	Woreda	Kebele	Sex	Age	Marital status	Occupation	Religion	Educational background	Residence period in the area	Socioeconomic activity
1	Abate Gebabaw	Gena Bosa	Mela Galeda	M	35	Married	Farmers	Catholic	7	Since birth	Mixed farming
2	Abate Goteto	Tocha	Boka	M	45	Married	Farmers	Protestant	4	Since birth	Mixed farming
3	Abate setegn Meshesha	Tocha	Boteri	M	65	Married	Farmers	Orthodox	0	Since birth	Mixed farming
4	Abayeneh Abete	Tocha	Medehanialem	M	25	Single	Farmers	Orthodox	12	Since birth	Mixed farming
5	Abayneh Beyene	Gena Bosa	Dilamo Mareka	M	25	Married	Farmers	Protestant	0	Since birth	Mixed farming
6	Abebe Gobena	Loma	Gesa Chare	M	18	Single	Farmers	Protestant	10	Since birth	Mixed farming
7	Abera Woja	Tocha	Boteri	M	77	Married	Farmers	Orthodox	0	Since birth	Mixed farming
8	Akalu Abay	Gena Bosa	Denba gena	M	40	Married	Farmers	Protestant	0	Since birth	Mixed farming
9	Alemu Tesema Abbeza	Tocha	Aba dahi	M	47	Married	Farmers	Protestant	0	Since birth	Mixed farming
10	Amachie Araro	Tocha	Gorika	M	20	Single	Farmers	Orthodox	0	Since birth	Mixed farming
11	Arebi Godato Kuyilu	Tocha	Boka	M	48	Married	Farmers	Protestant	8	Since birth	Mixed farming
12	Asefa Bancha Lorato	Loma	Ala Bacho	M	22	Married	Farmers	Protestant	4	Since birth	Mixed farming
13	Asefa Habete	Gena Bosa	Mela Galeda	M	45	Married	Farmers	Orthodox	8	Since birth	Mixed farming
14	Asfaw Choka	Gena Bosa	Mela Galeda	M	45	Married	Farmers	Protestant	8	Since birth	Mixed farming

(continued)

Table 1 (continued)

Serial No.	Name of informant	Woreda	Kebele	Sex	Age	Marital status	Occupation	Religion	Educational background	Residence period in the area	Socioeconomic activity
15	Assefa Mencho	Essera	Sengeti	M	35	Married	Farmers	Protestant	7	Since birth	Mixed farming
16	Ayele Alanche	Loma	Ala Bacho	M	50	Married	Farmers	Protestant	8	Since birth	Mixed farming
17	Ayele Atenafu	Gena Bosa	Bosa Shoga	M	45	Married	Farmers	Orthodox	5	Since birth	Mixed farming
18	Ayele Bedeke Barata	Loma	Zima Waruma	M	55	Married	Farmers	Catholic	6	Since birth	Mixed farming
19	Azalech Mekonnen	Gena Bosa	Bosa Shoga	F	19	Married	House wife	Protestant	10	Since birth	Mixed farming
20	Babulo Shungeto Atero	Tocha	Gibera	M	76	Married	Farmers	Orthodox	0	Since birth	Mixed farming
21	Bafa Adey	Loma	Lala ambie	M	40	Married	Farmers	Protestant	3	Since birth	Mixed farming
22	Balecha Tona Ayu	Loma	Zima Waruma	M	65	Married	Farmers	Catholic	3	Since birth	Mixed farming
23	Batisa Zema	Gena Bosa	Mela Galeda	m	62	Married	Farmers	Orthodox	0	Since birth	Mixed farming
24	Bekele Bala	Essera	Gudemo	M	46	Married	Farmers	Orthodox	0	Since birth	Mixed farming
25	Belayneh Babanto	Loma	Zima Waruma	M	60	Married	Farmers	Catholic	0	Since birth	Mixed farming
26	Belaynesh Debo	Gena Bosa	Bosa Shoga	M	40	Married	Farmers	Catholic	0	Since birth	Mixed farming
27	Beyene Gezemu	Gena Bosa	Chicho Hayu	M	32	Married	Gov. servant	Protestant	12 + 3	Since birth	Mixed farming
28	Beyenech Nigussie	Gena Bosa	Bosa Shoga	F	50	Married	House wife	Orthodox	0	Since birth	Mixed farming

29	Biramo Latibelu	Tocha	Boteri	M	45	Married	Farmers	Orthodox	0	Since birth	Mixed farming
30	Buntase Lugo	Loma	Ala Bacho	F	64	Married	House wife	Protestant	0	For the last 20 years	Mixed farming
31	Chaka chanaka	Tocha	Boteri	M	56	Married	Farmers	Protestant	0	Since birth	Mixed farming
32	Damene Kocho Machu	Tocha	Lala Genji	M	55	Married	Farmers	Protestant	0	Since birth	Mixed farming
33	Demessie Dara	Essera	Gudemo	M	56	Married	Farmers	Orthodox	0	Since birth	Mixed farming
34	Deneke Masana Chetu	Tocha	Lala Genji	M	26	Married	Farmers	Orthodox	2	Since birth	Mixed farming
35	Desta Bobicho Botore	Tocha	Gorika	M	28	Married	Farmers	Protestant	3	Since birth	Mixed farming
36	Desta Degefu	Gena Bosa	Chicho Hayu	M	50	Married	Farmers	Catholic	0	Since birth	Mixed farming
37	Duba Bakari	Tocha	Gibera	M	76	Married	Farmers	Orthodox	0	Since birth	Mixed farming
38	Dusha Chambure	Gena Bosa	Bosa Shoga	M	41	Married	Farmers	Catholic	0	Since birth	Mixed farming
39	Engidaw Cheneke	Gena Bosa	Chicho Hayu	M	45	Married	Farmers	Catholic	0	Since birth	Mixed farming
40	Esayas Beku	Loma	Ala Bacho	M	58	Married	Farmers	Protestant	4	Since birth	Mixed farming
41	Etenech Negash	Gena Bosa	Mela Galeda	F	30	Married	Farmers	Protestant	0	Since birth	Mixed farming
42	Filmon Choramo	Gena Bosa	Wozo Haylaxa	M	25	Married	Farmers	Protestant	0	Since birth	Mixed farming
43	Ganamo Gashaw	Gena Bosa	Mela Galeda	M	35	Married	Farmers	Protestant	3	Since birth	Mixed farming

(continued)

Table 1 (continued)

Serial No.	Name of informant	Woreda	Kebele	Sex	Age	Marital status	Occupation	Religion	Educational background	Residence period in the area	Socioeconomic activity
44	Getahun Kebede Gerero	Tocha	Medehanialem	M	29	Married	Farmers	Orthodox	12	Since birth	Mixed farming
45	Girma Fola	Loma	Ala Bacho	M	37	Married	Farmers	Protestant	0	Since birth	Mixed farming
46	Goba Barata	Essera	Sengeti	M	35	Married	Farmers	Cultural Christian	0	Since birth	Mixed farming
47	Hadaro Halabo	Tocha	Gorika	M	50	Married	Farmers	Orthodox	0	Since birth	Mixed farming
48	Haile Charku toga	Tocha	Boteri	M	55	Married	Farmers	Protestant	0	Since birth	Mixed farming
49	Haile Haringo	Essera	Ofa Wadi	M	70	Married	Farmers	Cultural Christian	8	Since birth	Mixed farming
50	Herano Sagaro	Gena Bosa	Wozo Haylaxa	M	40	Married	Farmers	Protestant	0	Since birth	Mixed farming
51	Jorga Aday	Loma	Lala ambie	M	60	Married	Farmers	Protestant	2	Since birth	Mixed farming
52	Kaleb Golu Godaro	Loma	Ala Bacho	M	38	Married	Farmers	Protestant	7	Since birth	Mixed farming
53	Kampasha Lembu	Gena Bosa	Mela Galeda	M	60	Married	Farmers	Protestant	0	Since birth	Mixed farming
54	Kampashe Karesso	Tocha	Gorika	M	45	Married	Farmers	Protestant	0	Since birth	Mixed farming
55	Kebede Kema	Gena Bosa	Chicho Hayu	M	75	Married	Farmers	Orthodox	0	Since birth	Mixed farming
56	Kebede Kerchu Ayanu	Tocha	Gibera	M	55	Married	Farmers	Cultural Christian	0	Since birth	Mixed farming
57	Kedir Goba Arimo	Tocha	Gorika	M	65	Married	Farmers	Orthodox	0	Since birth	Mixed farming

58	Konbara Muka Amba	Tocha	Gorika	M	35	Married	Farmers	Orthodox	0	Since birth	Mixed farming
59	Malalo Mengesha	Tocha	Gorika	M	18	Single	Farmers	Protestant	3	Since birth	Mixed farming
60	Mamo Lafebo Lekemengo	Essera	Neda	M	22	Married	Farmers	Protestant	8	Less than ten	Mixed farming
61	Mamo Maga	Tocha	Gibera	M	40	Married	Farmers	Protestant	0	Since birth	Mixed farming
62	Matiwos Mesele	Gena Bosa	Chicho Hayu	M	27	Married	Farmers	Catholic	Diploma	Since birth	Mixed farming
63	Mekedese Karesu	Tocha	Lala Genji	F	29	Married	House wife	Protestant	6	Since birth	Mixed farming
64	Mekuria Cheba	Gena Bosa	Mela Galeda	M	68	Married	Farmers	Orthodox	0	Since birth	Mixed farming
65	Mengistu Mogoro	Gena Bosa	Wozo Haylaxa	M	23	Single	Administrator	Protestant	10 + 3	Since birth	Mixed farming
66	Merkineh Mesele	Gena Bosa	Chicho Hayu	M	28	Married	Farmers	Catholic	0	Since birth	Mixed farming
67	Mitiku Ayene	Gena Bosa	Dilamo Mareka	M	35	Married	Farmers	Adventist	10	Since birth	Mixed farming
68	Mulatu Gebeyehu	Gena Bosa	Chicho Hayu	M	45	Married	School administrator	Protestant	Degree	Since birth	
69	Nega Tegegn	Tocha	Medehanialem	M	27	Married	Administrator	Orthodox	12	Since birth	Mixed farming
70	Okanto Oshu	Loma	Gesa Chare	M	35	Married	Farmers	Protestant	5	Since birth	Mixed farming
71	Oshu Uro	Tocha	Boteri	M	50	Married	Farmers	Orthodox	0	Since birth	Mixed farming
72	Shagire Bereda Bena	Tocha	Gibera	M	60	Married	Farmers	Orthodox	0	Since birth	Mixed farming

(continued)

Table 1 (continued)

Serial No.	Name of informant	Woreda	Kebele	Sex	Age	Marital status	Occupation	Religion	Educational background	Residence period in the area	Socioeconomic activity
73	Shamena Shasho	Essera	Ofa Wadi	M	87	Married	Farmers	Cultural Christian	0	Since birth	Mixed farming
74	Shirko Ashango	Loma	Denba bola	M	55	Married	Farmers	Catholic	0	Since birth	Mixed farming
75	Simion Debancho	Loma	Denba bola	M	55	Married	Farmers	Catholic	4	Since birth	Mixed farming
76	Tafese Chambura	Gena Bosa	Bosa Shoga	M	55	Married	Farmers	Orthodox	0	Since birth	Mixed farming
77	Tamerat Tanga	Essera	Neda	M	58	Married	Farmers	Protestant	0	Less than ten	Mixed farming
78	Taye Keshamo Adulo	Tocha	Gorika	M	23	Married	Farmers	Orthodox	6	Since birth	Mixed farming
79	Taye Shiferaw	Tocha	Boka	M	52	Married	Farmers	Protestant	0	Since birth	Mixed farming
80	Tayebela deneke	Gena Bosa	Chicho Hayu	M	28	Married	Farmers	Catholic	3	Since birth	Mixed farming
81	Tefera Belate	Essera	Gudemo	M	42	Married	Farmers	Orthodox	4	Since birth	Mixed farming
82	Tesema Otoro	Gena Bosa	Denba gena	M	78	Married	Farmers	Adventist	0	Since birth	Mixed farming
83	Tesema taddesse awashi	Tocha	Medehanialem	M	55	Married	Farmers	Orthodox	0	Since birth	Mixed farming
84	Teshome Ashenafi Anjulo	Loma	Zima Waruma	M	52	Married	Farmers	Catholic	0	Since birth	Mixed farming

85	Ute Anato	Loma	Tulema Tama	M	35	Married	Farmers	Protestant	4	Since birth	Mixed farming
86	Wajebo Walane	Tocha	Lala Genji	M	40	Married	Farmers	Protestant	0	Since birth	Mixed farming
87	Wodaje Minita	Essera	Sengeti	M	37	Married	Farmers	Orthodox	8	Since birth	Mixed farming
88	Woju Salato samara	Tocha	Medehanialem	M	62	Married	Farmers	Orthodox	0	Last 20 years	Mixed farming
89	Worabo Utha	Loma	Tulema Tama	M	35	Married	Farmers	Catholic	10	Since birth	Mixed farming
90	Yigezu Tona	Essera	Neda	M	92	Married	Farmers	Adventist	8	Last 10 years	Mixed farming
91	Zenebech Mandoye	Tocha	Gibera	F	35	Married	House wife	Protestant	0	Since birth	Mixed farming

Data Analysis

Facilities in MS Excel spreadsheet were utilized to make simple calculations, and determine proportions. Ethnobotanical data were entered into Excel spreadsheet and summarized using descriptive statistics. The spreadsheet data filter facility was employed to determine frequencies of citations so as to identify the most common ailments in the study area that popularly used medicinal plant species and multipurpose plant species, and to determine proportions of different variables like growth forms, source of collection, degree of scarcity, plant part used, methods of preparation, and threatening factors.

Medicinal Plants of the Study Area

Diversity of Medicinal Plants in the Study Area

A total of 216 medicinal plant species distributed in 69 families were collected and identified. *Asteraceae* is the most frequently used family containing 31 species and accounts for 44.93%. *Fabaceae* and *Lamiaceae* follow it by covering *39*.13% and 26.06% families having 27 and 18 species, respectively. The scientific name, families, growth form, and other information from the data gathered in the different sources are summarized in Table 2. Significant numbers of medicinal plants were documented in this ethnobotanical study when compared to similar studies in other parts of the country, Ethiopia. The most frequently and easily accessible known traditional medicinal plants were found in the family *Asteraceae*. There was a similar report in *Asteraceae* to Wonago (Mesfin 2007) and Wolayta (Seta 2013) and to Loma and Gena Bosa (Agize et al. 2013). It disagrees with a report (*Fabaceae)* to Ejaji (Amenu 2007) and Gimbi (Tolasa 2007). It may be due to the fact that collections and traditional healers interviewed were from highland and midland.

Out of 216 species 169 (78.24%) species were wild, while 36 (16.67%) and 11(5.09%) species were cultivated wild relatives and semi-wild, respectively (Table 3). Most of the traditional healers of the area collect their medicinal plants from the wild. Similar studies elsewhere, for example, Ejaji 78.7% (Amenu 2007), Loma and Gena Bosa area 57.9% (Agize et al. 2013), Konta 74.6% (Hailemariam 2007), Kafficho 74% (Awasa et al. 2007), Wonago 69.1% (Mesfin 2007), and Konso 55.7% (Gebre 2005), showed a similar trend. The significant number of medicinal plants documented may be due to either the large area covered (four woredas) or the diversity of indigenous knowledge to use different plants to treat different types of diseases. In other words, traditional practices and various cultural and seasonal restrictions of collecting medicinal plants have contributed to the management and conservation of diversified and rich medicinal plants compared to others.

Table 2 Medicinal plants of the study area

Scientific name	Family	Local name	Growth form	Source	Part used	Scarcity	Preparation and application	Route of administration	Disease treated	Treatment for	Voucher no.
Abrus precatoritus	Fabaceae	Badaluwa	Climber	Wild	Seed	Rare	Crushed and concocted with *Maesa lanceolata* and taken	Oral	Hepatitis/liver problem	Human	Dawro149
Acalypha villicaulis	Euphorbiaceae	Wak'ak'uwa	Herb	Wild	Root	Rare	Crushed and mixed with water and taken	Oral	Hepatitis/liver problem; babesios	Animal and human	Dawro171
Agarista salicifolia	Ericaceae	C'ank'uwa	Tree	Wild	Shoot	Rare	Crushed/decocted and mixed with water	Oral	Babesios	Animal	Dawro251
Ageratum conyzoides	Asteraceae		Shrub	Wild	Leaf	Rare	Crushed and applied through nose	Dermal	Eye disease	Human	Dawro173
Ajuga integrifolia var. alba	Lamiaceae	c'amiashiya	Herb	Wild	Leaf	Plenty	Powder mixed and a cup of it is taken	Oral	Colic pain	Human	Dawro139
Albiza schinperiana	Fabaceae	C'aattaa	Tree	Wild	Root	Less plenty	Crushed and taken	Oral	Snake bite	Human	Dawro205
Allium sativum	Alliaceae	Tumuwa	Herb	Cultivate	Leaf, stem	Plenty	Eat the parts with other food	Oral	Stomachache, malaria, and others	Human	Dawro26
Alysicarpus ferrugineus	Fabaceae	Warechiya	Climber	Wild	Root	Less plenty	Crushed/concocted with sura and taken	Oral	Snake bite	Human	Dawro339
Annona senegalensis	Annonaceae	Monok'uwa	Tree	Wild	Stem	Rare	Decocted and taken as two jugs	Oral	Abdominal pain	Animal	Dawro100

(continued)

Table 2 (continued)

Scientific name	Family	Local name	Growth form	Source	Part used	Scarcity	Preparation and application	Route of administration	Disease treated	Treatment for	Voucher no.
Artemisia absinthium	Asteraceae	Naatiruwa	Herb	Cultivate	All parts	Rare	Crushed and mixed/concocted with butter and taken	Oral	Removal of placenta during birth/for retained placenta	Human	Dawro35
Artemisia afra	Asteraceae	Agupiya	Herb	Cultivated	Root	Less plenty	Crushed and concocted with rue and wormwood	Oral	For different ailments	Animal	Dawro39
Arundincria alpina	Poaceae	Woosha	Shrub	Semi-wild	Leaf	Less plenty	Crush the leaf after drying and mix with water and take orally	Oral	Diarrhea	Animal	Dawro294
Asparagus flagellaris	Asparagaceae	Sereetiya	Climber	Wild	Root	Plenty	Crushed and decocted and mixed with butter and taken	Oral	For blocked urination	Animal	Dawro45
Astragalu membranasceus	Fabaceae	K'eeri-wusiwusiya	Shrub	Wild	Stem	Less plenty	Small pieces of it hanged on the neck	Dermal	Black leg	Human	Dawro198
Azadirachta indica	Meliaceae	Nimitriya	Tree	Cultivated	Leaf	Less plenty	Powdered, pressed, and applied	Nasal	Black leg	Animal	Dawro291
Basilicum polystachyon	Lamiaceae		Herb	Wild	Leaf	Less plenty	Crushed and applied on the skin	Dermal	Allergic reaction	Human	Dawro343
Becium obovatum	Lamiaceae	Gendiya	Climber	Wild	Root	Less plenty	Crushed and mixed with water and taken orally and the remaining rubbed on the skin	Dermal and oral	Tinea capitis	Human	Dawro220
Bersama abyssinica	Melianthaceae	Walasoniya	Tree	Wild	Leaf	Rare	Flesh part rubbed on it	Dermal	Tumor	Human	Dawro170

Biophytum umbraculum	Oxalidaceae	Dango/shidho	Herb	Wild	Root	Less plenty	Crushed/powdered and mixed with water a cup of it is taken	Oral	Anaphylactic shock; epilepsy	Human	Dawro299
Bothricline sp.	Asteraceae	Shankishasha	Shrub	Wild	Leaf	Plenty	Crushed and mixed/concocted with butter and taken	Oral	Pain	Human	Dawro268
Brachiaria brizontha	Poaceae	Shaalishattuwa	Herb	Wild	Root	Plenty	Crushed and taken orally for children and chewing the root by adults	Oral	Stomachache and anaphylactic shock; epilepsy	Human	Dawro129 and Dawro346
Brassica nigra	Brassicaceae	Sanafic'iya	Herb	Cultivate	Fruit	Plenty	Crushed and concocted with *Piper capense, Lepidium sativum*, and yoghurt	Oral	Abdominal pain	Animal and human	Dawro62
Bridelia scleroneura	Euphorbiaceae	Zuzia	Tree	Wild	Bark	Less plenty	Crushed and mixed with water and taken by mainly the pregnant women	Oral	Colic pain	Human	Dawro153
Brucea antidysenterica	Simaraubaceae	Shushaliya	Tree	Wild	Bark	Rare	Crushed and mixed with water and taken	Oral	Abdominal ache	Human	Dawro15
Brucea antidysentrica J.F.Mill	Solanaceae	Shureshuupiya	Tree	Wild	Root	Rare	Crushed and taken	Oral	Parasitic disease in children	Human	Dawro107
Buddleja polystachya	Luganiaceae	Kanfara	Tree	Wild	Leaf, root, bark	Rare	Dry the part and crush and take as a drink, smelling	Oral and nasal	Epilepsy; malaria; stomachache	Human	Dawro1

(continued)

Table 2 (continued)

Scientific name	Family	Local name	Growth form	Source	Part used	Scarcity	Preparation and application	Route of administration	Disease treated	Treatment for	Voucher no.
Capsicum frutescens	Solonaceae	Mis'imis'uwa	Herb	Cultivate	Stem	Plenty	Crushed and taken orally	Oral	Lymph adenitis	Human	Dawro298
Carex steudneri	Cyperaceae	–	Herb	Wild	Fruit	Rare	Decocted and taken	Oral	Snake bite	Human	Dawro245
Carica papaya	Caricaceae	Paapa	Shrub	Cultivate	Root	Less plenty	Crushed/decocted and taken for 3 days	Oral	Malaria	Human	Dawro286
Catha edulis	Celastraceae	Jimaa	Shrub	Cultivated and wild	Leaf	Plenty	Heated/decocted and taken	Oral	Liver problem	Human	Dawro283
Chamaecrista mimosoides	Fabaceae	Shosha entarsa/Bazo Mata/Deesha halakuwa	Shrub	Wild	Root, leaf, and stem	Plenty	Washed/crushed and rubbed on the skin; otherwise taken orally	Dermal and oral	Snake bite	Human	Dawro332;Dawro 201
Cissampelos mucronata	Menispermaceae	Bula Tura	Climber	Wild	Root	Plenty	Crushed/decocted and taken orally	Oral	Colic pain	Human	Dawro320
Cissus sp.	Vitaceae	Tussa	Climber	Wild	Leaf	Rare	Rubbed on the infected part of the skin	Dermal	Fire burning	Human	Dawro273
Citrus aurantifolia	Rutaceae	Lomiya	Tree	Cultivate	Fruit	Plenty	The fleshy part applied on the wound	Dermal	Wound	Human	Dawro103
Clausena anisata	Fabaceae	Zama	Tree	Wild	Root; bark	Less plenty	Crushed and mixed with water	Oral	Gall bladder	Animal	Dawro120
Clematis hirsuta	Ranunculaceae	Soguwa Tura	Climber	Wild	Leaf	Plenty	Powder and pressed/decocted and rubbed on the bite part	Dermal	Snake bite	Human	Dawro160
Clerodendrum cordifolium	Lamiaceae	Boye maata	Climber	Cultivate	Root	Less plenty	Crushed and applied orally	Oral	Anaphylactic shock	Animal	Dawro157

Clerodendrum myricoides	Lamiaceae	Alga	Shrub	Wild	Leaf	Less plenty	Crush 2/3 leaves and decoct half of a cup and apply orally; crushed and tied on the infected part	Dermal and oral	Chill and internal pain; severe abdominal cramp	Animal and human	Dawro77 and 229
Clerodendrum myricoides	Lamiaceae	Kareta Mata	Tree	Wild	Leaf	Rare	Crushed/decocted and taken	Oral	Evil eye	Human	Dawro210
Cluita lanceolata	Euphorbiaceae	Shosha D'aliya	Shrub	Wild	Leaf; root	Less plenty	Crushed/decocted and taken; concocted with *Entada abyssinica* and taken	Oral	Anaphylactic shock; snake bite	Animal and human	Dawro166 and Dawro 233
Coccinia abyssinica	Cucurbitaceae	Usik'iya/ ushushiya	Climber	Cultivate	Root	Less plenty	Crushed and mixed with the bark of korch and water taken	Oral	Gonorrhea; liver problem	Human	Dawro102
Combretum collinium	Combretaceae	Digisuwa	Tree	Wild	Leaf; root	Less plenty	Crush the leaf and root part and applied both dermally and orally	Dermal and oral	Black leg	Animal	Dawro165
Commelina africana	Commelinaceae	Gulbatiya	Herb	Wild	Leaf	Less plenty	Crushed, mixed with water, and taken	Oral	Vomiting in children	Human	Dawro106
Commelinia latifolia	Commelinaceae	Dal'isha/ Gassaa	Herb	Cultivate	Leaf	Plenty	Chewed the part and applied to the infected skin with insect for 3 days	Dermal	Insect bite	Human	Dawro20
Conyza pyrrhopappa	Asteraceae	D'oniya	Herb	Wild	Leaf	Less plenty	Crushed, decocted, a glass of it is taken	Oral	Diarrhea	Human	Dawro197

(continued)

Table 2 (continued)

Scientific name	Family	Local name	Growth form	Source	Part used	Scarcity	Preparation and application	Route of administration	Disease treated	Treatment for	Voucher no.
Conyza sp.	Asteraceae		Shrub	Wild	Leaf	Less plenty	Crushed and taken	Oral	Rheumatism	Human	Dawro347
Crassocephalum macropappum	Asteraceae	Botsa k'odhuwa	Climber	Wild	Leaf	Less plenty	Crushed and taken	Oral	Anthrax	Animal	Dawro335
Crepis achyrophoroides	Asteraceae	Mas'uwa-sawa	Herb	Wild	Leaf	Less plenty	Crushed/decocted and taken	Oral	Black leg	Animal	Dawro193
Crepis rueppellii	Asteraceae	Maas'uwa	Herb	Wild	All parts	Plenty	Decoction	Dermal	Evil eye	Human	Dawro48
Crepis xylorrchiza	Asteraceae	Shid'a-maas'oliya	Herb	Wild	Root	Less plenty	Crushed and taken	Oral	Anthrax/aba senga	Animal	Dawro246
Crotalaria rosenii	Fabaceae	Kishikisho	Shrub	Wild	Root	Less plenty	Cut the root into seven pieces and then crush/decoct and mix with milk and take	Oral	Colic pain	Human	Dawro235
Croton macrostachyus	Euphorbiaceae	Anka	Tree	Wild	Leaf	Plenty	Crush the leaf and apply on the infected skin	Dermal	Wound	Human	Dawro24
Cucumis ficifolium	Cucurbitaceae	Sukulo d'antha	Herb	Wild	Root	Less plenty	Crushed and taken orally and also rubbed on the skin	Dermal and oral	Snake bite	Human	Dawro232
Cuscuta reflexa	Convilvulaceae	Has'emamito	Climber	Wild	Leaf	Rare	Crushed/decocted	Oral	Internal parasites/worms	Animal	Dawro87
Cyathula cylindrical	Amaranthaceae	Gumpula/dorsa-k'arc'ocha	Herb	Wild	Leaf	Plenty	Crushed and mixed with water and taken	Oral	For swelling of abdomen	Animal	Dawro56
Cynodon spp	Poaceae	Sura	Runner	Wild	Shoot	Less plenty	Place the leaf on the injured area	Dermal	Fire burning	Human	Dawro341

Cynoglossum sp.	Boraginaceae	Shosha Taliya;S'ilkiya	Herb	Wild	Leaf; root	Less plenty	Crushed and mixed with water, taken orally, put the root on the pocket; crushed/decocted and taken	Oral	Snake bite; anthrax	Animal and human	Dawro253 and Dawuro 349
Cyperus articulata	Cypraceae	Bidaaraa	Herb	Wild	Root	Less plenty	Crushed and concocted with rue and wormwood	Oral	Stomachache	Human	Dawro38
Cyperus iria	Cyperceae	Bidara-mala	Herb	Wild	Root	Less plenty	Heat the root part and take	Oral	Facilitate digestion	Human	Dawro207
Cyphostemma niveum	Vitaceae	Shortiya Tura	Climber	Wild	Leaf	Less plenty	Crushed and taken orally, tied on the neck	Dermal and oral	Swelling on the neck/lymph adenitis; snake bite	Animal and human	Dawro261 and Dawro322
Cyphostemmo nivenum	Vitaceae	Banbari-Bachuwa	Shrub	Wild	Root	Rare	Crushed and taken	Oral	Trypanosomiasis	Animal	Dawro348
Datura stramonium	Solanaceae	Laflafuwa	Herb	Wild	Leaf	Plenty	Crushed with water and applied to skin	Dermal	Ringworm and skin disease	Animal	Dawro18
Dergea sp.	Asclepiadaceae	Ex'a d'aliya	Climber	Wild	Root	Less plenty	Washed/crushed/ decocted and taken	Oral	Abdominal pain	Human	Dawro203
Dichondra repens	Convolvulaceae	Ec'c'ere haytsa	Herb	Wild	Leaf; root	Less plenty	Crushed and concocted with *Tragia cinerea* and *Sida schimperiana*	Oral	Snake bite	Human	Dawro223
Dicliptera laxata	Acanthaceae	Toguwa	Herb	Cultivated	Shoot	Less plenty	Crush the part and apply on skin	Dermal	Eye disease/ allergic reaction	Human	Dawro32
Dicrocephula integrifolia	Asteraceae	Sa'a'-okata malaa	Herb	Wild	Leaf	Less plenty	Crushed/decocted and applied through nose and rubbed on the surface of head	Dermal and nasal	Anaphylactic shock; epilepsy	Human	Dawro228

(continued)

Table 2 (continued)

Scientific name	Family	Local name	Growth form	Source	Part used	Scarcity	Preparation and application	Route of administration	Disease treated	Treatment for	Voucher no.
Discopodium penninarvum	Solonaceae	C'oyd'a/A'inaa	Shrub	Wild	Leaf	Less plenty	Crushed and mixed with water	Oral	Black leg	Animal	Dawro194
Dombeya torrida	Sterculiaceae	Boshuwa/ loluwa	Tree	Wild	Leaf	Less plenty	Crushed/decocted and taken	Oral	Abdominal pain	Animal	Dawro292
Echinops amplexcaulis	Asteraceae	Wora bursa/ kashiya (zo'uwa gad'awa)	Shrub	Wild	Root	Plenty	Crushed/decocted and taken	Oral	For building	Animal	Dawro211
Echinops kebricho	Asteraceae	Bursa	Shrub	Wild	Root	Plenty	Smoked	Dermal	Headache, never reach snake around the area of smoking	Human	Dawro9
Ehretia cymosa	Boraginaceae	Etriwanjiya	Tree	Wild	Leaf	Less plenty	Crushed and tied on the infected part	Dermal	Severe abdominal cramp	Animal	Dawro78
Embelia schimperi	Myrsinaceae	K'uank'uula	Shrub	Wild	Fruit	Rare	Crushed and mixed with water and taken orally before breakfast	Oral	Tapeworm	Human	Dawro250
Ensete ventricosum	Musaceae	Utsa	Shrub	Cultivated	Root	Plenty	The root tied on neck	Dermal	Tumor	Human	Dawro72
Entada abyssinica	Fabaceae	Gelec'ec'a	Shrub	Wild	Leaf	Less plenty	Crushed and applied on the infected parts	Dermal	Wound	Animal and human	Dawro88
Erythrina abysinica	Fabaceae	Borttuwa-Gad'awa	Tree	Cultivated and wild	Bark	Rare	Crushed/decocted and taken	Oral	Evil eye	Animal and human	Dawro155

Erythrina brucci	Fabaceae	Bortuwa-Geziyawa	Tree	Cultivated and wild	Bark	Plenty	Crushed or powdered fresh and mixed with water and taken	Oral	Ascaris, stomachache	Human	Dawro3
Eucalyptus globules	Myrtaceae	Botha-Barzafiya	Tree	Cultivated	Leaf	Plenty	Smoking	Nasal	Common cold	Human	Dawro282
Euphorbia hirta	Euphorbiaceae	Shato-maataa	Herb	Wild	Root	Less plenty	Crushed and rubbed out to it	Dermal	Ringworm and skin disease	Human	Dawro176
Euphorbia inidica	Euphorbiaceae	Shato dhaliya	Climber	Wild	Sap	Less plenty	Cut and then apply the sap	Dermal	Skin disease	Human	Dawro324
Euphorbia tirucalli	Euphorbeaceae	Maxuwa-darawa/ S'aduwa	Shrub	Wild	Shoot	Rare	Decocted and taken	Oral	Swelling of stomach	Animal	Dawro27
Ficus thonningii	Moraceae	Shaynhiya	Tree	Wild	Bark	Less plenty	Crushed and decocted taken	Oral	Dysentery	Human	Dawro323
Ficus vasta	Moraceae	Esaa//etta	Tree	Wild	Bark	Rare	Crushed, decocted, and mixed/ concocted with *Rumex nepalensis, Erythrina brucei*, and *Solanum incanum* and taken	Oral	Stomach disorder	Human	Dawro154
Foeniculum vulgarie	Apiaceae	Shileria	Shrub	Wild	Leaf	Rare	Crushed and mixed with water and taken	Oral	Chill	Human	Dawro16
Galinirea coffeoides	Rubiaceae	Deesha loomiya	Shrub	Wild	Leaf; root	Rare	Crushed and mixed with milk and applied orally	Oral	Epilepsy	Human	Dawro307

(continued)

Table 2 (continued)

Scientific name	Family	Local name	Growth form	Source	Part used	Scarcity	Preparation and application	Route of administration	Disease treated	Treatment for	Voucher no.
Galinsoga paruifolra	Asteraceae	Ematiya/ bizdiya	Tree	Wild	Leaf; root	Less plenty	Crush the root and leaf together and take orally within 3 days' interval until healed	Oral	Hepatitis/liver problem	Human	Dawro222
Gallium aparinoides	Rubiaceae	Kangad'a	Herb	Wild	Shoot	Rare	Rubbing on the infected part	Dermal	Skin disease; impetigo	Human	Dawro259
Gardenia ternifolia	Rubiaceae	Genbbela	Tree	Wild	Leaf	Less plenty	Crushed/decocted and mixed with butter	Oral	Colic pain	Human	Dawro296
Geranium sp.	Geranaceae	Badinecho	Herb	Wild	Leaf	Less plenty	Rubbed on the infected part of the tooth	Dermal	Gum infection	Human	Dawro272
Gerbera piloselloides	Asteraceae	Sa-sheka	Herb	Wild	Leaf	Rare	Crushed/decocted and add little water when taken orally once	Oral	Severe abdominal cramp	Human	Dawro315
Girardinca bullosa	Urticaceae	Kona	Herb	Wild	Root	Rare	Crushed and taken	Oral	Constipation	Human	Dawro280
Glycine wightii var.longicauda	Fabaceae	Tooguwa Tura	Climber	Wild	Root	Less plenty	Crushed/decocted and can be concocted/mixed with milk and taken	Oral	Evil eye	Animal and human	Dawro231
Gnidia glauca	Thymelaeaceae	Migra	Shrub	Wild	Root	Less plenty	Crush and leave it overnight, mix with water, and take orally	Oral	Hepatitis/liver problem	Human	Dawro183

Gnidia stenophylla	Thymelaceae	K'uriya	Shrub	Cultivated	Root	Less plenty	Crushed/decocted and taken orally and then taken with milk	Oral	Abdominal pain	Human	Dawro351
Grewia bicolour Juss	Tiliaceae	Gumariya/ S'awayiya	Tree	Wild	Bark	Plenty	Crunched and added with water and taken	Oral	Swelling of stomach	Human	Dawro28
Guizotia scabra	Asteraceae	Tufaa	Herb	Semi-wild	Stem	Plenty	Crushed/decocted and applied on the skin	Dermal	Skin disease	Human	Dawro147
Habenstretia angolensis	Scrophuriaceae	Kayis'eriya	Shrub	Wild	Stem	Less plenty	The small pieces of it tied on the neck	Dermal	Tumor	Human	Dawro244
Hagnia abyssinica	Rosaceae	Soyid'uwa	Tree	Wild	Fruit	Rare	Crushed and mixed with water and taken	Oral	Tapeworm	Human	Dawro101
Helichrysum gerberifolium	Asteraceae		Herb	Wild	Shoot	Plenty	Crushed and taken	Oral	Meningitis	Human	Dawro168
Helichrysum sp.	Asteraceae	Samba lolo/ Yesamba mich	Herb	Wild	Leaf	Less plenty	Crushed and mixed with water and taken through eye, nose, and mouth and rubbed on the skin	Oral, nasal, dermal	Evil eye, pneumonia	Human	Dawro242
Hypericum peplidifolium	Guttiferaceae	Mali Mas'ino	Herb	Cultivated	Leaf	Rare	Crushed/decocted and applied on the skin	Dermal	Itching and scabies	Human	Dawro128
Hypericum revolutum	Guttiferaceae	K'irik'uwa	Shrub	Wild	Leaf	Plenty	Crushed/decocted and taken	Oral	Anaphylactic shock	Human	Dawro156
Hypoestes forskaolii	Acanthaceae	Ginginuwa malatiyawa	Herb	Wild	Root	Rare	Powdered and applied	Oral	Snake bite	Human	Dawro178

(continued)

Table 2 (continued)

Scientific name	Family	Local name	Growth form	Source	Part used	Scarcity	Preparation and application	Route of administration	Disease treated	Treatment for	Voucher no.
Indigofera arrecta	Fabaceae	Wusiwusiya	Shrub	Wild	Root	Plenty	Chew the root part	Oral	Abdominal ache; anthrax	Animal and human	Dawro54
Indigofera spicata	Fabaceae	Sheka/K'uriya/ Dangarsa d'oniya	Herb	Wild	Root	Less plenty	Crushed/decocted and mixed with water and taken orally	Oral	Snake bite; abdominal pain	Animal and human	Dawro328 and dawro98
Juniperus procera	Cuperssaceae	S'iida	Tree	Semi-wild	Leaf	Rare	Crushed and applied on the infected parts	Dermal	Wound	Animal	Dawro311
Justicia ladanoides	Acantaceae	Mulu muk'uwa	Herb	Wild	Leaf; root	Rare	Crushed, washed, rubbed on the skin, and also taken through nose; powdered, and liquid-filtered butter of a spoon given orally; crushed, mixed with butter, and rubbed on the part	Dermal and oral	Colic pain; hepatitis/liver problem; tumor	Animal and human	Dawro169 (175/187/196)
Laggera pterodonta	Asteraceae	Sesa/Gelesho tanbuwa	Shrub	Wild	Leaf	Rare	Crushed and applied orally	Oral	Evil eye	Human	Dawro59
Lannea fruticosa	Anacardiaceae	Dechi-marac'iya	Tree	Wild	Root	Rare	Crushed and rubbed on the infected part	Dermal	Wound; abdominal pain/ (Karishuwa)	Animal and human	Dawro99
Lantana tritolia	Verbensceae	Shanki-shasha	Herb	Wild	Leaf	Rare	Crushed and decocted and taken	Oral	Malaria	Human	Dawro97

Lantana viburnoides	Verbenaceae	Shanki-shasha malaa	Herb	Wild	Leaf	Less plenty	Crushed and applied through eye and nose	Nasal	Allergic reaction of eye	Human	Dawro192
Launea mtgbacea Jeffrcy	Asteraceae		Herb	Wild	Root	Less plenty	Crushed and mixed with water and taken	Oral	Abdominal pain	Human	Dawro234
Lepidium sativum	Brassicaceae	Sibika	Herb	Cultivated	Fruit	Plenty	Crushed and mixed/concocted with butter and taken	Oral	Abdominal pain and intestinal problem/ cramp (karshuwa)	Animal	Dawro137
Leucas abbyssinica	Lamiaceae	Kirikisa	Herb	Wild	Leaf	Less plenty	Crushed and mixed with water and taken one cup for adult and half cup for children	Oral	Diarrhea	Human	Dawro218
Leucas martinicensis	Lamiaceae	Gumpula	Herb	Semi-wild	Root	Less plenty	Powder and then the local areki is taken through mouth	Oral	Anthrax	Animal	Dawro146
Lippia adoensis var.koseret	Verbenaceae	Kosorotiya	Shrub	Semi-wild	Leaf	Less plenty	Concoction with milk	Oral	Appetizer	Human	Dawro68
Lobelia giberroa	Lobeliaceae	Ododiya	Shrub	Wild	Leaf	Rare	Crushed/concocted with tobacco and eucalyptus and taken	Oral	Trypanosomiasis	Animal	Dawro125
Lotus sp.	Fabaceae	Badanecha	Climber	Wild	Leaf; root	Less plenty	Crushed and rubbed on the surface of gum	Dermal	Gum infection	Human	Dawro338
Maerua oblongifolia	Capparidaceae	Sangana	Shrub	Wild	Root	Less plenty	Crushed and taken orally, smoked	Oral and nasal	Colic pain	Human	Dawro209

(continued)

Table 2 (continued)

Scientific name	Family	Local name	Growth form	Source	Part used	Scarcity	Preparation and application	Route of administration	Disease treated	Treatment for	Voucher no.
Maesa lanceolata	Myrrecenaceae	Gegec'uwa	Tree	Wild	Bark	Plenty	Crushed or powdered fresh and mixed with water and taken	Oral	Ascaris, stomachache	Human	Dawro5
Maytenus sega	Celastraceae	Putawuwa	Tree	Wild	Leaf; root	Less plenty	Crush the root and leaf/decoct and take	Oral	Evil eye	Animal and human	Dawro163
Milettia ferruginea	Fabaceae	Zagiya	Tree	Wild	Root	Less plenty	Crushed/decocted and applied	Oral	Trypanosomiasis	Animal	Dawro119
Momordica foetide	Cucuribitaceae	K'eca	Climber	Wild	Root	Less plenty	Crushed and taken orally with water	Oral and nasal	Rabies	Animal and human	Dawro304
Moringa stenopetala	Moringaceae	Halakuwa	Tree	Cultivated	Leaf	Less plenty	Heat the leaf and then eat until threatened	Oral	Malaria	Human	Dawro287
Mukia maderaspatana	Cucurbitaceae	Shosha Mata	Climber	Wild	Leaf	Less plenty	Crushed and mixed with water	Oral	Snake bite	Human	Dawro336
Musa x peradisacal	Musaceae	Muuziya	Shrub	Cultivated	All parts	Less plenty	Crushed/decocted and applied on the injured part	Dermal	Blood clotting	Human	Dawro305
Nephrolepis undulata	Oleandraceae	Bisa-gadhawa	Herb	Wild	Leaf	Less plenty	Crushed and mixed with water and taken	Oral	Snake bite	Human	Dawro213
Nicandra physaloides	Solanaceae	Puqaqiya (laflafuwa mala)	Herb	Wild	Leaf	Plenty	Crushed and a cup of it is taken	Oral	Hepatitis/liver problem	Human	Dawro290
Nicotiana tabacum	Solanaceae	Tambuwa	Herb	Cultivated	Leaf	Less plenty	Crushed/decocted and applied	Nasal	Leech	Animal and human	Dawro92

Ocimum americanum	Lamiaceae	Dunkiya-bunawa/Sa'a tusa	Shrub	Wild	Root	Less plenty	Crushed and pressed and mixed with the root of *Reichardia tingitana*/decocted and taken orally; crushed and mixed with water and a cup of it is taken	Oral	Anaphylactic shock	Human	Dawro161(326)
Ocimum basilicum	Lamiaceae	Dunkiya	Herb	Cultivated	Leaf	Rare	Crushed, pounded, and given	Oral	Stomachache	Human	Dawro12
Ocimum lamifolium	Lamiaceae	Damakesiya	Herb	Wild	Leaf	Plenty	Crushed and mixed/concocted with coffee and taken	Dermal and oral	Allergic reaction	Human	Dawro14
Ocimum utricifolium	Lamraceae	Guluuwa/ Desha-dunkiya	Shrub	Wild	Root	Less plenty	Crushed and mixed with water and taken orally	Oral	Colic pain	Human	Dawro195
Oncocalyx sp.	Loranthaceae	China Mita	Shrub	Wild	Bark	Rare	Crushed and mixed with butter and rubbed on the skin	Dermal	Skin disease	Human	Dawro185
Oxalis latifolia	Oxalidaceae	mac'igara	Climber	Wild	Leaf	Plenty	Powder taken	Oral	Colic pain	Human	Dawro144
Oxalis radicosa	Oxalidaceae	Shumachiya	Herb	Wild	Leaf	Plenty	Heated/crushed and mixed with water and taken	Oral	Appetizer	Human	Dawro217
Paspalum scrobiculatum	Poaceae	Gors'a-mala	Herb	Cultivated	Leaf	Less plenty	Rubbed on the infected part	Dermal	Snake bite	Human	Dawro74
Penisetum clandestinum	Poaceae	Gors'aa	Herb	Wild	Leaf	Rare	Crushed and mixed with water	Oral	Snake bite	Animal and human	Dawro136

(continued)

Table 2 (continued)

Scientific name	Family	Local name	Growth form	Source	Part used	Scarcity	Preparation and application	Route of administration	Disease treated	Treatment for	Voucher no.
Pentas lanceolata	Rubiaceae	Gergeda mitsa/ Dawridama mala	Shrub	Wild	Leaf; root; bark	Rare	Crushed/decocted and taken; crushed and taken orally, and the root cut into three pieces and tied on neck	Nasal	Breathing problem/ Tinnitus/tumor	Human	Dawro227 (256)
Pentas schimperiana	Rubiaceae	Dawuri Daamaa/ Dalbantsa	Shrub	Wild	Leaf	Rare	Crushed and mixed with water and taken with butter	Oral	Broken	Animal and human	Dawro138
Persea americana	Lauraceae	Abokatuwa	Tree	Cultivated	Bark	Less plenty	Crushed/decocted and taken	Oral	Colic pain	Animal and human	Dawro295
Phaulopsis imbricata	Acanthaceae	Umba	Herb	Wild	Root	Rare	Powdered and mixed with water and applied	Nasal	Black leg	Animal	Dawro177
Phoenix reclinta	Ariaceae	Zamba	Tree	Wild	Leaf	Rare	Chopped, powdered, and dropped into eye	Dermal	Eye disease	Human	Dawro288
Phragmanthera machosolen	Loranthaceae	Mitsa shapuwa	Climber	Wild	Bark	Rare	The plant part tied on the neck	Dermal	Lymph adenitis/ swelling of gland	Human	Dawro21
Phyllanthus maderaspatensis	Euphorbiaceae	Dal uwa	Herb	Wild	Root	Rare	Crushed and mixed with water and applied	Oral and nasal	For milk shortage	Animal	Dawro188
Phyllanthus reticulatus	Euphorbiaceae	Wusiwisiya mala	Shrub	Wild	Leaf; root	Less plenty	Crush both parts and take	Oral	Hepatitis/liver problem	Human	Dawro249
Phytolacca dodecandra	Phytolacaceae	Hanc'ic'iya	Shrub	Wild	Leaf, root, bark	Plenty	Crushed or powdered fresh and mixed with water and taken	Oral	Stomachache, malaria	Human	Dawro2

Pilea rivularis	Utricaceae	Hayitsa matta	Herb	Semi-wild	Leaf	Plenty	Crushed, pressed, and the solid is applied into ear	Dermal	Ear disease	Human	Dawro145
Pileatera phylla	Uritaceae		Herb	Wild	Leaf	Less plenty	Crush and smell it	Dermal	Headache, sweating	Human	Dawro240
Piper capense	Piperaceae	Tunja	Shrub	Cultivated	Fruit	Rare	Crushed and cococted with rue and *Echinops kebericho* and taken	Oral	Chill, stomachache, headache	Human	Dawro6
Plantago lanceolata	Plantagonaceae	Borada mala	Herb	Wild	Leaf	Rare	Crushed and rubbed on the skin	Dermal	Wound	Human	Dawro238
Plantago palmata	Plantagonaceae	Borodaa	Herb	Cultivated	Root	Rare	The root tied on neck	Dermal		Human	Dawro132
Plectranthus caninus	Lamiaceae	Mudha	Herb	Wild	Leaf	Plenty	Crushed and decocted	Dermal and oral	Allergic reaction and wound	Human	Dawro34
Plectranthus ornatus	Lamiaceae	Dissa	Herb	Wild	Leaf; root	Rare	Simply smell the leaf part but the root parts are chewed	Oral and nasal	Allergic reaction	Human	Dawro309
Plumbago zeylanica	Plumaginaceae		Herb	Wild	Root	Less plenty	Crushed/decocted and taken	Nasal	Snake bite	Human	Dawro190
Polygala persicarifolia	Polygonaceae	Gic'inda	Herb	Wild	Root	Less plenty	Powdered and applied	Nasal	Headache	Human	Dawro180
Pteris catoptera	Pteridaceae	Bisa	Herb	Wild	Root	Less plenty	Crushed and mixed with water and taken	Oral	Abdominal pain	Human	Dawro148
Pycnostachys abyssinica	Lamiaceae	Olomuwa	Shrub	Wild	Leaf	Less plenty	Crush the leaf and apply on the foot	Dermal	Athlete foot cracking/ maac'uwa	Human	Dawro306

(continued)

Table 2 (continued)

Scientific name	Family	Local name	Growth form	Source	Part used	Scarcity	Preparation and application	Route of administration	Disease treated	Treatment for	Voucher no.
Pycnostachys abyssinica	Lamiaceae	Olomuwa	Shrub	Wild	Leaf	Rare	The leaf placed on the eye	Dermal	Eye disease	Animal and human	Dawro58
Rhamnus prinoides	Rhamneceae	Geeshuwa	Shrub	Cultivated	Leaf	Plenty	Crushed, mixed with water, and heated	Dermal	Itching and scabies	Human	Dawro122
Rhoicissus revoilii	Vitaceae	Gegeluwa	Climber	Wild	Stem	Less plenty	Flesh of it tied on the neck	Dermal	Lymphadenitis	Human	Dawro159
Rhynchosia minima	Fabaceae	Galimentsuwa	Herb	Cultivated and wild	Leaf	Rare	Crushed and rubbed on the infected part	Dermal	Thorn toxic	Human	Dawro124
Rhynchosia orthobotrya	Fabaceae		Shrub	Wild	Leaf	Plenty	Crushed/decocted and mixed with little water	Oral	Allergic reaction and anaphylactic shock	Human	Dawro152
Richrdia tingetana	Asteraceae	Mas'oliya	Herb	Wild	Leaf	Less plenty	Crushed and taken orally	Oral	Gonorrhea and haemorrhage (kintarot)	Human	Dawro255
Rumex abyssinicus	Polygonaceae	C'olieya	Herb	Cultivated	Root	Plenty	Decocted and half a cup of it taken	Oral	Ascariasis	Human	Dawro57
Rumex nepalensis	Polyganaceae	Zans'ala	Herb	Wild	Root	Plenty	Crushed and tied on the infected part	Dermal	Insect bite (Sa'aba) disease	Human	Dawro80
Ruta chalopensis	Rutaceae	S'alotiya	Shrub	Cultivated	Fruit, leaf	Plenty	Chewed orally or mixed with water	Oral	Stomachache, chill	Human	Dawro8
Salvia nilotica	Lamiaceae	Sa'a Okata	Herb	Wild	Leaf	Plenty	Crushed and mixed with water and taken	Oral	Allergic reaction	Human	Dawro17

Satureja abyssinica	Lamiaceae	Wuta malaa	Herb	Wild	Leaf	Less plenty	Crushed and applied orally and dermally by simply smelling the leaf	Dermal and oral	Abdominal pain; epilepsy	Animal and human	Dawro118(241)
Satureja punctata	Lamiaceae		Herb	Wild	Leaf	Less plenty	Powdered and pressed and a cup of it is taken	Oral	Snake bite	Human	Dawro182
Satyrium aethiopicum	Orchidaceae	Ec'ere Hayitsa	Herb	Wild	Root	Less plenty	Crushed and mixed/decocted with lemon and taken orally	Oral	Anthrax	Animal	Dawro266
Scadoxus nutans	Maryllidaceae	Wara Mana	Herb	Wild	Leaf	Rare	Crushed and cococted with *Tragia cinerea* and taken	Oral	Snake bite	Human	Dawro95
Schrebera alata	Oleaceae	K'ara	Tree	Wild	Leaf	Less plenty	Crushed and mixed with water and taken	Oral	Wound	Human	Dawro212
Sclerocarya birrea	Anacardiaceae	Woshilachiya/ Tunk'aluwa	Tree	Wild	Stem	Less plenty	Five pieces of it hanged on the neck	Dermal	Lymphadenitis	Human	Dawro297
Senna peteriana	Fabaceae	Shosha enxarsa	Shrub	Wild	Leaf	Rare	Crushed/decocted and add little water and take once	Oral	Snake bite	Human	Dawro313
Sida rhombifolia	Malvaceae	Danduretsa	Herb	Wild	Root	Less plenty	Crushed and applied	Nasal	Dandreta	Animal and human	Dawro301
Sida schimperiana	Malraceae	Kindichuwa	Shrub	Cultivated	Leaf	Plenty	Crush the leaf and take, smelling the leaf	Dermal and oral	Evil eye	Human	Dawro73

(continued)

Table 2 (continued)

Scientific name	Family	Local name	Growth form	Source	Part used	Scarcity	Preparation and application	Route of administration	Disease treated	Treatment for	Voucher no.
Solanecio gigas	Asteraceae	Dook'a	Shrub	Cultivated	Leaf	Less plenty	Crushed and mixed with water and taken orally	Oral	Abdominal pain	Animal	Dawro134
Solanum capsicoides	Solanaceae	Karetsa buluwa/ Meetetiya buluwa	Herb	Wild	Shoot and flower	Plenty	Crushed and decocted and taken orally for children	Oral	Common cold, abdominal cramp	Human	Dawro47
Solanum incanum	Solanaceae	Wora buluwa	Shrub	Wild	Fruit	Less plenty	Crushed/decocted and taken orally	Oral	Snake bite	Animal and human	Dawro310
Solanum incanum L	Solanaceae	Buluwa/Wora buluwa	Shrub	Wild	Leaf	Plenty	Crushed, heated, and mixed with butter	Oral	Gastritis, allergic reaction	Human	Dawro63
Solanum sp.	Solanaceae	Puk'ek'iya	Shrub	Wild	Root	Rare	Washed/crushed, mixed with water, and taken orally three cups before breakfast for 3 days	Oral	Hepatitis/liver problem	Human	Dawro189
Solanum sp.	Solanaceae		Shrub	Cultivated	Leaf	Rare	Cut into small pieces and hung	Dermal	Lymphadenitis	Human	Dawro355
Sparmannia ricinocarpa	Tiliaceae	K'eri-cayshiya/ K'arc'ocha/ Baribacho	Herb	Wild	Leaf	Less plenty	Apply into nose/ drink	Oral and nasal	Snake bite	Human	Dawro181
Spilanthus mauritiana	Asteraceae	Aydamiya	Herb	Wild	Flower	Plenty	Chewed orally	Oral	Flatulence, for fattening	Human	Dawro13

Sporobolus pyramidalis	Poaceae	Gic'igiliya/ Gic'ariya	Herb	Wild	Leaf	Rare	Powdered and mixed with the leaf of *Conyza pyrrhopappa* a glass of it is taken	Oral	Diarrhea	Human	Dawro174
Sporobulus sp.	Poaceae	Sura mala	Herb	Wild	Leaf and stem	Less plenty	Crushed and applied on the skin	Dermal	Fire inflammation	Human	Dawro331
Stephenia abyssinica	Menispermaceae	Bazo Tura/ado Tura	Climber	Wild	Root	Rare	Crushed, decocted, and mixed with fresh milk and taken	Oral	Stomachache in children	Human	Dawro19
Syzygium guineense	Myrtaceae	Ocha	Tree	Wild	Bark	Rare	Crushed or powdered fresh and mixed with water decocted and taken	Oral	Ascaris, stomachache; abdominal pain	Human	Dawro4 (285)
Tagetes minuta	Asteraceae	Derek'a	Herb	Wild	Leaf; root	Less plenty	Crushed/decocted and mixed with yoghurt; crushed and given orally	Oral	Chill; sudden attack of digestive guts; rheumatism	Human	Dawro206 (334;352)
Tamarindus indica	Fabaceae	Koriya	Tree	Wild	Fruit	Rare	Crushed and mixed with water and taken	Oral	Diarrhea	Human	Dawro325
Tephrosia villosa	Fabaceae		Herb	Wild	Root	Rare	Crushed and added with water and taken	Oral	Anaphylactic shock	Animal and human	Dawro172
Terminalia schimperiana	Combretaceae	Ambiya	Tree	Wild	Bark	Plenty	Crushed and decocted and taken	Oral	Chill and stomachache	Human	Dawro43
Thalictrum rhynchocarpum	Ranunculaceae		Herb	Wild	Root	Less plenty	Crushed and mixed with milk and applied	Oral	Ascariasis	Human	Dawro257

(continued)

Table 2 (continued)

Scientific name	Family	Local name	Growth form	Source	Part used	Scarcity	Preparation and application	Route of administration	Disease treated	Treatment for	Voucher no.
Tragia cinerea	Euphorbiaceae	Kinklishuwa	Climber	Wild	Root	Less plenty	Flesh of it tied on the neck	Dermal	Lymph adenitis	Human	Dawro158
Tragia doryoges	Euphorbiaceae	Kinklishuwa	Herb	Wild	Leaf; root	Less plenty	Crush the root and leaf together concocted with the leaf and root of *Croton macrostachyus*, mix with water, and take orally	Oral	Snake bite	Human	Dawro237
Trichodeswa zeylanicum	Boraginaceae	Kontsotsuwa/ K'uro aguntsa/ Kachanchiliya	Herb	Wild	Leaf	Less plenty	Crushed, mixed with *Phytolacca dodecandra*, and applied	Dermal	Eye disease	Human	Dawro116
Trifolium decorum	Fabaceae	Azimiya	Herb	Cultivated	Leaf	Plenty	Crushed/decocted and applied on the skin	Dermal	Itching	Human	Dawro130
Trigonella foenumgraecum	Fabaceae	Shuk'uwa	Herb	Wild	Leaf	Less plenty	Crushed/ground and applied into the eye	Dermal	Eye disease	Human	Dawro289
Triticum polonicum	Poceae	K'anbara	Herb	Cultivated	Seed	Plenty	Pound and then ground and mixed with water and then heated	Oral	Building of body	Animal and human	Dawro64
Tropaeolum majus	Tropacolaceae	Faranjiya-Sibika	Climber	Cultivated	Fruit	Rare	Crushed and cococted with rue/*Ruta chalepensis* and taken orally	Oral	Stomachache	Human	Dawro10

Vepris danellii	Rutaceae	C'awula	Tree	Wild	Fruit	Rare	Crushed and mixed with water and taken	Oral	Stomachache, chill	Human	Dawro7
Verbena officinalis	Verbenaceae	Higisha D'aliya	Herb	Wild	Root	Rare	Crushed and taken orally	Oral	Evil eye	Human	Dawro350
Vernonia amygdalina	Astraceae	Garaa	Tree	Wild	Shoot (young)	Rare	Crush the root and concoct with the root of papaya	Oral	Malaria for human and swelling of abdomen for animals	Animal and human	Dawro23
Vernonia karaguensis	Asteraceae	Saguwa	Shrub	Wild	Leaf	Less plenty	Crushed, dried, and rubbed on the skin	Oral	Anthrax	Animal	Dawro262
Vernonia lasiopus	Asteraceae	Waramayiya	Shrub	Wild	Root	Less plenty	Crushed/decocted and taken orally	Oral	Dysentery	Animal	Dawro321
Vernonia sp.	Asteraceae	Yesheshuwa	Herb	Wild	Root	Less plenty	Powdered and press and then mixed/concocted with the root of *Abrus precatorius* and milk	Oral	Hepatitis/liver problem	Human	Dawro162
Vernonia sp.	Asteraceae	Kariya	Shrub	Cultivated	Leaf	Rare	Chewed concocted with *Syzygium guineense* taken through mouth	Oral	Anaphylactic shock	Human	Dawro358
Vernonia theophrastifolia	Asteraceae	Buuzuuwa	Shrub	Wild	Leaf	Rare	Crushed/concocted with tselotiya taken orally	Oral	Evil eye	Human	Dawro308
Vernonia urticifolia	Asteraceae	Zamuwa	Shrub	Semi-wild	Leaf	Rare	Take the tip of the leaf and crush/mix with water and take	Oral	Abdominal pain	Human	Dawro131

(continued)

Table 2 (continued)

Scientific name	Family	Local name	Growth form	Source	Part used	Scarcity	Preparation and application	Route of administration	Disease treated	Treatment for	Voucher no.
Vicia sp.	Fabaceae	Kishikishi mala	Shrub	Wild	Leaf	Less plenty	Crushed and applied on the skin	Dermal	Saba bite	Human	Dawro329
Vigna vexillata	Fabaceae	Tsoka	Climber	Wild	Leaf	Rare	Powdered and taken orally	Oral	Black leg	Animal	Dawro342
Zehneria scabra	Cucuribitaceae	Ecca	Climber	Wild	Root	Less plenty	Crushed and applied through mouth and nose	Oral and nasal	Gonorrhea	Human	Dawro303
Zernia pratensis	Fabaceae	X	Herb	Wild	Leaf	Less plenty	Crushed and rubbed on the bitten area of the skin at least three times within 3 days' interval	Dermal	Snake bite	Human	Dawro230
Zingiber officinale	Zingiberaceae	Yenjeluwa	Herb	Cultivated	Root	Plenty	Crushed and concocted/mixed with coffee for human and only with water for animals	Oral	Swelling of abdomen for human; trypanosomiasis for animals	Animal and human	Dawro22
Zornia glochidiato	Fabaceae	Korie	Herb	Wild	Root	Less plenty	Crushed and concocted/mixed with *Plumbago zeylanica* and taken through nose	Nasal	Snake bite	Human	Dawro191
Zornia partensis	Fabaceae		Shrub	Wild	Root	Less plenty	Crushed and taken orally	Oral	Abdominal pain	Animal and human	Dawro164

Table 3 Source of medicinal plants used

S. No.	Item	Number	%	Rank
1	Wild	169	78.24	1st
2	Cultivated	36	16.67	2nd
3	Cultivated wild relatives	11	5.09	3rd
Total		**216**	**100**	

Growth Forms and Parts of Medicinal Plant Species Used in the Study Area

Traditional healers in the study area predominantly use herbs 91(42.13%), followed by trees 40 (18.52%), shrubs 58 (26.85%), climbers 26 (12.04%), and creeper 1 (0.46%) (Table 4). People in the study area collect more herbs than trees and shrubs for medicinal purpose that is similar to the report for the Kafficho people in Kafa (Awassa et al. 2007) and Loma and Gena Bosa area (Agize et al. 2013) and a country report (Anon 2005b). This result differed from the report for the people of Bosat, Welenchiti area, who are said to use more of shrubs than herbs (Debela Hunde et al. 2004). This can be explained by the fact that Bosat is a degraded dryland area where the shrub elements take prominence over the herbs.

Leaves account for 89 (41.20%), roots 61 (28.24%), leaf and root parts 13 (6.02%), bark 13 (6.02%), and fruit 11 (5.09%), each to treat different types of diseases in that order. The other parts are used to a lesser percentage [shoot 7 (3.24%); stem 7 (3.24%); all parts 3 (1.39%); seed 2 (0.93%); fruit and leaf 1 (0.46%); leaf, root, and bark 1 (0.46%); leaf and stem 1 (0.46%) sap 1 (0.46%); and shoot and flower 1 (0.46%)] (Table 5). Traditional healers in the study area collect more of leaves than other parts of medicinal plants similar to a report of the Kafficho people in Kafa (Awassa et al. 2007); a report to Bosat, Welenchiti area (Debela Hunde et al. 2004); Loma and Gena Bosa area (Agize et al. 2013); and Konso (Gebre 2005) while it disagrees with the report that indicates the use of more of roots than leaves (Anon 2005b).

People who are using more of shrubs focus on root parts while those using herbs do focus on leaves of the plant. The most widely used life form of medicinal plants in the study area is herb followed by shrub due to their ease of availability. This has contribution to reduce the threat rate to medicinal plants. The use of leaves than roots, barks, stems, and whole plant minimizes the threat to the destruction of medicinal plants. This is because if it is harvested in sustainable manner, it gives opportunity for long life of the plant than others. However, the root part utilization is significant and threats the plant species in the study area.

Most of the medicinal plants are prepared alone and mixed with water, 190 (87.96%), while 26 (12.04%) are used in mixture (Table 6), prepared in combination with others. In this kind of indigenous knowledge, the chemicals in the mixture may dilute the toxicity of some dangerous plants that become suited to be used with others in the mixture. Most of the remedies in the study area depend on fresh plant

Table 4 Growth forms/habits of medicinal plants used in the preparation of remedies

S. No.	Growth forms	Number	%	Rank
1	Herbs	91	42.13	1st
2	Shrubs	58	26.85	2nd
3	Trees	40	18.52	3rd
4	Climbers	26	12.04	4th
5	Creepers	1	0.46	5th
Total		**216**	**100**	

Table 5 Parts of medicinal plants used in the preparation of remedies

S. No.	Parts	Number	%	Rank
1	Leaves	89	41.2	1st
2	Roots	61	28.24	2nd
3	Leaf and root	13	6.02	3rd
4	Bark	13	6.02	3rd
5	Fruit	11	5.09	4th
6	Shoot	7	3.24	5th
7	Stem	7	3.24	5th
8	All parts	3	1.39	6th
9	Seed	2	0.93	7th
10	Fruit and leaf	1	0.46	8th
11	Leaf, root, and bark	1	0.46	8th
12	Leaf and stem	1	0.46	8th
13	Sap	1	0.46	8th
14	Shoot and flower	1	0.46	8th
Total		**216**	**100**	

Table 6 Methods of preparation

S. No.	Preparation	Number	%	Rank
1	Prepared alone	190	87.96	1st
2	Mixed	26	12.04	2nd
Total		**216**	**100**	

material preparation; elsewhere, there are similar reports of Bosat area (Debela Hunde et al. 2004), Fentalle (Balemie et al. 2004), Ejaji area (Amenu 2007), Konta (Hailemariam 2007), and Loma and Gena Bosa area (Agize et al. 2013). Most of the time the fresh preparation is more threatened than dry preparation due to use and through mechanism of plant material that is not conservative. However, local people argue that fresh materials are effective in treatment as the contents are not lost before use compared to the dried one similar to Ejaji area (Amenu 2007) and Loma and Gena Bosa area (Agize et al. 2013).

Most of the medicinal plant prescription is orally administered 135 (62.5%). About 51 (23.61%) are applied on the skin/surface, 11 (5.09%) are through oral and

Table 7 Mode of administration of traditional medicine

S. No.	Application site	Number	%	Rank
1	Oral	135	62.5	1st
2	Skin/surface	51	23.6	2nd
3	Oral and dermal	11	5.09	3rd
4	Nasal	10	4.63	4th
5	Oral and nasal	7	3.24	5th
6	Dermal and nasal	1	0.46%	6th
Total		**216**	**100**	

dermal, 10 (4.63%) are nasally applied, 7 (3.24%) are through oral and nasal, and only 1 (0.46%) is administrated through dermal and nasal (Table 7). Most people in the study area take their preparations in liquid form. This is because it is easier to take and because of fear to taste some chemicals during chewing or masticating. However most of their prescription is orally administered, 135 (62.5%), similar to Fentalle (Balemie et al. 2004) and Chifra (Seifu et al. 2006) and Loma and Gena Bosa area (Agize et al. 2013).

Medicinal Plants and Indigenous Knowledge of Traditional Healers in the Study Area

Though there is a difference in know-how among healers due to the interest of individuals, age, education status, availability of plants, and occurrence of diseases, the traditional healers of the study area have accumulated traditional medicinal knowledge for a long period of time (for generations) due to their interaction with plants of their environment.

The traditional healers in the study area have knowledge of diseases and medicinal plants used to treat them (Table 1). Medicinal plants practiced in the area are more for treatment of human than cattle and other domestic animals. About 71.76% (155) of the medicinal plants are used to treat humans while 15.28% (33) are used to treat cattle disease only but about 12.96% (28) used for both human and cattle (Table 8).

Most traditional healers practice using the same plant species for treating different diseases of human and domestic animals. About 10 (4.63%) of them are used to treat three types, 30 (13.89%) of them are used to treat two types of diseases, and 176 (81.48%) of them are used to treat only one type of either human or animal diseases (Table 9).

Allium sativum, Artemisia afra, Brachiaria brizantha, Buddleja polystachya, Clerodendrum myricoides, Justicia ladanoides, Lannea fruticosa, Piper capense, Syzygium guineense, and *Tagetes minuta* are popular medicinal plants, each used to treat three types of diseases of either human, human, or animal. These and other plants are kept in the house or in the pocket for immediate usage of accidental illness. For this purpose, they are cultivated or allowed to grow in home gardens.

Table 8 Percentage of medicinal plants to treat human and domestic animals

S. No.	To be treated	Number	%	Rank
1	Human	155	71.76	1st
2	Domestic animals	33	15.28	2nd
3	Human and domestic animals	28	12.96	3rd
Total		**216**	**100**	

Table 9 Percentage of medicinal plants to treat ailments

S. No.	Aliments	Number	%	Rank
1	Three types	10	4.63	3rd
2	Two types	30	13.89	2nd
3	One type	176	81.48	1st
Total		**216**	**100**	

Table 10 The age distribution of respondents

		Male	Female	Total		Rank
S. No.	Age range	number	number	number	%	
1	16–25	10	1	11	12.09	5th
2	26–35	17	3	20	21.98	1st
3	36–45	19	_	19	20.88	2nd
4	46–55	17	1	18	19.78	3rd
5	56–65	13	1	14	15.38	4th
6	66–75	3	_	3	3.3	7th
7	76–85	4	_	4	4.4	6th
8	86–95	2	_	2	2.2	8th
Total (18–92)		**85**	**6**	**91**	**100**	

Healers that exercise traditional medicine (knowledge) are above 15 years of age and below 95 years' range (18–91 years) (Table 10). The traditional healers in the area developed indigenous knowledge to treat different types of diseases using different types of plant species at early ages (26–35 years). This is probably because of the availability of plants in the study area, encouragement/demand of local people, stress of diseases, efficacy in some treatment that cannot be healed using modern medicament, and absence of health centers around them at the time though there are recently established health posts and upgrading clinics of veterinary that are yet not well equipped with facilities and infrastructure for transportation unless the dry season roads. This also indicates that, there is transfer of knowledge at young age as some interviewees informed that because of modernization and less consideration of it/indigenous knowledge, they are willing to transfer their knowledge at earlier time of life. It is one's knowledge that determines the use of plants as medicinal value that others might use for treatment (Agize et al. 2013). In most cases, the tradition of conveying traditional medicinal knowledge to the next generation is at old ages to keep secret. In this process of transferring at the old age, most of indigenous

Table 11 Educational level of the respondents

S. No.	Educational level	Male number	Female number	Total number	%	Rank
1	0	46	4	50	54.95	1st
2	1–4	14	0	14	15.38	3rd
3	5–8	15	1	16	17.58	2nd
4	9–10	3	1	4	4.4	4th
5	11–12	3	0	3	3.3	6th
6	>12	4	0	4	4.4	4th
Total (18–92)		**85**	**6**	**91**	**100**	

knowledge and practices die out with the old knowledgeable individual because he/she loses memory and cannot walk to distant places to show/train important medicinal plants apart from the less ability of children to understand/memorize things.

About 54.95% (50) of them are not educated while 45.05% (41) are literate (Table 11). Educational level is considered as a factor to determine the inheritance of indigenous knowledge and conservation and sustainable use of medicinal plants, because most of the non-educated informants (traditional healers in the study area) harvest medicinal plants from the wild.

Indigenous Knowledge Associated to Conservation of Medicinal Plants

Home Garden Management

Home gardens are rich in species diversity than that of areas away from home due to hipping and spreading of household wastes that served as manure for their growth. Women regularly manure home garden plants with house wastes including cow dung. A similar pattern is reported in Kafa (Asfaw 2004). This traditional practice is used to conserve a lot of plant species with their associated knowledge.

Farmers deliberately preserve some native tree, shrub, and herb species for a variety of purposes specially for immediate access such as for remedy (medicinal), supply of fodder and fuel wood, food, collection of harvested crops, protection of crops and workers from the sun heat (for shade), improving soil fertility, etc. either in home garden or in scattered (on-farm tree) agroforestry system. The more the multiple uses a plant has for local people, the more conservation of that plant resource through cultivation and protection in and around home gardens and farm areas.

The indigenous knowledge of using plants to protect other plant species from disease, pests, and other harming agents has increased conservation value of these species, e.g., *Pycnostachys abyssinica* which is planted among *Ensete ventricosum* to destroy bacterial wilt spread; another option is to rotate the enset field with other crops such as taro and barley and also plant bacterial resistant variety of enset like "Mazya"—local name. Another IK in the community is planting enset and some other species when moon appears for propagation because they believe that species planted during moon were not decay. A similar finding is reported in Dawuro by Dea (1997) and Agize et al. (2013).

The Intercropping of Plants

The presence of on-farm agroforestry and home garden diversity with knowledge about their use is important for the conservation of plant diversity and environmental suitability. The intermixing of multipurpose plants in home gardens and in the farm field benefits the indigenous people. Such activity conserves plants of medicinal value with indigenous practices. It was observed that men are more knowledgeable than women regarding the use of trees and shrubs that were collected from the wild and planted in the home gardens. On the other hand, women are more knowledgeable than men regarding the usage, cultivation, and management of herbaceous species (root and tuber, vegetable crops, spices, condiments, and medicinal plants grown in home garden).

Tree and Shrub Management Practices of Indigenous Knowledge

In the study area home garden and on-farm tree and shrub management agroforestry practices were remarkable. The respondents informed that the tree species are managed by coppicing from the beginning of December up to the end of April or shortly before the rainy season. It was noted that coppice sprouts which are equivalent to straight stem are important consideration in the management of coppicing. The harvest from coppicing can be used to produce firewood and charcoal and other tree products. It was also noted that respondents in the present study site managed tree species by thinning. Respondents informed that the growth of seedling in the study areas is by watering. The other tree management practice further mentioned was pruning, also noted for the protection from splash erosion which would have destroyed the crop. In general the types of management employed vary from one agroecological zone to the other. It was mentioned that the cutting of trees in June, July, and August caused decay. It was preferred to cut trees near the ground at a height of 5–30 cm mainly to protect the sprouts from splitting by wind and to obtain more sprouts. It was noted that coppicing avoids the need to replant trees after harvesting. Generally, thinning, pruning, controlling lopping, watering, and coppicing are the most important farmer's indigenous known home garden and on-farm tree management practices in each agroecological zones of the study site.

Culture of Diversifying Income Generating and Food Security Plant Species

The present interest of community trend of planting various species on their home garden and on farmlands is for income generation and understanding the advantages of improving their livelihood status. On the other hand the culture of the community-obligating individuals planting various species and managing them on their home garden and on farm area has important contribution for biodiversity. The individual who cannot properly mange species diversity at home garden and on farmland is neglected/deprived from social works. The community stops cooperating with them on any social activities. As a result the nature of home gardens and farm area diversity in the study site are rich. A similar finding is reported in Kafa (Asfaw 2004).

Traditional Cultural Ceremony-Celebrating Places

There is indigenous knowledge of the community-preserving plant diversity around church, local tomb, and spiritual ceremonial places like forests or "kasha" (local name). After the death of community member, planting selective indigenous tree, shrub, and herb species on his/her grave/tomb is the common traditional practice. They believe that the type of species planted refers to dead person's strength on his/her lifetime and protecting the grave/tomb from replacement of others. They respect protected forests and big trees as the older men and cultural leaders. In addition to this they believe that "God" destroys them and whole community if they cut big trees (kasha) from culturally protected area. As result of culturally respecting the local protected areas in the study area, there was sustainability of species diversity and hence conservation of that type of plant species as well as others growing under or near the canopy of these plants. There are similar reports for indigenous knowledge of Loma and Gena Bosa woredas community of Dawuro (Agize et al. 2013).

Ochoa Kasha is another ceremony at which women in the neighborhood and relatives come together to the home of a newborn child. They wash the mother 3–4 days after child is born. They wash her after letting for an hour in the steam bath of spices and *Syzygium guineense*. According to their tradition, for this purpose, the spices used for the washing should be cultivated around the house, in the home garden. Therefore, this traditional practice encourages the growth of spices, medicinal plants, and other multipurpose species.

Selective Harvesting

Medicinal plants in the study area are collected in the morning only and not at midday in order to protect themselves from evil spirits that may be hiding in the plant. A similar report is from Chifra district (Seifu et al. 2006). They also claim traditional medicines to be effective only if the plants are collected at certain time of the day as reported in Der es Salaam and on the main island of Ukerewe, Tanzania (Gesseler et al. 1995 cited in Seifu et al. 2006). Medicinal plant collection and provision are also done in holidays (including Sunday, Wednesday, and Friday as God's day) as reported in Ejaji area, Chelya woreda in West Shoa (Amenu 2007), and Gimbi area, in western Wellega (Tolasa 2007). They think that it is effective especially on Sunday because it is their cultural belief that this is the day on which Christ became free from temptation challenges and even death and that might help them. The ritual and spiritual protected areas for celebration probably contain more plant diversity and have contribution to conservation activities in the area, a report similar to Ejaji area, Chelya woreda (Amenu 2007), and Gimbi area in western Wellega (Tolasa 2007). The selective harvesting of spices and herbs and taking of fresh roots and tubers, vegetable crops, spices, condiments, and medicinal plants to market are carried out by women rather than men which is one of the conservative practices and cultures.

The Most Popular Medicinal Plants of the Study Area

The traditional medicinal plants especially in the families *Asteraceae*, *Fabaceae*, and *Lamiaceae* played the most important role in curing illness of human and domestic animals for centuries. The most frequently and easily accessible known traditional medicinal plants are found in the family *Asteraceae*. The others *Allium sativum, Moringa stenopetala, Millettia ferruginea, Gnidia involucrate*, and *Indigofera spicata* are popularly used in the area. These and other plants are kept in the house or in the pocket for immediate usage in case of accidental illness. For this purpose, they are cultivated or allowed to grow in home gardens which agrees with research result of some areas of Dawuro (Agize et al. 2013).

Factors for Depletion of Medicinal Plants of the Study Area

There are different threats to medicinal plant availability and indigenous knowledge in the study area. Indigenous knowledge on medicinal plants is gradually disappearing due to secrecy, unwillingness of young generation to gain the knowledge, influence of modern education, and awareness factors besides other natural conditions and anthropogenic activities. Medicinal plants are being scarcer due to different factors such as agricultural expansion, overharvesting, overgrazing, uncontrolled bushfires, drought, disease and pests, introduction of modernization, and encouraging the new varieties and cultural shifts are also some of the threatening factors (Agize et al. 2013).

The death of old people with particular knowledge on cultural requirement and regarding medicinal uses of some plants that are more knowledgeable than younger individuals is one of the major threats to both medicinal plants and indigenous

Table 12 Time of the stay respondents in the study area

		Male	Female	Total		Rank
S. No.	Years age	number	number	number	%	
1	<10	2	_	2	2.2	2nd
2	10–19	1	_	1	1.1	3rd
3	20–29	1	–	1	1.1	3rd
4	Since birth	81	6	87	95.6	1st
Total		**85**	**6**	**91**	**100**	

knowledge as there is no documented IK of the use of traditional medicines. The acquisition and transfer are done verbally as top secret in the presence of only the healer and his inheritor with strong oath training. A similar study elsewhere in Kenya (Kokwaro 1979), Manyara, Tanzania (Efrem et al. 2004), Fentalle (Balemie et al. 2004), Konso (Gebre 2005), Gimbi (Tolasa 2007), and Loma and Gena Bosa area (Agize et al. 2013) in Ethiopia confirmed such a cultural threat to medicinal plants and IK on them. They treat their patients by disclosing medicinal plant name or never show it thinking that the medicine becomes powerless in curing patient if he/she knows (even if he/she knows that the patient is prohibited to call or show that plant until he/she gets safe).

The destructive practices in harvesting (collecting the whole plant and using one or some parts and throwing the rest unused parts) are the major threat to medicinal plants in specific and to plant diversity (taxa) in general. This may be due to the reason that most of them, about 54.95%, are none educated (Table 11) and 95.6% of them lived in remote area for a long time (Table 12). Uprooting and using roots for treatment result in threat of plant itself than relying on other parts. The heavy or continued exploitation risks the regeneration of the natural source population besides overharvesting and the destruction and conversion of their habitats to other purposes. This is reported as one of the medicinal plant threats (Frankel et al. 1995). The loss of plants causes the loss of traditional knowledge in turn. All informants agree that there is decrease in plant resources of medicinal value from nearby forests. Because of this, they travel long distances, even from one woreda to the other. The elders who cannot travel such a distance inform the threat in transferring their knowledge to the younger generation, as the medicinal plants are no longer available; even they themselves do not serve with/practice their knowledge. As plant resources decrease, indigenous knowledge and practices die out (Agize et al. 2013).

Conclusion and Recommendations

There are immense medicinal plant resources which are used for treatment of both human and livestock in the study area for generation. Traditional practices and various cultural and seasonal restrictions of collecting medicinal plants have contributed to the management and conservation of diversified and rich medicinal plants compared to others. Currently, these resources and indigenous knowledge on medicinal plants are gradually disappearing due to secrecy, unwillingness of young generation to gain the knowledge, influence of modern education, and awareness factors besides other natural conditions and anthropogenic activities like agricultural expansion, timber production, overharvesting for construction materials, and other purposes and overgrazing. The destructive practices in harvesting (collecting the whole plant and using one or some parts and throwing the rest) are the major threat to medicinal plants in specific and to plant diversity (taxa) in general. Most of the medicinal plants that were recorded and effectively used by the community are harvested from wild. For this purpose, they travel long distances even from one woreda to the other. Apart from this, all informants agree that there is a decrease in plant resources of medicinal value from nearby forests. In other words, they are cultivated or allowed to grow in the home garden if and only if they have multipurpose (medicinal, spices, etc.); otherwise they are uprooted by considering as weed. Therefore, training on conservation and sustainable usage, domesticating, and cultivating endangered multipurpose plants in specific and other plants in general should be given for traditional healers as well as for community. Distributing important information in the form of leaflets, brochures, posters, and other ways like media accessible in the area. The following are some suggestions and recommendations for the effective utilization of the medicinal plant species and the associated indigenous knowledge of the Dawuro people of the study area:

- Initiating pharmacological and biological activity testing of most popularly used traditional medicinal plants in the area
- Establishing botanical gardens and protected areas with community-based conservation
- Encouraging the use of home gardens for cultivation of multipurpose plants
- Area closure to allow regeneration of seed and establish community parks in each kebele for the sake of conservation too
- Assess the indigenous multipurpose trees and shrubs and associated indigenous knowledge and then later apply ex situ conservation (establish botanical garden, reforestation, field gene bank, etc.) before their loss

References

Abbink J (1995) Medicinal and ritual plants of Ethiopian southwest. An account of recent research. Indigenous Knowl Dev Monitor 3:6–8

Abebe D (1986) Traditional medicine in Ethiopia: the attempt bering made to promote it for effective and better utilization. SINET: Ethiopian J Sci 9 Suppl:61–69

Abebe D (2001a) Biodiversity conservation of medicinal plants: problems and prospects. In: Proceeding of the national workshop on biodiversity conservation and sustainable use of medicinal plants in Ethiopia, 28 April–01, may 1998. IBCR, Addis Ababa, pp 56–64

Abebe D (2001b) The role of medicinal plants in health care coverage of Ethiopia, the possible integration. In: Zewdu M, Demise A (eds) Proceeding of the national workshop on biodiversity conservation and sustainable use of medicinal plants in Ethiopia, 28 April–1 May 1999, IBCR, Addis Ababa, pp 6–21

Abebe D, Ayehu A (1993) Medicinal plants and enigmatic health practice of North Ethiopia. Berhanina Selam Printing Enterprise, Addis Ababa, p 341

Agize M, Demissew S, Asfaw Z (2013) Ethno botany of medicinal plants in Loma and Gena bosa districts (Woredas) of Dawuro zone, southern Ethiopia. Top Class J Herbal Med 2(9):194–212

Amenu E (2007) Use and management of medicinal plants by indigenous people of Ejaji Area, Chelya Woreda, West Shoa, Ethiopia: an ethnobotanical approach. M.Sc. Thesis, Addis Ababa University, p 102

Anon (2005a) The Gospel of Luke in Dawuro. The Bible Society of Ethiopia (BSE), Addis Abeba, Ethiopia, p 123

Anon (2005b) National biodiversity strategy and action plan. Institute of Biodiversity Conservation (IBC), Addis Ababa, Ethiopia, p 103

Asfaw Z (1999) Report on ethno botanical study nations of nationalities and people in Gambella and Benishangul Gumuz regional states. Progress Report to Research and Publication Office, Addis Ababa University, Addis Ababa

Asfaw Z (2001) The role of home garden in production and conservation of medicinal plants. In: Zewdu M, Demissie A (eds) Conservation and sustainable use of medicinal plants in Ethiopia. Proceeding of the national workshop on biodiversity conservation and sustainable use of medicinal plants in Ethiopia, 28 April–01, May 1998. IBCR, Addis Ababa, pp 76–91

Asfaw Z (2004) The Enset based home gardens of Ethiopia. In: Eyzaguirre PB, Linares OF (eds) Home gardens and agro biodiversity. Smithsonian Institution, Washington, pp 123–147

Awassa T, Asfaw Z, Nordal I, Demissew S (2007) Ethno botany of Berta and Gumuz People in Western Ethiopia. In: Plant diversity in Western Ethiopia: ecology, ethno botany and conservation, Doctor of Philosophy Dissertation, University of Oslo, Norway, pp 1–34

Balemie K, Kelbessa E, Asfaw Z (2004) Indigenous medicinal plant utilization, management and threats in Fentalle area, eastern Shewa, Ethiopia. Ethiopia J Biol Sci 3:37–58

Balick MJ, Cox PA (1996) Plant, people and culture: the science of ethno botany. Scientific American Library, New York, p 220

Bekelle E (2007) Study on actual situations of medicinal plants in Ethiopia prepared for Japan Association for international Collaboration of Agriculture and Forestry, pp 1–76

Birhanu A (2002) Use and conservation of traditional medicinal plants by indigenous people in Jabitehana Woreda, West Gojjam. M.Sc. Thesis, Addis Ababa University, Addis Ababa

Cotton CM (1996) Ethno botany: principles and applications. John Willey and Sons Ltd., Chichester, p 242

Cunningham AB (1993) African medicinal plants: setting priorities at the interface healthcare between conservation and primary health care. In: Sample A (ed) People and plants working paper. UNESCO, Paris, pp 1–50

Dea D (1997) Rural livelihoods and social stratification among the Dawuro, Southern Ethiopia. MSc. Thesis, Addis Ababa University, Addis Ababa, p 136

Farnsworth NR (1985) Plants and modern medicine: where science and folklore Meets. World Health Forum 6:76–80

Farnsworth NR, Soejarto DD (1991) Global importance of medicinal plants. In: Akerel O, Heywood N, Synge H (eds) Conservation of medicinal plants. Cambridge University Press, Cambridge

Frankel OH, Brown AHD, Burdon JJ (1995) The conservation of plant biodiversity. Cambridge University Press, Cambridge, p 299

Fullas F (2001) Ethiopian traditional medicine: common medicinal plants in perspective, Sioux City

Gebre T (2005) Ethno botanical study of medicinal plants in the Konso Special Woreda (SNNPR), Ethiopia, M.Sc. Thesis, Addis Ababa University, Addis Ababa

Getahun A (1976) Some common medicinal and poisonous plants used in Ethiopia folk medicine. Addis Ababa University, Ethiopia, p 63

Giday M (1999) An ethno botanical study of medicinal plants used by the Zay People in Ethiopia. M.Sc. Thesis, Uppsala, Sweden

Giday M (2007) Medicinal plants of the Bench, Meinit and Sheko socio-cultural groups in Ethiopia with emphasis on use diversity, distribution and abundance. Ph.D. Thesis (Unpublished). Addis Ababa University, Addis Ababa

Hailemariam T (2007) An ethno botanical study of medicinal plants used by local people in the lowlands of Konta Special Woreda, SNNPRs, Ethiopia. M.Sc. Thesis, Addis Ababa University, Addis Ababa, p 111

Hunde D (2001) Use and management of traditional medicinal plants by indigenous people in 'Bosat' Woreda 'Welenchiti' Area: an ethnobotanical approach. M.Sc. Thesis, Addis Ababa University, Addis Ababa, p 132

IUCN (1993) Guidelines on conservation of medicinal plants. Castlecary Press, UK

Jain SK (1986) Ethno botany. National Botanical Research Institute, Lucknow, India. Interdisciplinary Sci Rev 11(3):285–292

Jansen PCM (1981) Species, condiments and medicinal plants in Ethiopia, their taxonomy and agricultural significance. Center for Agricultural Publishing and Documentation, Wageningen, Netherlands, p 327

Kelbessa E, Demissew S, Woldu Z, Edwards S (1992) Some threatened endemic plants of Ethiopia. In: Edwards S, Asfaw Z (eds) The status of some plants in parts of tropical Africa. Botany 2000: East and Central Africa: NAPRECA Monograph Series No. 2. NAPRECA, Addis Ababa University, Addis Ababa, pp 33–55

Letchamo W, Craker LE (1996) Cultivating and breeding of medicinal and aromatic plants. J Herbs Spices Medicinal Plants 4:12

Lulekal E (2005) Ethno botanical Study of Medicinal Plants and Floristic Composition of the Manna Angatu Moist Montana Forest, Bale, M.Sc. Thesis, Addis Ababa, Addis Ababa

Martin GJ (1995) Ethno botany: a method manual. World wide Fund for Nature. Chapman and Hall, London, p 263

Mesfin F (2007) An ethno botanical study of medicinal plants in Wonago Woreda, SNNPR, Ethiopia. MSc. Thesis, Addis Ababa University, Addis Ababa, p 74

Pareek A, Trivedi PC (2011) Ethno botanical studies on medicinal plants of Kaladera region of Jaipur District. Indian J Fund App Life Sci 1:59–63

Pei SJ (1991) Conservation of biological diversity in temple yards and holly hills by the Dai ethnic minorities of China. Ethno Botany 3:27–35

Seifu T, Asres K, Gebre-Mariam T (2006) Ethno botanical and ethno pharmaceutical studies on medicinal plants of Chifra District, Afar region, North-Eastern Ethiopia. Ethiopian Pharm J 24:41–58

Seta T, Demissew S, Asfaw Z (2013) Home gardens of Wolayta, Southern Ethiopia: an ethno botanical profile. Acad J Med Plants 1(1):014–030

Sharma H, Kumar A (2011) Ethno botanical studies on medicinal plants of Rajasthan (India): a review. J Med Plant Res 5:1107–1112

Sirvastava R (2000) Studying the information needs of medicinal plant stake holders in Europe. TRAFFIC-International, UK

Taddese M (1986) Some medicinal plants of central Shewa and South Western Ethiopia. SINET: Ethiopian J Sci 9(Suppl)

Taddese M, Demissew S (1992) Medicinal Ethiopian plants: inventory, identification and classification. In: Edwards S, Asfaw Z (eds) Plants used in African traditional medicine as practiced in Ethiopia and Uganda, Botany 2000: East and Central Africa: NAPRECA Monograph Series No. 5. NAPRECA, Addis Ababa University, Addis Ababa, pp 1–19

Tamene B (2000) A floristic analysis and ethno botanical study of the semi wet land of Cheffa Area, South Wello, Ethiopia, M.Sc. Thesis, Addis Ababa University, Addis Ababa, p 139

Teklehaymanot T, Giday M (2007) Ethno botanical study of medicinal plants used by people in Zegie peninsula, north-western Ethiopia. J Ethno Biol Ethno Med 3:12

Tolasa E (2007) Use and conservation of traditional medicinal plants by indigenous people in Gimbi Woreda, Western Wellega, Ethiopia. M.Sc. Thesis, Addis Ababa University, Addis Ababa, p 111

Vanon T, Quyen D, Bich LD (2001) A survey of medicinal plants in Bavi National Park, Vietnam: methodology and implications for conservation and sustainable use. Biol Conserv 97:295–304

Venkataswamy R, Mohammad Mubarak H, Doss A, Ravi TK, Sukumar M (2010) Ethno botanical study of medicinal plants used by Malabar tribes in Coimbatore District of Tamil Nadu, South India. Asian J Exp Biol Sci 1:387–392

WCMC (1992) Global biodiversity status of the Earth's Living resources. World Conservation WHO (1991) Traditional medicine and modern health care: progress report by the director-general paper presented to the forty-fourth world health assembly. Monitoring centre. Chapman and Hall

WHO (2003) Traditional medicine. Fact sheet No. 134

World Health Organization (2002–2005).WHO traditional medical strategy, WHO, Geneva

Yineger H (2005) A study of ethno botany of medicinal plants and floristic composition of the dry Afromontane forest at Bale Mountains National Park. M.Sc. Thesis, Addis Ababa University, Addis Ababa

Medicinal Bryophytes Distributed in Turkey

Munir Ozturk, İsa Gökler, and Volkan Altay

Introduction

There are nearly 23,000 taxa of bryophytes distributed in the world. These are grouped under three: the mosses (14,000 species), the Marchantiophyta or liverworts (6000 species), and the Anthocerotophyta or hornworts (300 species). However, almost every year the number of taxa keeps changing due to reports of new species and synonymization of others (Sabovljević et al. 2016). Some authors consider these as the second largest group of land plants after angiosperms (Marko et al. 2001; Shaw and Renzaglia 2004). General concept about this group of plants is that these are the earliest plants. They are said to have evolved 472 million years ago. This depicts that they could be the ancestors of all land plants (Wellman et al. 2003; Asakawa and Ludwiczuk 2017). These plants seem to have played a great role in maintaining ecosystems because of their buffer system characteristics for other plants. On a global basis bryophytes form a major part of the biodiversity in moist environments, and are important components of the forests, wetlands, and high-altitude ecosystems (Hallingbäck and Hodgetts 2000; Chandra et al. 2017). They are found in all ecosystems except saline waters, but some species inhabit habitats rich in salts and are classified as halophytic bryophytes (*Riella* sp., *Entosthodon hungaricus*, *Hennediella heimii*). The reports show that *Physcomitrella patens*, a non-halophytic species, grows well in liquid media with 1.5% salt concentration

M. Ozturk (✉)
Vice President of the Islamic World, Academy of Sciences, Amann, Jordan

Department of Botany and Centre for Environmental Studies, Ege University, Izmir, Izmir, Turkey

İ. Gökler
Faculty of Science, Biology Department, Dokuz Eylül University, Izmir, Turkey

V. Altay
Faculty of Science & Arts, Biology Department, Hatay Mustafa Kemal University, Antakya, Hatay, Turkey

M. Ozturk, K. R. Hakeem (eds.), *Plant and Human Health, Volume 1*,
https://doi.org/10.1007/978-3-319-93997-1_8

under laboratory conditions (King et al. 2016; Sabovljević et al. 2016). This group of plants lacks seeds or flowers and is thus placed under the group of cryptogams, which are too small, characterized by dominant perennial gametophytic stages, with relatively small, unbranched sporophyte, which remains permanently attached to the maternal gametophyte (Cox et al. 2010). Most people do not have much information about these plants, due to their small size, and less biomass. The latter features have made these plants neglected for wide use (Harris 2008). On the other hand, the bryophytes have been used in packing, plugging, as well as decoration from very ancient times (Chandra et al. 2017).

The plant taxa belonging to this group are used as indicator species, for erosion control, as bioindicators of heavy metals in air pollution, as aquatic bioindicators, and as radioactivity indicators. They are also evaluated as material for seed beds, fuel, medicines, food sources, and pesticides; in nitrogen fixation, moss gardening, treatment of waste, constructions, clothing, furnishing, packing, and genetic engineering; and for soil conditioning and culturing (Saxena and Harinder 2004; Glime 2007). One of the important characteristics of bryophytes is that they serve as important environmental indicators. These plants have been used as predictors of past climate change to validate climate models and potential indicators of global warming (Gignac 2001; Azuelo et al. 2011).

The reports published sometime back have revealed that bryophytes are an important contributor to the total stream metabolism, nutrient cycling, food web interactions in streams, and a direct food source for some invertebrates (Azuelo et al. 2011). In particular the mosses like *Bryum, Hypnum*, and *Fissidens* grow in association with thallophytes in shallow waters of lakes, streams, and springs. They contain huge amounts of lime and act as rock builders. According to Chandra et al. (2017) the insoluble calcium carbonate precipitates along with mosses and their symbionts, thereby helping in the soil conservation. In industrialized countries like Sweden, West Germany, Finland, Poland, Ireland, and Soviet Union liverworts and mosses are evaluated as a source of fuel for generating natural gas, hydrogen, ethylene, and methanol. The peat moss is accepted as the best source of fuel for generation of heat and methane and this group of mosses shows rapid regeneration and low sulfur content, their heating value is greater than wood (Saxena and Harinder 2004; Chandra et al. 2017). The bryophytes in general lack leaf cuticle; this enables them to gain and lose water quickly. These plants are able to absorb minute quantities of available moisture from fog, mist, and dew. This characteristic is not generally found in the higher plants except *Welwitschia mirabilis*. Other uses of plants included here such as liverworts and mosses are regarded as good indicators of environmental conditions; they can act as growth regulators (auxins, gibberellins, cytokinins, and ethylene), whereas mosses are used as stuffing material (Sabovljevic et al. 2011a; Chandra et al. 2017).

The data published by Asakawa (1981, 1982, 1994, 1995, 2001, 2007), Pant and Tewari (1990), Asakawa et al. (2013), and Chandra et al. (2017) on the phytochemistry of bryophytes shows that they exhibit a greater range of biologically active compounds such as carbohydrates, lipids, proteins, steroids, polyphenols, terpenoids, organic acids, sugar alcohols, fatty acids, aliphatic compounds, acetogenins, phenyl-quinones, and aromatic and phenolic substances, all showing significant bioactivities. These active constituents are widely used as antibacterial, antifungal, cytotoxic, antitumor, and insecticidal agents (Asakawa 2007; Ucuncu et al. 2010). They are also used in the medicinal and agricultural areas (Saxena and Harinder 2004; Pant 1998). Some taxa are a great source of herbal medicines (Azuelo et al. 2011).

Bryophytes in the Traditional Medicine: Global Perspective

The use of bryophytes by various ethnic groups for their healthcare or other needs has not been investigated at length, because the feeling among the people is that these plants do not play a direct role in human health care (Alam et al. 2015; Chandra et al. 2017). Undoubtedly not much has been published on the ethnobotanical uses of plants from this group, when compared to the vascular plants. Their ethnobotanical importance has not attracted the attention of individuals in different cultures around the world. In all 136 taxa of bryophytes have been recorded up till now with their ethnobotanical use around the globe for different purposes (Harris 2006; Sabovljević et al. 2016). A half of these uses are of medical importance (Sabovljević et al. 2016) (Table 1; Fig. 1). These tiny creatures are widely used by different tribal groups in Nepal; different parts of South, North, and East India; Pakistan; Argentina; Poland; Africa; Australia; New Zealand; Japan; Taiwan; China; the USA, and Europe (Flowers 1957; Chandra et al. 2017).

The reasons for a restricted use of bryophytes as medicinal plants throughout the world are that they produce little biomass per single species per locality, and are small and hard to distinguish. However, in the areas like boreo-polar and tropical regions there are evidences of ethnobotanical use of bryophytes by the local people on a large scale, because their biomass in these areas is substantially higher. On the other hand, in the dryer areas ethnobotanical use of bryophytes decreases due to smaller biomass, in spite of high species diversity. Some big exceptions do exist, as in high-biomass-yielding species *Fontinalis antipyretica* and species of the genus *Polytrichum* the situation is different (Sabovljević et al. 2016).

An ancient way of looking at the medicinal value of plants is the concept of Paracelsus "doctrine of signatures," which deals with the resemblance of plant parts with the shape and structure of an organ in human or animal body for which it is remedial (Chandra et al. 2017). According to this philosophy, *Marchantia polymorpha* is used to cure hepatic disorders as its shape is like liver (Miller and Miller 1979). In the same way, the hair cup moss *Polytrichum commune* bears hairy calyptra and oil extracted from this moss has been used by the women in ancient time for their hair treatment (Glime 2007). Different ethnic groups around the world are using these tiny plants to cure various ailments in their daily lives. For example, people of Gaddi tribes of Himachal Pradesh in India are using *Plagiochasma appendiculatum* for treating skin diseases (Kumar et al. 2000). The thalloid *Targionia hypophylla* is used by Irular tribe of the Attappady valleys of Kerala state in India to cure skin diseases due to resemblance of thallus of this liverwort to the rough surface

Table 1 Medicinally important bryophytes used in the traditional medicine on global scale

Family/taxa	Medical uses	Sources[a]
Liver Worts		
Aneuraceae		
Riccardia sp.	Antileukemic activity	**1, 2**
Riccardia multifida	Antileukemic activity	**3, 4**
Aytoniaceae		
Plagiochasma sp.	Antimicrobial, cancer, thrombosis, muscle relaxation, burn infection	**5-7**
Plagiochasma appendiculatum	Skin diseases	**1, 8**
Plagiochasma intermedium	Antifungal	**9**
Reboulia hemisphaerica	Blotches, hemostasis, external wounds, bruises, antiplatelet, antiobesity	**9**
Conocephalaceae		
Conocephalum conicum	Antimicrobial, antifungal, antipyretic, cancer, antidotal activity, cuts, swollen tissue, scalds, burns, fractures, poisonous snake bites, gallstones	**1, 6, 10, 11**
Dumortieraceae		
Dumortiera hirsuta	Antimicrobial, cancer	**2, 12**
Herbertaceae		
Herbertus sp.	Antiseptics, antidiarrheal agents, expectorants, astringents	**1, 2**
Herbertus aduncus	Antifungal	**4, 13, 14**
Jubulaceae		
Frullania sp.	Cancer, antiseptic activity, nourishment of hair	**9, 15**
Frullania tamarisci	Antiseptic	**9**
Frullania ericoides	Anti-lice (in hair)	**15**
Jungermaniaceae		
Jungermannia sp.	Neurotrophic action, antituberculosis	**16, 17**
Lepidoziaceae		
Bazzania sp.	Antimicrobial, cancer	**18**
Lepidozia sp.	Cancer, antiplatelet	**19**
Lophocoleaceae		
Hepatostolonophora paucistipula	Cancer	**20**
Chiloscyphus rivularis	Cancer	**20**
Plicanthus hirtellus	Cancer	**4, 9**
Marchantiaceae		
Dumortiera hirsuta	Antibiotics	**2**
Marchantia sp.	Boils, abscesses	**21**

(continued)

Table 1 (continued)

Family/taxa	Medical uses	Sources[a]
Marchantia polymorpha	Diuretics, liver ailments, insect bites, boils, abscesses, pulmonary tuberculosis, cuts, fractures, poisonous snake bites, burns, scalds, open wounds, cardiovascular disease, antimicrobial, cancer, thrombosis, muscle relaxation, antiviral (flu), skin ailments, antileukemic	**2, 8, 22-26**
Marchantia convoluta	Hepatitis, fever, gastric intolerance	**27**
Marchantia palmata	Acute inflammation caused by the touch of fire and hot; boils, abscesses	**26, 28**
Marchantia paleacea	Skin tumefaction, hepatitis, antipyretic	**29**
Pallaviciniaceae		
Pallavicinia sp.	Antimicrobial, antifungal	**2, 4, 30, 31**
Pelliaceae		
Pellia endiviifolia	Antimicrobial, skin diseases	**32**
Plagiochilaceae		
Plagiochila sp.	Antileukemic, antimicrobial, cancer, neurotrophic action, antifeeding, antiviral (flu), burns, blisters, insecticidal	**1, 5, 9, 32–34**
Plagiochila beddomei	Wound healing	**35**
Porellaceae		
Porella sp.	Antimicrobial, cancer	**36**
Radulaceae		
Radula sp.	Antimicrobial, antiviral (flu), thrombosis	**4, 37**
Ricciaceae		
Riccia sp.	Ringworms (in children), antimicrobial	**8**
Scapaniaceae		
Diplophyllum sp.	Cancer	**38**
Targioniaceae		
Targionia hypophylla	Scabies, itches, and other skin diseases (in children)	**15**
Weisnerellaceace		
Wiesnerella denudata	Cancer	**1**
MOSSES		
Amblystegiaceae		
Cratoneuron filicinum	Heart disease	**9, 20, 39**
Leptodictyum riparium	Antipyretic	**20, 39**
Bartramiaceae		
Philonotis fontana	Pain, adenopharyngitis, antipyretic	**9, 39, 40**
Philonotis sp.	Burns, adenopharyngitis, antipyretic, antidotal	**9, 20**
Plagiopus oederianus	Sedative, epilepsy	**39**
Bryaceae		
Bryum sp.	Fever, body aches, healing wounds, burns, bruises, fungal infections	**2, 22, 41**
Bryum argenteum	Antidote, antipyretic, antifungal	**9, 42**

(continued)

Table 1 (continued)

Family/taxa	Medical uses	Sources[a]
Bryum capillare	Wounds, burns, bruises, fungal infections	**22, 41**
Rhodobryum giganteum	Cardiovascular problem, nervous prostration, angina, antihypoxia, diuretic, antipyretic, antihypertensive	**9, 10, 39, 43, 44**
Rhodobryum roseum	Nervous prostration, cardiovascular diseases	**9, 39, 43, 44**
Dicranaceae		
Dicranum scoparium	Antimicrobial	**4, 45**
Leucobryum bowringii	Body pain	**35**
Oreas martiana	Anodyne (pain), hemostasis, wounds, epilepsy, menorrhagia, neurasthenia (nervosism, nervous exhaustion)	**9**
Ditrichaceae		
Ceratodon purpureus	Antifungal	**42**
Ditrichum pallidum	Convulsions (in infants)	**9, 39**
Entodontaceae		
Entodon flavescens	Earache	**35**
Entodon myurus	Antibacterial activity	**6**
Fissidentaceae		
Fissidens adianthoides	Bandage wounds	**4**
Fissidens nobilis	Diuretic, swollen throats, hair growth stimulation	**2, 15, 39, 46, 47**
Funariaceae		
Funaria hygrometrica	Hemostasis, pulmonary tuberculosis, bruises, skin infection	**39**
Fontinalaceae		
Fontinalis antipyretica	Chest fever, antimicrobial	**48, 49**
Hypnaceae		
Hypnum cupressiforme	Antimicrobial, antifungal	**4, 50**
Taxiphyllum taxirameum	Wounds, hemostasis	**9, 39**
Meteoriaceae		
Aerobryum lanosum	Burns	**35**
Mniaceae		
Plagiomnium cuspidatum	Hemostasis, nose bleeding	**9, 39**
Mnium sp.	Pain of burns, bruises, and wounds; hemostasis; and nosebleed	**2, 6, 20**
Plagiomnium sp.	Infections and swellings	**2, 49**
Octoblepharaceae		
Octoblepharum albidum	Febrifuge and anodyne	**51**
Polytrichaceae		
Atrichum undulatum	Antimicrobial	**29, 52-54**

(continued)

Table 1 (continued)

Family/taxa	Medical uses	Sources[a]
Dawsonia superba	Diuretics, hair growth stimulation, cold	**2, 49**
Pogonatum macrophyllum	Inflammation, fever, detergent diuretic, laxative, hemostatic agent	**1, 2, 49**
Polytrichum sp.	Burns, bruises, wounds, fever, inflammation, antipyretic, antidotal, hemostasis, cuts, bleeding from gingivae, cold	**2, 22, 55**
Polytrichum commune	Hemostasis, wound, antipyretic, antidotal, dissolve kidney, gallbladder stones, to speed up labor process during child birth, burns and bruises, fever, inflammation, cuts, bleeding from gingivae, diuretic, laxative, and hemostatic agent	**8, 10, 22, 23, 39, 47, 55-58**
Polytrichum juniperinum	Prostate, urinary difficulties, skin ailments	**4, 55, 58-60**
Pottiaceae		
Barbula sp.	Antirheumatic febrifuge, colds, fever, body aches	**41**
Barbula unguiculata	Fever, cold, body aches	**2**
Barbula indica	Menstrual pain, intermittent fever	**35**
Hyophila attenuata	Cold, cough, neck pain	**35**
Weissia controversa	Cold, fever	**9, 39**
Sphagnaceae		
Sphagnum portoricense	Antimicrobial	**12**
Sphagnum sericeum	Dressing wounds, antimicrobial properties for skin ailments (insect bites, scabies, acne), hemorrhoids, and eye diseases	**2, 10, 21, 25**
Sphagnum strictum	Antimicrobial	**12**
Sphagnum teres	Eye diseases	**10, 59**
Thuidiaceae		
Haplocladium microphyllum	Cystitis, bronchitis, tonsillitis, pneumonia, fever	**10, 39**
HORNWORTS		
Ceratophyllaceae		
Ceratophyllum demersum	Purgative, astringent, constipating, antipyretic	**61**

[a]Sources: **1**: Alam 2012; **2**: Azuelo et al. 2011; **3**: Asakawa 1982; **4**: Sabovljević et al. 2016; **5**: Singh et al. 2006; **6**: Singh et al. 2011; **7**: Wigginton 2002; **8**: Shirsat 2008; **9**: Asakawa 2007; **10**: Ding 1982; **11**: Ando 1983; **12**: Madsen and Pates 1952; **13**: Matsuo et al. 1982; **14**: Matsuo et al. 1983; **15**: Remesh and Manju 2009; **16**: Kondoh et al. 2005; **17**: Grolle et al. 2005; **18**: Scher et al. 2004; **19**: Paliwal et al. 2014; **20**: Asakawa et al. 2013; **21**: Saxena and Harinder 2004; **22**: Beike et al. 2010; **23**: Hu 1987; **24**: Miller and Miller 1979; **25**: Bland 1971; **26**: Pant and Tewari 1989; **27**: Rao 2009; **28**: Tag et al. 2007; **29**: Sabovljevic et al. 2011a; **30**: Subhisha and Subramoniam 2005; **31**: Millar et al. 2007; **32**: Sharma et al. 2015; **33**: Kumar et al. 2000; **34**: Ramirez et al. 2010; **35**: Lubaina et al. 2014; **36**: Dey and Mukherjee 2015; **37**: Castle 1967; **38**: Hong 1980; **39**: Pant 1998; **40**: Flowers 1957; **41**: Sturtevant 1954; **42**: Frahm 2004; **43**: Wu 1977; **44**: Wu 1982; **45**: Pavletic and Stilinovic 1963; **46**: Harris 2002; **47**: Harris 2008; **48**: Drobnik and Stebel 2014; **49**: Chandra et al. 2017; **50**: Veljić et al. 2009; **51**: Singh 2011; **52**: McCleary and Walkington 1966; **53**: Sabovljevic et al. 2011b; **54**: Sabovljevic et al. 2010; **55**: Gulaban 1974; **56**: Turner et al. 1983; **57**: Sabovljevic et al. 2001; **58**: Hart 1981; **59**: Glime 2007; **60**: Belkin et al. 1952–1953; **61**: Pullaiah 2006

Fig. 1 Some of the medicinally important bryophytes in the world (**a**) *Conocephalum conicum* (www.bioref.lastdragon.org); (**b**) *Funaria hygrometrica* (www.uniprot.org); (**c**) *Haplocladium microphyllum* (www3.sfasu.edu); (**d**) *Marchantia polymorpha* (www.inaturalist.org); (**e**) *Polytrichum commune* (www.ohiomosslichen.org); (**f**) *Reboulia hemisphaerica* (www.bryophytes.plant.siu.edu); (**g**) *Rhodobryum giganteum* (www.bryophytes.plant.siu.edu); (**h**) *Ceratophyllum demersum* (www.gobotany.newenglandwild.org)

of the diseased part. The liverwort *Frullania ericoides* has a long stem and hairlike thallus; it is applied for hair-related afflictions by tribal people of South India (Remesh and Manju 2009; Chandra et al. 2017).

The Gasuite Indians in Utah, USA, are using several genera of bryophytes such as *Philonotis, Bryum, Mnium*, or some hypnaceous forms to alleviate the pain from burns. The reason behind this is the consequence of cooling the burnt skin; they also make a kind of paste and apply it as a poultice (Ando and Matsuo 1984; Sabovljevic et al. 2001, 2016). Cheyenne Indians in Montana, USA, and few other indigenous tribes from Alaska use the moss *Polytrichum juniperinum* for the same purpose (Ando and Matsuo 1984; Sabovljević et al. 2016). *Marchantia polymorpha* has been used as a diuretic in Europe. French liverworts were soaked with white liquor and patients drank the resulting mixture of liquor and extracts (Garnier et al. 1961).

In the traditional Chinese medicinal system bryophytes have been adequately used. According to Ding (1982) nearly 40 bryophyte species are evaluated as crude drugs in the Chinese traditional medicine. The thalloid liverworts *Conocephalum conicum* and *Marchantia polymorpha* are often mixed with vegetable oils and used as ointments for boils, eczema, cuts, bites, and burns. Peat moss *Sphagnum teres* is very popular among the Chinese for eye diseases. *Haplocladium microphyllum* moss is used for tonsillitis, bronchitis, cystitis, and tympanitis. *Polytrichum commune* is widely evaluated as a natural cure for many health disorders. They possess antipyretic, diuretic, and hemostatic features. The liverworts *Frullania tamarisci* and *Reboulia hemisphaerica* and mosses *Weissia controversa, Funaria hygrometrica, Bryum argenteum*, and *Climacium dendroides* are widely available and sold in the Chinese markets. Ding (1982) has published a detailed report on Chinese medicinal bryophytes with their Latin names, morphological features, habitats, distribution localities, pharmacological activity, and effects, together with their prescription uses. Several mosses are widely used medicinally in China, to treat burns, bruises, external wounds, snake bite, pulmonary tuberculosis, neurasthenia, fractures, convulsions, scald, uropathy, pneumonia, and neurasthenia, among other uses (Ding 1982; Asakawa 1999; Asakawa and Ludwiczuk 2017).

Use of Bryophytes in Clinical Practices: Global View

In many pharmaceutical laboratories, research institutes, and universities scientific research on medicinal uses of bryophytes is carried out on a large scale. These investigations involve work on the active ingredients of medicinally important bryophytes which are used in curing diseases such as hepatic disorders, skin diseases, cardiovascular diseases, and several other ailments. These research studies also deal with the discovery of new kinds of drugs from bryophytes, not explored so far. The drugs reported in Ayurveda, Unani, and Siddha system need to be further evaluated and validated scientifically (Ratra and Gupta 2015; Chandra et al. 2017). This new trend of evaluation and validation of traditional practices with modern knowledge provides significant opportunities for new drug discoveries; these will lead to an effective strategy for the improvement of human health care (Chandra et al. 2017).

The bryophytes are regarded as the earliest land plants. They are reported to have a capability to synthesize a number of secondary metabolites to combat against different kinds of stress. The phytochemicals from these plants can cope up with infection, predation, radiation, as well as temperature and salinity fluctuations. The pharmacological investigations, phytochemical evaluations, and clinical trials can be used for exploiting the diverse and novel nature of secondary constituents. Development of drug resistance in proliferative cells as well as in microbes can be controlled by using such novel natural products. The possible use of bryophytes as medicine may lead to cure of different ailments which have been difficult to treat by conventional medicine (Dey and Mukherjee 2015).

Anticancer Activity

Several useful anticancer agents are reported to occur in plants; there are a number of naturally derived compounds which act as an important source in this connection. *Polytrichum commune* is used in traditional Chinese medicine for curing lymphocytic leukemia as well as other diseases like fever, hemostatic and traumatic injury to pneumonia, and uterine prolapse (Zhonghua 1999). Moreover, the pharmacological exploration of acid and alcohol extract of *Polytrichum juniperinum* has been reported to exhibit inhibitory action against sarcoma (Asakawa 1982) implanted in CAF1 mice (Cheng et al. 2012). Similarly the compound ohioensin A isolated from *Polytrichum ohioense* is reported to show cytotoxicity against 9PS murine leukemia and MCF-7 human breast tumor cells. Benzonaphthoxanthenone and cinnamoyl bibenzyl derivatives isolated from ethanolic extract of *Polytrichum pallidisetum* are mentioned to significantly impede the growth of RPMI-7951 melanoma and U-251 glioblastoma multiforme (Zheng et al. 1993).

According to Asakawa (1982) bryophytes have a great potential for antileukemia activity; for example Marchantin A from *Marchantia paleacea*, *M. polymorpha*, and *M. tosana*; riccardin from *Riccardia multifida*; and perrottetin E from *Radula perrottetii* are reported to show cytotoxicity against the leukemic KB cells. The active constituent diplophyllin, an ent-eudesmanolide isolated from liverworts *Diplophyllum albicans* and *D. taxifolium*, has a-methylene lactone unit which shows significant anticancer activity against human epidermoid carcinoma (Ohta et al. 1977). Sesquiterpenoids costunolide and tulipinolide isolated from *Frullania monocera*, *Marchantia polymorpha*, *Porella japonica*, *Wiesnerella denudate*, *Conocephalum supradecomositum*, and *Plagiochila semidecurrens* are also recorded as to show anticancer activity against human carcinoma of the nasopharynx. The compound extracted from *Plagiochila fasciculata* is mentioned as to inhibit P388 cells (leukemia) (Asakawa 1981). The mosses are also mentioned as growth retarders for cancer cells in culture (Hallingbäck and Hodgetts 2000). The mechanism of how bryophytes inhibit growth of different cancer cell lines and exhibit anticancer activity has not been validated at length as yet (Chandra et al. 2017). More detailed studies are needed in this connection.

In the area of cancer chemotherapy, natural products derived from plants occupy an important place, because of their minimal side effects. For example, *Polytrichum commune* is reported to play a significant role in the therapy of lymphocytic leukemia. It has also been found that *P. ohioense* and *P. pallidisetum* show cytotoxicity against the 9PS murine leukemia and several other tumor cell lines. *Marchantia paleacea, M. polymorpha*, and *M. tosana; Riccardia multifida; and Radula perrottetii* are reported to show cytotoxicity against the leukemic KB cell (Chandra et al. 2017).

Other Potential Uses

Some of the liverworts are reported to produce hot-tasting substances, like capsaicin or α piperine, which are attributed to some sesquiterpene and diterpene dialdehydes. As against this, the species of mosses like *Fissidens* and *Rhodobryum* possess a strong sweet taste. These could be useful as spices for food or may be used as food additives (Asakawa and Ludwiczuk 2017). Some produce significant amounts of vitamins B2 and E and related compounds. These reports stress the fact that such bryophyte taxa are potentially important as food or spice; they can thus be exploited (Asakawa and Ludwiczuk 2017). The use of bryophytes as insecticidal agents has gained importance over the past few decades. Fatty acids derived from *Hypnum cupressiforme, Dicranum scoparium, Polytrichastrum formosum, Homalothecium lutescens*, and liverwort *Conocephalum conicum* have been used as insecticides against *Sitophilus granarius*. Similarly, the solution of powder from *Calymperes afzelii, Thuidium gratum, Bryum coronatum*, and *Barbula lambarenensis* is evaluated against maize stem borers (Abay et al. 2013; Ande et al. 2010).

Negative Impacts of Bryophytes

Generally, bryophytes are not damaged by bacteria and fungi, insect larvae and adults, snails, slugs, and small mammals. Some liverworts are allelopathic and do cause intense allergic contact dermatitis (Asakawa and Ludwiczuk 2017). Some taxa of *Frullania* like *F. dilatata, F. tamarisci, F. tamarisci* spp. *nisquallensis*, as well as *Chiloscyphus polyanthos* and *Schistochila appendiculata* are reported to be the main causative agents of potent allergic contact dermatitis, as these taxa contain a number of sesquiterpene lactones with α-methylene g-lacone functionality. The *Frullania* taxa cited above grow epiphytically on the bark of trees; as such they have been recognized as the cause of occupational contact dermatitis in forest workers, woodcutters, and olive pickers in some areas of Canada, the USA, Finland, and France (Mitchell 1986; Mitchell et al. 1970; Knoche et al. 1969). The allergens, (+)-frullanolide and (−)-frullanolide, isolated from *F. dilatata* and *F. tamarisci* ssp. *tamarisci*, respectively, cause very intense allergenic contact dermatitis. The allergens of the *Schistochila appendiculata* are long-chain alkylphenols, 3-undecyl, 6-undecyl, 3-tridecyl, 3-pentadecyl and 3-heptadecyl phenols; long-chain alkyl salicylic acids, 6-tridecyl, 6-pentadecyl salicylates; and their potassium salts, potassium 6-undecyl, 6-tridecyl, 6-pentadecyl salicylates, and 6-undecyl catechol. These too cause contact dermatitis (Asakawa 1994). *Marchantia polymorpha* and *Metzgeria furcata* also show allergenic contact dermatitis activity, but their allergens have not been isolated as yet (Asakawa 1982; Asakawa et al. 2013).

Studies on Bryophytes of Turkey

Bryofloristic Studies

Bryofloristic studies in Turkey by local bryologists started in 1980s (Gökler et al. 1984; Gökler and Öztürk 1986, 1987a, b, 1989). These have continued and much information has been pooled up on bryophyte diversity (Abay et al. 2016; Erdağ and Kürschner 2017a, b). The results have been published in the form of several provincial checklists. The first checklist of liverworts in Turkey was prepared by Gökler et al. (1985–1986), followed by the second detailed one on liverworts and hornworts (Çetin 1988a). Gökler and Öztürk (1991, 1992, 1994a, b, 1996) have published the list of Turkish liverworts. The first checklist of Turkish mosses was published by Çetin (1988b). The period between 1988 and 2004 is one of the active periods enlisting intensive bryological studies in the history of Turkish bryology (Abay et al. 2016). A major publication on the Turkish moss checklist was published by Uyar and Çetin (2004), followed by the second one for mosses, liverworts, and hornworts in Turkey (Kürschner and Erdağ 2005). However, the number of bryophytes in the country has kept on increasing during these years as many new species and new records have been added (Abay et al. 2016). In the year 2009 Özenoğlu Kiremit and Keçeli published a detailed list of the liverworts and hornworts of Turkey. Keeping in view all these published records, the bryophytes in the country are represented by a total of 960 species comprising approximately 6.5% (for now) of world's bryophytes. Among these, two species of bryophytes are endemic to Turkey (*Jungermannia caucasica* and *J. lignicola*). The mosses are represented by 773 species, liverworts 183 species, and hornworts 4 species in Turkey (Erdağ and Kürschner 2017a, b).

Bryo-Ecological and Bryo-Ecophysiological Studies

Bryo-ecological studies in Turkey by local bryologists were started in 1980s (Öztürk and Gökler 1988). These studies, although limited in number, continued during the following years (Gökler 1993; Gökler and Öztürk 1994b; Gökler and Özenoğlu

1999; Abay and Ursavaş 2009; Erkara 2017). However, during the last decade studies on the epiphytic bryophyte communities in Turkey have increased (Alataş et al. 2017). In bryosociological studies on epiphytic bryophytes, a total of 26 syntaxa have been reported from Turkey till to date (Alataş et al. 2017). The investigations on the bryophytes in Turkey were undertaken on a large scale on different ecophysiological features such as bioaccumulation and oxidative stress, biomonitoring of atmospheric heavy metal pollution, determination of heavy metal concentration, levels of cesium radionuclides, assays of 210Po and 210Pb, support materials, and upgrading of biomass materials as energy sources, pigment concentration, and allelopathic effects (Tonguç 1998; Baysal and Özdemir 1999; Demirbaş et al. 2000; Şahin et al. 2000; Tüzen et al. 2003; Uğur et al. 2003, 2004; Mendil et al. 2005, 2009; Sarı et al. 2005; Uyar et al. 2007a, b, 2009; Içel and Çobanoğlu 2009; Ezer et al. 2010, 2016; Batan et al. 2011; Belivermiş et al. 2016; Aydoğan et al. 2017; Demir et al. 2017; Türkyılmaz Ünal et al. 2017).

Bryo-Phytochemical Screening Studies

During the last few decades considerable number of studies have been carried out on the taxonomy, morphology, anatomy, ecology, and ecophysiology of bryophytes in Turkey. Some of these do focus on the phytochemical screening, such as essential oil composition, total phenol and luteolin contents, and some elemental levels (Yaylı et al. 2009; Cansu et al. 2010, 2013; Özdemir et al. 2010, 2013; Ucuncu et al. 2010; Gökbulut et al. 2012; Batan et al. 2013; Tosun et al. 2014; Aslanbaba et al. 2017; Çöteli et al. 2017; Yağlıoğlu et al. 2017; Yong Tan et al. 2017).

Medicinally Important Bryophytes of Turkey

In the reports on the traditional folk medicine of Turkey no information could be traced on the use of bryophytes. However, many species of bryophytes are widely used in traditional folk medicine in other countries in the world. These are well distributed in Turkey as well. The current research on bryophytes has been studied on antibacterial, antimicrobial and antifungal, antioxidative and antiproliferative, wound-healing activity, and insecticidal and cytotoxic potentials (Table 2; Fig. 2). The taxa belonging to the group of bryophytes are used in the treatment of several diseases such as hepatic disorders, cardiovascular diseases, and many other disorders in different parts of the world. But in Turkey no clinical studies have been undertaken till now in connection with the treatment of these diseases.

Table 2 Studies undertaken on the potential medicinal uses of bryophytes in Turkey

No.	Taxa	Medicinal activity	Source[a]
1	*Anomodon viticulosus*	Antimicrobial	**1, 2**
2	*Brachythecium campestre*	Antimicrobial	**3**
3	*Calliergonella cuspidata*	Antibacterial	**4, 5**
4	*Calliergonella lindbergii*	Antibacterial	**5**
5	*Cinclidotus riparius*	Antibacterial	**4**
6	*Cirriphyllum crassinervium*	Antibacterial	**4**
7	*Conocephalum conicum*	Insecticidal	**6**
8	*Corsinia coriandrina*	Wound healing	**7**
9	*Ctenidium molluscum*	Antimicrobial	**1, 8**
		Antioxidant	**8**
10	*Dicranum majus*	Antibacterial	**5**
11	*Dicranum scoparium*	Insecticidal	**6**
		Antiproliferative	**9**
12	*Diplophyllum taxifolium*	Antibacterial	**5**
13	*Eurhynchium angustirete*	Antimicrobial	**10**
14	*Eurhynchium pulchellum*	Antimicrobial	**3**
15	*Eurhynchium striatum*	Antimicrobial	**10**
		Antibacterial	**5**
16	*Eurhynchium striatulum*	Antimicrobial	**8**
		Antioxidant	**8**
17	*Fontinalis antipyretica*	Antimicrobial	**12**
		Antiproliferative	**12**
18	*Funaria hygrometrica*	Antibacterial	**11**
		Antifungal	**11**
19	*Grimmia alpestris*	Antibacterial	**5**
20	*Grimmia anodon*	Antibacterial	**13**
		Antifungal	**13**
21	*Grimmia orbicularis*	Antibacterial	**5**
22	*Hedwigia ciliata*	Antibacterial	**5**
23	*Homalothecium lutescens*	Antimicrobial	**8, 14**
		Antioxidant	**8**
24	*Homalothecium sericeum*	Antimicrobial	**2, 8, 15**
		Antioxidant	**8**
		Insecticidal	**6**
		Antiproliferative	**15**
25	*Hylocomium splendens*	Antimicrobial	**16**
26	*Hypnum cupressiforme*	Antimicrobial	**2, 8, 14**
		Antioxidant	**8**
		Antibacterial	**11**
		Antifungal	**11**
		Insecticidal	**6, 17**

(continued)

Table 2 (continued)

No.	Taxa	Medicinal activity	Source[a]
27	*Hypnum imponens*	Antibacterial	**11**
		Antifungal	**11**
28	*Isothecium alopecuroides*	Antibacterial	**5**
29	*Leucobryum glaucum*	Antibacterial	**5**
30	*Leucobryum juniperoideum*	Antibacterial	**4**
31	*Leucodon sciuroides*	Antimicrobial	**2, 8, 16**
		Antioxidant	**8**
32	*Mannia androgyna*	Wound healing	**7**
33	*Marchantia polymorpha*	Antioxidant	**18**
34	*Metzgeria conjugata*	Antibacterial	**5**
35	*Mnium stellare*	Antimicrobial	**19**
36	*Orthotrichum rupestre*	Antibacterial	**13**
		Antifungal	**13**
37	*Palustriella commutata*	Antimicrobial	**20**
38	*Plagiochasma rupestre*	Wound healing	**7**
39	*Plasteurhynchium meridionale*	Antimicrobial	**1**
40	*Platyhypnidium riparioides*	Antimicrobial	**1, 2**
		Antioxidant	**21**
41	*Pleurochaete squarrosa*	Antibacterial	**13**
		Antifungal	**13**
		Antimicrobial	**22**
42	*Pohlia nutans*	Antimicrobial	**14**
43	*Polytrichastrum formosum* (syn. *Polytrichum formosum*)	Antimicrobial	**1**
		Insecticidal	**6**
		Antibacterial	**5**
44	*Polytrichum commune*	Antibacterial	**5**
45	*Polytrichum juniperinum*	Antibacterial	**11**
		Antifungal	**11**
46	*Porella cordaeana*	Cytotoxic	**23**
		Wound healing	**7**
47	*Porella platyphylla*	Wound healing	**7**
48	*Pseudoscleropodium purum*	Antimicrobial	**10**
49	*Pterigynandrum filiforme*	Antimicrobial	**24**
50	*Ptychostomum capillare* (syn. *Bryum capillare*)	Antibacterial	**13**
		Antifungal	**13**
51	*Reboulia hemisphaerica*	Wound healing	**7**
52	*Rhytidiadelphus triquetrus*	Antiproliferative	**25**
		Cytotoxic	**25**
53	*Riccia fluitans*	Wound healing	**7**
54	*Schistidium papillosum*	Antibacterial	**5**

(continued)

Table 2 (continued)

No.	Taxa	Medicinal activity	Source[a]
55	*Schistidium trichodon*	Antibacterial	**5**
56	*Syntrichia calcicola*	Antibacterial	**5**
57	*Syntrichia laevipila*	Antibacterial	**5**
58	*Syntrichia montana* (Syn. *Syntricha intermedia*)	Antibacterial	**5**
59	*Syntrichia ruralis*	Antibacterial	**5, 13**
		Antifungal	**13**
60	*Syntrichia virescens*	Antibacterial	**5**
61	*Targionia hypophylla*	Wound healing	**7**
62	*Thamnobryum alopecurum*	Antibacterial	**4**
63	*Thuidium delicatulum*	Antimicrobial	**8, 26**
		Antioxidant	**8**
64	*Thuidium tamariscinum*	Antioxidant	**21**
65	*Tomentypnum nitens* (syn. *Homalothecium nitens*)	Antimicrobial	**8**
		Antioxidant	**8**
66	*Tortella humilis*	Antibacterial	**5**
67	*Tortella inclinata* var. *densa*	Antimicrobial	**22**
		Antibacterial	**5**
68	*Tortella tortuosa*	Antibacterial	**5, 11, 13**
		Antifungal	**11, 13**
		Antimicrobial	**22, 27**
		Antiproliferative	**25**
		Cytotoxic	**25**
69	*Tortula muralis*	Antimicrobial	**14**

[a]Sources: **1**: Dulger et al. 2009; **2**: Çolak et al. 2011; **3**: Yayintas and Yapıcı 2009; **4**: Uyar et al. 2016; **5**: Sevim et al. 2017; **6**: Abay et al. 2013; **7**: Tosun et al. 2016; **8**: Ertürk et al. 2015; **9**: Abay et al. 2015; **10**: Tosun et al. 2015; **11**: Savaroglu et al. 2011a; **12**: Savaroğlu et al. 2011b; **13**: Elibol et al. 2011; **14**: Ucuncu et al. 2010; **15**: Oztopcu-Vatan et al. 2011; **16**: Cansu et al. 2013; **17**: Abay et al. 2012; **18**: Gökbulut et al. 2012; **19**: Canli et al. 2015; **20**: Ilhan et al. 2006; **21**: Aslanbaba et al. 2017; **22**: Tosun et al. 2014; **23**: Yong Tan et al. 2017; **24**: Yetgin et al. 2017; **25**: Yağlıoğlu et al. 2017; **26**: Altuner and Çetin 2009; **27**: Altuner et al. 2010.

Conclusion

The bryophyte taxa are regarded as a "remarkable reservoir" of new natural products as well as secondary compounds, many of which show interesting biological activity (Abdel-Shafi et al. 2017). Out of approximately 23,000 taxonomically described taxa of bryophytes, only a few percent of this group of small photosynthesizing terrestrial green spore-forming plants have been investigated chemically, but nearly a thousand references are available (Ando and Matsuo 1984; Inoue 1988; Asakawa and Ludwiczuk 2017). The liverworts from this group of plants have been evaluated more comprehensively in terms of the chemistry and molecular biology as compared to the mosses and the hornworts. The plants from the latter groups lack

Fig. 2 Some examples of the potentially important medicinal bryophytes from Turkey. (**a**) *Homalothecium sericeum* (www.cisfbr.org.uk); (**b**) *Hypnum cupressiforme* (www.arcticatlas.org); (**c**) *Pleurochaete squarrosa* (www.inaturalist.org); (**d**) *Polytrichastrum formosum* (www.cisfbr.org.uk); (**e**) *Porella cordaeana* (www.inaturalist.org); (**f**) *Syntrichia ruralis* (www.cisfbr.org.uk); (**g**) *Thuidium delicatulum* (www.florafinder.org); (**h**) *Tortella tortuosa* (www.ohiomosslichen.org)

cell oil bodies found in the liverworts, from which are a source of more than 1000 secondary metabolites (Asakawa and Ludwiczuk 2017). Some mosses produce vitamin B2 and many kinds of unsaturated fatty acids and lipid (triglycerides)-containing unsaturated alkane moieties. In view of these findings, further strategies to focus on secondary metabolites that could be applied to the diets of cattle and cultured fish are needed. In liverworts the metabolites have potential drug use because of the presence of some biologically important compounds, particularly bis-bibenzyls, from the marchantin and isoplagiochin series; these are involved in the interference with the normal breakdown of microtubules during cell division of cancer cells, like the clinical drug paclitaxel (Asakawa and Ludwiczuk 2017). *Marchantia polymorpha* and *M. paleacea* ssp. *diptera* both produce a large amount of marchantin A (Yoshikawa et al. 2002) and its analogues, and are very easy to culture in the field and in the greenhouse (Asakawa and Ludwiczuk 2017). There are a large number of medicinally important bioactive compounds occurring in the bryophytes but with little information at present. They are used globally as drugs as remedies to cure various diseases (Bodade et al. 2008; Sabovljević et al. 2016). The medicinal importance of this group has not been exploited fully. They can be used as pharmaceutical products, in horticultural practices and for household purposes (Kumar et al. 1999). They have an ability to treat illness of cardiovascular system, tonsillitis, bronchitis, skin diseases, and burns. The bryophytes also possess anticancer and antimicrobial activity due to their unique chemical constituents (Banerjee and Sen 1979; Abdel-Shafi et al. 2017). *Plagiochila fasciculata* shows inhibitory effect on virus (*herpes simplex* type 1, Polio type 1) and bacteria (*Bacillus subtilis, E. coli, Candida albicans*, and *Cladosporium resinae*) (Lorimeres and Perry 1994). Both acetone and ethanol extracts of the bryophytes do inhibit the growth of *Escherichia coli, Bacillus cereus, Erwinia chrysanthemi*, and *Pseudomonas aeruginosa* on an agar plate (Kandpal et al. 2016). In fact, these plants have been proven to be potent, nontoxic, and broad-spectrum antibacterial substances (Lashin et al. 2015).

Bryophytes like ferns and other vascular plants embody immense potential in the development of allopathic medicines for evaluation in the treatment of some important diseases. The taxa known to us and used since ancient times in traditional folk medicinal system need to be followed in depth. These medicines can be produced for use by the commons and their supply arranged in order to meet the requirement at global scale. Such plants can be prescribed as herbal formulations to cure many diseases with a scope of economic earnings (Singh et al. 2010). It is strongly emphasized that there is an immediate need for all organic chemists and biochemists, molecular geneticists, pharmacy faculty researchers, and industries to concentrate on this group of plants to evaluate the tremendous genetic potential of bryophytes (as potential medicinal plants) in Turkey. The bryophytes could be an alternative solution for the treatment of several present-time diseases due to their secondary metabolites.

References

Abay G, Altun M, Karakoc OC, Gul F, Demirtas I (2013) Insecticidal activity of fatty acid-rich Turkish bryophyte extracts against *Sitophilus granarius* (Coleoptera: Curculionidae). Comb Chem High Throughput Screen 16:806–816

Abay G, Batan N, Özdemir T (2016) Bryophyte checklist of Rize, north-East Turkey. Arctoa 25(2):386–392

Abay G, Karakoç OC, Tufekçi AR, Koldaş S, Demirtaş İ (2012) Insecticidal activity of *Hypnum cupressiforme* (Bryophyta) against *Sitophilus granarius* (Coleoptera: Curculionidae). J Stored Prod Res 51:6–10

Abay G, Koldaş S, Tüfekçi AR, Demirtaş İ (2015) Determination of antiproliterative activities of volatile contents and HPLC profiles of *Dicranum scoparium* (Dicranaceae, Bryophyta). Comb Chem High Throughput Screen 18:453–463

Abay G, Ursavaş S (2009) The moss (*Musci*) flora and ecology of Çankırı Research Forest. Bartın Orman Fakültesi Dergisi 11(16):61–70

Abdel-Shafi S, Hussein Y, Lashin G, Abdel-Monaem A-S (2017) An evaluation of the antibacterial and antiviral activities of some bryophytes. Egypt J Microbiol 52(1):63–86

Alam A (2012) Some Indian bryophytes known for their biologically active compounds. Int Jappl Bio Pharma Tech 3:239–246

Alam A, Shrama V, Rawat KK, Verma PK (2015) Bryophytes-the ignored medicinal plants. SMU Med J 2:299–316

Alataş M, Batan N, Ezer T, Uyar G (2017) The epiphytic bryophyte flora and vegetation of Boraboy and Destek forests (Amasya, Turkey). Pak J Bot 49:1779–1786

Altuner EM, Çetin B (2009) Antimicrobial activity of *Thuidium delicatulum* (Bryopsida) extracts. Kafkas Üniv Fen Bil Enst Derg 2(2):85–92

Altuner EM, Çetin B, Çökmüş C (2010) *Tortella tortuosa* (Hedw.) Limpr. özütlerinin antimikrobiyal aktivitesi. Kafkas Üniv Fen Bil Enst Derg 10:111 116

Ande AT, Wahedi JA, Fatoba PO (2010) Biocidal activities of some tropical moss extracts against maize stem borers. Ethnobot Leafl 14:479–490

Ando H (1983) Use of bryophytes in China 2. Mosses indispensable to the production of Chinese gallnuts. Proc Bryol Soc Japan 3:124–125

Ando H, Matsuo A (1984) Applied bryology. In: Schultze-Motel W, Cramer J (eds) Advances in bryology, Vaduz, vol 2, pp 1–211

Asakawa Y (1981) Biologically active substances obtained from bryophytes. J Hattori Bot Lab 50:123–142

Asakawa Y (1982) Chemical constituents of the bryophytes. In: Herz et al (eds) Progress in the chemistry of organic natural products, vol 42. Springer, Vienna, pp 1–285

Asakawa Y (1994) Highlights in phytochemistry of hepaticae-biologically active terpenoids and aromatic compounds. Pure Appl Chem 66:2193–2196

Asakawa Y (1995) Chemical constituents of the bryophytes. In: Herz et al (eds) Progress in the chemistry of organic natural products, vol 65. Springer-Verlag, Wien, pp 1–562

Asakawa Y (1999) Phytochemistry of bryophytes. In: Romeo J (ed) Phytochemicals in human health protection, nutrition, and plant defense. Kluwer Academic/Plenum Publishers/Springer, New York, pp 319–342

Asakawa Y (2001) Pharmacologically active substances from oriental bryophytes and inedible mushrooms and fijian kava. World Conference on Medicinal and Aromatic Plants Book of Abstracts

Asakawa Y (2007) Biologically active compounds from bryophytes. Pure Appl Chem 79:557–580

Asakawa Y, Ludwiczuk A (2017) Chemical constituents of bryophytes: structures and biological activity. J Nat Prod. https://doi.org/10.1021/acs.jnatprod.6b01046

Asakawa Y, Ludwiczuk A, Nagashima F (2013) Phytochemical and biological studies of bryophytes. Phytochemistry 91:52–80

Aslanbaba B, Yılmaz S, Yayıntaş ÖT, Özyurt D, Öztürk BD (2017) Total phenol content and antioxidant activity of mosses from Yenice Forest (Ida Mountaın). J Scientific Perspect 1(1):1–12
Aydoğan S, Erdağ B, Aktaş L (2017) Bioacumulation and oxidative stress impact of Pb, Ni, cu and Cr heavy metals in two bryophyte species, *Pleurochaete squarrosa* and *Timmiella barbuloides*. Turk J Bot 41:464–475
Azuelo AG, Sariana LG, Pabualan MP (2011) Some medicinal bryophytes: their ethnobotanical uses and morphology. Asian J Biodivers 2(1):49–80
Banerjee RD, Sen SP (1979) Antibiotic activity of bryophytes. Bryologist 82(2):141–153
Batan N, Mendil D, Apaydin G, Cengiz E (2013) Some element levels in moss samples collected from the Igdir-Nahhicevan International Highway, Turkey. Asian J Chem 25(12):6741
Batan N, Özdemir T, Mendil D (2011) Determination heavy metal concentration of mosses in Değirmendere Valley of Trabzon province in Turkey. Asian J Chem 24(19):193–196
Baysal A, Özdemir T (1999) The levels of cesium radionuclides in mosses collected from Akçaabat, Maçka and Sürmene in the eastern Black Sea region of Turkey. Toxil Environ Chem 69:481–485
Beike AK, Decker E, Wolfgang F, Lang D, Verulit-Scheebaum M, Zimmer A, Reski R (2010) Appl Bryol Bryotechnol Trop Bryol 31:22–32
Belivermiş M, Kılıç Ö, Çayır A, Coşkun M, Coşkun M (2016) Assesment of 210Po and 210Pb in lichen, moss and soil around Çan coal - fired power plant, Turkey. J Radioanal Nucl Chem 307:523–531
Belkin M, Fitzgerald DB, Felix MD (1952–1953) Tumor-damaging capacity of plant materials. II. Plants used as diuretics. J Natl Cancer Inst 13: 741-744
Bland J (1971) Forests of Lilliput. Prentice-Hall, Inc., Englewood Cliffs, pp 1–210
Bodade RG, Borkar PS, Saiful MA, Khobragade CN (2008) In vitro screening of bryophytes for antimicrobial activity. J Med Plants 7:23–28
Canli K, Altuner EM, Akata I (2015) Antimicrobial screening of *Mnium stellare*. Bangladesh J Pharmacol 10(2):321–325
Cansu TB, Ucuncu O, Kahriman N, Ozdemir T, Yayli N (2010) Essential oil composition of *Grimmia trichophylla* grew and *G. decipiens* (Shultz) Lindb. Grown in Turkey. Asian J Chem 22(9):7280
Cansu TB, Yaylı B, Özdemir T, Batan N, Alpay Karaoğlu S, Yaylı N (2013) Antimicrobial activity and chemical composition of the essential oils of mosses (*Hylocomium splendens* (Hedw.) Schimp. And *Leucodon sciuroides* (Hedw.) Schwägr.) growing in Turkey. Turk J Chem 37:213–219
Castle H (1967) A revision of the genus radula. Part II. Subgenus Acroradula. Section 11. Complanatae Rev Bryol Lichénol 35:1–94
Çetin B (1988a) Checklist of the liverworts and hornworts of Turkey. Lindbergia 14:12–14
Çetin B (1988b) Checklist of the mosses of Turkey. Lindbergia 14:15–23
Chandra S, Chandra D, Barh A, Pandey RK, Sharma IP (2017) Bryophytes: hoard of remedies, an ethno-medicinal review. J Trad Complement Med 7(1):94–98
Cheng X, Xiao Y, Wang X et al (2012) Anti-tumor and pro-apoptotic activity of ethanolic extract and its various fractions from *Polytrichum commune* L. ex Hedw in L1210 cells. J Ethnopharmacol 143:49–56
Çolak E, Kara R, Ezer T, Çelik GY, Elibol B (2011) Investigation of antimicrobial activity of some Turkish pleurocarpic mosses. Afr J Biotechnol 10(60):12905–12908
Çöteli E, Alataş M, Batan N (2017) Comparing of glutathion ingradients of *Syntrichia ruralis and Syntrichia montana* (Pottiaceae) taxa. Anatolian Bryol 3(1):25–30
Cox CJ, Goffinet B, Wickett NJ, Boles SB, Shaw AJ (2010) Moss diversity: a molecular phylogenetic analysis of genera. Phytotaxa:175–195
Demir ME, Uygun-Aktaş D, Erdağ A, Akgöl S (2017) A new support material for Ig adsorption: *Syntrichia papillosissima* (Copp.) Loeske. Artif Cells Nanomed Biotechnol. https://doi.org/10.1080/21691401-2017-1296848

Demirbaş A, Özdemir T, Şahin B, Güllü D, Akdeniz F, Çağlar A (2000) Upgrading of biomass materials as energy sources: liquefaction of mosses from Turkey. Energy Sources 22:403–408
Dey A, Mukherjee A (2015) Therapeutic potential of bryophytes and derived compounds against cancer. J Acute Dis 4(3):236–248
Ding H (1982) Medicinal spore-bearing plants of China. Shanghai Science and Technology Press, Shanghai, pp 1–409
Drobnik J, Stebel A (2014) Medicinal mosses in pre-Linnaean bryophyte floras of Central Europe. An example from the natural history of Poland. J Ethnopharmacol 153:682–685
Dulger B, Hacioglu N, Uyar G (2009) Evaluation of antimicrobial activity of some mosses from Turkey. Asian J Chem 21(5):4093
Elibol B, Ezer T, Kara R, Yuvalı C, Colak E (2011) Antifungal and antibacterial effects of some acrocarpic mosses. Afr J Biotechnol 10(6):986–989
Erdağ A, Kürschner H (2017a) Türkiye Bitkileri Listesi (Karayosunları). Ali Nihat Gökyiğit Vakfı yayını, İstanbul, pp 1–181
Erdağ A, Kürschner H (2017b) A reference list of Turkish bryophytes. The state of knowledge from 1829 until 2017. Anatolian Bryol 3:81–102
Erkara IP (2017) Spore morphology, taxonomical and ecological importance of some Encalyptaceae Schimp. Species (Bryophyta) from Turkey. Bangladesh J Bot 46:139–145
Ertürk O, Sahın H, Ertürk EY, Hotaman HE, Koz B, Özdemır Ö (2015) The antimicrobial and antioxidant activities of extracts obtained from some moss species in Turkey. Herba Polonica 61:52–65
Ezer T, Kara R, Demir İ (2010) The comparison of pigment concentration in some aquatic and non-aquatic bryophytes: chlorophyll a/b and total carotenoid. Biyoloji Bilimleri Araştırma Dergisi 3:181–183
Ezer T, Yılmaz T, İşlek C, Türkyılmaz-Ünal B (2016) Impact of heavy metals (zinc and lead) on the photosynthetic pigment contents of *Bryum schleicheri* and *Plagiomnium undulatum* (Bryophyta). Biol Divers Conserv 9:122–127
Flowers S (1957) Ethnobryology of the Gosiute Indians of Utah. Bryologist 60:11–14
Frahm JP (2004) Recent developments of commercial products from bryophytes. Bryologist 107:277–283
Garnier G, Bézanger-Beauquesne L, Debranx G (1961) Resources médicinales de la flore française, vol 1. Vigot Freres, Paris, pp 1–1511
Gignac LD (2001) New frontiers in bryology and lichenology: bryophytes as indicators of climate change. Bryologist 104:410–420
Glime JM (2007) Economic and ethnic uses of bryophytes vol. 27. Fl North Am: Am Editorial Committee pp 14-41
Gökbulut A, Satilmiş B, Batçioğlu K, Cetin B, Şarer E (2012) Antioxidant activity and luteolin content of *Marchantia polymorpha* L. Turk J Biol 36(4):381–385
Gökler İ (1993) Studies on the taxonomy and ecology of some western Anatolian liverworts. Eğitim Bilimleri Dergisi 2:79–85
Gökler İ, Inoue H, Öztürk M (1984) A new record for Turkey, *Pellia neesiana* (Gott.) Limpr. J Faculty Sci Ege Univ B 7:85–89
Gökler İ, Özenoğlu H (1999) Taxonomy and ecology of liverworts of Kazdağı National Park and its environs. Çevre Koruma 8:22–26
Gökler İ, Öztürk M (1986) Taxonomical investigations on some liverworts (*Hepaticae*) distributed in Turkey. 1. *Jungermanniales Anacrogynae* and *J Acrogynae* Doğa, Türk Biyoloji Dergisi C 10: 163-170
Gökler İ, Öztürk M (1987a) A new record for Turkey, *Porella thuja* (dicks.) C. Jens. Doğa. Türk Botanik Dergisi 11:313–315
Gökler İ, Öztürk M (1987b) New concepts concerning the methods used in the identification of liverworts (Hepaticae). Doğa. Türk Botanik Dergisi C 11:306–312
Gökler İ, Öztürk M (1989) An investigation on the liverworts (Hepaticae) of Black Sea region Doğa. Türk Botanik Dergisi C 13:242–248

Gökler İ, Öztürk M (1991) Liverworts of Turkey and their position in south-West Asia. Candollea 46:359–366

Gökler İ, Öztürk M (1992) Liverworts (*Marchantiopsida*) from the Artvin province (A4, A5). XI. Ulusal Biyoloji Kongresi, 24–27 Haziran 1992, Elazig, pp. 185–194

Gökler İ, Öztürk M (1994a) Liverworts (*Marchantiopsida*) from the Istanbul province (A1). XII. Ulusal Biyoloji Kongresi, 6–8 Temmuz 1994, Edirne, pp. 174-178

Gökler İ, Öztürk M (1994b) Studies on the taxonomy and ecology of liverworts of the Kütahya province. Ege Üniversitesi Fen Fakültesi Dergisi Ser B 16:1525–1529

Gökler İ, Öztürk M (1996) Liverworts of Turkish Thrace. Bocconea 5:319–323

Gökler İ, Öztürk M, Kesercioğlu T (1985–1986) Checklist of liverworts (*Hepaticae*) recorded from Turkey. J Faculty Sci Ege Univ B 8: 1-10

Grolle R, Long D, McNeill J (2005) Nomenclatural problems in *Pleurozia* (Pleuroziaceae): the lectotypification of *Pleurozia* Dumort., *Jungermannia sphagnoides* Schwägr., *Pleurozia sphagnoides* Dumort., *P. purpurea* Lindb., *Mnium jungermannia* L. and *Jungermannia undulata* L.(Hepaticae). Taxon 54:503–508

Gulaban A (1974) Bryophytes as economic plants. Botanica 14:73–75

Hallingbäck T, Hodgetts N (2000) Status survey and conservation action plan for bryophytes: Mosses, Liverworts and Hornworts. Gland: IUCN/SSC Bryophyte Specialist Group, IUCN, pp 1–106

Harris E (2002) An examination of phylogenetic characters in mosses: Examples from *Fissidens* Hedw. (Fissidentaceae: Musci). In: Presentation and abstract presented at the annual meeting of the American Bryological and Lichenological Society, 26–27 July 2002, Storrs, CN, USA

Harris ES (2008) Ethnobryology: traditional uses and folk classification of bryophytes. Bryol 111:169–217

Harris ESJ (2006) Ethnobotany, evolution and chemistry of medicinal bryophytes: examples from the moss genus *Plagiomnium*. PhD thesis, Berkeley University of California, Berkeley

Hart JA (1981) The ethnobotany of the northern Cheyenne Indians of Montana. J Ethnopharmacol 4:1–55

Hong WS (1980) A study of the distribution of *Diplophyllum* in western North America. Bryologist 83:497–504

Hu R (1987) Bryology. Higher Education Press, Beijing

Içel Y, Çobanoglu G (2009) Biomonitoring of atmospheric heavy metal pollution using lichens and mosses in the city of Istanbul, Turkey. Fresen Environ Bull 18(11):2066–2071

Ilhan S, Savaroğlu F, Çolak F, İşçen CF, Erdemgil FZ (2006) Antimicrobial activity of *Palustriella commutata* (Hedw.) ochyra extracts (Bryophyta). Turk J Biol 30(3):149–152

Inoue H (1988) Bryophytes as an indicator of continental drift (Gondwana land). Kagaku-Asahi 48:116–121

Kandpal V, Chaturvedi P, Negi K, Gupta S, Sharma A (2016) Evaluation of the antibiotic and biochemical potential of bryophytes from Kumaun Hills and Taral Belt of Himalayas. Int J Pharmacy Pharmaceut Sci 8:65–69

King BC, Vavitsas K, Ikram NKBK, Schrøder J, Scharff LB, Hamberger B, Jensen PE, Simonsen HT (2016) In vivo assembly of DNA-fragments in the Mos, Physcomitrella patenss. Sci Rep 6:25030

Knoche H, Ourisson G, Perold GW, Foussereau J, Maleville J (1969) Allergenic component of a liverwort: a sesquiterpene lactone. Science 166:239–240

Kondoh M, Nagashima F, Suzuki I, Harada M, Fujii M, Asakawa Y, Watanabe Y (2005) Induction of apoptosis by new ent-kaurene-type diterpenoids isolated from the New Zealand liverwort *Jungermannia* species. Planta Med 71:1005–1009

Kumar K, Singh KK, Asthana AK, Nath V (1999) Ethno therapeutics of bryophyte *Plagiochasma appendiculatum* among the Gaddi tribes of Kangra Valley, Himachal Pradesh. Indian Pharmaceut Biol 37:1–4

Kumar K, Singh KK, Asthana AK, Nath V (2000) Ethnotherapeutics of bryophyte *Plagiochasma appendiculatum* among the Gaddi tribes of Kangra valley, Himachal Pradesh, India. Pharm Biol 38:353–356

Kürschner H, Erdağ A (2005) Bryophytes of Turkey: an annotated reference list of the species with synonyms from the recent literature and an annotated list of Turkish bryological literature. Turk J Bot 29:95–154

Lashin GMA, Abdel-Shafi S, Hussein Y, Osman A, Al A-M (2015) Efficient inhibition of pathogenic bacteria and potential toxicity by secondary metabolites of some bryophyte. Egypt J Bot Microbiol 3:475–497

Lorimeres SD, Perry NB (1994) Antifungal hydroxyl acetophenones from the New Zealand liverwort, *Plagiochila fasciculate*. Planta Med 60:386–387

Lubaina AS, Pradeep DP, Aswathy JM, Remya Krishnan MKV, Murugan K (2014) Traditional knowledge of medicinal bryophytes by the Kani tribes of Agasthiyarmalai biosphere reserve, southern western ghats. IAJPS 4:2116–2121

Madsen GC, Pates AL (1952) Occurrence of antimicrobial substances in chlorophyllose plants growing in Florida. Bot Gazette 113:293–300

Marko S, Aneta B, Dragoljub G (2001) Bryophytes as a potential source of medicinal compounds. Pregl Rev 21:17–29

Matsuo A, Yuki S, Nakayama M (1983) Herbertenediol and (−)-herbertenolide, two new sesquiterpenoids of the ent-herbertane class from the liverwort *Herberta adunca*. Chem Lett 12:1041–1042

Matsuo A, Yuki S, Nakayama M, Hayashi S (1982) Three new sesquiterpene phenols of the ent-herbertane class from the liverwort *Herberta adunca*. Chem Lett 11:463–466

McCleary JA, Walkington DL (1966) Moss and antibiosis. Rev Bryol Lichenol 34:309–314

Mendil D, Çelik F, Tuzen M, Soylak M (2009) Assessment of trace metal levels in some moss and lichen samples collected from near the motorway in Turkey. J Hazard Mater 166(2–3):1344–1350

Mendil D, Tuzen M, Sari H, Hasdemir E (2005) Trace metal levels in moss samples collected from the roadside in Tokat, Turkey. Trace Elem Electrol 22(1)

Millar KDL, Crandall-Stotler BJ, Ferreira JFS, Wood KV (2007) Antimicrobial properties of three liverworts in axenic culture: *Blasia pusilla, Pallavicinia lyellii* and *Radula obconica*. Cryptogam Bryol 28:197–210

Miller NG, Miller H (1979) Make ye the bryophytes. Horticulture 57:40–47

Mitchell JC (1986) *Frullania* (liverwort) phytodermatitis (woodcutter's eczema). Clin Dermatol 4:62–64

Mitchell JC, Fritig B, Singh B, Towers GHN (1970) Allergic contact dermatitis from *Frullania* and compositae. The role of sesquiterpene lactones. J Investig Dermatol 54:233–239

Ohta Y, Andersen NH, Liu CB (1977) Sesquiterpene constituents of two liverworts of genus *Diplophyllum*: novel eudesmanolides and cytotoxicity studies for enantiomeric methylene lactones. Tetrahedron 33:617–628

Özdemir T, Apaydın G, Mendil D, Bulut VN, Cengiz E, Gündoğdu A, Aylıkçı V (2010) Determination of some elements in moss samples from north eastern Anatolia, Turkey. Asian J Chem 22(1):346–352

Özdemir T, Batan N, Mendil D, Apaydın G, Cengiz E (2013) Some element levels in moss samples collected from the Iğdir-Nahhicevan international highway, Turkey. Asian J Chem 25:6741–6744

Özenoğlu Kiremıt H, Keçeli T (2009) An annotated check-list of the Hepaticae and Anthocerotae of Turkey. Cryptogamie Bryol 30:343–356

Oztopcu-Vatan P, Savaroglu F, Filik Iscen C, Kabadere S, Ilhan S, Uyar R (2011) Antimicrobial and antiproliferative activities of *Homalothecium sericeum* (Hedw.) Schimp. Extracts. Fresenius Environ Bull 20:461–466

Öztürk M, Gökler İ (1988) Ecology of West-Anatolian liverworts. Bryol Times 47:1–3

Paliwal A, Arjun M, Madhav NV, Murthy E, Aruna M (2014) Endangered treatment of traditional medicinal amphibian plants (Bryophytes). In: Proceedings of national seminar on traditional medicine & health practices, pp 61–75

Pant G, Tewari SD (1989) Various human uses of bryophytes in the Kumaun region of northwest Himalaya. Bryologist 92:120–122

Pant G, Tewari SD (1990) Bryophytes and mankind. Ethnobotany 2:97–103
Pant GP (1998) Medicinal uses of bryophytes. In: Chopra RN (ed) Topics in bryology. Allied Publisher Limited, New Delhi, pp 112–124
Pavletic Z, Stilinovic B (1963) Untersuchungen über die antibiotische Wirkung von Moosextrakten auf einige Bakterien. Acta Bot Croat 22:133–139
Pullaiah T (2006) Encyclopaedia of world medicinal plants, vol 1. Dayabooks, New Delhi, pp 513–514
Ramirez M, Kamiya N, Popich S, Asakawa Y, Bardon A (2010) Insecticidal constituents from the argentine liverwort *Plagiochila bursata*. Chem Biodivers 7:1855–1861
Rao M (2009) Microbes and non-flowering plants: impact and applications. Ane book Pvt. Ltd, New Delhi, pp 213–214
Ratra M, Gupta R (2015) Future prospects and aspects of herbal drug discovery in herbal medicines. Sciences 2:16–21
Remesh M, Manju CN (2009) Ethnobryological notes from Western Ghats, India. Bryol 112:532–537
Sabovljevic A, Sokovic M, Glamoclija J, Ciric A, Vujicic M, Pejin B, Sabovljevic M (2010) Comparison of extract bio-activities of in situ and in vitro grown selected bryophyte species. Afr J Microbiol Res 4:808–812
Sabovljevic A, Sokovic M, Glamoclija J, Ciric A, Vujicic M, Pejin B, Sabovljevic M (2011a) Bio-activities of extracts from some axenically farmed and naturally grown bryophytes. J Med Plants Res 5:565–571
Sabovljevic M, Bijelovic A, Grubisic D (2001) Bryophytes as a potential source of medicinal compounds. Lekovite Sirovine 21:17–29
Sabovljevic M, Sabovljevic A, Vujicic M, Ljaljevic-Grbic M, Rodda M, Girlanda M (2011b) Are there endobionts in bryophytes? The case study of peat-moss *Sphagnum palustre*. In: 19th symposium of the Serbian plant physiology society, Banja Vrujci, June 13–15, 2011, Book of Abstracts p 23
Sabovljević MS, Sabovljević AD, Ikram NKK, Peramuna A, Bae H, Simonsen HT (2016) Bryophytes-an emerging source for herbal remedies and chemical production. Plant Genet Resour 14(4):314–327
Şahin B, Doğan G, Demırbaş A, Özdemır T, Akdenız F, Aglar A (2000) Upgrading of biomass materials as energy sources: liquefaction of mosses from Turkey. Engery sources, part 1: recovery. Util Environ Effects 22:403–408
Sarı H, Mendil D, Tüzen M, Hasdemır E, Özdemir T (2005) AAS determination of trace metals in some moss samples from Trabzon, Turkey. Fresenius Environ Bull 14:473–477
Savaroglu F, Ilhan S, Filik-Iscen C (2011a) An evaluation of the antimicrobial activity of some Turkish mosses. J Med Plants Res 5(14):3286–3292
Savaroğlu F, İşçen CF, Vatan APÖ, Kabadere S, Ilhan S, Uyar R (2011b) Determination of antimicrobial and antiproliferative activities of the aquatic moss *Fontinalis antipyretica* Hedw. Turkish J Biol 35(3):361–369
Saxena DK, Harinder (2004) Uses of bryophytes. Resonance 9(6):56–65
Scher JM, Speakman JB, Zapp J, Becker H (2004) Bioactivity guided isolation of antifungal compounds from the liverwort *Bazzania trilobata* (L.) SF gray. Phytochemistry 65:2583–2588
Sevim E, Baş Y, Celik G, Pınarbaş M, Bozdeveci A, Özdemir T et al (2017) Antibacterial activity of bryophyte species against *Paenibacillus larvae* isolates. Turk J Vet Anim Sci 41(4):521–531
Sharma A, Slathia S, Gupta D, Handa N, Choudhary SP, Langer A, Bhardwaj R (2015) Antifungal and antioxidant profile of ethnomedicinally important liverworts (*Pellia endivaefolia* and *Plagiochasma appendiculatum*) used by indigenous tribes of district Reasi: north west Himalayas. Proc Natl Acad Sci India Sect B: Biol Sci 85:571–579
Shaw AJ, Renzaglia KJ (2004) Phylogeny and diversification of bryophytes. Am J Bot 91:1557–1581
Shirsat RP (2008) Ethnomedicinal uses of some common bryophytes and pteridophytes used by tribals of Melghat region (Ms), India. Ethnobot Leafl 1:92

Singh A (2011) Herbalism, phytochemistry and ethnopharmacology. CRC Press, New Delhi, pp 286–293

Singh AP, Rawat VK, Behera SK, Khare PB (2010) Perspectives of pteridophytes biodiversity: a source of economy elevation. In: National conference on biodiversity, development and poverty alleviation, 22nd May, India, pp 46–49

Singh M, Raghavan G, Nath V, Rawat AKS, Mehrotra S (2006) Antimicrobial, wound healing and antioxidant activity of *Plagiochasma appendiculatum* Lehm. & Lind. J Ethnopharmacol 107:67–72

Singh M, Singh S, Nath V, Sahu V, Rawat AKS (2011) Antibacterial activity of some bryophytes used traditionally for the treatment of burn infections. Pharm Biol 49:526–530

Sturtevant W (1954) The Mikasuki seminole: medical beliefs and practices. PhD Dissertation. Yale University

Subhisha S, Subramoniam A (2005) Antifungal activities of a steroid from *Pallavicinia lyellii*, a liverwort. Indian J Pharmacol 37:304–308

Tag H, Das AK, Loyi H (2007) Anti-inflammatory plants used by the Khamti tribe of Lohit district in eastern Arunachal Pradesh, India. Nat Prod Radiance 6:334–340

Tonguç Ö (1998) Determination of heavy metal levels in some moss species around thermic power stations. Turk J Biol 22:171–180

Tosun A, Süntar İ, Keleş H, Özenoğlu-Kiremit H, Asakawa Y, Küpeli-Akkol E (2016) Wound healing potential of selected liverworts growing in Turkey. Turk J Pharmacy 13(3):285–291

Tosun G, Yayli B, Batan N, Yayli N, Karaoglu SA (2014) Chemical composition and antimicrobial activity of essential oils from *Tortella inclinata* var. *densa*, *T. tortusa* and *Pleurochaete squarrosa*. Asian J Chem 26(7):2001

Tosun G, Yayli B, Özdemir T, Batan N, Bozdeveci A, Yayli N (2015) Volatiles and antimicrobial activity of the essential oils of the mosses *Pseudoscleropodium purum*, *Eurhynchium striatum*, and *Eurhynchium angustirete* grown in Turkey. Rec Nat Prod 9(2):237

Türkyılmaz Ünal B, İşlek C, Ezer T, Düzelten Z (2017) Allelopathic effects of *Cinclidotus pachylomoides* (Bryophyta) on pepper and corn plants. Anatol Bryol 3(2):58–67

Turner NJ, Thomas J, Carlson BF, Ogilvie RT (1983) Ethnobotany of the Nitinaht Indians of Vancouver Island. Victoria. In: British Columbia Provincial Museum Occasional Paper Number 24, Victoria. British Columbia, Canada

Tüzen M, Mendil D, Sarı H, Hasdemir E (2003) AAS determination of heavy metals in moss samples of Giresun—Turkey. Fresenius Environ Bull 12:1283–1286

Ucuncu O, Cansu TB, Ozdemir T, Alpaykaraoglu S, Yayli N (2010) Chemical composition and antimicrobial activity of the essential oils of mosses (*Tortula muralis* Hedw., *Homalothecium lutescens* (Hedw.) H. Rob., *Hypnum cupressiforme* Hedw, and *Pohlia nutans* (Hedw.) Lindb. from Turkey. Turk J Chem 34: 825–834

Uğur A, Özden B, Saç MM, Yener G (2003) Biomonitoring of 210Po and 210Pb using lichens and mosses around a uraniferous coal-field power plant in western Turkey. Atmos Environ 37:2237–2245

Uğur A, Özden B, Saç MM, Yener G, Altınbaş Ü, Kurucu Y, Bolca M (2004) Lichens and mosses for correlation between trace elements and 210Po in the areas near coal-fired power plant at Yatağan, Turkey. J Radioanal Nucl Chem 259:87–92

Uyar G, Avcıl E, Ören M, Karaca F, Öncel MS (2009) Determination of heavy metal pollution in Zonguldak (Turkey) by moss analysis (*Hypnum cupressiforme*). Environ Eng Sci 26(1):183–194

Uyar G, Çetin B (2004) A new check-list of the mosses of Turkey. J Bryol 26:203–220

Uyar G, Hacıoğlu Doğru N, Ören M, Çavuş A (2016) Determining antibacterial activity of some mosses [*Cinclidotus riparius* (host ex Brid.) Arn., *Calliergonella ciuspidata* (Hedw.) Loeske, *Thamnobryum alopecurum* (Hedw.) Gangulee, *Leucobryum juniperoideum* (Brid.) Müll. Hal., *Cirriphyllum crassinervium* (Taylor) Loeske & M.Fleisch.]. Anatol Bryol 2:1–8

Uyar G, Ören M, İnce M (2007b) Atmospheric heavy metal deposition in Düzce Province by using mosses as biomonitors. Fresenius Environ Bull 16:145–153

Uyar G, Ören M, Yıldırım Y, İnce M (2007a) Mosses as indicators of atmospheric heavy metal deposition around a coal-filed power plant in Turkey. Fresenius Environ Bull 16:182–192

Veljić M, Djurić A, Soković M, Ćirić A, Glamočlija J, Marin P (2009) Antimicrobial activity ofmethanol extracts of *Fontinalis antipyretica*, *Hypnum cupressiforme* and *Ctenidium molluscum*. Arch Biol Sci 61:225–229

Wellman CH, Osterloff PL, Mohiuddin U (2003) Fragments of the earliest land plants. Nature 425:282–285

Wigginton MJ (2002) Checklist and distribution of the liverwort and hornworts of sub-Saharan Africa, including the east African Islands. Trop Bryol Res Rep 3:1–88

Wu PC (1977) *Rhodobryum giganteum* (Schwaegr.) par can be used for curing cardiovascular disease. Acta Phytotaxonomica Sin 15:93

Wu PC (1982) Some uses of mosses in China. Bryol Times 13:5

Yağlıoğlu MŞ, Abay G, Demirtaş İ, Yağlıoğlu AŞ (2017) Phytochemical screening, antiproliferative and cytotoxic activities of the mosses *Rhytidiadelphus triquetrus* (Hedw.) Warnst. And *Tortella tortuosa* (Hedw.) Limpr. Anatol Bryol 3(1):31–42

Yayintas OT, Yapici BM (2009) In vitro antimicrobial activity of *Brachythecium campestre* and *Eurhynchium pulchellum* extracts. Asian J Chem 21(3):2193

Yayli N, Cansu TB, Volga C, Yayli N (2009) Essential oils in mosses (*Brachythecium salebrosum*, *Eurhynchium pulchellum* and *Plagiomnium undulatum*) grown in Turkey. Asian J Chem 21(7):5505

Yetgin A, Şenturan M, Benek A, Efe E, Canlı K (2017) *Pterigynandrum filiforme* Hedw. türünün antimikrobiyal aktivitesinin belirlenmesi. Anatol Bryol 3(1):43–47

Yong Tan C, Inagaki M, Chaia HB, Karadeniz Lambrechtsc M, Önder A, Özenoglu-Kiremit H, Harinantenaina-Rakontondraibea L (2017) Phytochemical and cytotoxic investigations of pinguisanoids from liverwort *Porella cordeana*. Phytochem Lett 19:77–82

Yoshikawa H, Ichiki Y, Sakakibara KD, Tamura H, Suıko M (2002) The biological and structural similarity between lunularic acid and abscisic acid. Biosci Biotechnol Biochem 66(4):840–846

Zheng GQ, Chang CJ, Stout TJ, Clardy J, Ho DK, Cassady JM (1993) Ohioensins: novel benzonaphthoxanthenones from *Polytrichum ohioense*. J Org Chem 58:366–372

Zhonghua B (1999) State administration of traditional Chinese medicine of the People's Republic of China. Shanghai Science Technology Press, Shanghai, pp 22–23

A Comparative Analysis of the Medicinal Pteridophytes in Turkey, Pakistan, and Malaysia

Munir Ozturk, Volkan Altay, Abdul Latıff, Tabinda Salman, and Iqbal Choudhry

Introduction

The pteridophytes first appear in the fossil records of the Carboniferous but many of the current families and species did not appear until roughly the late Cretaceous (after flowering plants came to dominate many environments). The origin of this group of plants is uncertain, but it is believed that they are more closely related to algae rather than mosses and may have originated from them. The highest diversity of species is found at lower altitudes. They are most common in the understory of humid temperate and tropical forests. However, pteridophytes are widely distributed, and also grow in the Arctic and alpine tundra, saline mangrove swamps, semi-arid deserts, and coastal rocks swept by salt spray. They provide the bulk of biomass in some tropical forests and dominate the understories of some temperate coniferous forests. Only a relatively small number is adapted to dry and sunny conditions by having either reduced surfaces or a covering of hairs or scales to prevent loss of moisture. The greatest number of the species is found in the tropics of both hemispheres (Jahns 1983; Rost et al. 2006; Umi Kalsom 2010).

M. Ozturk (✉)
Vice President of the Islamic World, Academy of Sciences, Amann, Jordan

Department of Botany and Centre for Environmental Studies, Ege University, Izmir, Izmir, Turkey

V. Altay
Faculty of Science & Arts, Biology Department, Hatay Mustafa Kemal University, Antakya, Hatay, Turkey

A. Latıff
Faculty of Science & Technology, Universiti Kebangsaan Malaysia, Bangi, Selangor, Malaysia

T. Salman · I. Choudhry
International Center for Chemical and Biological Sciences, University of Karachi, Karachi, Sindh, Pakistan

M. Ozturk, K. R. Hakeem (eds.), *Plant and Human Health, Volume 1*,
https://doi.org/10.1007/978-3-319-93997-1_9

Currently the pteridophytes, distributed in the tropical, subtropical, and remote tropical islands and in different habitats or even in deeper rainforests, play a definitive role to constitute the strong carpet flora and undercover vegetation (Goswami et al. 2016). A strong network of their roots is highly rich in organic compounds, showing a great soil-binding capacity. These serve as the home for a large number of smaller mammals, rodents, as well as amphibians, invertebrates, even arthropods, and insects (Balick et al. 1978; Cooper-Drive 1978; Gerson, 1979). The pteridophytes are ecologically strong natural companions of a rich forest (Goswami 2009; Mehltreter et al. 2010). Many among these are too hardy, colonizing, and adapted to survive and reproduce in disturbed or barren land with the modest possibility of moisture and drain water (Goswami et al. 2016). This group of plants is also able to adapt to xeric environments, and appear as frequent colonizers following disturbances such as hurricanes, landslides, fires, floods, and even volcanic eruptions (Mehltreter et al. 2010). The five major environments where these plants flourish are tropical mesic, tropical xeric epiphytic, tropical xeric terrestrial, tropical alpine, and temperate (Page 1979). *Acrostichum aureum* grows in the saline waters; *Pteris vittata* often occupies the soil near the water drainage pipelines, on walls, and even on cement-calcareous waste matter. A large number of ecologists have referred them to be able to be disseminated on barren lands. These plants multiply aggressively by vegetatively; an open area can be better utilized for their inhabitation before nasty invasive species occupy the place. The tree fern *Dicksonia antarctica* has become invasive in São Miguel Island (Arosa et al. 2012). This group of plants can survive drought and even occupy the man-made disturbed habitats due to huge construction works with cement concrete residue, but these genera may be different from the original populations (Goswami et al. 2016).

Nearly 12,000 taxa of pteridophytes are reported to be distributed on earth in different ecological niches. Out of these, 10,000 taxa are ferns and the rest fern allies (Goswami et al. 2016). On the basis of morphological and gene sequencing techniques pteridophytes are regarded as paraphyletic assemblage of lycophytes and filicophytes (Smith et al. 2006). Although pteridophytes are the second largest group, they represent only 5–7% of the total vascular plants, but still play an important ecological role particularly in the tropical vegetation (Page 1979; Goswami et al. 2016). The number of taxa of pteridophytes included within *The Plant List* belongs to 48 families and 587 genera. This list includes 47,439 scientific plant names of species rank for the pteridophytes. Of these 10,620 are accepted species names (www.plantlist.com). In South East Asia around 4400 species are known and 1165 species have been recorded from the tropical rainforest of Malaysia (Roos 1996; Parris and Latiff 1997; Umi Kalsom 2010). In Turkey, 89 species belonging to 19 families and 34 genera have been recorded (Güner et al. 2012). A total of 133 species belonging to 41 genera and 9 families from West Pakistan and Kashmir have been reported (Stewart 1972). Majority of these are growing in mixed coniferous forests in mountainous regions forming a substantial component of terrestrial plant communities. The list of pteridophytes published by Nakaike and Malik (1992) includes 82 species belonging to 30 genera and 18 families along with their distribution in Pakistan.

This group of plants is long known for its medicinal and therapeutic uses. In ancient times they were prescribed as herbal extract for the cure of several diseases. Theophrastus (327–287 BC) and Dioscorides (50 AD) have listed many pteridophytes as a potential herbal formulation to cure more deadly disorders (Puri 1970; Parihar et al. 2004; Goswami et al. 2016). However, Singh et al. (2010) in their monumental contribution on the medicinal attributes of this plant group have enormously mentioned the utility of *Marsilea minuta, Adiantum capillus-veneris*, and few others. Few of the prescriptions based on the doctrine of signature reveal that plants of particular shape were recommended for the cure of certain organs resembling to it; these include the lower plants as well (Singh et al. 2010; Goswami et al. 2016). In nineteenth century first attempt has been made by Caius (1935) who presented the medicinal uses of ferns in India. He is recognized as the first researcher to take this initiative in pteridophytes. Recently, tremendous work has been done to determine the potentiality of pteridophytes in relation to their chemical composition and other aspects. The pteridophytes are rich in glycosides, flavonoids, terpenoids, alkaloids, and several primary vis-à-vis secondary metabolites. These are used for the preparation of expectorant and formulations too are advised as supplement of aphrodisiac, appetizer, and stimulants. As against this, certain species are used as diuretic, for treatment of ulcer as well as stomach disorders. Not very many taxa from this group are historically in practice in homeopathy and Ayurvedic system of medicines. *Selaginella bryopteris* and *Lycopodium clavatum* are well-known examples used in homeopathic system of medicine; the former is prescribed for the cure of neurological disorders and heat stroke effects, whereas the latter is recommended to the patients suffering from splinted bones. *Helminthostachys zeylanica* is a part of the well-known herbal Ayurvedic formulation used to enhance the sexual efficiency and as a source of stimulant and aphrodisiac. Few of the pteridophytes have been and are being screened out chemically and numbers of active novel new chemical compounds are validated. The marsiline isolated from *Marsilea minuta* has great importance because it is used in psychopathy, diarrhea, cough, skin diseases, dyspepsia, fever, and insomnia. Many other taxa have been extensively explored and found to exhibit great economic value. *Pteris vittata*, "the Bracken fern," is reported to show antimicrobial activities against a number of gastrointestinal bacterial strains. These findings stress the fact that pteridophytes possess great importance due to their vast medicinal scope; they can prove as biological resources for the upliftment of humans (Singh et al. 2010).

Interestingly great majority of indigenous people have little information on the uses of pteridophytes, the reason being that they are not easily available like flowering plants. The pteridophytes have an important role in the global biodiversity. Although both economic and medicinal values of higher plants have been and are being investigated thoroughly, unfortunately pteridophytes have been ignored. The reason may be the problem with the collection of large quantities of material needed to find out the chemical compounds in these plants. Not much information is available in the literature related to the medicinal importance of these plants except a few studies (Caius 1935; Manandhar 1996; Kumar and Kaushik 1999; Benjamin and Manickam 2007; Karthik et al. 2011). It is a must that we document and evaluate the medicinal uses of pteridophytes which were used by the indigenous people. The indigenous ethnomedicinal knowledge of these plants may show a great potential for research and as a source of new materials for the discovery of new drugs by the phytopharmaceutical industry to fight the diseases (Karthik et al. 2011). There is an urgent need for the conservation of traditional ethnobotanical knowledge which is important for sustainable biological and cultural diversity (Sujarwo et al. 2014). Our main aim here has been to analyze the distribution of the traditional medicinal knowledge about the pteridophytes in Turkey, Pakistan, and Malaysia.

Study Areas

Turkey

Turkey is a meeting place of different phytogeographical regions showing great variation in plant diversity with different types of ecosystems, occupying different habitats and showing varying vegetational characteristics (Ozturk et al. 2006, 2012a, b; Ozturk et al. 2017a, b). Among the countries in southwest Asia and the Mediterranean basin, as well as whole of Europe, the richest flora has been reported for the Anatolian peninsula. Phytogeographically the country is a meeting place for two centers of diversity and origin: the Near East and the Mediterranean (Fig. 1). The number of flowering plant taxa distributed in the country is estimated to be more than 11,000, which is very near to the number recorded from the whole of Europe (Güner et al. 2012). A great variety of geomorphological, topographical, and climatic features are responsible for its widespread habitat and plant diversities as well as endemism, with 3035 taxa of endemics, comprising 31.12% of the total flora and confined to narrow and restricted ecological niches (Güner et al. 2012; Ozturk et al. 2016).

Fig. 1 Map of Turkey

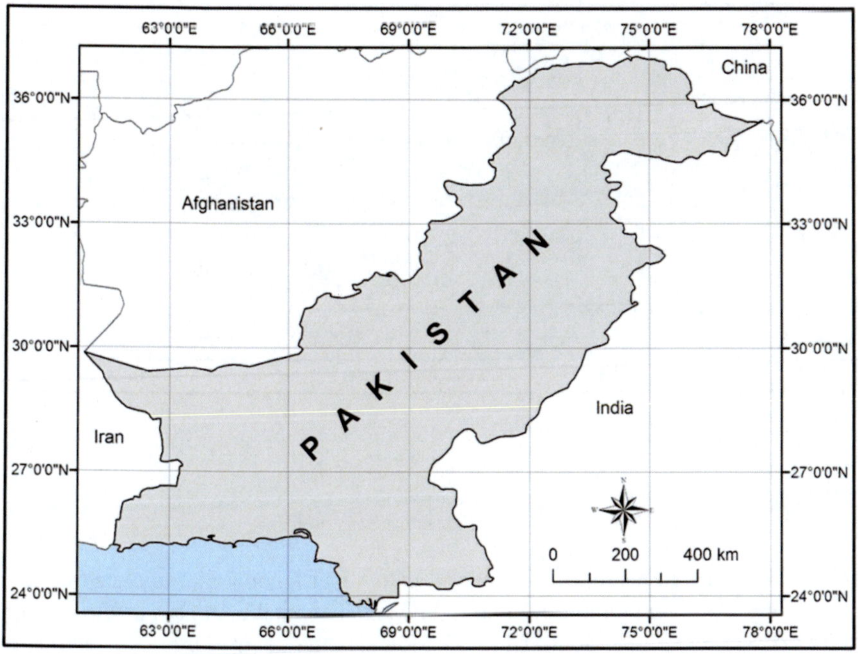

Fig. 2 Map of Pakistan

Pakistan

Pakistan too has a peculiar geographical position (Fig. 2), and harbors a great diversity of flora. More than 6000 vascular plant species are reported to occur in this region (Shinwari and Shinwari 2010), out of which 5600 species have been described to date in the Flora of Pakistan, representing 22 families and about 150 genera (Nasir and Ali 1970–1995). Among the lower plants, there are at least 189 pteridophytes (ferns and their allies), of which 153 are Sino-Japanese elements and 36 Euro-Siberian (Nasir and Ali 1970–1995; Shinwari and Shinwari 2010). Four monotypic genera of flowering plants (*Douepia, Suleimania, Spiroseris, Wendelboa*) and around 400 species (7.8%) are endemic to Pakistan. Most endemics are Irano-Turanian and Sino-Japanese (Shinwari and Shinwari 2010).

A short analysis of plant diversity in Pakistan reveals that the number of species per genus is much lower than the global average, indicating a high rate of diversity at the gene level (Ali and Qaiser 1986). The plant wealth of the country includes elements of six phytogeographical regions, namely the Mediterranean, Saharo-Sindian, Euro-Siberian, Irano-Turanian, Sin-Japanese, and Indian.

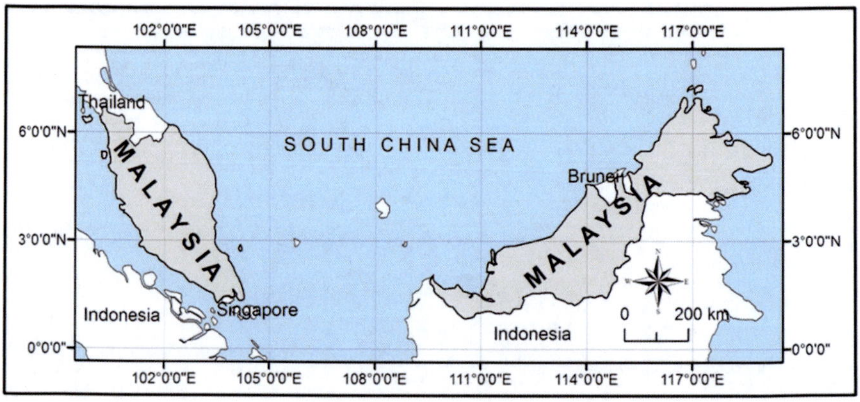

Fig. 3 Map of Malaysia

Malaysia

Malaysia is located just north of equator, with a warm equatorial climate but little seasonality, hot, and humid throughout the year. The total landmass of the country is 329.847 km^2 (Fig. 3). South China Sea divides the country into two regions of similar size: Peninsular Malaysia connected to mainland Asia on the western half and the states of Sabah and Sarawak on the island of Borneo on the east (Saw and Chung 2015). These two halves of the country pose interesting challenges towards documenting the Malaysian flora. Peninsular Malaysia, part of the Malay Peninsula (here includes Singapore and Peninsular Thailand), contains the floristic elements of the Sunda Shelf and also of the mainland Asiatic species from seasonal climates (Wong 1998; Saw and Chung 2015).

Up till now no comprehensive and up-to-date checklist for the Malaysian flora has been published (Saw and Chung 2015). There are several checklists prepared by different workers due to different botanical history of the two main regions of Malaysia. The work of Ridley (1922–1926) for Peninsular Malaysia is an outdated one. But it provided the first complete enumeration of the vascular plants of the Malay Peninsula. The angiosperms have been published in five volumes during 1922–1925. Subsequently, a separate checklist of ferns was published (Ridley 1926). “A Catalogue of the Vascular Plants of Malaya” (Turner 1997) serves as the most recent checklist for the Peninsular Malaysia based on an existing literature survey. This catalogue enumerates 8198 species. Parris and Latiff (1997) have published a further update on the ferns and fern allies with some additions and nomenclatural changes. In this checklist, ferns and fern allies of Sabah and Sarawak were included to provide the first complete checklist of the group for Malaysia (Parris and Latiff 1997; Saw and Chung 2015).

Pteridophytes and Traditional Medicine

Turkey

The total number of pteridophytes evaluated in the traditional medicine includes 19 species from 8 families and 11 genera (Table 1). These are given alphabetically with their botanical name, part used, ailment treated, and information on the preparations used (Table 1). The following families are represented by the largest number of species: Equisetaceae (six species), Aspleniaceae (four species), Dryopteridaceae (three species), and Pteridaceae (two species). Species/genera ratio is higher in Equisetaceae (6.0), Aspleniaceae (2.0), Dryopteridaceae (1.5), and Pteridaceae (1.0) (Table 2).

Pakistan

The total number used in the traditional medicine includes 59 species belonging to 18 families and 28 genera, listed alphabetically with their botanical name, part used, ailment treated, and information on the preparations used (Table 3). The following families are represented by the largest number of species: Pteridaceae (17 species), Dryopteridaceae (8 species), Equisetaceae (5 species), Ophioglossaceae (5 species), Athyriaceae (4 species), and Aspleniaceae (3 species). Species/genera ratio is higher in Dryopteridaceae (4.0), Pteridaceae (3.4), Aspleniaceae (3.0), Equisetaceae (2.5), Ophioglossaceae (2.5), and Athyriaceae (1.3) (Table 4).

Malaysia

Total number of plants used in the traditional medicine includes 40 species spread over 17 families and 30 genera, arranged alphabetically with their botanical name, part used, ailment treated, and information on the preparations used in Malaysia (Table 5). The families with largest number of taxa are Polypodiaceae (nine species), Pteridaceae (five species), Lycopodiaceae (four species), and Tectariaceae (three species). Species/genera ratio is higher in Lygodiaceae (4.0), Polypodiaceae (1.8), Tectariaceae (1.5), and Pteridaceae (1.0) (Table 6).

Table 1 Pteridophytes used in the traditional medicine in Turkey

No	Taxa	Family	Part used	Preparation	Ailments	Resource
1	*Adiantum capillus-veneris* L.	Pteridaceae	AP, LE, UP	AT, BO, DE, JU, IN, PN, PU	Appetizing, shortness of breath, chest tightness, expectorant, anti-dandruff, menstrual regulatory, gastrointestinal diseases, bronchitis, urinary disorders, kidney stones and sand, diarrhea, kidney problems, cough, for veterinary purposes (diarrhea in calves), male aphrodisiac, diuretic, stomachache, swollen testicles, prostate, tonic.	**1–21**
2	*Asplenium adiantum-nigrum* L.	Aspleniaceae	AP, FS, LE, WP	BO, DE, IN	Hemorrhoid, menstrual regulatory, kidney stone, stomachache, shortness of breath, diuretic, eczema, sinusitis, tension regulator	**21–27**
3	*Asplenium scolopendrium* L.	Aspleniaceae	AP, LE, WP	AT, DE, IN	Diuretic, chest tightness, constipation, hemorrhoid, stomachache, arteriosclerosis, anemia, diabetes, wounds, burns	**5, 26, 28**
4	*Asplenium trichomanes* L.	Aspleniaceae	AP, WP	BO, CO	Stomachache, hemorrhoid, eczema	**26, 28**
5	*Ceterach officinarum* Willd.	Aspleniaceae	AP, LE, SP, WP	BO, DE, IN, PU	Abdominal pain, kidney stone and sand, diuretic, constipation, hemorrhoid, cold, bronchitis, stomachache, shortness of breath, urinary tract disorders, urinary antiseptic, gallbladder diseases, gallstones, wounds, kidney diseases, stomach diseases, ulcer, gastrointestinal diseases, gynecology diseases	**6, 21, 22, 24, 25, 29–43**

(continued)

Table 1 (continued)

No	Taxa	Family	Part used	Preparation	Ailments	Resource
6	*Dryopteris filix-max* (L.) Schott	Dryopteridaceae	AP, RH	DE	Intestinal parasites, stomachache	**5, 7, 28**
7	*Dryopteris raddeana* (Fomin) Fomin	Dryopteridaceae	LE, WP	IN	Kidney stone	**13, 21**
8	*Equisetum arvense* L.	Equisetaceae	AP, LE, SO, ST, WP	AT, BO, DE, IN	Diuretic, kidney stone and sand, rheumatism, urinary tract diseases, wounds, prostate diseases, kidney diseases, skin diseases, eczema, hemorrhoid, internal bleeding, nose bleeds, blood purification, edema, circulatory disorders, hair straighteners, gum inflammations, tonsillitis, arteriosclerosis, tachycardia, calcification, cancer, gallbladder diseases, nephralgia, gastrointestinal diseases, hypertension, menstrual regulatory, cystitis, bronchitis, ovaritis, antiperspirant, liver diseases, weakens, lung injury, tuberculosis, shortness of breath, cough, diabetes	**5, 7, 10, 12, 26, 28, 35, 37, 39–62**
9	*Equisetum fluviatile* L.	Equisetaceae	AP	BO	Tension, kidney diseases	**56**
10	*Equisetum hyemale* L.	Equisetaceae	AP	BO, DE	Kidney stone, clean stomach, diuretic	**63–65**
11	*Equisetum palustre* L.	Equisetaceae	AP, WP	DE, IN	Hemorrhoids, kidney stones, peptic ulcer	**38, 66**

(continued)

Table 1 (continued)

No	Taxa	Family	Part used	Preparation	Ailments	Resource
12	*Equisetum ramosissimum* Desf.	Equisetaceae	AP, LE, SO, ST, WP	BO, DE, IN	Kidney stone and sand, wounds, diuretic, urinary tract diseases, stomachache, anti-leptotrichia, anticoagulant, arteriosclerosis, cardiovascular diseases, cancer, vulnerary, stomach disorders, skin diseases, tuberculosis, diabetes, hemostatic at inner bleeding, gum inflammations, tonsillitis, eczema, rheumatism, mouth and throat inflammation	**6, 7, 9, 10, 13, 14, 19, 21, 27, 36, 40, 46, 67–76**
13	*Equisetum telmateia* Ehrh.	Equisetaceae	AP, LE, WP	BO, DE, IN	Kidney stone and sand, stomachache, diuretic, kidney diseases, prostate diseases, pain, vasodilating, urinary tract diseases, acne, rheumatism, pain in former broken bones, expectorant, strengthen hair, skin and nails, infections in the mouth, chronic eczema, antifungal, wounds, internal bleeding, shortness of breath, cystitis, nephritis, cardiac deficiency, arteriosclerosis	**5, 14, 19, 21, 24, 43, 66, 68, 77–85**
14	*Lycopodium clavatum* L.	Lycopodiaceae	AP, LE, SP, ST	DE, IN	For irritated body region, carminative (in infants), skin diseases, intertrigo, liver diseases, hepatitis, rheumatism	**28, 55, 63**
15	*Osmunda regalis* L.	Osmundaceae	RH	IN	Diuretic, constipation, Roborant	**5**
16	*Polypodium vulgare* L.	Polypodiaceae	AP, RO	DE	Kidney stones, gallstones, headache, tonsillitis, carminative, abdominal pain	**31**

(continued)

Table 1 (continued)

No	Taxa	Family	Part used	Preparation	Ailments	Resource
17	*Polystichum aculeatum* (L.) Roth	Dryopteridaceae	AP		Abdominal pain	**86**
18	*Pteridium aquilinum* (L.) Kuhn.	Dennstaedtiaceae	AP, LE, RO	BO, DE, IN	Eczema, rheumatism, toothache, intestinal parasites, diuretic, constipation, kidney diseases, wounds	**16, 21, 22, 52, 79, 85, 87, 89**
19	*Pteris cretica* L.	Pteridaceae	AP	BO	Hemorrhoid	**88**

Part used: *AP* Aerial parts, *FS* fronds, *LE* leaves, *RH* rhizome, *RO* root; *SO* shoot; *SP* spores; *ST* stem, *UP* underground parts, *WP* whole plants

Preparation: *AT* As tea; *BO* boiled; *CO* cooked; *CR* crushed; *DE* decoction; *JU* juice; *IN* infusion; *PA* paste; *PU* poultice, *PN* pounded; *PO* powder; *SY* syrup

Resource: 1: Gençay 2007; **2**: Özçelik et al. 1990; **3**: Kilic and Bagci 2013; **4**: Sürmeli et al. 2000; **5**: Baytop 1984; **6**: Demirci and Özhatay 2012; **7**: Fakir et al. 2009; **8**: Altan and Alçıtepe 2000; **9**: Mart 2006; **10**: Altundag and Ozturk 2011; **11**: Topaloğlu 1987; **12**: Çömlekçioğlu and Karaman 2008; **13**: Aslan et al. 2007; **14**: Koçak 1999; **15**: Ertug 1999; **16**: Sargin 2015; **17**: Sezik et al. 1991; **18**: Sargin et al. 2015; **19**: Koçyiğit and Özhatay 2006; **20**: Hayta et al. 2014; **21**: Tuzlacı 2006; **22**: Bulut and Tuzlacı 2015; **23**: Başer et al. 1986; **24**: Tuzlacı and Aymaz 2001; **25**: Tuzlacı and Aymaz 2001; **26**: Sağıroğlu et al. 2012; **27**: Uysal et al. 2010; **28**: Saraç et al. 2013; **29**: Tuzlacı 2005; **30**: Tuzlacı and Erol 1999; **31**: Polat and Satil 2012; **32**: Yeşilada et al. 1993; **33**: Alpınar 1999; **34**: Ertuğ et al. 2004; **35**: Uysal et al. 2012; **36**: Tuzlacı and Sadıkoğlu 2007; **37**: Sarı et al. 2010; **38**: Honda et al. 1996; **39**: Yücel 2014; **40**: Bulut and Tuzlacı 2013; **41**: Sadıkoğlu 2003; **42**: Gürdal and Kültür 2013; **43**: Tuzlacı and Tolon 2000; **44**: Akan et al. 2005; **45**: Ozgokce and Ozcelik 2004; **46**: Korkmaz et al. 2011; **47**: Toksoy et al. 2010; **48**: Yiğit 2014; **49**: Sargin et al. 2013; **50**: Vural et al. 1997; **51**: Akdoğan and Akgün 2006; **52**: Güler et al. 2015; **53**: Cakilcioglu et al. 2011; **54**: Polat et al. 2011; **55**: Akbulut and Bayramoglu 2014; **56**: Kaval et al. 2014; **57**: Korkmaz and Karakurt 2014; **58**: Everest and Ozturk 2005; **59**: Kargioğlu et al. 2008; **60:** Özüdoğru et al. 2011; **61**: Ugurlu and Secmen 2008; **62**: Tetik 2011; **63**: Yeşilada et al. 1999; **64**: Bulut 2006; **65**: Ozçelik and Balabanlı 2005; **66**: Yeşilada et al. 1995; **67**: Bulut and Tuzlacı 2005; **68**: Genç and Özhatay 2006; **69**: Öztürk and Ölçücü 2011; **70**: Çakılcıoğlu et al. 2007; **71**: Tümen et al. 2005; **72**: Öz Aydın et al. 2005; **73**: Polat et al. 2013; **74**: Arıcan et al. 2013; **75:** Karataş 2007; **76**: Doğanoğlu et al. 2006; **77**: Eryaşar 1998; **78**: Fenercioğlu 1997; **79**: Bulut 2011; **80**: Uzun et al. 2004; **81**: Polat et al. 2015; **82**: Sadıkoğlu and Alpınar 2000; **83**: Ertuğ 2004; **84**. Ezer and Arisan 2006; **85**: Kültür 2007; **86**: Fujita et al. 1995; **87**: Mart and Türkmen 2008; **88**: Saraç 2013; **89**: Sağiroğlu et al. 2013.

Table 2 Genera and species with highest numbers (Turkey)

Family	Genera	Species	Species/genus ratio
Equisetaceae	1	6	6.0
Aspleniaceae	2	4	2.0
Dryopteridaceae	2	3	1.5
Pteridaceae	2	2	1.0

Table 3 Pteridophytes used in the traditional medicine in Pakistan

	Taxa	Familia	Part used	Preparation	Ailments	Source
1	*Actiniopteris australis* link	Pteridaceae	RH	PA, PO	Snake bite	**35**
2	*Adiantum capillus-veneris* L.	Pteridaceae	AP, FS, LE, WP	AT, BO, CR, DE, JU, IN, PA, SY	Gastrointestinal diseases, bronchial disorders, cough, diuretic, cold, flu, asthma, jaundice, hair tonic, scorpion bites, expectorant, laxative, emetic, anti skin allergy, chest pain, measles, removing spleen stones, emollient, febrifuge, snake bite, dog bite, pneumonia, catarrh, sore throat, demulcent, female sterility	**1–19, 35**
3	*Adiantum caudatum* L.	Pteridaceae	FS		Wound healing	**35**
4	*Adiantum chilense* Kaulf. var. *sulphureum* (Kaulf.) Kuntze ex Hicken	Pteridaceae	WP	PU	Congestion	**35**
5	*Adiantum incisum* Forsk.	Pteridaceae	FS, LE, WP	JU	Skin diseases, febrifuge, cough, diabetes, cardiovascular diseases, headache, diaphoretic, chronic diarrhea, dysentery, jaundice, menstrual regulatory, diuretic, bronchitis, general body weakness, malaria, fever, bronchial diseases, falling hairs	**1, 3, 20–25, 35**
6	*Adiantum myriosorum* baker	Pteridaceae	WP	DE	Chronic catarrhs	**35**
7	*Adiantum raddianum* C. Presl.	Pteridaceae	LE	PA	Antidote in snake bite	**12**

(continued)

Table 3 (continued)

	Taxa	Familia	Part used	Preparation	Ailments	Source
8	*Adiantum venustum* D. Don	Pteridaceae	FS, LE, SO, WP	BO, JU, PA, PO	Scorpion bites, expectorant, emetic, diuretic, hair growth, laxative, febrifuge, eye disorders, cold, cough, headache, snake bite, lung disorders, aphrodisiac, stomachache, backache, blood purifier	**5, 9, 10, 14, 20–23, 26–28, 35**
9	*Allantodia aspera* (Blume) Ching	Athyriaceae	FS	EX	Demulcent, hypotensive, tonic, antiviral, antibacterial	**35**
10	*Allantodia squamigera* (Mett.) Ching	Athyriaceae	LE, RH	PA	Tonic	**35**
11	*Asplenium adiantum-nigrum* L.	Aspleniaceae	FS		Scorpion bites	**21**
12	*Asplenium dalhousiae* hook.	Aspleniaceae	LE, RH	DE, JU	Snake bite, gonorrhea, hepatitis	**14, 35**
13	*Asplenium septentrionale* (L.) Hoffm.	Aspleniaceae	LE		For veterinary purposes (cattle oral cavity infection)	**3**
14	*Botrychium lunaria* (L.) Sw.	Ophioglossaceae	WP	IN	Fever	**35**
15	*Botrychium virginianum* (L.) Sw.	Ophioglossaceae	RO		Hypertension, wounds	**35**
16	*Cheilanthes acrostica* (Balb.) Tod.	Pteridaceae	LE	DE	Bleeding, skin diseases	**35**
17	*Cheilanthes albomarginata* C.B. Clarke	Pteridaceae	LE	PO	Body pain	**35**
18	*Cheilanthes bicolor* Fraser-Jenk.	Pteridaceae	WP		Weakness	**35**
19	*Cheilanthes farinosa* (Forssk.) Kaulf.	Pteridaceae	RO, RH	BO, PA	Tonic, gout, rheumatism	**35**
20	*Cheilanthes pteridioides* C. Chr.	Pteridaceae	LE	JU, PO	Dysentery, leprosy, skin diseases	**35**

(continued)

Table 3 (continued)

	Taxa	Familia	Part used	Preparation	Ailments	Source
21	*Christella dentata* (Forssk.) Brownsey & Jermy	Thelypteridaceae	LE, RH, RO	EX	Spermatorrhea, rheumatism, antidiabetic	**35**
22	*Deparia japonica* (Thunb.) M. Kato	Athyriaceae	FS, RH		Weakness	**35**
23	*Diplazium esculentum* (Retz.) Sw.	Athyriaceae	FS, RH		Asthma, cold, cough, tonic	**35**
24	*Dryopteris cochleata* (D. Don) C. Chr.	Dryopteridaceae	FS, RH	PA, PO	Epilepsy, leprosy, pain	**35**
25	*Dryopteris erythrosora* (D.C.Eaton) Kuntze	Dryopteridaceae	FS		Stimulant	**23**
26	*Dryopteris juxtaposita* Christ	Dryopteridaceae	FS, YSO		Digestive power, aphrodisiac, febrifuge	**10, 20, 26, 27**
27	*Dryopteris nigropaleacea* (Fraser-Jenk.) Fraser-Jenk.	Dryopteridaceae	RH	DE	Snake bite, rheumatism, leprosy	**35**
28	*Dryopteris raddeana* (Fomin) Fomin	Dryopteridaceae	WP		Chronic dysentery	**29**
29	*Dryopteris ramosa* (C. Hope) C. Chr.	Dryopteridaceae	FS, LE, ST	JU	Gastric ulcer, constipation, stomachache, aphrodisiac	**11, 29, 30**
30	*Equisetum arvense* L.	Equisetaceae	AP, CN, SO, WP	AT, DE, PA, PO, PU	Kidney stones, urinary tract diseases, wound healing, prostatitis, hair tonic, cystitis, diabetes, diarrhea, gallstone, against lice, cleaning and shinning of teeth, cholera	**2, 10, 16, 21, 31–35**
31	*Equisetum debile* Roxb. Ex Vaucher	Equisetaceae	AP	JU	Jaundice, intestinal inflammation	**17, 19**
32	*Equisetum palustre* L.	Equisetaceae	CN	DE	Stomach disorders	**35**

(continued)

Table 3 (continued)

	Taxa	Familia	Part used	Preparation	Ailments	Source
33	*Equisetum ramosissimum* Desf.	Equisetaceae	SO, WP		Diuretic, gonorrhea, anti-lice, tonic, renal disorders, scabies, itches, skin infections, bone fracture, female infertility, wound healing	**3, 28, 35**
34	*Hippochaete debilis* (Roxb. Ex Vaucher) Ching	Equisetaceae	WP	DE, PA, PO	Fracture, dislocation of bones, stomach disorder, fertility	**35**
35	*Hypodematium crenatum* (Forssk.) Kuhn	Hypodematiaceae	FS, RH, SP	DE, PA	Aphrodisiac, febrifuge, infertility, wounds, snake, scorpion and dog bites	**26, 35**
36	*Lygodium japonicum* (Thunb.) Sw.	Lygodiaceae	LE	IN	Burns as cooling agent, chest congestion	**35**
37	*Lygodium scandens* (L.) Sw.	Lygodiaceae	LE		Female infertility	**35**
38	*Marsilea minuta* L.	Marsileaceae	LE-P		Cough, insomnia, expectorant	**35**
39	*Marsilea quadrifolia* L.	Marsileaceae	LE, LE-P	BO, CR, JU, PA	Flu, migraine, nerve relaxant, infantile diarrhea	**23, 35**
40	*Microsorum membranaceum* (D. Don) Ching	Polypodiaceae	LE	PA	Purgative, diuretic, wound healing	**35**
41	*Nephrolepis cordifolia* (L.) C. Presl	Nephrolepidaceae	FS	PA	Bleeding	**35**
42	*Ophioglossum capense* Sw.	Ophioglossaceae	LE	PA	Menstrual disorders	**35**
43	*Ophioglossum petiolatum* hook.	Ophioglossaceae	LE	DE	Dysentery	**35**
44	*Ophioglossum vulgatum* L.	Ophioglossaceae	LE	DE	Joint pain	**35**
45	*Osmunda regalis* L.	Osmundaceae	WP	DE, PA	Rickets, rheumatism	**35**
46	*Pellaea calomelanos* (Sw.) link	Pteridaceae	FS		Cold, cough	**35**
47	*Polystichum lonchitis* (L.) Roth	Dryopteridaceae	SP	PO	Wound healing	**3**

(continued)

Table 3 (continued)

	Taxa	Familia	Part used	Preparation	Ailments	Source
48	*Polystichum squarrosum* (D. Don) Fée	Dryopteridaceae	FS, SP	PO	Wound healing	**35**
49	*Pseudophegopteris levingei* (C.B. Clarke) Ching	Thelypteridaceae	RO	DE	Spermatorrhea	**35**
50	*Psilotum nudum* (L.) P. Beauv.	Psilotaceae	SP		Diarrhea	**35**
51	*Pteridium aquilinum* (L.) Kuhn.	Dennstaedtiaceae	FS, RH	DE, IN, SY	Dysentery, pain, earache, mouth sores, sore throat, vomiting, diarrhea, blood purifier, gingivitis, scabies, stomach cramps, against worms, abortifacient	**18, 33, 35**
52	*Pteridium revolutum* (Blume) Nakai	Dennstaedtiaceae	RH	DE	Belly worms	**35**
53	*Pteris cretica* L.	Pteridaceae	FS, LE	DE, PA, PO	Wound healing	**23, 35**
54	*Pteris quadriaurita* Retz.	Pteridaceae	FS, RH	EX	Wounds	**35**
55	*Pteris vittata* L.	Pteridaceae	LE	DE, PA, PO	Wounds, burns, infections bacterial	**35**
56	*Salvinia adnata* Desv.	Salviniaceae	WP	DE	Antifungal	**35**
57	*Salvinia auriculata* Aubl.	Salviniaceae	WP		Malaria	**35**
58	*Schizaea dichotoma* (L.) J. Sm.	Schizaeaceae	RH	EX	Chest congestion	**35**
59	*Woodwardia radicans* (L.) Sm.	Blechnaceae	FS	EX	Worms	**35**

Part used: *AP* Aerial parts, *CN* cones, *FS* fronds, *LE* leaves, *LE-P* leaves with petioles, *RH* rhizome; *RO* root, *SO* shoot, *SP* spores, *ST* stem, *WP* whole plants

Preparation: *AT* As tea, *BO* boiled, *CR* crushed, *DE* decoction, *EX* extract, *JU* juice, *IN* infusion, *PA* paste, *PU* poultice, *PO* powder, *SY* syrup

Source: 1: Murad et al. 2011; **2**: Shah et al. 2013; **3**: Shah et al. 2015; **4**: Shinwari et al. 2006; **5**: Kayani et al. 2014; **6**: Bibi et al. 2014; **7**: Ullah et al. 2013; **8**: Shinwari and Shinwari 2010; **9**: Shah 2015; **10**: Alam et al. 2011; **11**: Ahmed et al. 2013; **12**: Ali et al. 2015; **13**: Shaheen et al. 2010; **14**: Butt et al. 2015; **15**: Hussain et al. 2012; **16**: Ahmad et al. 2011; **17**: Abbasi et al. 2009; **18**: Abbasi et al. 2010; **19**: Abbasi et al. 2011; **20**: Hamayun et al. 2007; **21**: Shinwari et al. 2003; **22**: Hamayun 2007; **23**: Ali 2014; **24**: Arshad and Ahmad 2005; **25**: Mahmood et al. 2011; **26**: Barkatullah et al. 2015; **27**: Sher et al. 2015; **28**: Awan et al. 2011; **29**: Arshad and Ahmad 2004; **30**: Ahmad and Habib 2014; **31**: Khan 2014; **32**: Nisar and Ali 2012; **33**: Mahmood et al. 2012; **34**: Khan et al. 2015; **35**: Gul et al. 2016.

Table 4 Genera and species with highest numbers (Pakistan)

Family	Genera	Species	Species/genus ratio
Pteridaceae	5	17	3.4
Dryopteridaceae	2	8	4.0
Equisetaceae	2	5	2.5
Ophioglossaceae	2	5	2.5
Athyriaceae	3	4	1.3
Aspleniaceae	1	3	3.0

Table 5 Pteridophytes used in the traditional medicine in Malaysia (Umi Kalsom 2010; Latiff, 2016)

No	Taxa	Family	Part used	Preparation	Ailments
1	*Acrostichum aureum* L.	Pteridaceae	FS, RH		Anthelmintic, vulnerary, antifungal, ulcer
2	*Allantodia aspera* (Blume) Ching	Athyriaceae	FS, LE, WP		To treat body odor, swellings, pain
3	*Angiopteris evecta* (G. Forst.) Hoffm.	Marattiaceae	FS, LE, SP		To treat beriberi, dysentery, leprosy, skin diseases
4	*Angiopteris palmiformis* (Cav.) C. Chr.	Marattiaceae	FS, LE		Stomachache, abdominal problems
5	*Aspidium blechnoides* Sm.	Tectariaceae	LE		Treatment after childbirth
6	*Aspidium crenatus* (Cav.) Ching	Tectariaceae	FS, RH, WP	BO	Gonorrhea
7	*Asplenium nidus* L.	Aspleniaceae	FS, LE, WP	LO	To enrich the hair growth, depurative, sedative, febrifuge, pain
8	*Blechnum orientale* L.	Blechnaceae	FS		Urinary bladder complaints, intestinal wounds, anthelmintic
9	*Ceratopteris thalictroides* (L.) Brongniart	Pteridaceae	FS	PU	Skin diseases
10	*Cheilanthes tenuifolia* Swartz.	Pteridaceae	RH, RO		Hair tonic
11	*Cheilosoria tenuifolia* (Burnm. f.) Trev	Pteridaceae	FS		To enrich the hair growth
12	*Cibotium barometz* (L.) J. Sm.	Cibotiaceae	RH, RO		Styptic, lumbago, itching
13	*Dicranopteris linearis* (Burnm. f.) Underw.	Gleicheniaceae	LE	BO, PU	Febrifuge, ulcer, wounds

(continued)

Table 5 (continued)

No	Taxa	Family	Part used	Preparation	Ailments
14	*Diplazium esculentum* (Retz.) Sw.	Athyriaceae	LE		Tonic after childbirth
15	*Drynaria quercifolia* (L.) J. Sm.	Polypodiaceae	RH		Cough, febrifuge
16	*Drynaria sparsisora* (Desv.) T. Moore	Polypodiaceae	FS, RH	BO	To treat eye sore, gonorrhea
17	*Equisetum debile* Roxb. Ex Vaucher	Equisetaceae	ST		Joint problems
18	*Helminthostachys zeylanica* (L.) hook.	Ophioglossaceae	FS, LE, RH, WP		Dysentery, aphrodisiac, tonic to treat syphilis, whooping cough
19	*Lycopodiella cernua* (L.) pic. Serm.	Lycopodiaceae	FS, RH, WP		Cough, chronic asthma
20	*Lycopodium carinatum* Poir.	Lycopodiaceae	FS, RH	CR	To enrich the hair growth
21	*Lygodium circinatum* (Burnm. f.) Sw.	Lygodiaceae	FS, RH	PU	Insect bites, wounds
22	*Lygodium flexuosum* (L.) Sw.	Lygodiaceae	RH		Gonorrhea, scabies, rheumatism, eczema
23	*Lygodium japonicum* (Thunb.) Sw.	Lygodiaceae	LE		Expectorant
24	*Lygodium microphyllum* (Cav.) R. Br.	Lygodiaceae	FS, LE	CR, PU	Dysentery, skin diseases, swelling, cough
25	*Microsorum rubidium* (J.Sm.) Copell	Polypodiaceae	RH	PU	Wound
26	*Nephrolepis auriculata* (L.) Trimen	Nephrolepidaceae	FS		Hypertension
27	*Ophioglossum pendulum* L.	Ophioglossaceae	FS		Hair tonic
28	*Osmunda regalis* L.	Osmundaceae	WP		Rickets, rheumatism, intestinal gripping, tonic, styptic
29	*Phymatosorus longissimus* (Blume) pic. Serm.	Polypodiaceae	FS, RH		Insect bites
30	*Pityrogramma calomelanos* (L.) link	Pteridaceae	RH, WP		Kidney diseases, dysentery
31	*Platycerium holttumii* Joncheere & Hennipman	Polypodiaceae	LE		Swellings
32	*Platycerium wallichii* hook.	Polypodiaceae	FS		To treat inflammation

(continued)

Table 5 (continued)

No	Taxa	Family	Part used	Preparation	Ailments
33	*Pleocnemia irregularis* (C. Presl) Holttum	Tectariaceae	FS, LE, RH		Diarrhea, skin diseases
34	*Pyrrosia lanceolata* (L.) Farw.	Polypodiaceae	WP		Cold, sore throat
35	*Pyrrosia nummulariifolia* (Sw.) Ching	Polypodiaceae	FS, LE, WP		Cough, anticancer properties
36	*Pyrrosia piloselloides* (L.) M.G. Price	Polypodiaceae	FS, LE, WP		Cough, gonorrhea, constipation, skin diseases
37	*Schizaea dichotoma* (L.) J. Sm.	Schizaeaceae	FS, RH, WP		Cough, troubles of the throat, aphrodisiac
38	*Selaginella plana* (Desv. Ex Pior.) Hieron.	Selaginellaceae	FS, RH, WP	CR	Styptic
39	*Selaginella willdenowii* (Desv. Ex Poir.) baker	Selaginellaceae	FS, RH, WP		Ringworm, skin itches
40	*Stenochlaena palustris* (Burnm. f.) Bedd.	Blechnaceae	FS, LE, RH, WP		Constipation, febrifuge

Part used: *FS* Fronds, *LE* leaves, *RH* rhizome, *RO* root, *SP* spores, *ST* stem, *WP* whole plants
Preparation: *BO* Boiled, *CR* crushed, *LO* lotion, *PU* poultice

Table 6 Genera and species with highest numbers (Malaysia)

Family	Genera	Species	Species/genus ratio
Polypodiaceae	5	9	1.8
Pteridaceae	5	5	1.0
Lygodiaceae	1	4	4.0
Tectariaceae	2	3	1.5

Comparative Evaluation of Pteridophyte Taxa on the Basis of Disease Groups

A total of 103 taxa of pteridophytes are used in traditional medicine. These are listed in Table 7. The taxa determined in three countries show the following distribution: 19 taxa in Turkey, 59 taxa in Pakistan, and 40 taxa in Malaysia. A total of nine taxa are common among Turkey-Pakistan, six between Pakistan and Malaysia, and only one between Turkey and Malaysia (Table 7; Figs. 4 and 5). Only one taxon (*Osmunda regalis* L.) is common in Turkey-Pakistan-Malaysia (Table 7; Figs. 4 and 5).

Table 7 Medicinal pteridophytes distributed in three countries, a general list

	Taxa	Turkey	Pakistan	Malaysia
1	*Acrostichum aureum* L.			x
2	*Actiniopteris australis* link		x	
3	*Adiantum capillus-veneris* L.	x	x	
4	*Adiantum caudatum* L.		x	
5	*Adiantum chilense* Kaulf. Var. *sulphureum* (Kaulf.) Kuntze ex Hicken		x	
6	*Adiantum incisum* Forsk.		x	
7	*Adiantum myriosorum* baker		x	
8	*Adiantum raddianum* C. Presl.		x	
9	*Adiantum venustum* D. Don		x	
10	*Allantodia aspera* (Blume) Ching		x	x
11	*Allantodia squamigera* (Mett.) Ching		x	
12	*Angiopteris evecta* (G. Forst.) Hoffm.			x
13	*Angiopteris palmiformis* (Cav.) C. Chr.			x
14	*Aspidium blechnoides* Sm.			x
15	*Aspidium crenatus* (Cav.) Ching			x
16	*Asplenium adiantum-nigrum* L.	x	x	
17	*Asplenium dalhousiae* hook.		x	
18	*Asplenium nidus* L.			x
19	*Asplenium scolopendrium* L.	x		
20	*Asplenium septentrionale* (L.) Hoffm.		x	
21	*Asplenium trichomanes* L.	x		
22	*Blechnum orientale* L.			x
23	*Botrychium lunaria* (L.) Sw.		x	
24	*Botrychium virginianum* (L.) Sw.		x	
25	*Ceratopteris thalictroides* (L.) Brongniart			x
26	*Ceterach officinarum* Willd.	x		
27	*Cheilanthes acrostica* (Balb.) Tod.		x	
28	*Cheilanthes albomarginata* C.B. Clarke		x	
29	*Cheilanthes bicolor* Fraser-Jenk.		x	
30	*Cheilanthes farinosa* (Forssk.) Kaulf.		x	
31	*Cheilanthes pteridioides* C. Chr.		x	
32	*Cheilanthes tenuifolia* Swartz.			x
33	*Cheilosoria tenuifolia* (Burm. f.) Trev.			x
34	*Christella dentata* (Forssk.) Brownsey & Jermy		x	
35	*Cibotium barometz* (L.) J.Sm.			x
36	*Deparia japonica* (Thunb.) M. Kato		x	
37	*Dicranopteris linearis* (Burm.f.) Underw.			x
38	*Diplazium esculentum* (Retz.) Sw.		x	x
39	*Drynaria quercifolia* (L.) J. Sm			x
40	*Drynaria sparsisora* (Desv.) T. Moore			x

(continued)

Table 7 (continued)

	Taxa	Turkey	Pakistan	Malaysia
41	*Dryopteris cochleata* (D. Don) C. Chr.		x	
42	*Dryopteris erythrosora* (D.C.Eaton) Kuntze		x	
43	*Dryopteris filix-max* (L.) Schott	x		
44	*Dryopteris juxtaposita* Christ		x	
45	*Dryopteris nigropaleacea* (Frase-Jenk.) Fraser-Jenk.		x	
46	*Dryopteris raddeana* (Fomin) Fomin	x	x	
47	*Dryopteris ramosa* (C. Hope) C. Chr.		x	
48	*Equisetum arvense* L.	x	x	
49	*Equisetum debile* Roxb. Ex Vaucher		x	x
50	*Equisetum fluviatile* L.	x		
51	*Equisetum hyemale* L.	x		
52	*Equisetum palustre* L.	x	x	
53	*Equisetum ramosissimum* Desf.	x	x	
54	*Equisetum telmateia* Ehrh.	x		
55	*Helminthostachys zeylanica* (L.) hook.			x
56	*Hippochaete debilis* (Roxb. Ex Vaucher) Ching		x	
57	*Hypodematium crenatum* (Forssk.) Kuhn		x	
58	*Lycopodiella cernua* (L.) pic. Serm.			x
59	*Lycopodium carinatum* Poir.			x
60	*Lycopodium clavatum* L.	x		
61	*Lygodium circinatum* (Burm. f.) Sw			x
62	*Lygodium flexuosum* (L.) Sw.			x
63	*Lygodium japonicum* (Thunb.) Sw.		x	x
64	*Lygodium microphyllum* (Cav.) R. Br.			x
65	*Lygodium scandens* (L.) Sw.		x	
66	*Marsilea minuta* L.		x	
67	*Marsilea quadrifolia* L.		x	
68	*Microsorum membranaceum* (D. Don) Ching		x	
69	*Microsorum rubidium* (J.Sm.) Copell.			x
70	*Nephrolepis auriculata* (L.) Trimen			x
71	*Nephrolepis cordifolia* (L.) C. Presl		x	
72	*Ophioglossum capense* Sw.		x	
73	*Ophioglossum pendulum* L.			x
74	*Ophioglossum petiolatum* hook.		x	
75	*Ophioglossum vulgatum* L.		x	
76	***Osmunda regalis* L.**	x	x	x
77	*Pellaea calomelanos* (Sw.) link		x	
78	*Phymatosorus longissimus* (Blume) pic. Serm.			x
79	*Pityrogramma calomelanos* (L.) link			x
80	*Platycerium holttumii* Joncheere & Hennipman			x
81	*Platycerium wallichii* hook.			x

(continued)

Table 7 (continued)

	Taxa	Turkey	Pakistan	Malaysia
82	*Pleocnemia irregularis* (C. Presl) Holttum			x
83	*Polypodium vulgare* L.	x		
84	*Polystichum aculeatum* (L.) Roth	x		
85	*Polystichum lonchitis* (L.) Roth		x	
86	*Polystichum squarrosum* (D. Don) fee		x	
87	*Pseudophegopteris levingei* (C.B. Clarke) Ching		x	
88	*Psilotum nudum* (L.) P. Beauv.		x	
89	*Pteridium aquilinum* (L.) Kuhn.	x	x	
90	*Pteridium revolutum* (Blume) Nakai		x	
91	*Pteris cretica* L.	x	x	
92	*Pteris quadriaurita* Retz.		x	
93	*Pteris vittata* L.		x	
94	*Pyrrosia lanceolata* (L.) Farw.			x
95	*Pyrrosia nummulariifolia* (Sw.) Ching			x
96	*Pyrrosia piloselloides* (L.) M.G. Price			x
97	*Salvinia adnata* Desv.		x	
98	*Salvinia auriculata* Aubl.		x	
99	*Schizaea dichotoma* (L.) J. Sm.		x	x
100	*Selaginella plana* (Desv. Ex Pior.) Hieron.			x
101	*Selaginella willdenowii* (Desv. Ex Poir.) baker			x
102	*Stenochlaena palustris* (Burm. f.) Bedd.			x
103	*Woodwardia radicans* (L.) Sm.		x	

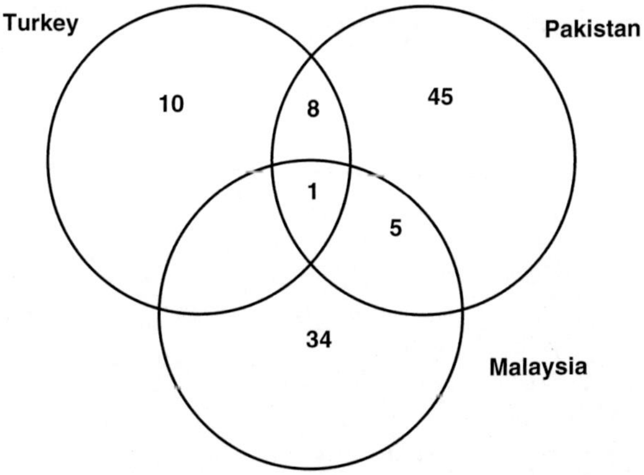

Fig. 4 Venn diagram representing the overlap of medicinally important pteridophytes cited from three countries

Fig. 5 Some medicinal pteridophytes distributed in three countries. **1**: *Osmunda regalis* (www.floraitaliae.actaplantorum.org); **2**: *Pteridium aquilinum* (www.freenatureimages.eu); **3**: *Diplazium esculentum* (www.onlineplantguide.com); **4**: *Asplenium adiantum-nigrum* (www.en.wikipedia.org); **5**: *Equisetum arvense* (www.botanologio.com); **6**: *Equisetum debile* (www.macaubiodiversity.org); **7**: *Pteris cretica* (www.nzpcn.org.nz); **8**: *Adiantum capillus-veneris* (www.luirig.altervista.org)

Digestive System

The use of plants for medicinal purposes is categorized as follows: stomachache, stomach disorders, abdominal pain, gastrointestinal diseases, intestinal parasites/anthelmintic/against worms, ulcer, gallstones, gallbladder diseases, carminative, appetizing, dysentery, emetic/vomiting, digestive powder, stomach cramps, intestinal inflammation, intestinal wounds, intestinal gripping, and abdominal problems. These are seen in all three countries. Their use in stomachache, intestinal parasites/anthelmintic/against worms, and ulcer disease group is outlined in traditional folk medicine (Table 8). If evaluated on a nationality basis, the most common disease group is stomachache in Turkey; dysentery, emetic/vomiting, and intestinal parasites/anthelmintic/against worms in Pakistan; and dysentery in Malaysia.

Dermal System

Use as medicine is categorized as wounds, eczema, skin diseases, hair straighteners/hair tonic, vulnerary, anti-dandruff, strengthen skin and nails, burns, acne, skin allergy, scabies, leprosy, skin itches/itching, skin infection, falling hairs, and ringworm that are seen in all three countries. The use in wounds, skin diseases, and hair straighteners/hair tonic disease groups is mentioned in traditional folk medicine at length (Table 9). An evaluation on a nationality basis reveals that the most common disease groups are wounds and eczema in Turkey; wounds in Pakistan; and skin diseases and hair straighteners/hair tonic in Malaysia.

Respiratory System

Evaluation in the treatment of different diseases is categorized as shortness of breath/asthma, bronchitis, cough, expectorant, tuberculosis, chest tightness, cold, lung injury, flu, emollient, chest pain, lung disorders, pneumonia, catarrh, chest congestion, and whooping cough that are seen in all three research areas. In these three countries, the use of pteridophytes in shortness of breath/asthma, cough, expectorant, and cold disease groups is mentioned in the traditional folk medicine of all three countries (Table 10). If evaluated on a nationality basis, the most common disease groups are shortness of breath/asthma in Turkey; cough and cold in Pakistan; and cough in Malaysia.

Table 8 Number of pteridophytes used in the digestive system disorders in three countries

Medicinal use categories	Turkey	Pakistan	Malaysia
Stomachache	**8**	**2**	**1**
Stomach disorders	3	2	–
Abdominal pain	3	–	–
Gastrointestinal diseases	3	1	–
Intestinal parasites/anthelmintic/against worms	**2**	**3**	**2**
Ulcer	**2**	**1**	**2**
Gallstones	2	1	–
Gallbladder diseases	2	–	–
Carminative	2	–	–
Appetizing	1	–	–
Dysentery	–	5	4
Emetic/vomiting	–	3	–
Digestive powder	–	1	–
Stomach cramps	–	1	–
Intestinal inflammation	–	1	–
Intestinal wounds	–	–	1
Intestinal gripping	–	–	1
Abdominal problems	–	–	1

Table 9 Number of pteridophytes used in the dermal system disorders in three countries

Medicinal use categories	Turkey	Pakistan	Malaysia
Wounds	**6**	**11**	**3**
Eczema	6	–	1
Skin diseases	**4**	**3**	**5**
Hair straighteners/hair tonic	**2**	**3**	**5**
Vulnerary	1	–	1
Anti-dandruff	1	–	–
Strengthen skin and nails	1	–	–
Burns	1	2	–
Acne	1	–	–
Skin allergy	–	1	–
Scabies	–	2	1
Leprosy	–	3	1
Skin itches/itching	–	1	2
Skin infection	–	1	–
Falling hairs	–	1	–
Ringworm	–	–	1

Table 10 Number of pteridophytes used in the respiratory system disorders in three countries

Medicinal use categories	Turkey	Pakistan	Malaysia
Shortness of breath/asthma	**5**	**2**	**1**
Bronchitis	3	3	–
Cough	**2**	**6**	**6**
Expectorant	**2**	**3**	**1**
Tuberculosis	2	–	–
Chest tightness	2	–	–
Cold	**1**	**4**	**1**
Lung injury	1	–	–
Flu	–	2	–
Emollient	–	1	–
Chest pain	–	1	–
Lung disorders	–	1	–
Pneumonia	–	1	–
Catarrh	–	2	–
Chest congestion	–	2	–
Whooping cough	–	–	1

Urogenital System

The medicinal evaluation is categorized as diuretic, kidney stone, hemorrhoids, kidney diseases/renal disorders, urinary tract diseases, kidney sand, constipation, prostate diseases, cystitis/urinary bladder complaints, swollen testicles, urinary antiseptic, nephritis, diarrhea/diarrhoea, nephralgia, gonorrhea, laxative, spermatorrhea, and purgative that have been recorded from all three countries. In all these countries, use of medicinal pteridophytes in kidney diseases/renal disorders, constipation, and cystitis/urinary bladder complaints disease groups is outlined in traditional folk medicine (Table 11). On a nationality basis, the most common disease groups are diuretic, kidney stone, hemorrhoids, kidney diseases/renal disorders, urinary tract diseases, kidney sand, and constipation in Turkey; diuretic and diarrhea/diarrhoea in Pakistan; and gonorrhea in Malaysia.

Cardiovascular System

The pteridophytes are used in the following categories: arteriosclerosis, tension regulator, stop internal bleeding, anticoagulant, hemostatic in inner bleeding, circulatory disorders, vasodilating, anemia, cardiac deficiency, tachycardia, hypertension, blood purification, cardiovascular diseases, bleeding, malaria, congestion, hypotensive, and styptic in all three countries. Use of plants from this group in hypertension is mentioned in traditional folk medicine as well (Table 12). When evaluated on a nationality basis, the most common disease groups are arteriosclerosis, tension regulator, and stop internal bleeding in Turkey; blood purification, bleeding, and malaria in Pakistan; and styptic in Malaysia.

Table 11 Number of pteridophytes used in the urogenital system disorders in three countries

Medicinal use categories	Turkey	Pakistan	Malaysia
Diuretic	10	5	–
Kidney stone	10	1	–
Hemorrhoids	7	–	–
Kidney diseases/renal disorders	**6**	**1**	**1**
Urinary tract diseases	5	1	–
Kidney sand	5	–	–
Constipation	**4**	**1**	**2**
Prostate diseases	3	1	–
Cystitis/urinary bladder complaints	**2**	**1**	**1**
Swollen testicles	1	–	–
Urinary antiseptic	1	–	–
Nephritis	1	–	–
Diarrhea/diarrhoea	1	5	1
Nephralgia	1	–	–
Gonorrhea	–	2	4
Laxative	–	2	–
Spermatorrhea	–	2	–
Purgative	–	1	–

Table 12 Number of pteridophytes used in the cardiovascular disorders in three countries

Medicinal use categories	Turkey	Pakistan	Malaysia
Arteriosclerosis	4	–	–
Tension regulator	2	–	–
Internal bleeding	2	–	–
Anticoagulant	1	–	–
Hemostatic at inner bleeding	1	–	–
Circulatory disorders	1	–	–
Vasodilating	1	–	–
Anemia	1	–	–
Cardiac deficiency	1	–	–
Tachycardia	1	–	–
Hypertension	**1**	**1**	**1**
Blood purification	1	2	–
Cardiovascular diseases	1	1	–
Bleeding	–	2	–
Malaria	–	2	–
Congestion	–	1	–
Hypotensive	–	1	–
Styptic	–	–	3

Table 13 Number of pteridophytes used in the skeletal-muscular disorders in three countries

Medicinal use categories	Turkey	Pakistan	Malaysia
Rheumatism	**5**	**4**	**2**
Edema	1	–	–
Calcification	1	–	–
Pain in former broken bones	1	–	–
Bone fracture	–	2	–
Backache	–	1	–
Dislocation of bones	–	1	–
Joint pain	–	1	–
Gout	–	1	–
Rickets	–	1	1
Lumbago	–	–	1
Joint problems	–	–	1

Table 14 Number of pteridophytes used in the gynecological disorders in three countries

Medicinal use category	Turkey	Pakistan	Malaysia
Menstrual regulatory	3	2	–
Female sterility/infertility	–	4	–
Fertility	–	1	–
Abortifacient	–	1	–
Ovaritis	1	–	–
Gynecology diseases	1	–	–
Treatment (or tonic) after childbirth	–	–	2

Skeletal-Muscular System

The medicinal uses of pteridophytes in three countries can be categorized as rheumatism, edema, calcification, pain in former broken bones, bone fracture, backache, dislocation of bones, joint pain, gout, rickets, lumbago, and problems related to joints. Their use in rheumatism is mentioned in traditional folk medicine at length (Table 13). If evaluated on a nationality basis, the most common disease groups are rheumatism in Turkey and Malaysia, but rheumatism and bone fracture in Pakistan.

Gynecological System

Pteridophytes are medicinally used in the treatment of; menstrual regulatory, female sterility/infertility, fertility, abortifacient, ovaritis, gynecology diseases, and treatment (or tonic) after childbirth are seen in all these countries. On a nationality basis, the most common disease groups are menstrual regulatory in Turkey; female sterility/infertility and menstrual regulatory in Pakistan; and treatment (or tonic) after childbirth in Malaysia (Table 14).

Table 15 Number of pteridophytes used in the ear, nose, and throat disorders in three countries

Medicinal use category	Turkey	Pakistan	Malaysia
Tonsillitis	3	–	–
Throat inflammation	1	–	–
Sinusitis	1	–	–
Nose bleeds	1	–	–
Sore throat	–	2	1
Earache	–	1	–
Troubles of the throat	–	–	1

Table 16 Number of pteridophytes used in the neurological and psychological disorders in three countries

Medicinal use category	Turkey	Pakistan	Malaysia
Headache	1	2	–
Demulcent	–	2	–
Stimulate	–	1	–
Insomnia	–	1	–
Epilepsy	–	1	–
Migraine	–	1	–
Nerve relaxant	–	1	–
Sedative	–	–	1

Ear, Nose, and Throat System

The medicinal uses can be categorized as tonsillitis, throat inflammation, sinusitis, stop nose bleeding, sore throat, earache, and throat troubles met with in all three countries (Table 15). In an evaluation on a nationality basis, the most common disease groups are tonsillitis in Turkey; sore throat in Pakistan; and sore throat and throat troubles in Malaysia (Table 15).

Neurological and Psychological System

The use for medicinal purposes can be categorized as headache, demulcent, insomnia, epilepsy, migraine, nerve relaxant, and sedative as seen in all three countries. When evaluated on a nationally basis, the most common disease groups are headache in Turkey; headache and demulcent in Pakistan; and sedative in Malaysia (Table 16).

Table 17 Number of pteridophytes used in the mouth and teeth ailments in three countries

Medicinal use category	Turkey	Pakistan	Malaysia
Gum inflammations	2	–	–
Mouth inflammation	2	–	–
Toothache	1	–	–
Mouth sores	–	1	–
Gingivitis	–	1	–
For cleaning and shinning of teeth	–	1	–

Mouth and Teeth Ailments

The medicinal evaluation can be categorized as gum inflammations, mouth inflammation, toothache, mouth sores, gingivitis, and teeth cleaning and shinning as observed in all three countries (Table 17). On a nationality basis, the most common disease groups are gum inflammations and mouth inflammation in Turkey, and mouth sores, gingivitis, and cleaning and shinning of teeth in Pakistan (Table 17).

Other Ailments

These can be categorized as diabetes, liver diseases, cancer, irritated body region, tonic, aphrodisiac, pain, antifungal, veterinary purposes, hepatitis, general body weakness, anti-leptotrichia, antiperspirant, intertrigo, roborant, febrifuge/fever, snake bites, scorpion bites, jaundice, against lice, antibacterial, dog bite, cholera, spleen stones, diaphoretic, antiviral, measles, eye disorders, swellings, insect bites, inflammation, body odor, beriberi, tonic to treat syphilis, and depurative that are seen in all three countries. Their use as tonic, aphrodisiac, in pains, and antifungal are mentioned in traditional folk medicine as well (Table 18). On a nationality basis, the most common disease groups are diabetes, liver diseases, and cancer in Turkey; febrifuge/fever, snake bites, tonic, aphrodisiac, and scorpion bites in Pakistan; and febrifuge/fever and swellings in Malaysia.

An evaluation of the data presented above depicts that, on the basis of diseases major number of taxa are used for urogenital system disorders (29.69%), followed by digestive system disorders (14.58%) and dermal system disorders (11.98%) in Turkey. In Pakistan this distribution is as follows: for respiratory disorders and dermal system disorders (14.14% each), followed by urogenital system disorders (11.62%). The data for Malaysia reveals that use for dermal system disorders (22.99%) is followed by digestive system disorders (13.79%) and respiratory disorders (11.49%) (Table 19; Fig. 6). In all three countries, diseases related to the herbal treatments of digestive, respiratory, dermal, and urogenital systems top the list (Table 19; Fig. 6); other ailments in Pakistan (27.27%) and Malaysia (25.29%) stand at high level on the basis of treatment with pteridophytes, whereas in Turkey the ratio is medium (9.89%) (Table 19; Fig. 6). The neurological and psychological

Table 18 Number of pteridophytes used in other ailments in three countries

Medicinal use category	Turkey	Pakistan	Malaysia
Diabetes	3	3	–
Liver diseases	2	–	–
Cancer	2	–	1
Irritated body region	1	–	–
Tonic	**1**	**5**	**1**
Aphrodisiac	**1**	**4**	**2**
Pain	**1**	**3**	**2**
Antifungal	**1**	**1**	**1**
Veterinary purposes	1	1	–
Hepatitis	1	1	–
General body weakness	1	3	–
Anti-leptotrichia	1	–	–
Antiperspirant	1	–	–
Intertrigo	1	–	–
Roborant	1	–	–
Febrifuge/fever	–	7	4
Snake bites	–	7	–
Scorpion bites	–	4	–
Jaundice	–	3	–
Against lice	–	2	–
Antibacterial	–	2	–
Dog bite	–	2	–
Cholera	–	1	–
Spleen stones	–	1	–
Diaphoretic	–	1	–
Antiviral	–	1	–
Measles	–	1	–
Eye disorders	–	1	1
Swellings	–	–	3
Insect bites	–	–	2
Inflammation	–	–	1
Body odor	–	–	1
Beriberi	–	–	1
Tonic to treat syphilis	–	–	1
Depurative	–	–	1

system disorders in Pakistan (4.55%) stand at medium level on the basis of treatment with medicinal pteridophytes, whereas in Malaysia the ratio is low (1.15%) and in Turkey very very low (0.52%) (Table 19; Fig. 6). In addition, the disorders related to the mouth and teeth ailments in Turkey (2.60%) and Pakistan (1.51%) stand at low level on the basis of treatment with medicinal pteridophytes, whereas in Malaysia the ratio is 0% (Table 19; Fig. 6).

Table 19 Therapeutic uses of pteridophytes in three countries (%)

Medicinal use category	Turkey	Pakistan	Malaysia
Digestive system disorders	**14.58**	10.61	**13.79**
Respiratory system disorders	9.38	**14.14**	**11.49**
Urogenital system disorders	**29.69**	**11.62**	10.34
Dermal system disorders	**11.98**	**14.14**	**22.99**
Cardiovascular system disorders	9.38	5.05	4.60
Skeletal-muscular system disorders	6.25	5.56	5.75
Gynecological system disorders	2.60	4.04	2.30
Ear, nose, and throat system disorders	3.13	1.51	2.30
Neurological and psychological system	0.52	4.55	1.15
Mouth and teeth ailment system	2.60	1.51	–
Other ailments	9.89	27.27	25.29

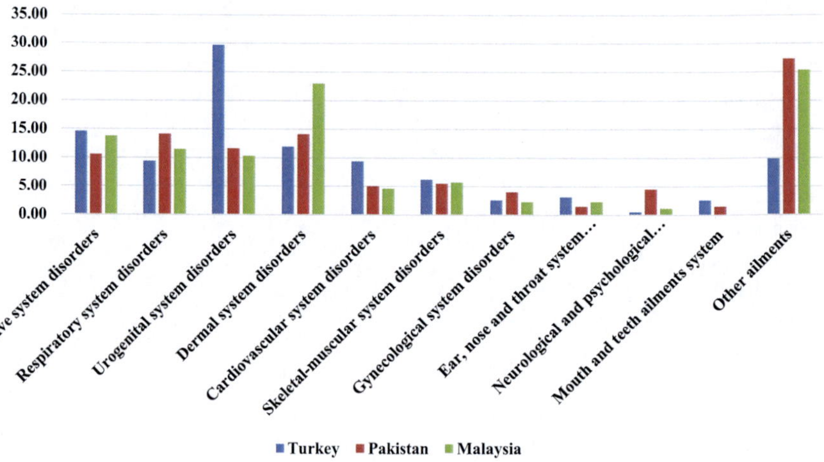

Fig. 6 Comparison of therapeutic uses of pteridophytes in three countries (%)

On the basis of top three diseases in each of these countries, maximum use is as follows: diuretic (10 taxa), kidney stone (10 taxa), and stomachache (8 taxa) in Turkey; wound (11 taxa), febrifuge (7 taxa), and snake bites (7 taxa) in Pakistan; and cough (6 taxa), skin diseases (5 taxa), and hair tonic (5 taxa) in Malaysia (Table 20).

In Turkey and Pakistan five taxa are widely distributed and used almost for the same applications: *Adiantum capillus-veneris* (for cough, gastrointestinal diseases, diuretic, shortness of breath/asthma, expectorant, bronchitis, and chest tightness/chest pain), *Equisetum arvense* (for kidney stone, urinary tract diseases, wounds, prostate diseases, hair straighteners/hair tonic, cystitis, diabetes, and gallstone/gall-bladder diseases), *Equisetum ramosissimum* (for diuretic, skin diseases, kidney stone and sand, and wounds), *Osmunda regalis* (for rickets and rheumatism), and *Pteridium aquilinum* (for intestinal parasites/against worms) (Table 21). For

Table 20 Pteridophytes used for the most common ten diseases in three countries

No	Turkey	Number of taxa	Pakistan	Number of taxa	Malaysia	Number of taxa
1	Diuretic	**10**	Wound	**11**	Cough	**6**
2	Kidney stone	**10**	Febrifuge	**7**	Skin diseases	**5**
3	Stomachache	**8**	Snake bites	**7**	Hair tonic	**5**
4	Hemorrhoids	**7**	Cough	**6**	Dysentery	**4**
5	Wound	**6**	Diuretic	**5**	Gonorrhea	**4**
6	Kidney diseases	**6**	Dysentery	**5**	Febrifuge/ fever	**4**
7	Eczema	**6**	Tonic	**5**	Wounds	**3**
8	Shortness of breath	**5**	Rheumatism	**4**	Swellings	**3**
9	Urinary tract diseases	**5**	Aphrodisiac	**4**	Styptic	**3**
10	Rheumatism	**5**	Female sterility	**4**	Aphrodisiac	**2**

Table 21 Same and/or similar uses of pteridophytes in three countries

	Taxa	Turkey	Pakistan	Malaysia
1	*Adiantum capillus-veneris* L.	Cough	Cough	
		Gastrointestinal diseases	Gastrointestinal diseases	
		Diuretic	Diuretic	
		Shortness of breath	Asthma	
		Expectorant	Expectorant	
		Bronchitis	Bronchitis	
		Chest tightness	Chest pain	
2	*Diplazium esculentum* (Retz.) Sw.		Tonic	Tonic
3	*Equisetum arvense* L.	Kidney stone	Kidney stone	
		Urinary tract diseases	Urinary tract diseases	
		Wounds	Wounds	
		Prostate diseases	Prostate diseases	
		Hair straighteners	Hair tonic	
		Cystitis	Cystitis	
		Diabetes	Diabetes	
		Gallstone	Gallbladder diseases	
4	*Equisetum ramosissimum* Desf.	Diuretic	Diuretic	
		Skin diseases	Skin infections	
		Kidney stone and sand	Renal disorders	
		Wounds	Wound healing	
5	*Osmunda regalis* L.	Rickets	Rickets	
		Rheumatism	Rheumatism	
6	*Pteridium aquilinum* (L.) Kuhn.	Intestinal parasites	Against worms	

Pakistan and Malaysia only one taxa is used for the same applications, *Diplazium esculentum* (for tonic) (Table 21). There are no common taxa between Turkey and Malaysia for the same applications (Table 21).

Conclusion

Pteridophytes embody immense potential in the development of allopathic medicines and their evaluation in the treatment of some dangerous and epidemic diseases. The taxa known to us and used since ancient times in homeopathic and Ayurvedic systems need to be followed in depth. These medicines can be produced in large quantities for use by the commons and their supply arranged in order to meet its requirement at a global scale. Most important information we have is that the herbal medicines hardly show any side effects. Such plants can be prescribed as herbal formulations to cure many diseases with a great scope of economic earnings. Several uses of pteridophytes have been recorded by locals in the treatment of health problems, but we need proper screening, testing, characterization, and validation of compounds together with their theraupeutical values (Singh et al. 2010).

The decorative ferns are in great demand at the international markets which can be cultivated at a large scale engaging indigenous communities to maintain their better livelihood (Singh et al. 2010). Several researchers have stressed the need for evaluation of pteridophytes and a detailed information has been put forth by Goswami et al. (2016). These workers have strongly emphasized the fact that there is an immediate need for all organic chemists and biochemists, evolutionary and molecular geneticists, and pharmaceutical faculty researchers and industries to concentrate on this group of plants to evaluate the tremendous genetic potentials of pteridophytes. This should be not only for producing useful drugs but also to conserve health of our ecosystems in particular forests. The pteridophytes commonly known as seedless vascular plants are used by tribal, isolated populations in different regions of the world, for different ailments or disorders (Goswami et al. 2016).

All through the history of humans these seedless plants have been used on a large scale as ornamentals, in domestic utensils, in handicrafts, as components of cosmetic formulations and foodstuffs, and for medicinal purposes (Morais-Braga et al. 2012). The therapeutic effectiveness, as well as scientific curiosity and the need for new drugs, has prompted several groups to conduct pharmacological research on these plants (Cao et al. 2017). Studies undertaken on the pharmacological have shown that substances in this group of lower plants exhibit diverse pharmacological effects such as cytotoxicity (Radhika et al. 2010), hepatoprotective activity (Wills and Asha 2006), antihyperglycemic activity (Zheng et al. 2011a; Zheng et al. 2011b), leishmanicidal activity (Socolsky et al. 2016), trypanocidal activity (Morais-Braga et al. 2013a; Morais-Braga et al. 2013b), anti-nociceptive activity, anti-inflammatory activity (Yonathan et al. 2006), immunomodulatory activity (Wu et al. 2005), and chemopreventive effects (Wills and Asha 2009). There is a great need for new medicines with such activities. The pteridophytes together with other lower plants could be a solution for the treatment of several present-time diseases due to their secondary metabolites (Cao et al. 2017).

References

Abbasi AM, Khan MA, Ahmad M, Zafar M (2011) Medicinal plant biodiversity of lesser Himalayas-Pakistan. Springer Science & Business Media

Abbasi AM, Khan MA, Ahmad M, Zafar M, Jahan S, Sultana S (2010) Ethnopharmacological application of medicinal plants to cure skin diseases and in folk cosmetics among the tribal communities of North-West Frontier Province, Pakistan. J Ethnopharmacol 128(2):322–335

Abbasi AM, Khan MA, Ahmad M, Zafar M, Khan H, Muhammad N, Sultana S (2009) Medicinal plants used for the treatment of jaundice and hepatitis based on socio-economic documentation. Afr J Biotechnol 8(8):1643–1650

Ahmad I, Ibrar M, Barkatullah AN (2011) Ethnobotanical Study of Tehsil Kabal, Swat District, KPK, Pakistan. J Bot:1–9

Ahmad KS, Habib S (2014) Indigenous knowledge of some medicinal plants of Himalaya region, Dawarian Village, Neelum Valley, Azad Jammu and Kashmir, Pakistan. Univ J Plant Sci 2(2):40–47

Ahmed E, Arshad M, Saboor A, Qureshi R, Mustafa G, Sadiq S, Chaudhari SK (2013) Ethnobotanical appraisal and medicinal use of plants in Patriata, New Murree. J Ethnobiol Ethnomed 9(1):1–11

Akan H, M.M.B. Aslan (2005) GAP yöresindeki tıbbi ve aromatik bitkiler. TUBITAK Proje No: TBAG/Ç. SEK 22 (103-T009)

Akbulut S, Bayramoglu MM (2014) Reflections of socio-economic and demographic structure of urban and rural on the use of medicinal and aromatic plants: the sample of Trabzon province. Ethno Med 8(1):89–100

Akdoğan H, Akgün B (2006) Göksun (Kahramanmaras) çevresinde halk ilacı olarak kullanılan bazı bitkisel gıdalar. Türkiye 9. Gıda Kongresi (Food Congress), 24–26 May, Bolu, Turkey, pp 183–186

Alam N, Shinwari ZK, Ilyas M, Ullah Z (2011) Indigenous knowledge of medicinal plants of Chagharzai valley, District Buner. Pakistan Pak J Bot 43(2):773–780

Ali S (2014) Eco-taxonomic study of district Buner (KPK) Pakistan. Ph D thesis. University of Karachi, Pakistan

Ali S, Perveen A, Qaiser M (2015) Vegetation structure, edaphalogy and ethnobotany of Mahaban and Malka (district Buner) KPK, Pakistan. Pak J Bot 47(SI):15–22

Ali SI, Qaiser M (1986) A phytogeographic analysis of the phanerogams of Pakistan and Kashmir. Proc Royal Soc Edinburgh 89B:89–101

Alpınar K (1999) Ayvalık (Balıkesir) ve yakınındaki adaların floristik ve etnobotanik açıdan değerlendirilmesi. TBAG-1407

Altan Y, Alçıtepe E (2000) Kuşadası (Aydın)'nın Etnobotanik Özellikleri. Geçmişten Geleceğe Kuşadası Sempozyumu, 23–26 Şubat 2000, Meta Basım, Izmir, pp 499–506

Altundag E, Ozturk M (2011) Ethnomedicinal studies on the plant resources of east Anatolia, Turkey. Procedia Soc Behav Sci 19:756–777

Arıcan YE, Yeşil Y, Genç GE (2013) A preliminary ethnobotanical survey of Kumluca (Antalya). J Fac Pharm Istanbul 43(2):95–102

Arosa ML, Ceia RS, Quintanilla LG, Ramos JA (2012) The tree fern Dicksonia antarctica invades two habitats of European conservation priority in São Miguel Island, Azores. Biol Invasions 14(7):1317–1323

Arshad M, Ahmad M (2004) Medico-botanical investigation of medicinally important plants from Galiat areas-NWFP (Pakistan). In: Internationl symposium medicinal plants: linkages beyond national boundaries, September 7–9, 2004, Pakistan, pp 240–249
Arshad M, Ahmad M (2005) Ethnobotanical study of Galliyat for botanical demography and bio-ecological diversification. Ethnobot Leaflets 1(4):1–30
Aslan A, Mat A, Özhatay N, Sarıyar G (2007) A contribution to traditional medicine in West Anatolia. J Fac Pharm Istanbul 39:73–83
Awan MR, Iqbal Z, Shah SM, Jamal Z, Jan G, Afzal M et al (2011) Studies on traditional knowledge of economically important plants of Kaghan Valley, Mansehra District. Pakistan J Med Plants 5(16):3958–3967
Balick MJ, Furth DG, Cooper-Driver G (1978) Biochemical and evolutionary aspects of arthropod predation on ferns. Oecologia 89:55–89
Barkatullah IM, Rauf A, Ben Hadda T, Mubarak MS, Patel S (2015) Quantitative ethnobotanical survey of medicinal flora thriving in Malakand Pass Hills, Khyber Pakhtunkhwa, Pakistan. J Ethnopharmacol 169:335–346
Başer KHC, Honda G, Mıkı W (1986) Herb drugs and herbalists in Turkey. Institue for the study of languages and cultures of Asia and Africa. Studia Culturae Islamicae. 27, Tokyo
Baytop T (1984) Türkiye'de Bitkiler ile Tedavi. Istanbul. İstanbul Üniversitesi Yayınları No: 3255, Eczacılık Fakültesi No: 40, İstanbul
Benjamin A, Manickam VS (2007) Medicinal pteridophytes from the Western Ghats. Indian J Trad Knowl 6(4):611–618
Bibi S, Sultana J, Sultana H, Malik RN (2014) Ethnobotanical uses of medicinal plants in the highlands of Soan Valley, Salt Range, Pakistan. J Ethnopharmacol 155(1):352–361
Bulut EG, Tuzlacı E (2005) An ethnobotanical study in Bozcaada (Çanakkale-Turkey). IVth International Congress of Ethnobotany (ICEB, 2005) pp 581–583
Bulut G (2011) Folk medicinal plants of Silivri (İstanbul-Turkey). Marmara Pharmaceut J 1(15):25–29
Bulut G, Tuzlaci E (2013) An ethnobotanical study of medicinal plants in Turgutlu (Manisa-Turkey). J Ethnopharmacol 149(3):633–647
Bulut G, Tuzlacı E (2015) An ethnobotanical study of medicinal plants in Bayramîç. Marmara Pharm J 19(3):268–282
Bulut Y (2006) Manavgat (Antalya) yöresinin faydalı bitkileri. Msc thesis. Suleyman Demirel Uninersity, Isparta
Butt MA, Ahmad M, Fatima A, Sultana S, Zafar M, Yaseen G, Ashraf MA, Shinwari ZK, Kayani S (2015) Ethnomedicinal uses of plants for the treatment of snake and scorpion bite in Northern Pakistan. J Ethnopharmacol 168:164–181
Caius JF (1935) The medicinal and poisonous ferns of India. J Bombay Nat Hist Soc 38:341–361
Çakilcioglu U, Khatun S, Turkoglu I, Hayta S (2011) Ethnopharmacological survey of medicinal plants in Maden (Elazig-Turkey). J Ethnopharmacol 137(1):469–486
Çakılcıoğlu U, Türkoğlu İ, Kürşat M (2007) Harput (Elazığ) ve çevresının etnobotanik özellikleri. Doğu Anadolu Bölgesi Araştırmaları:22–28
Cao H, Chai TT, Wang X, Morais-Braga MFB, Yang JH, Wong FC et al (2017) Phytochemicals from fern species: potential for medicine applications. Phytochem Rev 16(3):379–440
Çömlekçioğlu N, Karaman Ş (2008) Kahramanmaraş şehir merkezindeki aktarlarda bulunan tıbbi bitkiler. KSU J Sci Eng 11(1):23–32
Cooper-Drive G (1978) Insect-fern associations. Entomol Exp Appl 24:310–316
Demırcı S, Özhatay N (2012) Wild plants used for medicinal purpose in Andırın, Kahramanmaraş Turkish. J Pharm Sci 9(1):75–92
Doğanoğlu Ö, Gezer A, Yücedağ C (2006) Studies on some important medicinal and aromatic plant taxa in Yenişarbademli locality of Lakes District. Süleyman Demirel Üniversitesi Fen Bilimleri Enstitüsü Dergisi 10(1):66–73

Ertug F (1999) Plant, animal and human relationships in the Folk Medicine of Turkey. In: Herbs, humans and animals proceedings of the international seminar, Coreglia (Tuscany), Italy. Koln: Experiences, pp 45–63

Ertuğ F (2004) Bodrum yöresinde halk tıbbında yararlanılan bitkiler. Retrieved from documents. anadolu.edu. tr/bihat/e-kitap/fertugpdf.pdf E

Ertuğ F, Tümen G, Çelik A, Dirmenci T (2004) Buldan (Denizli) etnobotanik alan araştırması 2003. TÜBA Kültür Envanteri Dergisi 2

Eryaşar P (1998) Gönen (Balıkesir) yöresinin geleneksel halk ilacı olarak kullanılan bitkileri. Marmara Üniversitesi Sağlık Bilimleri Enstitüsü Farmasötik Botanik Anabilim Dalı. Yüksek Lisans Tezi, İstanbul

Everest A, Ozturk E (2005) Focusing on the ethnobotanical uses of plants in Mersin and Adana provinces (Turkey). J Ethnobiol Ethnomed 1:1–6

Ezer N, Arisan OM (2006) Folk medicines in Merzifon (Amasya, Turkey). Turk J Bot 30(3):223–230

Fakir H, Korkmaz M, Güller B (2009) Medicinal plant diversity of Western Mediterranean Region in Turkey. J Appl Biol Sci 3(2):3–4

Fenercioğlu E (1997) Şile (İstanbul) yöresinin geleneksel halk ilacı olarak kullanılan bitkileri. Msc thesis. Marmara University, Istanbul

Fujita T, Sezik E, Tabata M, Yeşilada E, Honda G, Takeda Y, Tanaka T, Takaishi Y (1995) Traditional medicine in Turkey VII. Folk medicine in middle and west Black Sea Regions. Econ Bot 49:406–422

Genç EG, Özhatay N (2006) An ethnobotanical study in Çatalca (European part of Istanbul) II. Turk J Pharmaceut Sci 3(2):73–89

Gençay A (2007) Cizre (Şırnak)'Nin Etnobotanik Özellikleri. Msc thesis. Yüzüncü Yıl University, Van

Gerson U (1979) The associations between pteridophytes and arthropods. Fern Gazette 12:29–45

Goswami HK (2009) Non angiospermic plants are also ancient medicinal plants: conserve and explore them. Bionature 29:95–107

Goswami HK, Sen K, Mukhopadhyay R (2016) Pteridophytes: evolutionary boon as medicinal plants. Plant Genet Resour: Charact Util 14(4):328–355

Gul A, Alam J, Ahmad H, Irfan M (2016) An updated checklist of pteridophytes of district Mansehra, Khyber Pukhtunkhwa-Pakistan. Int J Biosci 9(5):116–133

Güler B, Manav E, Uğurlu E (2015) Medicinal plants used by traditional healers in Bozüyük (Bilecik-Turkey). J Ethnopharmacol 173:39–47

Güner A, Aslan S, Ekim T, Vural M, Babaç MT (2012) Türkiye Bitkileri Listesi (Damarlı Bitkiler). Istanbul-Türkiye. Nezahat Gökyiğit Botanik Bahçesi ve Flora Araştırmaları Derneği Yayını

Gürdal B, Kültür Ş (2013) An ethnobotanical study of medicinal plants in Marmaris (Muğla, Turkey). J Ethnopharmacol 146(1):113–126

Hamayun M (2007) Traditional uses of some medicinal plants of Swat Valley. Pakistan Indian J Traditional Knowl 6(4):636–641

Hamayun M, Khan MA, Chudhary MF, Ahmad H (2007) Studies on traditional knowledge of medicinal herbs of Swat Kohistan, District Swat, Pakistan. J Herbs Spices Med Plants 12(4):11–28

Hayta S, Polat R, Selvi S (2014) Traditional uses of medicinal plants in Elazığ (Turkey). J Ethnopharmacol 154(3):613–623

Honda G, Yeşilada E, Tabata M, Sezik E, Fujita T, Takeda Y et al (1996) Traditional medicine in Turkey VI. Folk medicine in West Anatolia: Afyon, Kutahya, Denizli, Mugla, Aydin provinces. J Ethnopharmacol 53(2):75–87

Hussain W, Hussain J, Ali R, Hussain S, Khan MA, Khan I et al (2012) Phytomedicinal studies of kurram agency in the federally administered tribal areas (FATA) of Pakistan. J Appl Pharmaceut Sci 2(10):81–85

Jahns HM (1983) Collins guide to the Ferns, Mosses and Lichens of Britain and North and Central Europe. William Collins Sons and Co. Ltd, London

Karataş H (2007) Ilgaz (Çankırı) ilçesi ve çevresinin etnobotaniği. MSc thesis. Gazi University, Ankara
Kargioğlu M, Cenkci S, Serteser A, Evliyaoğlu N, Konuk M, Kök MŞ, Bağci Y (2008) An ethnobotanical survey of inner-west Anatolia, Turkey. Hum Ecol 36(5):763–777
Karthik V, Raju K, Ayyanar M, Gowrishankar K, Sekar T (2011) Ethnomedicinal uses of pteridophytes in Kolli Hills, Eastern Ghats of Tamil Nadu, India 1(2): 50–55
Kaval I, Behçet L, Cakilcioglu U (2014) Ethnobotanical study on medicinal plants in Geçitli and its surrounding (Hakkari-Turkey). J Ethnopharmacol 155(1):171–184
Kayani S, Ahmad M, Zafar M, Sultana S, Khan MPZ, Ashraf MA et al (2014) Ethnobotanical uses of medicinal plants for respiratory disorders among the inhabitants of Gallies- Abbottabad, Northern Pakistan. J Ethnopharmacol 156:47–60
Khan KU (2014) Ethnobotanical and ethnopharmacognostic exploration of Folk medicines of Deosai Plateau Ph D thesis. Hazara University, Pakistan
Khan MT, Hashim S, Ayub S, Jan A, Marwat KB (2015) A case study of ethnobotany and biodiversity conservation from tehsil Barawal, Upper Dir, Khyber Pakhtunkhwa. Pak J Bot 47:7–13
Kilic O, Bagci E (2013) An ethnobotanical survey of some medicinal plants in Keban (Elazg-Turkey). J Med Plant Res 7(23):1675–1684
Koçak, S. (1999) Karaman yöresinde etnobotanik bir araştırma. Msc thesis. Istanbul University, Istanbul
Koçyiğit M, Özhatay N (2006) Wild plants used as medicinal purpose in Yalova (Northwest Turkey). Turk J Pharmaceut Sci 3(2):91–103
Korkmaz M, Fakir H, Guller B (2011) Consumer preferences for medicinal and aromatic plant products: surveys of urban consumers and sellers in western mediterranean region of Turkey. J Med Plant Res 5(10):2054–2063
Korkmaz M, Karakurt E (2014) Medicinal plants sold in the herbal markets in Kelkit (Gümüşhane). Suleyman Demirel Univ J Nat Appl Sci 18(3):60–80
Kültür Ş (2007) Medicinal plants used in Kirklareli Province (Turkey). J Ethnopharmacol 111(2):341–364
Kumar A, Kaushik P (1999) Antibacterial activity of *Adiantum capillus-veneris*. Indian Fern J 16:1–2
Latiff A (2016) Medicinal and aromatic ferns and lycophytes of Malaysia. Unpublished personal report, Malaysia
Mahmood A, Mahmood A, Malik RN (2012) Indigenous knowledge of medicinal plants from Leepa valley, Azad Jammu and Kashmir, Pakistan. J Ethnopharmacol 143(1):338–346
Mahmood A, Qureshi RA, Mahmood A, Sangi Y, Shaheen H, Ahmad I, Nawaz Z (2011) Ethnobotanical survey of common medicinal plants used by people of district Mirpur, AJK. Pakistan J Med Plants Res 5(18):4493–4498
Manandhar PN (1996) Ethnobotanical observations on ferns and fern allies of Nepal. J Econ Tax Bot 12:414–422
Mart S (2006) Bahçe ve Hasanbeyli (Osmaniye) halkının kullandığı doğal bitkilerin etnobotanik yönden araştırılması. Msc thesis. Çukurova University, Adana
Mart S, Türkmen N (2008) Bahçe ve Hasanbeyli (Osmaniye) bölgesinin etnobotanik kültürü. Ot Sistematik Botanik Dergisi 15(2):137–150
Mehltreter K, Walker IR, Sharpe JM (2010) Fern ecology. Cambridge University Press, Cambridge
Morais-Braga MF, Souza TM, Santos KK, Guedes GM, Andrade JC, Tintino SR (2012) Phenolic compounds and interaction between aminoglycosides and natural products of *Lygodium venustum* SW against multiresistant bacteria. Chemotherapy 58:337–340
Morais-Braga MFB, Souza TM, Santos KKA et al (2013a) Citotocixidade e atividade antiparasita'ria de *Lygodium venustum* SW. Acta Toxicol Argos 21:1–12
Morais-Braga MFB, Souza TM, Santos KKA et al (2013b) Phenol composition, cytotoxic and anti-kinetoplastidae activities of *Lygodium venustum* SW. (Lygodiaceae). Exp Parasitol 134:178–182

Murad W, Ahmad A, Gilani SA, Khan MA (2011) Indigenous knowledge and folk use of medicinal plants by the tribal communities of Hazar Nao Forest, Malakand District. North Pakistan J Med Plants Res 5(7):1072–1086

Nakaike T, Malik S (1992) A list of pteridophytes collected from Pakistan in 1990. Cryptogamic Flora of Pakistan

Nasir E, Ali SI (1970–1995) Flora of Pakistan. Islamabad and Karachi: National Herbarium/NARC and Department of Botany, University of Karachi

Nisar M, Ali Z (2012) Ethnobotanical wealth of Jandool Valley, Dir Lower, Khyber Pakhtunkhwa (Kpk), Pakistan. Int J Phytomed 4(3):351

Öz Aydın S, Dirmenci T, Tümen G, Başer KHC (2005) Plants used as analgesic in the folk medicine of Turkey. In: Proceeding of the IVth international congress of ethnobotany (ICEB 2005), pp 197–171

Özçelik H, Ay G, Öztürk M (1990) Doğu ve Güneydoğu Anadolu'nun ekonomik yönden önemli bazı bitkileri. X. Ulusal Biyoloji Kongresi, 18–20 Temmuz 1990, Erzurum pp 1–10

Ozçelik H, Balabanlı C (2005) Burdur ilinin tıbbi ve aromatik bitkileri. In: I. Burdur Sempozyumu. Burdur, Turkiye pp 1127–1136

Ozgokce F, Ozcelik H (2004) Ethnobotanical aspects of some taxa in east Anatolia, Turkey. Econ Bot 58(4):697–704

Öztürk F, Ölçücü C (2011) Ethnobotanical features of some plants in the district Şemdinli (Hakkari-Turkey). Int J Acad Res 3(1):117–121

Ozturk M, Altay V, Aksoy A (2016) Ecology of endangered endemic plant taxa of Turkiye in relation to climate change. In: Conservation and sustainable use of genepool of plant World in Eur Secur at the present stage International Scientific Conference within Day of Kazakhstan. September 3, EXPO-2016, Antalya, Turkiye

Ozturk M, Altay V, Gonenç TM (2017b) Herbal from High Mountains in the East Mediterranean. In: Bhojraj S et al (eds) Drug discovery from herbs - approaches and applications Centre for Scinece & Technology of the non-aligned and other developing countries (NAM S & T Centre). DAYA Publishing House, New Delhi., Chapter 24, pp 327–367

Ozturk M, Altay V, Gücel S, Altundağ E (2017a) Plant diversity of the Drylands in southeast Anatolia-Turkey: role in human health and food security. In: Ansari AA, Gill SS (eds) Plant biodiversity: monitoring, assessment and conservation. CABI, pp 83–124

Ozturk M, Altundağ E, Gücel S (2012b) Medicinal and aromatic plants (Turkey). Ethnopharmacol Encyclopedia of Life Support Systems (EOLSS). Retrieved from http://www.eolss.net/sample-chapters/c03/e6-79-48.pdf

Ozturk M, Gucel S, Altundag E, Mert T, Gork C, Gork G, Akcicek E (2012a) An overview of the medicinal plants of Turkey. In: Singh R (ed) Genetic resources, chromosome engineering and crop improvement: medicinal plants. CRC Press, LLC, Taylor & Francis, USA., Chapter 7, pp 181–206

Ozturk M, Guvensen A, Aksoy A, Beyazgul M (2006) An overview of the Soil and sustainable land use in Turkiye. In: Proceedings of the fifth international GAP engineering congress, Sanlıurfa, Turkey, pp 1548–1555

Özüdoru B, Akaydin G, Erik S, Yesilada E (2011) Inferences from an ethnobotanical field expedition in the selected locations of Sivas and Yozgat provinces (Turkey). J Ethnopharmacol 137(1):85–98

Page CN (1979) The diversity of ferns, an ecological perspective. In: Dyer AF (ed) The experimental biology of ferns. Academic Press, London, pp 9–56

Parihar P, Parihar L, Bohra A (2004) Antifungal activity of *Cheilanthes albomarginata* Clarke and *Marsilea minuta* Linn. against Aspergillus flavus. Indian Fern J 21:140–143

Parris BS, Latiff A (1997) Towards a pteridophyte flora of Malaysia: a provisional checklist of taxa. Malayan Nat J 50:235–280

Polat R, Cakilcioglu U, Kaltalioğlu K, Ulusan MD, Türkmen Z (2015) An ethnobotanical study on medicinal plants in Espiye and its surrounding (Giresun-Turkey). J Ethnopharmacol 163:1–11

Polat R, Cakilcioglu U, Satil F (2013) Traditional uses of medicinal plants in Solhan (Bingöl-Turkey). J Ethnopharmacol 148(3):951–963
Polat R, Satil F (2012) An ethnobotanical survey of medicinal plants in Edremit Gulf (Balikesir-Turkey). J Ethnopharmacol 139(2):626–641
Polat R, Satıl F, Çakılcıoğlu U (2011) Medicinal plants and their use properties of sold in herbal market in Bingöl (Turkey) district. Biol Divers Conserv J 4(3):25–35
Puri HS (1970) Indian pteridophytes used in folk remedies. Am Fern J 60:137–142
Radhika NK, Sreejith PS, Asha VV (2010) Cytotoxic and apoptotic activity of *Cheilanthes farinosa* (Forsk.) Kaulf. Against human hepatoma, Hep3B cells. J Ethnopharmacol 128(1):166–171
Ridley HN (1922–1925) The flora of the Malay peninsula. L. Reeve & Co. Ltd, London vol. 1: 918., 2: 672, 3: 405, 4: 383, 5: 470
Ridley HN (1926) The ferns of the Malay peninsula. Journal of the Malayan Branch, Royal Asiatic. Society 4:1–121
Roos M (1996) Mapping the world's pteridophyte diversity—systematic and floras. In: Camus JM et al (eds) Pteridology in perspective. Royal Botanic Gardens, Kew, pp 29–42
Rost TL, Barbour MG, Stockinga CR, Murphy TM (2006) Plant biol Thomson Brooks/Cole, Toronto
Sadıkoğlu E (2003) Koçarlı (Aydın) yöresinin geleneksel halk ilacı olarak kullanılan bitkileri. Msc thesis. Marmara University, Istanbul
Sadıkoğlu N, Alpınar K (2000) Bartın: from an ethnobotanical point of view. In: XIII th meeting on plant originated crude drugs proceeding book. Marmara University Press, pp 87–100
Sağıroğlu M, Olgaç E, Ertürk B, Turna M (2012) An ethnobotanical survey from Şile (Istanbul) and Karasu (Sakarya). Ot Sistematik Botanik Dergisi 19(2):93–104
Sağiroğlu M, Topuz T, Ceylan K, Turna M (2013) An ethnobotanical survey from Yahyalı (Kayseri) and tarsus (Mersin). SAÜ Fen Edebiyat Dergisi 2:13–37
Saraç DU (2013) Rize ili etnobotanik özellikleri. Msc thesis. Karadeniz Teknik University, Trabzon
Saraç DU, Özkan ZC, Akbulut S (2013) Ethnobotanic features of Rize/Turkey province. Biol Divers Conserv 6(3):57–66
Sargin SA (2015) Ethnobotanical survey of medicinal plants in Bozyazi district of Mersin, Turkey. J Ethnopharmacol 173:105–126
Sargin SA, Akçicek E, Selvi S (2013) An ethnobotanical study of medicinal plants used by the local people of Alaşehir (Manisa) in Turkey. J Ethnopharmacol 150(3):860–874
Sargin SA, Selvi S, Büyükcengiz M (2015) Ethnomedicinal plants of Aydincik District of Mersin, Turkey. J Ethnopharmacol 174:200–216
Sarı AO, Oğuz B, Güvensen A, Bilgiç A, Şenol SG, Tort N (2010) Ege ve Güney Marmara Bölgelerinde halk ilacı olarak kullanılan bitkiler. Anadolu J AARI 20(2):1–21
Saw LG, Chung RCK (2015) The flora of Malaysia projects. Rodriguesia 66(4):947–960
Sezik E, Tabata M, Yeşilada E, Honda G, Goto K, Ikeshiro Y (1991) Traditional medicine in Turkey I. Folk medicine in North-east Anatolia. J Ethnopharmacol 35:191–196
Shah A, Marwat SK, Gohar F, Khan A, Bhatti KH, Amin M et al (2013) Ethnobotanical study of medicinal plants of semi-tribal Area of Makerwal & amp; Gulla Khel (Lying between Khyber Pakhtunkhwa and Punjab Provinces), Pakistan. Am J Plant Sci 4(1):98–116
Shah AH (2015) Cultural drivers of plant biodiversity of district Tor Ghar. Ph.D thesis. Hazara University, Pakistan
Shah AH, Khan SM, Shah AH, Mehmood A, Rahman IU, Ahmad H (2015) Cultural uses of plants among Basikhel tribe of district tor Ghar, Khyber Pakhtunkhwa. Pakistan Pak J Bot 47:23–41
Shaheen H, Qureshi R, Zehra I, Arshad M, Chaudhry AK, Iqbal S (2010) Ethnobotanical survey of Santh Saroola, Kotli Sattian, Rawalpindi, Pakistan. Proceeding of 2nd International Conference "Biodiversity is our life" (IC Biour Life) December 2010, Pakistan pp 137–153
Sher H, Aldosari A, Ali A, De Boer HJ (2015) Indigenous knowledge of folk medicines among tribal minorities in Khyber Pakhtunkhwa, northwestern Pakistan. J Ethnopharmacol 166:157–167
Shinwari MI, Shinwari MI (2010) Botanical diversity in Pakistan; past present and future. World Environ:85–104

Shinwari ZK, Khan AA, Nakaike T (2003) Medicinal and other useful plants of District Swat, Pakistan. WWF, Pakistan 68

Shinwari ZK, Watanabe T, Rehman M, Youshikawa T (2006) A pictorial guide to Medicinal Plants of Pakistan. KUST. Kohat, Pakistan

Singh AP, Rawat VK, Behera SK, Khare PB (2010) Perspectives of Pteridophytes biodiversity: a source of economy elevation. National Conference on biodiversity, development and poverty alleviation, 22nd may, India pp 46–49

Smith AR, Pryer KM, Schuettpelz E, Korall P, Schneider H, Wolf PG (2006) A classification for extant ferns. Taxon 55(3):705–731

Socolsky C, Salamanca E, Giménez A, Borkosky SA, Bardón A (2016) Prenylated Acylphloroglucinols with Leishmanicidal activity from the Fern *Elaphoglossum lindbergii*. J Nat Prod 79(1):98–105

Stewart RR (1972) An annotated catalogue of the vascular plants of West Pakistan and Kashmir. Fakhri Print. Press

Sujarwo W, Lugrayasa N, Caneva G (2014) Ethnobotanical study of edible ferns used in Bali Indonesia. J Sustain Food Energy 2(2):1–4

Sürmeli B, Sakcali S, Ozturk M, Serin M (2000) Preliminary studies on the medicinal plants of Kilis. XIII. Plant Raw Materials Meeting. Marmara University, Istanbul, Turkey pp 211–220

Tetik F (2011) Malatya ilinin etnobotanik değeri olan bitkileri üzerine bir araştırma. Msc thesis. Fırat University, Elazığ

Toksoy D, Bayramoglu M, Hacisalihoglu S (2010) Usage and the economic potential of the medicinal plants in Eastern Black Sea Region of Turkey. J Environ Biol/Acad Environ Biol India 31(5):623–628

Topaloğlu M (1987) Antakya bölgesinin gıda ve ilaç olarak kullanılan bitkiler üzerinde araştırmalar. Msc thesis. Istanbul University, Istanbul

Tümen G, Malyer H, Başer KHC, Öz Aydın S (2005) Plants used in Anatolia for wound healing. Proceeding of the IVth International Congress of Ethnobotany (ICEB 2005) pp 217–221

Turner IM (1997) A catalogue of the vascular plants of Malaya. Gardens' Bull Singapore 47:1–757

Tuzlacı E (2005) Geçmişten günümüze Bodrum'da Bitkiler ve Yaşam. Güzel Sanatlar Matbaası A.Ş., İstanbul

Tuzlacı E (2006) Şifa Niyetine Türkiye'Nin Bitkisel Halk İlaçları. Alfa Yayınları, Istanbul

Tuzlacı E, Aymaz P (2001) Turkish Folk Medicinal Plants, Part IV: Gönen (Balıkesir-Turkey). Fitoterapia 72:323–343

Tuzlacı E, Erol MK (1999) Turkish folk medicinal plants. Part II: Eğirdir (Isparta). Fitoterapia 70(6):593–610

Tuzlacı E, Sadıkoğlu E (2007) Turkish Folk Medicinal Plants, Part VI: Koçarlı (Aydın). J Fac Pharm İstanbul 39:25–37

Tuzlacı E, Tolon E (2000) Turkish folk medicinal plants, part III: Şile (İstanbul). Fitoterapia 71:673–685

Ugurlu E, Secmen O (2008) Medicinal plants popularly used in the villages of Yunt Mountain. Fitoterapia 79(2):126–131

Ullah M, Khan MU, Mahmood A, Malik RN, Hussain M, Wazir SM et al (2013) An ethnobotanical survey of indigenous medicinal plants in Wana district south Waziristan agency, Pakistan. J Ethnopharmacol 150(3):918–924

Umi Kalsom Y (2010) Ferns of Malaysian rain Forest—a journey through the fern world. Universiti Putra Malaysia Press, Serdang

Uysal I, Gücel S, Tütenocakli T, Öztürk M (2012) Studies on the medicinal plants of Ayvacik-Çanakkale in Turkey. Pak J Bot 44:239–244

Uysal I, Onar S, Karabacak E, Çelik S (2010) Ethnobotanical aspects of Kapıdağ peninsula (Turkey). Biol Divers Conserv 3:15–22

Uzun E, Sariyar G, Adsersen A, Karakoc B, Ötük G, Oktayoglu E, Pirildar S (2004) Traditional medicine in Sakarya province (Turkey) and antimicrobial activities of selected species. J Ethnopharmacol 95(2–3):287–296

Vural M, Karavelioğulları FA, Polat H (1997) Çiçekdağı (Kırşehir) ve çevresinin etnobotanik özellikleri. Ot Sistematik Botanik Dergisi 4(1):117–124
Wills PJ, Asha V (2006) Protective effect of *Lygodium flexuosum* (L.) Sw. Extract against carbon tetrachloride-induced acute liver injury in rats. J Ethnopharmacol 108(3):320–326
Wills PJ, Asha V (2009) Chemopreventive action of *Lygodium flexuosum* extract in human hepatoma PLC/PRF/5 and Hep 3B cells. J Ethnopharmacol 122(2):294–303
Wong KM (1998) Patterns of plant endemism and rarity in Borneo and the Malay peninsula. In: Peng C-I, Lowry II (eds) Rare, threatened, and endangered floras of Asia and the Pacific Rim, pp 139–169
Wu MJ, Weng CY, Wang L, Lian TW (2005) Immunomodulatory mechanism of the aqueous extract of sword brake fern (*Pteris ensiformis* Burm.). J Ethnopharmacol 98(1–2):73–81
Yeşilada E, Honda G, Sezik E, Tabata M, Fujita T, Tanaka T et al (1995) Traditional medicine in Turkey V: folk medicine in the inner Taurus Mountains. J Ethnopharmacol 46(3):133–152
Yesilada E, Honda G, Sezik E, Tabata M, Goto K, Ikeshiro Y (1993) Traditional medicine in Turkey IV. Folk medicine in the Mediterranean subdivision. J Ethnopharmacol 39(1):31–38
Yeşilada E, Sezik E, Honda G, Takaishi Y, Takeda Y, Tanaka T (1999) Traditional medicine in Turkey IX: folk medicine in north-west Anatolia. J Ethnopharmacol 64(3):195–210
Yiğit SŞ (2014) Gaziantep ili aktarlarında satılan bitkiler ve etnobotanik özellikleri. Msc thesis. Gaziantep University, Gaziantep
Yonathan M, Asres K, Assefa A, Bucar F (2006) In vivo anti-inflammatory and anti-nociceptive activities of *Cheilanthes farinosa*. J Ethnopharmacol 108(3):462–470
Yücel E (2014) Growing in Turkey guide to medicinal plants Türmatsan Organize Matbaacılık, Eskişehir
Zheng XK, Li YJ, Zhang L, Feng WS, Zhang X (2011a) Antihyperglycemic activity of *Selaginella tamariscina* (Beauv.) spring. J Ethnopharmacol 133(2):531–537
Zheng XK, Zhang L, Wang WW, Wu YY, Zhang QB, Feng WS (2011b) Anti-diabetic activity and potential mechanism of total flavonoids of *Selaginella tamariscina* (Beauv.) spring in rats induced by high fat diet and low dose STZ. J Ethnopharmacol 137(1):662–668

Medicinal Shrubs and Trees from the Nara Desert, Pakistan

Rahmatullah Qureshi

Introduction

The Study Area

The Nara desert is situated between 26°–28° N and 68°–70° E (elevation 50–115 m) in the Sindh province, Pakistan (Fig. 1). The approximately 23,000 km^2 desert is semi-arid, receiving most of its 88–135 mm of annual rainfall episodically during the monsoon season. Mean minimum and maximum temperature averages are 20 °C and 45 °C, respectively (Qureshi and Bhatti 2005a). Topography is distinctly marked with sandy hills, steep slopes, and vast low-lying areas locally known as *Patt*. The vegetation, typical of arid regions, consists mainly of xerophytes that are adapted to extreme temperature fluctuations and a wide variety of edaphic conditions. Perennial vegetation is sparse, comprised mainly of stunted, thorny, or prickly shrubs and herbs capable of drought resistance. Trees are few and scattered. Ephemerals emerge during the rainy season, complete their life cycle before the advent of summer, and the bulk of the area is once more transformed into an open sandy plain, desolate and barren (Qureshi 2012).

This desert sustains a relatively high human and livestock populations, i.e., 1.05 and 1.25 million, respectively (Anonymous 1992). The language of the inhabitants is *Sindhi*. Commonly the people of this area are called *Thari* people. The majority of the population consists of nomads who follow the distribution of rainfall and resultant forage. Livestock keeping is the only livelihood of the inhabitants of the study area. Agriculture is in practice on both the flanks, the right and left sides of Nara canal of this desert. It lies in the same topographical region, but the land under cultivation has been physically modified to receive water for irrigation purpose from

R. Qureshi (✉)
Department of Botany, Pir Mehr Ali Shah Arid Agriculture University, Rawalpindi, Pakistan
e-mail: rahmatullahq@uaar.edu.pk

M. Ozturk, K. R. Hakeem (eds.), *Plant and Human Health, Volume 1*,
https://doi.org/10.1007/978-3-319-93997-1_10

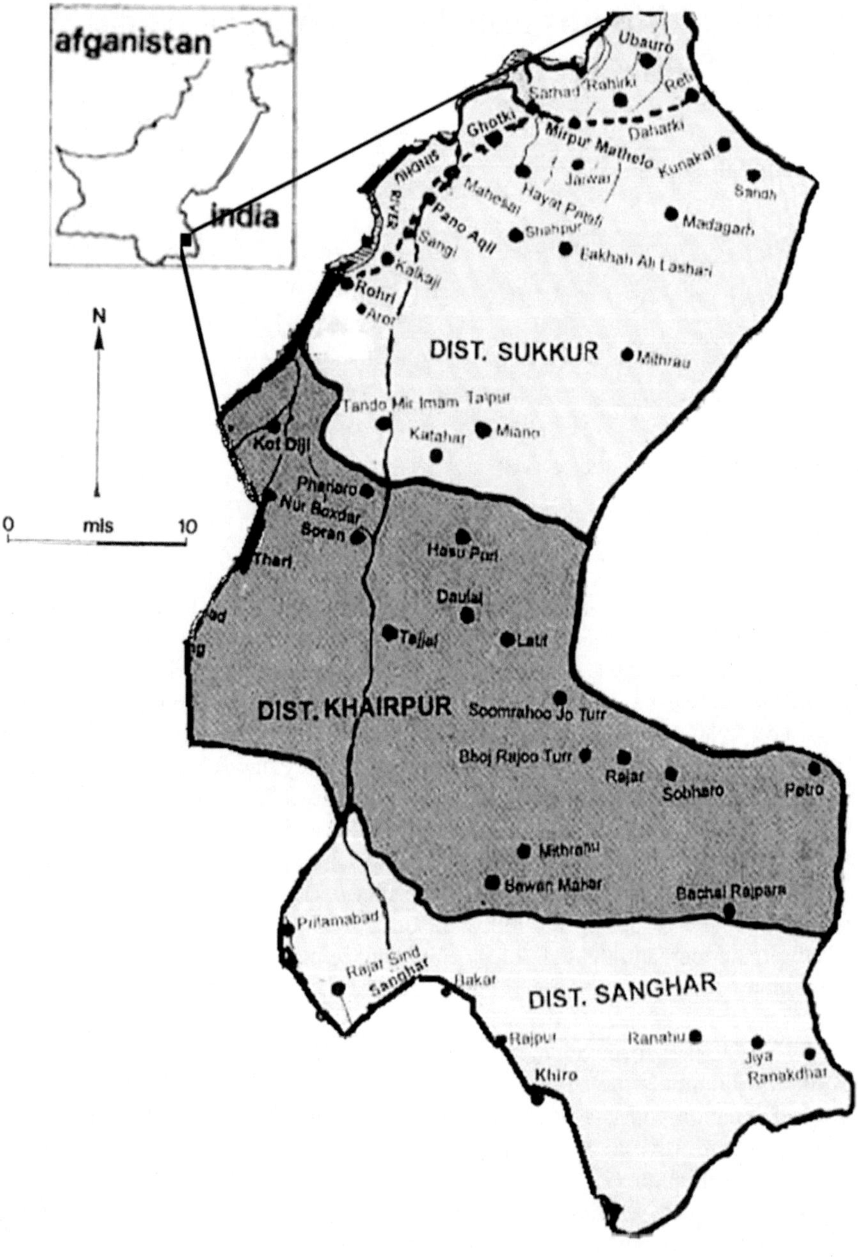

Fig. 1 Location map of the Nara desert, Sindh, Pakistan

the Nara canal. The outermost boundaries are intermingled with desert habitat. Cotton and *Guar* are the major crops of Rabi seasons whereas wheat, *Brassica*, and *Alfalfa* are cultivated in Kharif season (Qureshi and Bhatti 2005b).

A review of the literature reveals that much of the Nara desert is unexplored with respect to the medicinal value of its flora and very little information is available with respect to ethnobotany. Previously Qureshi and Bhatti (2008) and Qureshi et al. (2010) have written medicinal aspects of herbs from the Nara desert, Pakistan. The aim of this study is to document medicinal uses of shrubs and trees which are being used by the local inhabitants including herbalists (*Hakeems*) and elderly known people of the study area.

Materials and Methods

Ethnobotanical Data Collection

The study area was surveyed during September 1998 to December 2001 and 2004–2007 to record medicinal uses of tree and shrubs from the study area. In all, 100 informants including herbal practitioners (*Hakeems*), women, religious teachers (*Pesh Imam*), and elders were interviewed. A semi-structured questionnaire was designed to collect information about native trees/shrubs, local names, parts used, method of preparation, mode of administration, and ailments treated (Qureshi and Bhatti 2008). In order to cross-check data, different age and gender classes were interviewed to seek homogeneity of the information. The folk literature on medicinal uses (Ambasta 1986; Kritikar and Basu 1918; Nadkarni 1954; Dastur 1962; Dymock et al. 1972; Baquar and Tasnif 1984; Murray 1989; Asolkar et al. 1992; Sivarajan and Balachandran 1996) was matched to confirm their uses and any novelty use.

The fidelity level (FL%) is used to calculate the most frequent diseases coupled with the use of a certain plant for the same major purpose by using the following formula:

$$\mathrm{FL}\% = \frac{N_{\mathrm{p}}}{N} \times 100$$

where N_{p} is the number of informants who reported a use of a plant species to treat a particular disease, and N is the number of informants who used the plants as a medicine to treat any disease (Alexiades 1996).

Plant Specimen Collection and Preservation

Plant specimens were also collected, pressed, dried, and identified through various floras (Jafri 1966; Nasir and Ali 1970–1989; Batanouny 1981; Ali and Nasir 1990–1991; Ali and Qaiser 1993–1995, 2000–2004; Matthew 1981–83; Bhandari 1990; Boulos 1991) and deposited in the Shah Latif University Botanic Garden and Herbarium, Khairpur, Pakistan, for record.

Results

Result of the survey is presented in Table 1. For each taxon, the information includes botanical name, plant family, local names, part(s) used, method of preparation, and ailments treated. Fifty-two indigenous trees and shrubs belonged to 44 genera and 27 families are being used for curing various diseases by the inhabitants of the study area. Among them, 16 species are reported containing novel medicinal uses compared with medicinal literature. Fabaceae, Mimosaceae, and Solanaceae were leading families which jointly contributed 12 species, followed by Amaranthaceae, Boraginaceae, Capparidaceae, Moraceae, and Poaceae each having 3 species.

Twelve different parts of plant species were used by the people of the Nara desert as crude medicine (Fig. 2). For preparing various recipes, leaves were highly utilized part (34%), followed by fruits (20%), and roots and flowers (7.37% each), while rest of the parts were scarcely used (Fig. 3).

With reference to different recipe formations (Fig. 3), powder was the most preferred medicament for which 20 species were used, followed by juice, paste (12 spp. each), decoction, poultice (7 spp.), and potherb (4 spp.), while others were less used. In all, oral way of administration was preferred (60.53%), followed by external one (39.47%).

Fifty-two medicinal plant species were used for treating 49 different diseases (Table 2). Joint pain was treated by the highest number of plant species (9 spp., 7.44%), followed by cold, cough and wounds (8 spp., 6.61% each), asthma, spermatorrhea, and as cooling agent (6 spp., 4.96% each), rheumatic pain (5 spp., 4.13%), and diarrhea and inflammation (4 spp., 3.31% each), whereas rest of the diseases were treated by less number of species.

Fidelity level (FL%) determines the level of significance of any species which is preferably used in treating any particular disease. The fidelity level of novel use of plants is provided in Table 3. *Leptadenia pyrotechnica* was the most important medicinal plant in the study area which was particularly used to treat eczema/ringworm with 115 use reports and 92% fidelity, followed by *Tamarix aphylla* (89.6% FL for boils and wounds), *Tephrosia falciformis* (88.8 FL% for rheumatic pain), *Calligonum polygonoides* (88% FL for cooling agent), *Calotropis procera* (87.2% FL for earache, sciatic pain and pneumonia, painful joints, and chest pain), *Iphiona grantioides* (78.4% Fl for sores), and *Schweinfurthia papilionacea* (73.6% FL for typhoid fever; skin impurities, chronic ulcers, and cancerous wounds).

Table 1 Medicinal use of trees and shrubs recorded from the Nara desert, Pakistan

S. No.	Family	Botanical name and voucher specimen	Local names	Parts used	Preparations (administrations)	Uses/ailments treated
1	Amaranthaceae	*Achyranthes aspera* L. (325)	Ubat Kandri	Whole plant	Root powder with honey (I)	Asthma, cough, cold, pneumonia, and joints pain
				Leaves	Paste (E)	Cuts and wounds
2		*Aerva javanica* var. *bovei* Webb. (56)	Booh	Whole plant	Decoction (I)	Toothache
			Booh	Leaves, flowers	Paste (E)	Wounds
3		*Aerva javanica* var. *javanica* (Burm. f.) Juss ex J. A. Shultes. (57)	Booh	Leaves, flowers	Paste (E)	Wounds
4	Anacardiaceae	*Mangifera indica* L. (112)	Amb	Seeds	Dried under shade and powdered (I)	Diarrhea, dysentery
5	Asclepiadaceae	*Calotropis procera* (Willd.) R. Br. (918)	Ak	Whole plant	Latex (E)	To expel spine from the body parts[a]
				Leaves	Paste (E)	Burnt injury and swellings
				Yellow leaves	Juice obtained by slightly warming and compression (E)	Otalgia and earache[a]
				Leaves	Juice burnt in sesame oil (E)	Applied over inflamed joints, and for sciatic pain and pneumonia[a]
				Leaves	Coated with sesame oil and slightly warmed (E)	Painful joints and chest pain[a]
				Leaves	Powder (E)	Wounds[a]
				Whole plant	Ash with honey (I)	Asthma and cough[a]

(continued)

Table 1 (continued)

S. No.	Family	Botanical name and voucher specimen	Local names	Parts used	Preparations (administrations)	Uses/ailments treated
6		*Leptadenia pyrotechnica* (Forssk.) Decne. (2, 44)	Khipp	Stem	Latex (E)	Ringworm[a]
7	Asteraceae	*Iphiona granticoides* (Boiss.) Anderb. (898)	Gandraf	Leaves	Paste (E)	Sores[a]
8		*Pluchea lanceolata* (DC.) Oliv. & Hiern. (208)	Phaar/ Resham Buti	Whole plant	Crushed in water combined with candy (I)	Cooling agent[a]
9		*Xanthium indicum* Koenig ex Roxb. (823)	Bhurt	Roots	Paste (E)	Cancerous wounds
10	Moraceae	*Ficus bengalensis* L. (222)	Barr	Latex	Eaten with candy (I)	Spermatorrhea
11	Bignoniaceae	*Tecomella undulata* (Sm.) Seem. (719)	Lohero	Bark of stem	Paste (E)	Eczema
12	Boraginaceae	*Cordia myxa* L. (725)	Lesuro	Leaves	Burnt on fire and powdered (I)	Flu and cough
13		*Cordia gharaf* (Forssk.) Ehren. ex Asch.	Lesuro	Leaves	Burnt on fire and powdered (I)	Flu and cough
14		*Heliotropium crispum* Desf. (111)	Kharsan	Whole plant	Grinded in water with candy (I)	Cooling agent
15	Burseraceae	*Commiphora stocksiana* (Engl.) Engl.	Gugur	Resin	Mixed with herbs and pill formation (I)	Piles
16	Caesalpiniaceae	*Caesalpinia bonduc* (L.) Roxb. (122)	Pahar Wal	Seeds	Powder with black pepper (I)	Malarial and intermittent fever[a]
17		*Senna italica* Mill. (139)	Ghorawal	Leaflets	1 gm powder along with honey	Backache, sciatic, joints pain, colic pain, laxative
18	Capparidaceae	*Cadaba farinosa* Forssk. (619)	Khabri Wal	Leaves	Infusion (I)	Cold and cough
				Leaves	Paste (E)	Sores
19		*Capparis decidua* (Forssk.) Edgew. (22)	Kirrar	Bark	Coal of old plant macerated with sesame oil (E)	Muscular injuries and wounds

				Stem	Coal of old stem mixed with honey (I)	Asthma, cough, and joint pain
				Tender leaves/twigs	Paste (E)	Boils, pain, and inflammation
				Flower bud/ unripe fruit	Boiled and cooked (I)	Rheumatic pain
20		*Capparis spinosa* L. (1252)	Kalvari/ Golaro	Roots	Paste (E)	Sores
				Leaves	Hot poultice (E)	Gout
				Fruits	Potherb (I)	Rheumatic pain
				Fruits	Pickled (I)	Rheumatic pain[a]
21	Cucurbitaceae	*Citrullus colocynthis* (L.) Schrad. (51, 167)	Trooh	Roots	Tooth stick (E)	Toothache
				Leaves	Juice (E)	Baldness
				Fruits	Black grams soaked in the juice thrice and made into powder	250 mg dose in diabetes
				Fruits	Powder (i)	Leukorrhea[a]
				Seeds	Oil (E)	Loss of hairs
22	Euphorbiaceae	*Ricinus communis* L.	Heran buti	Leaves	Coated with sesame oil and slightly warmed (E)	Painful joints and chest pain[a]
23	Fabaceae	*Alhagi maurorum* Medic. (979)	Kandero	Whole plant	Decoction (E)	Piles
24		*Crotalaria burhia* Ham. ex Benth.	Chagg	Stem	Soaked in water and rectified	Cooling agent

(continued)

Table 1 (continued)

S. No.	Family	Botanical name and voucher specimen	Local names	Parts used	Preparations (administrations)	Uses/ailments treated
25		*Dalbergia sissoo* L. (231)	Tari	Fallen leaves	Soaked overnight in water and drunk (I)	Spermatorrhea[a]
26		*Tephrosia falciformis* Romaswami. (106)	Drebbar	Seeds	Seeds (I)	Rheumatic pain[a]
27	Malvaceae	*Abutilon indicum* (L.) Sweet (25)	Pat Teer	Leaves	Decoction (E)	To clean wounds and as gargle for tender gums and toothache
28	Meliaceae	*Azadirachta indica* Juss. (980)	Nim	Leaves	Soaked and then grinded in water (I)	Skin diseases
29	Menispermaceae	*Cocculus hirsutus* (L.) Diels. (517)	Fareed Buti	Leaves	Poultice (E)	Rheumatic pain
				Leaves	Juice (I)	Skin diseases
				Leaves	Juice with candy (I)	Cooling agent for gonorrhea and micturation
30	Mimosaceae	*Acacia nilotica* (L.) Delile (1178)	Sindhi Babur	Bark	Gargle (bark soaked in water with potash alum	Pyorrhea
				Bark	Decoction (I)	Diarrhea and dysentery[a]
				Leaves, flower, gum, pods	Mixed and powdered (I)	Spermatorrhea
31		*Acacia senegal* (L.) Willd. (1243)	Khaunr	Gum	Mixed with wheat flour, sugar and roasted in desi ghee (I)	Tonic[a]
32		*Albizia lebbeck* (L.) Willd. (1921)	Sarinh	Seeds	Powder (I)	Spermatorrhea
33		*Prosopis cineraria* (L.) Druce (1264)	Kandi	Leaves	Paste (E)	Skin affection[a]
				Unripe pods	Cooked as vegetable (I)	Diarrhea[a]
				Ripened fruits	Powder (I)	Congestion of chest[a]

34	Moraceae	*Morus alba* L. (231)	Achho toot	Fruit	Eaten (I)	Constipation
35		*Morus nigra* L. (891)	Karo Toot	Fruit	Syrup (I)	Throat pain, cough
36	Moringaceae	*Moringa oleifera* Lamk. (1256)	Suhanjaro	Fresh flowers	Cooked as vegetable (I)	Joint pain and inflammation
37	Myrtaceae	*Syzygium cumini* (143)	Jamun	Seeds	Powder (I)	Diabetes
38	Poaceae	*Cymbopogon jwarancusa* (Jones) Schult. (96)	Kattan	Leaves, flower, roots	Decoction (I)	Seasonal fevers and catarrhal complaints
39		*Desmostachya bipinnata* (L.) Stapf. (221)	Drabh	Roots	Poultice (E)	Carbuncle
40		*Saccharum spontaneum* L. (46)	Booro	Bark of stem	Decoction (E)	Mouth gargle to relieve laryngitis and phthisis
				Bark of stem	Smoke (I)	Hiccough[a]
41	Polygonaceae	*Calligonum polygonoides* L. (29)	Phog	Fallen twigs	Soaked in water in earthen pot for 36 h (I)	Cooling agent[a]
42	Rhamnaceae	*Ziziphus mauritiana* Lam. (109)	Ber	Ripened fruits	Eaten (I)	Constipation
43		*Ziziphus nummularia* (Burm. f.) Wt. & Arn. (270)	Jhangoori Ber	Unripe fruit	Dried under shade and powdered (I)	Spermatorrhea[a]
44	Salvadoraceae	*Salvadora oleoides* Decne. (309)	Jaar	Ripened fruits	Raw fruit chewed (I)	Carminative and purgative
45	Scrophulariaceae	*Anticharis glandulosa* Aschers. (149)	Gaamesh	Leaves, flowers	Boiled in sesame oil (E)	Earache
46		*Schweinfurthia papilionacea* (Burm. f.) Boiss. (668)	Akri/Paneer Wal	Leaves; fruits	Powder (I)	Chronic typhoid fever; skin impurities[a]
				Leaves	Juice (I and E)	Chronic ulcers and cancerous wounds[a]

(continued)

Table 1 (continued)

S. No.	Family	Botanical name and voucher specimen	Local names	Parts used	Preparations (administrations)	Uses/ailments treated
47	Solanaceae	*Datura fastuosa* L. (1258)	Charyo Dhaaturo	Leaves	Coated with sesame oil and slightly warmed (E)	Carbuncle and boils
48		*Solanum surattense* Burm.f. (165)	Kanderi Wal	Ripened fruits	Decoction of the fruit along with honey (I)	Bronchial asthma
				Ripened fruits	Powder (E)	Snuffed for headache and migraine
				Ripened fruits	Powder mixed with honey (I)	Cough and asthma
				Ripened fruits	Pills made with equal quantity of old molasses	Joint pain
49		*Withania coagulans* (L.) Dunal. (1341)	Paneer	Fruit	Soaked in water (I)	Blood purifier/cooling agent
50		*Withania somnifera* (L.) Dunal (225)	Akri/Aksan Buti	Fresh roots	Paste (E)	Scrupulous and glandular swellings
				Dried roots	Powder combined with candy (I)	Joint pain, spermatorrhea, and as a nerve tonic
51	Tamaricaceae	*Tamarix aphylla* (L.) Karst. (609)	Lao	Leaves	Poultice (E)	Boils and wounds[a]
52		*Tamarix indica* Willd. (21)	Lai	Galls	Powder (I)	Diarrhea and dysentery

Way of administration: (E) External use. (I) Internal use
[a]Not previously reported in the Indo-Pak literature of medicinal plants

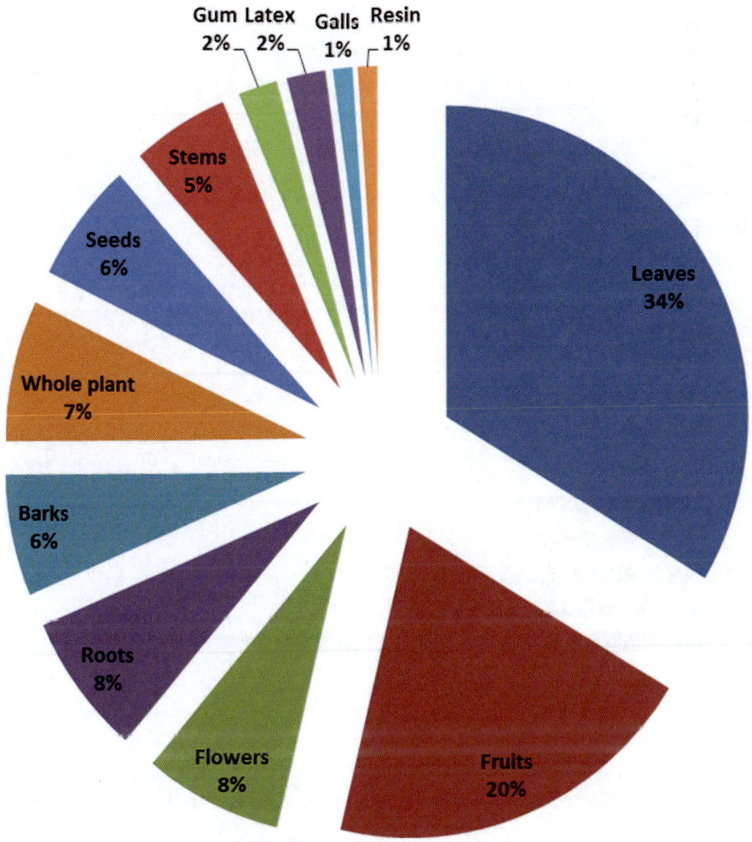

Fig. 2 Various plant parts used in preparing herbal preparation

Discussion

Medicinal plants are of special interest to the inhabitants of this desert due to little or sometimes no access to the modern healthcare facilities. The availability of effective herbal drug resources at the doorstep associated with the generation-blocked traditional knowledge retains the traditional communities to depend for more than 95% of their primary healthcare medicine from phytomedicine. Qureshi and Bhatti (2008) reported 51 plant species distributed across 43 genera and 28 families which are being used by the inhabitants for treating various diseases.

The study was aimed to collect information about the medicinal uses of native plants, methods of preparation of herbal drugs, and part of plants used in various phytotherapies. The ethnobotanical survey of medicinal trees and shrubs was conducted from the Nara desert, Pakistan. The study revealed that the dwellers mostly depended on the wild plant resources to fulfill their basic health needs as reported by the author in previous communications (Qureshi and Bhatti 2008; Qureshi et al. 2010).

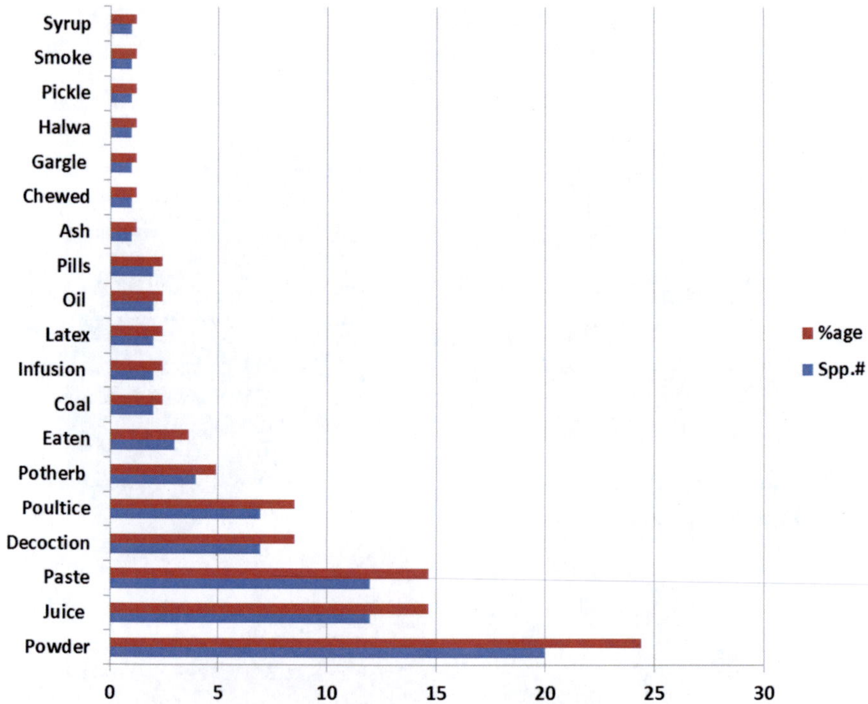

Fig. 3 Medicaments of medicinal plants from the Nara desert, Pakistan

Comparing Indo-Pak medicinal plant literature (Ambasta 1986; Kritikar and Basu 1918; Nadkarni 1954; Dastur 1962; Dymock et al. 1972; Baquar and Tasnif 1984; Murray 1989; Asolkar et al. 1992; Sivarajan and Balachandran 1996), 17 species such as *Acacia nilotica, Acacia senegal, Caesalpinia bonduc, Calligonum polygonoides, Calotropis procera, Capparis spinosa, Citrullus colocynthis, Dalbergia sissoo, Iphiona grantioides, Leptadenia pyrotechnica, Pluchea lanceolata, Prosopis cineraria, Saccharum spontaneum, Schweinfurthia papilionacea, Tamarix aphylla, Tephrosia falciformis*, and *Ziziphus nummularia* had new uses (Table 1). Besides, *Achyranthes aspera, Citrullus colocynthis*, and *Solanum surattense* possessed some additional uses not previously known, while rest of the species were previously known.

It is worthwhile to mention that the dwellers commonly stated the use of a species to treat two or more diseases. For example *Achyranthes aspera* was used to treat asthma, cough, cold, pneumonia, joint pain, cuts, and wounds. *Capparis decidua* is reported in asthma, cough, joint pain, inflammation, boils, muscular injuries, and wounds. *Cocculus hirsutus* is prescribed in rheumatic pain and skin diseases, and as a cooling agent for gonorrhea and micturation. *Senna italica* is given in backache, sciatic, joint pain, colic pain, and constipation. *Withania somnifera* is used in scrofulous and glandular swellings, joint pain, and spermatorrhea and as a nerve tonic. These diseases are distinctive to the Eastern Mediterranean region (Aburjai et al. 2007).

Table 2 Diseases treated by number of species along with their proportions

#	Diseases	No. of species	Percentage
1	Joint pain	9	7.44
2	Cold	8	6.61
3	Cough	8	6.61
4	Wounds	8	6.61
5	Asthma	6	4.96
6	Cooling agent	6	4.96
7	Spermatorrhea	6	4.96
8	Rheumatic pain	5	4.13
9	Diarrhea	4	3.31
10	Inflammation	4	3.31
11	Boils	3	2.48
12	Dysentery	3	2.48
13	Skin diseases	3	2.48
14	Sores	3	2.48
15	Toothache	3	2.48
16	Blood purifier	2	1.65
17	Carbuncle	2	1.65
18	Constipation	2	1.65
19	Diabetes	2	1.65
20	Earache	2	1.65
21	Piles	2	1.65
22	Pneumonia	2	1.65
23	Sciatic pain	2	1.65
24	Abdominal worms	1	0.83
25	Backache	1	0.83
26	Baldness	1	0.83
27	Burnt injury	1	0.83
28	Carminative	1	0.83
29	Colic pain	1	0.83
30	Eczema	1	0.83
31	Gout	1	0.83
32	Hair fall	1	0.83
33	Headache and migraine	1	0.83
34	Hiccough	1	0.83
35	Intermittent fever	1	0.83
36	Laryngitis	1	0.83
37	Laxative	1	0.83
38	Leukorrhea	1	0.83
39	Malaria	1	0.83
40	Nerve tonic	1	0.83
41	Otalgia	1	0.83

(continued)

Table 2 (continued)

#	Diseases	No. of species	Percentage
42	Pain	1	0.83
43	Phthisis	1	0.83
44	Purgative	1	0.83
45	Pyorrhea	1	0.83
46	Ringworm	1	0.83
47	Throat pain	1	0.83
48	Tonic	1	0.83
49	Typhoid fever	1	0.83
		121	100.00

Table 3 Novel medicinal uses recorded from the Nara desert, Pakistan

Plant species	Diseases treated	Use report	FL%
Acacia nilotica (L.) Delile	Pyorrhea, diarrhea, and dysentery	105	84.00
Acacia senegal (L.) Willd.	Tonic	79	63.20
Caesalpinia bonduc (L.) Roxb.	Malarial and intermittent fever	78	62.40
Calligonum polygonoides L.	Cooling agent	110	88.00
Calotropis procera (Willd.) R. Br.	Otalgia and earache, sciatic pain, pneumonia, painful joints, and chest pain	109	87.20
Capparis spinosa L.	Rheumatic pain	49	39.20
Citrullus colocynthis (L.) Schrad.	Leukorrhea	15	12.00
Dalbergia sissoo L.	Spermatorrhea	79	63.20
Iphiona grantioides (Boiss.) Anderb.	Sores	98	78.40
Leptadenia pyrotechnica (Forssk.) Decne.	Eczema, ringworm	115	92.00
Pluchea lanceolata (DC.) Oliv. & Hiern.	Cooling agent	69	55.20
Prosopis cineraria (L.) Druce	Skin affection, diarrhea, congestion of chest	64	51.20
Saccharum spontaneum L.	Mouth gargle to relieve laryngitis and phthisis, and hiccough	43	34.40
Schweinfurthia papilionacea (Burm. f.) Boiss.	Chronic typhoid fever; skin impurities, chronic ulcers, and cancerous wounds	92	73.60
Tamarix aphylla (L.) Karst.	Boils and wounds	112	89.60
Tephrosia falciformis Romaswami.	Rheumatic pain	111	88.80
Ziziphus nummularia (Burm. f.) Wt. & Arn.	Spermatorrhea	45	36.00

Some of the species possessed novel use not mentioned in the literature. For instance, *Prosopis cineraria* was given in skin affection, diarrhea, and congestion of chest. *Schweinfurthia papilionacea* is utilized in chronic typhoid fever, skin impurities, chronic ulcers, and cancerous wounds. The ethnopharmacology for newly reported plants is sporadically known (Rastogi and Mehrotra 1993; Said et al. 1986). It is therefore suggested that the reported species should be screened in terms of phytochemicals as well as biological assays in order to establish their scientific ground for the anecdotal uses reported in this communication.

It is worthwhile to mention that some of the toxic species such as *Calotropis procera* (Asclepiadaceae), *Datura fastuosa* (Solanaceae), and *Ricinus communis* (Euphorbiaceae) are popularly used; however their use is known as either topical or mixed with other herbs. The milky latex of former species is externally applied as well as the powder of leaves is dusted on wounds to heal. The leaf powder of *Datura fastuosa* is smoked and inhaled to treat cough and asthma. The fresh leaves of *Ricinus communis* are coated with sesame oil and slightly warmed over fire which is tied on swelled joints to relieve pain and inflammation. According to Baydoun et al. (2015), informants consider that the native species may always guarantee the harmlessness and their safer use. There is no doubt that adequate knowledge about the potential toxicity of species is necessary to prevent the ingestion of such toxic plants or plant parts.

Mixing of herbs for making compounds culturally varies from area to area as well as region. From this area, powder formation was common (Fig. 3) for which 20 species were used, followed by juice, paste (12 spp. each), decoction, poultice (7 spp.), etc. Some of the studies reported the same trend (Qureshi and Bhatti 2008; Rauf et al. 2012; Qureshi and Shaheen 2013).

Fidelity level (FL%) of medicinal plants depicts the preference/reliance of key informants most used in treating particular ailments (Table 3). Some of the important species include *Leptadenia pyrotechnica* (92% FL for eczema/ringworm), *Tamarix aphylla* (89.6% FL for boils and wounds), *Tephrosia falciformis* (88.8 FL% for rheumatic pain), *Calligonum polygonoides* (88% FL for cooling agent), *Calotropis procera* (87.2% FL for earache, sciatic pain, pneumonia, painful joints, and chest pain), *Iphiona grantioides* (78.4% FL for sores), and *Schweinfurthia papilionacea.* The most widely used species by the dwellers always possess higher FL% than that of less used ones. This index is intended to measure the significance of the species for a particular use (Baydoun et al. 2015).

The people of the study area are well aware about the use of plant. Nevertheless, none of the informants in the study was aware of the growing recent controversial scientific evidence regarding the effects of the plant in terms of the plant parts used, way of preparation, dosage, and activity of the separate constituents of extract. Only beneficial effects were actually reported with no mentioning of any side effects (El Thaher et al. 2001; Zanoli et al. 2005).

Conclusion

This ethnobotanical inventory includes a wide range of therapeutic uses of 52 medicinal trees and shrubs cited for 49 illnesses. Besides, this study revealed some of the novel/additional uses of medicinal plants not previously known in the medicinal literature. The method of preparations recorded for the local popular medicine of the studied area is mostly based on the use of a single species instead of mixtures. Regardless of the wide range of their use by the dwellers, scientific validation and authentication are required. In vitro and in vivo activities should be undertaken in order to authenticate the claimed uses of native and endemic species. Besides, there is need for identifying of active compounds that may be used in drug discovery program.

Acknowledgment The author gratefully acknowledges the local people and herbalists (Hakeems) who provided valuable information on medicinal uses of plants.

References

Aburjai T, Hudaib M, Tayyem R (2007) Ethnopharmacological survey of medicinal herbs in Jordan, the Ajloun Heights region. J Ethnopharmacol 110(2):294–304

Alexiades MN (1996) Collecting ethnobotanical data: an introduction to basic concepts and techniques. In: Alexiades MN (ed) Selected guidelines for ethnobotanical research: a field manual. The New York Botanical Garden, New York, pp 53–94

Ali SI, Nasir YJ (Eds.) (1990–1991) Flora of Pakistan, Islamabad, Karachi

Ali SI, Qaiser M (Eds.) (1993–1995) Flora of Pakistan, Islamabad, Karachi

Ambasta SP (1986) The useful plants of India, CSIR, New Delhi

Anonymous (1992) Pak. Swiss Range and Livestock Improvement Project

Asolkar LV, Kakkar KK, Chakre OJ (1992) Secondary supplement to glossary of Indian medicinal plants with active principles, Publications and Information Directorate (CSIR), Dr. K.S. Krishnan Marg, New Delhi

Baquar SR, Tasnif M (1984) Medicinal plants of Southern West Pakistan, Periodical Expert Book Agency, D-42, Vivek Vihar, Delhi-113334

Batanouny KH (1981) Ecology and Flora of Qatar. Centre for scientific and applied Research, University of Qatar, P.O. Box 2713, Doha

Baydoun S, Lamisb C, Helenaa D, Nellya A (2015) Ethnopharmacological survey of medicinal plants used in traditional medicine by the communities of Mount Hermon, Lebanon. J Ethnopharmacol 173:139–156

Bhandari MM (1990) Flora of the Indian desert. M.P.S. Repros., Rajasthan

Boulos L (1991) Flora of Egypt, vol 1. Al Hadara Publishing, Cairo, Egypt

Dastur JF (1962) Medicinal plants of India and Pakistan. D.B. Taraporevala, Bombay

Dymock W, Warden CJH, Hooper D (1972) Pharmacographia Indica, vol I–IV. Hamdard Foundation, Nazimabad, Karachi

El-Thaher TS, Matalka KZ, Taha HA, Badwan AA (2001) Ferula hermonis 'zallouh' and enhancing erectile function in rats: efficacy and toxicity study. Int J Impot Res 13:247–251

Jafri SMH (1966) The Flora of Karachi. The Book Corporation, Karachi, Pakistan

Kritikar KR, Basu BD (1918) Indian medicinal plants, vol II–III. Indian Press, Allahabad

Matthew KM (1981 1983) Flora of Tamilnadu Carnatic. The Rapinat Herbarium, St. Joseph's College, Tiruchirapalli, India, vol 1–3.

Murray JA (1989) The plants and drugs of Sindh, Karachi. Indus Publications

Nadkarni AK (1954) Indian materia medica, vol 1. Popular Prakashan Pvt. Ltd., Bombay, pp 237–242

Nasir E, Ali SI (Eds) (1970–1989) Flora of Pakistan, Islamabad, Karachi.

Queshi R, Bhatti GR (2008) Ethnobotany of plants used by the Thari people of Nara Desert, Pakistan. Fitoterapia 79:468–473

Qureshi R (2012) The Flora of Nara Desert, Pakistan, 1st edn. Nova Science Publishers, New York
Qureshi R, Bhatti GR (2005a) Nara Desert, Pakistan. Part I: Soils, climate and vegetation. Rangelands 27(5):27–31
Qureshi R, Bhatti GR (2005b) Nara Desert, Pakistan. Part II: Human life. Rangelands 27(5):32–35
Qureshi R, Bhatti GR, Memon RA (2010) Ethnomedicinal uses of herbs from Nara Desert, Pakistan. Pak J Bot 42(2):839–851
Qureshi R, Shaheen H (2013) The Ethnobotanical profile of Tehsil Kotli Sattian, Rawalpindi, Pakistan. Nova Science Publishers Inc., Hauppauge, New York, pp 1–165
Rastogi RP, Mehrotra BN (1993) Compendium of Indian medicinal plants, vol 1–3. CDRI, Luknow and PID, New Delhi
Rauf F, Qureshi R, Shaeen H (2012) Folk medicinal uses of indigenous plant species of Barroha, Bhara Kahu and Maanga in Islamabad, Pakistan. J Med Pl Res 6(11):2061–2070
Said HM, Ahmad VD, Rahman AU (1986) Pakistan encyclopedia planta medica, Hamdard Foundation Press, Hamdard Centre, Nazimabad, Karachi
Sivarajan VV, Balachandran I (1996) Ayurvedic drugs and plants resources. Oxford and IBH Publishing Co. Pvt. Ltd., New Delhi
Zanoli P, Rivasi M, Zavatti M, Brusiani F, Vezzalini F, Baraldi M (2005) Activity of single components of Ferula hermonis on male rat sexual behavior. Int J Impot Res 17:513–518

A Comparative Analysis of the Medicinal Plants Used for Diabetes Mellitus in the Traditional Medicine in Turkey, Pakistan, and Malaysia

Munir Ozturk, Volkan Altay, Abdul Latiff, M. Asad Ziaee, M. Iqbal Choudhry, Farzana Shaheen, and Cenk Durmuşkahya

Introduction

One of the fast-growing and complex and medical problem at global scale in both developed and developing countries is diabetes mellitus. It is prevalent equally well in both developed as well as developing countries, being the most common endocrine disorder in the world and as a serious global health problem (Tripathi et al. 2011; Kumar et al. 2013; Tiwari and Rana 2015). The herbal therapeutic remedies are used with success to treat this disorder and its ramifications. The drugs derived from the herbs and preparations made from these are generally accepted as less toxic and free from side effects. However, this subject has not been validated scientifically in order to move ahead for their substitution for the current therapeutics (Arulselvan et al. 2014).

M. Ozturk (✉)
Vice President of the Islamic World, Academy of Sciences, Amann, Jordan

Department of Botany and Centre for Environmental Studies, Ege University, Izmir, Izmir, Turkey

V. Altay
Faculty of Science and Arts, Department of Biology, Hatay Mustafa Kemal University, Hatay, Turkey

A. Latiff
Faculty of Science and Technology, Universiti Kebangsaan Malaysia, Selangor, Malaysia

M. Asad Ziaee
Fujian Institute of Research on Structure of Matter, University of Chinese Academy of Sciences, Beijing, People's Republic of China

M. Iqbal Choudhry · F. Shaheen
International Center for Chemical and Biological Sciences, University of Karachi, Karachi City, Sindh, Pakistan

C. Durmuşkahya
Faculty of Forestry, Katip Çelebi University, Izmir, Turkey

M. Ozturk, K. R. Hakeem (eds.), *Plant and Human Health, Volume 1*,
https://doi.org/10.1007/978-3-319-93997-1_11

Traditional antidiabetic plants used in the traditional medicine can serve as a useful source for developing new oral hypoglycemic compounds. These may be evaluated as simple dietary adjuncts to existing therapies. Such investigations can offer a natural key to unlock future diabetologist's pharmacy, because it is one of the major chronic diseases and healthcare challenges. At present it globally affects over 370 million people. Several reports published show that more than 4.5 million deaths have occurred due to diabetes in 2012. The latest estimates from the "International Diabetes Federation" mentions that more than 550 million people will be facing this disorder by 2030, and more than 75% of diabetic patients will be from developing countries (International Diabetes Federation 2013; Nowbandegani et al. 2015), especially the people with ages between 45 and 64 years (Roglic 2004; Trojan-Rodrigues et al. 2012).

It is a multifarious group of disorders affiliated to the disturbance in the metabolism of carbohydrates, fat, and protein and results in a shortage or lack of insulin secretion and/or reduced sensitivity of the tissue to insulin (Arulselvan et al. 2014). Despite advances in understanding and management of this metabolic disorder, the rate of morbidity and mortality due to this disorder is increasing every year (Leite et al. 2013; Arulselvan et al. 2014). The number of cases is rapidly increasing all over the globe, and its complications are a major cause of disability and hospitalization, posing a significant financial burden (Arulselvan et al. 2014).

Although tremendous information has been gathered in the understanding of pathophysiology of diabetes and management of diabetes, the disease and disease-associated complications are increasing. Much information is available on the presence of known antidiabetic medicine on the pharmaceutical market. Therapeutic remedies from herbs are used with success to treat this disorder and its ramifications.

Among the populations undergoing modernization of lifestyle, a striking emergence of non-insulin-dependent diabetes mellitus as a major health problem is recorded now both in developing nations and in rural areas of developed countries (Bennett 1983; Bransome 1992; WHO 1985; Gohdes 1986; Schraer et al. 1988; Marles and Farnsworth 1995). Alternate strategies for the prevention and treatment of diabetes need be developed due to enormous costs of modern treatments. Nearly 90% of the residents in rural areas of developing countries still rely on traditional medicines for their primary healthcare. The scientific investigations of traditional medicines have led to the discovery of at least 88 drugs now in professional use worldwide (Soejarto and Farnsworth 1989; Marles and Farnsworth 1995). A rationally designed interdisciplinary research program could lead to the development of indigenous, renewable, medicinal plant resources as practical and cost-efficient alternatives. The synthesis of local traditional and modern knowledge as well as techniques for the management of diabetes should be feasible (Marles and Farnsworth 1995).

Earliest recorded treatments for this disorder involve the use of plants, such as the Papyrus Ebers of 1550 BC which show a high-fiber diet of wheat grains and ochre. Large number of herbs, spices, and other plant materials have been recorded for the treatment of diabetes globally starting with by ancient cultures (Ajgaonkar

1979; Day and Bailey 1988a, b; Bailey and Day 1989; Marles and Farnsworth 1995; Soumyanath 2005; Durmuşkahya and Öztürk 2013; Subramoniam 2016a, b). The availability of insulin, folklore medicines for diabetes have almost disappeared in occidental societies, but continue to be the cornerstone of therapy in underdeveloped countries. Much attention to alternative medicines and natural therapies has stimulated a new wave of research interest in traditional practices, and "WHO" expert committee on diabetes has listed it as one of its recommendations that traditional methods of treatment for diabetes should be further investigated (Bailey and Day 1989).

Ethnopharmacologically, more than 1000 taxa of organisms have been used to experimentally treat the symptoms of this disease. These belong to more than 700 genera from 180 families. This list extends phylogenetically all the way from marine algae and fungi to advanced plants. The large and widely distributed families are Fabaceae, Asteraceae, Lamiaceae, Liliaceae, Poaceae, and Euphorbiaceae. The large number of taxa reported to have been used traditionally or experimentally for the treatment of diabetes may be coincidental. The phylogenetic distance between the selected groups of families is a strong indication of the varied nature of the active constituents. While chemotaxonomic investigations are useful in the discovery of new plants with biologically active constituents, it will be necessary to learn more about particular groups of hypoglycemic natural products and their mechanisms of action before this method of drug discovery can be successfully used (Marles and Farnsworth 1995).

More than 1/3 of all plant-derived drugs come from tropical rainforest plant taxa. This number may rise substantially if we learn more about the phytochemistry and pharmacology of tropical plants. Even with their greater accessibility and longer history of study, most temperate plants have not been exhaustively studied for therapeutic usefulness (Soejarto and Farnsworth 1989; Marles and Farnsworth 1995).

The destruction of tropical rainforests is resulting in the loss of a tremendous natural resource for potential new drugs. Their germ plasm would be necessary for genetic improvement of cultivated varieties and tissue cultures and their constituents which may serve as new drugs or prototypes for synthetic drug research. The economic potential of novel drugs derived from primary rainforest plant taxa could serve as an incentive for preservation of rainforest and its management as a renewable resource rather than just a source of land for mineral and agricultural exploitation (Marles and Farnsworth 1995).

If the same or a closely related plant is used traditionally for the same at more than one place, it suggests either cultural contact between the countries or independent discovery. As such, the conservation of such traditional use indicates a higher probability that the traditional practitioners found the remedy to be effective (Marles and Farnsworth 1995). The best strategy will involve the study of traditional antidiabetic plants, and discovery of locally available alternative medicines to treat diabetics in developing countries, as well as commercial development of new botanical hypoglycemic agents and adjuncts to antidiabetic therapy (Marles and Farnsworth 1995).

Our main aim here has been to analyze the distribution of the traditional medicinal plant knowledge used for diabetes mellitus in Turkey, Pakistan, and Malaysia. The specific aims are to compare the composition and richness of medicinal ethnofloras.

Study Areas

Turkey is a meeting place of different phytogeographical regions showing great variation in plant diversity with different types of ecosystems, occupying different habitats and showing varying vegetational characteristics (Ozturk et al. 2006, 2012a, b, 2017a, b). Among the countries in the Southwest Asia and the Mediterranean basin, as well as whole of Europe, the richest flora has been reported for the Anatolian peninsula. Phytogeographically, the country is a meeting place for two centers of diversity and origin: the Near East and the Mediterranean (Fig. 1). The number of flowering plant taxa distributed in the country is estimated to be more than 11.000, which is very near to the number recorded from the whole of Europe (Güner et al. 2012). A great variety of geomorphological, topographical, and climatic features are responsible for its widespread habitat and plant diversities as well as endemism, with 3035 taxa of endemics, comprising 31.12% of the total flora and confined to narrow and restricted ecological niches (Güner et al. 2012; Ozturk et al. 2016).

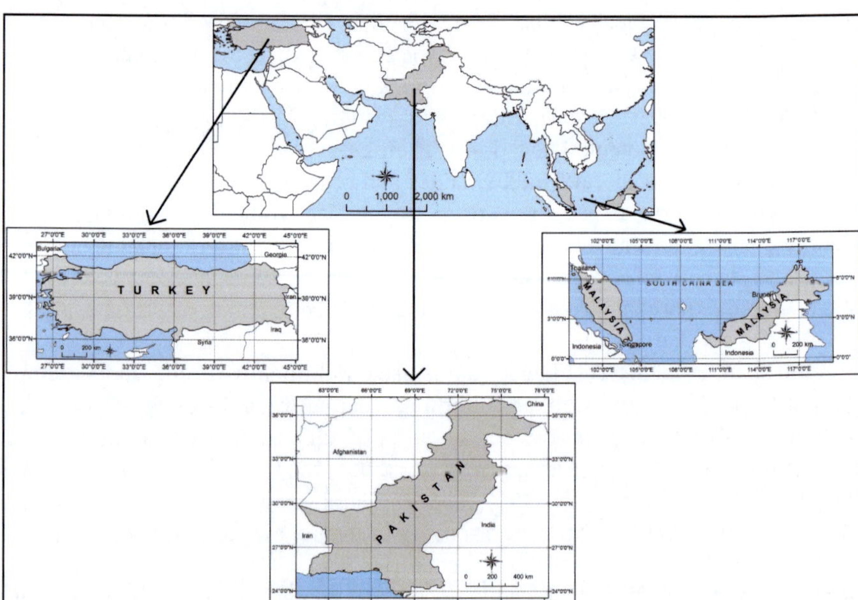

Fig. 1 The map showing the countries included in this study

Nearly 1280 plant taxa belonging to 458 genera and 114 plant families have been recorded to be used in Turkish folk medicine for food and health benefits. Most frequently referred plant genera for healing potentials are as follows: *Hypericum*, *Origanum*, *Sideritis*, *Rosa*, *Urtica*, *Sambucus*, *Plantago*, *Ecballium*, *Equisetum*, *Cotinus*, *Thymus*, *Malva*, *Helichrysum*, *Polygonum*, *Mentha*, *Achillea*, *Teucrium*, and *Allium* (Yeşilada 2013).

Pakistan has a unique biodiversity, having nine major ecological zones and a peculiar geographical position (Fig. 1) and harbors a great diversity of flora. More than 6000 vascular plant species are reported to occur in this region (Shinwari and Shinwari 2010), out of which 5.600 species have been described to date in the flora of Pakistan, representing 22 families and about 150 genera (Nasir and Ali 1970–1995). Among the lower plants, there are at least 189 pteridophytes (ferns and their allies), of which 153 are Sino-Japanese elements and 36 Euro-Siberian (Nasir and Ali 1970–1995; Shinwari and Shinwari 2010). Four monotypic genera of flowering plants (*Douepia*, *Sulaimania*, *Spiroseris*, *Wendelboa*) and around 400 species (7.8%) are endemic to Pakistan. Most endemics are Irano-Turanian and Sino-Japanese (Shinwari and Shinwari 2010). A short analysis of plant diversity in Pakistan reveals that the number of species per genus is much lower than the global average, indicating a high rate of diversity at the gene level (Ali and Qaiser 1986). The plant wealth of the country includes elements of six phytogeographical regions, namely, the Mediterranean, Saharo-Sindian, Euro-Siberian, Irano-Turanian, Sin-Japanese, and Indian. Nearly 400–600 plant taxa are medicinally important (Rehman et al. 2015). These are mainly used in Tibbi Dawakhan (Industries of herbal medicines) with very little attention being paid to the ethnobotanical research (Hamayun 2003; Rehman et al. 2015). The studies on the traditional knowledge of medicinal plants has increased in the country during the last three decades; consequently significant knowledge has been documented (Rehman et al. 2015).

Malaysia is located just north of equator, with a warm equatorial climate but little seasonality, hot and humid throughout the year. The total landmass of the country is 329.847 km^2 (Fig. 1). South China Sea divides the country into two regions of similar size: Peninsular Malaysia connected to mainland Asia on the western half and the states of Sabah and Sarawak on the island of Borneo on the east (Saw and Chung 2015). These two halves of the country pose interesting challenges toward documenting the Malaysian flora. Peninsular Malaysia, part of the Malay Peninsula (here includes Singapore and Peninsular Thailand), contains the floristic elements of the Sunda Shelf and also of the mainland Asiatic species from seasonal climates (Wong 1998; Saw and Chung 2015).

Up till now no comprehensive and up-to-date checklist for the Malaysian flora has been published (Saw and Chung 2015). There are several checklists prepared by different workers due to different botanical history of the two main regions of Malaysia. The work of Ridley (1922–1925) for Peninsular Malaysia is an outdated one. But, it provided the first complete enumeration of the vascular plants of the Malay Peninsula. The angiosperms have been published in five volumes during 1922–1925. Subsequently, a separate checklist of ferns was published (Ridley 1926). "A Catalogue of the Vascular Plants of Malaya" (Turner 1997) serves as the most recent checklist for the Peninsular Malaysia based on an existing literature survey. This catalogue enumerates 8.198 species. Parris and Latiff (1997) have published a further update on the ferns and fern allies with some additions and nomenclatural changes.

The plant diversity of Malaysia is among some of the most diverse, being one of the world's richest in the world and most varied biophysical resources (Premilla 2002; Adnan and Othman 2012). The rainforest of the country turns out to be the perfect place for luxuriant plant growth. Furthermore, it is also set apart with vast of resources of plants either medicinal or with other usage (Adnan and Othman 2012). Ethnobotany in this multiracial country is widely used in a broad manner. Every ethnic group or race practices and uses its ethnobotanical knowledge in its own way and depends on their beliefs (Adnan and Othman 2012). These communities commonly use plants for dietary purposes (food consumption). The utilities such as materials for construction, medicinal, ritual, dyes, and cosmetics are also common. The studies on the relationship between plants and the Malay culture help to preserve the integrity of the cultures and at the same time protect the natural heritage and its knowledge (Adnan and Othman 2012).

Malaysian traditions embody an exceptional vast knowledge and practices derived from Malay, Chinese, Indian, and indigenous people's traditions with regard to plants. Malaysians have inherited multiple traditional medicine and ways of healing. It has started with the indigenous people called Orang Asli since this group of people embodied a wonderful knowledge and respect for plant life together with the nature and environment that they inherited from their ancestors (Adnan and Othman 2012).

Comparative Evaluation on Country Basis

Turkey

The total number of medicinal plants evaluated for diabetes in the traditional medicine includes 340 taxa. These belong to 66 families and 185 genera and are given alphabetically with their botanical name, part used, ailment treated, and information on the preparations used (Appendix 1). The following families are represented by the largest number of taxa: Rosaceae (49), Lamiaceae (48), Asteraceae (45), Apiaceae (20), and Fabaceae (17). The genera with highest number of taxa are *Prunus* (11 taxa), *Thymus* (9 taxa), *Quercus* (7 taxa), and *Juniperus*, *Hypericum*, *Crataegus*, *Pyrus*, *Rubus*, and *Salvia* (6 taxa each) (Appendix 1).

The most commonly used taxa among these are *Rosa canina*, *Teucrium polium*, *Urtica dioica*, *Juglans regia*, *Rubus sanctus*, *Viscum album* ssp. *album*, *Olea europaea*, *Morus nigra*, *Prunus dulcis*, *Thymbra spicata*, and *Prunus spinosa* ssp. *dasphylla* (Appendix 1).

On the basis of parts mainly used, we found the numbers as follows: leaves used (107 taxa), aerial parts (104 taxa), fruits (77 taxa), and flower (55 taxa) (Table 1). The most common preparations used are decoction (177 taxa), followed by infusion (138 taxa), fresh (65 taxa), and raw (25 taxa). Other uses and their taxa numbers are given in Table 2.

Table 1 The parts of MAPS used in three countries studied with the number of taxa

Part used	Code	Turkey	Pakistan	Malaysia
Leaves	L	107	87	35
Aerial parts	AP	104	11	2
Fruit	FR	77	54	12
Fruit shell	FR-S	2	–	–
Fruit juice	FR-J	6	6	3
Flower	FL	55	20	5
Root	R	40	26	22
Seed	S	39	41	7
Shoot	SH	18	6	1
Stem	ST	13	10	5
Herb	H	10	–	–
Stem bark	ST-B	7	15	1
Cones	CO	7	–	–
Tuber	T	6	1	1
Whole plants	WP	5	51	3
Bulb	B	5	3	2
Branches	BR	5	2	–
Buds	BD	4	1	–
Rhizome	RH	3	6	2
Oil	OO	2	1	–
Pine tar	PT	2	–	–
Resin	RS	2	2	–
Capitulum	CA	1	1	–
Latex	LA	1	5	–
Cupula	CU	1	–	–
Fronds	FO	–	2	1
Receptacles	RE	–	1	–
Immature pods	IP	–	1	–

Pakistan

The total number of medicinal plants evaluated for diabetes in the traditional medicine includes 281 taxa from 76 families and 209 genera. These are presented here alphabetically with their botanical name, part used, ailment treated, and information on the preparations used (Appendix 2). The following families are represented by the largest number of taxa: Asteraceae (27), Fabaceae (25), Lamiaceae (14), Cucurbitaceae (12), Apocynaceae (12), and Moraceae (11). The genera with highest number of taxa are *Ficus* (9 taxa), *Artemisia* (6 taxa), *Phyllanthus* (5 taxa), and *Ziziphus*, *Solanum*, and *Fagonia* (4 taxa each) (Appendix 2). The most commonly used taxa among these are *Syzygium cumini*, *Allium sativum*, *Momordica charantia*, *Justicia adhatoda*, *Trigonella foenum-graecum*, *Ficus benghalensis*, and *Olea ferruginea* (Appendix 2).

Table 2 The preparations used in the studied countries with the number of taxa

Preparations	Code	Turkey	Pakistan	Malaysia
Decoction	DE	177	42	32
Infusion	IN	138	13	2
Eaten fresh	EF	65	13	7
Raw	R	25	3	5
Cooked	C	20	15	1
Boiled	BO	15	41	13
Mash	MA	9	–	–
Crushed	CR	6	1	–
Pickled	PI	5	–	–
Dried	DR	5	17	2
Powdered	PW	5	62	3
Poultice	PU	4	1	–
Syrup	SY	4	–	–
Jam	JA	2	2	–
Pounded	PN	2	–	3
Cataplasm	CP	2	–	–
Medicinal oil	MO	2	1	–
As tea	AT	1	1	–
Extract	EX	1	72	21
Gum	G	1	6	–
Chewing	CH	1	4	–
Maceration	MC	1	–	–
One drop oil added, a glass of water	DOA	1	–	–
Juice	JU	–	17	2
As vegetable	AV	–	10	1
Pulp	PL	–	5	–
Roasted	RA	–	2	–

On the basis of parts mainly used, we found the numbers as follows: leaves used (87 taxa), fruits (54 taxa), whole plants (51 taxa), and seed (41 taxa) (Table 1). The most common preparations used are; extract (72 taxa), followed by powder (62 taxa), decoction (42 taxa), and boiled (41 taxa). Other uses and their taxa numbers are given in Table 2.

Malaysia

The total number of medicinal plants evaluated for diabetes in the traditional medicine includes 76 taxa from 43 families and 65 genera. Appendix 3 includes these alphabetically with their botanical name, part used, ailment treated, and information on the preparations used. The following families are represented by the largest number of taxa: Fabaceae (8), Acanthaceae (5), Malvaceae and Phyllanthaccac (4 each), and Annonaceae and Zingiberaceae (3 each). The genera with highest number of taxa are *Phyllanthus* (3 taxa) and *Piper*, *Orthosiphon*, *Sindora*, *Allium*, *Annona*, *Terminalia*, and *Cnestis* (2 taxa each) (Appendix 3). The most commonly used taxa among these are *Parkia speciosa*, *Andrographis paniculata*, *Averrhoa bilimbi*, *Momordica charantia*, and *Morinda citrifolia* (Appendix 3).

On the basis of parts mainly used, we found the numbers as follows: leaves used (35 taxa), root (22 taxa), and fruits (12 taxa) (Table 1). The most common preparations used are decoction (32 taxa), followed by extract (21 taxa) and boiled (13 taxa). Other uses and their taxa number are given in Table 2.

In all three countries, for each of the MAPS scientific name, preparations, part used and its use has been recorded (Appendices 1, 2, and 3). The information recorded has been symbolized as follows from the three countries in order to follow the appendices easily:

Part used: **AP**, aerial parts; **BD**, buds; **BR**, branches; **B**, bulb; **CA**, capitulum; **CO**, cones; **CU**, cupula; **FL**, flowers; **FO**, fronds; **FR**, fruit; **FR-J**, fruit juice; **FR-S**, fruit shell; **H**, herb; **IP**, immature pods; **LA**, latex; **L**, leaves; **OO**, oil; **PT**, pine tar; **RE**, receptacles; **RH**, rhizome; **RS**, resin; **R**, root; **S**, seed; **SH**, shoots; **ST**, stem; **ST-B**, stem bark; **T**, tuber; **WP**, whole plants

Preparations: **AT**, as tea; **AV**, as vegetable; **BO**, boiled; **C**, cooked; **CH**, chewing; **CP**, cataplasm; **CR**, crushed; **DE**, decoction; **DOA**, one drop oil added, a glass of water; **DR**, dried; **EF**, eaten fresh; **EX**, extract; **G**, gum; **IN**, infusion; **JA**, jam; **JU**, juice; **MA**, mash; **MC**, maceration; **MO**, medicinal oil; **PI**, pickled; **PL**, pulp; **PU**, poultice; **PN**, pounded; **PW**, powdered; **R**, raw; **RA**, roasted; **SY**, syrup

The taxa determined in three countries show the following distribution: 340 taxa in Turkey, 281 taxa in Pakistan, and 76 taxa in Malaysia. A total of 42 taxa are common among Turkey and Pakistan, 11 between Pakistan-Malaysia, and 6 between Turkey-Malaysia (Table 3). A comparison on the basis of MAPS reveals that in all these countries, five taxa are commonly used for the diabetes mellitus: *Allium cepa*, *Allium sativum*, *Momordica charantia*, *Nigella sativa*, and *Punica granatum* (Table 3).

If the data presented above is taken into consideration, similarity ratios can be calculated, using Jaccard similarity index, ethnoflora used for diabetes mellitus (Table 4) in three countries.

This index allows the percentage of similarity to be calculated in each area. The formula to calculate this index is as follows:

$$\text{Index of Jaccard} = 100 \times C / \left(A + B - C\right)$$

where A is the number of species of the sample A, B the number of species of the sample, and C is the number of species common to A and B (González-Tejero et al. 2008).

Table 3 The medicinal plant taxa used in three countries for diabetes mellitus treatment

Taxa	Turkey	Pakistan	Malaysia
Abelmoschus esculentus	x	x	
Allium cepa	x	x	x
Allium sativum	x	x	x
Aloe vera		x	x
Artemisia absinthium	x	x	
Capparis spinosa	x	x	
Catharanthus roseus		x	x
Centaurea iberica	x	x	
Cichorium intybus	x	x	
Coriandrum sativum		x	x
Cucumis sativus	x	x	
Cynara scolymus	x	x	
Daucus carota	x	x	
Elaeagnus angustifolia	x	x	
Equisetum arvense	x	x	
Eruca vesicaria	x	x	
Ficus carica	x	x	
Foeniculum vulgare	x	x	
Gentiana olivieri	x	x	
Hedera helix	x	x	
Hordeum vulgare	x	x	
Juniperus communis	x	x	
Juniperus excelsa	x	x	
Lamium amplexicaule	x	x	
Malus domestica	x	x	
Malva neglecta	x	x	
Mangifera indica		x	x
Mentha longifolia	x	x	
Momordica charantia	x	x	x
Morus alba	x	x	
Morus nigra	x	x	
Myrtus communis	x	x	
Nigella sativa	x	x	x
Ocimum basilicum	x	x	
Olea europaea	x	x	
Origanum vulgare	x	x	
Phyllanthus acidus		x	x
Phyllanthus amarus		x	x
Plantago major	x		x
Portulaca oleracea	x	x	
Prunus dulcis	x	x	

(continued)

Table 3 (continued)

Taxa	Turkey	Pakistan	Malaysia
Prunus persica	x	x	
Punica granatum	x	x	x
Raphanus raphanistrum	x	x	
Sesamum indicum	x	x	
Taraxacum campylodes	x	x	
Trigonella foenum-graecum	x	x	
Viscum album	x	x	
Ziziphus jujuba	x	x	

Table 4 Jaccard similarity index related to diabetes mellitus in three countries studied

	Turkey-Pakistan (%)	Pakistan-Malaysia (%)	Turkey-Malaysia (%)
Diabetes mellitus disease group	13.58	6.16	2.88

Conclusions

Antidiabetic plants used in the traditional medicine do provide useful information on the source of new oral hypoglycemic compounds for development as pharmaceutical entities or as simple dietary adjuncts to existing therapies (Bailey and Day 1989). For a range of diabetic presentations in the world, traditional plant medicines are used widely and investigating such medicines can offer a natural key to unlock a diabetologist's pharmacy for the future (Bailey and Day 1989).

The recommendations put forth by WHO reveal that antihyperglycemic agents of natural plant origin used in traditional medicine are important. Antidiabetic potential of herbals is due to their ability to restore the function of pancreatic tissues by causing an increase in concentration of insulin. The research for alternate herbal remedies for diabetes mellitus is continuing widely all over the globe, because this disease poses many challenges not only to the physician but also to the researcher (Arulselvan et al. 2014).

In view of the increase in the number of diabetic patients, expenses related to this are increasing and effecting our health economy. Therefore, there is urgent need for higher number of more economical and effective alternative cheaper treatments. Moreover, treatment of diabetes is still a challenging issue for the health specialists. This chronic disease requires "lifestyle modifications" like exercise, balanced nutrition, and weight control as well as regular drug use. It also requires discipline and regular monitoring and follow-up.

The percentage of diabetic patients in the world who achieve treatment targets (including Turkey) is unfortunately still around 43%. The reason for this low rate can be listed as problems related to healthcare professionals and health system, insufficiency of present therapies, and patients' poor compliance with the recommended treatments. Additionally, like many other chronic diseases, diabetes has a degenerative and progressive character, which may raise feelings of despair, helplessness, and even tiredness in patients (Parildar et al. 2011).

In the treatment of type 2 diabetes mellitus, herbal treatments have been used since ancient times. The historical documents 2000 years old depict this fact clearly as recorded on the Egyptian Papyruses, in the works of Hippocrates and Chinese medicine and Ayurveda texts. These sources also mention about herbal antidiabetic treatments. But, there is only one antidiabetic drug generated from a plant: "metformin" which originates from French lilac (*Galega officinalis* L.). Having gained gradual importance in recent years, metformin has been successfully used in diabetes treatment for the last five decades (Parildar et al. 2011).

The phytotherapy treatments of type 2 diabetes mellitus are attracting much attention lately and are expected to increase gradually during the coming years. Such treatments can be effective in the prevention of diabetes and its complications as well as optimization of the treatment and life standards. The mechanism here is closely related with several chronic metabolic diseases. In particular diabetes shows closeness to oxidative stress and inflammation in the body. Because of the antioxidant features of herbs, these are considered for both prevention and treatment of diabetes (Parildar et al. 2011).

The herbs constitute a part of culture at global scale. This is a well-known fact that for quite a long time, herbs are like nutrients and have beneficial effects on health. It is quite important to utilize nature's riches in the most efficient way. Special attention need be paid to the herbal treatments, which have been traditionally used by people for thousands of years. Undertaking more scientific investigations on these treatments will open new horizons in the treatment of diabetes. Currently, scientific evidence and findings support the efficiency of herbal remedies routinely in clinical practice, but the applications are insufficient. However, adverse effects and/or drug interactions in these widely used therapies need to be definitely taken into consideration by the healthcare professionals (Parildar et al. 2011).

In addition to this, we should not forget that phytotherapy is not an alternative but a complementary and supportive treatment to the conventional diabetes therapy and should be accompanied by diet and exercise treatment. The most important support and information source for reliability, efficiency, and/or side effects of herbs and herbal preparations should be health professionals who can undertake leadership role effectively by improving and updating themselves in this field, as has to be in all other fields (Parildar et al. 2011).

Appendix 1: Herbs Used for Diabetes Mellitus in the Traditional Medicine in Turkey

No	Family/taxa	Parts used	Preparation	Resources[a]
	Adoxaceae			
1	*Sambucus ebulus*	S		124
2	*Sambucus nigra*	FL	DE	1, 44
3	*Viburnum lantana*	FR	DE	10
4	*Viburnum opulus*	FR	DE	57
	Amaranthaceae			
5	*Beta vulgaris*	L	EX	150
	Amaryllidaceae			
6	*Allium akaka*	B	EF	17
7	*Allium ampeloprasum*	B	DE, R	43, 57
8	*Allium cepa*	B	DE	47, 72
9	*Allium sativum*	B, FL, L	EF, R	16, 43, 103, 104
	Anacardiaceae			
10	*Cotinus coggygria*	L	DE	1–3
11	*Pistacia terebinthus*	L	DE	4
12	*Pistacia vera*	FR, S	EF	147
13	*Rhus coriaria*	L	CR, DE	147
	Apiaceae			
14	*Ammi visnaga*	L	IN	133
15	*Apium graveolens*	R	DE	6
16	*Chaerophyllum bulbosum*	RH	R	19
17	*Daucus carota*	AP	IN	7
18	*Diplotaenia cachrydifolia*	AP	DE	142, 146
19	*Echinophora tenuifolia* ssp. *sibthorpiana*	AP	DE	8
20	*Eryngium campestre* var. *virens*	ST	EF	9
21	*Ferula caspica*	AP	DE	10
22	*Ferula orientalis*	AP	C	11
23	*Ferula rigidula*	H	PU	10
24	*Foeniculum vulgare*	S	IN	12, 13
25	*Heracleum persicum*	WP	DE	146
26	*Laser trilobum*	FR	EF	14
27	*Petroselinum crispum*	AP, L, R	EF, IN	15, 16
28	*Peucedanum longifolium*	H	PI	10

No	Family/taxa	Parts used	Preparation	Resources[a]
29	*Prangos ferulacea*	**SH, ST**	**BO, C**	**42, 127**
30	*Prangos pabularia*	**R**	**PI**	**135**
31	*Scandix pecten-veneris*	**AP**	**IN**	**149**
32	*Smyrnium connatum*	**RH**	**EF, C**	**14**
33	*Zosima absinthifolia*	**L**	**DE**	**10**
	Araceae			
34	*Arum rupicola* var. *virescens*	**L, R, T**	**BO, DE, DR, IN**	**10, 17, 18, 19**
	Araliaceae			
35	*Hedera helix*	**L, ST**	**DE**	**3, 20–22**
	Asparagaceae			
36	*Asparagus acutifolius*	**FR, R**	**DE, IN**	**49**
37	*Polygonatum multiflorum*	**L**	**IN**	**12, 13**
	Aspleniaceae			
38	*Asplenium scolopendrium*	**AP**	**AT, DE, IN**	**151**
39	*Ceterach officinarum*	**AP**	**DE**	**23**
	Asteraceae			
40	*Achillea arabica*	**FL**	**IN**	**24**
41	*Achillea millefolium*	**FL**	**IN**	**24**
42	*Achillea schischkinii*	**FL**	**DE**	**25**
43	*Achillea tenuifolia*	**L**	**IN**	**10**
44	*Anthemis cotula*	**AP**	**DE**	**11**
45	*Artemisia absinthium*	**AP, L, FL**	**DE, IN**	**3, 5, 10, 11, 14, 26–29**
46	*Artemisia annua*	**AP**	**CR**	**114, 128**
47	*Artemisia chamaemelifolia*	**FL, H**	**DE**	**10**
48	*Artemisia vulgaris*	**AP**	**DE**	**22**
49	*Carduus acanthoides*	**AP**	**DE**	**3**
50	*Carduus nutans* ssp. *leiophyllus*	**AP**	**DE**	**3**
51	*Carduus pycnocephalus*	**AP**	**EF**	**9**
52	*Centaurea benedicta*	**AP, L, ST**	**DE, IN**	**12, 13, 32, 33**
53	*Centaurea iberica*	**AP**	**DE, EF**	**23, 30**
54	*Centaurea virgata*	**FL**	**DE**	**129**
55	*Cichorium intybus*	**AP, SH**	**C, EF**	**30, 31**
56	*Cirsium hypoleucum*	**FL**	**R**	**16**
57	*Cirsium vulgare*	**R, ST**	**BO**	**34**
58	*Cota austriaca*	**AP**	**IN**	**24**
59	*Cota tinctoria*		**DE**	**22**

No	Family/taxa	Parts used	Preparation	Resources[a]
60	*Cota wiedemanniana*	**BR**	**IN**	**19**
61	*Cynara scolymus*	**WP**	**EF**	**35, 36**
62	*Filago arvensis*	**AP**	**BO, MA**	**43, 141**
63	*Gundelia tournefortii*	**R, ST**	**DE**	**10, 37**
64	*Helianthus tuberosus*	**FR, ST, T**	**C, DE, EF**	**4, 10, 11, 14, 16, 19, 38–41**
65	*Helichrysum armenium*	**AP**	**IN**	**146**
66	*Helichrysum plicatum* ssp. *plicatum*	**AP, FL, H**	**DE, IN**	**10, 19, 24, 26, 37, 146**
67	*Helichrysum plicatum* ssp. *pseudoplicatum*	**AP**	**DE**	**146**
68	*Helichrysum rubicundum*	**FL**	**DE**	**42**
69	*Jurinea moschus* ssp. *pinnatisecta*	**H**	**DE**	**10**
70	*Matricaria chamomilla*	**FL**	**IN**	**44**
71	*Onopordum acanthium*	**AP**	**EF**	**9**
72	*Onopordum tauricum*	**AP, FL**	**DE, IN**	**10, 25, 37**
73	*Scolymus hispanicus*	**AP, R**	**DE**	**6, 16**
74	*Scorzonera cinerea*	**T**	**R**	**19**
75	*Scorzonera mollis* ssp. *szovitzii*	**T**	**R**	**19**
76	*Scorzonera semicana*	**L**	**EF, C**	**32**
77	*Tanacetum aureum*	**AP**	**IN**	**142**
78	*Tanacetum polycephalum* ssp. *argyrophyllum*	**AP**	**IN**	**142**
79	*Taraxacum campylodes*	**AP, FL, L, R, SH**	**DE, IN, R**	**15, 47–49**
80	*Taraxacum farinosum*	**L**	**DE**	**14**
81	*Taraxacum macrolepium*	**L, R**	**DE**	**11**
82	*Taraxacum stevenii*	**FL**	**BO**	**46**
83	*Tripleurospermum parviflorum*	**CA**	**DE**	**136**
84	*Xanthium strumarium*	**FR**		**15**
	Berberidaceae			
85	*Berberis crataegina*	**L, FR, R**	**DE, EF, IN**	**10, 12, 13, 16, 41, 50–52**
86	*Berberis integerrima*	**FR**	**EF**	**42**
87	*Berberis vulgaris*	**FR**	**DE**	**10**
88	*Bongardia chrysogonum*	**T**	**IN**	**129**

No	Family/taxa	Parts used	Preparation	Resources[a]
	Boraginaceae			
89	*Anchusa azurea*	**AP, R**	**DE**	**19**
90	*Anchusa undulata* ssp. *hybrida*	**L**	**BO**	**137**
	Brassicaceae			
91	*Brassica elongata*	**FL**	**DE**	**14**
92	*Brassica oleracea*	**AP**	**C**	**16**
93	*Brassica nigra*	**AP**	**DE**	**23**
94	*Capsella bursa-pastoris*	**AP**	**DE, IN**	**10, 16, 39**
95	*Eruca vesicaria*	**L**	**EF**	**45**
96	*Lepidium sativum*	**L**	**R**	**133**
97	*Nasturtium officinale*	**AP, FL, L, SH**	**EF, IN**	**9, 49, 53**
98	*Raphanus raphanistrum*	**AP**	**C, EF**	**54**
99	*Sinapis alba*	**FL**	**IN**	**36, 55**
100	*Sinapis arvensis*	**FL, SH**	**EF, IN**	**30, 36, 55**
	Capparaceae			
101	*Capparis spinosa*	**AP**	**DE**	**56**
	Caprifoliaceae			
102	*Dipsacus laciniatus*	**R**	**DE**	**10**
	Caryophyllaceae			
103	*Dianthus carmelitarum*	**FL**	**IN**	**16**
	Cistaceae			
104	*Cistus creticus*	**L**	**DE, IN**	**35, 36, 43**
105	*Cistus laurifolius*	**BD, BR, FL, L, R, SH**	**DE, IN, MA**	**6, 12, 13, 26, 35, 36, 43, 51, 58**
106	*Cistus salviifolius*	**BR**	**IN**	**36**
	Cornaceae			
107	*Cornus mas*	**FR, L**	**DE, EF**	**1, 2, 59–61**
	Cucurbitaceae			
108	*Cucumis sativus*	**FR**	**EF**	**62, 63**
109	*Cucurbita moschata*	**S**	**EF**	**29**
110	*Ecballium elaterium*			**64**
111	*Momordica charantia*	**FR**	**PW**	**12, 13, 15**
	Cupressaceae			
112	*Cupressus sempervirens*	**CO, S**	**DE**	**16, 34, 64, 65**
113	*Juniperus communis*	**FR**	**DE**	**62**
114	*Juniperus drupacea*	**FR, R**		**15**
115	*Juniperus excelsa*	**CO**	**DE**	**45, 62**
116	*Juniperus foetidissima*	**CO**	**EF, IN**	**16, 66**

No	Family/taxa	Parts used	Preparation	Resources[a]
117	*Juniperus oxycedrus*	BR, CO, L, S	DE, IN	14, 16, 33, 65–70
118	*Juniperus sabina*	SH	DE	42
	Dioscoreaceae			
119	*Dioscorea communis*	FL, R	BO, R	43
	Ebenaceae			
120	*Diospyros kaki*	FR	EF	139
	Elaeagnaceae			
121	*Elaeagnus angustifolia*	FR, L	DE, IN	16, 43
	Equisetaceae			
122	*Equisetum arvense*	AP, L		15
123	*Equisetum ramosissimum*	AP	BO, DE, IN	10, 151
	Ericaceae			
124	*Arbutus andrachne*	FR	DE	71
125	*Arbutus unedo*	FR	EF	144
126	*Vaccinium myrtillus*	FR, L	DR, EF, IN	12, 13, 72, 73
	Euphorbiaceae			
127	*Euphorbia rigida*	LA		123
128	*Ricinus communis*	S	EF, DE	12, 13
	Fabaceae			
129	*Astracantha gummifera*	R	DE, IN	10, 37, 146
130	*Astragalus brevicalyx*	R	R	25
131	*Astragalus bustillosii*	R	DE	19
132	*Astragalus ceramicus* var. *filifolius*	R	IN	146
133	*Ceratonia siliqua*	FR	DE	14, 16
134	*Galega officinalis*	FL	IN	15
135	*Glycyrrhiza glabra*	R	DE	16
136	*Lathyrus sativus*	S	DE	74
137	*Lathyrus tuberosus*	L	C	11
138	*Lupinus albus*	S	CR, EF	13, 27
139	*Lupinus angustifolius* ssp. *angustifolius*	S	CR	21
140	*Lupinus pilosus*	FR	C, DR, EF, PW	26
141	*Phaseolus vulgaris*	AP	DE	133
142	*Robinia pseudoacacia*	FL	R	128
143	*Trigonella foenum-graecum*	S	DE, IN, PN, PW	12, 13, 18, 72, 75
144	*Vicia ervilia*	S	DE	52
145	*Vicia faba*	FR, S	EF	74, 75

No	Family/taxa	Parts used	Preparation	Resources[a]
	Fagaceae			
146	*Castanea sativa*	**FL**	**IN**	**31, 76**
147	*Quercus brantii*	**FR**	**EF**	**30, 56**
148	*Quercus cerris*	**FR**	**DE**	**24**
149	*Quercus coccifera*	**BR, CU, FR, R**	**DE, EF**	**14, 16, 43, 66, 77**
150	*Quercus infectoria* ssp. *veneris*	**FR**	**DE, EF**	**77, 78**
151	*Quercus ithaburensis* ssp. *macrolepis*	**FR**	**EF**	**30**
152	*Quercus petraea* ssp. *pinnatiloba*	**FR**	**DE, IN**	**19**
153	*Quercus robur*	**FR, ST**	**DE**	**24, 78**
	Gentianaceae			
154	*Centaurium erythraea*	**AP**	**DE, IN, MO**	**43**
155	*Gentiana olivieri*	**FL**	**IN**	**78**
	Geraniaceae			
156	*Erodium cedrorum* ssp. *salmoneum*	**AP**	**IN**	**149**
157	*Erodium cicutarium*	**AP**	**IN**	**149**
158	*Erodium gruinum*	**AP**	**IN**	**149**
159	*Erodium pelargoniflorum*	**AP**	**IN**	**149**
160	*Geranium robertianum*	**AP**	**DE, IN**	**12, 13, 71**
161	*Geranium tuberosum*	**AP**	**DE**	**17**
162	*Pelargonium graveolens*	**L**	**DE**	**133**
	Hypericaceae			
163	*Hypericum adenotrichum*	**AP**	**BO**	**76**
164	*Hypericum empetrifolium*	**AP**	**IN**	**54**
165	*Hypericum perforatum*	**AP, FL, L**	**DE, IN**	**3, 16, 44, 64, 75, 79, 80**
166	*Hypericum scabrum*	**FL**	**DE**	**16**
167	*Hypericum tetrapterum*	**AP**	**DE**	**80**
168	*Hypericum triquetrifolium*	**AP**	**DE**	**16, 56, 80**
	Juglandaceae			
169	*Juglans regia*	**FR, FR-S, L, S, ST-B**	**DE, IN, R, MC**	**1, 12–14, 16, 18, 19, 24, 25, 31, 32, 40, 43, 64, 72, 75, 81–84**
	Lamiaceae			

No	Family/taxa	Parts used	Preparation	Resources[a]
170	*Ajuga chamaepitys* ssp. *chia*	**L**	**DE**	**71**
171	*Lamium amplexicaule*	**WP**	**DE**	**85**
172	*Lavandula stoechas*	**AP, FL, L**	**DE**	**15, 16, 54, 75, 86**
173	*Melissa officinalis*	**AP, FL, L, SH**	**DE, IN**	**3, 15, 16, 49, 60, 87, 88**
174	*Mentha aquatica*	**AP**	**EF**	**9**
175	*Mentha longifolia*	**L**	**DE**	**74**
176	*Mentha pulegium*	**AP**	**DE**	**54**
177	*Mentha spicata*	**L**	**DE**	**89**
178	*Micromeria cristata* ssp. *orientalis*	**AP**	**IN**	**135**
179	*Micromeria juliana*	**AP**	**IN**	**76**
180	*Ocimum basilicum*	**L**	**CP, IN, MA**	**43**
181	*Origanum majorana*	**AP**	**IN**	**16, 90**
182	*Origanum onites*	**AP, FL, L, ST**	**DE, EF, IN**	**6, 12, 13, 16, 35, 36, 54, 80**
183	*Origanum vulgare* ssp. *gracile*	**AP**	**IN**	**39**
184	*Origanum vulgare* ssp. *hirtum*	**AP, FL, L**	**IN**	**7, 16, 60, 88, 91**
185	*Origanum vulgare* ssp. *vulgare*	**AP**	**DE**	**1, 2**
186	*Phlomis linearis*	**AP**	**DE**	**24**
187	*Rosmarinus officinalis*	**AP, FL, L**	**DE, IN**	**15, 43, 49, 62, 80, 92**
188	*Salvia cyanescens*	**AP**	**DE**	**136**
189	*Salvia fruticosa*	**L**	**IN**	**133**
190	*Salvia hydrangea*	**H**	**IN**	**10**
191	*Salvia multicaulis*	**AP**	**IN**	**33**
192	*Salvia officinalis*	**L**	**IN**	**133**
193	*Salvia virgata*			**93**
194	*Satureja cuneifolia*	**AP, FL, L, ST**	**IN**	**12, 13, 24**
195	*Satureja thymbra*	**AP**	**IN**	**149**
196	*Sideritis congesta*	**L**	**IN**	**74**
197	*Sideritis lanata*	**FL**	**IN**	**133**
198	*Sideritis perfoliata*			**92**
199	*Stachys annua*	**AP**	**IN**	**94**
200	*Stachys cretica*	**AP**		**141**
201	*Teucrium chamaedrys* ssp. *chamaedrys*	**FL**	**IN**	**49**
202	*Teucrium chamaedrys* ssp. *lydium*	**AP**	**IN**	**16**

No	Family/taxa	Parts used	Preparation	Resources[a]
203	*Teucrium chamaedrys* ssp. *sinuatum*	**S, AP**	**DE, IN**	**32, 27**
204	*Teucrium chamaedrys* ssp. *tauricola*	**FL, L**	**IN**	**130**
205	*Teucrium polium*	**AP, FL, L, WP**	**DE, IN, EF**	**2, 8, 10, 12, 13, 16, 19, 25, 32, 33, 35–37, 45–47, 52, 61, 62, 66, 78, 95–100**
206	*Thymbra capitata*	**AP, OO**	**DOA, IN**	**6, 149**
207	*Thymbra spicata*	**AP, FL, L**	**CP, IN**	**7, 16, 35, 36, 43, 60, 76, 88, 92, 96**
208	*Thymus cilicicus*	**L**	**IN**	**96**
209	*Thymus fallax*	**AP**	**DE, IN**	**10, 52**
210	*Thymus kotschyanus* ssp. *kotschyanus*	**AP, H, L**	**DE, IN**	**10, 22, 46, 52**
211	*Thymus leucostomus*	**AP**	**IN**	**88**
212	*Thymus longicaulis* ssp. *chaubardii*	**AP, L**	**DE, IN**	**3, 16, 31, 48, 60, 88**
213	*Thymus praecox* ssp. *jankae*	**AP**	**DE, IN**	**16, 101**
214	*Thymus nummularius*	**AP**	**IN**	**16**
215	*Thymus sipyleus*	**AP**	**DE, IN**	**10, 20, 47, 52, 88**
216	*Thymus zygioides*	**AP**	**IN**	**1, 7, 26, 67, 68**
217	*Vitex agnus-castus*	**S**	**BO**	**76, 98**
	Lauraceae			
218	*Cinnamomum zeylanicum*	**ST-B**	**DE**	**27**
219	*Laurus nobilis*	**FR, L**	**DE, EF**	**14, 16, 89, 102**
	Linaceae			
220	*Linum hirsutum*	**S**	**PW**	**67**
221	*Linum usitatissimum*	**S**	**IN**	**21**
222	*Linum tenuifolium*	**AP**	**DE**	**3**
	Lythraceae			
223	*Punica granatum*	**FL, FR-J, FR**	**DE, IN, SY**	**4, 16, 23, 35, 36, 43, 62, 110**
	Malvaceae			
224	*Abelmoschus esculentus*	**S**	**IN**	**27, 28, 107**
225	*Malva neglecta*	**FL, L**	**C, IN, MA**	**43, 108**
226	*Malva sylvestris*	**AP, FL, L, R**	**DE, IN, MA**	**23, 31, 43**
	Moraceae			
227	*Ficus carica*			**144**
228	*Morus alba*	**L, FR**	**DE, SY**	**2, 16, 22, 31, 59, 64, 89, 94**

No	Family/taxa	Parts used	Preparation	Resources[a]
229	*Morus nigra*	**FR, L, ST-B**	**DE, IN**	**3, 4, 15, 16, 32, 72, 74, 94, 96, 108, 109**
230	*Morus rubra*	**FR-J, FR, L**	**IN, R, SY**	**43, 66**
	Myrtaceae			
231	*Eucalyptus camaldulensis*	**L**	**DE, IN**	**12, 13**
232	*Myrtus communis*	**FR, L**	**DE, EF, IN**	**6, 12–14, 16, 23, 29, 45, 54, 62, 74, 110**
	Oleaceae			
233	*Jasminum officinale*	**FL**	**DE**	**133**
234	*Olea europaea*	**BD, FL, FR, FR-J, L, OO, S**	**DE, IN, MA, PU, R**	**6, 13, 15, 16, 40, 43, 45, 49, 54, 64, 72, 96, 97, 90**
235	*Phillyrea latifolia*	**FR, L**	**EF, IN**	**16, 60**
	Orchidaceae			
236	*Orchis simia*	**B**	**DE**	**111**
	Paeoniaceae			
237	*Paeonia arietina*	**AP**	**IN**	**10, 39**
	Papaveraceae			
238	*Fumaria officinalis*	**AP**	**C**	**61**
239	*Papaver argemone*	**AP**	**IN**	**80**
240	*Papaver dubium* ssp. *lecoqii*	**AP**	**IN**	**80**
241	*Papaver rhoeas*	**AP**	**IN**	**22, 80**
	Pedaliaceae			
242	*Sesamum indicum*	**FR, S**	**EF**	**12, 13**
	Pinaceae			
243	*Cedrus libani*	**R**	**DE**	**126**
244	*Pinus brutia*	**CO, L, PT, RS, SH, ST-B**	**CH, DE, IN, PU, R**	**6, 14, 43, 110**
245	*Pinus nigra* ssp. *pallasiana*	**CO, PT, RS**	**BO, G**	**34, 43, 112**
246	*Pinus sylvestris*	**CO**	**BO, DE**	**113**
	Plantaginaceae			
247	*Plantago lanceolata*	**L**	**DE**	**46, 50**
248	*Plantago major* ssp. *major*	**L**	**DE, IN**	**2, 3, 16, 44**
249	*Plantago major* ssp. *intermedia*	**L**	**DE**	**67, 114**
	Platanaceae			

No	Family/taxa	Parts used	Preparation	Resources[a]
250	*Platanus orientalis*	**FR, L**	**DE, IN**	**25, 32, 60, 102**
	Poaceae			
251	*Elymus repens*	**AP, RH**	**DE**	**15, 22, 33**
252	*Avena barbata*	**L, SH**	**IN**	**49**
253	*Avena sativa*	**FR**	**DE**	**133**
254	*Cynodon dactylon*	**AP, L, R**	**DE, IN**	**31, 46, 60, 138**
255	*Hordeum vulgare*	**S**	**C**	**73**
	Polygonaceae			
256	*Polygonum cognatum*	**L**	**C, EF, IN**	**33, 96**
257	*Rheum ribes*	**R, SH, ST**	**DE, IN, R**	**10, 17, 18, 39, 42, 104, 127, 132, 142**
258	*Rumex acetosella*	**L**	**EF**	**23, 39, 63**
259	*Rumex crispus*	**L**	**DE, IN, C**	**25, 42, 94**
260	*Rumex pulcher*	**L**	**R**	**94, 134**
261	*Rumex scutatus*	**H**	**DE**	**42**
262	*Rumex tuberosus*	**H, L, R**	**C, EF, IN, R**	**25, 42, 115**
	Portulacaceae			
263	*Portulaca oleracea*	**AP, L**	**C, DE, IN, MA, PU, R**	**19, 25, 30, 33, 43, 45**
	Ranunculaceae			
264	*Nigella sativa*	**S**	**DE, EF, PW**	**12, 13, 62, 110**
265	*Nigella segetalis*	**S, SH**	**DE**	**10, 18, 21**
	Rhamnaceae			
266	*Paliurus spina-christi*	**R, S**	**CR, DE**	**2, 43, 44, 54, 62**
267	*Rhamnus lycioides* ssp. *oleoides*	**FR**	**EF**	**24**
268	*Ziziphus jujuba*	**FR**	**EF, DE, DR**	**12, 13, 16, 35, 36, 116**
	Rosaceae			
269	*Alchemilla compactilis*	**FL, L**	**DE, IN**	**148**
270	*Prunus dulcis*	**FR, S**	**EF**	**12, 13, 16, 27, 43, 54, 62, 67, 75, 94, 144**
271	*Cotoneaster nummularius*	**FR, L**	**DE, PN, R**	**19, 24, 113**
272	*Crataegus azoralus* var. *aronia*	**FL, FR**	**DE**	**10, 20**
273	*Crataegus meyeri*	**FR, R**	**DE, EF**	**10, 52**
274	*Crataegus monogyna*	**SH**	**DE**	**14**
275	*Crataegus orientalis* ssp. *orientalis*	**FR, L, S**	**DE, IN, MA, R**	**43**
276	*Crataegus orientalis* ssp. *szovitsii*	**FL, L, ST-B, SH**	**DE, IN**	**134**
277	*Crataegus tanacetifolia*	**FL, SH**	**DE**	**113**

No	Family/taxa	Parts used	Preparation	Resources[a]
278	*Cydonia oblonga*	**FR, L**	**DE, EF, IN**	**16, 19, 35, 36, 44, 103, 110**
279	*Eriobotrya japonica*	**L**	**DE, IN**	**149**
280	*Fragaria vesca*	**FR, R**	**DE**	**16**
281	*Fragaria viridis*	**FR**	**R**	**140**
282	*Malus domestica*	**FR**	**DE**	**35**
283	*Malus pumila*	**FL**	**IN**	**62**
284	*Malus sylvestris* ssp. *orientalis var. orientalis*	**FR, L**	**DE, EF**	**3, 10, 38, 106**
285	*Mespilus germanica*	**FR, L**	**EF, IN**	**31, 103**
286	*Prunus armeniaca*	**FR, S**	**R**	**25**
287	*Prunus avium*			**135, 145**
288	*Prunus cerasus*	**S**	**MO**	**125**
289	*Prunus cocomilia*	**FR**	**DE**	**3**
290	*Prunus divaricata* var. *divaricata*	**FR**	**DE, R, SY**	**43, 61**
291	*Prunus laurocerasus*	**FR, L, S**	**DE, EF**	**16, 20, 31, 60, 73**
292	*Prunus mahaleb*	**FR, S**	**EF, IN**	**10, 12, 13, 22, 27, 47, 56, 61**
293	*Prunus orientalis*	**S**	**BO, EF**	**63, 94**
294	*Prunus persica*	**S**	**EF**	**30**
295	*Prunus spinosa* ssp. *dasphylla*	**FR**	**BO, DE, DR, EF, JA, IN**	**1–3, 15, 16, 31, 44, 62, 79, 103**
296	*Pyrus amygdaliformis* var. *amygdaliformis*	**FL, FR**	**DE**	**6, 92, 117**
297	*Pyrus amygdaliformis* var. *lanceolata*	**FR**	**PI**	**131**
298	*Pyrus bulgarica*	**FR**	**EF, PI**	**1**
299	*Pyrus communis* ssp. *caucasica*	**FR**	**DE**	**33**
300	*Pyrus communis* ssp. *communis*	**FR, R, ST-B**		**15**
301	*Pyrus elaeagnifolia* ssp. *elaeagnifolia*	**FL, FR, L**	**DE, IN**	**3, 29, 62**
302	*Rosa boissieri*	**FR-J**		**135**
303	*Rosa canina*	**FL, FR, FR-S, L, R, S**	**DE, EF, JA, IN**	**2, 4, 10, 14, 16, 17, 31, 33, 35–38, 43, 46, 47, 50, 51, 58–60, 71, 74, 81, 82, 85, 96, 98, 100, 103, 108, 109, 114, 115, 118, 119**
304	*Rosa mollis*	**FR-J**		**135**
305	*Rosa phoenicia*	**FR**	**DE**	**35, 36**
306	*Rubus caesius*	**R**	**DE**	**23**
307	*Rubus canescens*	**BD, FR, L, R, SH**	**DE, IN**	**1, 7, 10, 16, 21, 30, 45, 75**
308	*Rubus hirtus*	**L, R**	**DE, IN**	**10, 16, 57**

No	Family/taxa	Parts used	Preparation	Resources[a]
309	*Rubus ibericus*	**L, R**	**DE, IN**	**1, 16, 33**
310	*Rubus idaeus*	**L, R**	**DE, IN**	**16, 62**
311	*Rubus sanctus*	**BD, FR, L, R, SH**	**IN**	**2, 7, 12–14, 16, 19, 22, 49, 57, 76, 79, 92, 94, 101, 114, 117, 118**
312	*Sarcopoterium spinosum*	**AP, R, ST, ST-B**	**DE, IN**	**12, 13, 72, 104**
313	*Sorbus aucuparia*	**FR**	**EF**	**60**
314	*Sorbus domestica*	**FR, L**	**DE, EF, PI**	**3, 12, 13, 16, 72, 91, 106**
315	*Sorbus umbellata* var. *cretica*	**FR, L**	**IN**	**49**
316	*Sorbus umbellata* var. *umbellata*	**FR, L**	**IN**	**49**
317	*Sorbus torminalis*	**L**	**DE**	**3**
	Rubiaceae			
318	*Galium aparine*	**AP**	**DE**	**6, 33**
	Rutaceae			
319	*Citrus maxima*	**FR-J**	**EF**	**6**
	Salicaceae			
320	*Populus tremula*	**L**	**IN**	**1, 116**
321	*Salix alba*	**BR, FL, L**	**DE**	**14, 16, 60**
	Santalaceae			
322	*Viscum album* ssp. *abietis*	**FR, L**	**DE**	**149**
323	*Viscum album* ssp. *album*	**AP, H, L**	**DE, IN**	**6, 10, 14, 16, 24, 26, 31, 38, 39, 51, 61, 94, 98, 100, 101, 106**
324	*Viscum album* ssp. *austriacum*	**L**	**DE**	**43**
	Scrophulariaceae			
325	*Verbascum macrurum*	**FL**	**IN**	**16**
326	*Verbascum stenostachyum*	**FL, L**	**EF, IN**	**38**
	Simaroubaceae			
327	*Quassia amara*	**ST**	**IN**	**103**
	Smilacaceae			
328	*Smilax excelsa*	**SH**	**BO**	**105**
	Solanaceae			
329	*Physalis alkekengi*	**FR, S**	**DE**	**34, 74**
330	*Solanum tuberosum*	**T**	**R**	**133, 143**
	Styracaceae			
331	*Styrax officinalis*	**L, S**	**DE, EF**	**14, 26**
	Urticaceae			
332	*Parietaria judaica*	**AP**	**DE**	**100**

No	Family/taxa	Parts used	Preparation	Resources[a]
333	*Parietaria officinalis*	AP	DE	133
334	*Urtica dioica*	AP, L, S, WP	C, DE, IN	2, 4, 6, 11, 16, 22, 23, 29, 31–34, 37, 41, 45, 57, 60, 68, 78, 92, 101, 103, 109, 113, 115, 119–122
335	*Urtica pilulifera*	AP, S	C, DE	14, 80, 81
336	*Urtica urens*	AP, L, R, S	CR, DE	9, 16, 80
	Vitaceae			
337	*Vitis vinifera*	L		150
	Xanthorrhoeaceae			
338	*Asphodelus aestivus*	L	C	63
339	*Eremurus spectabilis*	AP, R	DE, MA	19, 39
	Zygophyllaceae			
340	*Tribulus terrestris*	AP	DE	23, 96

[a]**Resources**: **1**: (Genç and Özhatay 2006); **2**: (Alparslan and Tuzlacı 2006); **3**: (Kültür 2007); **4**: (Sezik et al. 2001); **5**: (Çelik et al. 2008); **6:** (Ertuğ 2002); **7**: (Bulut 2008); **8**: (Balos 2007); **9**: (Sarper et al. 2009); **10**: (Altundag and Ozturk 2011); **11**: (Güneş and Özhatay 2011); **12**: (Durmuşkahya and Öztürk 2013); **13**: (Durmuşkahya and Oztürk 2013); **14**: (Güneş 2010); **15**: (Sarı et al. 2010); **16**: (Tuzlacı 2006); **17**: (Öztürk and Ölçücü 2011); **18**: (Özgokce and Ozcelik 2004); **19**: (Polat et al. 2013); **20**: (Koyuncu 2005); **21**: (Ugulu et al. 2009); **22**: (Çakilcioglu et al. 2011); **23**: (Kıran 2006); **24**: (Keskin 2011); **25**: (Tetik et al. 2013); **26**: (Oral 2007); **27**: (Çömlekçioğlu and Karaman 2008); **28:** (Uysal 2010); **29**: (Uysal et al. 2010); **30**: (Gençay 2007); **31**: (Kızılaslan 2008); **32**: (Çakılcıoğlu et al. 2007); **33**: (Çakilcioglu and Turkoglu 2010); **34**: (Sağıroğlu et al. 2012a); **35**: (Fujita et al. 1995); **36**: (Polat and Satil 2012); **37**: (Çakılcıoğlu et al. 2010); **38**: (Kargioğlu et al. 2008); **39**: (Tuzlacı and Dogan 2010); **40**: (Kahraman and Tatlı 2004); **41**: (Savran et al. 2009); **42**: (Özgen et al. 2012); **43**: (Sargin et al. 2013); **44**: (Tuzlacı et al. 2010); **45**: (Akaydın et al. 2013); **46**: (Doğan and Bağcı 2011); **47**: (Öztürk 2006); **48**: (Uysal et al. 2012); **49**: (Fakir et al. 2009); **50**: (Gençler and Koyuncu 2005); **51**: (Vural 2008); **52**: (Yeşil and Akalın 2009); **53**: (Polat et al. 2012); **54**: (Gürdal and Kültür 2013); **55**: (Polat and Satil 2012); **56**: (Akgül 2008); **57**: (Sezik et al. 1997); **58**: (Unsal et al. 2010); **59**: (Yeşilada et al. 1999); **60**: (Koçyiğit and Özhatay 2006); **61**: (Demirci and Özhatay 2012); **62**: (Metin 2009); **63**: (Tekin 2011); **64**: (Başer et al. 1986); **65**: (Emre 2003); **66**: (Tuzlacı and Erol 1999); **67**: (Deniz et al. 2010); **68**: (Akçiçek and Vural 2003); **69**: (Alparslan 2003); **70**: (Sadıkoğlu 2003); **71**: (Saday 2009); **72**: (Baytop 1984); **73**: (Sağıroğlu et al. 2012b); **74**: (Everest and Ozturk 2005); **75**: (Akan et al. 2005); **76**: (Ertuğ et al. 2004); **77**: (Keskin and Alpınar 2002); **78**: (Şığva and Seçmen 2009); **79**: (Bulut 2011); **80**: (Alpınar 1999); **81**: (Türkan et al. 2006); **82**: (Sayar et al. 1995); **83**: (Tabata et al. 1988); **84**: (Vural 2004); **85**: (Akyol and Altan 2013); **86**: (Malyer et al. 2004) **87**: (Yücel and Tülükoğlu 2000); **88**: (Koyuncu et al. 2010); **89**: (Yazıcıoğlu and Tuzlacı 1996); **90**: (Uzun et al. 2004); **91**: (Tuzlacı and Aymaz 2001); **92**: (Tümen and Sekendiz 1990); **93**: (Ahiskalı et al. 2012); **94**: (Tuzlacı and Şenkardeş 2011); **95**: (Mart 2006); **96**: (Özçelik and Balabanlı 2005); **97:** (Çubukçu and Özhatay 1987); **98**: (Koçak 1999); **99:** (Özdemir 2005); **100**: (Öztürk and Dinç 2005); **101**: (Ezer and Arisan 2006); **102**: (Genç 2003); **103**: (Tuzlacı and Tolon 2000); **104**: (Yıldırım et al. 2008); **105**: (Koca and Yıldırımlı 2010); **106**: (Aktan 2011); **107**: (İlçim and Varol 1996); **108**: (Ezer and Avcı 2004); **109**: (Çakılcıoğlu and Türkoğlu 2009); **110**: (Polat et al. 2011); **111**: (Yapıcı et al. 2009); **112**: (Honda et al. 1996); **113**: (Özüdoru et al. 2011); **114**: (Akalın and Alpınar 1994); **115**: (Elçi and Erik 2006); **116**: (Ugurlu and Secmen 2008); **117**: (Bulut and Tuzlacı 2005); **118**: (Aslan 2002); **119**: (Şimsek et al. 2004); **120**: (Akan et al. 2008); **121**: (Gümüş 1994); **122**: (Şimşek et al. 2001); **123**: (Keklik Koçoğlu et al. 1996); **124**: (Sadıkoğlu and Alpınar 2000);

125: (Karaman and Kocabas 2001); **126**: (Yeşilada et al. 1995); **127**: (Özgen and Coşkun 2000); **128**: (Akalın 1998); **129**: (Özuslu 2005); **130**: (Çubukcu et al. 1994); **131**: (Duran et al. 2001); **132**: (Güneş and Özhatay 2011); **133**: (Tümen and Selvi 2011); **134**: (Şenkardeş 2014); **135**: (Korkmaz and Karakurt 2014); **136**: (Han and Bulut 2015); **137**: (Arı et al. 2015); **138**: (Yeşilyurt et al. 2017); **139**: (Şenkardeş and Tuzlaci 2014); **140**: (Polat et al. 2015); **141**: (Sargin et al. 2015a); **142**: (Kaval et al. 2014); **143**: (Güler et al. 2015); **144**: (Sargin et al. 2015b); **145**: (Günbatan et al. 2016); **146**: (Mükemre et al. 2015); **147**: (Bulut et al. 2017); **148**: (Akbulut and Bayramoglu 2014); **149**: (Sargin 2015); **150**: (Wong et al. 2011); **151**: (Oztürk et al. 2018)

Appendix 2: Medicinal Plants Used in the Treatment of Diabetes Mellitus in the Traditional Medicine in Pakistan

No	Family/taxa	Parts used	Preparations	Resources[a]
	Acanthaceae			
1	*Justicia adhatoda*	**BD, FR, L, WP**	**DE, EF, EX, JU**	**1–8**
	Adoxaceae			
2	*Viburnum grandiflorum*	**L**	**EX**	**9**
	Aizoaceae			
3	*Trianthema triquetra*	**WP**	**DE, IN, PW**	**10**
	Alismataceae			
4	*Alisma plantago-aquatica*	**L, RH**	**PW**	**8, 11**
	Amaranthaceae			
5	*Achyranthes aspera*	**S**	**PW**	**12, 13**
6	*Aerva javanica*		**DE**	**14, 15**
7	*Chenopodium album*			**14, 15**
8	*Chenopodium foliosum*	**R**	**DE**	**16**
9	*Chenopodium murale*			**14, 15**
10	*Dysphania botrys*	**AP**		**17**
11	*Salsola imbricata*		**DE**	**15**
	Amaryllldaceae			
12	*Allium cepa*	**B**	**G**	**1, 2, 8, 18, 19**
13	*Allium sativum*	**B, L**	**C, EF**	**1, 2, 8, 11, 19, 20–24**
	Asparagaceae			
14	*Asparagus officinalis*	**S**	**PW**	**25**
	Anacardiaceae			
15	*Mangifera indica*	**L, S**	**DR, EX**	**24, 26**
	Annonaceae			
16	*Polyalthia longifolia*	**L, R, ST**	**BO**	**8, 27**
	Apiaceae			
17	*Carum carvi*	**S**	**BO, EX**	**28**
18	*Coriandrum sativum*	**OO, S**	**CH**	**24, 29, 30**

No	Family/taxa	Parts used	Preparations	Resources[a]
19	*Cuminum nigrum*	FR		31
20	*Daucus carota*		JU	2, 15, 32
21	*Ferula narthex*	WP	PW	8
22	*Foeniculum vulgare*	L, S	BO, EX, PW	19, 33
23	*Narthex asafoetida*	RS		34
	Apocynaceae			
24	*Calotropis procera*	FL	BO, EX	8
25	*Caralluma adscendens* var. *fimbriata*	ST		35
26	*Caralluma edulis*	AP, WP	C, EX, PW	1, 2, 6, 8, 36, 37
27	*Caralluma tuberculata*	FR, R, ST, WP	AV, C, EX	6, 36, 38, 39
28	*Carissa spinarum*	L, R	DE	4
29	*Catharanthus roseus*	L, WP	JU	1, 8, 11, 40, 41
30	*Gymnema sylvestre*	L	BO, IN	22, 42
31	*Nerium oleander*	L, LA, S	IN, PW	19
32	*Pergularia tomentosa*			8
33	*Rauvolfia serpentina*	L	EX	43, 44
34	*Rhazya stricta*	L, WP	BO, PW	6, 8, 45, 47, 126
35	*Tylophora hirsuta*	BR, L	EX, JU	2, 8
	Araceae			
36	*Anthurium* sp.			14
37	*Monstera deliciosa*	FR	PL	9
38	*Pistia stratiotes*	L, ST	EX	48
	Araliaceae			
39	*Hedera helix*	B, L	EX	5, 8, 49, 50
40	*Hedera nepalensis*	L, FR	EX, PW	7, 11, 19, 49, 51, 52
	Arecaceae			
41	*Nannorrhops ritchiana*			14
42	*Phoenix sylvestris*	S	DE	53
	Asteraceae			
43	*Achillea santolinoides* ssp. *wilhelmsii*	WP	DE, IN	33
44	*Arctium lappa*	L, R, ST	DE	8, 54
45	*Artemisia absinthium*	FL, L, WP	BO, DE, EX	8, 16, 54
46	*Artemisia gmelinii*			55
47	*Artemisia herba-alba*	WP	PW	56
48	*Artemisia indica*	AP	EX	56, 57
49	*Artemisia roxburghiana*	L	EX	9
50	*Artemisia scoparia*	FR	EF	17
51	*Baccharoides anthelmintica*	S	PW	52
52	*Centaurea iberica*	L	EX	9

No	Family/taxa	Parts used	Preparations	Resources[a]
53	*Chrysanthemum indicum*	FL	EX	52
54	*Cichorium intybus*	R, L, SH	DE, PW	1, 2, 58, 59
55	*Erigeron bonariensis*			14
56	*Cousinia thomsonii*	CA	DR, PW	16
57	*Cynara scolymus*	RE		11
58	*Galinsoga parviflora*			60
59	*Lactuca sativa*	L	JU	11
60	*Launaea procumbens*	L, WP	AV, C, EX	8, 27
61	*Parthenium hysterophorus*	FL	PW	8, 37, 61
62	*Seriphidium quettense*	L		33
63	*Silybum marianum*	WP	DE	4
64	*Sonchus asper*	L, R	AV, C	7, 52
65	*Stevia rebaudiana*	L	EX	62
66	*Tanacetum artemisioides*	WP	BO	8, 54
67	*Taraxacum campylodes*	L, R, WP	AV, DE, PW	1, 2, 8, 9, 16, 37
68	*Tricholepis furcata*			8
69	*Vernonia cinerea*	FR		39
	Berberidaceae			
70	*Berberis brandisiana*	R, ST	PW	8
71	*Berberis lycium*	FR, R, ST-B, WP	DE, EX, PW	8, 52, 63, 64, 65
72	*Berberis orthobotrys*	FR, L, R		16
	Betulaceae			
73	*Alnus nitida*	L		8
	Boraginaceae			
74	*Arnebia benthamii*	FL, L	DE	16
75	*Onosma echioides*	R	EX	12
	Brassicaceae			
76	*Brassica cretica*	FL	AV	66
77	*Lepidium didymum*			15
78	*Eruca vesicaria*	S, WP	AV, C, MO	4, 52
79	*Farsetia stylosa*	WP	DE, PW	10, 67
80	*Raphanus raphanistrum* ssp. *sativus*	R	BO	8
	Cactaceae			
81	*Opuntia dillenii*	FR, FR-J, LA	DR, EF, JU	8, 68, 69
82	*Opuntia monacantha*	FR	DR	8
	Capparaceae			
83	*Capparis cartilaginea*	L	PW	8
84	*Capparis decidua*	FR, S	EX	8, 70, 71
85	*Capparis spinosa*	L	BO, EX	6, 8

No	Family/taxa	Parts used	Preparations	Resources[a]
	Cleomaceae			
86	*Cleome scaposa*	**L**	**BO, DE**	**10**
	Caricaceae			
87	*Carica papaya*	**FR**		**8**
	Caryophyllaceae			
88	*Stellaria media*	**WP**	**IN**	**72**
	Convolvulaceae			
89	*Convolvulus arvensis*			**14**
90	*Convolvulus prostratus*	**L**	**BO, DE**	**10**
91	*Cuscuta campestris*	**WP**	**BO, EX**	**8**
92	*Cuscuta reflexa*	**SH, WP**	**EX**	**8, 33, 39**
	Cucurbitaceae			
93	*Coccinia grandis*	**L, R**	**JU**	**8, 73–75**
94	*Cucumis melo*	**FR**	**PL**	**17**
95	*Cucumis sativus*	**FR-J**		**8**
96	*Cucurbita maxima*	**FR**	**R**	**33**
97	*Cucurbita pepo*	**FR**		**8, 14**
98	*Luffa acutangula*	**FR**	**C**	**17, 76**
99	*Momordica balsamina*	**FR**	**EX, PW**	**8, 77**
100	*Momordica charantia*	**FR, L**	**C, DR, EX, JU, PW, R**	**1–3, 8, 15, 22, 33, 78–80**
101	*Momordica dioica*	**FR**	**DR, EF, JU, PW**	**8, 45, 81**
102	*Mukia maderaspatana*			**14**
103	*Citrullus colocynthis*	**FR, R, S**	**EF, PW**	**6, 8, 10, 24, 38, 45**
104	*Citrullus lanatus*	**FR**	**EF**	**8, 14**
	Cupressaceae			
105	*Juniperus communis* var. *saxatilis*		**IN**	**82**
106	*Juniperus excelsa*	**FR**	**BO**	**8**
	Cyperaceae			
107	*Cyperus rotundus*	**AP, R, T**	**BO, DE**	**2, 10, 83, 84**
	Dioscoreaceae			
108	*Dioscorea deltoidea*			**60**
	Elaeagnaceae			
109	*Elaeagnus angustifolia*	**FR**	**R**	**33**
110	*Elaeagnus rhamnoides*	**FR, S**	**JA**	**8, 85**
	Equisetaceae			
111	*Equisetum arvense*	**AP, WP**	**AT, DE, PU**	**125, 127**
	Euphorbiaceae			
112	*Acalypha wilkesiana*			**86**
113	*Euphorbia helioscopia*	**FL, L**	**EX**	**9**
114	*Euphorbia hirta*	**L, WP**	**BO, EX**	**8**
115	*Euphorbia thymifolia*	**WP**	**PW**	**8, 87**
116	*Manihot esculenta*			**86**

No	Family/taxa	Parts used	Preparations	Resources[a]
	Fabaceae			
117	*Acacia modesta*	**FL, S**	**DE, G, PW**	**8**
118	*Acacia nilotica*	**L, S, ST**	**BO, DE, EF, G, IN**	**6, 8, 74, 79, 88**
119	*Acacia senegal*		**G**	**74, 11**
120	*Argyrolobium roseum*	**WP**	**IN**	**6, 8**
121	*Alhagi maurorum*	**R**	**PW**	**8, 73, 74**
122	*Albizia lebbeck*	**S**	**PW**	**6, 8, 89**
123	*Albizia procera*			**60**
124	*Bauhinia purpurea*	**WP**	**AV**	**8, 41**
125	*Bauhinia variegata*	**L**	**IN**	**9**
126	*Butea monosperma*	**FL, LA, ST-B**	**G, PW**	**8, 53, 59**
127	*Cajanus cajan*	**S**	**C**	**2, 8, 90**
128	*Cassia fistula*	**L, S**	**DR, PL**	**8, 12, 65**
129	*Cicer arietinum*	**FR, S**	**DE, RA**	**2, 11, 14, 31**
130	*Dalbergia sissoo*		**G, PW**	**19**
131	*Glycine max*	**S**	**EX**	**11, 91**
132	*Medicago monantha*	**WP**	**AV**	**92**
133	*Medicago polymorpha*			**14**
134	*Melilotus albus*			**14**
135	*Mucuna pruriens*	**S**	**DE**	**12, 62**
136	*Parkinsonia aculeata*	**FL, ST-B**		**92**
137	*Prosopis cineraria*			**14**
138	*Senna obtusifolia*	**FL**	**EX**	**88**
139	*Trigonella foenum-graecum*	**L, S**	**DE, EX**	**1, 2, 8, 22, 78, 93, 94**
140	*Vigna mungo*	**S**	**C**	**2**
141	*Vigna unguiculata*	**IP**	**AV**	**2**
	Fagaceae			
142	*Quercus ilex*			**95**
	Gentianaceae			
143	*Gentiana olivieri*	**AP**	**EX**	**96**
144	*Gentiana tianschanica*	**FL, L**	**BO, EX, IN**	**8, 16, 58**
145	*Swertia chirata*	**WP**	**DR, EX**	**8, 52**
146	*Swertia petiolata*	**WP**	**DE**	**16**
	Lamiaceae			
147	*Ajuga integrifolia*	**WP**	**BO**	**8**
148	*Ajuga parviflora*	**AP**	**EX**	**97**
149	*Ajuga reptans*	**WP**		**39**
150	*Clerodendrum phlomidis*			**8**
151	*Lamium amplexicaule*	**L**	**JU**	**59**
152	*Mentha x piperita*	**L, S**	**BO, IN, PW**	**8, 19**

No	Family/taxa	Parts used	Preparations	Resources[a]
153	*Mentha longifolia*	L	BO, DE, EX	8, 33
154	*Ocimum basilicum*	L	PW	8
155	*Ocimum tenuiflorum*	L	PW	8
156	*Origanum vulgare*	L	EX	8, 79
157	*Pseudocaryopteris bicolor*	FL, L	PW	8
158	*Salvia coccinea*	L	EX	9
159	*Teucrium stocksianum*	SH, WP	BO, CR, DE, EX	8, 17, 33, 98
160	*Vitex negundo*	FL	PW	53
	Limeaceae			
161	*Limeum obovatum*	WP	BO, IN	10
	Lythraceae			
162	*Lawsonia inermis*	L	EX	78, 99
163	*Punica granatum*	FR	DR, EF, PW	5, 51, 69
164	*Woodfordia fruticosa*	FL	PW	53
	Malvaceae			
165	*Abelmoschus esculentus*	R	DR	32
166	*Bombax ceiba*	L, WP	EX	19, 100
167	*Grewia asiatica*	FR	EF	12, 17, 34, 38, 101, 102
168	*Malva neglecta*	L	EX, JU	8, 61
169	*Sida cordata*	WP	EX	19
	Meliaceae			
170	*Azadirachta indica*	FL, FR, L	DR, PW	2, 8
171	*Cedrela serrata*	L, ST	DE, EX, JU	51, 53, 104
172	*Melia azedarach*	FR, L, WP	DE, DR, JU	1, 4, 18, 32, 39, 45
173	*Toona ciliata*	L	PW	105
	Menispermaceae			
174	*Cissampelos pareira*	S	PW	19
175	*Cocculus hirsutus*	AP	EX	96
	Moraceae			
176	*Ficus benghalensis*	BR, FR, L, LA, R, ST-B		2, 3, 6, 8, 24, 106, 107
177	*Ficus carica*	L	BO, EX	8, 24, 42
178	*Ficus hispida*	ST-B	EX	8
179	*Ficus lacor*	FR	DR, PW	8
180	*Ficus microcarpa*	FR, L, ST-B	PW	8, 106
181	*Ficus palmata*	FR		8, 81, 89
182	*Ficus racemosa*	ST-B	BO, EX	8, 12
183	*Ficus religiosa*	ST-B		8, 106
184	*Ficus virens*	L		8
185	*Morus alba*	L, R	BO	8, 19, 65, 78, 108

No	Family/taxa	Parts used	Preparations	Resources[a]
186	*Morus nigra*	**FR, L, ST-B**		**8, 52**
	Moringaceae			
187	*Moringa oleifera*	**FL, FR, S**	**BO, C**	**8, 11**
	Musaceae			
188	*Musa acuminata*	**FL**	**PW, RA**	**38**
	Myrtaceae			
189	*Eucalyptus globulus*	**ST-B**	**BO**	**19**
190	*Eucalyptus obliqua*			**15**
191	*Myrtus communis*	**L**	**EX**	**109**
192	*Psidium guajava*	**L, ST-B**	**EX**	**1, 2, 8, 14, 52**
193	*Syzygium cumini*	**FR, S**	**BO, PW**	**1, 2, 8, 11, 12, 14, 22, 24, 37, 38, 40, 42, 52, 61, 78**
	Nitrariaceae			
194	*Peganum harmala*	**WP**	**JU**	**92**
	Nyctaginaceae			
195	*Boerhavia diffusa*	**L**	**EX**	**19**
	Orchidaceae			
196	*Dactylorhiza hatagirea*	**RH**	**BO**	**16**
	Oleaceae			
197	*Fraxinus excelsior*	**S**	**PW**	**19**
198	*Olea europaea*	**FR**	**EF, DE, DR**	**4, 6, 8**
199	*Olea ferruginea*	**FR**	**BO, DE, DR, EF**	**1, 2, 8, 17, 38, 74, 75**
	Oxalidaceae			
200	*Oxalis corniculata*	**WP**	**EX**	**19, 33, 65**
	Papaveraceae			
201	*Fumaria indica*	**WP**	**JU**	**8, 18, 52, 100**
202	*Fumaria parviflora*	**WP**	**DE**	**92**
203	*Papaver somniferum*	**FL, FR, LA, S**	**BO, EX**	**8, 52**
	Pedaliaceae			
204	*Sesamum indicum*			**8**
	Phyllanthaceae			
205	*Leptopus cordifolius*	**L**	**BO, EX**	**8, 49**
206	*Phyllanthus acidus*			**86**
207	*Phyllanthus amarus*			**86**
208	*Phyllanthus emblica*	**FR**	**DR, JA, PW**	**8, 65, 86**
209	*Phyllanthus maderaspatensis*			**86**
210	*Phyllanthus urinaria*			**86**
	Pinaceae			
211	*Abies pindrow*	**L**	**EX**	**9**
212	*Pinus roxburghii*	**RS**		**17**
	Plantaginaceae			

No	Family/taxa	Parts used	Preparations	Resources[a]
213	*Kickxia incana*	WP	PW	92
214	*Nanorhinum ramosissimum*	WP	PW	1, 2, 75, 92
215	*Picrorhiza kurroa*	R	PW	61
	Poaceae			
216	*Avena fatua*			14
217	*Hordeum vulgare*	S, WP	BO, PW	2, 24
218	*Oryza sativa*	S	PW	2
219	*Ochthochloa compressa*			14
220	*Pennisetum glaucum*	S		8
221	*Phalaris minor*			15
222	*Sporobolus ioclados*			14
223	*Triticum aestivum*	FR, S		2, 81, 110
224	*Zea mays*	FL	EX	2, 8, 111
	Polygonaceae			
225	*Fagopyrum esculentum*	L, S	C, PW	8, 85
226	*Polygonum plebeium*			14
227	*Rumex acetosa*	SH		39
228	*Rumex hastatus*	WP		4
	Portulacaceae			
229	*Portulaca oleracea*	L	C	38
	Primulaceae			
230	*Anagallis arvensis*			14
	Pteridaceae			
231	*Adiantum capillus-veneris*	ST-B	BO, DE	8, 22, 112
232	*Adiantum incisum*	FO	JU	8, 59, 104, 112, 113
	Ranunculaceae			
233	*Aconitum chasmanthum*	RH		11
234	*Delphinium brunonianum*	R	PW	16
235	*Nigella sativa*	L, S, WP	BO, EX, PW	8, 78, 114
236	*Ranunculus muricatus*			15
	Rhamnaceae			
237	*Ziziphus jujuba*	FR, L, WP	CH, DE, EX, PW	8, 17, 35, 79, 115
238	*Ziziphus nummularia*	FR, L, R, S, ST-B	DE, EF, PW	8, 116, 120
239	*Ziziphus oxyphylla*	L, R	DE, EX	7, 117, 118, 119
240	*Ziziphus spinosa*	L		8, 14
	Rosaceae			
241	*Fragaria indica*	WP	PW	8
242	*Fragaria nubicola*	FR		11
243	*Prunus dulcis*	FR, L		8

No	Family/taxa	Parts used	Preparations	Resources[a]
244	*Prunus persica*	**FR**	**EF**	**8, 11**
245	*Malus domestica*	**FR-J**		**8**
246	*Rosa x damascena*	**S**	**PW**	**8**
247	*Rubus ellipticus*	**FR-J**		**8, 69, 120**
	Rubiaceae			
248	*Spermacoce articularis*			**60**
	Rutaceae			
249	*Aegle marmelos*	**L**	**EX**	**96**
250	*Citrus limon*	**FR-J**		**8**
251	*Citrus medica*	**FR-J**		**8**
252	*Citrus reticulata* cv. Murcott	**L**	**CH**	**37**
253	*Zanthoxylum armatum*			**60**
	Salicaceae			
254	*Flacourtia indica*	**FR**	**EF, PL**	**4, 7**
255	*Populus alba*	**ST-B**	**DE**	**60**
256	*Salix babylonica*	**L**	**EX**	**115**
	Santalaceae			
257	*Viscum album*	**L, SH**	**BO, EX**	**121**
	Sapindaceae			
258	*Dodonaea viscosa*	**L, ST-B**	**CH, DE**	**1, 2, 4, 8**
	Saxifragaceae			
259	*Bergenia ciliata*	**R, RH**	**BO, PW**	**7, 72, 122**
260	*Bergenia pacumbis*	**RH**	**BO**	**8, 9**
	Solanaceae			
261	*Atropa belladonna*	**S, WP**	**EX**	**19**
262	*Datura innoxia*			**8**
263	*Solanum americanum*	**AP, L, ST**	**C, JU**	**1, 8, 61, 81**
264	*Solanum incanum*	**AP**	**AV**	**2, 8, 116**
265	*Solanum melongena*	**FR**	**C**	**76**
266	*Solanum surattense*	**FR**	**DE, DR, PW**	**8, 81, 98**
267	*Withania coagulans*	**FR, S**	**PW**	**1, 2, 8, 81, 123**
268	*Withania somnifera*	**L, R**	**DE**	**6, 10, 15**
	Tamaricaceae			
269	*Tamarix aphylla*	**FR**	**BO**	**8**
	Thelypteridaceae			
270	*Christella dentata*	**FO, RH**	**EX**	**125, 127**
	Verbenaceae			
271	*Lantana camara*	**FR, WP**	**DE, EX**	**8, 53**
272	*Phyla nodiflora*	**S**	**BO**	**123**
	Xanthorrhoeaceae			
273	*Aloe vera*	**L**	**PL**	**2, 6, 18, 52, 124**
274	*Asphodelus tenuifolius*			**14**
	Zingiberaceae			

No	Family/taxa	Parts used	Preparations	Resources[a]
275	*Elettaria cardamomum*	**S**	**PW**	**2**
	Zygophyllaceae			
276	*Balanites aegyptiaca*	**FR**	**EX**	**96**
277	*Fagonia bruguieri*	**WP**	**IN**	**33**
278	*Fagonia cretica*	**WP**	**EX**	**35, 116**
279	*Fagonia indica*	**AP, L, ST, WP**	**EX**	**2, 81, 89**
280	*Fagonia olivieri*	**SH**	**EX**	**98**
281	*Tribulus pentandrus*			**14**

[a]**Resources: 1**: (Ahmad et al. 2004); **2**: (Ahmad et al. 2009); **3**: (Ahmed et al. 2007); **4**: (Shaheen et al. 2010); **5**: (Saqib et al. 2014); **6**: (Shah et al. 2013); **7**: (Ahmed et al. 2013); **8**: (Yaseen et al. 2015); **9**: (Hussain et al. 2004); **10**: (Ahmad et al. 2014a); **11**: (Shinwari et al. 2006); **12**: (Akhtar 1992); **13**: (Qureshi and Bhatti 2009); **14**: (Ahmed et al. 2014); **15**: (Ahmed et al. 2015); **16**: (Khan 2014); **17**: (Khan et al. 2015); **18**: (Abbasi et al. 2010a); **19**: (Shah and Khan 2006); **20**: (Ashraf et al. 2011); **21**: (Begum et al. 2015); **22**: (Fatima et al. 2005); **23**: (Hussain et al. 2009); **24**: (Marwat et al. 2011); **25**: (Hafizur et al. 2012); **26**: (Khattak and Khattak 2011); **27**: (Hussain et al. 2010a); **28**: (Sadiq et al. 2010); **29**: (Waheed et al. 2006); **30**: (Sahib et al. 2013); **31**: (Ahmad et al. 2000); **32**: (Malik et al. 2004); **33**: (Bibi et al. 2014); **34**: (Zia-Ul-Haq et al. 2012a); **35**: (Barkatullah et al. 2015); **36**: (Mahmood et al. 2010); **37**: (Mahmood et al. 2013); **38**: (Adnan et al. 2014b); **39**: (Shah et al. 2015); **40**: (Arshad et al. 2011); **41**: (Kanwal et al. 2011); **42**: (Ishtiaq et al. 2015); **43**: (Azmi and Qureshi 2012); **44**: (Harisaranraj et al. 2009); **45**: (Ahmad 2006); **46**: (Gilani et al. 2007); **47**: (Qureshi 2012); **48**: (Khan et al. 2014); **49**: (Alam et al. 2011); **50**: (Ibrar et al. 2004); **51**: (Awan et al. 2011); **52**: (Sabeen and Ahmad 2009); **53**: (Rashid et al. 2015); **54**: (Khan and Khatoon 2008), **55**: (Khan and Khatoon 2004); **56**: (Hayat et al. 2009); **57**: (Ahmad et al. 2014b); **58**: (Pushparaj et al. 2007); **59**: (Ali 2014); **60**: (Shah 2015); **61**: (Mahmood et al. 2012); **62**: (Khan et al. 2008); **63**: (Arshad and Ahmad 2004); **64**: (Ahmad and Alamgeer 2009); **65**: (Husain et al. 2008); **66**: (Munir and Qureshi 2018); **67**: (Hayat et al. 2014); **68**: (Mahmood et al. 2011); **69**: (Abbasi et al. 2013); **70**: (Zia-Ul-Haq et al. 2011); **71**: (Rathee et al. 2010); **72**: (Ali et al. 2011); **73**: (Panhwar and Abro 2007); **74**: (Panhwar 2013); **75**: (Hussain 2013); **76**: (Khan et al. 2013); **77**: (Zammurad and Qaiser 2011); **78**: (Arayne et al. 2007); **79**: (Akhtar and Begum 2009); **80**: (Singh et al. 2011); **81**: (Ahmad et al. 2010); **82**: (Kayani et al. 2015); **83**: (Ardestani and Yazdanparast 2007); **84**: (Raut and Gaikwad 2006); **85**: (Hussain et al. 2011); **86**: (Khalil et al. 2014); **87**: (Qureshi and Bhatti 2008), **88**. (Wadood et al. 1989); **89**: (Mushtaq et al. 2012); **90**: (Amalraj and Ignacimuthu 1998); **91**: (Khushk et al. 2010); **92**: (Rehman et al. 2015); **93**: (Zia et al. 2001); **94**: (Abdel-Barry et al. 1997); **95**: (Hussain et al. 2012); **96**: (Wong et al. 2011); **97**: (Khan et al. 1999); **98**: (Ibrar and Hussain 2009); **99**: (Chaudhary et al. 2010); **100**: (Saleem et al. 1999); **101**: (Zia-Ul-Haq et al. 2012b); **102**: (Mesaik et al. 2013); **103**: (Shah and Khan 2014); **104**: (Shinwari et al. 2003); **105**: (Abbasi et al. 2010b); **106**: (Khan et al. 2011); **107**: (Khan et al. 2012); **108**: (Singab et al. 2005); **109**: (Choudhary et al. 2013); **110**: (MacFarlane et al. 2003); **111**: (Miao et al. 2008); **112**: (Hamayun et al. 2006); **113**: (Hamayun 2007); **114**: (Ahmad et al. 2013); **115**: (Haq et al. 2011); **116**: (Murad et al. 2013); **117**: (Nisar et al. 2007); **118**: (Nisar et al. 2010); **119**: (Choudhary et al. 2011); **120**: (Zada Khan et al. 2015); **121**: (Adnan et al. 2014a); **122**: (Ali et al. 2015); **123**: (Ullah et al. 2013); **124**: (Jafri et al. 2011); **125**: (Oztürk et al. 2018); **126**: (Hussain et al. 2010b); **127**: (Gul et al. 2016).

Appendix 3: Medicinal Plants Used for the Treatment of Diabetes Mellitus in the Traditional Medicine in Malaysia

No	Family/taxa	Part used	Preparation	Resources[a]
	Acanthaceae			
1	*Acanthus ilicifolius*	FR	PN	1
2	*Andrographis paniculata*	AP, L, R, ST, WP	DE, EX, IN, R	1–7
3	*Barleria lupulina*	AP	EX	2
4	*Clinacanthus nutans*	L		8
5	*Strobilanthes crispus*	L	EX	2, 3
	Amaryllidaceae			
6	*Allium cepa*	B	JU	2
7	*Allium sativum*	B	JU	2, 3
	Anacardiaceae			
8	*Anacardium occidentale*	L, S	EX	2, 6
9	*Mangifera indica*	L	EX	2
	Annonaceae			
10	*Annona muricata*	FR-J		5
11	*Annona squamosa*	L		3
12	*Polyalthia bullata*	FL, L, R	DE, PN	1, 7, 9
	Apiaceae			
13	*Centella asiatica*	L, ST	DE	1
14	*Coriandrum sativum*	S	PW	2, 3
	Apocynaceae			
15	*Catharanthus roseus*	L, R, WP	DE, EX	2–4
	Arecaceae			
16	*Nypa fruticans*	FR	DR	1
17	*Sanguis draxonis*			2
	Asteraceae			
18	*Cosmos caudatus*	FL, L		6
19	*Gynura procumbens*	L	EF	5, 6
	Blechnaceae			
20	*Stenochlaena palustris*	FO	C	1
	Bromeliaceae			
21	*Ananas comosus*	L	EX	2
	Cactaceae			
22	*Hylocereus lemairei*	FR	EF	2
23	*Pereskia bleo*	L	DE	1
	Combretaceae			
24	*Terminalia catappa*	FR	BO, EX	2, 3
25	*Terminalia chebula*	FR	EX	2
	Connaraceae			
26	*Cnestis* sp.	R, ST	DE	1, 7

No	Family/taxa	Part used	Preparation	Resources[a]
27	*Cnestis palala*	**R**	**DE**	**12**
	Convolvulaceae			
28	*Ipomea batatas*	**L**		**2**
	Crassulaceae			
29	*Bryophyllum pinnatum*	**L**	**BO**	**3**
	Cucurbitaceae			
30	*Momordica charantia*	**FR, FR-J**	**AV, BO, EF, EX, R**	**1–3, 9, 11**
	Dilleniaceae			
31	*Tetracera indica*	**R**	**DE**	**1**
	Fabaceae			
32	*Archidendron bubalinum*	**R, S**	**DE, R**	**1**
33	*Archidendron jiringa*	**R, S**	**DE, EF, R**	**1, 4, 7**
34	*Leucaena leucocephala*	**R**	**DE**	**1**
35	*Mimosa pudica*	**L**	**EX**	**2**
36	*Parkia speciosa*	**FR, R, S**	**BO, DE, EF, EX, R**	**1–3, 5, 7, 11–13**
37	*Pongamia pinnata*	**FL**	**EX**	**2**
38	*Sindora coriacea*	**FR**	**DE**	**9**
39	*Sindora wallichii*	**FR**	**DE**	**9**
	Lamiaceae			
40	*Orthosiphon aristatus*	**L**	**DE**	**1, 5, 13**
41	*Orthosiphon stamineus*	**L**		**6**
	Lythraceae			
42	*Lagerstroemia speciosa*	**L**		**2**
43	*Punica granatum*	**FL**	**BO**	**3**
	Malvaceae			
44	*Ceiba pentandra*	**ST-B**	**EX**	**2, 3**
45	*Durio zibethinus*	**R**	**DE**	**1, 7**
46	*Hibiscus rosa-sinensis*	**FL**	**EX**	**2**
47	*Theobroma cacao*		**EX**	**2**
	Menispermaceae			
48	*Tinospora crispa*	**ST**	**DE, EX**	**1, 2, 14**
	Moraceae			
49	*Ficus deltoidea*	**L, R**		**6**
	Muntingiaceae			
50	*Muntingia calabura*	**L**	**DE**	**1, 7**
	Myrtaceae			
51	*Psidium guajava*	**SH**	**BO, EF**	**3**
	Oleaceae			
52	*Jasminum sambac*	**R**	**BO, IN**	**14**
	Ophioglossaceae			
53	*Helminthostachys zeylanica*	**R**	**DE**	**10**
	Oxalidaceae			
54	*Averrhoa bilimbi*	**FR-J, L**	**BO, DE**	**3–5, 9, 11**
55	*Averrhoa carambola*	**L, R**	**DE, DR**	**1, 7**

No	Family/taxa	Part used	Preparation	Resources[a]
	Pandanaceae			
56	*Pandanus amaryllifolius*	R	EX	2
	Phyllanthaceae			
57	*Bridelia stipularis*	R	PN	15
58	*Phyllanthus acidus*	L	DE	5
59	*Phyllanthus amarus*	L, S	BO, DE	1, 3
60	*Phyllanthus niruri*	L		6
	Piperaceae			
61	*Piper betle*	L	EX	2
62	*Piper sarmentosum*	L, R	DE	5, 6
	Plantaginaceae			
63	*Plantago major*	L, R, WP	BO, DE	9, 16, 17
	Poaceae			
64	*Imperata cylindrica*	R	DE	1
	Ranunculaceae			
65	*Nigella sativa*	S		3
	Rubiaceae			
66	*Morinda citrifolia*	FR, L	BO, DE, EF	1, 3, 4, 9
	Rutaceae			
67	*Murraya koenigii*	FR, L	EX	2
	Simaroubaceae			
68	*Eurycoma longifolia*	R, ST	BO, DE	1, 3, 7
	Smilacaceae			
69	*Smilax myosotiflora*	T	DE	1
	Solanaceae			
70	*Physalis minima*	FR, L, R	DE	17
	Urticaceae			
71	*Leucosyke capitellata*	L	DE	17
	Vitaceae			
72	*Leea indica*	L	DE	10
	Xanthorrhoeaceae			
73	*Aloe vera*	L		18
	Zingiberaceae			
74	*Alpinia galanga*	RH	EX, PW	2
75	*Curcuma longa*	RH	BO, PW	3
76	*Zingiber officinale*			19

[a]**Resources: 1**: (Ong and Azliza 2015); **2**: (Wong et al. 2011); **3**: (Latiff 2016); **4**: (Ong et al. 2011a); **5**: (Ong et al. 2011b); **6**: (Sekar et al. 2014); **7**: (Azliza et al. 2012); **8**: (Kassim et al. 2016); **9**: (Ong and Nordiana 1999); **10**: (Mohammad et al. 2012); **11**: (Alsarhan et al. 2012); **12**: (Ong et al. 2012); **13**: (Ramli et al. 2015); **14**: (Samuel et al. 2010); **15**: (Kulip 2003); **16**: (Kulip 1997); **17**: (Ahmad and Holdsworth 2003); **18**: (Mustaffa et al. 2011); **19**: (Alsarhan et al. 2014)

References

Abbasi AM, Khan M, Ahmad M, Zafar M, Jahan S, Sultana S (2010a) Ethnopharmacological application of medicinal plants to cure skin diseases and in folk cosmetics among the tribal communities of North-West Frontier Province, Pakistan. J Ethnopharmacol 128(2):322–335

Abbasi AM, Khan MA, Ahmed M, Zafar M (2010b) Herbal medicines used to cure various ailments by the inhabitants of Abbottabad district, North West Frontier Province, Pakistan. Indian J Tradit Knowl 9:175–183

Abbasi AM, Khan MA, Khan N, Shah MH (2013) Ethnobotanical survey of medicinally important wild edible fruits species used by tribal communities of Lesser Himalayas-Pakistan. J Ethnopharmacol 148(2):528–536

Abdel-Barry JA, Abdel-Hassan IA, Al-Hakiem MH (1997) Hypoglycaemic and antihyperglycaemic effects of Trigonella foenum-graecum leaf in normal and alloxan induced diabetic rats. J Ethnopharmacol 58(3):149–155

Adnan N, Othman N (2012) The relationship between plants and the Malay Culture. Proc Soc Behav Sci 42:231–241

Adnan M, Bibi R, Mussarat S, Tariq A, Shinwari ZK (2014a) Ethnomedicinal and phytochemical review of Pakistani medicinal plants used as antibacterial agents against Escherichia coli. Ann Clin Microbiol Antimicrob 13(40):1–18

Adnan M, Ullah I, Tariq A, Murad W, Azizullah A, Khan AL, Ali N (2014b) Ethnomedicine use in the war affected region of Northwest Pakistan. J Ethnobiol Ethnomed 10(16):1–16

Ahiskalı M, Arı Ç, Selvi S (2012) Edible wild plants and their consumption during winter in a rural village on Kazdağı (Mount Ida). Bocconea 24:195–198

Ahmad M (2006) Checklist of medicinal flora of Tehsil Isakhel, District Mianwali-Pakistan. Ethnobot Leaflets 10:41–48

Ahmad M, Alamgeer ST (2009) A potential adjunct to insulin: Berberis lycium Royle. Diabetol Croat 38:13–18

Ahmad FB, Holdsworth DK (2003) Medicinal plants of Sabah, East Malaysia-Part I. Pharm Biol 41(5):340–346

Ahmad M, Akhtar MS, Malik T, Gilani AH (2000) Hypoglycaemic action of the flavonoid fraction of Cuminum nigrum seeds. Phytother Res 14(2):103–106

Ahmad M, Khan MA, Arshad M, Zafar M (2004) Ethnophytotherapical approaches for the treatment of diabetes by the local inhabitants of district Attock (Pakistan). Ethnobot Leaflets 2004:7

Ahmad M, Qureshi R, Arshad M, Khan MA, Zafar M (2009) Traditional herbal remedies used for the treatment of diabetes from district Attock (Pakistan). Pak J Bot 41(6):2777–2782

Ahmad M, Khan MA, Awan AG, Zafar M, Sultana S (2010) Floral biodiversity of medicinal plants in salt range (Punjab) Pakistan: Current threats and conservation measures. Proceedings of 2nd International Conference "Biodiversity is our life" (IC Biour Life) December 2010 pp 41–59

Ahmad A, Husain A, Mujeeb M, Khan SA, Najmi AK, Siddique NA, Damanhouri ZA, Anwar F (2013) A review on therapeutic potential of Nigella sativa: a miracle herb. Asian Pac J Trop Biomed 3(5):337–352

Ahmad S, Wariss HM, Alam K, Anjum S, Mukhtar M (2014a) Ethnobotanical studies of plant resources of Cholistan desert, Pakistan. IJSR 3(6):1781–1788

Ahmad W, Khan I, Khan MA, Ahmad M, Subhan F, Karim N (2014b) Evaluation of antidiabetic and antihyperlipidemic activity of Artemisia indica linn (aeriel parts) in Streptozotocin induced diabetic rats. J Ethnopharmacol 151(1):618–623

Ahmed M, Khan MA, Zafar M, Sultana S (2007) Treatment of common ailments by plant-based remedies among the people of district Attock (Punjab) of northern Pakistan. Afr J Trad Complement Alternat Med 4:112–120

Ahmed E, Arshad M, Saboor A, Qureshi R, Mustafa G, Sadiq S, Chaudhari SK (2013) Ethnobotanical appraisal and medicinal use of plants in Patriata, New Murree, evidence from Pakistan. J Ethnobiol Ethnomed 9(1):1–11

Ahmed N, Mahmood A, Tahir SS, Bano A (2014) Ethnomedicinal knowledge and relative importance of indigenous medicinal plants of Cholistan desert, Punjab Province, Pakistan. J Ethnopharmacol 155(2):1263–1275
Ahmed N, Mahmood A, Mahmood A, Sadeghi Z, Farman M (2015) Ethnopharmacological importance of medicinal flora from the district of Vehari, Punjab province, Pakistan. J Ethnopharmacol 168:66–78
Ajgaonkar SS (1979) Herbal drugs in the treatment of diabetes: a review. IDF Bull 24:10–17
Akalın E (1998) Tekirdağ İli halk ilaçları ve gıda olarak kullanılan yabani bitkiler. Geleneksel ve Folklorik Droglar Dergisi 5(1):1–98
Akalın E, Alpınar K (1994) Tekirdağ'ın tıbbi ve yenen bitkileri hakkında bir araştırma. Ege Üniversitesi Eczacılık Fak Derg 1(2):1–11
Akan H, Aslan M, Balos MM (2005) GAP yöresindeki tıbbi ve aromatik bitkiler. TUBITAK Proje No: TBAG/Ç. SEK 22 (103-T009)
Akan H, Korkut MM, Balos MM (2008) Arat Dağı ve çevresinde (Birecik, Şanlıurfa) etnobotanik bir araştırma. Fırat Üniversitesi Fen ve Mühendislik Bilimleri Dergisi 20(1):67–81
Akaydın G, Şimşek I, Arıtuluk ZC, Yeşilada E (2013) An ethnobotanical survey in selected towns of the Mediterranean subregion (Turkey). Turk J Biol 37:230–247
Akbulut S, Bayramoglu MM (2014) Reflections of socio-economic and demographic structure of urban and rural on the use of medicinal and aromatic plants: the sample of Trabzon province. Ethno Med 8(1):89–100
Akçiçek E, Vural M (2003) Kumalar dağı (Afyon) ve çevresindeki bazı bitkilerin yöresel adları ve etnobotanik özellikleri. Ot Sistematik Botanik Dergisi 12(2):151–162
Akgül A (2008) Ethnobotany at Midyat (Mardin-Turkey). Msc thesis. Ege University, Izmir
Akhtar MS (1992) Hypoglycaemic activities of some indigenous medicinal plants traditionally used as antidiabetic drugs. J Pak Med Assoc 42:271
Akhtar N, Begum S (2009) Ethnopharmacological important plants of Jalala, district Mardan, Pakistan. Pak J Pl Sci 15(2):95–100
Aktan T (2011) Ethnobotany of Yenişehir Villages (Bursa-Turkey). Msc thesis. Celal Bayar University, Manisa
Akyol Y, Altan Y (2013) Ethnobotanical studies in the Maldan Village (Province Manisa, Turkey). Marmara Pharmaceut J (17):21–25
Alam N, Shinwari Z, Ilyas M, Ullah Z (2011) Indigenous knowledge of medicinal plants of Chagharzai valley, District Buner, Pakistan. Pak J Bot 43:773–780
Ali S (2014) Eco-Taxonomic study of District Buner (KPK) Pakistan. PhD thesis. University of Karachi, Karachi, Pakistan
Ali SI, Qaiser M (1986) A phytogeographic analysis of the phanerogams of Pakistan and Kashmir. Proc R Soc Edinb B 89:89–101
Ali H, Sannai J, Sher H, Rashid A (2011) Ethnobotanical profile of some plant resources in Malam Jabba valley of Swat, Pakistan. J Med Plant Res 5:4676–4687
Ali H, Uddin S, Jalal S (2015) Chemistry and biological activities of Berberis lycium Royle. JBAPN 5:295–312
Alparslan DF (2003) Babaeski (Kırklareli) yöresinin geleneksel halk ilacı olarak kullanılan bitkileri. Msc thesis. Marmara University, Marmara, İstanbul
Alparslan DF, Tuzlacı E (2006) The folk medicinal plants of the European part of Turkey. Proceedings of the IV. Intenational Congress of Ethnobotany (ICEB 2005), Istanbul pp. 201–204
Alpınar K (1999) Ayvalık (Balıkesir) ve yakınındaki adaların floristik ve etnobotanik açıdan değerlendirilmesi. TUBITAK, TBAG-1407, Turkey
Alsarhan A, Sultana N, Kadir MRA, Aburjai T (2012) Ethnopharmacological survey of medicinal plants in Malaysia, the Kangkar Pulai region. Int J Pharmacol 8(8):679–686
Alsarhan A, Sultana N, Al-khatib A, Rafiq M, Kadir A (2014) Review on some Malaysian traditional medicinal plants with therapeutic properties. Aust J Basic Appl Sci 10:149–159
Altundag E, Ozturk M (2011) Ethnomedicinal studies on the plant resources of east Anatolia, Turkey. Proc Soc Behav Sci 19:756–777

Amalraj T, Ignacimuthu S (1998) Hypoglycemic activity of *Cajanus cajan* (seeds) in mice. Indian J Exp Biol 36(10):1032–1033

Arayne MS, Sultana N, Mirza AZ, Zuberi MH, Siddiqui FA (2007) In vitro hypoglycemic activity of methanolic extract of some indigenous plants. Pak J Pharm Sci 20(4):268–273

Ardestani A, Yazdanparast R (2007) Cyperus rotundus suppresses AGE formation and protein oxidation in a model of fructose-mediated protein glycoxidation. Int J Biol Macromol 41(5):572–578

Arı S, Temel M, Kargıoğlu M, Konuk M (2015) Ethnobotanical survey of plants used in Afyonkarahisar-Turkey. J Ethnobiol Ethnomed 11(1):84

Arshad M, Ahmad M (2004) Medico-botanical investigation of medicinally important plants from Galiat areas-NWFP (Pakistan). Internationl Symposium Medicinal Plants: Linkages Beyond National Boundaries, September 7–9, 2004, Pakistan, pp 240–249

Arshad M, Nisar MF, Majeed A, Ismail S, Ahmad M (2011) Ethnomedicinal flora in district Sialkot, Punjab, Pakistan. Middle-East J Sci Res 9:209–214

Arulselvan P, Amynurliyana H, Ghofar A, Firdaus M, Halim A, Syafiq M, Ghafar A (2014) Antidiabetic therapeutics from natural source: a systematic review. Biomed Prev Nutr 4(4):607–617

Ashraf R, Khan RA, Ashraf I (2011) Garlic (*Allium sativum*) supplementation with standard antidiabetic agent provides better diabetic control in type 2 diabetes patients. Pak J Pharm Sci 24(4):565–570

Aslan A (2002) Ege Bölgesi bazı halk ilaçları üzerinde etnofarmakognozik bir değerlendirme. Msc thesis. İstanbul University, İstanbul

Awan MR, Iqbal Z, Shah SM, Jamal Z, Jan G, Afzal M, Majid A, Gul A (2011) Studies on traditional knowledge of economically important plants of Kaghan Valley, Mansehra District, Pakistan. J Med Plant Res 5:3958–3967

Azliza MA, Ong HC, Vikineswary S, Noorlidah A, Haron NW (2012) Ethno-medicinal resources used by the Temuan in Ulu Kuang Village. Ethno-Med 6(1):17–22

Azmi MB, Qureshi SA (2012) Methanolic root extract of *Rauwolfia serpentina* improves the glucose tolerance in wister mice. J Food Drug Anal 20(2)

Bailey CJ, Day C (1989) Traditional plant medicines as treatments for diabetes. Diabetes Care 12(8):553–564

Balos MM (2007) The flora and ethobotany of the region between Zeytinbahce and Akarçay (Birecik-Urfa-Turkey). Harran University, Şanlıurfa, Msc thesis

Barkatullah IM, Rauf A, Ben Hadda T, Mubarak MS, Patel S (2015) Quantitative ethnobotanical survey of medicinal flora thriving in Malakand Pass Hills, Khyber Pakhtunkhwa, Pakistan. J Ethnopharmacol 169:335–346

Başer KHC, Honda G, Mıkı W (1986) Herb drugs and herbalists in Turkey, Institute for the study of languages and cultures of Asia and Africa. Studia Culturae Islamicae 27, Tokyo

Baytop T (1984) Türkiye'de Bitkiler ile Tedavi. İstanbul Üniversitesi Yayınları No: 3255, Istanbul

Begum HA, Hamayun M, Zaman K, Hussain A, Ruaf M (2015) Phytochemical Evaluation of Ethnobotanically Selected Medicinal Plants of Mardan, Pakistan. J Adv Bot Zool 3:1–5

Bennett PH (1983) Diabetes in developing countries and unusual populations. In: Mann JI, Pyorala K, Teuscher A (eds) Diabetes in epidemiological perspective. Churchill Livingstone, Edinburgh, pp 43–57

Bibi S, Sultana J, Sultana H, Malik RN (2014) Ethnobotanical uses of medicinal plants in the highlands of Soan Valley, Salt Range, Pakistan. J Ethnopharmacol 155(1):352–361

Bransome ED Jr (1992) Financing the care of diabetes mellitus in the U.S. Diabetes Care 15(Supp. 1): 1–5

Bulut GE (2008) Bayramiç (Çanakkale) yöresinde etnobotanik araştırmalar. PhD thesis. Marmara Üniversitesi, İstanbul

Bulut G (2011) Folk Medicinal Plants of Silivri (İstanbul-Turkey). Marmara Pharmaceut J (15):25–29

Bulut, EG, Tuzlacı E (2005) An ethnobotanical study in Bozcaada (Çanakkale-Turkey). IVth International Congress of Ethnobotany (ICEB, 2005), pp 581–583

Bulut G, Korkmaz A, Tuzlacı E (2017) The ethnobotanical notes from Nizip (Gaziantep-Turkey). Istanbul J Pharm 47(2):57–62

Çakılcıoğlu U, Türkoğlu İ (2009) Çitli Ovası (Elazığ) ve çevresinin etnobotanik özellikleri. Ecolog Life Sci 4(2):81–85

Çakilcioglu U, Turkoglu I (2010) An ethnobotanical survey of medicinal plants in Sivrice (Elazıg-Turkey). J Ethnopharmacol 132:165–175

Çakılcıoğlu U, Türkoğlu İ, Kürşat M (2007) Harput (Elazığ) ve çevresinin etnobotanik özellikleri. Doğu Anadolu Bölgesi Araştırmaları:22–28

Çakılcıoğlu U, Şengün MT, Türkoğlu I (2010) An ethnobotanical survey of medicinal plants of Yazıkonak and Yurtbaşı districts of Elazığ province, Turkey. J Med Plant Res 4(7):567–572

Çakilcioglu U, Khatun S, Turkoglu I, Hayta S (2011) Ethnopharmacological survey of medicinal plants in Maden (Elazig-Turkey). J Ethnopharmacol 137(1):469–486

Çelik S, Karabacak E, Uysal I (2008) Plants have been collected from Mythological Kazdağı (Mt. Ida) National Park, West Turkey by Turkmens and their folk, cultural and social uses. Eur J Sci Res 19(4):835–843

Chaudhary G, Goyal S, Poonia P (2010) *Lawsonia inermis* L.: a phytopharmacological review. Int J Pharm Sci Drug Res 2(2):91–98

Choudhary MI, Adhikari A, Rasheed S, Marasini BP, Hussain N, Kaleem WA (2011) Cyclopeptide alkaloids of *Ziziphus oxyphylla* Edgew as novel inhibitors of α-glucosidase enzyme and protein glycation. Phytochem Lett 4:404–406

Choudhary MI, Khan N, Ahmad M, Yousuf S, Fun HK, Soomro S, Asif M, Mesaik MA, Shaheen F (2013) New inhibitors of ROS generation and T-cell proliferation from *Myrtus communis*. Org Lett 15(8):1862–1865

Çömlekçioğlu N, Karaman Ş (2008) Kahramanmaraş şehir merkezindeki aktar'larda bulunan tıbbi bitkiler. KSU J Sci Eng 11(1):23–32

Çubukçu B, Özhatay N (1987) Anadolu halk ilaçları hakkından bir araştırma. III. Milletlerarası Türk Folklor Kongresi Bildirileri, Kültür ve Turizm Bakanlığı Milli Folklor Araştırma Dairesi Yayınları: 86, Başbakanlık Basımevi, Ankara

Çubukcu B, Atay M, Sarıyar G, Özhatay N (1994) Aydın ili halk ilaçları. Geleneksel ve Folklorik Droglar Dergisi 1(1):1–55

Day C, Bailey CJ (1988a) Hypoglycemic agents from traditional plant treatments for diabetes. Int Ind Biotechnol 8:5–8

Day C, Bailey CJ (1988b) A Diabetologist's Herbal: Current Medical Literature on Diabetes. London, R. Soc. Med. 5

Demirci S, Özhatay N (2012) An ethnobotanical study in Kahramanmaraş (Turkey); wild plants used for medicinal purpose in Andırın, Kahramanmaraş. Turk J Pharmaceut Sci 9(1):75–92

Deniz L, Serteser A, Kargıoğlu M (2010) Uşak Üniversitesi ve yakın çevresindeki bazı bitkilerin mahalli adları ve etnobotanik özellikleri. AKÜ Fen Bilimleri Dergisi 1:57–72

Doğan G, Bağcı E (2011) The plants benefit to public in some residential area of Elazığ (Turkey) using traditional ecological knowledge and their ethnobotanical features. Fırat Univ Fen Bilimleri Dergisi 23(2):77–86

Duran A, Satıl F, Tümen G (2001) The edible wild fruits in Balıkesir and their ethnobotanical features. Ot Sistematik Botanik Dergisi 8(1):87–94

Durmuşkahya C, Öztürk M (2013) Ethnobotanical survey of medicinal plants used for the treatment of diabetes in Manisa, Turkey. Sains Malaysiana 42(10):1431–1438

Elçi B, Erik S (2006) Güdül (Ankara) ve çevresinin etnobotanik özellikleri. Hacettepe Üniversitesi Eczacılık Fak Dergisi 26:57–64

Emre G (2003) Ezine (Çanakkale) yöresinin geleneksel halk ilacı olarak kullanılan bitkileri. Msc thesis. Marmara University, İstanbul

Ertuğ F (2002) Bodrum yöresinde halk tıbbında yararlanılan bitkiler. 14. Bitkisel ilaç hammaddeleri toplantısı. 29–31 Mayıs 2002, Eskişehir

Ertuğ F, Tümen G, Çelik A, Dirmenci T (2004) Buldan (Denizli) etnobotanik alan araştırması 2003. TÜBA Kültür Envanteri Dergisi 2
Everest A, Ozturk E (2005) Focusing on the ethnobotanical uses of plants in Mersin and Adana provinces (Turkey). J Ethnobiol Ethnomed 1:1–6
Ezer N, Arisan OM (2006) Folk medicines in Merzifon (Amasya, Turkey). Turk J Bot 30(3):223–230
Ezer N, Avcı K (2004) Çerkeş (Çankırı) yöresinde kullanılan halk ilaçları. Hacettepe Üniversitesi Eczacılık Fakültesi Dergisi 24(2):67–80
Fakir H, Korkmaz M, Güller B (2009) Medicinal plant diversity of western Mediterranean region in Turkey. J Appl Biol Sci 3(2):3–4
Fatima N, Maqsood Z, Khan B (2005) Study of some micronutrients in selected medicinal plants. Sci Iran 12:269–273
Fujita T, Sezik E, Tabata M, Yeşilada E, Honda G, Takeda Y, Tanaka T, Takaishi Y (1995) Traditional medicine in Turkey VII. Folk medicine in middle and west Black Sea regions. Econ Bot 49:406–422
Genç EG (2003) Çatalca yöresinde etnobotanik bir çalışma. Msc thesis. Istanul University, İstanbul
Genç EG, Özhatay N (2006) An ethnobotanical study in Çatalca (European part of Istanbul) II. Turk J Pharmaceut Sci 3(2):73–89
Gençay A (2007) Cizre (Şırnak)'nin Etnobotanik Özellikleri. Msc thesis. Yüzüncü Yıl University, Van
Gençler ÖAM, Koyuncu M (2005) Traditional medicinal plants used in Pınarbaşı area (Kayseri-Turkey). Turk J Pharm Sci 2(2):63–82
Gilani SA, Kikuchi A, Shinwari ZK, Khattak ZI, Watanabe KN (2007) Phytochemical, pharmacological and ethnobotanical studies of *Rhazya stricta* Decne. Phytother Res 21(4):301–307
Gohdes DM (1986) Diabetes in American Indians: a growing problem. Diabetes Care 9:609–613
González-Tejero MR, Casares-Porcel M, Sánchez-Rojas CP, Ramiro-Gutiérrez JM, Molero-Mesa J et al (2008) Medicinal plants in the Mediterranean area: synthesis of the results of the project Rubia. J Ethnopharmacol 116(2):341–357
Gul A, Alam J, Ahmad H, Irfan M (2016) An updated checklist of Pteridophytes of district Mansehra, Khyber Pakhtunkhwa-Pakistan. Int J Biosci 9(5):116–133
Güler B, Manav E, Uğurlu E (2015) Medicinal plants used by traditional healers in Bozüyük (Bilecik-Turkey). J Ethnopharmacol 173:39–47
Gümüş İ (1994) Ağrı yöresinde yetişen bazı faydalı bitkilerin yerel adları ve kullanılışları. Turk J Bot 18:107–112
Günbatan T, Gürbüz İ, Özkan AMG (2016) The current status of ethnopharmacobotanical knowledge in Çamlıdere (Ankara, Turkey). Turk J Bot 40(3):241–249
Güner A, Aslan S, Ekim T, Vural M, Babaç MT (2012) Türkiye Bitkileri Listesi (Damarlı Bitkiler). Istanbul-Türkiye. Nezahat Gökyiğit Botanik Bahçesi ve Flora Araştırmaları Derneği Yayını
Güneş S (2010) Karaisalı (Adana-Turkey) and villages natural plants used by the ethnobotanical respect investigation. Msc thesis. Nigde University, Nigde
Güneş F, Özhatay N (2011) An ethnobotanical study from Kars (Eastern) Turkey. Biol Divers Conserv 4(1):30–41
Gürdal B, Kültür Ş (2013) An ethnobotanical study of medicinal plants in Marmaris (Muğla, Turkey). J Ethnopharmacol 146(1):113–126
Hafizur RM, Kabir N, Chishti S (2012) *Asparagus officinalis* extract controls blood glucose by improving insulin secretion and β-cell function in streptozotocin-induced type 2 diabetic rats. Br J Nutr 108(9):1586–1595
Hamayun M (2003) Ethnobotanical studies of some useful shrubs and trees of District Buner, NWFP, Pakistan. Ethnobot Leaflets 12
Hamayun M (2007) Traditional uses of some medicinal plants of Swat Valley, Pakistan. Indian J Tradit Knowl 6(4):636–641
Hamayun M, Khan SA, Sohn EY, Lee I-J (2006) Folk medicinal knowledge and conservation status of some economically valued medicinal plants of district swat, Pakistan. Lyonia 11:101–113

Han MI, Bulut G (2015) The folk-medicinal plants of Kadişehri (Yozgat-Turkey). Acta Soc Bot Pol 84(2):237–248

Haq F, Ahmad H, Alam M (2011) Traditional uses of medicinal plants of Nandiar Khuwarr catchment (District Battagram), Pakistan. J Med Plant Res 5:39–48

Harisaranraj R, Suresh K, Saravanababu S, Achudhan VV (2009) Phytochemical based strategies for pathogen control and antioxidant capacities of *Rauwolfia serpentina* extracts. Recent Res Sci Technol 1(2)

Hayat MQ, Khan MA, Ashraf M, Jabeen S (2009) Ethnobotany of the genus *Artemisia* L.(Asteraceae) in Pakistan. Ethnobot Res Appl 7:147–162

Hayat MM, Sarwar S, Anjum S, Uzair M, Rasheed HMF, Jabeen Q, Choudhary BA, Ashraf M (2014) Anti-diabetic and spasmolytic potential of *Farsetia hamiltonii* Royle from Cholistan desert. J Ethnopharmacol 156:347–352

Honda G, Yeşilada E, Tabata M, Sezik E, Fujita T, Takeda Y et al (1996) Traditional medicine in Turkey VI. Folk medicine in West Anatolia: Afyon, Kutahya, Denizli, Mugla, Aydin provinces. J Ethnopharmacol 53(2):75–87

Husain SZ, Malik RN, Javaid M, Bibi S (2008) Ethonobotanical properties and uses of medicinal plants of Morgah biodiversity park, Rawalpindi. Pak J Bot 40:1897–1911

Hussain MI (2013) Studies on plant biodiversity conservation and sustainable utilization of medicinally important plant species of Johi & Gorakh Hill. PhD thesis. University of Karachi, Karachi, Pakistan

Hussain Z, Waheed A, Qureshi RA, Burdi DK, Verspohl EJ, Khan N, Hasan M (2004) The effect of medicinal plants of Islamabad and Murree region of Pakistan on insulin secretion from INS-1 cells. Phytother Res 18:73–77

Hussain J, Khan AL, Rehman N, Hamayun M, Shah T, Nisar M, Bano T, Shinwari ZK, Lee I (2009) Proximate and nutrient analysis of selected vegetable species: a case study of Karak region, Pakistan. Afr J Biotechnol 8

Hussain K, Nisar MF, Majeed A, Nawaz K, Bhatti KH (2010a) Ethnomedicinal survey for important plants of Jalalpur Jattan, district Gujrat, Punjab, Pakistan. Ethnobot Leaflets 2010:11

Hussain J, Ullah R, Rehman N, Khan AL, Muhammad Z, Khan FU, Hussain ST, Anwar S (2010b) Endogenous transitional metal and proximate analysis of selected medicinal plants from Pakistan. J Med Plant Res 4:267–270

Hussain I, Bano A, Ullah F (2011) Traditional drug therapies from various medicinal plants of central karakoram national park, Gilgit-Baltistan Pakistan. Pak J Bot 43:79–84

Hussain W, Hussain J, Ali R, Hussain S, Khan MA, Khan I, Shinwari ZK, Lopes WA, Nascimento IA (2012). Phytomedicinal Studies of Kurram Agency in the Federally Administered Tribal Areas (FATA) of Pakistan

Ibrar M, Hussain F (2009) Ethnobotanical studies of plants of Charkotli hills, Batkhela district, Malakand, Pakistan. Front Biol China 4(4):539

Ibrar M, Ilahi I, Hussain F (2004) Hypoglycemic activity of *Hedera helix* L. leaves and the possible mechanism of action. Pak J Bot 35:805–810

İlçim A, Varol Ö (1996) Hatay ve K. Maraş (Türkiye) illerindeki bazı bitkilerin etnobotanik özellikleri. Ot Sistematik Botanik Dergisi 3(1):69–74

International Diabetes Federation (IDF) (2013) IDF diabetes atlas, 6th edn. International Diabetes Federation, Brussels, Belgium

Ishtiaq M, Mahmood A, Maqbool M (2015) Indigenous knowledge of medicinal plants from Sudhanoti district (AJK), Pakistan. J Ethnopharmacol 168:201–207

Jafri SA, Hasan SS, Nadeem A, Iqbal J (2011) Hypoglycemic effect of *Aloe vera* Extract in alloxan-induced diabetic Albino rats. Med J Islamic World Acad Sci 19:127–130

Kahraman A, Tatlı A (2004) Umarbaba dağı (Eşme-Uşak) ve çevresindeki bazı bitkilerin mahalli adları ve etnobotanik özellikleri. Ot Sistematik Botanik Dergisi 11(2):147–154

Kanwal AMM, Shaukat S, Javed R, Ilyas R (2011) Exploration of ethnomedicinal values of imperative plants of District Gujrat, Pakistan. Middle-East J Sci Res 7:397–400

Karaman S, Kocabas YZ (2001) Traditional medicinal plants of K. Maras (Turkey). Science 1(3):125–128

Kargioğlu M, Cenkci S, Serteser A, Evliyaoğlu N, Konuk M, Kök MŞ, Bağci Y (2008) An ethnobotanical survey of inner-west Anatolia, Turkey. Hum Ecol 36(5):763–777

Kassim DHA, Raduan SZ, Aziz MA, Chelum A, Morni AAM, Wahab RA (2016) Indigenous knowledge of medicinal plants used and its ımplication towards health-seeking behavior among the Melanau in Pulau Bruit, Sarawak, Malaysia. J Adv Res Soc Behav Sci 4(2):136–145

Kaval I, Behçet L, Cakilcioglu U (2014) Ethnobotanical study on medicinal plants in Geçitli and its surrounding (Hakkari-Turkey). J Ethnopharmacol 155(1):171–184

Kayani S, Ahmad M, Sultana S, Khan Z (2015) Ethnobotany of medicinal plants among the communities of alpine and sub-alpine regions of Pakistan. J Ethnopharmacol 164:186–202

Keklik Koçoğlu T, Çubukcu B, Özhatay N (1996) Konya ve Karaman illeri halk ilaçları. Geleneksel ve Folklorik Droglar Dergisi 3(1):1–70

Keskin L (2011) Ethnobotanical features of plants in Kadinhani (Konya-Turkey) and enviroments. Msc thesis. Selcuk University, Konya

Keskin M, Alpınar K (2002) Kışlak (Yayladağı-Hatay) hakkında etnobotanik bir araştırma. Ot Sistematik Botanik Dergisi 9:91–100

Khalil AT, Shinwari ZK, Qaiser M, Marwat KB (2014) Phyto-therapeutic claims about euphorbeaceous plants belonging to Pakistan; an ethnomedicinal review. Pak J Bot 46(3):1137–1144

Khan KU (2014) Ethnobotanical and ethnopharmacognostic exploration of folk medicines of Deosai Plateau. PhD thesis. Hazara University, Hazara, Pakistan

Khan SW, Khatoon S (2004) Ethnobotanical studies in Haramosh and Bugrote valleys (Gilgit). Int J Biol Biotechnol 1(4):585–589

Khan SW, Khatoon S (2008) Ethnobotanical studies on some useful herbs of Haramosh and Bugrote valleys in Gilgit, northern areas of Pakistan. Pak J Bot 40:43

Khan PM, Nawaz HR, Ahmad S, Malik A (1999) Ajugins C and D, new withanolides from *Ajuga parviflora*. Helv Chim Acta 82(9):1423–1426

Khan SA, Khan L, Hussain I, Marwat KB, Akhtar N (2008) Profile of heavy metals ın selected medicinal plants. PJWSR 14:101–110

Khan KY, Khan MA, Niamat R, Munir M, Fazal H, Mazari P, Seema N, Bashir T, Kanwal A, Ahmed SN (2011) Element content analysis of plants of genus Ficus using atomic absorption spectrometer. Afr J Pharm Pharmacol 5:317–321

Khan KY, Khan MA, Niamat R, Shah GM, Fazal H, Seema N, Hussain I, Ahmad I, Inayat H, Jan G (2012) Elemental content of some anti-diabetic ethnomedicinal species of genus Ficus Linn using atomic absorption spectrophotometry technique. J Med Plant Res 6:2136–2140

Khan SM, Page S, Ahmad H, Shaheen H, Ullah Z, Ahmad M, Harper DM (2013) Medicinal flora and ethnoecological knowledge in the Naran Valley, Western Himalaya, Pakistan. J Ethnobiol Ethnomed 9(4):2–13

Khan MA, Marwat KB, Gul B, Wahid F, Khan H, Hashim S (2014) *Pistia stratiotes* L.(Araceae): Phytochemistry, use ın medicines, phytoremediation, biogas and management options. Pak J Bot 46:851–860

Khan AA, Ali F, Ihsan M, Hayat K, Nabi G (2015) Ethnobotanical study of the medicinal plants of Tehsil Charbagh, District Swat, Khyber Pakhtunkhwa, Pakistan. Am Eurasian Agric Environ Sci 15:1464–1474

Khattak MI, Khattak MI (2011) Study of heavy trace metals in some medicinal–herbal plants of Pakistan. Pak J Bot 43:2003–2009

Khushk I, Dahot M, Baloach S, Bhutto M (2010) The evaluation of soybean extracts in alloxan-induced diabetic rabbits. World Appl Sci J 8:22–25

Kıran Ö (2006) Kozan yöresi florasındaki tıbbi bitkiler ve bunların halk tıbbında kullanılışı. Msc thesis. Çukurova Üniversity, Adana

Kızılaslan Ç (2008) An ethnobotanical survey in the South Part of Izmit Gulf. Msc thesis. Istanbul University, İstanbul

Koca AD, Yıldırımlı Ş (2010) Ethnobotanical proporties of Akçakoca District in Düzce (Turkey). Hacettepe J Biol Chem 38(1):63–69
Koçak S (1999) Karaman yöresinde etnobotanik bir araştırma. Msc thesis. İstanbul University, İstanbul
Koçyiğit M, Özhatay N (2006) Wild plants used as medicinal purpose in Yalova (Northwest Turkey). Turk J Pharmaceut Sci 3(2):91–103
Korkmaz M, Karakurt E (2014) Medicinal plants sold in the herbal markets in Kelkit (Gümüşhane). Suleyman Demirel Univ J Nat Appl Sci 18(3):60–80
Koyuncu O (2005) Investigation of floristic and ethnobotanical aspect of Geyve (Sakarya) and its environs. PhD thesis. Osmangazi University, Eskişehir
Koyuncu O, Yaylacı ÖK, Öztürk D, Erkara İP, Savaroğlu F, Akçoşkun Ö, Ardıç M (2010) Risk categories and ethnobotanical features of the Lamiaceae taxa growing naturally in Osmaneli (Bilecik/Turkey) and environs. Biol Divers Conserv 3(3):31–45
Kulip J (1997) A preliminary survey of traditional medicinal plants in the west coast and interior of Sabah. J Trop For Sci 10(2):271–274
Kulip J (2003) An ethnobotanical survey of medicinal and other useful plants of Muruts in Sabah, Malaysia. Telopea 10(1):81–98
Kültür Ş (2007) Medicinal plants used in Kirklareli Province (Turkey). J Ethnopharmacol 111(2):341–364
Kumar A, Goel MK, Jain RB, Khanna P, Chaudhary V (2013) India towards diabetes control: key issues. Australias Med J 6(10):524–531
Latiff A (2016) Antidiabetic plants of Malaysia. Unpublished personal report, Malaysia
Leite RS, Marlow NM, Fernandes JK (2013) Oral health and type 2 diabetes. Am J Med Sci 345(4):271–273
MacFarlane AJ, Burghardt KM, Kelly J, Simell T, Simell O, Altosaar I, Scott FW (2003) A type 1 diabetes-related protein from wheat (*Triticum aestivum*) cDNA clone of a wheat storage globulin, Glb1, linked to islet damage. J Biol Chem 278(1):54–63
Mahmood T, Muhammad S, Shinwari ZK (2010) Molecular and morphological characterization of *Caralluma* species. Pak J Bot 42:1163–1171
Mahmood A, Mahmood A, Tabassum A (2011) Ethnomedicinal survey of plants from District Sialkot, Pakistan. J Appl Pharmacol 2:212–220
Mahmood A, Mahmood A, Malik RN (2012) Indigenous knowledge of medicinal plants from Leepa valley, Azad Jammu and Kashmir, Pakistan. J Ethnopharmacol 143(1):338–346
Mahmood A, Rashid S, Naseem R (2013) Determination of toxic heavy metals in indigenous medicinal plants used in Rawalpindi and Islamabad cities, Pakistan. J Ethnopharmacol 148(1):158–164
Malik RN, Rafique A, Husain SZ (2004) Ethnobotanical resources of District Sialkot. International Symposium Medicinal Plants: Linkages Beyond National Boundaries, September 7–9, 2004 pp 42–49
Malyer H, Aydın SÖ, Tümen G, Er S (2004) Tekirdağ ve çevresindeki aktarlarda satılan bazı bitkiler ve tıbbi kullanım özellikleri. Dumlupınar Üniversitesi Fen Bilimleri Enstitüsü Dergisi 7:103–112
Marles RJ, Farnsworth NR (1995) Antidiabetic plants and their active constituents 1. Phytomedicine 2(2):137–189
Mart S (2006) Bahçe ve Hasanbeyli (Osmaniye) halkının kullandığı doğal bitkilerin etnobotanik yönden araştırılması. Msc thesis. Çukurova University, Adana
Marwat SK, Mak F-U-R, Ahmad M, Zafar M, Ghulam S (2011) Medicinal folk recipes used as traditional phytotherapies in district Dera Ismail Khan, KPK, Pakistan. Pak J Bot 43:1453–1462
Mesaik MA, Ahmed A, Khalid AS, Jan S, Siddiqui AA, Perveen S, Azim MK (2013) Effect of *Grewia asiatica* fruit on *Glycemic index* and phagocytosis tested in healthy human subjects. Pak J Pharm Sci 26(1)
Metin A (2009) Ethnobotanical features of plants in Mut (Mersin) and its environments. Msc thesis. Selçuk University, Konya

Miao MS, Zhang GL, Miao YY, Shi JJ, Liu HL (2008) Influence of *Zea mays* L. saponin (ZMLS) on ultrastructure of kidney and pancreas in diabetes rats induced by streptozocin. China J Chin Materia Medica 33(10):1179–1183
Mohammad NS, Milow P, Ong HC (2012) Traditional medicinal plants used by the Kensiu tribe of Lubuk Ulu Legong, Kedah, Malaysia. Ethno-Med 6(3):149–153
Mükemre M, Behçet L, Çakılcıoğlu U (2015) Ethnobotanical study on medicinal plants in villages of Çatak (van-Turkey). J Ethnopharmacol 166:361–374
Munir M, Qureshi R (2018) Antidiabetic plants of Pakistan. (In Press)
Murad W, Azizullah A, Adnan M, Tariq A, Khan KU, Waheed S, Ahmad A (2013) Ethnobotanical assessment of plant resources of Banda Daud Shah, District Karak, Pakistan. J Ethnobiol Ethnomed 9:77
Mushtaq A, Muhammad Z, Ajab KM, Shazia S, Mujtaba SG, Jan G (2012) Ethnomedicinal investigation of phytomedicines among the local communities of. Arid Areas of Pakistan 11(July):436–446
Mustaffa F, Indurkar J, Ali NIM, Hanapi A, Shah M, Ismail S, Mansor SM (2011) A review of Malaysian medicinal plants with potential antidiabetic activity. J Pharm Res 4(11):4217–4224
Nasir E, Ali SI (1970–1995) Flora of Pakistan. National Herbarium/NARC and Department of Botany, University of Karachi, Islamabad/Karachi
Nisar M, Adzu B, Inamullah K, Bashir A, Ihsan A, Gilani A (2007) Antinociceptive and antipyretic activities of the *Ziziphus oxyphylla* Edgew. Leaves. Phytother Res 21:693–695
Nisar M, Kaleem WA, Adhikari A, Ali Z, Hussain N, Khan I, Qayum M, Choudhary MI (2010) Stereochemistry and NMR data assignment of cyclopeptide alkaloids from *Ziziphus oxyphylla*. Nat Prod Commun 5(8):1205–1208
Nowbandegani AS, Kiumarcy S, Rahmani F, Dokouhaki M, Khademian S, Zarshenas MM, Faridi P (2015) Ethnopharmacological knowledge of Shiraz and Fasa in Fars region of Iran for diabetes mellitus. J Ethnopharmacol 172:281–287
Ong HC, Azliza MA (2015) Medicinal plants for diabetes by the Orang Asli in Selangor, Malaysia. Ethno-Med 9(1):77–84
Ong HC, Nordiana M (1999) Malay ethno-medico botany in Machang, Kelantan, Malaysia. Fitoterapia 70(5):502–513
Ong HC, Ruzalila BN, Milow P (2011a) Traditional knowledge of medicinal plants among the Malay villagers in Kampung Tanjung Sabtu, Terengganu, Malaysia. Indian J Tradit Knowl 10(3):460–465
Ong HC, Zuki RM, Milow P (2011b) Traditional knowledge of medicinal plants among the Malaysia. Ethno-Med 5(3):175–185
Ong HC, Faezah AW, Milow P (2012) Traditional knowledge and usage of medicinal plants among the Semai Orang Asli at Kampung Batu 16, Tapah, Perak, Malaysia. Ethno-Med 6(3):207–211
Oral DÇ (2007) Konya ilinde kullanılan halk ilaçları üzerinde etnobotanik araştırmalar. Msc thesis. Gazi University, Ankara
Özçelik H, Balabanlı C (2005) Burdur ilinin tıbbi ve aromatik bitkileri. I. Burdur Sempozyumu, Burdur/Turkiye
Özdemir E (2005) Niğde-Aladağlar'ın batısında etnobotanik bir araştırma. Msc thesis. Istanbul University, Istanbul
Özgen U, Coşkun M (2000) Ilıca (Erzurum) ilçesine bağlı köylerde halk ilacı olarak kullanılan bitkiler. XIII. Bitkisel İlaç Hammaddeleri Toplantısı, 20–22 Eylül 2000, Istanbul pp 135–143
Özgen U, Kaya Y, Houghton P (2012) Folk medicines in the villages of Ilıca District (Erzurum, Turkey). Turk J Biol 36:93–106
Özgokce F, Ozcelik H (2004) Ethnobotanical aspects of some taxa in East Anatolia, Turkey. Econ Bot 58(4):697–704
Öztürk M (2006) Nizip bölgesi (Aksaray) florası ve etnobotanik özellikleri. Msc thesis. Selçuk University, Konya
Öztürk M, Dinç M (2005) Nizip (Aksaray) bölgesinin etnobotanik özellikleri. Ot Sistematik Botanik Dergisi 12(1):93–102

Öztürk F, Ölçücü C (2011) Ethnobotanical features of some plants in the district Şemdinli (Hakkari-Turkey). Int J Acad Res 3(1):117–121
Ozturk M, Guvensen A, Aksoy A, Beyazgul M (2006) An overview of the soils and sustainable land use in Turkiye. Proceedings of the Fifth International GAP Engineering Congress, Sanlıurfa, Turkey, pp 1548–1555
Ozturk M, Gucel S, Altundag E, Mert T, Gork C, Gork G, Akcicek E (2012a) Chapter 7: An overview of the medicinal plants of Turkey. In: Singh R (ed) Genetic Resources, chromosome engineering and crop improvement: medicinal plants. CRC Press, LLC, Taylor & Francis, USA, pp 181–206
Ozturk M, Altundağ E, Gücel S (2012b) Medicinal and Aromatic Plants (Turkey). Ethnopharmacology, Encyclopedia of Life Support Systems (EOLSS). http://www.eolss.net/sample-chapters/c03/e6-79-48.pdf
Ozturk M, Altay V, Aksoy A (2016) Ecology of endangered endemic plant taxa of Turkiye in relation to climate change. In: Conservation and Sustainable use of genepool of plant World in Eurasia at the present stage International Scientific Conference within Day of Kazakhstan. September 3, EXPO-2016, Antalya, Turkiye
Ozturk M, Altay V, Gonenç TM (2017a) Chapter 24: Herbal from High Mountains in the East Mediterranean. In: Bhojraj S et al (eds) Drug Discovery from Herbs-Approaches and Applications Centre for Scinece & Technology of the Non-aligned and other Developing Countries (NAM S & T Centre). DAYA Publishing House, New Delhi, India, pp 327–367
Ozturk M, Altay V, Gücel S, Altundağ E (2017b) Plant Diversity of the drylands in Southeast Anatolia-Turkey: role in human health and food security. In: Ansari AA, Gill SS (eds) Plant biodiversity: monitoring, assessment and conservation. CABI, UK, pp 83–124
Oztürk M, Altay V, Latiff A, Salman T, Choudhry I (2018) A comparative analysis of the medicinal pteridophytes in Turkey, Pakistan and Malaysia. (In press)
Özüdoru B, Akaydin G, Erik S, Yesilada E (2011) Inferences from an ethnobotanical field expedition in the selected locations of Sivas and Yozgat provinces (Turkey). J Ethnopharmacol 137(1):85–98
Özuslu E (2005) Sof Dağı (Gaziantep) yöresindeki bazı bitkilerin etnobotanik özellikleri ve mahalli adları. Kırsal Çevre Yıllığı:7–22
Panhwar AQ (2013) Ethnobotanical and taxonomic studies of selected plant species from Mahal Kohistan region of Sindh, Pakistan. PhD thesis. University of Sindh Jamshoro, Pakistan
Panhwar AQ, Abro H (2007) Ethnobotanical studies of Mahal Kohistan (Khirthar National Park). Pak J Bot 39:2301–2315
Parildar H, Serter R, Yesilada E (2011) Diabetes mellitus and phytotherapy in Turkey. J Pak Med Associat 61:1116–1120
Parris BS, Latiff A (1997) Towards a pteridophyte flora of Malaysia: a provisional checklist of taxa. Malay Nat J 50:235–280
Polat R, Satil F (2012) An ethnobotanical survey of medicinal plants in Edremit Gulf (Balikesir-Turkey). J Ethnopharmacol 139(2):626–641
Polat R, Satıl F, Çakılcıoğlu U (2011) Medicinal plants and their use properties of sold in herbal market in Bingöl (Turkey) district. Biol Divers Conserv J 4(3):25–35
Polat R, Selvi S, Çakılcıoğlu U, Açar M (2012) Investigations of ethnobotanical aspect of wild plants sold in Bingöl (Turkey) local markets. Biol Divers Conserv J 5(3):155–161
Polat R, Cakilcioglu U, Satil F (2013) Traditional uses of medicinal plants in Solhan (Bingöl-Turkey). J Ethnopharmacol 148(3):951–963
Polat R, Cakılcıoglu U, Kaltalioglu K, Ulusan MD, Türkmen Z (2015) An ethnobotanical study on medicinal plants in Espiye and its surrounding (Giresun-Turkey). J Ethnopharmacol 163:1–11
Premilla M (2002) Green Malaysia: Rainforest Encounters Editions Didier Millet Sdn Bhd: Kuala Lumpur, Malaysia
Pushparaj PN, Low HK, Manikandan J, Tan BKH, Tan CH (2007) Anti-diabetic effects of *Cichorium intybus* in streptozotocin-induced diabetic rats. J Ethnopharmacol 111(2):430–434

Qureshi R (2012) Medicinal flora of Hingol National Park, Baluchistan, Pakistan. Pak J Bot 44:725–732

Qureshi R, Bhatti GR (2008) Ethnobotany of plants used by the Thari people of Nara Desert, Pakistan. Fitoterapia 79(6):468–473

Qureshi R, Bhatti G (2009) Folklore uses of Amaranthaceae family from Nara desert, Pakistan. Pak J Bot 41:1565–1572

Ramli MR, Milow P, Chooi OH (2015) Traditional knowledge of a practitioner in medicinal plants of Masjid Ijok village, Perak, Malaysia. Ethno-Med 9(1):59–66

Rashid S, Ahmad M, Zafar M, Sultana S, Ayub M, Khan MA, Yaseen G (2015) Ethnobotanical survey of medicinally important shrubs and trees of Himalayan region of Azad Jammu and Kashmir, Pakistan. J Ethnopharmacol 166:340–351

Rathee S, Mogla OP, Sardana S, Vats M, Rathee P (2010) Antidiabetic activity of Capparis decidua Forsk Edgew. J Pharm Res 3:231–234

Raut NA, Gaikwad NJ (2006) Antidiabetic activity of hydro-ethanolic extract of *Cyperus rotundus* in alloxan induced diabetes in rats. Fitoterapia 77(7):585–588

Rehman K, Mashwani Z, Ali M, Ullah Z (2015) An ethno botanical perspective of traditional medicinal plants from the Khattak tribe of Chonthra Karak, Pakistan. J Ethnopharmacol 165:251–259

Ridley HN (1922–1925) The flora of the Malay Peninsula. L. Reeve & Co. Ltd, London1: 918, 2: 672, 3: 405, 4: 383, 5: 470

Ridley HN (1926) The ferns of the Malay Peninsula. J Malayan Branch Roy Asiatic Soc 4:1–121

Roglic G (2004) Global prevalence of diabetes: estimates for the year 2000 and projections for 2030. Diabetes Care 27:10147–11053

Sabeen M, Ahmad SS (2009) Exploring the folk medicinal flora of Abbotabad city, Pakistan. Ethnobot Leaflets 2009:1

Saday H (2009) Ethnobotanical properties of Guzeloluk Village and its vicinity (Erdemli-Mersin). Msc thesis. Selçuk University, Konya

Sadıkoğlu E (2003) Koçarlı (Aydın) yöresinin geleneksel halk ilacı olarak kullanılan bitkileri. Msc thesis. Marmara University, Istanbul

Sadıkoğlu N, Alpınar K (2000) Bartın: from an ethnobotanical point of view. XIII th meeting on Plant Originated Crude Drugs Proceeding Book. Marmara University press pp 87–100

Sadiq S, Nagi AH, Shahzad M, Zia A (2010) The reno-protective effect of aqueous extract of *Carum carvi* (black zeera) seeds in streptozotocin induced diabetic nephropathy in rodents. Saudi J Kidney Dis Transpl 21:1058

Sağıroğlu M, Olgaç E, Ertürk B, Turna M (2012a) An ethnobotanical survey from Şile (Istanbul) and Karasu (Sakarya). Ot Sistematik Botanik Dergisi 19(2):93–104

Sağıroğlu M, Aslantürk A, Akdemir ZK, Turna M (2012b) An ethnobotanical survey from Hayrat (Trabzon) and Kalkandere (Rize/Turkey). Biol Divers Conserv 5(1):31–43

Sahib NG, Anwar F, Gilani AH, Hamid AA, Saari N, Alkharfy KM (2013) Coriander (*Coriandrum sativum* L.): a potential source of high-value components for functional foods and Nutraceuticals-a review. Phytother Res 27(10):1439–1456

Saleem R, Ahmad M, Hussain SA, Qazi AM, Ahmad SI, Qazi MH, Ali M, Faizi S, Akhtar S, Husnain SN (1999) Hypotensive, hypoglycaemic and toxicological studies on the flavonol C-glycoside shamimin from *Bombax ceiba*. Planta Med 65(4):331–334

Samuel AJSJ, Kalusalingam A, Chellappan DK, Gopinath R, Radhamani S, Husain HA et al (2010) Ethnomedical survey of plants used by the Orang Asli in Kampung Bawong, Perak, West Malaysia. J Ethnobiol Ethnomed 6(1):5

Saqib Z, Mahmood A, Malik RN, Mahmood A, Syed JH, Ahmad T (2014) Indigenous knowledge of medicinal plants in Kotli Sattian, Rawalpindi district, Pakistan. J Ethnopharmacol 151:820–828

Sargin SA (2015) Ethnobotanical survey of medicinal plants in Bozyazi district of Mersin, Turkey. J Ethnopharmacol 173:105–126

Sargin SA, Akçicek E, Selvi S (2013) An ethnobotanical study of medicinal plants used by the local people of Alaşehir (Manisa) in Turkey. J Ethnopharmacol 150(3):860–874
Sargin SA, Selvi S, López V (2015a) Ethnomedicinal plants of Sarigöl district (Manisa), Turkey. J Ethnopharmacol 171:64–84
Sargin SA, Selvi S, Büyükcengiz M (2015b) Ethnomedicinal plants of Aydincik District of Mersin, Turkey. J Ethnopharmacol 174:200–216
Sarı AO, Oğuz B, Güvensen A, Bilgiç A, Şenol SG, Tort N (2010) Ege ve Güney Marmara Bölgelerinde halk ilacı olarak kullanılan bitkiler. Anadolu J AARI 20(2):1–21
Sarper F, Akaydın G, Şimşek I, Yeşilada E (2009) An ethnobotanical field survey in the Haymana District of Ankara Province in Turkey. Turk J Biol 33:79–88
Savran A, Bağcı Y, Kargıoğlu M (2009) Gemerek (Sivas) ve çevresindeki bazı bitkilerin yerel adları ve etnobotanik özellikleri. Afyon Kocatepe Üniversitesi Fen Bilimleri Dergisi 8(1):313–321
Saw LG, Chung RCK (2015) The flora of Malaysia projects. Rodriguesia 66(4):947–960
Sayar A, Güvensen A, Özdemir F, Öztürk M (1995) Muğla (Türkiye) ilindeki bazı türlerin etnobotanik özellikleri. Ot Sistematik Botanik Dergisi 2(1):151–160
Schraer CD, Lanier AP, Boyko E, Gohdes D, Murphy NJ (1988) Prevalence of diabetes mellitus in Alaskan Eskimos, Indians and Aleuts. Diabetes Care 11:693–700
Sekar M, Zulhilmi M, Hamdi AY, Nabila N, Zahida Z, Shafiq M (2014) Ten commonly available medicinal plants in Malaysia used for the treatment of diabetes-a review. Asian J Pharm Clin Res 7(1):1–5
Şenkardeş I (2014) Nevşehir'in güney ilçelerinde (Acıgöl, Derinkuyu, Gülşehir, Nevşehir-merkez, Ürgüp) etnobotanik araştırmalar. PhD thesis. Marmara University, Istanbul
Şenkardeş I, Tuzlaci E (2014) Some ethnobotanical notes from Gundogmus District (Antalya/Turkey). Marmara Üniversitesi Saglik Bilimleri Enstitüsü Dergisi 4(2):63–75
Sezik E, Yeşilada E, Tabata M, Honda G, Takaishi Y, Fujita T, Tanaka T, Takeda Y (1997) Traditional medicine in Turkey VIII. Folk medicine in East Anatolia: Erzurum, Erzincan, Ağrı, Kars, Iğdır Provinces. Econ Bot 51(3):195–211
Sezik E, Yeşilada E, Honda G, Takaishi Y, Takeda Y, Tanaka T (2001) Traditional medicine in Turkey X: folk medicine in central Anatolia. J Ethnopharmacol 75:95–115
Shah AH (2015) Cultural drivers of plant biodiversity of district Tor Ghar. PhD thesis. Hazara University, Pakistan
Shah GM, Khan MA (2006) Checklist of Medicinal Plants of Siran Valley, Mansehra, Pakistan. Ethnobot Leaflets 10:63–71
Shah NA, Khan MR (2014) Antidiabetic effect of *Sida cordata* in alloxan induced diabetic rats. Biomed Res Int 2014:15
Shah A, Marwat SK, Gohar F, Khan A, Bhatti KH, Amin M, Din NU, Ahmad M, Zafar M (2013) Ethnobotanical study of medicinal plants of semi-tribal area of Makerwal & Gulla Khel (lying between Khyber Pakhtunkhwa and Punjab Provinces), Pakistan. Am J Plant Sci 4:98–116
Shah AH, Khan SM, Shah AH, Mehmood A, Rahman IU, Ahmad H (2015) Cultural uses of plants among Basikhel tribe of District Tor Ghar, Khyber Pakhtunkhwa, Pakistan. Pak J Bot 47:23–41
Shaheen H, Qureshi R, Zehra I, Arshad M, Chaudhry AK, Iqbal S (2010) Ethnobotanical survey of Santh Saroola, Kotli Sattian, Rawalpindi, Pakistan. Proceeding of 2nd International Conference "Biodiversity is our life" (IC Biour Life) December 2010, Pakistan pp 137–153
Shinwari MI, Shinwari MI (2010) Botanical diversity in Pakistan; past present and future. World Environ:85–104
Shinwari ZK, Khan AA, Nakaike T (2003) Medicinal and other useful plants of District Swat, Pakistan. WWF, Pakistan 68
Shinwari ZK, Watanabe T, Rehman M, Youshikawa T (2006) A pictorial guide to medicinal plants of Pakistan. KUST, Kohat, Pakistan
Şığva HÖ, Seçmen Ö (2009) Ethnobotanic survey of Işıklı (Çarpın) Dağdancık and Tokdemir in Gaziantep (Turkey). IUFS J Biol 68(1):19–26
Şimşek I, Aytekin F, Yeşilada E, Yıldırımlı Ş (2001) Ankara, Gölbaşı'nda yabani bitkilerin kullanılış amaçları ve şekilleri üzerinde bir araştırma. Ot Sistematik Botanik Dergisi 8(2):105–120

Şimsek I, Aytekin F, Yesilada E, Yıldırımlı Ş (2004) An ethnobotanical survey of the Beypazarı, Ayas and Güdül District towns of Ankara Province (Turkey). Econ Bot 58(4):705–720

Singab ANB, El-Beshbishy HA, Yonekawa M, Nomura T, Fukai T (2005) Hypoglycemic effect of Egyptian *Morus alba* root bark extract: effect on diabetes and lipid peroxidation of streptozotocin-induced diabetic rats. J Ethnopharmacol 100(3):333–338

Singh J, Cumming E, Manoharan G, Kalasz H, Adeghate E (2011) Medicinal chemistry of the anti-diabetic effects of *Momordica charantia*: active constituents and modes of actions. Open Med Chem J 5(Supp. 2):70

Soejarto DO, Farnsworth NR (1989) Tropical rain forests: potential source of new drugs? Perspect Biol Med 32:244–256

Soumyanath A (2005) Traditional medicines for modern times: antidiabetic plants. CRC Press, Boca Raton, FL

Subramoniam A (2016a) Anti-diabetes mellitus plants: active principles, mechanisms of action and sustainable utilization. CRC Press, Boca Raton, FL

Subramoniam A (2016b) Plants with anti-diabetes mellitus properties. CRC Press, Boca Raton, FL

Tabata M, Honda G, Sezik E (1988) A report on traditional medicine and medicinal plants in Turkey. Faculty of Pharmaceutical Sciences Kyoto University, Kyoto

Tekin S (2011) Ethnobotanic aspects of Üzümlü town (Erzincan-Turkey). Msc thesis. Erzincan University, Erzincan

Tetik F, Civelek S, Cakilcioglu U (2013) Traditional uses of some medicinal plants in Malatya (Turkey). J Ethnopharmacol 146:331–346

Tiwari R, Rana CS (2015) Phytomedicine for the diabetes: a traditional approach. Ann Phytomed 4(1):108–110

Tripathi AK, Bhoyar PK, Baheti JK, Biyanim DM, Khaliquem M, Kothmirem MS, Yogesh M, Amgaonkar YM, Bhanarkar AB (2011) Herbal antidiabetics-a review. Int J Res Pharm Sci 2(1):30–37

Trojan-Rodrigues M, Alves TLS, Soares GLG, Ritter MR (2012) Plants used as antidiabetics in popular medicine in Rio Grande do Sul, southern Brazil. J Ethnopharmacol 139(1):155–163

Tümen G, Sekendiz O (1990) Balıkesir ve merkez köylerinde halk ilacı olarak kullanılan bitkiler. VIII. Bitkisel İlaç Hammaddeleri Toplantısı, İstanbul, 19–21 Mayıs 1989, İstanbul Üniversitesi Yayınları No: 3733, Eczacılık Fakültesi Yayınları No: 64, İstanbul, pp 347–354

Tümen G, Selvi S (2011) Türkiye'de yaygın görülen tip II diyabet tedavisinde kullanılan tıbbi bitkiler üzerine etnobotanik bir araştırma. Merkez Efendi Tıp günleri, 16–20 Mayıs 2011. Denizli Belediyesi Kültür Yayınları, Denizli, pp 68–81

Türkan Ş, Malyer H, Aydın SÖ, Tümen G (2006) Ordu ili ve çevresinde yetişen bazı bitkilerin etnobotanik özellikleri. Süleyman Demirel Üniversitesi Fen Bilimleri Enstitüsü Dergisi 10(2):162–166

Turner IM (1997) A catalogue of the vascular plants of Malaya. Gardens' Bulletin Singapore 47:1–757

Tuzlacı E (2006) Şifa niyetine Türkiye'nin bitkisel halk ilaçları. Alfa Yayınları, Istanbul

Tuzlacı E, Aymaz P (2001) Turkish Folk Medicinal Plants, Part IV: Gönen (Balıkesir-Turkey). Fitoterapia 72:323–343

Tuzlacı E, Dogan A (2010) Turkish folk medicinal plants, IX Ovacık (Tunceli Turkey). Marmara Pharmaceut J 14:136–143

Tuzlacı E, Erol MK (1999) Turkish folk medicinal plants. Part II: Eğirdir (Isparta). Fitoterapia 70(6):593–610

Tuzlacı E, Şenkardeş I (2011) Turkish folk medicinal plants, X: Ürgüp (Nevşehir-Turkey). Marmara Pharmaceut J 15:58–68

Tuzlacı E, Tolon E (2000) Turkish folk medicinal plants, part III: Şile (İstanbul). Fitoterapia 71:673–685

Tuzlacı E, Alparslan İDF, Bulut G (2010) Turkish folk medicinal plants, VIII. Lalapaşa (Edirne) 14:47–52

Ugulu I, Baslar S, Yorek N, Dogan Y (2009) The investigation and quantitative ethnobotanical evaluation of medicinal plants used around Izmir province, Turkey. J Med Plant Res 3(5):345–367

Ugurlu E, Secmen O (2008) Medicinal plants popularly used in the villages of Yunt Mountain (Manisa-Turkey). Fitoterpia 79:126–131

Ullah M, Khan MU, Mahmood A, Malik RN, Hussain M, Wazir SM et al (2013) An ethnobotanical survey of indigenous medicinal plants in Wana district south Waziristan agency, Pakistan. J Ethnopharmacol 150(3):918–924

Unsal Ç, Vural H, Sarıyar G, Özbek B, Ötük G (2010) Traditional mecine in Bilecik province (Turkey) and antimicrobial activities selected species. Turk J Pharm Sci 7(2):139–150

Uysal I (2010) An overwiev of plant diversity of Kazdagi Forest National Park, Turkey. J Enviroment Biol 31:141–147

Uysal I, Onar S, Karabacak E, Çelik S (2010) Ethnobotanical aspects of Kapıdağ Peninsula (Turkey). Biol Divers Conserv 3:15–22

Uysal I, Gücel S, Tütenocakli T, Öztürk M (2012) Studies on the medicinal plants of Ayvacik-Çanakkale in Turkey. Pak J Bot 44(Supp. 1):239–244

Uzun E, Sariyar G, Adsersen A, Karakoc B, Ötük G, Oktayoglu E, Pirildar S (2004) Traditional medicine in Sakarya province (Turkey) and antimicrobial activities of selected species. J Ethnopharmacol 95(2–3):287–296

Vural H (2004) Bilecik ili bazı halk ilaçları üzerine farmakognozik araştırmalar. Msc thesis. Istanbul University, Istanbul

Vural G (2008) Ethnobotanical features some of the wild plants on the Honaz Mountain and its environment. Msc thesis. Afyon Kocatepe University, Afyon

Wadood A, Wadood N, Shah S (1989) Effects of *Acacia arabica* and *Caralluma edulis* on blood glucose levels of normal and alloxan diabetic rabbits. J Pak Med Assoc 39:208–212

Waheed A, Miana GA, Ahmad SI, Khan MA (2006) Clinical investigation of hypoglycemic effect of *Coriandrum sativum* in type-2 (NIDDM) diabetic patients. Pak J Pharmacol 23(1):7–11

WHO (1985) Diabetes Mellitus: Report of a WHO Study Group. WHO Technical Report Series 727. WHO, Geneva

Wong KM (1998) Patterns of plant endemism and rarity in Borneo and the Malay Peninsula. In: Peng C-I., Lowry II (eds) Rare, threatened, and endangered floras of Asia and the Pacific Rim. Academia Sinica Monograph Series No 16, pp 139–169

Wong TW, Choo CY, Ilyanie Y (2011) Antidiabetic plants Monographs-Outcome of scientific evaluation of antidiabetic plants. LAP LAMPERT Academic Publishing pp 1–472

Yapıcı İÜ, Hoşgören H, Saya Ö (2009) Ethnobotanical features of Kurtalan (Siirt) District. Dicle Üniversitesi Ziya Gökalp Eğitim Fakültesi Dergisi 12:191–196

Yaseen G, Ahmad M, Zafar M, Sultana S, Kayani S, Cetto AA, Shaheen S (2015) Traditional management of diabetes in Pakistan: Ethnobotanical investigation from traditional health practitioners. J Ethnopharmacol 174:91–117

Yazıcıoğlu A, Tuzlacı E (1996) Folk medicinal plants of Trabzon (Turkey). Fitoterapia 67(4):307–318

Yeşil Y, Akalın E (2009) Folk Medicinal Plants in Kürecik Area (Akçadağ Malatya-Turkey). Turk J Pharm Sci 6(3):207–220

Yeşilada E (2013) An overview of Turkish folk medicine; past and present. Curr Drug Deliv 10(1):92–95

Yeşilada E, Honda G, Sezik E, Tabata M, Fujita T, Tanaka T et al (1995) Traditional medicine in Turkey V: folk medicine in the inner Taurus Mountains. J Ethnopharmacol 46(3):133–152

Yeşılada E, Sezık E, Honda G, Takaishi Y, Takeda Y, Tanaka T (1999) Traditional medicine in Turkey IX: folk medicine in north-west Anatolia. J Ethnopharmacol 64(3):195–210

Yeşilyurt EB, Şimşek I, Akaydin G, Yeşilada E (2017) An ethnobotanical survey in selected districts of the Black Sea region (Turkey). Turk J Bot 41(1):47–62

Yıldırım B, Terzioğlu Ö, Özgökçe F, Türközü D (2008) Ethnobotanical and pharmacological uses of some plants in the districts of Karpuzalan and Adigüzel (Van-Turkey). J Anim Vet Adv 7(7):873–878

Yücel E, Tülükoğlu A (2000) Gediz (Kütahya) çevresinde halk ilacı olarak kullanılan bitkiler. Ekoloji Çevre ve Doğa Dergisi 36:12–14

Zada Khan MP, Ahmad M, Zafar M, Sultana S, Ali MI, Sun H (2015) Ethnomedicinal uses of edible wild fruits (EWFs) in Swat Valley, Northern Pakistan. J Ethnopharmacol 173:191–203

Zammurad S, Qaiser M (2011) Pakistan'daki Nagar Parkar-Sindh yöresi bitkilerin geleneksel tedavide kullanım alanları. I. Merkez Efendi Tıp Günleri. Denizli Belediyesi Kültür Yayınları, Denizli, pp 63–67

Zia T, Hasnain SN, Hasan SK (2001) Evaluation of the oral hypoglycaemic effect of *Trigonella foenum-graecum* L. (Methi) in normal mice. J Ethnopharmacol 75(2):191–195

Zia-Ul-Haq M, Ćavar S, Qayum M, Imran I, Feo VD (2011) Compositional studies: antioxidant and antidiabetic activities of *Capparis decidua* (Forsk.) Edgew. Int J Mol Sci 12(12):8846–8861

Zia-Ul-Haq M, Raza Shah M, Qayum M, Ercisli S (2012a) Biological screening of selected flora of Pakistan. Biol Res 45:375–379

Zia-Ul-Haq M, Shahid SA, Muhammed S, Qayum M, Khan I, Ahmad S (2012b) Antimalarial, anti-emetic and antidiabetic potential of *Grewia asiatica* L. leaves. J Med Plant Res 6:3087–3092

Antidiabetic Plants of Pakistan

Mubashrah Munir and Rahmatullah Qureshi

Introduction

Diabetes mellitus (DM) is a chronic illness, and every year, millions of people suffer from this disease (Ezuruike and Prieto 2014). This is a metabolic disorder in which blood sugar level remains high over a prolonged period triggered by the inequity among insulin supply and demand (Andrade-Cetto and Heinrich 2005). DM is classified into many types among which type 1 and 2 are most common (Group 2007). Diabetic patients suffer from hyperglycemia that damages many systems of their body (Mukherjee et al. 2006). DM has caused approximately 1.5 million mortalities across the globe, 80% of which are in the low- and middle-income countries. DM would rank seventh among the causes of the deaths by 2030 (WHO 2016).

The intensity of this disease is very rapidly increasing, and according to a recent survey, 371 million people were recorded in 2012 (Song et al. 2014), which may reach up to 552 million by 2030 (Oputa and Chinenye 2012). This disease is the most prevalent (80%) in underdeveloped and developing countries (Oputa and Chinenye 2012) that caused about 3.2 million deaths per year. This situation necessitates in developing useful antidiabetic drugs. Many drugs are available to control blood glucose level which poses serious side effects (Currie et al. 2009; Mukherjee et al. 2006), while plants are good source of antidiabetic medicine with least side effects (Riya et al. 2015; Venkatesh et al. 2003) and low cost (Sunmonu and Afolayan 2013). Therefore, it is worthwhile to find out traditional medicinal plants for control of diabetes (Suba et al. 2004). Increasing intensity of side effects and reduced efficacy of antidiabetic allopathic drugs caught the attention of patients, physicians, and researchers toward searching for an alternative source of plant-based medicament.

M. Munir · R. Qureshi (✉)
Department of Botany, Pir Mehr Ali Shah Arid Agriculture University, Rawalpindi, Pakistan
e-mail: rahmatullahq@uaar.edu.pk

M. Ozturk, K. R. Hakeem (eds.), *Plant and Human Health, Volume 1*,
https://doi.org/10.1007/978-3-319-93997-1_12

There it is the need of hours to investigate the potential use of plants and herbal therapies used in diabetes control. Presently, medicinal plants are targeted in searching novel antidiabetic agents due to their efficacy in human clinical trials and the less adverse effects as well as easy availability (Suba et al. 2004).

The prevalence of DM is increasing in almost every country. Pakistan occupies the seventh position in the list of Middle East and North African countries with large population of diabetics (Whiting et al. 2011). The country has a rich heritage in using a wide range of plants and their compounds in treating diabetes. Worldwide, there is reawakening in the use of herbs for treating such noxious diseases. Several native plants have been evaluated for their effectiveness in diabetes. Regardless of remarkable advancement in medicine during the recent past, still there is no cure and effective prevention/treatment of utmost importance to prevent a potential rise in the disease. Plants possess commendable drug sources, and many of the existing drugs have been obtained from them. Today, herbal products are believed to be used safely compared to the synthetics (Haridasan 2001; Jain 1989). Many folk medicinal and scientific investigations dealing with the antidiabetic plants of Pakistan have been published during the last few decades (Ahmad et al. 2009; Yaseen et al. 2015). However, no comprehensive evidence-based review is available. Therefore this review was aimed to document the antidiabetic plants, their effective parts and folk recipes, and their distribution across the country.

Methodology

Various literature resources such as online bibliographic databases and libraries (MEDLINE, CAM-PubMed, HealthSTAR and the Cochrane Library) were searched from 1989 to 2015 for the information pertinent to antidiabetic plants, which are commonly used by the traditional health practitioners (THPs) of Pakistan. Moreover, semi-structured ethnobotanical questionnaires also gathered traditional information from herbal practitioners. For the validation of information, we contacted experts in the field to validate the information as well as handsearched references of key articles.

Data analysis

Data obtained on antidiabetic plants from various sources were analyzed, and the antidiabetic plant families of medicinal plants were ranked based on number of species contributed. Percentage (%age) of mode of administration, usage of plant parts, and life forms was ranked. Frequency of citation (FC) of each plant was computed based on the number of times a plant is reported in the literature as antidiabetic remedy.

Results and Discussion

Antidiabetic Plants and Their Families' Contribution

A total of 209 antidiabetic plant species belonging to 74 families were reported, of which 182 species (Table 1) are used individually and 27 are used in herbal mixtures (Table 2). Moreover, six herbal mixtures are new to science for the treatment of DM (Table 2). Three species, viz., *Astragalus gummifer*, *Lactuca sativa*, and *Santalum album*, are first time recorded having antidiabetic potential in the herbal mixtures. Besides, 38 species such as *Adiantum incisum*, *Alnus nitida*, *Andrachne cordifolia*, *Artemisia roxburghiana*, *Berberis brandisiana*, *Capparis cartilaginea*, *Caryopteris odorata*, *Centaurea iberica*, *Chrysanthemum indicum*, *Cleome scaposa*, *Convolvulus prostratus*, *Cuscuta campestris*, *Euphorbia helioscopia*, *Farsetia hamiltonii*, *Ferula narthex*, *Ficus virens*, *Fragaria indica*, *Gentianodes tianschanica*, *Hedera nepalensis*, *Kickxia ramosissima*, *Limeum indicum*, *Malva neglecta*, *Mentha longifolia*, *Onosma echioides*, *Opuntia monacantha*, *Oxalis corniculata*, *Papaver somniferum*, *Prunus amygdalus*, *P. persica*, *Pyrus malus*, *Salix babylonica*, *Stellaria media*, *Tamarix aphylla*, *Tanacetum artemisioides*, *Taraxacum officinale*, *Tylophora hirsuta*, *Ziziphus oxyphylla*, and *Z. spinosa* are ethnobotanically used to treat diabetes, which are required to be scientifically evaluated for in vitro and in vivo antidiabetic activity. Some of the species like *Alisma plantago-aquatica*, *Caralluma edulis*, *Cedrela serrata*, *Fagonia cretica*, *Juniperus excelsa*, *Launaea procumbus*, *Salvia coccinea*, *Sonchus asper*, and *Trianthema triquetra* are sporadically known for the scientific validation of hypoglycemic activity.

The most dominant family was Fabaceae which contributed 18 species (8.6%), followed by Asteraceae (16 spp., 7.66%); Moraceae (11 spp., 5.26%); Lamiaceae (10 spp., 4.78%); Cucurbitaceae (9 spp., 4.30%); Rosaceae (8 spp., 3.82%); Apocynaceae, Solanaceae, and Apiaceae (7 spp., 3.35%); Rhamnaceae (6 spp., 2.87%); and Poaceae (5 spp., 2.39%), while Brassicaceae, Capparidaceae, Euphorbiaceae, and Meliaceae were represented by four species each (1.19% each). The rest of the 59 families contributed species from 3 to 1 (Fig. 1).

Over 200 species are used in the treatment of DM worldwide as traditional medicine (Simpson and Morris 2014). The present endeavor reported 209 plant species from Pakistan (Tables 1 and 2) by adding 89 species to the antidiabetic plants of Pakistan. Previously, Yaseen et al. (2015) reported 120 species from Pakistan. They indicated Moraceae as the leading family which contributed 11 species; however, present investigation reported Fabaceae as the most dominant family. Marles and Farnsworth (1995) and Patel et al. (2012b) also reported that Fabaceae contributes the largest taxa for antidiabetic activity.

Table 1 Single herb recipes used for treatment of diabetes mellitus in Pakistan

Sr. #	Botanical name (local name)/ family/habit	Parts used	Recipe	Phytochemicals reported	Pharmacological activity	Ethnobotanical studies
1.	*Justicia adhatoda* L. (Baikar)/ Acanthaceae/herb	Fruits, fresh leaves, vegetative buds	Vegetative buds are eaten fresh. One cup of juice of freshly ground leaves and fruit is used	Vasicoline, vasicolinone, vasicinone, vasicine, adhatodine, anisotine, deoxyvasicine, adhatodine, adhatonine, adhavasinon E, anisotine, hydroxypeg anine glucosides, kaempferol, essential oils, resins, gum, vitamin "C," phenols, tannins, anthraquinone, saponins, betaines, and steroids (Jha et al. 2012; Yaseen et al. 2015)	Antidiabetic (Gulfraz et al. 2011), antimicrobial (Pa and Mathew 2012)	Ahmad et al. (2004, 2009), Ahmed et al. (2007), Shah et al. (2013a), Saqib et al. (2014), Yaseen et al. (2015)
2.	*Viburnum foetens* Dcne./Ghar Meva/ Adoxaceae/shrub	Leaves	Leaf extract is used	Flavanoids, coumarins, and tannins (Bibi et al. 2010)	Insulin secretagogue activity (Hussain et al. 2004; Bibi et al. 2010)	Hussain et al. (2004)
3.	*Limeum indicum* Stocks ex T. Anderson/Lonri/ Aizoaceae/herb	Whole plant	Plant material is boiled for 5 min and infusion is used	–	–	Ahmad et al. (2014a)
4.	*Trianthema triquetra* Rottl. & Willd./Choti Ulwaiti/Aizoaceae/ herb	Whole plant	Powder, infusion, or decoction is used	–	Antioxidant (Chitra and Nithyanandhi 2007)	Ahmad et al. (2014a)
5.	*Alisma plantago-aquatica* L./Jabai/ Alismataceae/herb	Leaves	10 g powder of dried leaves is administered three times a day	Alisolide, alisol O, alisol P, and triterpene (Zhao et al. 2008)	Radical scavenging activity (Kim et al. 2007)	Yaseen et al. (2015)

6.	*Aloe vera* (L.) Burm.f./Kunwar Gandal/Alliaceae/ herb	Leaves	Leaf pulp is used with sugar	Lophenol, 24-methyl-lophenol, 24-ethyl-lophenol, cycloartanol, and 24-methylene-cycloartanol (Tanaka et al. 2006)	In vivo hypoglycemic and antioxidant activity (Rajasekaran et al. 2005) and antidiabetic (Tanaka et al. 2006)	Sabeen and Ahmad (2009), Abbasi et al. (2010a), Jafri et al. (2011), Shah et al. (2013a)
7.	*Achyranthes aspera* L./Ubat Kandri/ Amaranthaceae/ herb	Seeds	Seed powder is used. Also tested on diabetic rabbit to decrease blood glucose	Saponins, oleanolic acid, dihydroxy ketones, alkaloids (Srivastav et al. 2011)	Hypoglycemic effects on rabbits (Akhtar and Iqbal 1991), free radical scavenging and antihyperglycemic activity (Zohura Talukder et al. 2012)	Akhtar (1992), Qureshi and Bhatti (2009)
8.	*Allium cepa* L./ Piaz/ Amaryllidaceae/ herb	Bulb, gum, root, bark	Fresh juice of bulb is taken or used in cooked form. Pieces of bulb are soaked in water overnight and one cup is drunk	*S*-methyl cysteine sulfoxide (Kumari and Augusti 2002), isoallin (Corzo-Martínez et al. 2007)	In vivo antioxidant activity in rats (Kumari and Augusti 2002) and antihyperglycemic activity (El-Demerdash et al. 2005; Eldin et al. 2010)	Ahmad et al. (2004, 2009), Shah and Khan (2006), Abbasi et al. (2010a), Yaseen et al. (2015)
9.	*Allium sativum* L./ Lahsan/ Amaryllidaceae/ herb	Rhizome, leaf, cloves	Blub is used in fresh form or cooked with vegetable and used daily twice. Fresh leaves are also eaten	Alliin, cysteine, allylcysteine, cycloalliin, sulfoxides, glutamyl peptides, thiosulfinates (Sendl 1995), sulfur containing amino (Shori and Baba 2014)	Antidiabetic activity in rats (Eidi et al. 2006), improved glucose tolerance and renal function in rats (Liu et al. 2006)	Ahmad et al. (2004, 2009), Fatima et al. (2005), Shah and Khan (2006), Hussain et al. (2009), Marwat et al. (2011), Begum et al. (2015), Yaseen et al. (2015)

(continued)

Table 1 (continued)

Sr. #	Botanical name (local name)/ family/habit	Parts used	Recipe	Phytochemicals reported	Pharmacological activity	Ethnobotanical studies
10.	*Mangifera indica* L./Aam/ Anacardiaceae/tree	Leaves, seeds	Water extract of leaves is taken. Dried or fresh seed is consumed daily	Polyphenols, mangiferin (Núñez Sellés et al. 2002; Rastogi et al. 2007)	Hypoglycemic activity in mice (Aderibigbe et al. 2001), antihyperglycemic in rats (Bhowmik et al. 2009; Muruganandan et al. 2005), antidiabetic and antioxidant (Shah et al. 2010)	Khattak and Khattak (2011), Marwat et al. (2011)
11.	*Polyalthia longifolia* (Sonnerat) Thwait/ Ulta Ashok/ Annonaceae/tree	Leaves, root, stem	Plant parts are boiled in half liter of water for 10 min. One cup of this tea is used thrice a day	Steroids, alkaloids, diterpenoids, carbohydrates, amino acids, essential oil, phenolics, and flavonoids (Ghosh et al. 2008)	In vivo hypoglycemic and antihyperglycemic and antioxidant (Nair et al. 2007; Ghosh et al. 2010; Sivashanmugam and Chatterjee 2013)	Hussain et al. (2010d), Yaseen et al. (2015)
12.	*Coriandrum sativum* L./Dahnia, Kashneez Khushk/ Apiaceae/herb	Seeds, essential oil	Oil is taken orally, or seeds are chewed daily	Mostly alcohols and aldehyde. 2*E*-decenal, decanal, 2*E*-decen-1-ol, and *n*-decanol (Matasyoh et al. 2009), linalool, geranyl acetate, and γ-terpinene (Bhuiyan et al. 2009)	In vivo hypoglycemic (Aissaoui et al. 2011), hypoglycemic, hypolipidemic, and antioxidant (Sreelatha and Inbavalli 2012)	Waheed et al. (2006), Marwat et al. (2011)
13.	*Foeniculum vulgare* Mill./Sonf/ Apiaceae/herb	Leaves, seeds	Leaves and seeds are boiled in water and extract is used	Methyl chavicol, fenchone, anethole, and fencho (García-Jiménez et al. 2000), linoleic acid, palmitic acid, and oleic acid (Singh et al. 2006a), phenol and phenolic glycosides (Rather et al. 2012)	In vivo hypoglycemic (El-Soud et al. 2011; Badgujar et al. 2014), antidiabetic (Rather et al. 2012)	Shah and Khan (2006)
14.	*Ferula asafoetida* L./Hing/Apiaceae/ herb	Resin	Ground resin is used	Resin, gum essential oil, phenolic, and tannins (Iranshahy and Iranshahi 2011; Iranshahi and Alizadeh 2012)	In vitro hypoglycemic and hyperinsulinemic (Helal et al. 2005), in vivo antidiabetic (Abu-Zaiton 2010), antihyperglycemic (Iranshahi and Alizadeh 2012)	Zia-Ul-Haq et al. (2012a)

15.	*Ferula narthex* Boiss./Jangli sonf/ Apiaceae/herb	Whole plant	1–10 g of plant powder is consumed three times a day	Luteolin and pinene (Shinwari and Gilani 2003)	–	Yaseen et al. (2015)
16.	*Catharanthus roseus* (L.) G. Don./Sada Bahar/ Apocynaceae/shrub	Leaves	Juice of fresh leaves is used after each meal	Flavanoids and alkaloids (Islam et al. 2009), monoterpenoid indole alkaloid (Jaleel et al. 2009)	In vivo antidiabetic and hypolipidemic (Nammi et al. 2003; Islam et al. 2009; Rasineni et al. 2010; Zhang et al. 2016)	Ahmad et al. (2004), Arshad et al. (2011), Kanwal et al. (2011a), Yaseen et al. (2015)
17.	*Caralluma edulis* (L.) Benth ex Hook. F./Chugain/ Apocynaceae/herb	Aerial parts	Aerial parts are cooked and used. Flowers are placed in water overnight and their extract is used early in the morning. Powder of dried whole plant is also used once a day	Fatty acid, hydrocarbon, and sterols (Rizwani et al. 1993a)	Radical scavenging activity (Ansari et al. 2005)	Ahmad et al. (2004, 2009), Mahmood et al. (2010), Shah et al. (2013a), Yaseen et al. (2015)
18.	*Caralluma tuberculata* N. E. Br./Pamanay or Pawoona/ Apocynaceae/herb	Whole plant, roots, fruit	Used as vegetable in cooked form. Young plant extract is also used	Steroids, terpenoids, reducing sugars, tannins, betacyanin, and amino acid (Rauf et al. 2013), pregnane glycosides (Rizwani et al. 1993b)	Antidiabetic effect in rats (Abdel-Sattar et al. 2013), antihyperglycemic and antilipidemic (Abdel-Sattar et al. 2011)	Mahmood et al. (2010), Shah et al. (2013a), Adnan et al. (2014b)
19.	*Gymnema sylvestre* R. Br./Gurmar buti/ Apocynaceae/herb	Leaves	Leaves are boiled in water and infusion is taken twice per day	Gymnemic acid (Saneja et al. 2010), gymnemasaponins and gurmarin (Porchezhian and Dobriyal 2003)	Enhance endogenous insulin (Shanmugasundaram et al. 1990), increase insulin release in vitro (Persaud et al. 1999)	Fatima et al. (2005)

(continued)

Table 1 (continued)

Sr. #	Botanical name (local name)/ family/habit	Parts used	Recipe	Phytochemicals reported	Pharmacological activity	Ethnobotanical studies
20.	*Nerium oleander* L./Kaner/ Apocynaceae/shrub	Leaf, latex, seed	Tea of leaves is taken. Powder of seed and latex is also taken twice a day	Oleanderol (Siddiqui et al. 1988), neriine, digitoxigenin, amorphane, 1.8-cineole, α-pinene, calarene, limonene, B-phellandrene, terpinene-4-ol, sabinene, isoledene, 3-carene, humulene, β-pinene, and cymen-8-ol (Derwich et al. 2010)	Antidiabetic in rats along with glimepiride (Yassin and Mwafy 2007), in vivo and in vitro improved fat and glucose metabolism (Bas et al. 2012)	Shah and Khan (2006)
21.	*Rhazya stricta* Decne./Winra, Verian/ Apocynaceae/herb	Leaves, whole plant	One spoon of powder of dried leaves is used after each meal. Sometimes aerial parts are boiled for 30 min and decoction is taken. Also used with *Withania somnifera*	Alkaloids, flavonoids, glycosides, peptides, triterpenes, fatty esters (Ullah 2012)	In vivo antidiabetic [−] (Wasfi et al. 1994; Ali 1997)	Ahmad (2006), Hussain et al. (2010c), Qureshi (2012), Shah et al. (2013a), Yaseen et al. (2015)
22.	*Tylophora hirsuta* (Wall.) Wight/Akri/ Apocynaceae/shrub	Branches and leaves	Fresh juice is used three times a day	Alkaloids, flavonoids, tannins, saponins, glycosides, terpenoids, sterols, and carbohydrates (Ali 2013), phenanthroindolizidine alkaloids (Bhutani et al. 1984)	–	Yaseen et al. (2015)
23.	*Monstera deliciosa* Liebm./Araceae/ climber	Fruit pulp	Paste of fruit is used three times a day	Tannins, steroids, flavonoids, alkaloids, and saponins (Rao et al. 2015)	In vivo insulin secretagogue activity (Lim 2012), in vivo antihyperglycemic and antihyperlipidemic (Abo-Ellil 2014)	Hussain et al. (2004)

24.	*Pistia stratiotes* L./ Jalkumbhi/ Araceae/herb	Leaves and stem	Extract of leaves and stem is used	Alkaloids, glycosides, flavonoids, and phytosterols (Khan et al. 2014a)	Antidiabetic and diuretic in rats (Tripathi 2011), antidiabetic activity in rats (Khan et al. 2014a)	Khan et al. (2014a)
25.	*Hedera helix* L./ Payo Zelani, Bilari/ Araliaceae/climber	Leaves, bulb	Half cup of extract of 5–10 leaves soaked overnight is used	Triterpene and steroid saponins (Facino et al. 1995; Bedir et al. 2000), falcarinol and didehydrofalcarinol (Hausen et al. 1987), alkaloids, saponins, terpenoids, and tannins (Uddin et al. 2011)	Hypoglycemic activity in rats (Ibrar et al. 2004)	Ibrar et al. (2004), Alam et al. (2011), Saqib et al. (2014), Yaseen et al. (2015)
26.	*Hedera nepalensis* K. Koch/Arbumbal, Phalol, Bilari, Albomor, Zailai/ Araliaceae/climber	Leaves	Leaves (5–10) are soaked whole night and half cup of extract is used before meal thrice a day	Terpenoids, flavonoids, alkaloids, steroids, tannins, glycosides and saponins (Kanwal et al. 2011b)	–	Shah and Khan (2006), Sabeen and Ahmad (2009), Alam et al. (2011), Awan et al. (2011)
27.	*Calotropis procera* (Aiton.) W. T. Aiton/Akk/ Asclepiadaceae/ shrub	Flowers	Flowers are boiled for 30 min and half cup of extract is used before meal thrice a day	Alkaloids, flavonoids, tannins, saponins, cardiac glycosides, balsams and volatile oil, and steroids (Mainasara et al. 2012)	In vivo antihyperglycemic and antioxidant activity [+] (Roy et al. 2005; Bhaskar and Ajay 2009; Neto et al. 2013)	Yaseen et al. (2015)
28.	*Artemisia absinthium* L./ Kakamush, Afsanteen/ Asteraceae/herb	Whole plant, flower, and leaves	Half cup of juice of whole plant is used early in the morning. Extract of flower and leaves is used after boiling in water three times a day	Hydroxybenzoic acids, hydroxycinnamic acids, flavonols, salicylic acid, myricetin, caffeic acid, gallic acid and ferulic acid (Lee et al. 2013), essential oil has chrysanthenyl acetate (Wani et al. 2014)	In vivo antidiabetic [+] (Daradka et al. 2014), in vitro antioxidant [+] (Singh et al. 2012; Wani et al. 2014)	Khan and Khatoon (2008), Yaseen et al. (2015)

(continued)

Table 1 (continued)

Sr. #	Botanical name (local name)/ family/habit	Parts used	Recipe	Phytochemicals reported	Pharmacological activity	Ethnobotanical studies
29.	*Artemisia herba-alba* Asso./ Mushki afsantheen/ Asteraceae/herb	Whole plant	Plant powder is used	Sesquiterpene lactones, flavonoid, phenols, wax (Mohamed et al. 2010), camphor, *trans*-pinocarveol, chrysanthenone and β-thujone, and monoterpenoids (Dob and Benabdelkader 2006)	In vivo antidiabetic [+] (Iriadam et al. 2006; Taştekin et al. 2006)	Hayat et al. (2009)
30.	*Artemisia indica* Willd./Nagdona/ Asteraceae/herb	Aerial parts	Aerial parts extract	Artemisia ketone, germacrene B, borneol, and *cis*-chrysanthenyl acetate (Rashid et al. 2013b), eucalyptol, camphor, borneol, terpineol, cubebene, caryophyllene, caryophyllene oxide, and germacrene D (Wu et al. 2008)	Antidiabetic and antihyperlipidemic in rats [+] (Ahmad et al. 2014b)	Hayat et al. (2009), Ahmad et al. (2014b)
31.	*Artemisia roxburghiana* Wall. Ex Besser/ Garrotra/ Asteraceae/herb	Leaves	Extract of leaves is used	A-thujone and β-thujone (Mathela et al. 1994), thujane, camphane and 1,8-cineole, monoterpenes, caryophyllene, and cadinene (Thakur et al. 1990; Bicchi et al. 1998)	–	Hussain et al. (2004)
32.	*Arctium lappa* Linn./Chero/ Asteraceae/herb	Leaves, seeds, and roots	Half cup of juice of leaves and roots is recommended. A cup of decoction is taken after every meal	Neoarctin B (VI), arctigenin (II), arctiin (III), matairesinol (IV), and lappaol F (V) (Wang and Yang 1992), fructan (Kardošová et al. 2003), tannin, arctigenin, arctiin, beta-eudesmol, caffeic acid, chlorogenic acid, inulin, trachelogenin 4, sitosterol-beta-D-glucopyranoside, lappaol, and diarectigenin (Chan et al. 2011)	In vivo antidiabetic [+] (Cao et al. 2012), in vivo and in vitro antioxidant activity [+] (Liu et al. 2014)	Khan and Khatoon (2008), Yaseen et al. (2015)

33.	*Centaurea iberica* Trevir. & Spreng./ Kandiyara/ Asteraceae/herb	Leaves	Extract of leaves is used	–	–	Hussain et al. (2004)
34.	*Chrysanthemum indicum* L./ Gul-e-daudi/ Asteraceae/herb	Flower	Flower extract is prepared by soaking in water for few hours and administered	8-Cineole, camphor, borneol, and bornyl acetate (Shunying et al. 2005), volatiles, flavonoids, (quercitrin, myricetin, and luteolin-7-glucoside), and flavonoid glycosides (Wu et al. 2010)	–	Sabeen and Ahmad (2009)
35.	*Cichorium intybus* L./Kasni/ Asteraceae/herb	Roots	Before meal powder of roots is taken twice a day	Chicoric acid (Benalla et al. 2010), saccharides, methoxycoumarin cichorine, flavonoids, essential oils, anthocyanins, octane, n-nonadecane, pentadecanone, and hexadecane (Street et al. 2013)	In vivo and in vitro antihyperglycemic [+] (Azay-Milhau et al. 2013; Street et al. 2013), antioxidant [+] (Samarghandian et al. 2013)	Ahmad et al. (2004, 2009)
36.	*Launaea procumbus* Roxb/ Bhatter/Asteraceae/ herb	Whole plant, leaves	Water extract is consumed or leaves are cooked as vegetable and taken as per need	Flavonoids and phenolics (myricetin, catechin, vitexin, orientin, hyperoside, and rutin) (Khan et al. 2012b), flavanols, flavanones, flavones, and isoflavones (Mishra et al. 2012)	In vitro antioxidant [+]	Hussain et al. (2010d), Yaseen et al. (2015)
37.	*Parthenium hysterophorus* L./ Thandan, Partha/ Asteraceae/herb	Aerial parts	Powder of plant is used twice or three times a day	Saponins and hysterolactone (Shah et al. 2009), histamine, saponin, glucosides, and triterpene (sesquiterpene) (Khushk et al. 2010), flavonoids, tannins, phenolic compounds, steroids, terpenoids, and alkaloids (Sinha and Paul 2014)	In vivo hypoglycemic [−] (Patel et al. 2008), in vitro antioxidant [+] (Sinha and Paul 2014)	Mahmood et al. (2012), Yaseen et al. (2015)

(continued)

Table 1 (continued)

Sr. #	Botanical name (local name)/ family/habit	Parts used	Recipe	Phytochemicals reported	Pharmacological activity	Ethnobotanical studies
38.	*Sonchus asper* (L.) Hill/Hund/ Asteraceae/herb	Leaves and roots	Cooked as vegetable	Alkaloids, saponins, flavonoids, phenols, saponins, and tannins (Hussain et al. 2010a), tannic acid, quercetin, and catechin (Khan et al. 2012c)	In vitro antioxidant [+] (Xu et al. 2011)	Sabeen and Ahmad (2009)
39.	*Stevia rebaudiana* (Bertoni) Bertoni/ Methi booti/ Asteraceae/shrub	Leaves	Fresh leaves or their extract is recommended twice a day	Diterpene glycosides (Brandle et al. 1998), stevioside (Abou-Arab et al. 2010)	Recovers islet β-cell dysfunction [+] (Kosta et al. 2015), antidiabetic activity [–] (Kujur et al. 2010), antihyperglycemic, insulinotropic, and glucagonostatic actions [+] (Jeppesen et al. 2002; Gregersen et al. 2004)	Khan et al. (2008)
40.	*Tanacetum artemisioides* Schultz-Bip.ex Hook.f.	Whole plant	Tea is prepared by boiling plant for 15 min and once cup is taken three times a day	Ceramide, tanacetamide D, 5-demethylnobiletin, and 5-hydroxy-3,6,7,8,3′,4′-hexamethoxyflavone (Hussain et al. 2010b)	–	Khan and Khatoon (2008), Yaseen et al. (2015)
41.	*Taraxacum officinale* Weber./ Paloyo Zoon/ Asteraceae/herb	Leaves, roots	Used as vegetable	Luteolin 7-glucoside and two luteolin 7-diglucosides, hydroxycinnamic acids, chicoric acid, monocaffeyltartaric acid, and chlorogenic acid (Williams et al. 1996)	–	Ahmad et al. (2004, 2009), Hussain et al. (2004), Yaseen et al. (2015)
42.	*Vernonia anthelmintica* Willd./Kali Zeri/ Asteraceae/herb	Seeds	Powder of seed is taken twice a day	Steroid: vernoanthelsteron (Hua et al. 2012), oil, resin, myristic acid, palmitic acid, stearic acid, oleic acid, linoleic acid, vernolic acid, and methyl vernolate (Manvar and Desai 2012)	Antidiabetic and antihyperlipidemic [+] (Fatima et al. 2010; Manvar and Desai 2012)	Sabeen and Ahmad (2009)

43.	*Berberis brandisiana* Ahrendt/Sumbloo/ Berberidaceae/ shrub	Root and stem	Powder of roots and stem is mixed with butter oil and taken with milk three times a day	Berberine, berbamine, palmitine, jatrorrhizine, and isotetrandrine (Srivastava et al. 2015)	–	Yaseen et al. (2015)
44.	*Berberis lycium* Rnoyl./Sumblu/ Berberidaceae/ shrub	Root, bark, whole plant, fruit	Roots roasted powder is consumed with milk. This powder is also mixed with butter and used with milk. Fruit and leaves decoction is used. Antidiabetic potential of root bark was tested on rabbits	Alkaloids, cardioactive glycosides, saponins, tannins, anthocyanins, vitamins, carbohydrates, proteins, lipids, fiber content, β-carotene, cellulose, phytic acid, and phytate phosphorous (Shabbir et al. 2012), alkaloids: berberine, tannin, flavonoids, phenols, terpenoids, fat, and resin (Gupta et al. 2015)	Antihyperglycemic and antilipidemic (Ali et al. 2015b), hypoglycemic [+] (Ahmad and Alamgeer 2009; Gulfraz et al. 2011)	Husain et al. (2008), Ahmad and Alamgeer (2009), Sabeen and Ahmad (2009), Yaseen et al. (2015)
45.	*Alnus nitida* Endl./ Gheray/Betulaceae/ tree	Leaves	Leaves are soaked overnight and half cup is taken before breakfast	–	–	Yaseen et al. (2015)
46.	*Bombax ceiba* Linn./Sambal/ Bombacaceae/tree	Leaves, whole plant	Plant extract is prepared and used daily	Shamimin (Saleem et al. 1999), triterpenoid (Bhavsar and Talele 2013), mangiferin, stigma-5-en-3-*O*-glucoside, and amyrin (Faizi et al. 2012), quercetin (Verma et al. 2014)	In vivo hypoglycemic [+] (Saleem et al. 1999; Bhavsar and Talele 2013)	Shah and Khan (2006)

(continued)

Table 1 (continued)

Sr. #	Botanical name (local name)/ family/habit	Parts used	Recipe	Phytochemicals reported	Pharmacological activity	Ethnobotanical studies
47.	*Onosma echioides* L./Rattanjot// Boraginaceae/herb	Roots	Roots are soaked in water overnight and water extract is taken in the morning	Alkannin, shikonin, and naphthoquinone derivatives (Papageorgiou et al. 1999, 2006; Sagratini et al. 2008)	–	Akhtar (1992)
48.	*Brassica olerecea* var. *botrytis*/Phul gobi/Brassicaceae/ herb	Flower	Used as vegetable	Saponins, tannins, triterpenes, alkaloids, and flavonoids (Asadujjaman et al. 2011)	In vivo antihyperglycemic and antioxidant [+] (Kataya and Hamza 2008; Asadujjaman et al. 2011; Haque et al. 2013), hypoglycemic and hypolipidemic [+] (Assad et al. 2014)	Khan et al. (2013a, 2013b)
49.	*Eruca sativa* Lamk/ Tara Mera/ Brassicaceae/herb	Whole plant	Plant is cooked as vegetable and used three times a day	Carotenoids, vitamin C, fibers, flavonoids, and glucosinolates (Barillari et al. 2005a), 4-methylthiobutylisothiocyanate (60.13%) and 5-methylthiopentanonitrile (Miyazawa et al. 2002), erucin (Melchini and Traka 2010)	Antidiabetic [+] (El-Missiry and El Gindy 2000; Melchini and Traka 2010)	Sabeen and Ahmad (2009)
50.	*Raphanus sativus* L./Mooli/ Brassicaceae/herb	Root	Two to three fresh roots are boiled in 1 L water for 10–15 min. Take one cup of this water thrice a day	Glucosinolates (Barillari et al. 2005b; Jing et al. 2012), sinapinic acid esters and flavonoids (Takaya et al. 2003)	In vivo hypoglycemic [+] (Taniguchi et al. 2006; Shukla et al. 2011), antioxidants [+] (Takaya et al. 2003; Papi et al. 2008)	Yaseen et al. (2015)

51.	*Farsetia hamiltonii* Royle/Fareed buti or Lathia/ Brassicaceae/shrub	Whole plant	Plant powder or decoction is used	Flavonoids (Shahat et al. 2005)	–	Ahmad et al. (2014a), Hayat et al. (2014)
52.	*Opuntia dillenii* (Ker Gawler) Haworth/Chattar Thoar, Zaqoom/ Cactaceae/herb	Fruit, latex	Fruit juice is taken or fruit paste is used three times per day. Shade-dried latex is taken two times a day	Polysaccharide (Zhao et al. 2011), tannins, saponins, and mucilages (Nougbodé et al. 2013)	In vivo antihyperglycemic [+] (Zhao et al. 2011), in vivo antidiabetic [+] (Gao et al. 2015)	Mahmood et al. (2011), Yaseen et al. (2015)
53.	*Opuntia monacantha* Haworth/Thooer/ Cactaceae/herb	Fruit	Dried fruit three doses are administered three times per day	Alkaloids, tannins, saponins, flavonoids, and polysaccharides (Hussain et al. 2010a)	–	Yaseen et al. (2015)
54.	*Capparis cartilaginea* Decne./Kirip/ Capparidaceae/ shrub	Leaves	Paste of leaves is used three times. Powder of dried leaves is also consumed.	Isothiocyanates (Hammed et al. 2007)	–	Yaseen et al. (2015)
55.	*Capparis decidua* Pax/Kair/ Capparidaceae/ shrub	Seeds and fruits	Plant material is soaked in water overnight and extract is used	Carbohydrates, lipids and proteins, tocopherols, and glucosinolates (Zia-Ul-Haq et al. 2011), tannin, flavonoid, alkaloid, phenol, and steroid (Shad et al. 2014)	Hypoglycemic [+] (Rathee et al. 2010), in vivo antidiabetic [+] (Sharma et al. 2010)	Zia-Ul-Haq et al. (2011), Yaseen et al. (2015)

(continued)

Table 1 (continued)

Sr. #	Botanical name (local name)/ family/habit	Parts used	Recipe	Phytochemicals reported	Pharmacological activity	Ethnobotanical studies
56.	*Capparis spinosa* L./Kavir, Karein/ Capparidaceae/ shrub	Leaves	Leaves are boiled in water when half of it is evaporated and water extract is used two times a day	Β-sitosterylglucoside-6-octadecanoate, −methyl-2-butenyl-β-glucoside, sorhamnitine-3-*O*-rutinoside, 1-tetradecanol, *p*-hydroxybenzaldehyde, 6,10,14-trimethyl-2-pentadecanone, ursolic acid, glycerol monotetracostanoate, 4-coumaric acid, nicotinamide, methyl hexadecanoate, −sitosterol, −sitosteryl glucoside, cadabicine, octadecanoic acid, rutin, and stachydrine (Khanfar et al. 2003), methyl isothiocyanate, benzyl isothiocyanate, sesquiterpenes, and monoterpenes (Romeo et al. 2007), glucosinolates, fatty acid, sterols, and tocopherol (Matthäus and Özcan 2005)	In vivo hypolipidemic [+] (Eddouks et al. 2005), antioxidant [+] (Yang et al. 2010), antihyperglycemic [+] (Huseini et al. 2013)	Shah et al. (2013a), Yaseen et al. (2015)
57.	*Cleome scaposa* DC./Kastoori Buti/ Capparidaceae/herb	Leaves	Leaves are boiled for 30 min and decoction is used once a day	–	–	Ahmad et al. (2014a)
58.	*Carica papaya* L./ Papita/Caricaceae/ tree	Fruit	Unripe fruit paste is taken four to five times a day	Saponins and cardenolides, sodium, calcium, iron, phosphorus, zinc, copper, magnesium, and manganese (Oloyede 2005)	In vivo antihyperglycemic and hypolipidemic [+] (Sasidharan et al. 2011; Maniyar and Bhixavatimath 2012), hypoglycemic and antioxidant [+] (Juárez-Rojop et al. 2012)	Yaseen et al. (2015)

59.	*Stellaria media* L./ Oulalai/ Caryophyllaceae/ herb	Whole plant	Tea of plant is used daily	–	–	Ali et al. (2011a)
60.	*Cuscuta campestris* Yunck./Baypaari/ Convolvulaceae/ herb	Whole plant	Fresh material is boiled for 20 min and extract is used three times in one day	Flavonoid, quercetin, and polysaccharide (Lee et al. 2011)	–	Yaseen et al. (2015)
61.	*Cuscuta reflexa* Roxb./Neeli taar/ Convolvulaceae/ herb	Whole plant	Extract is used three times a day	7-(3,4-dihydroxyphenyl)-*N*-[(4-methoxyphenyl)ethyl]propenamide; 7-(4-hydroxy,3-methoxyphenyl)-*N*-[(4-butylphenyl)ethyl]propenamide; 6,7-dimethoxy-2*H*-1-benzopyran-2-one; 3-(3,4-dihydroxyphenyl)-2-propen-1-ethanoate; 6,7,8-trimethoxy-2*H*-1-benzopyran-2-one; 3-(4-*O*-β-D-glucopyranoside-3,5-dimethoxyphenyl)-2-propen-1-ol; and 2-(3-hydroxy-4-methoxyphenyl)-3,5-dihydroxy-7-*O*-β-D-glucopyranoside-4*H*-1-benzopyrane-4-one (Anis et al. 2002)	Glucoside inhibitory [+] (Anis et al. 2002), in vivo hypoglycemic (Rahmatullah et al. 2010)	Mahmood et al. (2013), Yaseen et al. (2015)
62.	*Convolvulus prostratus* Forssk./ Hiran Buti/ Convolvulaceae/ climber	Leaves	Leaves are boiled for 30 min and decoction is used once a day	–	–	Ahmad et al. (2014a)

(continued)

Table 1 (continued)

Sr. #	Botanical name (local name)/ family/habit	Parts used	Recipe	Phytochemicals reported	Pharmacological activity	Ethnobotanical studies
63.	*Coccinia grandis* (L.) Voigt./Golaru, Kanduri/ Cucurbitaceae/ climber	Root, leaf	Fresh juice of root and leaves is used in the morning	Heptacosane, cephalandrol, β-sitosterol, alkaloids cephalandrins A and B, β-amyrin acetate, lupeol, cucurbitacin B, taraxerone, taraxerol, β-carotene, lycopene, cryptoxanthin, xyloglucan, β-sitosterol, and stigma-7-en-3-one. Resin, alkaloids, starch, fatty acids, carbonic acid, triterpenoid, saponin coccinoside, flavonoid glycoside, lupeol, β-amyrin, β-sitosterol, and taraxerol (Pekamwar et al. 2013)	Antihyperglycemic [+] (Gunjan et al. 2010), in vivo hypoglycemic [+] (Pekamwar et al. 2013)	Panhwar and Abro (2007), Yaseen et al. (2015)
64.	*Cucumis sativus* L./ Khira/ Cucurbitaceae/herb	Fruit	Fresh fruit juice is taken three times with black salt	Phytosterols, amyrin, multiflorenol, 24-methylenecycloartanol, cycloartenol, tirucallol, and isopentenyl adenosine trialcohol (Yaseen et al. 2015)	Antidiabetic and antihyperlipidemic [+] (Karthiyayini et al. 2009), hypoglycemic (Sharmin et al. 2012)	Yaseen et al. (2015)
65.	*Cucumis melo* L./ Kerkunda/ Cucurbitaceae/herb	Fruit	Pulp is used	Phenolic compounds and flavonoids (Ismail et al. 2010)	In vitro antidiabetic potential [+] (Chen and Kang 2013)	Khan et al. (2015)
66.	*Cucurbita pepo* Wall./Kadu/ Cucurbitaceae/herb	Fruit	Paste of fresh fruit is used three times a day. Also cooked as vegetable	Fiber, protein, −carotene, carbohydrates, minerals, and fatty acids (Badr et al. 2011), tocopherol (Bharti et al. 2013)	In vivo hypoglycemic [+] (Sedigheh et al. 2011)	Yaseen et al. (2015)
67.	*Luffa acutangula* (L.) Roxb./Tori/ Cucurbitaceae/ climber	Fruit	Cooked as vegetable and taken two times a day	Triterpene, saponins, acutosides A–G (Nagao et al. 1991)	Hypoglycemic [+] (Quanico et al. 2008), antidiabetic [+] (Raj et al. 2012), antihyperglycemic [+] (Akther et al. 2014)	Khan et al. (2015, 2013a)

68.	*Momordica charantia* L./ Kerala/ Cucurbitaceae/ climber	Fruit	Fresh fruit extract is used or one teaspoon of shade-dried fruit powder is consumed three times a day	Lipids, nonpolar lipids, phospholipids, and glycolipids (Yuwai et al. 1991), charantin, vicine, and polypeptide-p (Krawinkel and Keding 2006), *trans*-nerolidol, apiole, *cis*-dihydrocarveol, and germacrene D (Braca et al. 2008)	Hypoglycemic [+] (Sarkar et al. 1996; Miura et al. 2001; Trakoon-osot et al. 2013), in vivo antidiabetic [+] (Xu et al. 2015)	Akhtar (1992), Ahmad et al. (2004, 2009), Fatima et al. (2005), Ahmed et al. (2007), Akhtar and Begum (2009), Yaseen et al. (2015)
69.	*Momordica dioica* Roxb. Ex Willd./ Jungli Kerala/ Cucurbitaceae/ climber	Fruit	Fruit is used fresh or dried to make powder for use with water daily	Steroids, fatty acids and proteins, saponin glycosides, and triterpenes (Ilango et al. 2012)	Antihyperglycemic and antilipidemic (Ilango et al. 2009), antidiabetic [+] (Sharma and Singh 2014)	Ahmad (2006), Yaseen et al. (2015)
70.	*Citrullus colocynthis* (L.) Schrad./Tumba/ Cucurbitaceae/herb	Roots, fruits, and seeds	One seed is swallowed early in the morning. Seed powder (one spoon) is also used three times a day. Paste of fresh fruit is also recommended. It is used as vegetable as well	Glycosides (Hatam et al. 1989), glucosides (Tannin-Spitz et al. 2007), tannins, saponins, alkaloids, flavonoids, and glycosides (Najafi et al. 2010)	Clinical investigation-proved hypoglycemic activity (Huseini et al. 2009), antidiabetic (Gurudeeban and Ramanathan 2010), hypoglycemic [+] (Agarwal et al. 2012)	Ahmad (2006), Marwat et al. (2011), Shah et al. (2013a), Adnan et al. (2014b), Ahmad et al. (2014a), Yaseen et al. (2015)
71.	*Citrullus vulgaris* Schrad/Tarbooz/ Cucurbitaceae/herb	Fruit	Fresh fruit is taken	Violaxanthin, luteoxanthin, lycopene, β-carotene, sucrose, fructose, citric acid, malic acid, and limonene (Liu et al. 2012)	Antidiabetic and protects pancreatic cells in mice [+] (Ahn et al. 2011; Simpson and Morris 2014)	Yaseen et al. (2015)

(continued)

Table 1 (continued)

Sr. #	Botanical name (local name)/ family/habit	Parts used	Recipe	Phytochemicals reported	Pharmacological activity	Ethnobotanical studies
72.	*Juniperus excelsa* Webb ex Parl/Chili/ Cupressaceae/tree	Fruit	Fruit is boiled to make tea for 15 min and taken in the morning	Diterpenes (Topçu et al. 1999), α-pinene (Ehsani et al. 2012), β-myrcene (6.99%), (E,E)-farnesol (4.66%), and β-pinene (Bakkour et al. 2013)	Antioxidant [−] (Bakkour et al. 2013)	Yaseen et al. (2015)
73.	*Cyperus rotundus* L./Deela or Murki, Nagar Mooth/ Cyperaceae/herb	Roots and tubers	Roots and tubers are boiled for 30 min and decoction is used once a day	Sesquiterpene hydrocarbons, isorotundene, cypera-2,4(15)-diene, norrotundene, ketone cyperadione (Sonwa and König 2001), α-cyperone, myrtenol, caryophyllene oxide, and β-pinene (Lawal and Oyedeji 2009), 7-isopropenyl-1,4a-5,6,7,8-hexahydro-3*H*-naphthalen-2-one, zierone, and (+)-*cis*-longipinan (Surendran and Vijayalakshmi 2011)	Antihyperglycemic and antioxidants (Raut and Gaikwad 2006)	Ahmad et al. (2014a)
74.	*Hippophae rhamnoides* L./Soq, Boo hay/ Elaeagnaceae/herb	Fruit and seeds	Jams and jellies are effective for DM. Juice of equal weight of fruit and seed is drunk daily	Flavonoids (Qunhua et al. 2003; Cao et al. 2005), flavones, quercetin, and isorhamnetin (Lan et al. 2004), vitamins A, B1, B12, C, E, K, and P, flavonoids, lycopene, carotenoids, and phytosterols (Patel et al. 2012a)	Hypoglycemic effect in diabetic rats (Zhang et al. 2010), antihyperglycemic [+] and antioxidant (Sharma et al. 2011)	Hussain et al. (2011), Yaseen et al. (2015)

75.	*Andrachne cordifolia* Mull. Arg./Chagzip Panra/ Euphorbiaceae/ shrub	Leaves	In clay pot, water is taken and leaves are placed for whole night and half cup is taken in the morning. Leaf extract can also be prepared by boiling for 5 min, taken once in a day	Glut-5(10)-en-3-one (Mukherjee et al. 1986; Mukherjee and Bhattacharjee 1987)	–	Alam et al. (2011), Yaseen et al. (2015)
76.	*Euphorbia helioscopia* L./ Chattri dodak/ Euphorbiaceae/ herb	Leaves and flowers	Extract of parts is administered daily	Triterpenoids, diterpenoids, flavonoids, tannins, steroids, and lipids (Zhang and Guo 2006), jatrophane diterpene ester, lupane derivatives, and triterpenoids	–	Hussain et al. (2004)
77.	*Euphorbia hirta* L./ Kaazi Dustaar/ Euphorbiaceae/ herb	Leaves, whole plant	It is boiled in water for 30 min and one cup extract is taken three times. Paste or juice is also used	Phenols and flavonoids (Yi et al. 2012), steroids, terpenoids, saponins, tannins, phenol, and quinone (Gopinath et al. 2012)	In vitro antidiabetic and antioxidant [+] (Kumar et al. 2010b), in vivo antidiabetic [+] (Subramanian et al. 2011b), in vivo antioxidant (Widharna et al. 2010)	Yaseen et al. (2015)
78.	*Euphorbia thymifolia* L./Kheer Wal/ Euphorbiaceae/ herb	Whole plant	Powder of plant material is taken twice a day	Flavonoids: luteonin (EL01) and quercetin-3-*O*-arabinofuranoside (Quyen 2013), tannins (Lee et al. 1990)	Antihyperglycemic [+] (Rahmatullah et al. 2012)	Qureshi and Bhatti (2008), Yaseen et al. (2015)
79.	*Alhagi maurorum* Medic./Kas Kundero/Fabaceae/ herb	Roots	Powder of roots is taken twice a day	Tannins, unsaturated sterols, triterpenes, flavonoids, flavanone glycosides (Samejo et al. 2012)	In vivo antihyperglycemic, antihyperlipidemic, antioxidant [+] (Sheweita et al. 2016)	Panhwar and Abro (2007), Yaseen et al. (2015)

(continued)

Table 1 (continued)

Sr. #	Botanical name (local name)/ family/habit	Parts used	Recipe	Phytochemicals reported	Pharmacological activity	Ethnobotanical studies
80.	*Acacia arabica* Lam./Babul/ Fabaceae/tree	Seeds, leaves	Leaves and seeds are boiled for 20 min sand and decoction is used	Phenolic compound (Sundaram and Mitra 2007; Yasir et al. 2010), linoleic acid and oleic acid (Maity and Mandal 1990)	Antioxidant [+] (Sundaram and Mitra 2007), in vivo hypoglycemic and antihyperglycemic [+] (Modak et al. 2007; Yasir et al. 2010; Patil et al. 2011, 2010)	Wadood et al. (1989), Shah et al. (2013a)
81.	*Acacia modesta* Wall./Phulai/ Fabaceae/tree	Gum, flower bark, seed/ powder,	Gum is roasted, mixed with oil, eggs, almonds, poppy seed, and dates, and used three times a day. Juice or decoction can be used	Alkaloids, terpenoids, flavonoids, and tannin (Bukhari et al. 2010)	In vivo hypoglycemic [+] (Singh et al. 1975)	Yaseen et al. (2015)
82.	*Acacia nilotica* (L.) Delile/Desi Kikar/ Fabaceae/tree	Leaves, gum, pods	Leaf infusion is taken or gum is taken twice a day. Fresh soft pods are eaten as vegetable	Tannin (Sotohy et al. 1997), phytosterols, fixed oils, fats, phenolic compounds, flavonoids, and saponins, crude protein, crude fiber, crude fat, ash, carbohydrates, potassium, phosphorus, magnesium, iron and manganese, cystine, methionine, threonine, lysine, tryptophan, and oleic and linoleic acids (Siddhuraju et al. 1996a), kaempferol (Singh et al. 2008)	In vivo antidiabetic and hypolipidemic [+] (Ahmad et al. 2008), in vivo hypoglycemic [+] (Asad et al. 2011)	Akhtar and Begum (2009), Shah et al. (2013a), Yaseen et al. (2015)

83.	*Bauhinia purpurea* Dc. Ex walp/ Kachnar/Fabaceae/ tree	Whole plant	Used as vegetable	Albumins and globulins and potassium (Rajaram and Janardhanan 1991), crude protein, crude fiber, crude fat, ash, carbohydrates, potassium, phosphorus, iron, cystine, methionine, threonine, lysine and tryptophan, and oleic and linolenic acid (Vijayakumari et al. 1997), glycolipids, phospholipids, and tocopherols (Ramadan et al. 2006)	In vitro antidiabetic [+] (Gupta et al. 2012a, 2013)	Kanwal et al. (2011a), Yaseen et al. (2015)
84.	*Bauhinia variegata* (L.) Benth/ Kachnar/Fabaceae/ tree	Leaves	Leaf infusion	Flavonoids (Reddy et al. 2003)	In vivo antidiabetic [+] (Koti et al. 2009; Thiruvenkatasubramaniam and Jayakar 2010; Kumar et al. 2012)	Hussain et al. (2004)
85.	*Butea monosperma* L./Palas/Fabaceae/ tree	Flower, latex	Powder of flower taken three times a day/small amount of dried latex	Butrin, butein, butin, flavonoids butrin, and isobutrin (Choedon et al. 2010)	In vivo antidiabetic [+] (Somani et al. 2006), in vivo antihyperglycemic [+] (Bavarva and Narasimhacharya 2008)	Yaseen et al. (2015)
86.	*Cassia fistula* L./ Amaltas/Fabaceae/ tree	Leaves, seed, and pulp	Juice of fresh or dried leaves is used two to three times a day. Confection of pulp and seed is also used	Ca, Fe, Mn, aspartic acid, glutamic acid, and lysine (Barthakur et al. 1995), hydrocarbons, sterols, triterpene, anthraquinone, coumarins, and chromones (Lee et al. 2001), anthraquinones and flavonoids (Bahorun et al. 2005)	In vivo antihyperglycemic and antilipidemic [+] (Nirmala et al. 2008), in vivo antidiabetic (Ratnasooriya et al. 2004; Khan et al. 2010b)	Akhtar (1992), Husain et al. (2008), Yaseen et al. (2015)
87.	*Cassia obtusifolia* L./Chaund/ Fabaceae/herb	Flowers	Extract of flowers taken in the morning daily	Naphthopyrones (Tianaka and takido 1988; Jiang et al. 2005), emodin (Yang et al. 2003)	In vitro antioxidant and antidiabetic (Vadivel et al. 2012)	Yaseen et al. (2015)
88.	*Cajanus cajan* (L.) Druce/Arar ke dal/ Fabaceae/shrub	Seed	Pulse cocked	Hydrocyanic acid and vicianine (Yaseen et al. 2015)	In vivo antihyperglycemic (Jaiswal et al. 2008; Anwar et al. 2010)	Yaseen et al. (2015)

(continued)

Table 1 (continued)

Sr. #	Botanical name (local name)/ family/habit	Parts used	Recipe	Phytochemicals reported	Pharmacological activity	Ethnobotanical studies
89.	*Dalbergia sissoo* Roxb./Talhi/ Fabaceae/shrub	Gum	Gum roasted for a while, powdered, and used three times	Polysaccharides (Rana et al. 2012)	In vivo antidiabetic [+] (Niranjan et al. 2010)	Shah and Khan (2006)
90.	*Glycine max* (L.) Merr./Soybean/ Fabaceae/herb	Seed	Extract is used once a day	Kaempferol glycoside (Zang et al. 2011), beta-sitosterol (Tang et al. 2008)	In vivo antidiabetic [+] (Badole and Bodhankar 2009; Zang et al. 2011)	Khushk et al. (2010)
91.	*Mucuna pruriens* DC. Ex Pammel/ Kowanch/ Fabaceae/herb	Seeds	Seeds decoction is prepared and used twice a day	Crude protein, crude fiber, crude fat, ash, and carbohydrates; potassium, phosphorus, and calcium; globulins and albumins; valine and tryptophan; oleic and linoleic acids; and palmitic acid (Siddhuraju et al. 1996b)	In vivo antihyperglycemic [+] (Bhaskar et al. 2008; Murugan and Reddy 2009; Majekodunmi et al. 2011)	Akhtar (1992), Khan et al. (2008)
92.	*Trigonella foenum-graecum* L./Mathi/Fabaceae/ herb	Leaves, seeds	Seeds (21 g) are taken orally for 21 days	4-Hydroxyisoleucine (Haeri et al. 2012), galactomannans (Kamble et al. 2013), furostanolic saponins (Swaroop et al. 2014)	Decrease in insulin resistance in clinical trials [+] (Gupta et al. 2001), in vivo hypoglycemic [+] (Baquer et al. 2011; Haeri et al. 2012), in vivo antihyperglycemic [+] (Kamble et al. 2013), in vivo and in vitro antidiabetic [+] (Swaroop et al. 2014)	Zia et al. (2001), Ahmad et al. (2004, 2009), Fatima et al. (2005), Yaseen et al. (2015)
93.	*Vigna sinensis* (L.) Savi ex Hassk./ Safed Lobia/ Fabaceae/herb	Immature pods	Soft immature pods are used as vegetable	Protein similar to bivine insulin (Venâncio et al. 2003), saponin, thiamine, riboflavin, niacin, vitamin B6, pantothenic acid and foliate (Chandrasekaran et al. 2015).	In vitro antioxidant (Zia-Ul-Haq et al. 2013a), in vivo antihyperglycemic [+] (Tazin et al. 2014), hypolipidemic and hypoglycemic [+] (Weththasinghe et al. 2014)	Ahmad et al. (2009)

94.	*Gentianodes tianschanica* (Rupr. ex Kusn.) Omer, Ali & Qaiser/ Kamalay Char, Jangli boti/ Gentianaceae/herb	Leaves	Extract of leaves is prepared by boiling in water for 20 min and a cup is used two times a day.	–	–	Khan and Khatoon (2008), Yaseen et al. (2015)
95.	*Swertia chirayita* (Roxb ex. Fleming) H. Karst./Choriata, Karaita/ Gentianaceae/herb	Whole plant	Fresh water extract is advised thrice a day. Dried plant is ground to make powder. Two teaspoons are taken two times a day	Chiratin, xanthones, flavonoids, terpenoids, iridoids, glycosides carbonates, and phosphates of calcium, potassium, and magnesium (Phoboo et al. 2010)	In vitro antidiabetic [+] (Phoboo et al. 2013)	Sabeen and Ahmad (2009), Yaseen et al. (2015)
96.	*Ajuga bracteosa* Benth/Hari booti/ Lamiaceae/herb	Whole plant	Plant is boiled in water, placed in dew whole night, and taken before breakfast	Bracteonin-A (1) 6 f-acetoxy; 15 (R&S)-methoxy; 18-neoclerodane; 14,15-dihydroajugapitin; 14-hydro-15-hydroxy-ajugapitin; g-sitosterol; and stigmasterol (Verma et al. 2002), 1,2-benzenedicarboxylic acid bis (2*S*-methyl heptyl) ester; ajugarin-I; reptoside; 8-*O*-acetyl harpagide; and linalyl acetate (Singh et al. 2006b)	Reduce blood glucose level [+] (Gupta et al. 2008), clinically proved antidiabetic [+] (Chaudhary et al. 2015)	Yaseen et al. (2015)
97.	*Caryopteris odorata* B. L. Rob./ Path gar/ Lamiaceae/shrub	Leaves, flowers	One teaspoon powder of equal amount of leaves and flowers is taken thrice a day	Glucosides (Abbasi et al. 2014), 8-*O*-*trans*-cinnamoyl caryoptoside, 8-*O*-*trans*-cinnamoyl shanzhiside methylester, 8-*O*-*trans*-cinnamoyl mussaenoside, and 8-*O*-cafeoyl massenoside (Shahzadi et al. 2013)	-	Yaseen et al. (2015)

(continued)

Table 1 (continued)

Sr. #	Botanical name (local name)/ family/habit	Parts used	Recipe	Phytochemicals reported	Pharmacological activity	Ethnobotanical studies
98.	*Mentha piperita* L. Pro spec. & Hylander/Podina/ Lamiaceae/herb	Seeds, leaves	Seeds and leaves are boiled for 5 min and infusion is used twice. Powder of leaves (two teaspoons) is used after each meal	Menthon, menthol, menthyl acetate, and menthofuran (Andoğan et al. 2002), tannin and flavonoid (Pramila et al. 2012)	In vivo antihyperglycemic [−] (Alarcon-Aguilara et al. 1998), in vitro antioxidant and antiglycation activity [+] (Ramkissoon et al. 2013), in vivo hypoglycemic [+] (Barbalho et al. 2011)	Shah and Khan (2006), Yaseen et al. (2015)
99.	*Mentha longifolia* Host/Podina/ Lamiaceae/herb	Leaves	Leaves are boiled in water. Half cup of this extract is taken three times	Pulegone, isomenthone, 1,8-cineole, borneol, and piperitenone oxide (Mkaddem et al. 2009)	–	Yaseen et al. (2015)
100.	*Salvia coccinea* Juss. Ex Murr./ Lamiaceae/herb	Leaves	Leaves are soaked in water and extract is used daily	Salviacoccin (Savona et al. 1982), linolenic acid, linoleic acid, oleic acid, and stearic acid (Delange et al. 2012)	In vitro antioxidant [+] (Yadav and Mukundan 2011)	Hussain et al. (2004)
101.	*Ocimum album* L./ Chiti Tulsi/ Lamiaceae/herb	Leaves	10 g of powder of dried leaves is used two times daily	Flavonoids, phenylpropanoids, and anthocyanins (Yaseen et al. 2015)	In vivo hypoglycemic effect [−] (Yaseen et al. 2015)	Yaseen et al. (2015)
102.	*Ocimum sanctum* L./Tulsi, Niazbo/ Lamiaceae/herb	Leaves	10 g of powder of dried leaves is used two times daily	Essential oil and eugenol (Kumar et al. 2010a), stearic, palmitic, oleic, linoleic, and linolenic acids (Singh et al. 1996), eugenol, euginal, ursolic acid, carvacrol, linalool, limatrol, caryophyllene, methyl chavicol, sitosterol, sugars, and anthocyanins (Pattanayak et al. 2010)	In vivo hypoglycemic [+] (Rai et al. 1997; Gupta et al. 2002; Vats et al. 2002, 2004), in vivo antihyperglycemic [+] (Hannan et al. 2006)	Yaseen et al. (2015)

103.	*Origanum vulgare* L./Shamake/ Lamiaceae/herb	Leaves	Extract of leaves soaked in clay pot for whole night is taken in the morning	Phenolics (Şahin et al. 2004), origanol A, origanol B along with ursolic acid, oleanolic acid, β-sitosterol, and i(Rao et al. 2011)	In vivo antioxidant, immunomodulator and in an antiapoptotic manner, protect from diabetes development (Vujicic et al. 2015)	Akhtar and Begum (2009), Yaseen et al. (2015)
104.	*Teucrium stocksianum* Boiss/ Aspa botay/ Lamiaceae/herb	Shoots	Plant is boiled for 15 min and one cup of extract is taken three times per day	Tannin, flavonoids, sterols, and saponins (Ali et al. 2011b)	In vivo antidiabetic [+] (Rashid et al. 2013a)	Yaseen et al. (2015)
105.	*Viscum album* L. Kishmish kaabuli/ Loranthaceae/shrub	Leaves and twigs	One cup of leaves and twigs extract is prepared after boiling. Half cup is administered daily one time	Flavonoid, tannins, alkaloids, and carbohydrate (Oguntoye et al. 2008), phenols (Oluwaseun and Ganiyu 2008)	In vivo antidiabetic [+] (Orhan et al. 2005; Adaramoye et al. 2012; Ibegbulem and Chikezie 2013)	Adnan et al. (2014a)
106.	*Malva neglecta* Wall./Sonchal/ Malvaceae/herb	Leaves	Leaves are soaked along with small quantity of table salt for whole night and half cup of extract is used in the morning	Alkaloids, flavonoids, tannins, and saponins (Mojab et al. 2010)	–	Yaseen et al. (2015)
107.	*Sida cordata* (Burm. F.) Waalkes/Bhuinii/ Malvaceae/herb/	Whole plant	Extract is used	Flavonoid (Shah et al. 2013b)	In vivo antidiabetic [+] (Srinivasan et al. 2013)	Shah and Khan (2014)

(continued)

Table 1 (continued)

Sr. #	Botanical name (local name)/ family/habit	Parts used	Recipe	Phytochemicals reported	Pharmacological activity	Ethnobotanical studies
108.	*Azadirachta indica* A. Juss./Neem/ Meliaceae/tree	Leaf, bark, fruit, and flower	Half tablespoon of dried fruit powder is used. Fresh juice of leaf is taken. Tea of flowers may be used three times. All plant parts are placed in water overnight, and one cup of extract is taken in the morning	Salannin and azadirachtin (Johnson et al. 1996), tetranortriterpenoid and odoratone (Siddiqui et al. 2003), flavonoids, tannins, saponins, polyphenols, and alkaloids (Atangwho et al. 2009), triterpenoids: neemfruitins A and B (Chianese et al. 2010)	In vivo antidiabetic [+] (Akter et al. 2013), in vivo hypoglycemic [+] (Ojiako et al. 2014)	Ahmad et al. (2009), Yaseen et al. (2015)
109.	*Cedrela serrata* Royle/Darawa/ Meliaceae/tree	Leaves and wood	Extract is made by soaking plant material in water and taken	Phenols (Ahmad et al. 2013b)	Antioxidant in vitro [+] (Ahmad et al. 2013b)	Awan et al. (2011)
110.	*Cedrela toona* Roxb. Ex Willd./ Nem/Meliaceae/ tree	Leaves	Leaf powder is taken along with table salt daily	Cedrelone, 1,2-dihydrocedrelone, bergapten, and β-sitosterol (Chatterjee et al. 1971; Modey et al. 1996), astrin, antiper, toonafolin, and toonacillin (Kumari and Kakkar 2008)	In vitro antioxidants [−] (Kumari and Kakkar 2008), lipid peroxidation inhibitory potential [+] (Asif 2015), in vivo antihyperglycemic (Rana et al. 2016)	Abbasi et al. (2010b)
111.	*Melia azedarach* L./Dharek/ Meliaceae/tree	Fruits	Half tablespoon of dried ground pericarp is used for 1 month	Flavonoids and phenols (Italo et al. 2009; Safithri and Sari 2016)	In vivo antidiabetic [+] (Chasturvedi et al. 2005), in vitro antidiabetic [+] (Khan et al. 2014b), in vitro antihyperglycemic [+] (Safithri and Sari 2016)	Ahmad et al. (2004), Ahmad (2006), Abbasi et al. (2010a),

112.	*Cissampelos pareira* L./Ghora Sum/ Menispermaceae/ herb	Seed	Seed powder is used twice a day after meal	Cissampareine (Kupchan et al. 1965), cissampeloflavone (Ramírez et al. 2003), alkaloids (Bafna and Mishra 2010)	In vivo antidiabetic [+] (Yadav et al. 2013; Basumata 2016)	Shah and Khan (2006)
113.	*Albizia lebbeck* (L.) Benth./Shrin/Kala Shareen/ Mimosaceae/tree	Seed	One spoon of powder of seed is used daily in the morning or twice a day	Saponins (Barbosa 2014), phenols, flavonoids, steroids, phytosterol, triterpenoid, and tannin (Jeeva et al. 2011), 5-deoxyflavone (geraldone), luteolin, and isookanin (Ahmed et al. 2014a)	In vivo antihyperglycemic and antihyperlipidemic [+] (Ahmed et al. 2014b), in vitro antidiabetic [+] (Ahmed et al. 2014a)	Shah et al. (2013a), Yaseen et al. (2015)
114.	*Ficus benghalensis* L./Burgad, Bohr/ Moraceae/tree	Latex, bark, roots, fruit, prop root, leaves, and branches	Latex mixed with honey is used. Fresh milk latex is poured in water and used three times a day	Carbohydrates, flavonoids, amino acids, steroids, saponins, and tannin (Uma et al. 2009; Joseph and Raj 2010)	In vivo antidiabetic [+] (Shukla et al. 1994; Singh et al. 2009)	Ahmed et al. (2007), Ahmad et al. (2009), Khan et al. (2011, 2012a), Marwat et al. (2011), Shah et al. (2013a), Yaseen et al. (2015)
115.	*Ficus carica* L./ Anjeer/Moraceae/ shrub	Leaf	10–15 leaves are boiled and extract is taken	Pyrogallic acid, ferulic acid, coumaric acid, galangin, cinnamic acid, quercetin, and pinostobin (El-Shobaki et al. 2010), aldehydes, alcohols, ketone, monoterpenes, sesquiterpenes, oxalic, citric, malic, quinic, shikimic, and fumaric acid (Oliveira et al. 2010), coumarins (Lazreg-Aref et al. 2012)	In vivo hypoglycemic and antioxidant [+] (El-Shobaki et al. 2010), in vivo antidiabetic [+] (Stalin et al. 2012)	Marwat et al. (2011), Yaseen et al. (2015)

(continued)

Table 1 (continued)

Sr. #	Botanical name (local name)/ family/habit	Parts used	Recipe	Phytochemicals reported	Pharmacological activity	Ethnobotanical studies
116.	*Ficus hispida* Roxb. Ex wall./ Botaye/Moraceae/ shrub	Bark	The fresh bark is soaked in water whole night and half cup of extract is used in the morning	Alkaloids, phenolics, tannins, and cyanogenic glycosides (Mahajan and Badgujar 2008), hispidacine and hispiloscine (Yap et al. 2015)	In vivo antidiabetic [+] (Ghosh et al. 2004; Shahreen et al. 2012)	Yaseen et al. (2015)
117.	*Ficus lacor* Buch. Ham./Anjeer/ Moraceae/tree	Fruit	Dried fruit is ground and powder is used three times a day	Flavonoids, carbohydrates, saponins, phenolic compounds, sterols, and amino acids (Sindhu and Arora 2013)	In vivo antihyperglycemic [−] (Shahreen et al. 2012)	Yaseen et al. (2015)
118.	*Ficus microcarpa* Hort. Berol. Ex Walp/Anjeer/ Moraceae/tree	Leaf, bark, fruit	Equal quantity of all parts is taken to make powder and two teaspoons are used three times a day	Phenolic compound and catechin (Kalaskar and Surana 2011), oleanolic acid, betulinic acid, lupeol, β-sitosterol, catechin, and gallic acid (Kalaskar and Surana 2012), polysaccharide (Jiang et al. 2014)	In vivo hypoglycemic [+] (Kumar et al. 2007).	Khan et al. (2011), Yaseen et al. (2015)
119.	*Ficus racemosa* Roxb./Oormal, Bara Anjeer/ Moraceae/tree	Bark	Bark is boiled in water for 30 min and extract is used daily	Tannin (Velayutham et al. 2012), phenolic compounds (Ahmed and Urooj 2009)	In vivo antidiabetic [+] (Bhaskara Rao et al. 2002; Sachan et al. 2009; Veerapur et al. 2012), in vitro hypoglycemic [−] (Ahmed and Urooj 2010)	Akhtar (1992), Yaseen et al. (2015)
120.	*Ficus palmata* Forssk./Anjeer/ Moraceae/tree	Fruit	Two spoons of paste of fruit are administered twice after breakfast and dinner	Alkaloids, flavonoids, tannins, unsaturated sterols/triterpenes, resins, and phenolic compounds (Saklani and Chandra 2012)	In vivo antidiabetic [+] (Singh et al. 2014)	Yaseen et al. (2015)

121.	*Ficus religiosa* L./ Peepal/Moraceae/ tree	Bark	Two spoons of bark juice are taken three times a day	Phytosterols, amino acids, furanocoumarins, phenolic components, hydrocarbons, aliphatic alcohols, volatile components (Singh et al. 2011)	In vivo hypoglycemic [+] (Pandit et al. 2010; Parikh et al. 2014)	Khan et al. (2011), Yaseen et al. (2015)
122.	*Ficus virens* Aiton./ Jangli peepal/ Moraceae/tree	Leaves	Five to eight fresh leaves are soaked in water overnight and half cup of this water is drunk three times a day	Phenols and flavonoids (Abdel-Hameed 2009; Chen and Kang 2013)	–	Yaseen et al. (2015)
123.	*Morus alba* Bureau/Shahtoot, Toot/Moraceae/tree	Leaf, root	Paste of fresh fruit is used. Root is boiled in water and one cup is used in the morning	Moracin, Steppogenin-4′-*O*-β-D-glucoside, mulberroside (Zhang et al. 2009; Naik et al. 2015), benzyl alcohol, ethyl benzoate, *t*-cinnamic acid, *p*-hydroxyacetophenone, *t*-coniferyl alcohol, and synapil alcohol (Hunyadi et al. 2013)	In vivo hypoglycemic [+] (Zhang et al. 2009), in vivo antidiabetic [+] (Sarikaphuti et al. 2013)	Shah and Khan (2006), Husain et al. (2008), Yaseen et al. (2015)
124.	*Morus nigra* L./ Kala Toot/ Moraceae/tree	Fruit, leaves, and bark	Bark is used to make juice. Half cup of this juice is taken orally two times a day. Fresh fruit is used to make paste and two to three spoons are taken two times a day. Root is boiled in water for 10–15 min. And one cup is drunk daily early in the morning	Flavonoids (Abd et al. 2010), total phenolics, flavonoids, and anthocyanins (Kamiloglu et al. 2013)	In vivo antidiabetic [+] (Abd et al. 2010)	Sabeen and Ahmad (2009), Yaseen et al. (2015)

(continued)

Table 1 (continued)

Sr. #	Botanical name (local name)/ family/habit	Parts used	Recipe	Phytochemicals reported	Pharmacological activity	Ethnobotanical studies
125.	*Moringa oleifera* Lam./Sohanjana/ Moringaceae/tree	Fruit, flower, seed	Plant is cooked as vegetable and taken twice a day. Half cup of juice of leaves may be taken twice a day. Seeds are boiled in water and extract is taken three times	Tannins, saponin, flavonoids, steroids, terpenoids, and glycosides (Nepolean et al. 2009)	In vivo hypoglycemic [+] (Ara et al. 2008; Aja et al. 2015; Olayaki et al. 2015), in vivo antidiabetic [+] (Gupta et al. 2012b; Edoga et al. 2013; Al-Malki and El Rabey 2015)	Kanwal et al. (2011a), Yaseen et al. (2015)
126.	*Musa acuminata Colla.*/Kela/ Musaceae/tree	Flower	Flowers are roasted and powdered and one spoon is used daily with water	Dietary fibers, fructan, campesterol, β-sitosterol, stigmasterol, and polyphenol (Menezes et al. 2011)	In vivo hypoglycemic [+] (Liyanage et al. 2015)	Adnan et al. (2014b)
127.	*Eucalyptus globulus Labill.*/ Gond/Myrtaceae/ tree	Bark	Bark extract is made after boiling and one cup is used daily	Phenolic compound and tannins (Boulekbache-Makhlouf et al. 2013)	Reduce glucose diffusion in vitro [+] (Gallagher et al. 2003), in vivo and in vitro antidiabetic [+] (Jouad et al. 2004; Mahmoudzadeh-Sagheb et al. 2010; Yen et al. 2015)	Shah and Khan (2006)
128.	*Psidium guajava* L./Amrud/ Myrtaceae/shrub	Dried leaves, bark	Extract is administered once a day	Quercetin, quercetin-3-*O*-α-L-arabinofuranoside, quercetin-3-*O*-β-D-arabinopyranoside, quercetin-3-*O*-β-D-glucoside, and quercetin-3-*O*-β-D-galactoside (Metwally et al. 2010), carotenoids, vitamin C, and polyphenols and triterpenoid (Barbalho et al. 2012)	In vitro and in vivo antidiabetic [+] (Huang et al. 2011; Basha and Kumari 2012), in vivo antidiabetic and antidiarrheal activities [+] (Mazumdar et al. 2015)	Ahmad et al. (2004, 2009), Sabeen and Ahmad (2009), Yaseen et al. (2015)

129.	*Syzygium cumini* Skeels/Jaman/ Myrtaceae/tree	Fresh fruits, seeds	Seed powder is administered three times a day. Besides, 5–15 seeds are boiled in water and taken three times a day	Anthocyanins, glucoside, ellagic acid, isoquercetin, kaempferol and myricetin, alkaloid, jambosine, and glycoside jambolin or antimellin (Ayyanar and Subash-Babu 2012), triterpenes/steroids, glycosides, carbohydrates, alkaloids, flavonoids, saponins, tannins, and amino acids (Tripathi and Kohli 2014)	In vivo antidiabetic [+] (Kumar et al. 2008; Tripathi and Kohli 2014), in vitro antidiabetic [+] (De Bona et al. 2014).	Akhtar (1992), Ahmad et al. (2004, 2009), Fatima et al. (2005), Sabeen and Ahmad (2009), Arshad et al. (2011), Marwat et al. (2011), , Adnan et al. 2014b, Yaseen et al. (2015)
130.	*Boerhavia diffusa* L. Nom.cons/Itsit/ Nyctaginaceae/herb	Leaf	Leaf extract is taken two times in a day	Alkaloids, flavonoids, steroids, terpenoids, reducing sugars, saponins, tannins, cardiac glycosides, and anthraquinones (Apu et al. 2012)	In vivo antidiabetic [+] (Pari and Satheesh 2004; Nalamolu et al. 2007; Dora et al. 2015), in vivo antioxidant [+] (Apu et al. 2012)	Shah and Khan (2006)
131.	*Fraxinus excelsior* L./Sum/Oleaceae/ tree	Seed	One spoon of seed powder is administered with water daily	Glucose, sorbitol, galactose, mannotriose and stachyose, mannose, sucrose, mannitol and sorbitol, elenolic acid, tyrosol, homovanillic acid, dopaol, pinoresinol and fraxetin, albeit, esculetin, *p*-hydroxybenzoic acid, 4-hydroxyphenacetic acid, 3,4-hydroxybenzoic acid, hydroxypinoresinol, medioresinol and syringaresinol (Caligiani et al. 2013)	In vivo hypoglycemic [+] (Eddouks and Maghrani 2004; Maghrani et al. 2004), hepatoprotective in diabetes [+] (Gomez-Garcia et al. 2015)	Shah and Khan (2006)

(continued)

Table 1 (continued)

Sr. #	Botanical name (local name)/ family/habit	Parts used	Recipe	Phytochemicals reported	Pharmacological activity	Ethnobotanical studies
132.	*Olea europaea* subsp. *europaea* L./ Zaitoon//Oleaceae/ tree	Fruit	Fruits are eaten fresh and dried for winter	Oleuropein (Al-Azzawie and Alhamdani 2006), phenolic compound (Pereira et al. 2007; Vogel et al. 2015), maslinic acid (Sánchez-González et al. 2013), oleanolic acid (Nazaruk and Borzym-Kluczyk 2015), ursolic acids, erythrodiol, and uvaol (Giménez et al. 2015)	In vivo antidiabetic [+] (Eidi et al. 2009; Chandak and Shrangare 2010; Sangi et al. 2015)	Shah et al. (2013a), Yaseen et al. (2015)
133.	*Olea ferruginea* (Sol.) Steud./ Zaiton, Khan, Khau/Oleaceae/tree	Fruit	Fruits are eaten fresh or dried for use year round. Leaves are boiled for few minutes and decoction is taken at night	Oleanolic acid (Sultana and Ata 2008), tetrahydroxyflavanone (Hashmi et al. 2014), flavonoid, ursane-type triterpene, and seco-iridoids (Hashmi et al. 2015)	In vitro insulin secretagogue activity [−] (Hussain et al. 2004)	(Adnan et al. (2014b), Ahmad et al. (2004, 2009), Khan et al. (2015), Yaseen et al. (2015)
134.	*Oxalis corniculata* L./Khatti buti, Khat kurla/Oxalidaceae/ herb	Whole plant	Whole plant extract is used daily	Flavonoids (Mizokami et al. 2008), carbohydrates and glycosides, phytosterols, phenolic compounds/ tannins, flavonoids, proteins and amino acids, and volatile oils (Raghavendra et al. 2006; Sakat et al. 2012)	–	Shah and Khan (2006), Husain et al. (2008)
135.	*Fumaria indica* (Haussk) Pugsely/ Papra/ Papaveraceae/herb	Whole plant	Half cup of juice of plant material is taken daily	Alkaloids, flavonoids, glycosides, tannins, saponins, steroids, and triterpenoids (Rao et al. 2007)	In vivo hypoglycemic [+] (Gilani et al. 2005)	Sabeen and Ahmad (2009), Abbasi et al. (2010a), Yaseen et al. (2015)

136.	*Papaver somniferum* L./ Posht, Khas Khas, Khashkash Safeed/ Papaveraceae/herb	Fruit, seed, flowers, and Latex	Flowers are soaked in water and extract is used. Fruit and seed are boiled for 15 min in water. One cup of this extract is taken three times a day	Papaverine, noscapine, sanguinarine, morphine, codeine, and thebaine, and L-tyrosine (Kraml and Dicosmo 1993), morphinan alkaloids (Larkin et al. 2007), fatty acid, tocopherol, and sterol (Erinç et al. 2009)	–	Sabeen and Ahmad (2009), Yaseen et al. (2015)
137.	*Argyrolobium roseum* (Cambess.) Jaub. & Spach/ Makhan Booti/ Papilionaceae/herb	Whole plant	Infusion is prepared and used twice a day	Flavonoid glycoside (Gupta et al. 2005)	In vitro and in vivo hypoglycemic [+] (Gupta et al. 2005, 2011b), stimulation of insulin secretion in vivo and in vitro [+] (Ahmed et al. 2008)	Shah et al. (2013a), Yaseen et al. (2015)
138.	*Phyllanthus emblica* L./Amla/ Phyllanthaceae/ shrub	Fruit	Fresh fruit, its jam, and dried plant powder are used three times a day	Cisplatin, 5-flurouracil, phyllaemblic acid, phyllaemblicin A, phyllaemblicin B, phyllaemblicin C, L-malic acid 2-*O*-gallate, mucic acid 2-*O*-gallate, hydrolyzable tannins, flavonoids, and condensed tannins (Zhang et al. 2004), isomallotusinin (Luo et al. 2012), gallic acid (Sawant et al. 2012),	In vivo antidiabetic [+] (Krishnaveni et al. 2010)	Husain et al. (2008), Yaseen et al. (2015)
139.	*Abies pindrow* (Royle ex D. Don) Royle/Partal/ Pinaceae/tree	Leaves	Leaves are soaked overnight and extract is used	Pindrolactone (Tripathi et al. 1996), pinitol (Singh et al. 2001)	Insulin secretagogue activity [+] (Bhushan et al. 2010; Chawla et al. 2013)	Hussain et al. (2004)
140.	*Pinus roxburghii* Sargent/Nakhtar/ Pinaceae/tree	Resin	Resin is used	Flavonoids (Kaushik et al. 2015b)	In vivo antihyperglycemic and antihyperlipidemic [+] (Kaushik et al. 2015a)	Khan et al. (2015)

(continued)

Table 1 (continued)

Sr. #	Botanical name (local name)/ family/habit	Parts used	Recipe	Phytochemicals reported	Pharmacological activity	Ethnobotanical studies
141.	*Kickxia ramosissima* (Wall) Janchen/Wal/ Plantaginaceae/ herb	Whole plant	One spoon of powder of whole plant is taken twice a day	Iridoids, *p*-hydroxy-coumaric acid methyl ester, and *p*-hydroxy-benzoic acid methyl ester (Amin et al. 2015), tannins, flavonoids, alkaloids, coumarins, cardiac glycosides, saponins, terpenoids, and phlobatannins (Jan and Khan 2016)	–	Ahmad et al. (2004, 2009)
142.	*Picrorhiza kurroa* Royl ex Benth./ Kore/ Plantaginaceae/ herb	Roots	One spoon of powder of roots is taken twice per day	Cucurbitacin glycosides (Stuppner and Moller 1993), picrosides (Upadhyay et al. 2013; Kumar et al. 2015)	In vivo antidiabetic [+] (Hurakadle et al. 2009; Husain et al. 2009, 2014; Naveen 2010)	Mahmood et al. (2012)
143.	*Hordeum vulgare* L./Bajara/Poaceae/ herb	Seeds, whole plant	Seeds are boiled in water till they become soft and stained and water is drunk daily	Hydroxycinnamic acids, ferulic acid (Verardo et al. 2008), volatile oils, alkaloids, saponins, terpenoids (Semwal et al. 2007)	In vivo antidiabetic [+] (Minaiyan et al. 2014), in vivo antioxidant [+] (Yu et al. 2008)	Marwat et al. (2011)
144.	Zea mays L./ Makai/Poaceae/ herb	Flower	Flowers are soaked in water for whole night and half cup of extract is taken in the morning	Flavonoids (Zhang et al. 2015)	In vitro α-glucosidase inhibitory activity [+] (Nile and Park 2014), in vivo antidiabetic [+] (Huang et al. 2015; Zhang et al. 2015)	Yaseen et al. (2015)

145.	*Fagopyrum esculentum* Moench./Ghiawas/ Polygonaceae/herb	Leaf, seed	Leaf is used as vegetable and flour is used to make bread and eaten twice. Besides, powder of plant is used three times a day. Flour is used to make bread taken twice per day	Polyphenol (Bystricka et al. 2014), flavonoids and rutin (Bae et al. 2015).	Hypoglycemic (Giménez-Bastida and Zieliński 2015), in vivo hypoglycemic [+] (Hong et al. 2011; Li et al. 2016).	Hussain et al. (2011), Yaseen et al. (2015)
146.	*Portulaca oleracea* L./Kulfa/ Portulacaceae/herb	Leaves	Leaves are cooked as vegetable and taken two times	Polysaccharide (Li and Yu 2011), alkaloids (Xiang et al. 2005), fatty acids, organic acids, and phenolic compound (Oliveira et al. 2009)	In vivo hypoglycemic [+] (Li and Yu 2011; Singh and Kori 2014; Gu et al. 2015), in vitro hypoglycemic [+] (Gu et al. 2015)	Adnan et al. (2014b)
147.	*Adiantum capillus-veneris* L./ Persayoshayon, Sumbal/ Pteridaceae/herb	Bark	Bark is boiled in water and half cup of decoction is used daily	Isoadiantone, isoadiantol B, 3-methoxy-4-hydroxyfilicane, 3,4-dihydroxyfilicane, quercetin, quercetin-3-*O*-glucoside, and quercetin-3-*O*-rutinoside (rutin) (Ibraheim et al. 2011), steroids, flavonoids, terpenoids, fats, tannins, and phenolic compounds (Ranjan and Vats 2016)	In vivo hypoglycemic [+] (Ibraheim et al. 2011; Jiang et al. 2011)	Fatima et al. (2005), Hamayun et al. (2006), Yaseen et al. (2015)
148.	*Adiantum incisum* Forssk/Sumbal/ Pteridaceae/herb	Fronds	Fresh fronds half-cup juice is taken once in a day	Triterpenes (Hayat et al. 2002)	–	Hamayun et al. (2006), Yaseen et al. (2015)

(continued)

Table 1 (continued)

Sr. #	Botanical name (local name)/ family/habit	Parts used	Recipe	Phytochemicals reported	Pharmacological activity	Ethnobotanical studies
149.	*Punica granatum* L./Anar, Dani. Druna/Punicaceae/ tree	Fruit, peel	Fresh fruit is eaten daily once	Anthocyanin (Al-Muammar and Khan 2012)	In vivo antidiabetic [+] (Khalil 2004; Parmar and Kar 2007; Bhaskar and Kumar 2012; Das and Barman 2012)	Awan et al. (2011), Saqib et al. (2014)
150.	*Nigella sativa* L./ Kalongi/ Ranunculaceae/ herb	Whole plant, leaves, seed	Seed are powdered and half spoon is taken three times a day. Leaves are boiled to get extract to use. Moreover, whole plant is soaked at night and extract is taken before breakfast for 1 week	Linoleic acid, oleic acid, eicosenoic acid, palmitic acid, stearic acid, myristic acid, lauric acid, arachidic acid, palmitoleic acid, linolenic acid, and thymoquinone (Bamosa 2015; Heshmati et al. 2015)	In vivo hypoglycemic [+] (Abdelmeguid et al. 2010; Sultan et al. 2014; Kaatabi et al. 2015), improves serum level of insulin (Omar and Atia 2012)	Yaseen et al. (2015)
151.	*Ziziphus mauritiana* Lam./ Beri/Rhamnaceae/ tree	Fruit	Dried fruit powder is eaten early morning time daily	Flavonoid and phenolic compounds (Ashraf et al. 2015), ceanothenic acid, zizymauritic acids A–C, and ceanothic acid (Grishko et al. 2015)	In vivo hypoglycemic [+] (Jarald et al. 2009; Bhatia and Mishra 2010)	Yaseen et al. (2015)
152.	*Ziziphus nummularia* (Burm. F.) Wight and Arn./Jhar beri, Karkanra/ Rhamnaceae/tree	Young leaves, bark, and root	Leaves are eaten fresh. Powder of roots and bark (5–8 g) is taken daily. Similarly, leaves are soaked in water at nighttime and one cup of that water is drunk early in the morning	Alkaloid, saponins phenolic compound, flavonoid, and tannin (Gupta et al. 2011a; Dureja and Dhiman 2012).	In vivo hypoglycemic [+] (Rajasekaran et al. 2013)	Murad et al. (2013), Yaseen et al. (2015)

153.	*Ziziphus sativa* Gaertn./Bari/Unab, Markhanai/ Rhamnaceae/tree	Leaves	Leaves extract is used. Four to five leaves may be chewed or decoction is taken twice per day	Jujubosides A and B (Otsuka et al. 1978), cyclopeptide alkaloid (Shah et al. 1985), flavonoids: quercetin, kaempferol, and phloretin derivatives (Pawlowska et al. 2009), zizyberanalic acid, zizyberenalic acid, and zizyberanal acid (Grishko et al. 2015)	In vivo hypoglycemic [−] (Anand et al. 1989), in vivo antihyperglycemic [+] (Hussein et al. 2006)	Khan et al. (2015)
154.	*Ziziphus spinosa* (Bunge) Hu ex F. H. Chen/Mada Bera/Rhamnaceae/ shrub	Leaves	Leaves are soaked in water at nighttime. One cup of it is taken early in the morning.	Triterpenoid, saponins, jujuboside B, jujuboside A phenolic acid, ferulic acid, and flavonoid (Zeng et al. 1987)	–	Yaseen et al. (2015)
155.	*Ziziphus vulgaris* Lam./Markha-nai/ Rhamnaceae/tree	Whole plant	Whole plant is soaked in water for whole night and one cup of extract is taken in the morning	3-*O*-robinobioside, quercetin 3-*O*-rutinoside, 3-*O*-α-L-arabinosyl-(1 → 2)-α-L-rhamnoside, 3-*O*-β-D-xylosyl-(1 → 2)-α-L-rhamnoside, 3′,5′-di-C-β-D-glucosylphloretin, 3-*O*-β-D-xylosyl-(1 → 2)-α-L-rhamnoside-4′-*O*-α-L-rhamnoside (Adnan et al. 2014a)	In vivo prevented serum insulin decrease (Goli-malekabadi et al. 2014)	Haq et al. (2011)
156.	*Ziziphus oxyphylla* Edgew/Ber Maloki/ Rhamnaceae/shrub	Leaf	Extract of leaves is taken in morning	Alkaloids, terpenoids, and flavonoids (Mazhar et al. 2015), flavonoid glycosides (Ahmad et al. 2016)	–	Choudhary et al. (2011), Nisar et al. (2007)
157.	*Fragaria indica* Andrews/Jangli booti/Rosaceae/ shrub	Fruit, leaves, whole plant	Paste, juice, or powder of whole plant material is used two times a day	Phenols, flavonoids, anthocyanins, and vitamin C (Badhani et al. 2011)	–	Yaseen et al. (2015)

(continued)

Table 1 (continued)

Sr. #	Botanical name (local name)/ family/habit	Parts used	Recipe	Phytochemicals reported	Pharmacological activity	Ethnobotanical studies
158.	*Prunus amygdalus* stokes var. Amara CD./Nashpatia/ Rosaceae/shrub	Fruit, leaves	Leaves are soaked in clay pot overnight and water extract is taken early in the morning. Moreover, paste obtained from fresh fruit is also used	Anthocyanidin and procyanidin, flavonol glycosides, flavanone glycosides, flavonol aglycones, and phenolic compounds (Esfahlan et al. 2010)	–	Yaseen et al. (2015)
159.	*Prunus persica* Stokes/Aroo/ Rosaceae/shrub	Fruit	Fresh fruit is consumed three times a day	Phenolic compounds (Survay et al. 2010)	–	Yaseen et al. (2015)
160.	*Pyrus malus* L./ Saeb/Rosaceae/tree	Fruit	Fruit juice is used three times a day	Potassium, metabisulphite, and citric acid (Muhammad et al. 2011)	–	Yaseen et al. (2015)
161.	*Rosa damascena* Mill./Gulab, GulSurak/ Rosaceae/herb	Seed	10 g seed powder is used three times a day	Citronellol, geraniol, and nonadecane (Mahboubi 2015)	In vivo hypoglycemic [+] (Gholamhoseinian and Fallah 2009)	Yaseen et al. (2015)
162.	*Rubus ellipticus* Smith/Akha/ Rosaceae/shrub	Fruit	One cup of fresh fruit juice is taken daily	Alkaloids (Lawrence and Gunasekaran 2014)	In vivo antidiabetic [+] (Sharma and Kumar 2011), inhibition of α-glucosidase activity (Latha et al. 2015)	Yaseen et al. (2015)
163.	*Citrus limon* (L.) Burm. F./Lemoo/ Rutaceae/tree	Fruit	One glass of juice is taken three times a day with black salt	Coumarin (Miyake et al. 1999), flavonoids (Del Rio et al. 2004), phenolic acid (González-Molina et al. 2010), carotenoids and vitamin C (Khosa et al. 2011)	In vivo antidiabetic [+] (Naim et al. 2012; Youssef et al. 2013), antihyperglycemic [+] (Shen et al. 2012), wound healing in diabetic rats [+] (Ahmad et al. 2013a)	Yaseen et al. (2015)

164.	*Citrus medica* L./ Sangtra/Rutaceae/ tree	Fruit	One glass of juice is taken three to five times a day with black salt	Flavonoids (Menichini et al. 2011), carbohydrate, protein, and amino acids (Nagaraju et al. 2012)	In vitro antioxidant and inhibition of α-amylase and αglucosidase activity [+] (Menichini et al. 2011),	Yaseen et al. (2015)
165.	*Citrus reticulata* Blanco Cv. Murcot (Honey)/Malta/ Rutaceae/tree	Leaves and rind	Leaves are chewed. Essential oil of rind is used	Phenols (Jayaprakasha et al. 1997), limonoids (Zhang et al. 2014), reticulataursenoside, citrusteryl arachidate, and citruslanosteroside (Khan et al. 2010a)	In vivo antidiabetic [+] (Mehmood et al. 2013)	Mehmood et al. (2013)
166.	*Salix babylonica* Linn./Asela ola/ Salicaceae/tree	Leaves	Extract of leaves is taken daily two times	Phenolics and saponins (Salem et al. 2011a), tritetracontane, octadecenoic acid-1,2,3-propanetriyl ester, hexadecanoic acid-methyl ester (10.5%), and 1,3-dioxane-4-(hexadecyloxy)-2-pentadecyl (Salem et al. 2011b)	–	Haq et al. (2011)
167.	*Dodonaea viscosa* (L.) Jacq./Sanatha/ Sapindaceae/shrub	Leaves	Two to three leaves are chewed daily	Diterpenes (Ortega et al. 2001), kaempferol methyl esters (Teffo et al. 2010), quercetin (Veerapur et al. 2010b)	In vivo antidiabetic [+] (Veerapur et al. 2010a, 2010b; Meenu et al. 2011; Muthukumran et al. 2011)	Ahmad et al. (2004, 2009), Yaseen et al. (2015)
168.	*Bergenia himalaica* Boiss./Badmia/ Saxifragaceae/herb	Rhizome	Rhizomes are boiled in water for 30 min and one cup is used three times a day	Bergenicin and bergelin (Siddiqui et al. 2014), O–H alcoholic/acid, C–H alkyl and amp; aromatic ring, carbonyl, and C–O–C groups (Mohani et al. 2014)	In vivo antihyperglycemic [+] (Siddiqui et al. 2014)	Hussain et al. (2004), Yaseen et al. (2015)
169.	*Bergenia ciliata* (Haw.) Stermb./ Maknar path/ Saxifragaceae/herb	Rhizome	Rhizomes are boiled in water and half cup of extract is used daily	Bergenin, catechin, gallicin, and gallic acid (Dhalwal et al. 2008; Dharmender et al. 2012), steroid, flavonides, and tannins (Uddin et al. 2012)	In vivo hypoglycemic [+] (Islam et al. 2002)	Ali et al. (2011a)

(continued)

Table 1 (continued)

Sr. #	Botanical name (local name)/ family/habit	Parts used	Recipe	Phytochemicals reported	Pharmacological activity	Ethnobotanical studies
170.	*Atropa belladonna* L./Cheelalubar/ Solanaceae/herb	Whole plant, seed	Extract	Atropine, hyoscyamine (Shetty 1997; Chaubey and Pandeya 2011; Biradar 2015), calystegines and glycosides (Asano et al. 2000)	–	Shah and Khan (2006)
171.	*Solanum incanum* L./Mhokri, Marongay/ Solanaceae/shrub	Leaves	Aerial parts are used as vegetable	Flavonoid, saponins, and oxalate (Auta et al. 2011), steroid glycosides (Manase et al. 2012), steroidal alkaloids, glycoalkaloids, flavonoids and chlorogenics, and saponins (Mwonjoria et al. 2014)	In vivo hypoglycemic [+] (Okolie et al. 2009; Wambugu et al. 2014)	Ahmad et al. (2009), Murad et al. (2013), Yaseen et al. (2015)
172.	*Solanum melongena* L./ Bangan, Bengarah/ Solanaceae/herb	Fruit	Cooked as vegetable and used three times a day	Polyamines (Rodriguez et al. 1999), phenolics (Kwon et al. 2008), caffeic acid (Lo Scalzo et al. 2010), caffeic acid methyl ester (Rani and Devanand 2013)	In vitro antioxidant and antihyperglycemic [+] (Kwon et al. 2008)	Khan et al. (2013a, 2013b)
173.	*Solanum nigrum* Lesch.ex Dunal/ Mako/Solanaceae/ herb	Aerial parts	Cooked as vegetable and three times taken in each meal	Alkaloids, flavonoids, steroids, tannins, and phenols (Sridhar et al. 2011; Gogoi and Islam 2012)	In vivo antihyperglycemic [+] (Maharana et al. 2010), in vivo hypoglycemic [+] (Meonah et al. 2012), in vivo hypoglycemic and hypolipidemic [+] (Sengottaiyan et al. 2012)	Ahmad et al. (2004), Yaseen et al. (2015)
174.	*Solanum surattense* Burm.f./Maragho one, Kandiari/ Solanaceae/herb	Fruit	Decoction of fresh fruit is used. Dried fruit powder is taken three times a day	Tannins and phenols, gum, and mucilage (Shahiladevi et al. 2006), β-sitosterol (Gupta et al. 2011c), steroidal alkaloid saponins and steroidal saponins (Lu et al. 2011)	In vivo antidiabetic and antioxidant [+] (Sridevi et al. 2007; Gupta et al. 2011c)	Yaseen et al. (2015)

175.	*Withania coagulans* (L.) Dunal./Paneer/Solanaceae/shrub	Seeds, fruit	Water-soaked seeds (15 g) are eaten before breakfast. Powder of fruit and seed is mixed with wheat flour, butter oil, and sugar and used daily	Steroidal lactones (Ali et al. 2015a; Zhang and Timmermann 2016), withacoagulin G, withacoagulin H, and withacoagulin I (Youn et al. 2013)	In vivo hypoglycemic [+] (Hemalatha et al. 2004; Jaiswal et al. 2009; Yasir et al. 2012).	Ahmad et al. (2004, 2009), Yaseen et al. (2015)
176.	*Withania somnifera* (L.) Dunal/Asghanh, Verian/Solanaceae/herb	Leaves and root	Decoction is used once daily	B-sitosterol, stigmasterol, β-sitosterol glucoside, stigmasterol glucoside, and α + β glucose (Misra et al. 2008), steroidal alkaloids and steroidal lactones (Verma and Kumar 2011)	Hypoglycemic effect on human [+] (Andallu and Radhika 2000), in vivo hypoglycemic and hypolipidemic [+] (Udayakumar et al. 2009)	Ahmad et al. (2014a), Shah et al. (2013a)
177.	*Tamarix aphylla* (L.) Karst./Rukhh/Tamaraceae/tree	Fruit	Fruit extract is prepared after boiling for few minutes in water and one cup is taken three times a day	Phenolics (Souliman et al. 1991), polyphenols (Mahfoudhi et al. 2014), flavonoids (Shafaghat 2010)	–	Yaseen et al. (2015)
178.	*Grewia asiatica* L./Falsa/Tiliaceae/shrub	Leaves	Fresh fruit is eaten daily for 1 month	Polyphenols (Siddiqi et al. 2011), alkaloids, carbohydrates, glycosides, proteins and amino acids, saponins, steroids, acids, mucilage, fixed oils, and fats (Zia-Ul-Haq et al. 2013b), alkaloids, tannins, anthraquinones, glycosides, saponins, flavonoids, steroids, coumarins, and resins (Sharma and Patni 2013)	In vivo antihyperglycemic [+] (Parveen et al. 2012; Khattab et al. 2015)	Zia-Ul-Haq et al. (2012a, 2012b), Akhtar (1992), Adnan et al. (2014b), Khan et al. (2015)

(continued)

Table 1 (continued)

Sr. #	Botanical name (local name)/ family/habit	Parts used	Recipe	Phytochemicals reported	Pharmacological activity	Ethnobotanical studies
179.	*Carum carvi* L./ Kala Zeera/ Apiaceae/herb	Seeds	Seeds are boiled in water and extract is used once a day	Carvone, limonene, β-myrcene, *trans*-dihydrocarvone, and *trans*-carveol (Raal et al. 2012), (R)-carvone, D-limonene, α-pinene, *cis*-carveol, and β-myrcene (Agrahari and Singh 2014)	In vivo hypoglycemic activity [+] (Eddouks et al. 2004; Eidi et al. 2010)	Sadiq et al. (2010)
180.	*Lantana camara* L./Panch phulli/ Verbenaceae/shrub	Fruit	Water extract of fresh fruit is prepared by soaking throughout the night and taken in the morning	Lantadene D (Sharma et al. 1990), alkaloids, tannin, flavonoids, and triterpenoids (Patel et al. 2011), Ursolic acid stearoyl glucoside (Kazmi et al. 2012)	In vivo antidiabetic [+] (Dash et al. 2001; Kazmi et al. 2012)	Yaseen et al. (2015)
181.	*Phyla nodiflora* (L.) Greene/ Bakaanrah/ Verbenaceae/herb	Seed	Seeds are boiled and extract is used daily one time	Alkaloids, flavonoids, tannin, steroids, terpenoids, saponins, and anthraquinones (Priya and Ravindhran 2015)	In vivo antidiabetic, hypolipidemic, and antioxidant [+] (Balamurugan et al. 2011; Balamurugan and Ignacimuthu 2011; Subramanian et al. 2011a)	Ullah et al. (2014)
182.	*Fagonia cretica* Burm.f./Spelaghza, Dha mana/ Zygophyllaceae/ herb	Whole plant	Plant extract is used once in the morning daily	Triterpenoid compounds: saponin I and saponin II (Saeed and Sabir 1999, Khalik et al. 2000)	In vitro potent dipeptidyl peptidase-4 (DPP-4) inhibitory activity [+] (Saleem et al. 2014)	Murad et al. (2013)

Table 2 Herbal formulations used for treatment of diabetes mellitus in Pakistan

Sr. #	Botanical name	Parts used	Recipe	References
1.	*Aloe vera* Nill + *Fagonia indica* L. + *Tinospora cordifolia* (Thunb.) Miers.	Leaf pulp of *A. vera*, aerial parts of *F. indica*, branches of *T. cordifolia*	Extracts of parts of three plants in equal amount are mixed, and one teaspoon is taken three times a day	Ahmed et al. (2007)
2.	*Aloe vera* Nill + *Fagonia indica* L. + *Tylophora hirsuta* L.	Leaf pulp of *A. vera*, aerial parts of *F. indica*, branches of *T. hirsuta*	Extracts of parts of three plants in equal amount are mixed, and one teaspoon is taken three times a day	Ahmad et al. (2004, 2009), Yaseen et al. (2015)
3.	*Tylophora hirsuta* (Wall.)Wight + *Trigonella foenum-graecum* L. + *Fumaria indica* (Haussk) Pugsely	*T. hirsuta* leaves, *T. foenum-graecum* seeds, and aerial parts of *F. indica*	The equal quantity of each plant is ground and made into powder. One tablespoon is taken twice daily	Ahmad et al. (2004, 2009)
4.	*Rhazya stricta Decne* + *Withania somnifera* (L.) Dunal	*R. stricta* leaves and whole plant and *W. somnifera* leaves and roots	Parts of both plants are boiled, and decoction is used once a day	Ahmad (2006), Hussain et al. (2010c), Qureshi (2012), Shah et al. (2013a), Yaseen et al. (2015)
5.	*Capparis spinosa* L. + *Rhazya stricta* Decne.	*C. spinosa* leaves, whole plant of *R. stricta*	Extract of both plants are mixed, and two spoons are taken twice a day	Shah et al. (2013a), Yaseen et al. (2015)
6.	*Cajanus cajan* (L.) Druce and *Vigna mungo* (Burm. f.) Walp.	Seed	Both pulses are mixed in equal quantity, cooked, and taken two times a day.	Ahmad et al. (2004, 2009)
7.	*Cicer arietinum* L. + *Daucus carota* L. + *Hordeum vulgare* L. + *Oryza sativa* L. + *Triticum aestivum* L. + and *Zea mays* L.	Seeds	Seeds are dried, and their flour is mixed in equal quantity to prepare antidiabetic bread at home	Ahmad et al. (2009)
8.	*Ocimum sanctum* L. + Ocimum *album* L.	Leaves	1 gram dried leaf powder is taken twice a day	Ahmad et al. (2004, 2009)
9.	*Melia azedarach* A. Juss., *Phyllanthus emblica* L., and *Terminalia chebula* Retz.	Dry fruit	Paste of all herbs is prepared manually, mixed together, and used orally daily	Abbasi et al. (2010a)
10.	*Syzygium cumini* Skeels + *Momordica charantia* L. + *Cyperus rotundus* L. + *Rosa alba* L.	Seeds of *S. cumini*, fruit of *M. charantia*, seeds of *C. Rotundus*, and seeds of *Rosa alba* L.	*S. cumini* (25 g), *M. charantia* (12 g), *C. rotundus* (12 g), and *R. alba* (12 g) are made into powder for use one time daily	Ahmad et al. (2009)

(continued)

Table 2 (continued)

Sr. #	Botanical name	Parts used	Recipe	References
11.	*Hordeum vulgare* L. + *Cicer arietinum* L. + *Elettaria cardamomum* Maton	Seeds	125 g roasted seeds of *H. vulgare* and 50 g seeds of each *C. arietinum* and *E. cardamomum* are mixed, and half spoon is prescribed three times a day	Ahmad et al. (2004, 2009), Yaseen et al. (2015)
12.	*Fragaria nubicola* Lindl. ex Lacaita	Fruit	Fresh fruit is consumed. The juice of the fruit is mixed with bark powder of *Berberis lycium* and used three times a day	Khan et al. (2013a, b), Yaseen et al. (2015)
13.	*Bergenia ligulata* (Wall.) Engl. + *Asteracantha longifolia* (L.) Nees in Wall + *Argyreia speciosa* (Linn. f.) Sweet + *Cinnamomum cassia* (Nees T. Nees) J. Presl	*B. ligulata* (roots) + *A. longifolia* (seeds) + *A. speciosa* (roots) + *C. cassia* (bark)	Equal weight of these plant parts are mixed in powder form and used early in the morning daily	Akhtar (1992)
14.	*Syzygium cumini* Skeels + *Papaver somniferous* L.	Seeds of *S. cumini* and opium water extract	Seeds (11.66 g) of *S. cumini* are ground in the opium extract, and 32 tablets are made and are used in the morning and evening	New record
15.	*Santalum album* L. + *Astragalus gummifer* Labill. + *Lactuca sativa* L. + *Portulaca oleracea* L. + *Acacia Arabica* Lam. + *Punica granatum* L. + *Rhus coriaria* L. + *Quercus baloot* Griffith	Bark of *S. album* (3 2.91 g), gum of *A. gummifer* (10.63 g), seeds of *L. sativa* (10.63 g), seeds of *P. oleracea* (10.63 g), gum of *A. arabica* (17. 39 g), flower of *P. granatum* (17. 39 g), fruit of *Rhus coriaria* L. (17. 39 g), fruit of *Quercus baloot* Griffith (17. 39 g)	All ingredients are made into powder and mixed with (17. 39 g) natural aluminum silicate containing iron oxide. On spoon is taken twice with water	New record
16.	*Tinospora cordifolia* (Thunb.) Miers. + (*Gymnema sylvestre* R. Br.)	Flower of *T. cordifolia* (11.66 g), whole plant of *G. sylvestre* (11.66 g))	The plant parts are ground and mixed with brown sugar. Two tablespoon of this powder is soaked in water whole night and stained in the morning, and extract is taken with pomegranate juice	New record

17.	*Chlorophytum borivilianum* Sant et. Fernand., + *Asparagus racemosus* Willd. + *Phyllanthus emblica* L. + *Vitis venefera* L. + *Pistacio lentiscus* L. + *Tribulus terrestris* Linn. + *Tinospora cordifolia* (Thunb.) Miers. + *Bumboosa vulgaris* Schrad ex J.C. Wendl. + *Coleus barbatus* Benth. + *Centella asiatica* (L.) Urban	Roots of *C. borivilianum*, roots and stem of *A. racemosus* Willd., fruit of *P. emblica*, fruit of *V. vinifera*, Mustagi of *P. lentiscus*, seed of *T. terrestris*, leaves and branches of *T. cordifolia*, Tabshir/bamboo resin of *B. vulgaris* Schrad ex J.C. Wendl., Pakhan Baid leaves and roots of *C. barbatus*, Gul mandi flowers of *C. asiatica* (L.) Urban	All the herbs are taken in equal quantity, or 5.82 g of each herb is ground separately and mixed with 5.82 g of Salajit and 34.98 g of old bricks powder of a well. It is used in the morning daily with water	New record
18.	*Gossypium herbaceum* L.	Seed	11.66 g seeds are ground to powder and 23.32 g of raw sugar is mixed into it. Mixture is boiled till it becomes thick. It is eaten once in a day	New record
19.	*Cyperus rotundus* L. + *Boswellia glabra* Roxb. + *P. lentiscus* L. + *Quercus incana* W. Bartram + *Papaver somniferous* L., *Bambusa vulgaris* Schrad ex J.C. Wendl. + *Coriandrum sativum* + *P. granatum* + *Rosa damascene* + *Cinnamomum cassia*	Roots and tubers of *C. rotundus* L., gum of *B. glabra* Roxb., fruit of *P. lentiscus*, fruit of *Q. incana* W. Bartram, seed *P. somniferous* L., Tabashir/bamboo resin of *B. vulgaris* Schrad ex J.C. Wendl., seeds of coriander (*C. sativum*), flower of *P. granatum*, flower of *Rosa damascena*, bark of *C. cassia*.	2.91 g of each herb is dried, ground, and mixed together along with 2.91 g natural aluminum silicate containing iron oxide and brown sugar. Two tablespoons of mixture are taken with water daily	New record

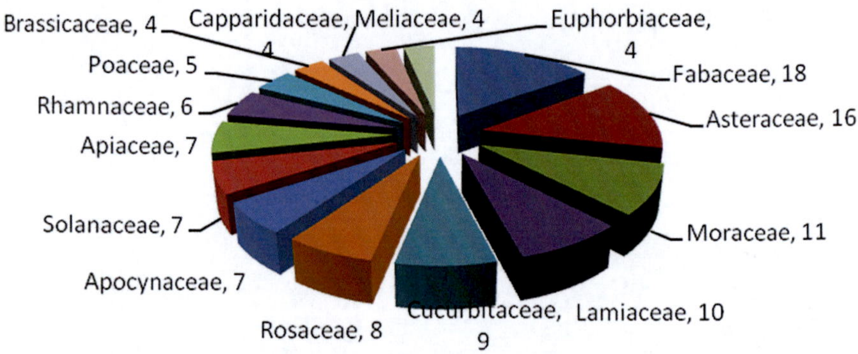

Fig. 1 Contribution of top 15 families for treatment of diabetes mellitus

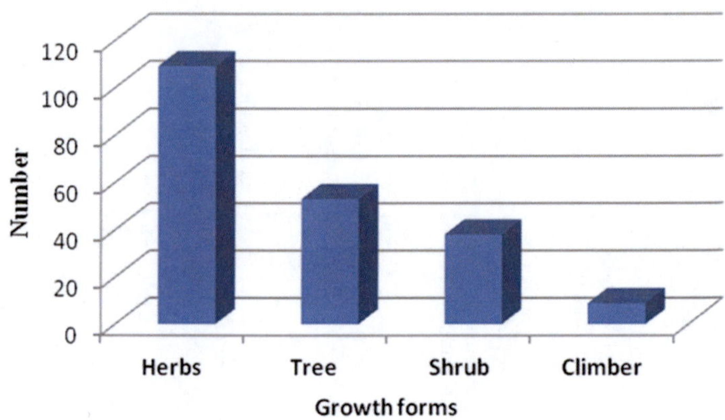

Fig. 2 Growth forms of plants used for treatment of diabetes mellitus

Life Forms

The antidiabetic plant species exhibited four growth forms in which herbs (109) were highly utilized, followed by trees (53), shrub (38), and climbers (9) as shown in Fig. 2. The preference of herbs as antidiabetic drug may be attributed due to their easy availability from the localities (Ayyanar and Ignacimuthu 2005; Giday et al. 2009; Ragupathy et al. 2008; Sanz-Biset et al. 2009; Shrestha and Dhillion 2003; Tabuti et al. 2012; Uniyal et al. 2006) and easy harvesting (Giday et al. 2003). Besides, they may possess certain active ingredients (Giday et al. 2009; Lulekal et al. 2013; Mesfin et al. 2009; Teklehaymanot 2009).

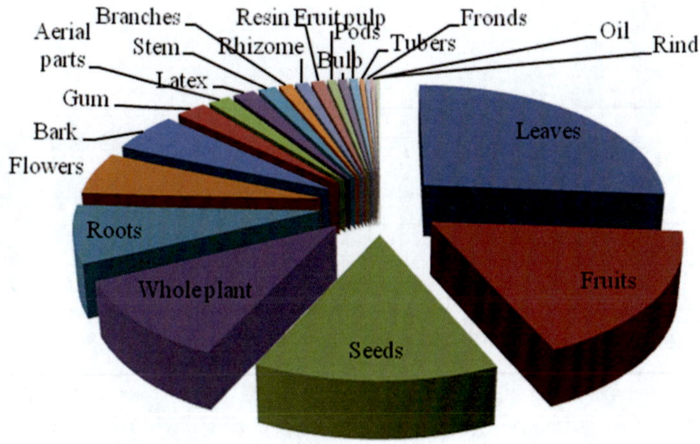

Fig. 3 Parts of plants used for treatment of diabetes mellitus

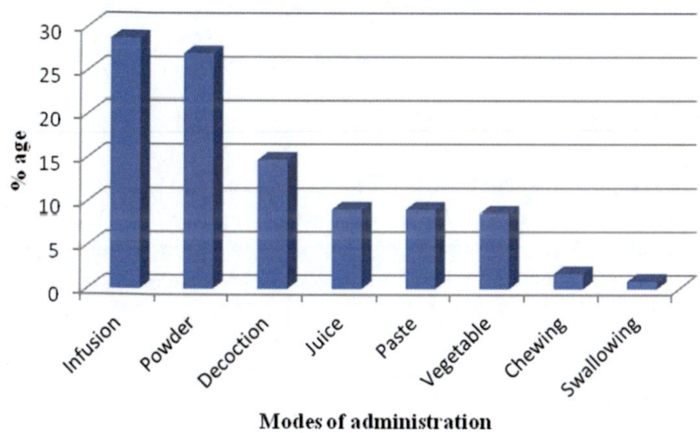

Fig. 4 Traditional modes of administration of plant-based recipes for DM treatment

Plant Parts and Mode of Administration

Eighteen various plant parts were exploited in treating DM by the THPs, in which leaves were the most frequently used (78 reports), followed by fruit, seed, whole plant, and root (Fig. 3). The high usage of leaves in the folk medicines clearly indicates their easy accessibility and availability to the local communities. The harvesting of leaves does not harm the plant growth compared to roots, rhizome, or bulb. Besides, gathering of leaves sustainably helps conserve plant diversity. It is suggested that the frequent use of the leaves is due to their high healing power (Mootoosamy and Mahomoodally 2014). Moreover, leaves of antidiabetic plants like walnut possess α-glucosidase inhibitor that reduces glucose level (Said et al. 2008).

Figure 4 represents the traditional modes of administration for recipes of DM. The most popular mode of preparation of recipe was infusion (28.70%), followed

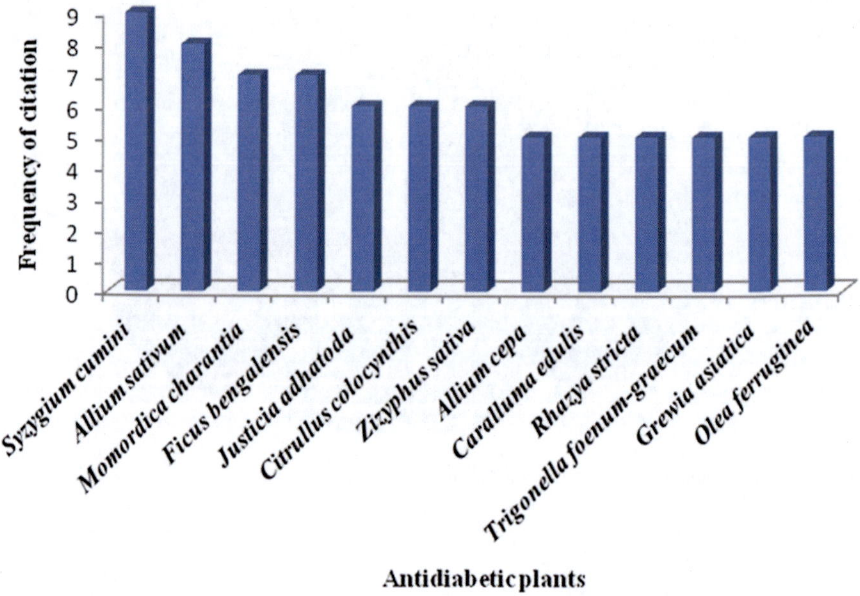

Fig. 5 Most frequently used plant species

by powder (26.96%), decoction (14.78%), juice/paste (9.13% each), cooking vegetable (8.70%), chewing (1.74%), and swallowing (0.87%). Most of the study reported decoction as the formal mode of preparation of the herbal recipe for treating diabetes (Diallo et al. 2012; Kadir et al. 2012; Nowbandegani et al. 2015).

Frequency of Citation (FC)

In all, 13 species were frequently quoted for treating diabetes as shown in Fig. 5. *Syzygium cumini* had highest frequency of citation (FC = 9), followed by *Allium sativum* (FC = 8); *Momordica charantia* and *Ficus benghalensis* (FC = 7 each); *Justicia adhatoda*, *Citrullus* colocynthis, and *Ziziphus sativa* (FC = 6 each); and *Allium cepa*, *Caralluma edulis*, *Rhazya stricta*, *Trigonella foenum-graecum*, *Grewia asiatica*, and *Olea ferruginea* (FC = 5 each). The rest of the species possessed FC in the range of 4–1 (Table 1). Of them, except *C. edulis* and *O. ferruginea*, the rest of the 11 species are scientifically proved to have antidiabetic potential such as *S. cumini* (Prince et al. 2003, 2004), *A. sativum* (Islam and Choi 2008; Thomson et al. 2007), *M. charantia* (Ahmed et al. 2001), *F. benghalensis* (Achrekar et al. 1991), *J. adhatoda* (Gulfraz et al. 2011), *C. colocynthis* (Gurudeeban and Ramanathan 2010), *A. cepa* (Augusti 1996; Kumari and Augusti 2002), *R. stricta* (Tanira et al. 1996), *T. foenum-graecum* (Vats et al. 2002), and *G. asiatica* (Parveen et al. 2012) as reflected in the Table 1. It is interesting to note that a large number of species are yet to be verified through scientific endeavor and are promising candidates for scientific validation based on this survey.

Conclusion

This study presents a first comprehensive review-cum-survey of antidiabetic plants used traditionally solely or in mixture by the traditional healers of Pakistan. Through this endeavor, efforts were undertaken to identify plant resources which may lead to the development of indigenous and inexpensive sources of new antidiabetic medicaments. The frequency of citation revealed some potential species (e.g., *Caralluma edulis*) as candidate which can be utilized in drug discovery program. Pharmacological studies of 39 species and phytochemistry of 8 species used solo are not carried out previously. Their further clinical studies are strongly recommended to evaluate long-term efficacy and toxicity in diabetic patients to scientifically validate their use in traditional medicines.

References

Abbasi AM, Khan M, Ahmad M, Zafar M, Jahan S, Sultana S (2010a) Ethnopharmacological application of medicinal plants to cure skin diseases and in folk cosmetics among the tribal communities of North-West Frontier Province, Pakistan. J Ethnopharmacol 128(2):322–335

Abbasi AM, Khan MA, Ahmed M, Zafar M (2010b) Herbal medicines used to cure various ailments by the inhabitants of Abbottabad district, North West Frontier Province, Pakistan. Indian J Tradit Knowl 9:175–183

Abbasi MA, Shahzadi T, Ahmad VU (2014) New Iridoid Glucosides from *Caryopteris Odorata* with Suitable Antioxidant Potential. Chem Nat Compd 50:836–841

Abd E-MA, Mohamed KM, Mostafa AM (2010) Induction of biologically active flavonoids in cell cultures of *Morus nigra* and testing their hypoglycemic efficacy. Sci Pharm 79:951–961

Abdel-Hameed E-SS (2009) Total phenolic contents and free radical scavenging activity of certain Egyptian Ficus species leaf samples. Food Chem 114:1271–1277

Abdelmeguid NE, Fakhoury R, Kamal SM, Al Wafai RJ (2010) Effects of *Nigella sativa* and thymoquinone on biochemical and subcellular changes in pancreatic β-cells of streptozotocin-induced diabetic rats. J Diabetes 2:256–266

Abdel-Sattar E, Harraz FM, Ghareib SA, Elberry AA, Gabr S, Suliaman MI (2011) Antihyperglycaemic and hypolipidaemic effects of the methanolic extract of *Caralluma tuberculata* in streptozotocin-induced diabetic rats. Nat Prod Res 25:1171–1179

Abdel-Sattar EA, Abdallah HM, Khedr A, Abdel-Naim AB, Shehata IA (2013) Antihyperglycemic activity of *Caralluma tuberculata* in streptozotocin-induced diabetic rats. Food Chem Toxicol 59:111–117

Abo-Ellil MEA (2014) Phytochemical and biological study of some plants belonging to family araceae. PhD thesis. Mansoura University, Mansoura

Abou-Arab AE, Abou-Arab AA, Abu-Salem MF (2010) Physico-chemical assessment of natural sweeteners steviosides produced from *Stevia rebaudiana* Bertoni plant. Afr J Food Sci 4:269–281

Abu-Zaiton AS (2010) Anti-diabetic activity of Ferula assafoetida extract in normal and alloxan-induced diabetic rats. Pak J Biol Sci 13:97

Achrekar S, Kaklij G, Pote M, Kelkar S (1991) Hypoglycemic activity of *Eugenia jambolana* and *Ficus benghalensis*: mechanism of action. In Vivo 5:143

Adaramoye O, Amanlou M, Habibi-Rezaei M, Pasalar P, Ali M-M (2012) Methanolic extract of African mistletoe (*Viscum album*) improves carbohydrate metabolism and hyperlipidemia in streptozotocin-induced diabetic rats. APJTM 5:427–433

Aderibigbe A, Emudianughe T, Lawal B (2001) Evaluation of the antidiabetic action of *Mangifera indica* in mice. Phytother Res 15:456–458

Adnan M, Bibi R, Mussarat S, Tariq A, Shinwari ZK (2014a) Ethnomedicinal and phytochemical review of Pakistani medicinal plants used as antibacterial agents against *Escherichia coli*. Ann Clin Microbiol Antimicrob 13(40):1–18

Adnan M, Ullah I, Tariq A, Murad W, Azizullah A, Khan AL, Ali N (2014b) Ethnomedicine use in the war affected region of Northwest Pakistan. J Ethnobiol Ethnomed 10(16):1–16

Agarwal V, Sharma AK, Upadhyay A, Singh G, Gupta R (2012) Hypoglycemic effects of *Citrullus colocynthis* roots. Acta Pol Pharm 69:75–79

Agrahari P, Singh DK (2014) A review on the pharmacological aspects of *Carum carvi*. J Biol Earth Sci 4:M1–M13

Ahmad M (2006) Checklist of medicinal flora of Tehsil Isakhel, District Mianwali-Pakistan. Ethnobot Leaflets 10:41–48

Ahmad M, Alamgeer ST (2009) A potential adjunct to insulin: *Berberis lycium* Royle. Diabetol Croat 38:13–18

Ahmad M, Khan MA, Arshad M, Zafar M (2004) Ethnophytotherapical approaches for the treatment of diabetes by the local inhabitants of district Attock (Pakistan). Ethnobot Leaflets 2004:7

Ahmad M, Zaman F, Sharif T, Ch MZ (2008) Antidiabetic and hypolipidemic effects of aqueous methanolic extract of *Acacia nilotica* pods in alloxan-induced diabetic rabbits. Scand J Lab Anim Sci 35:29–34

Ahmad M, Qureshi R, Arshad M, Khan MA, Zafar M (2009) Traditional herbal remedies used for the treatment of diabetes from district Attock (Pakistan). Pak J Bot 41:2777–2782

Ahmad M, Ansari M, Alam A, Khan T (2013a) Oral dose of citrus peel extracts promotes wound repair in diabetic rats. Pak J Biol Sci 16:1086

Ahmad R, Upadhyay A, Ahmad M, Pieters L (2013b) Antioxidant, antliglycation and antimicrobial activities of *Ziziphus oxyphylla* and *Cedrela serrata* extracts. Eur J Med Plants 3(4):520–529

Ahmad S, Wariss HM, Alam K, Anjum S, Mukhtar M (2014a) Ethnobotanical studies of plant resources of Cholistan desert, Pakistan. IJSR 3(6):1781–1788

Ahmad W, Khan I, Khan MA, Ahmad M, Subhan F, Karim N (2014b) Evaluation of antidiabetic and antihyperlipidemic activity of *Artemisia indica* linn (aeriel parts) in Streptozotocin induced diabetic rats. J Ethnopharmacol 151(1):618–623

Ahmad R, Ahmad N, Naqvi AA, Exarchou V, Upadhyay A, Tuenter E, Foubert K, Apers S, Hermans N, Pieters L (2016) Antioxidant and Antiglycating Constituents from Leaves of *Ziziphus oxyphylla* and *Cedrela serrata*. Antioxid 5(1):1–9

Ahmed F, Urooj A (2009) Antioxidant activities, of various extracts of *Ficus racemosa* stem bark. Nat Life Sci 6:69–74

Ahmed F, Urooj A (2010) In vitro studies on the hypoglycemic potential of *Ficus racemosa* stem bark. J Sci Food Agric 90:397–401

Ahmed I, Lakhani M, Gillett M, John A, Raza H (2001) Hypotriglyceridemic and hypocholesterolemic effects of anti-diabetic *Momordica charantia* (karela) fruit extract in streptozotocin-induced diabetic rats. Diabetes Res Clin Pract 51:155–161

Ahmed M, Khan MA, Zafar M, Sultana S (2007) Treatment of common ailments by plant-based remedies among the people of district Attock (Punjab) of northern Pakistan. Afr J Trad Complement Alternat Med 4:112–120

Ahmed Z, Bhagat A, Gupta OP, Gupta KK, Ram G, Qazi GN (2008) Insulin secretagogue fraction of *Argyrolobium roseum*. Diabetol Croat 37:3–12

Ahmed D, Kumar V, Sharma M, Verma A (2014a) Target guided isolation, in-vitro antidiabetic, antioxidant activity and molecular docking studies of some flavonoids from *Albizzia Lebbeck* Benth. bark. BMC Complement Altern Med 14:1

Ahmed D, Kumar V, Verma A, Gupta PS, Kumar H, Dhingra V, Mishra V, Sharma M (2014b) Antidiabetic, renal/hepatic/pancreas/cardiac protective and antioxidant potential of methanol/dichloromethane extract of *Albizzia Lebbeck* Benth. stem bark (ALEx) on streptozotocin induced diabetic rats. BMC Complement Altern Med 14(243):1–17

Ahn J, Choi W, Kim S, Ha T (2011) Anti-diabetic effect of watermelon (*Citrullus vulgaris* Schrad) on Streptozotocin-induced diabetic mice. Food Sci Biotechnol 20:251–254

Aissaoui A, Zizi S, Israili ZH, Lyoussi B (2011) Hypoglycemic and hypolipidemic effects of *Coriandrum sativum* L. in Meriones shawi rats. J Ethnopharmacol 137:652–661

Aja P, Igwenyi I, Okechukwu PU, Orji O, Alum E (2015) Evaluation of anti-diabetic effect and liver function indices of ethanol extracts of *Moringa oleifera* and *Cajanus cajan* leaves in alloxan induced diabetic albino rats. Glob Vet 14:439–447

Akhtar MS (1992) Hypoglycaemic activities of some indigenous medicinal plants traditionally used as antidiabetic drugs. J Pak Med Assoc 42:271–271

Akhtar N, Begum S (2009) Ethnopharmacological important plants of Jalala, district Mardan, Pakistan. Pak J Sci 15:95–100

Akhtar MS, Iqbal J (1991) Evaluation of the hypoglycaemic effect of *Achyranthes aspera* in normal and alloxan-diabetic rabbits. J Ethnopharmacol 31:49–57

Akter R, Mahabub-Uz-Zaman M, Rahman Md S, Khatun M, Abdullah A, Ahmed N, Islam F (2013) Comparative studies on antidiabetic effect with phytochemical screening of *Azadirachta indica* and *Andrographis paniculata*. J Pharm Biol Sci 5:2278–3008

Akther F, Rahman A, Proma JJ, Kabir MZ, Paul PK, Rahmatullah M (2014) Methanolic extract of *Luffa cylindrica* fruits show antihyperglycemic potential in Swiss albino mice. ANAS 8:62–66

Alam N, Shinwari Z, Ilyas M, Ullah Z (2011) Indigenous knowledge of medicinal plants of Chagharzai valley, District Buner, Pakistan. Pak J Bot 43:773–780

Alarcon-Aguilara F, Roman-Ramos R, Perez-Gutierrez S, Aguilar-Contreras A, Contreras-Weber C, Flores-Saenz J (1998) Study of the anti-hyperglycemic effect of plants used as antidiabetics. J Ethnopharmacol 61:101–110

Al-Azzawie HF, Alhamdani M-SS (2006) Hypoglycemic and antioxidant effect of oleuropein in alloxan-diabetic rabbits. Life Sci 78:1371–1377

Ali B (1997) The effect on plasma glucose, insulin and glucagon levels of treatment of diabetic rats with the medicinal plant *Rhazya stricta* and with glibenclamide, alone and in combination. J Pharm Pharmacol 49.1003–1007

Ali N (2013) Brine shrimp cytotoxicity of crude methanol extract and antispasmodic activity of α-amyrin acetate from Tylophora hirsuta Wall. BMC Complement Altern Med 13:135

Ali H, Sannai J, Sher H, Rashid A (2011a) Ethnobotanical profile of some plant resources in Malam Jabba valley of Swat, Pakistan. J Med Plant Res 5:4676–4687

Ali N, Shah SW, Shah I, Ahmed G, Ghias M, Khan I (2011b) Cytotoxic and anthelmintic potential of crude saponins isolated from *Achillea Wilhelmsii* C. Koch and *Teucrium Stocksianum* boiss. BMC Complement Altern Med 11:106

Ali A, Maher S, Khan SA, Chaudhary MI, Musharraf SG (2015a) Sensitive quantification of six steroidal lactones in *Withania coagulans* extract by UHPLC electrospray tandem mass spectrometry. Steroids 104:176–181

Ali H, Uddin S, Jalal S (2015b) Chemistry and biological activities of *Berberis lycium* Royle. JBAPN 5:295–312

Al-Malki AL, El Rabey HA (2015) The antidiabetic effect of low doses of *Moringa oleifera* Lam. seeds on streptozotocin induced diabetes and diabetic nephropathy in male rats. Biomed Res Int 2015:1–13

Al-Muammar MN, Khan F (2012) Obesity: the preventive role of the pomegranate (*Punica granatum*). J Nutr 28:595–604

Amin A, Cos P, Maes L, Apers S, Exarchou V, Pieters L (2015) Phytochemical and pharmacological investigation of *Kickxia ramosissima*. Planta Med 81(16):YRW_04

Anand K, Singh B, Grand D, Chandan B, Gupta V (1989) Effect of *Zizyphus sativa* leaves on blood glucose levels in normal and alloxan-diabetic rats. J Ethnopharmacol 27:121–127

Andallu B, Radhika B (2000) Hypoglycemic, diuretic and hypocholesterolemic effect of winter cherry (*Withania somnifera*, Dunal) root. Indian J Exp Biol 38:607–609

Andoğan BC, Baydar H, Kaya S, Demirci M, Özbaşar D, Mumcu E (2002) Antimicrobial activity and chemical composition of some essential oils. Arch Pharm Res 25:860–864

Andrade-Cetto A, Heinrich M (2005) Mexican plants with hypoglycaemic effect used in the treatment of diabetes. J Ethnopharmacol 99:325–348

Anis E, Anis I, Ahmed S, Mustafa G, Malik A, Afza N, Hai SMA, Shahzad-ul-hussan S, Choudhary MI (2002) ALPHA.-Glucosidase Inhibitory Constituents from *Cuscuta reflexa*. Chem Pharm Bull 50:112–114

Ansari N, Houlihan L, Hussain B, Pieroni A (2005) Antioxidant activity of five vegetables traditionally consumed by south-Asian migrants in Bradford, Yorkshire, UK. Phytother Res 19:907–911

Anwar MM, Kalpana MA, Bhadra B, Rahman S, Sarker S, Chowdhury MH, Rahmatullah M (2010) Antihyperglycemic activity and brine shrimp lethality studies on methanol extract of *Cajanus cajan* (L.) Millsp. leaves and roots. Adv Nat Appl Sci 4:311–317

Apu AS, Liza MS, Jamaluddin A, Howlader MA, Saha RK, Rizwan F, Nasrin N (2012) Phytochemical screening and in vitro bioactivities of the extracts of aerial part of *Boerhavia diffusa* Linn. Asian Pac J Trop Biomed 2:673–678

Ara N, Rashid M, Amran MS (2008) Comparison of *Moringa oleifera* leaves extract with atenolol on serum triglyceride, serum cholesterol, blood glucose, heart weight, body weight in adrenaline induced rats. Saudi J Biol Sci 15:253–258

Arshad M, Nisar MF, Majeed A, Ismail S, Ahmad M (2011) Ethnomedicinal flora in district Sialkot, Punjab, Pakistan. Middle-East J Sci Res 9:209–214

Asad M, Aslam M, Munir TA, Nadeem A (2011) Effect of *Acacia nilotica* leaves extract on hyperglycaemia, lipid profile and platelet aggregation in streptozotocin induced diabetic rats. J Ayub Med Coll Abbottabad 23:3–7

Asadujjaman M, Hossain M, Khan M, Anisuzzaman A, Ahmed M, Islam A (2011) Antihyperglycemic and glycogenesis effects of different fractions of *Brassica oleracea* in Alloxan induced diabetic rats. IJPSR 2:1436

Asano N, Nash RJ, Molyneux RJ, Fleet GW (2000) Sugar-mimic glycosidase inhibitors: natural occurrence, biological activity and prospects for therapeutic application. Tetrahedron Asymmetry 11:1645–1680

Ashraf A, Sarfraz RA, Anwar F, Shahid SA, Alkharfy KM (2015) Chemical composition and biological activities of leaves of *Ziziphus mauritiana* L. native to Pakistan. Pak J Bot 47:367–376

Asif M (2015) Chemistry and antioxidant activity of plants containing some phenolic compounds. Chem Int 1:35–52

Assad T, Khan RA, Feroz Z (2014) Evaluation of hypoglycemic and hypolipidemic activity of methanol extract of *Brassica oleracea*. Chin J Nat Med 12:648–653

Atangwho I, Ebong P, Eyong E, Williams I, Eten M, Egbung G (2009) Comparative chemical composition of leaves of some antidiabetic medicinal plants: *Azadirachta indica*, *Vernonia amygdalina* and *Gongronema latifolium*. Afr J Biotechnol 8(18):4685–4689

Augusti K (1996) Therapeutic values of onion (*Allium cepa* L.) and garlic (*Allium sativum* L.). Indian J Exp Biol 34:634–640

Auta R, James S, Auta T, Sofa E (2011) Nutritive value and phytochemical composition of processed *Solanum incanum* (Bitter garden egg). Sci World J 6:5–6

Awan MR, Iqbal Z, Shah SM, Jamal Z, Jan G, Afzal M, Majid A, Gul A (2011) Studies on traditional knowledge of economically important plants of Kaghan Valley, Mansehra District, Pakistan. J Med Plant Res 5:3958–3967

Ayyanar M, Ignacimuthu S (2005) Traditional knowledge of kani tribals in Kouthalai of Tirunelveli hills, Tamil Nadu, India. J Ethnopharmacol 102:246–255

Ayyanar M, Subash-Babu P (2012) *Syzygium cumini* (L.) Skeels: A review of its phytochemical constituents and traditional uses. Asian Pac J Trop Biomed 2:240–246

Azay-Milhau J, Ferrare K, Leroy J, Aubaterre J, Tournier M, Lajoix A-D, Tousch D (2013) Antihyperglycemic effect of a natural chicoric acid extract of chicory (*Cichorium intybus* L.): a comparative in vitro study with the effects of caffeic and ferulic acids. J Ethnopharmacol 150:755–760

Badgujar SB, Patel VV, Bandivdekar AH (2014) *Foeniculum vulgare* Mill: a review of its botany, phytochemistry, pharmacology, contemporary application, and toxicology. Biomed Res Int 2014:32

Badhani A, Sakalani S, Mishra A (2011) Variation in Biochemical's and antioxidant activity of some wild edible fruits of uttarakhand. Report and Opinion 3:1–10
Badole SL, Bodhankar SL (2009) Investigation of antihyperglycemic activity of *Glycine max* (L.) Merr. on serum glucose level in diabetic mice. J Complement Integr Med 6(1):1553
Badr SE, Shaaban M, Elkholy YM, Helal MH, Hamza AS, Masoud MS, El Safty MM (2011) Chemical composition and biological activity of ripe pumpkin fruits (*Cucurbita pepo* L.) cultivated in Egyptian habitats. Nat Prod Res 25:1524–1539
Bae IY, Choi AS, Lee HG (2015) Impact of buckwheat flavonoids on *in vitro* starch digestibility and noodle making property. Cereal Chem 93(3):299–305
Bafna A, Mishra S (2010) Antioxidant and immunomodulatory activity of the alkaloidal fraction of *Cissampelos pareira* Linn. Sci Pharm 78:21
Bahorun T, Neergheen VS, Aruoma OI (2005) Phytochemical constituents of *Cassia fistula*. Afr J Biotechnol 4(13):1530–1540
Bakkour Y, El-Achi N, Tabcheh M, El-Nakat H, El Omar F (2013) Chemical composition and antioxidant activities of the essential oils from green and ripe berries of *Juniperus excelsa* growing in Lebanon. Int J Pharma Life Sci 4(2):2362–2367
Balamurugan R, Ignacimuthu S (2011) Antidiabetic and hypolipidemic effect of methanol extract of *Lippia nodiflora* L. in streptozotocin induced diabetic rats. Asian Pac J Trop Biomed 1:S30–S36
Balamurugan R, Duraipandiyan V, Ignacimuthu S (2011) Antidiabetic activity of γ-sitosterol isolated from *Lippia nodiflora* L. in streptozotocin induced diabetic rats. Eur J Pharmacol 667:410–418
Bamosa AO (2015) A review on the hypoglycemic effect of *Nigella sativa* and thymoquinone. Saudi J Med Med Sci 3(1):2–3
Baquer NZ, Kumar P, Taha A, Kale R, Cowsik S, McLean P (2011) Metabolic and molecular action of *Trigonella foenum-graecum* (fenugreek) and trace metals in experimental diabetic tissues. J Biosci 36:383–396
Barbalho SM, Damasceno DC, Spada APM, VSd S, Martuchi KA, Oshiiwa M, Machado F, Mendes CG (2011) Metabolic profile of offspring from diabetic Wistar rats treated with *Mentha piperita* (peppermint). Evid Based Complement Alternat Med 11:1–6
Barbalho S, Farinazzi-Machado F, de Alvares Goulart R, Brunnati AS, Ottoboni A, Nicolau C (2012) *Psidium Guajava* (Guava): a plant of multipurpose medicinal applications. Med Aromat Plants 1:2167
Barbosa ADP (2014) Pharmacologically active saponins from the genus Albizia (Fabaceae). Int J Pharm Pharm Sci 6(11):32–36
Barillari J, Canistro D, Paolini M, Ferroni F, Pedulli GF, Iori R, Valgimigli L (2005a) Direct antioxidant activity of purified glucoerucin, the dietary secondary metabolite contained in rocket (Eruca sativa Mill.) seeds and sprouts. J Agric Food Chem 53:2475–2482
Barillari J, Cervellati R, Paolini M, Tatibouët A, Rollin P, Iori R (2005b) Isolation of 4-methylthio-3-butenyl glucosinolate from *Raphanus sativus* sprouts (Kaiware Daikon) and its redox properties. J Agric Food Chem 53:9890–9896
Barthakur N, Arnold N, Alli I (1995) The Indian laburnum (*Cassia fistula* L.) fruit: an analysis of its chemical constituents. Plant Foods Hum Nutr 47:55–62
Bas AL, Demirci S, Yazihan N, Uney K, Ermis Kaya E (2012) *Nerium oleander* distillate improves fat and glucose metabolism in high-fat diet-fed streptozotocin-induced diabetic rats. Int J Endocrinol 2012:10
Basha SK, Kumari VS (2012) In vitro antidiabetic activity of *Psidium guajava* leaves extracts. Asian Pac J Trop Dis 2:S98–S100
Basumata C (2016) Investigation on antidiabetic properties of C*issampelos pareira* Linn a common medicinal plant of Assam. PhD Thesis. Gauhati University, Gauhati
Bavarva J, Narasimhacharya A (2008) Preliminary study on antihyperglycemic and antihyperlipaemic effects of *Butea monosperma* in NIDDM rats. Fitoterapia 79:328–331
Bedir E, Kırmızıpekmez H, Sticher O, Çalış İ (2000) Triterpene saponins from the fruits of *Hedera helix*. Phytochemistry 53:905–909

Begum HA, Hamayun M, Zaman K, Hussain A, Ruaf M (2015) Phytochemical Evaluation of Ethnobotanically Selected Medicinal Plants of Mardan, Pakistan. J Adv Bot Zool 3:1–5
Benalla W, Bellahcen S, Bnouham M (2010) Antidiabetic medicinal plants as a source of alpha glucosidase inhibitors. Curr Diabetes Rev 6:247–254
Bharti SK, Kumar A, Sharma NK, Prakash O, Jaiswal SK, Krishnan S, Gupta AK, Kumar A (2013) Tocopherol from seeds of *Cucurbita pepo* against diabetes: Validation by in vivo experiments supported by computational docking. J Formos Med Assoc 112:676–690
Bhaskar V, Ajay SS (2009) Evaluation of antihperglycemic activity of extracts of *Calotropis procera* (Ait.) R. Br on streptozotocin induced diabetic rats. Glob J Pharmacol 3:95–98
Bhaskar A, Kumar A (2012) Antihyperglycemic, antioxidant and hypolipidemic effect of *Punica granatum* L flower extract in streptozotocin induced diabetic rats. Asian Pac J Trop Biomed 2:S1764–S1769
Bhaskar A, Vidhya V, Ramya M (2008) Hypoglycemic effect of *Mucuna pruriens* seed extract on normal and streptozotocin-diabetic rats. Fitoterapia 79:539–543
Bhaskara Rao R, Murugesan T, Sinha S, Saha B, Pal M, Mandal SC (2002) Glucose lowering efficacy of *Ficus racemosa* bark extract in normal and alloxan diabetic rats. Phytother Res 16:590–592
Bhatia A, Mishra T (2010) Hypoglycemic activity of *Ziziphus mauritiana* aqueous ethanol seed extract in alloxan-induced diabetic mice. Pharm Biol 48:604–610
Bhavsar C, Talele GS (2013) Potential anti-diabetic activity of *Bombax ceiba*. Bangladesh J Pharmacol 8:102–106
Bhowmik A, Khan LA, Akhter M, Rokeya B (2009) Studies on the antidiabetic effects of *Mangifera indica* stem-barks and leaves on nondiabetic, type 1 and type 2 diabetic model rats. Bangladesh J Pharmacol 4:110–114
Bhuiyan MNI, Begum J, Sultana M (2009) Chemical composition of leaf and seed essential oil of *Coriandrum sativum* L. from Bangladesh. Bangladesh J Pharmacol 4:150–153
Bhushan MS, Rao C, Ojha S, Vijayakumar M, Verma A (2010) An analytical review of plants for anti diabetic activity with their phytoconstituent & mechanism of action. Int J Pharm Sci Res 1:29–46
Bhutani K, Ali M, Atal C (1984) Alkaloids from Tylophora hirsuta. Phytochemistry 23:1765–1769
Bibi Y, Nisa S, Waheed A, Zia M, Sarwar S, Ahmed S, Chaudhary MF (2010) Evaluation of *Viburnum foetens* for anticancer and antibacterial potential and phytochemical analysis. Afr J Biotechnol 9:5611
Bicchi C, Rubiolo P, Marschall H, Weyerstahl P, Laurent R (1998) Constituents of *Artemisia roxburghiana* Besser essential oil. Flavour Fragr J 13:40–46
Biradar D (2015) Medicinal plants and Phytomedicines. Ann Phytomed 4:1–5
Boulekbache-Makhlouf L, Slimani S, Madani K (2013) Total phenolic content, antioxidant and antibacterial activities of fruits of *Eucalyptus globulus* cultivated in Algeria. Ind Crop Prod 41:85–89
Braca A, Siciliano T, D'Arrigo M, Germanò MP (2008) Chemical composition and antimicrobial activity of *Momordica charantia* seed essential oil. Fitoterapia 79:123–125
Brandle J, Starratt A, Gijzen M (1998) *Stevia rebaudiana*: its agricultural, biological, and chemical properties. Can J Plant Sci 78:527–536
Bukhari IA, Khan RA, Gilani AH, Ahmed S, Saeed SA (2010) Analgesic, anti-inflammatory and antiplatelet activities of the methanolic extract of *Acacia modesta* leaves. Inflammopharmacology 18:187–196
Bystricka J, Musilova J, Tomas J, Vollmannova A, Lachman J, Kavalcova P (2014) Changes of Polyphenolic Substances in the Anatomical Parts of Buckwheat (*Fagopyrum esculentum* Moench.) during Its Growth Phases. Foods 3:558–568
Caligiani A, Tonelli L, Palla G, Marseglia A, Rossi D, Bruni R (2013) Looking beyond sugars: Phytochemical profiling and standardization of manna exudates from Sicilian *Fraxinus excelsior* L. Fitoterapia 90:65–72
Cao Q, Qu WJ, Niu W, Deng YX, Wang YW, Xie JJ (2005) The antihyperglycemic effect of flavonoids from *Hippophae rhamnoides* L. on diabetic rats induced by streptozotocin. Acta Nutrimenta Sinica 2:016

Cao J, Li C, Zhang P, Cao X, Huang T, Bai Y, Chen K (2012) Antidiabetic effect of burdock (*Arctium lappa* L.) root ethanolic extract on streptozotocin-induced diabetic rats. Afr J Biotechnol 11:9079

Chan Y-S, Cheng L-N, Wu J-H, Chan E, Kwan Y-W, Lee SM-Y, Leung GP-H, Yu PH-F, Chan S-W (2011) A review of the pharmacological effects of *Arctium lappa* (burdock). Inflammopharmacology 19:245–254

Chandak R, Shrangare G (2010) Evalution of antidiabetic activity of *Olea-europea* in wistar albino rats. Int J App Biol Pharma Technol 1(3):952–956

Chandrasekaran S, Rajkishore VB, Ramalingam R (2015) *Vigna unguiculata*-an overall review. Res J Pharmacog Phytochem 7:219–222

Chasturvedi P, Raseroka B, Batisane B (2005) Acute effects of some organic extracts of *Melia azedarach* fruit on blood glucose levels in albino rats. J Appl Zool Res 16:238–241

Chatterjee A, Chakrabortty T, Chandrasekharan S (1971) Chemical investigation of *Cedrela toona*. Phytochemistry 10:2533–2535

Chaubey A, Pandeya S (2011) Pyridine: a versatile nucleuse in pharmaceutical field. Asian J Pharm Clin Res 4:5–8

Chaudhary V, Johri S, Rana AK (2015) Evaluation of comparative efficacy of neelkanthi (*Ajuga bracteosa*), tejapatra (*Cinnamomum tamala*) and methika beeja (*Trigonella foenum graecum*) churna in the management of diabetes mellitus. Int J Ayurveda Pharma Res 3

Chawla R, Thakur P, Chowdhry A, Jaiswal S, Sharma A, Goel R, Sharma J, Priyadarshi SS, Kumar V, Sharma RK (2013) Evidence based herbal drug standardization approach in coping with challenges of holistic management of diabetes: a dreadful lifestyle disorder of 21st century. J Diabetes Metabo Disord 12:1

Chen L, Kang Y-H (2013) In vitro inhibitory effect of oriental melon (*Cucumis melo* L. var. makuwa Makino) seed on key enzyme linked to type 2 diabetes: Assessment of anti-diabetic potential of functional food. J Funct Foods 5:981–986

Chianese G, Yerbanga SR, Lucantoni L, Habluetzel A, Basilico N, Taramelli D, Fattorusso E, Taglialatela-Scafati O (2010) Antiplasmodial triterpenoids from the fruits of neem, *Azadirachta indica*. J Nat Prod 73:1448–1452

Chitra M, Nithyanandhi K (2007) Radical scavenging activity of *Trianthema triquetra* in male albino rats intoxicated with CCl~ 4. J Environ Biol 28:283

Choedon T, Shukla SK, Kumar V (2010) Chemopreventive and anti-cancer properties of the aqueous extract of flowers of *Butea monosperma*. J Ethnopharmacol 129:208–213

Choudhary MI, Adhikari A, Rasheed S, Marasini BP, Hussain N, Kaleem WA (2011) Cyclopeptide alkaloids of *Ziziphus oxyphylla* Edgw as novel inhibitors of α-glucosidase enzyme and protein glycation. Phytochem Lett 4:404–406

Corzo-Martínez M, Corzo N, Villamiel M (2007) Biological properties of onions and garlic. Trends Food Sci Technol 18:609–625

Currie CJ, Poole CD, Gale E (2009) The influence of glucose-lowering therapies on cancer risk in type 2 diabetes. Diabetologia 52:1766–1777

Daradka HM, Abas MM, Mohammad MA, Jaffar MM (2014) Antidiabetic effect of *Artemisia absinthium* extracts on alloxan-induced diabetic rats. Comp Clin Pathol 23:1733–1742

Das S, Barman S (2012) Antidiabetic and antihyperlipidemic effects of ethanolic extract of leaves of *Punica granatum* in alloxan-induced non-insulin-dependent diabetes mellitus albino rats. Indian J Pharm 44:219

Dash GK, Suresh P, Ganapaty S (2001) Studies on hypoglycaemic and wound healing activities of *Lantana camara* Linn. J Nat Rem 1:105–110

De Bona KS, Bonfanti G, Bitencourt PE, Cargnelutti LO, da Silva PS, da Silva TP, Zanette RA, Pigatto AS, Moretto MB (2014) *Syzygium cumini* is more effective in preventing the increase of erythrocytic ADA activity than phenolic compounds under hyperglycemic conditions in vitro. J Physiol Biochem 70:321–330

Del Rıo J, Fuster M, Gómez P, Porras I, Garcıa-Lidón A, Ortuño A (2004) *Citrus limon*: a source of flavonoids of pharmaceutical interest. Food Chem 84:457–461

Delange DM, Rico CLM, Canavaciolo VLG, Pérez RS, Leyes EAR (2012) Fatty acid composition of seed oil from *Salvia coccinea* grown in Cuba. Anal Chem Lett 2:114–117

Derwich E, Benziane Z, Boukir A (2010) Antibacterial activity and chemical composition of the essential oil from flowers of *Nerium oleander*. Electro J Environ Agric Food Chem 9:1074–1084
Dhalwal K, Shinde V, Biradar Y, Mahadik K (2008) Simultaneous quantification of bergenin, catechin, and gallic acid from *Bergenia ciliata* and *Bergenia ligulata* by using thin-layer chromatography. J Food Compos Anal 21:496–500
Dharmender R, Madhavi T, Reena A, Sheetal A (2012) Simultaneous quantification of bergenin,(+)-catechin, gallicin and gallic acid; and quantification of β-sitosterol using HPTLC from *Bergenia ciliata* (Haw.) Sternb. Forma ligulata Yeo (Pasanbheda). Pharm Anal Acta 1(1):1–9
Diallo A, Traore MS, Keita SM, Balde MA, Keita A, Camara M, Van Miert S, Pieters L, Balde AM (2012) Management of diabetes in Guinean traditional medicine: an ethnobotanical investigation in the coastal lowlands. J Ethnopharmacol 144:353–361
Dob T, Benabdelkader T (2006) Chemical composition of the essential oil of *Artemisia herba-alba* Asso grown in Algeria. J Essent Oil Res 18:685–690
Dora BB, Gupta S, Sital S, Pastore A (2015) Punarnava (Boerhavia diffusa): A promising indigenous herbal drug and its effect on different disease conditions. Res Rev: J Herb Sci 4:21–24
Dureja AG, Dhiman K (2012) Free radical scavenging potential and total phenolic and flavonoid content of *Ziziphus mauritiana* and *Ziziphus nummularia* fruit extracts. Int J Green Pharm (IJGP) 6(3):187–192
Eddouks M, Maghrani M (2004) Phlorizin-like effect of *Fraxinus excelsior* in normal and diabetic rats. J Ethnopharmacol 94:149–154
Eddouks M, Lemhadri A, Michel J-B (2004) Caraway and caper: potential anti-hyperglycaemic plants in diabetic rats. J Ethnopharmacol 94:143–148
Eddouks M, Lemhadri A, Michel J-B (2005) Hypolipidemic activity of aqueous extract of *Capparis spinosa* L. in normal and diabetic rats. J Ethnopharmacol 98:345–350
Edoga C, Njoku O, Amadi E, Okeke J (2013) Blood sugar lowering effect of *Moringa oleifera* Lam. in albino rats. Int J Sci Technol 3:88–90
Ehsani E, Akbari K, Teimouri M, Khadem A (2012) Chemical composition and antibacterial activity of two Juniperus species essential oils. Afr J Microbiol Res 6:6704–6710
Eidi A, Eidi M, Esmaeili E (2006) Antidiabetic effect of garlic (*Allium sativum* L.) in normal and streptozotocin-induced diabetic rats. Phytomedicine 13:624–629
Eidi A, Eidi M, Darzi R (2009) Antidiabetic effect of *Olea europaea* L. in normal and diabetic rats. Phytother Res 23:347–350
Eidi A, Eidi M, Haeri Rohani A, Basati F (2010) Hypoglycemic effect of ethanolic extract of *Carum carvi* L. seeds in normal and streptozotocin-induced diabetic rats. J Med Plant 3:106–113. فصلنامه علمی پژوهشی گیاهان دارویی
El-Demerdash F, Yousef M, El-Naga NA (2005) Biochemical study on the hypoglycemic effects of onion and garlic in alloxan-induced diabetic rats. Food Chem Toxicol 43:57–63
Eldin IMT, Ahmed EM, HM AE (2010) Preliminary study of the clinical hypoglycemic effects of *Allium cepa* (red onion) in type 1 and type 2 diabetic patients. Environ Health Insights 4:71
El-Missiry M, El Gindy A (2000) Amelioration of alloxan induced diabetes mellitus and oxidative stress in rats by oil of *Eruca sativa* seeds. Ann Nutr Metab 44:97–100
El-Shobaki F, El-Bahay A, Esmail R, El-Megeid A, Esmail N (2010) Effect of figs fruit (*Ficus carica* L.) and its leaves on hyperglycemia in alloxan diabetic rats. World J Dairy Food Sci 5:47–57
El-Soud N, El-Laithy N, El-Saeed G, Wahby M, Khalil M, Morsy F, Shaffie N (2011) Antidiabetic activities of *Foeniculum vulgare* mill. Essential oil in streptozotocin-induced diabetic rats. Maced J Med Sci 4:139–146
Erinç H, Tekin A, Musa Özcan M (2009) Determination of fatty acid, tocopherol and phytosterol contents of the oils of various poppy (*Papaver somniferum* L.) seeds. Grasas Aceites 60:375–381
Esfahlan AJ, Jamei R, Esfahlan RJ (2010) The importance of almond (*Prunus amygdalus* L.) and its by-products. Food Chem 120:349–360
Ezuruike UF, Prieto JM (2014) The use of plants in the traditional management of diabetes in Nigeria: Pharmacological and toxicological considerations. J Ethnopharmacol 155:857–924
Facino RM, Carini M, Stefani R, Aldini G, Saibene L (1995) Anti-Elastase and Anti-Hyaluronidase Activities of Saponins and Sapogenins from *Hedera helix*, *Aesculus hippocastanum*, and

Ruscus aculeatus: Factors contributing to their efficacy in the treatment of venous insufficiency. Arch Pharm 328:720–724

Faizi S, Zikr-ur-Rehman S, Naz A, Versiani MA, Dar A, Naqvi S (2012) Bioassay-guided studies on *Bombax ceiba* leaf extract: isolation of shamimoside, a new antioxidant xanthone C-glucoside. Chem Nat Compd 48:774–779

Fatima N, Maqsood Z, Khan B (2005) Study of some micronutrients in selected medicinal plants. Sci Iran 12:269–273

Fatima SS, Rajasekhar MD, Kumar KV, Kumar MTS, Babu KR, Rao CA (2010) Antidiabetic and antihyperlipidemic activity of ethyl acetate: isopropanol (1: 1) fraction of *Vernonia anthelmintica* seeds in streptozotocin induced diabetic rats. Food Chem Toxicol 48:495–501

Gallagher A, Flatt P, Duffy G, Abdel-Wahab Y (2003) The effects of traditional antidiabetic plants on in vitro glucose diffusion. Nutr Res 23:413–424

Gao J, Han Y-L, Jin Z-Y, Xu X-M, Zha X-Q, Chen H-Q, Yin Y-Y (2015) Protective effect of polysaccharides from *Opuntia dillenii* Haw. fruits on streptozotocin-induced diabetic rats. Carbohydr Polym 124:25–34

García-Jiménez N, Péerez-Alonso MJ, Velasco-Negueruela A (2000) Chemical composition of fennel oil, *Foeniculum vulgare* Miller, from Spain. J Essent Oil Res 12:159–162

Gholamhoseinian A, Fallah H (2009) Inhibitory effect of methanol extract of *Rosa damascena* Mill. flowers on α-glucosidase activity and postprandial hyperglycemia in normal and diabetic rats. Phytomedicine 16:935–941

Ghosh R, Sharatchandra K, Rita S, Thokchom I (2004) Hypoglycemic activity of *Ficus hispida* (bark) in normal and diabetic albino rats. Indian J Pharm 36:222

Ghosh A, Das BK, Chatterjee SK, Chandra G (2008) Antibacterial potentiality and phytochemical analysis of mature leaves of *Polyalthia longifolia* (Magnoliales: Annonaceae). The South Pac J Nat Sci 26:68–72

Ghosh G, Kar DM, Subudhi BB, Mishra SK (2010) Antihyperglycemic and antioxidant activity of stem bark of *Polyalthia longifolia* var. angustifolia. Der Pharm Lett 2:206–216

Giday M, Asfaw Z, Elmqvist T, Woldu Z (2003) An ethnobotanical study of medicinal plants used by the Zay people in Ethiopia. J Ethnopharmacol 85:43–52

Giday M, Asfaw Z, Woldu Z (2009) Medicinal plants of the Meinit ethnic group of Ethiopia: an ethnobotanical study. J Ethnopharmacol 124:513–521

Gilani AH, Bashir S, Janbaz KH, Khan A (2005) Pharmacological basis for the use of *Fumaria indica* in constipation and diarrhea. J Ethnopharmacol 96:585–589

Giménez E, Juan ME, Calvo-Melià S, Barbosa J, Sanz-Nebot V, Planas JM (2015) Pentacyclic triterpene in *Olea europaea* L: A simultaneous determination by high-performance liquid chromatography coupled to mass spectrometry. J Chromatogr A 1410:68–75

Giménez-Bastida JA, Zieliński H (2015) Buckwheat as a functional food and its effects on health. J Agric Food Chem 63:7896–7913

Gogoi P, Islam M (2012) Phytochemical Screening of *Solanum nigrum* L and *S. myriacanthus* Dunal from Districts of Upper Assam, India. IOSR J Pharma 2(3):455–459

Goli-malekabadi N, Asgary S, Rashidi B, Rafieian-Kopaei M, Ghannadian M, Hajian S, Sahebkar A (2014) The protective effects of *Ziziphus vulgaris* L. fruits on biochemical and histological abnormalities induced by diabetes in rats. J Complement Integr Med 11:171–177

Gomez-Garcia F, Flanagan J, García-Molina O, Vilaplana-Vivo V, García-Carrillo N, Berthon PF, Bily A, Roller M, Ortega VV, Issaly N (2015) Preventive Effect of a *Fraxinus Excelsior* L Seeds/Fruits Extract on Hepatic Steatosis in Obese Type 2 Diabetic Mice. J Diabetes Metab 6(4):1–5

González-Molina E, Domínguez-Perles R, Moreno D, García-Viguera C (2010) Natural bioactive compounds of *Citrus limon* for food and health. J Pharm Biomed Anal 51:327–345

Gopinath S, Rakesh C, PATIL GA, Dayananda K (2012) Preliminary phytochemical evaluation of leaf extracts of *Euphorbia hirta*, *Syzygium cumini* of siddarabetta, Tumkur district, Karnataka. Int J Pharm Bio Sci 3:431–435

Gregersen S, Jeppesen PB, Holst JJ, Hermansen K (2004) Antihyperglycemic effects of stevioside in type 2 diabetic subjects. Metabolism 53:73–76

Grishko V, Tolmacheva I, Pereslavtseva A (2015) Triterpenoids with a five-membered a-Ring: distribution in nature, transformations, synthesis, and biological activity. Chem Nat Compd 51:1–21

Group UHS (2007) Risk of hypoglycaemia in types 1 and 2 diabetes: effects of treatment modalities and their duration. Diabetologia 50:1140–1147

Gu J-F, Zheng Z-Y, Yuan J-R, Zhao B-J, Wang C-F, Zhang L, Xu Q-Y, Yin G-W, Feng L, Jia X-B (2015) Comparison on hypoglycemic and antioxidant activities of the fresh and dried *Portulaca oleracea* L. in insulin-resistant HepG2 cells and streptozotocin-induced C57BL/6J diabetic mice. J Ethnopharmacol 161:214–223

Gulfraz M, Ahmad A, Asad MJ, Afzal U, Imran M, Anwar P, Zeenat A, Abbasi KS, Maqsood S, Qureshi RU (2011) Antidiabetic activities of leaves and root extracts of *Justicia adhatoda* Linn against alloxan induced diabetes in rats. Afr J Biotechnol 10:6101–6106

Gunjan M, Jana GK, Jha A, Mishra U (2010) Pharmacognostic and antihyperglycemic study of *Coccinia indica*. Int J Phytomed 2:36–40

Gupta A, Gupta R, Lal B (2001) Effect of *Trigonella foenum-graecum* (fenugreek) seeds on glycaemic control and insulin resistance in type 2 diabetes mellitus: a double blind placebo controlled study. J Assoc Physicians India 49:1057–1061

Gupta S, Prakash J, Srivastava S (2002) Validation of traditional claim of Tulsi, *Ocimum sanctum* Linn. as a medicinal plant. Indian J Exp Biol 40:765–773

Gupta OP, Ahmed Z, Bhagat A, Gupta KK, Handa SS (2005) Method of treating diabetes using plant *Argyrolobium roseum* extract, and a process for the isolation of extract from the said plant. Google Patents

Gupta S, Sood S, Singh N, Gupta M (2008) Effect of Neelkanthi (*Ajuga bracteosa*) and Plakhar (Ficus lacor) on the energy expenditure of diabetic subjects. Pantnagar J Res 6:286–290

Gupta D, Mann S, Jain I, Gupta RK (2011a) Phytochemical, nutritional and antioxidant activity evaluation of fruits of *Ziziphus nummularia* burm. F Int J Pharma Bio Sci 2(4):630–638

Gupta O, Malhotra P, Kudyar S, Singh A, Gupta G (2011b) Experimental model proposed to deduce pharmacological evidence of the beta-cell neogenesis activity of *Argyrolobium roseum* vis-a-vis an evaluation of its hypoglycemic activity. Int J Appl Basic Med Res 1:97

Gupta R, Sharma AK, Dobhal M, Sharma M, Gupta R (2011c) Antidiabetic and antioxidant potential of β-sitosterol in streptozotocin-induced experimental hyperglycemia. J Diabetes 3:29–37

Gupta D, Chandrashekar K, Lobo R, Nayak Y, Gupta N (2012a) In-vitro Antidiabetic activity of stem bark of *Bauhinia purpurea* Linn. Der Pharma Lett 4:614–619

Gupta R, Mathur M, Bajaj VK, Katariya P, Yadav S, Kamal R, Gupta RS (2012b) Evaluation of antidiabetic and antioxidant activity of *Moringa oleifera* in experimental diabetes. J Diabetes 4:164–171

Gupta D, Kondongala SC, Pai GK (2013) *In vitro* Antidiabetic activity of Pentacyclic tritrpenoids and fatty acid esters from *Bauhinia Purpurea*. Int J Pharma Pharma Technol 2:25–28

Gupta M, Singh A, Joshi HC (2015) *Berberis Lycium* multipotential medicinal application: An overview. IJCS 3:10–13

Gurudeeban S, Ramanathan T (2010) Antidiabetic effect of *Citrullus colocynthis* in alloxon-induced diabetic rats. Inventi Rapid: Ethnopharmacol 1:112

Haeri MR, Limaki HK, White CJB, White KN (2012) Non-insulin dependent anti-diabetic activity of (2S, 3R, 4S) 4-hydroxyisoleucine of fenugreek (*Trigonella foenum graecum*) in streptozotocin-induced type I diabetic rats. Phytomedicine 19:571–574

Hamayun M, Khan SA, Sohn EY, Lee I-J (2006) Folk medicinal knowledge and conservation status of some economically valued medicinal plants of District Swat, Pakistan. Lyonia 11:101–113

Hammed A, Abdel-Azim N, Ismail S, Hammouda F (2007) Chemical investigation of some Capparis species growing in Egypt and their antioxidant activity. Planta Med 73:380

Hannan J, Marenah L, Ali L, Rokeya B, Flatt P, Abdel-Wahab Y (2006) *Ocimum sanctum* leaf extracts stimulate insulin secretion from perfused pancreas, isolated islets and clonal pancreatic β-cells. J Endocrinol 189:127–136

Haq F, Ahmad H, Alam M (2011) Traditional uses of medicinal plants of Nandiar Khuwarr catchment (District Battagram), Pakistan. J Med Plant Res 5:39–48

Haque S, Naznine T, Ali M, Azad TT, Morshed T, Afsana NA, Ahmed I, Rahmatullah M (2013) Antihyperglycemic activities of leaves of *Brassica oleracea*, *Centella asiatica* and *Zizyphus mauritiana*: Evaluation through oral glucose tolerance tests. Adv Nat Appl Sci 7:519–526

Haridasan K (2001). Ethnobotanical studies in Arunachal Pradesh - A Status Report, In: Ethnomedicine of North East India, NISCAIR, New Delhi, p. 54–169

Hashmi MA, Khan A, Ayub K, Farooq U (2014) Spectroscopic and density functional theory studies of 5, 7, 3′, 5′-tetrahydroxyflavanone from the leaves of *Olea ferruginea*. Spectrochim Acta Part A: Mol Biomol Spectrosc 128:225–230

Hashmi MA, Shah HS, Khan A, Farooq U, Iqbal J, Ahmad VU, Perveen S (2015) Anticancer and alkaline phosphatase inhibitory effects of compounds isolated from the leaves of *Olea ferruginea* royle. Rec Nat Prod 9:164–168

Hatam NA, Whiting DA, Yousif NJ (1989) Cucurbitacin glycosides from *Citrullus colocynthis*. Phytochemistry 28:1268–1271

Hausen B, Bröhas J, König W, Faasch H, Hahn H, Bruhn G (1987) Allergic and irritant contact dermatitis from falcarinol and didehydrofalcarinol in common ivy (*Hedera helix* L.). Contact Dermatitis 17:1–9

Hayat S, Rahman A-U, Choudhary MI, Khan KM, Latif H, Bayer E (2002) Two new triterpenes from fern *Adiantum incisum*. Z Naturforsch B 57:233–238

Hayat MQ, Khan MA, Ashraf M, Jabeen S (2009) Ethnobotany of the genus Artemisia L.(Asteraceae) in Pakistan. Ethnobot Res Appl 7:147–162

Hayat MM, Sarwar S, Anjum S, Uzair M, Rasheed HMF, Jabeen Q, Choudhary BA, Ashraf M (2014) Anti-diabetic and spasmolytic potential of *Farsetia hamiltonii* Royle from Cholistan desert. J Ethnopharmacol 156:347–352

Helal E, Mostafa AM, MhMood AF, Kahwash AA (2005) Hypoglycemic and hyperinsulinemic effects of *Ferula assafoetida* on diabetic male albino rats. Egypt J Hosp Med 21:95–108

Hemalatha S, Wahi A, Singh P, Chansouria J (2004) Hypoglycemic activity of *Withania coagulans* Dunal in streptozotocin induced diabetic rats. J Ethnopharmacol 93:261–264

Heshmati J, Namazi N, Memarzadeh M-R, Taghizadeh M, Kolahdooz F (2015) *Nigella sativa* oil affects glucose metabolism and lipid concentrations in patients with type 2 diabetes: A randomized, double-blind, placebo-controlled trial. Food Res Int 70:87–93

Hong Z, Gang L, Xiao Yu Z, Chun Ping H, Min K, Shan Cai T (2011) Hyperglycemic effect of black buckwheat powder on alloxan-induced diabetic mice. Med Plant 2(40-41):44

Hua L, Li Y, Wang F, Lu D-F, Gao K (2012) Biologically active steroids from the aerial parts of *Vernonia anthelmintica* Willd. Fitoterapia 83:1036–1041

Huang C-S, Yin M-C, Chiu L-C (2011) Antihyperglycemic and antioxidative potential of *Psidium guajava* fruit in streptozotocin-induced diabetic rats. Food Chem Toxicol 49:2189–2195

Huang B, Wang Z, Park JH, Ryu OH, Choi MK, Lee J-Y, Kang Y-H, Lim SS (2015) Anti-diabetic effect of purple corn extract on C57BL/KsJ db/db mice. Nutr Res Pract 9:22–29

Hunyadi A, Veres K, Danko B, Kele Z, Weber E, Hetenyi A, Zupko I, Hsieh TJ (2013) In vitro Anti-diabetic Activity and Chemical Characterization of an Apolar Fraction of *Morus alba* Leaf Water Extract. Phytother Res 27:847–851

Hurakadle P, Patil M, Jalalpure S, Manvi F (2009) Effect of *Picrorhiza kurroa* roots on blood glucose level on alloxan induced diabetic rats. Afr J Traditional, Complement Altern Med 6:439

Husain SZ, Malik RN, Javaid M, Bibi S (2008) Ethonobotanical properties and uses of medicinal plants of Morgah biodiversity park, Rawalpindi. Pak J Bot 40:1897–1911

Husain GM, Singh PN, Kumar V (2009) Antidiabetic activity of standardized extract of *Picrorhiza kurroa* in rat model of NIDDM. Drug Discov Ther 3:88–92

Husain GM, Rai R, Rai G, Singh HB, Thakur AK, Kumar V (2014) Potential mechanism of anti-diabetic activity of *Picrorhiza kurroa*. TANG [Humanitas Medicine] 4:e27

Huseini HF, Darvishzadeh F, Heshmat R, Jafariazar Z, Raza M, Larijani B (2009) The clinical investigation of *Citrullus colocynthis* (L.) schrad fruit in treatment of Type II diabetic patients: a randomized, double blind, placebo-controlled clinical trial. Phytother Res 23:1186–1189

Huseini HF, Hasani-Rnjbar S, Nayebi N, Heshmat R, Sigaroodi FK, Ahvazi M, Alaei BA, Kianbakht S (2013) *Capparis spinosa* L.(Caper) fruit extract in treatment of type 2 diabetic patients: a randomized double-blind placebo-controlled clinical trial. Complement Ther Med 21:447–452

Hussain Z, Waheed A, Qureshi RA, Burdi DK, Verspohl EJ, Khan N, Hasan M (2004) The effect of medicinal plants of Islamabad and Murree region of Pakistan on insulin secretion from INS-1 cells. Phytother Res 18:73–77

Hussain J, Khan AL, Rehman N, Hamayun M, Shah T, Nisar M, Bano T, Shinwari ZK, Lee I (2009) Proximate and nutrient analysis of selected vegetable species: A case study of Karak region, Pakistan. Afr J Biotechnol 8

Hussain J, Muhammad Z, Ullah R, Khan FU, Ullah I, Khan N, Ali J, Jan S (2010a) Evaluation of the chemical composition of *Sonchus eruca* and *Sonchus asper*. J Am Sci 6:231–235

Hussain J, Munir M, Hassan Z, Bano N, Arshad S, Ahmad VU (2010b) Tanacetamide D: a new ceramide from *Tanacetum artemisioides*. Helv Chim Acta 93:350–353

Hussain J, Ullah R, Rehman N, Khan AL, Muhammad Z, Khan FU, Hussain ST, Anwar S (2010c) Endogenous transitional metal and proximate analysis of selected medicinal plants from Pakistan. J Med Plant Res 4:267–270

Hussain K, Nisar MF, Majeed A, Nawaz K, Bhatti KH (2010d) Ethnomedicinal survey for important plants of Jalalpur Jattan, district Gujrat, Punjab, Pakistan. Ethnobot Leaflets 2010:11

Hussain I, Bano A, Ullah F (2011) Traditional drug therapies from various medicinal plants of central karakoram national park, Gilgit-Baltistan Pakistan. Pak J Bot 43:79–84

Hussein HM, El-Sayed EM, Said AA (2006) Antihyperglycemic, antihyperlipidemic and antioxidant effects of *Zizyphus spina christi* and *Zizyphus jujuba* in alloxan diabetic rats. Int J Pharm 2:563–570

Ibegbulem C, Chikezie P (2013) Hypoglycemic properties of ethanolic extracts of *Gongronema latifolium*, *Aloe perryi*, *Viscum album* and *Allium sativum* administered to alloxan-induced diabetic albino rats (Rattus norvegicus). Phcog Commn 3:12

Ibraheim ZZ, Ahmed AS, Gouda YG (2011) Phytochemical and biological studies of *Adiantum capillus-veneris* L. Saudi Pharma J 19:65–74

Ibrar M, Ilahi I, Hussain F (2004) Hypoglycemic activity of *Hedera helix* L. leaves and the possible mechanism of action. Pak J Bot 35:805–810

Ilango K, Maharajan G, Narasimhan S (2009) Hypoglycemic and Hypolipidemic Activities of *Momordica dioica* Roxb Fruit Pulp Extracts on alloxan-induced diabetic rats. Int J Health Res 2(2):195–199

Ilango K, Maharajan G, Narasimhan S (2012) Preliminary phytochemical screening and antibacterial activity of fruit pulp of *Momordica dioica* Roxb.(Cucurbitaceae). Afr J Basic Appl Sci 4:12–15

Iranshahi M, Alizadeh M (2012) Antihyperglycemic effect of Asafoetida (*Ferula assafoetida* Oleo-Gum-Resin) in streptozotocin-induced diabetic rats. World Appl Sci J 17:157–162

Iranshahy M, Iranshahi M (2011) Traditional uses, phytochemistry and pharmacology of asafoetida (*Ferula assa-foetida* oleo-gum-resin)—A review. J Ethnopharmacol 134:1–10

Iriadam M, Musa D, Gumushan H, Baba F (2006) Effects of two Turkish medicinal plants *Artemisia herba-alba* and *Teucrium polium* on blood glucose levels and other biochemical parameters in rabbits. J Cell Mol Biol 5:19–24

Islam MS, Choi H (2008) Comparative effects of dietary ginger (*Zingiber officinale*) and garlic (*Allium sativum*) investigated in a type 2 diabetes model of rats. J Med Food 11:152–159

Islam M, Azhar I, Usmanghani K (2002) Bioactivity evaluation of *Bergenia ciliata*. Pak J Pharm Sci 15:15–33

Islam M, Akhtar MA, Islam M, Hossain M, Alam M, Wahed M, Rahman B, Anisuzzaman A, Shaheen S, Ahmed M (2009) Antidiabetic and hypolipidemic effects of different fractions of *Catharanthus roseus* (Linn.) on normal and streptozotocin-induced diabetic rats. J Sci Res 1:334–344

Ismail HI, Chan KW, Mariod AA, Ismail M (2010) Phenolic content and antioxidant activity of cantaloupe (*Cucumis melo*) methanolic extracts. Food Chem 119:643–647

Italo CG, Amanda HF, Diego LR (2009) Physical and chemical characterization of *Melia azedarach* L. fruit and leaf for use as botanical insecticide. Chil J Agr Res 69(1):38–45

Jafri SA, Hasan SS, Nadeem A, Iqbal J (2011) Hypoglycemic effect of *Aloe vera* Extract in alloxan-induced diabetic Albino rats. Med J Islamic World Acad Sci 19:127–130

Jain SK (1989). Medicinal Plants. National Book Trust, New Delhi
Jaiswal D, Rai PK, Kumar A, Watal G (2008) Study of glycemic profile of *Cajanus cajan* leaves in experimental rats. IJCB 23:167–170
Jaiswal D, Rai PK, Watal G (2009) Antidiabetic effect of *Withania coagulans* in experimental rats. Indian J Clin Biochem 24:88–93
Jaleel CA, Gopi R, Gomathinayagam M, Panneerselvam R (2009) Traditional and non-traditional plant growth regulators alters phytochemical constituents in *Catharanthus roseus*. Process Biochem 44:205–209
Jan S, Khan MR (2016) Antipyretic, analgesic and anti-inflammatory effects of *Kickxia ramosissima*. J Ethnopharmacol 182:90–100
Jarald E, Joshi S, Jain D (2009) Antidiabetic activity of extracts and fraction of *Zizyphus mauritiana*. Pharm Biol 47:328–334
Jayaprakasha G, Singh R, Pereira J, Sakariah K (1997) Limonoids from *Citrus reticulata* and their moult inhibiting activity in mosquito Culex quinquefasciatus larvae. Phytochemistry 44:843–846
Jeeva S, Johnson M, Aparna J, Irudayaraj V (2011) Preliminary phytochemical and antibacterial studies on flowers of selected medicinal plants. Int J Med Arom Plants 1:107–114
Jeppesen PB, Gregersen S, Alstrup K, Hermansen K (2002) Stevioside induces antihyperglycaemic, insulinotropic and glucagonostatic effects *in vivo*: studies in the diabetic Goto-Kakizaki (GK) rats. Phytomedicine 9:9–14
Jha DK, Panda L, Lavanya P, Ramaiah S, Anbarasu A (2012) Detection and confirmation of alkaloids in leaves of *Justicia adhatoda* and bioinformatics approach to elicit its anti-tuberculosis activity. Appl Biochem Biotechnol 168:980–990
Jiang TF, Lv ZH, Wang YH (2005) Separation and determination of anthraquinones in *Cassia obtusifolia* (Leguminosae) by micellar electrokinetic capillary electrophoresis. J Sep Sci 28:2225–2229
Jiang M-Z, Yan H, Wen Y, Li X-M (2011) In vitro and in vivo studies of antioxidant activities of flavonoids from *Adiantum capillus-veneris* L. Afr J Pharm Pharmacol 5:2079–2085
Jiang C, Li X, Jiao Y, Jiang D, Zhang L, Fan B, Zhang Q (2014) Optimization for ultrasound-assisted extraction of polysaccharides with antioxidant activity in vitro from the aerial root of *Ficus microcarpa*. Carbohydr Polym 110:10–17
Jing P, Zhao S-J, Ruan S-Y, Xie Z-H, Dong Y, Yu LL (2012) Anthocyanin and glucosinolate occurrences in the roots of Chinese red radish (*Raphanus sativus* L.), and their stability to heat and pH. Food Chem 133:1569–1576
Johnson S, Morgan ED, Peiris CN (1996) Development of the major triterpenoids and oil in the fruit and seeds of neem (*Azadirachta indica*). Ann Bot 78:383–388
Joseph B, Raj SJ (2010) Phytopharmacological and phytochemical properties of three Ficus species-an overview. Int J Pharm Bio Sci 1:246–253
Jouad H, Maghrani M, Hassani RAE, Eddouks M (2004) Hypoglycemic activity of aqueous extract of *Eucalyptus globulus* in normal and streptozotocin-induced diabetic rats. J Herbs Spices Med Plants 10:19–28
Juárez-Rojop IE, Díaz-Zagoya JC, Ble-Castillo JL, Miranda-Osorio PH, Castell-Rodríguez AE, Tovilla-Zárate CA, Rodríguez-Hernández A, Aguilar-Mariscal H, Ramón-Frías T, Bermúdez-Ocaña DY (2012) Hypoglycemic effect of *Carica papaya* leaves in streptozotocin-induced diabetic rats. BMC Complement Altern Med 12:1
Kaatabi H, Bamosa AO, Badar A, Al-Elq A, Abou-Hozaifa B, Lebda F, Al-Khadra A, Al-Almaie S (2015) *Nigella sativa* improves glycemic control and ameliorates oxidative stress in patients with type 2 diabetes mellitus: placebo controlled participant blinded clinical trial. PLoS One 10:e0113486
Kadir MF, Sayeed MSB, Shams T, Mia M (2012) Ethnobotanical survey of medicinal plants used by Bangladeshi traditional health practitioners in the management of diabetes mellitus. J Ethnopharmacol 144:605–611
Kalaskar MG, Surana SJ (2011) Free radical scavenging and hepatoprotective potential of *Ficus microcarpa* L. fil. bark extracts. J Nat Med 65:633–640

Kalaskar MG, Surana SJ (2012) Pharmacognostic and phytochemical studies on *Ficus Microcarpa* L. fil. Anc Sci Life 32:107
Kamble H, Kandhare AD, Bodhankar S, Mohan V, Thakurdesai P (2013) Effect of low molecular weight galactomannans from fenugreek seeds on animal models of diabetes mellitus. Biomed Aging Pathol 3:145–151
Kamiloglu S, Serali O, Unal N, Capanoglu E (2013) Antioxidant activity and polyphenol composition of black mulberry (*Morus nigra* L.) products. J Berry Res 3:41–51
Kanwal AMM, Shaukat S, Javed R, Ilyas R (2011a) Exploration of ethnomediciiial values of imperative plants of District Gujrat, Pakistan. Middle-East J Sci Res 7:397–400
Kanwal S, Ullah N, Haq I, Afzal I, Mirza B (2011b) Antioxidant, antitumor activities and phytochemical investigation of *Hedera nepalensis* K. Koch, an important medicinal plant from Pakistan. Pak J Bot 43:85–89
Kardošová A, Ebringerova A, Alföldi J, Nosál'ová G, Fraňová S, Hřibalová V (2003) A biologically active fructan from the roots of *Arctium lappa* L., var. Herkules. Int J Bio Macromolec 33:135–140
Karthiyayini T, Kumar R, Kumar K, Sahu R, Roy A (2009) Evaluation of antidiabetic and hypolipidemic effect of *Cucumis sativus* fruit in streptozotocin-induced-diabetic rats. Biomed Pharmacol J 2:351–355
Kataya HA, Hamza AA (2008) Red cabbage (*Brassica oleracea*) ameliorates diabetic nephropathy in rats. J Evid Based Complement Altern Med 5:281–287
Kaushik D, Kaushik P, Khokra S (2015a) Evaluation of antidiabetic potential of *Pinus roxburghii* bark extract in alloxan induced diabetic rats. J Pharmacog Nat Prod 1(1):2–5
Kaushik P, Singh G, Khokra SL, Kaushik D (2015b) Bioassay Guided Fractionation and Î±-Amylase Inhibitory Activity of Flavanoid Isolated from *Pinus roxburghii* Sarg. Nat Prod Chem Res 3(4):2–3
Kazmi I, Rahman M, Afzal M, Gupta G, Saleem S, Afzal O, Shaharyar MA, Nautiyal U, Ahmed S, Anwar F (2012) Anti-diabetic potential of ursolic acid stearoyl glucoside: a new triterpenic gycosidic ester from *Lantana camara*. Fitoterapia 83:142–146
Khalik SA, Miyase T, El-Ashaal HA, Melek F (2000) Triterpenoid saponins from *Fagonia cretica*. Phytochemistry 54:853–859
Khalil EA (2004) Antidiabetic effect of an aqueous extract of Pomegranate (*Punica granatum* L.) peels in normal and alloxan diabetic rats. EJHM 16:92–99
Khan SW, Khatoon S (2008) Ethnobotanical studies on some useful herbs of Haramosh and Bugrote valleys in Gilgit, northern areas of Pakistan. Pak J Bot 40:43
Khan SA, Khan L, Hussain I, Marwat KB, Akhtar N (2008) Profile of heavy metals in selected medicinal plants. PJWSR 14:101–110
Khan ZI, Nahar B, Jakaria A, Rahman S, Chowdhury MH, Rahmatullah M (2010) Oxid Med Cell Longev 2015
Khan MA, Ali M, Alam P (2010a) Phytochemical investigation of the fruit peels of *Citrus reticulata* Blanco. Nat Prod Res 24:610–620
Khan KY, Khan MA, Niamat R, Munir M, Fazal H, Mazari P, Seema N, Bashir T, Kanwal A, Ahmed SN (2011) Element content analysis of plants of genus Ficus using atomic absorption spectrometer. Afr J Pharm Pharmacol 5:317–321
Khan KY, Khan MA, Niamat R, Shah GM, Fazal H, Seema N, Hussain I, Ahmad I, Inayat H, Jan G (2012a) Elemental content of some anti-diabetic ethnomedicinal species of genus Ficus Linn. using atomic absorption spectrophotometry technique. J Med Plant Res 6:2136–2140
Khan RA, Khan MR, Sahreen S, Ahmed M (2012b) Assessment of flavonoids contents and in vitro antioxidant activity of *Launaea procumbens*. Chem Central J 6:43
Khan RA, Khan MR, Sahreen S, Ahmed M (2012c) Evaluation of phenolic contents and antioxidant activity of various solvent extracts of *Sonchus asper* (L.) Hill. Chem Cent J 6:12
Khan SM, Page S, Ahmad H, Shaheen H, Ullah Z, Ahmad M, Harper DM (2013a) Medicinal flora and ethnoecological knowledge in the Naran Valley, Western Himalaya, Pakistan. J Ethnobiol Ethnomed 9(4):2–13

Khan SU, Khan RU, Mehmood S, Sherwani SK, Muhammad A, Bokhari TZ, Khan A, Ullah I (2013b) Medicinally important underground fruit and leafy vegetables of frontier regions of Bannu, Khyber Pakhtunkhwa, Pakistan. Eur Acad Res 1(7):1613–1623

Khan MA, Marwat KB, Gul B, Wahid F, Khan H, Hashim S (2014a) *Pistia stratiotes* L.(Araceae): Phytochemistry, use in medicines, phytoremediation, biogas and management options. Pak J Bot 46:851–860

Khan MF, Rawat AK, Pawar B, Gautam S, Srivastava AK, Negi DS (2014b) Bioactivity-guided chemical analysis of *Melia azedarach* L.(Meliaceae), displaying antidiabetic activity. Fitoterapia 98:98–103

Khan AA, Ali F, Ihsan M, Hayat K, Nabi G (2015) Ethnobotanical study of the medicinal plants of Tehsil Charbagh, District Swat, Khyber Pakhtunkhwa, Pakistan. Am Eurasian Agric Environ Sci 15:1464–1474

Khanfar MA, Sabri SS, Abu Zarga MH, Zeller K-P (2003) The chemical constituents of *Capparis spinosa* of Jordanian origin. Nat Prod Res 17:9–14

Khattab HA, El-Shitany NA, Abdallah IZ, Yousef FM, Alkreathy HM (2015) Antihyperglycemic potential of Grewia asiatica fruit extract against streptozotocin-induced hyperglycemia in rats: anti-inflammatory and antioxidant mechanisms. J Chem Soc Pak 33(2):188–192

Khattak MI, Khattak MI (2011) Study of heavy trace metals in some medicinal–herbal plants of Pakistan. Pak J Bot 43:2003–2009

Khosa MK, Hussain A, Riaz H, Aslam K, Chatha S, Zia K (2011) Spectrophotometric quantification of antioxidant phytochemicals in juices from four different varieties of *Citrus limon*, Indigenous to Pakistan. .An evaluation of antihyperglycemic and antinociceptive effects of methanol extract of *Cassia fistula* l.(Fabaceae) leaves in Swiss albino mice. Adv Nat Appl Sci 4:305–311

Khushk I, Dahot M, Baloach S, Bhutto M (2010) The evaluation of soybean extracts in alloxan-induced diabetic rabbits. World Appl Sci J 8:22–25

Kim S, Rhyu D, Yokozawa T, Park J (2007) Antioxidant Effect of *Alisma plantago-aquatica* var. orientale and its main component. Korean J Pharmacogn 38(4):372–375

Kosta S, Dangi C, Shekhar C, Kaur M, Malviya S, Tiwari M, Khan SK, Kapoor S (2015) Novel anti-diabetic formula: Recover Islet β-Cell dysfunction with the implementation of molecular approach. Int J Adv Sci Res 1:71–79

Koti B, Biradar S, Karadi R, Taranalli A, Benade V (2009) Effect of *Bauhinia variegate* bark extract on blood glucose level in normal and alloxanised diabetic rats. J Nat Remedies 9:27–34

Kraml MM, Dicosmo F (1993) A rapid high performance liquid chromatographic method for the separation of the alkaloid precursor L-tyrosine and six tetrahydroisoquinoline alkaloids of *Papaver somniferum*. Phytochem Anal 4:103–104

Krawinkel MB, Keding GB (2006) Bitter gourd (*Momordica Charantia*): a dietary approach to hyperglycemia. Nutr Rev 64:331–337

Krishnaveni M, Mirunalini S, Karthishwaran K, Dhamodharan G (2010) Antidiabetic and anti-hyperlipidemic properties of *Phyllanthus emblica* Linn.(Euphorbiaceae) on streptozotocin induced diabetic rats. Pak J Nutr 9:43–51

Kujur R, Singh V, Ram M, Yadava HN, Singh K, Kumari S, Roy B (2010) Antidiabetic activity and phytochemical screening of crude extract of *Stevia Rebaudiana* in alloxan-induced diabetiis rats. Pharm J 2:27–32

Kumar KA, Maheshwari M, Sivashanmugam A, Devi VS, Prasanth N, Ravi T (2007) Hypoglycemic effect of *Ficus microcarpa* leaves (Chinese banyan) on alloxan-induced diabetic rats. J Biol Sci 7:321–326

Kumar A, Ilavarasan R, Jayachandran T, Deecaraman M, Aravindan P, Padmanabhan N, Krishan M (2008) Anti-diabetic activity of *Syzygium cumini* and its isolated compound against streptozotocin-induced diabetic rats. J Med Plant Res 2:246–249

Kumar A, Shukla R, Singh P, Dubey NK (2010a) Chemical composition, antifungal and antiaflatoxigenic activities of *Ocimum sanctum* L. essential oil and its safety assessment as plant based antimicrobial. Food Chem Toxicol 48:539–543

Kumar S, Malhotra R, Kumar D (2010b) Antidiabetic and free radicals scavenging potential of *Euphorbia hirta* flower extract. Indian J Pharm Sci 72:533

Kumar P, Baraiya S, Gaidhani S, Gupta M, Wanjari MM (2012) Antidiabetic activity of stem bark of *Bauhinia variegata* in alloxan-induced hyperglycemic rats. J Pharmacol Pharmacother 3:64

Kumar V, Chauhan RS, Sood H, Tandon C (2015) Cost effective quantification of picrosides in *Picrorhiza kurroa* by employing response surface methodology using HPLC-UV. J Plant Biochem Biotechnol 24:376–384

Kumari K, Augusti K (2002) Antidiabetic and antioxidant effects of S-methyl cysteine sulfoxide isolated from onions (*Allium cepa* Linn) as compared to standard drugs in alloxan diabetic rats. Indian J Exp Biol 40:1005–1009

Kumari A, Kakkar P (2008) Screening of antioxidant potential of selected barks of Indian medicinal plants by multiple in vitro assays. Biomed Environ Sci 21:24–29

Kupchan SM, Patel A, Fujita E (1965) Tumor inhibitors VI. Cissampareine, new cytotoxic alkaloid from *Cissampelos pareira*. Cytotoxicity of bisbenzylisoquinoline alkaloids. J Pharm Sci 54:580–583

Kwon Y-I, Apostolidis E, Shetty K (2008) In vitro studies of eggplant (*Solanum melongena*) phenolics as inhibitors of key enzymes relevant for type 2 diabetes and hypertension. Bioresour Technol 99:2981–2988

Lan K, X-h J, Ren J (2004) Determination of total flavones, quercetin and isorhamnetin of *Hippophae rhamnoides* sustained release capsules. West China J Pharm Sci 19:41–43

Larkin PJ, Miller JA, Allen RS, Chitty JA, Gerlach WL, Frick S, Kutchan TM, Fist AJ (2007) Increasing morphinan alkaloid production by over-expressing codeinone reductase in transgenic *Papaver somniferum*. Plant Biotechnol J 5:26–37

Latha R, Jansy S, Sarkar T (2015) Evaluation of antimicrobial efficiency and αlpha-glucosidase inhibition of *Rubus ellipticus* smith. leaf extracts and its phytochemical analysis. Asian J Pharm Clin Res:8

Lawal OA, Oyedeji AO (2009) Chemical composition of the essential oils of *Cyperus rotundus* L. from South Africa. Molecules 14:2909–2917

Lawrence T, Gunasekaran S (2014) Isolation, molecular structural estimation and characterized by FTIR, UV, Visible and GCMS Data of (9-Benzyl-9H-carbazole-3, 6-diyl) bis (2, 1-7ethynediyl-1, 5-cyclohexadien-1-yl-4-ylidene)-dimethanone alkaloid from *Rubus ellipticus* and with Theoretical Conformational Analysis. J Chem Biol Phys Sci 4:2445

Lazreg-Aref H, Mars M, Fekih A, Aouni M, Said K (2012) Chemical composition and antibacterial activity of a hexane extract of *Tunisian caprifig* latex from the unripe fruit of Ficus carica. Pharm Biol 50:407–412

Lee S-H, Tanaka T, G-i N, Nishioka I (1990) Hydrolysable tannins from *Euphorbia thymifolia*. Phytochemistry 29:3621–3625

Lee CK, Lee PH, Kuo YH (2001) The chemical constituents from the aril of *Cassia fistula* L. J Chin Chem Soc 48:1053–1058

Lee M-S, Chen C-J, Wan L, Koizumi A, Chang W-T, Yang M-J, Lin W-H, Tsai F-J, Lin M-K (2011) Quercetin is increased in heat-processed *Cuscuta campestris* seeds, which enhances the seed's anti-inflammatory and anti-proliferative activities. Process Biochem 46:2248–2254

Lee Y-J, Thiruvengadam M, Chung I-M, Nagella P (2013) Polyphenol composition and antioxidant activity from the vegetable plant *Artemisia absinthium* L. Aust J Crop Sci 7:1921

Li F-L, Yu L (2011) Hypoglycemic and hypolipidemic effects of polysaccharides form *Portulaca oleracea* L.[J]. China Food Addit 1:008

Li J, Gong F, Li F (2016) Hypoglycemic and hypolipidemic effects of flavonoids from tatary buckwheat in type 2 diabetic rats. Biomed Res

Lim T (2012) Monstera deliciosa. Edible medicinal and non-medicinal plants. Springer, Berlin, pp 252–256

Liu C-T, Wong P-L, Lii C-K, Hse H, Sheen L-Y (2006) Antidiabetic effect of garlic oil but not diallyl disulfide in rats with streptozotocin-induced diabetes. Food Chem Toxicol 44:1377–1384

Liu C, Zhang H, Dai Z, Liu X, Liu Y, Deng X, Chen F, Xu J (2012) Volatile chemical and carotenoid profiles in watermelons [*Citrullus vulgaris* (Thunb.) Schrad (Cucurbitaceae)] with different flesh colors. Food Sci Biotechnol 21:531–541

Liu W, Wang J, Zhang Z, Xu J, Xie Z, Slavin M, Gao X (2014) In vitro and in vivo antioxidant activity of a fructan from the roots of *Arctium lappa* L. Int J Biol Macromol 65:446–453

Liyanage R, Rizliya V, Jayathilake C, Jayawardana B, Vidanarachchi J (2015) Hypolipidemic activity and hypoglycemic effects of banana blossom (*Musa acuminate* Colla) incorporated experimental diets in Wistar rats. Cholestrol:1–6

Lo Scalzo R, Fibiani M, Mennella G, Rotino GL, Dal Sasso M, Culici M, Spallino A, Braga PC (2010) Thermal treatment of eggplant (*Solanum melongena* L.) increases the antioxidant content and the inhibitory effect on human neutrophil burst. J Agric Food Chem 58:3371–3379

Lu Y, Luo J, Kong L (2011) Steroidal alkaloid saponins and steroidal saponins from *Solanum surattense*. Phytochemistry 72:668–673

Lulekal E, Asfaw Z, Kelbessa E, Van Damme P (2013) Ethnomedicinal study of plants used for human ailments in Ankober District, North Shewa Zone, Amhara region, Ethiopia. J Ethnobiol Ethnomed 9:1

Luo W, Wen L, Zhao M, Yang B, Ren J, Shen G, Rao G (2012) Structural identification of isomallotusinin and other phenolics in *Phyllanthus emblica* L. fruit hull. Food Chem 132:1527–1533

Maghrani M, Zeggwagh N-A, Lemhadri A, El Amraoui M, Michel J-B, Eddouks M (2004) Study of the hypoglycaemic activity of *Fraxinus excelsior* and *Silybum marianum* in an animal model of type 1 diabetes mellitus. J Ethnopharmacol 91:309–316

Mahajan R, Badgujar S (2008) Phytochemical investigations of some laticiferous plants belonging to Khandesh region of Maharashtra. Ethnobot Leaflets 2008:151

Maharana L, Pattnaik S, Kar DM, Sahu PK, Si SC (2010) Anti-hyperglycemic potential of aqueous extract of leaves of *Solanum nigrum* Linn. Der Pharma Lettre 2

Mahboubi M (2015) *Rosa damascena* as holy ancient herb with novel applications. J Trad Complement Med 6(1):10–16

Mahfoudhi A, Prencipe FP, Mighri Z, Pellati F (2014) Metabolite profiling of polyphenols in the Tunisian plant *Tamarix aphylla* (L.) Karst. J Pharm Biomed Anal 99:97–105

Mahmood T, Muhammad S, Shinwari ZK (2010) Molecular and morphological characterization of Caralluma species. Pak J Bot 42:1163–1171

Mahmood A, Mahmood A, Tabassum A (2011) Ethnomedicinal survey of plants from District Sialkot, Pakistan. J Appl Pharmacol 2:212–220

Mahmood A, Mahmood A, Malik RN (2012) Indigenous knowledge of medicinal plants from Leepa valley, Azad Jammu and Kashmir, Pakistan. J Ethnopharmacol 143:338–346

Mahmood A, Mahmood A, Malik RN, Shinwari ZK (2013) Indigenous knowledge of medicinal plants from Gujranwala district, Pakistan. J Ethnopharmacol 148:714–723

Mahmoudzadeh-Sagheb H, Heidari Z, Bokaeian M, Moudi B (2010) Antidiabetic effects of *Eucalyptus globulus* on pancreatic islets: a stereological study. Folia Morphol (Warsz) 69:112–118

Mainasara M, Aliero B, Aliero A, Yakubu M (2012) Phytochemical and antibacterial properties of root and leaf extracts of *Calotropis procera*. Nigerian J Basic Appl Sci 20:1–6

Maity C, Mandal B (1990) Chemical and nutritional studies on the seed oil of *Acacia arabica*. J Am Oil Chem Soc 67:433–434

Majekodunmi SO, Oyagbemi AA, Umukoro S, Odeku OA (2011) Evaluation of the anti–diabetic properties of *Mucuna pruriens* seed extract. Asian Pac J Trop Med 4:632–636

Manase MJ, Mitaine-Offer A-C, Pertuit D, Miyamoto T, Tanaka C, Delemasure S, Dutartre P, Mirjolet J-F, Duchamp O, Lacaille-Dubois M-A (2012) *Solanum incanum* and *S. heteracanthum* as sources of biologically active steroid glycosides: Confirmation of their synonymy. Fitoterapia 83:1115–1119

Maniyar Y, Bhixavatimath P (2012) Antihyperglycemic and hypolipidemic activities of aqueous extract of *Carica papaya* Linn. leaves in alloxan-induced diabetic rats. J Ayur Integr Med 3:70

Manvar MN, Desai T (2012) *Vernonia anthelmintica* Willd: an overview on phytopharmacological properties. Inventi Rapid: Ethnopharmacol 2012(4):1–4

Marles R, Farnsworth N (1995) Antidiabetic plants and their active constituents. Phytomedicine 2:137–189

Marwat SK, Fazal-Ur-Rehman M, Ahmad M, Zafar M, Ghulam S (2011) Medicinal folk recipes used as traditional phytotherapies in district Dera Ismail Khan, KPK, Pakistan. Pak J Bot 43:1453–1462

Matasyoh J, Maiyo Z, Ngure R, Chepkorir R (2009) Chemical composition and antimicrobial activity of the essential oil of *Coriandrum sativum*. Food Chem 113:526–529

Mathela C, Kharkwal H, Shah G (1994) Essential oil composition of some Himalayan Artemisia species. J Essent Oil Res 6:345–348

Matthäus B, Özcan M (2005) Glucosinolates and fatty acid, sterol, and tocopherol composition of seed oils from *Capparis spinosa* Var. spinosa and *Capparis ovata* Desf. Var. canescens (Coss.) Heywood. J Agric Food Chem 53:7136–7141

Mazhar F, Khanum R, Ajaib M, Jahangir M (2015) Potent AChE enzyme inhibition activity of *Zizyphus oxyphylla*: A new source of antioxidant compounds. Pak J Pharm Sci 1:2053–2059

Mazumdar S, Akter R, Talukder D (2015) Antidiabetic and antidiarrhoeal effects on ethanolic extract of *Psidium guajava* (L.) bat. leaves in wister rats. Asian Pac J Trop Biomed 5:10–14

Meenu J, Sunil S, Manoj K (2011) Evaluation of antihyperglycemic activity of *Dodonaea viscosa* leaves in normal and STZ-diabetic rats. Int J Pharm Pharm Sci 3:69–74

Mehmood F, Zaheer-ud-din K, Shahzadi P, Yaseen T, Mughal TA, Raza SH, Qasim M (2013) A comparative study of in vitro total antioxidant, in vivo antidiabetic and antimicrobial activities of essential oils from leaves and rind of Citrus reticulata Blanco cv. Murcot (Honey). Pak J Bot 45:1571–1576

Melchini A, Traka MH (2010) Biological profile of erucin: a new promising anticancer agent from cruciferous vegetables. Toxins 2:593–612

Menezes EW, Tadini CC, Tribess TB, Zuleta A, Binaghi J, Pak N, Vera G, Dan MCT, Bertolini AC, Cordenunsi BR (2011) Chemical composition and nutritional value of unripe banana flour (*Musa acuminata*, var. Nanicão). Plant Foods Hum Nutr 66:231–237

Menichini F, Loizzo MR, Bonesi M, Conforti F, De Luca D, Statti GA, de Cindio B, Menichini F, Tundis R (2011) Phytochemical profile, antioxidant, anti-inflammatory and hypoglycemic potential of hydroalcoholic extracts from *Citrus medica* L. cv Diamante flowers, leaves and fruits at two maturity stages. Food Chem Toxicol 49:1549–1555

Meonah SS, Palaniswamy M, Keerthy S, Rajkumar LP, Nandhini RU (2012) Pharmacognostical and hypoglycemic activity of different parts of *Solanum nigrum* Linn plant. Int J Pharm Pharm Sci 4:221–224

Mesfin F, Demissew S, Teklehaymanot T (2009) An ethnobotanical study of medicinal plants in Wonago Woreda, SNNPR, Ethiopia. J Ethnobiol Ethnomed 5(28):1–18

Metwally A, Omar A, Harraz F, El Sohafy S (2010) Phytochemical investigation and antimicrobial activity of *Psidium guajava* L. leaves. Pharmacogn Mag 6:212

Minaiyan M, Ghannadi A, Movahedian A, Hakim-Elahi I (2014) Effect of *Hordeum vulgare* L.(Barley) on blood glucose levels of normal and STZ-induced diabetic rats. Res Pharmaceut Sci 9:173

Mishra GJ, Reddy M, Rana JS (2012) Isolation of flavonoid constituent from *Launaea procumbens* Roxb. by preparative HPTLC method. IOSR J Pharm 2:5–11

Misra L, Mishra P, Pandey A, Sangwan RS, Sangwan NS, Tuli R (2008) Withanolides from *Withania somnifera* roots. Phytochemistry 69:1000–1004

Miura T, Itoh C, Iwamoto N, Kato M, Kawai M, Park SR, Suzuki I (2001) Hypoglycemic activity of the fruit of the *Momordica charantia* in type 2 diabetic mice. J Nutr Sci Vitaminol 47:340–344

Miyake Y, Murakami A, Sugiyama Y, Isobe M, Koshimizu K, Ohigashi H (1999) Identification of coumarins from lemon fruit (*Citrus limon*) as inhibitors of in vitro tumor promotion and superoxide and nitric oxide generation. J Agric Food Chem 47:3151–3157

Miyazawa M, Maehara T, Kurose K (2002) Composition of the essential oil from the leaves of *Eruca sativa*. Fla Frag J 17:187–190

Mizokami H, Tomita-Yokotani K, Yoshitama K (2008) Flavonoids in the leaves of *Oxalis corniculata* and sequestration of the flavonoids in the wing scales of the pale grass blue butterfly, Pseudozizeeria maha. J Plant Res 121:133–136

Mkaddem M, Bouajila J, Ennajar M, Lebrihi A, Mathieu F, Romdhane M (2009) Chemical composition and antimicrobial and antioxidant activities of Mentha (longifolia L. and viridis) essential oils. J Food Sci 74:M358–M363

Modak M, Dixit P, Londhe J, Ghaskadbi S, Devasagayam TPA (2007) Indian herbs and herbal drugs used for the treatment of diabetes. J Clin Biochem Nutr 40:163–173

Modey WK, Mulholland DA, Mahomed H, Raynor MW (1996) Analysis of extracts from *Cedrela toona* (meliaceae) by on-line and off-line supercritical fluid extraction-capillary gas chromatography. J Microcol Separat 8:67–74

Mohamed A, El-Sayed MA, Hegazy ME, Helaly SE, Esmail AM, Mohamed NS (2010) Chemical constituents and biological activities of *Artemisia herba-alba*. Rec Nat Prod 4:1–25

Mohani N, Ahmad M, Jahan N (2014) Evaluation of phytoconstituents of three plants *Acorus calamus* Linn. *Artemisia absinthium* Linn and *Bergenia himalaica* boriss by FTIR spectroscopic analysis. Pak J Pharm Sci 27(6):2251–2255

Mojab F, Kamalinejad M, Ghaderi N, Vahidipour HR (2010) Phytochemical screening of some species of Iranian plants. Iran J Pharm Res 2:77–82

Mootoosamy A, Mahomoodally MF (2014) Ethnomedicinal application of native remedies used against diabetes and related complications in Mauritius. J Ethnopharmacol 151:413–444

Muhammad A, Ayub M, Zeb A, Durrani Y, Ullah J, Afridi S (2011) Physicochemical analysis of apple pulp from Mashaday variety during storage. Agric Biol J N Am 2(2):192–196

Mukherjee K, Bhattacharjee P (1987) Glut-5 (10)-en-3-one from *Andrachne cordifolia*. Phytochemistry 26:1539–1540

Mukherjee K, Bhattacharjee P, Mukherjee R, Ghosh P (1986) A triterpenoid of *Andrachne cordifolia*. Phytochemistry 25:2669–2670

Mukherjee PK, Maiti K, Mukherjee K, Houghton PJ (2006) Leads from Indian medicinal plants with hypoglycemic potentials. J Ethnopharmacol 106:1–28

Murad W, Azizullah A, Adnan M, Tariq A, Khan KU, Waheed S, Ahmad A (2013) Ethnobotanical assessment of plant resources of Banda Daud Shah, District Karak, Pakistan. J Ethnobiol Ethnomed 9:77

Murugan M, Reddy C (2009) Hypoglycemic and hypolipidemic activity of leaves of *Mucuna pruriens* in alloxan induced diabetic rats. Int J Curr Res 3(6):323–327

Muruganandan S, Srinivasan K, Gupta S, Gupta P, Lal J (2005) Effect of mangiferin on hyperglycemia and atherogenicity in streptozotocin diabetic rats. J Ethnopharmacol 97:497–501

Muthukumran P, Begumand VH, Kalaiarasan P (2011) Anti-diabetic activity of *Dodonaea viscosa* (L) leaf extracts. Int J Pharm Tech Res 3:1

Mwonjoria JK, Ngeranwa JJ, Kariuki HN, Githinji CG, Sagini MN, Wambugu SN (2014) Ethno medicinal, phytochemical and pharmacological aspects of *Solanum incanum* (lin.). Int J Pharmacol Toxicol 2:17–20

Nagao T, Tanaka R, Iwase Y, Hanazono H, Okabe H (1991) Studies on the constituents of *Luffa acutangula* Roxb. I. Structures of acutosides A--G, oleanane-type triterpene saponins isolated from the herb. Chem Pharm Bull 39:599–606

Nagaraju B, Anand S, Ahmed N, Chandra JNS, Ahmed F, Padmavathi G (2012) Antiulcer activity of aqueous extract of *Citrus medica* Linn. fruit against ethanol induced ulcer in rats. Adv Biol Res 6:24–29

Naik R, Harmalkar DS, Xu X, Jang K, Lee K (2015) Bioactive benzofuran derivatives: Moracins A–Z in medicinal chemistry. Eur J Med Chem 90:379–393

Naim M, Amjad FM, Sultana S, Islam SN, Hossain MA, Begum R, Rashid MA, Amran MS (2012) Comparative study of antidiabetic activity of hexane-extract of lemon peel (*Limon citrus*) and glimepiride in alloxan-induced diabetic rats. Bangladesh Pharmaceut J 15:131–134

Nair R, Shukla V, Chanda S (2007) Assessment of *Polyalthia longifolia* var. pendula for hypoglycemic and antihyperglycemic activity. J Clin Diagn Res 1:1–3

Najafi S, Sanadgol N, Nejad BS, Beiragi MA, Sanadgol E (2010) Phytochemical screening and antibacterial activity of *Citrullus colocynthis* (Linn.) Schrad against Staphylococcus aureus. J Med Plant 4:2321–2325

Nalamolu RK, Boini KM, Nammi S (2007) Effect of chronic administration of *Boerhaavia diffusa* Linn. leaf extract on experimental diabetes in rats. Trop J Pharm Res 3:305–309

Nammi S, Boini MK, Lodagala SD, Behara RBS (2003) The juice of fresh leaves of *Catharanthus roseus* Linn. reduces blood glucose in normal and alloxan diabetic rabbits. BMC Complement Altern Med 3:1

Naveen A (2010) Antidiabetic activity of polyherbal formulation in alloxan-induced diabetes mellitus. J Pharm Res 3:1119–1120

Nazaruk J, Borzym-Kluczyk M (2015) The role of triterpenes in the management of diabetes mellitus and its complications. Phytochem Rev 14:675–690

Nepolean P, Anitha J, Emilin R (2009) Isolation, analysis and identification of phytochemicals of antimicrobial activity of *Moringa oleifera* Lam. Current Biotica 3:33–37

Neto L, Mário C, de Vasconcelos CF, Thijan VN, Caldas GF, Araújo AV, Costa-Silva JH, Amorim EL, Ferreira F, de Oliveira AF (2013) Evaluation of antihyperglycaemic activity of *Calotropis procera* leaves extract on streptozotocin-induced diabetes in Wistar rats. Rev Bras Farm 23:913–919

Nile SH, Park SW (2014) Antioxidant, α-Glucosidase and Xanthine Oxidase Inhibitory Activity of Bioactive Compounds From Maize (Zea mays L.). Chem Bio Drug Des 83:119–125

Niranjan SP, Dharmendra S, Kiran P (2010) Antidiabetic activity of ethanolic extract of *Dalbergis sissoo* L. Leaves in Alloxan-Induced diabetic rats. Int J Curr Pharm Res 2(2):24–27

Nirmala A, Eliza J, Rajalakshmi M, Priya E, Daisy P (2008) Effect of hexane extract of *Cassia fistula* barks on blood glucose and lipid profile in streptozotocin diabetic rats. Int J Pharm 4:292–296

Nisar M, Adzu B, Inamullah K, Bashir A, Ihsan A, Gilani A (2007) Antinociceptive and antipyretic activities of the *Zizyphus oxyphylla* Edgew. leaves. Phytother Res 21:693–695

Nougbodé YAEI, Agbangnan CP, Koudoro AY, Dèdjiho CA, Aïna MP, Mama D, Sohounhloué DCK (2013) Evaluation of the *Opuntia dillenii* as natural coagulant in water clarification: case of treatment of highly turbid surface water. J Water Resource Prot 5:1242

Nowbandegani AS, Kiumarcy S, Rahmani F, Dokouhaki M, Khademian S, Zarshenas MM, Faridi P (2015) Ethnopharmacological knowledge of Shiraz and Fasa in Fars region of Iran for diabetes mellitus. J Ethnopharmacol 172:281–287

Núñez Sellés AJ, Vélez Castro HT, Agüero-Agüero J, González-González J, Naddeo F, De Simone F, Rastrelli L (2002) Isolation and quantitative analysis of phenolic antioxidants, free sugars, and polyols from mango (*Mangifera indica* L.) stem bark aqueous decoction used in Cuba as a nutritional supplement. J Agric Food Chem 50:762–766

Oguntoye S, Olatunji G, Kolawole O, Enonbun K (2008) Phytochemical screening and antibacterial activity of *Viscum album* (Mistletoe) extracts. Plant Sci Res 1:44–46

Ojiako AO, Chidoka C, Chidoka P (2014) Comparative proximate composition and hypoglycemic properties of three medicinal plants (*Verononia amygdalina*, *Azadirachta indica* and *Moringa oleifera*). Phcog Commn 4:40

Okolie VU, Okeke EC, Ehiemere OI, Ezenduka OP (2009) Investigation of the effect of *Solanum incanum* on postprandial blood glucose concentration of normoglycemic Nigerians. Pak J Nutr 8:1631–1635

Olayaki LA, Irekpita JE, Yakubu MT, Ojo OO (2015) Methanolic extract of *Moringa oleifera* leaves improves glucose tolerance, glycogen synthesis and lipid metabolism in alloxan-induced diabetic rats. J Basic Clin Physiol Pharmacol 26:585–593

Oliveira I, Valentão P, Lopes R, Andrade PB, Bento A, Pereira JA (2009) Phytochemical characterization and radical scavenging activity of *Portulaca oleraceae* L. leaves and stems. Microchem J 92:129–134

Oliveira AP, Silva LR, Ferreres F, Guedes de Pinho P, Pc V, Silva BM, Pereira JA, Andrade PB (2010) Chemical assessment and in vitro antioxidant capacity of *Ficus carica* latex. J Agric Food Chem 58:3393–3398

Oloyede O (2005) Chemical profile of unripe pulp of *Carica papaya*. Pak J Nutr 4:379–381

Oluwaseun AA, Ganiyu O (2008) Antioxidant properties of methanolic extracts of mistletoes (*Viscum album*) from cocoa and cashew trees in Nigeria. Afr J Biotechnol 7(17):3138–3142

Omar NM, Atia GM (2012) Effect of *Nigella sativa* on pancreatic β-cell damage in streptozotocin-induced diabetic rats: histological and immunohistochemical study. Egypt J Histol 35:106–116

Oputa R, Chinenye S (2012) Diabetes mellitus: a global epidemic with potential solutions. Afr J Diabetes Med 2(2):33–35

Orhan DD, Aslan M, Sendogdu N, Ergun F, Yesilada E (2005) Evaluation of the hypoglycemic effect and antioxidant activity of three *Viscum album* subspecies (European mistletoe) in streptozotocin-diabetic rats. J Ethnopharmacol 98:95–102

Ortega A, García PE, Cárdenas J, Mancera C, Marquina S, del Carmen Garduño ML, Maldonado E (2001) Methyl dodonates, a new type of diterpenes with a modified clerodane skeleton from *Dodonaea viscosa*. Tetrahedron 57:2981–2989

Otsuka H, Akiyama T, Kawai K-I, Shibata S, Inoue O, Ogihara Y (1978) The structure of jujubosides A and B, the saponins isolated from the seeds of Zizyphus jujuba. Phytochemistry 17:1349–1352

Pa R, Mathew L (2012) Antimicrobial activity of leaf extracts of *Justicia adhatoda* L. in comparison with vasicine. Asian Pac J Trop Biomed 2:S1556–S1560

Pandit R, Phadke A, Jagtap A (2010) Antidiabetic effect of *Ficus religiosa* extract in streptozotocin-induced diabetic rats. J Ethnopharmacol 128:462–466

Panhwar AQ, Abro H (2007) Ethnobotanical studies of Mahal Kohistan (Khirthar National Park). Pak J Bot 39:2301–2315

Papageorgiou VP, Assimopoulou AN, Couladouros EA, Hepworth D, Nicolaou K (1999) The chemistry and biology of alkannin, shikonin, and related naphthazarin natural products. Angew Chem Int Ed 38:270–301

Papageorgiou VP, Assimopoulou AN, Samanidou V, Papadoyannis I (2006) Recent advances in chemistry, biology and biotechnology of alkannins and shikonins. Curr Org Chem 10:2123–2142

Papi A, Orlandi M, Bartolini G, Barillari J, Iori R, Paolini M, Ferroni F, Grazia Fumo M, Pedulli GF, Valgimigli L (2008) Cytotoxic and antioxidant activity of 4-methylthio-3-butenyl isothiocyanate from *Raphanus sativus* L. (Kaiware Daikon) sprouts. J Agric Food Chem 56:875–883

Pari L, Satheesh MA (2004) Antidiabetic activity of *Boerhaavia diffusa* L.: effect on hepatic key enzymes in experimental diabetes. J Ethnopharmacol 91:109–113

Parikh NH, Parikh PK, Kothari C (2014) Indigenous plant medicines for health care: treatment of Diabetes mellitus and hyperlipidemia. Chin J Nat Med 12:335–344

Parmar HS, Kar A (2007) Antidiabetic potential of *Citrus sinensis* and *Punica granatum* peel extracts in alloxan treated male mice. Biofactors 31:17–24

Parveen A, Irfan M, Mohammad F (2012) Antihyperglycemic activity in *Grewia asiatica*, a comparative investigation. Int J Pharm Pharm Sci 4:210–213

Patel VS, Chitra V, Prasanna PL, Krishnaraju V (2008) Hypoglycemic effect of aqueous extract of *Parthenium hysterophorus* L. in normal and alloxan induced diabetic rats. Indian J Pharm 40(4):183–185

Patel J, Kumar G, Deviprasad S, Deepika S, Qureshi MS (2011) Phytochemical and anthelmintic evaluation of *Lantana camara* (L.) Var. Aculeate leaves against pheretima posthuma. JGTPS 2:11–20

Patel CA, Divakar K, Santani D, Solanki HK, Thakkar JH (2012a) Remedial prospective of *Hippophaë rhamnoides* Linn. (sea buckthorn). ISRN Pharmacol 2012:436857

Patel D, Prasad S, Kumar R, Hemalatha S (2012b) An overview on antidiabetic medicinal plants having insulin mimetic property. Asian Pac J Trop Biomed 2:320–330

Patil RN, Patil RY, Ahirwar D (2010) Study of some medicinal plants for antidiabetic activity in alloxan induced diabetes. Pharmacol Ther 1:53–60

Patil RN, Patil RY, Ahirwar B, Ahirwar D (2011) Evaluation of antidiabetic and related actions of some Indian medicinal plants in diabetic rats. Asian Pac J Trop Med 4:20–23

Pattanayak P, Behera P, Das D, Panda SK (2010) *Ocimum sanctum* Linn. A reservoir plant for therapeutic applications: An overview. Pharmacogn Rev 4:95

Pawlowska AM, Camangi F, Bader A, Braca A (2009) Flavonoids of *Zizyphus jujuba* L. and Zizyphus spina-christi (L.) Willd (Rhamnaceae) fruits. Food Chem 112:858–862

Pekamwar S, Kalyankar T, Kokate S (2013) Pharmacological activities of *Coccinia grandis*: Review. J Appl Pharmaceut Sci 3:114–119

Pereira AP, Ferreira IC, Marcelino F, Valentão P, Andrade PB, Seabra R, Estevinho L, Bento A, Pereira JA (2007) Phenolic compounds and antimicrobial activity of olive (*Olea europaea* L. Cv. Cobrançosa) leaves. Molecules 12:1153–1162

Persaud S, Al-Majed H, Raman A, Jones P (1999) *Gymnema sylvestre* stimulates insulin release in vitro by increased membrane permeability. J Endocrinol 163:207–212

Phoboo S, Pinto MDS, Bhowmik PC, Jha PK, Shetty K (2010) Quantification of major phytochemicals of *Swertia chirayita*, a medicinal plant from Nepal. Ecoprint 17:59–68

Phoboo S, Pinto MDS, Barbosa ACL, Sarkar D, Bhowmik PC, Jha PK, Shetty K (2013) Phenolic-linked biochemical rationale for the anti-diabetic properties of *Swertia chirayita* (Roxb. ex Flem.) Karst. Phytother Res 27:227–235

Porchezhian E, Dobriyal R (2003) An overview on the advances of *Gymnema sylvestre*: chemistry, pharmacology and patents. Die Pharmazie-An Int J Pharm Sci 58:5–12

Pramila D, Xavier R, Marimuthu K, Kathiresan S, Khoo M, Senthilkumar M, Sathya K, Sreeramanan S (2012) Phytochemical analysis and antimicrobial potential of methanolic leaf extract of peppermint (*Mentha piperita*: Lamiaceae). J Med Plant Res 6:331–335

Prince PSM, Kamalakkannan N, Menon VP (2003) *Syzygium cumini* seed extracts reduce tissue damage in diabetic rat brain. J Ethnopharmacol 84:205–209

Prince PSM, Kamalakkannan N, Menon VP (2004) Antidiabetic and antihyperlipidaemic effect of alcoholic *Syzygium cumini* seeds in alloxan induced diabetic albino rats. Jl Ethnopharmacology 91:209–213

Priya SE, Ravindhran R (2015) Phytochemical Analysis and antimicrobial properties of extracts from aerial parts of *Phyla nodiflora* (L) Greene. Int J Curr Microbiol App Sci 4:347–358

Quanico JP, Amor EC, Perez GG (2008) Analgesic and hypoglycemicactivities of *Bixa orellana*, *Kyllinga monocephala* and *Luffa acutangula*. Philipp J Sci 137:69–76

Qunhua C, Weijing Q, Yunxia D, Zhicai Z, Wei N, Yifeng P (2003) Effect of flavonoids from the seed and fruit residue of *Hippophae rhamnoides* L. on Glycometabolism in Mice. J Chin Med Mater 10:020

Qureshi R (2012) Medicinal flora of Hingol National Park, Baluchistan, Pakistan. Pak J Bot 44:725–732

Qureshi R, Bhatti GR (2008) Ethnobotany of plants used by the Thari people of Nara Desert, Pakistan. Fitoterapia 79:468–473

Qureshi R, Bhatti G (2009) Folklore uses of Amaranthaceae family from Nara desert, Pakistan. Pak J Bot 41:1565–1572

Quyen VTL (2013) Evaluation of pharmacological activity of flavonoids in *Euphorbia thymifolia* Linn. International University HCMC, Vietnam

Raal A, Arak E, Orav A (2012) The content and composition of the essential oil Found in *Carum carvi* L. commercial fruits obtained from different countries. J Essent Oil Res 24:53–59

Raghavendra M, Satish S, Raveesha K (2006) Phytochemical analysis and antibacterial activity of *Oxalis corniculata*; a known medicinal plant. Myscience 1:72–78

Ragupathy S, Steven NG, Maruthakkutti M, Velusamy B, Ul-Huda MM (2008) Consensus of the 'Malasars' traditional aboriginal knowledge of medicinal plants in the Velliangiri holy hills, India. J Ethnobiol Ethnomed 4:8

Rahmatullah M, Sultan S, Toma T, Lucky S, Chowdhury M, Haque W, Annay E, Jahan R (2010) Effect of *Cuscuta reflexa* stem and *Calotropis procera* leaf extracts on glucose tolerance in glucose-induced hyperglycemic rats and mice. Afr Tradit Complement Altern Med 7(2):109–112

Rahmatullah M, Hasan S, Ali Z, Rahman S, Jahan R (2012) Antihyperglycemic and antinociceptive activities of methanolic extract of *Euphorbia thymifolia* L. whole plants. Zhong Xi Yi Jie He Xue Bao 10:228–232

Rai V, Iyer U, Mani U (1997) Effect of Tulasi (*Ocimum sanctum*) leaf powder supplementation on blood sugar levels, serum lipids and tissues lipids in diabetic rats. Plant Foods Hum Nutr 50:9–16

Raj SM, Mohammed S, Kumar SV, Debnath S (2012) Antidiabetic effect of *Luffa acutangula* fruits and histology of organs in streptozotocin induced diabetic in rats. Res J Pharmacogn Phytochem 4:64–69

Rajaram N, Janardhanan K (1991) Chemical composition and nutritional potential of the tribal pulses *Bauhinia purpurea*, *B. racemosa* and *B. vahlii*. J Sci Food Agric 55:423–431

Rajasekaran S, Sivagnanam K, Subramanian S (2005) Antioxidant effect of Aloe vera gel extract in streptozotocin-induced diabetes in rats. Pharmacol Rep 57:90–96

Rajasekaran S, Jaykar B, Anandan R (2013) Anti-diabetic activity of leaves of *Zizyphus nummularia* by dexamethasone induced diabetic rat model. Int J Pharm Tech 5(2):544–551

Ramadan MF, Sharanabasappa G, Seetharam Y, Seshagiri M, Moersel J-T (2006) Characterisation of fatty acids and bioactive compounds of kachnar (*Bauhinia purpurea* L.) seed oil. Food Chem 98:359–365

Ramírez I, Carabot A, Meléndez P, Carmona J, Jimenez M, Patel AV, Crabb TA, Blunden G, Cary PD, Croft SL (2003) Cissampeloflavone, a chalcone-flavone dimer from *Cissampelos pareira*. Phytochemistry 64:645–647

Ramkissoon J, Mahomoodally M, Ahmed N, Subratty A (2013) Antioxidant and anti–glycation activities correlates with phenolic composition of tropical medicinal herbs. Asian Pac J Trop Med 6:561–569

Rana V, Kumar V, Soni P (2012) Structural characterization of an acidic polysaccharide from *Dalbergia sissoo* Roxb. leaves. Carbohydr Polym 90:243–250

Rana M, Kumar S, Rana M, Dhatwalia V (2016) Role of *Toona ciliata* extract in diabetes against streptozotocin–nicotinamide induced diabetic rats. J Ayu Herb Med 2:6–10

Rani PU, Devanand P (2013) Bioactivities of caffeic acid methyl ester (methyl-(E)-3-(3, 4-dihydroxyphenyl) prop-2-enoate): a hydroxycinnamic acid derivative from *Solanum melongena* L. fruits. J Pest Sci 86:579–589

Ranjan V, Vats M (2016) Pharmacognostical and physico-chemical standardisation of whole plant of *Adiantum capillus veneris* Linn. Int J Pharm Sci Res 7:773

Rao C, Verma A, Gupta P, Vijayakumar M (2007) Anti-inflammatory and anti-nociceptive activities of *Fumaria indica* whole plant extract in experimental animals. Acta Pharma 57:491–498

Rao GV, Mukhopadhyay T, Annamalai T, Radhakrishnan N, Sahoo M (2011) Chemical constituents and biological studies of *Origanum vulgare* Linn. Pharm Res 3:143

Rao VU, Viteesha V, Suma K, Nagababu P (2015) Evaluation of phytochemical constituents, antibacterial and antioxidant activities of *Monstera deliciosa* Liebm. stem extracts. World J Pharm Pharm Sci 4(11):1422–1433

Rashid M, Bashir S, Mushtaq MN, Khan HU, Malik MNH, Qayyum A, ur Rahaman MS (2013a) Comparative hypoglycemic activity of different extracts of *Teucrium stocksianum* in diabetic rabbits. Bangladesh J Pharm 8:186–193

Rashid S, Rather MA, Shah WA, Bhat BA (2013b) Chemical composition, antimicrobial, cytotoxic and antioxidant activities of the essential oil of *Artemisia indica* Willd. Food Chem 138:693–700

Rasineni K, Bellamkonda R, Singareddy SR, Desireddy S (2010) Antihyperglycemic activity of *Catharanthus roseus* leaf powder in streptozotocin-induced diabetic rats. Pharm Res 2:195

Rastogi S, Pandey M, Rawat A (2007) A new, convenient method for determination of mangiferin, an anti-diabetic compound, in *Mangifera indica* L. J Planar Chromat 20:317–320

Rathee S, Mogla O, Sardana S, Vats M, Rathee P (2010) Antidiabetic activity of *Capparis decidua* Forsk Edgew. J Pharm Res 3:231–234

Rather MA, Dar BA, Sofi SN, Bhat BA, Qurishi MA (2012) *Foeniculum vulgare*: a comprehensive review of its traditional use, phytochemistry, pharmacology, and safety. Arab J Chem 9(2):S1574–S1583

Ratnasooriya W, Jayakody J, Hettiarachchi H (2004) *Cassia fistula* and hypoglycaemia. AJMH 16:8

Rauf A, Jan M, Rehman W, Muhammad N (2013) Phytochemical, phytotoxic and antioxidant profile of *Caralluma tuberculata* NE Brown. Wudpecker J Pharm Pharmacol 2:21–25

Raut NA, Gaikwad NJ (2006) Antidiabetic activity of hydro-ethanolic extract of *Cyperus rotundus* in alloxan induced diabetes in rats. Fitoterapia 77:585–588

Reddy MV, Reddy MK, Gunasekar D, Caux C, Bodo B (2003) A flavanone and a dihydrodibenzoxepin from *Bauhinia variegata*. Phytochemistry 64:879–882

Riya MP, Antu KA, Pal S, Chandrakanth KC, Anilkumar KS, Tamrakar AK, Srivastava AK, Raghu KG (2015) Antidiabetic property of *Aerva lanata* (L.) Juss. ex Schult. is mediated by inhibition of alpha glucosidase, protein glycation and stimulation of adipogenesis. J Diabetes 7(4):548–561

Rizwani GH, Aslam M, Ahmad M, Usmanghani K, Ahmad V (1993a) Lipid components of *Caralluma edulis* (edgew.) hook. Pak J Pharm Sci 6:19–27

Rizwani GH, Usmanghani K, Ahmad M, Ahmad VU (1993b) Pregnane glycosides from *Caralluma tuberculata*. Nat Prod Lett 2:97–104

Rodriguez SC, López B, Chaves AR (1999) Changes in polyamines and ethylene during the development and ripening of eggplant fruits (*Solanum melongena*). J Agric Food Chem 47:1431–1434

Romeo V, Ziino M, Giuffrida D, Condurso C, Verzera A (2007) Flavour profile of capers (*Capparis spinosa* L.) from the Eolian Archipelago by HS-SPME/GC–MS. Food Chem 101:1272–1278

Roy S, Sehgal R, Padhy B, Kumar V (2005) Antioxidant and protective effect of latex of *Calotropis procera* against alloxan-induced diabetes in rats. J Ethnopharmacol 102:470–473

Sabeen M, Ahmad SS (2009) Exploring the folk medicinal flora of Abbotabad city, Pakistan. Ethnobot Leaflets 2009:1

Sachan NK, Kumar Y, Pushkar S, Thakur R, Gangwar SS, Kalaichelvan V (2009) Antidiabetic potential of alcoholic and aqueous extracts of *Ficus racemosa* Linn. bark in normal and alloxan induced diabetic rats. Int J Pharm Sci Drug Res 1:24–27

Sadiq S, Nagi AH, Shahzad M, Zia A (2010) The reno-protective effect of aqueous extract of *Carum carvi* (black zeera) seeds in streptozotocin induced diabetic nephropathy in rodents. Saudi J Kidney Dis Transpl 21:1058

Saeed MA, Sabir A (1999) Effects of *Fagonia cretica* L. constituents on various endocrinological parameters in rabbits. Turk J Biol 23:187–198

Safithri M, Sari YP (2016) Inhibition of α-glucosidase activity by ethanolic extract of *Melia azedarach* L. leaves. IOP Conference Series: Earth and Environmental Science. IOP Publishing, Bristol, p 012025

Sagratini G, Cristalli G, Giardinà D, Gioventù G, Maggi F, Ricciutelli M, Vittori S (2008) Alkannin/shikonin mixture from roots of *Onosma echioides* (L.) L.: Extraction method study and quantification. J Sep Sci 31:945–952

Şahin F, Güllüce M, Daferera D, Sökmen A, Sökmen M, Polissiou M, Agar G, Özer H (2004) Biological activities of the essential oils and methanol extract of *Origanum vulgare* ssp. *vulgare* in the Eastern Anatolia region of Turkey. Food Control 15:549–557

Said O, Fulder S, Khalil K, Azaizeh H, Kassis E, Saad B (2008) Maintaining a physiological blood glucose level with 'glucolevel', a combination of four anti-diabetes plants used in the traditional Arab herbal medicine. Evid Based Complement Alternat Med 5:421–428

Sakat S, Tupe P, Juvekar A (2012) Gastroprotective effect of *Oxalis corniculata* (whole plant) on experimentally induced gastric ulceration in Wistar rats. Indian J Pharm Sci 74:48

Saklani S, Chandra S (2012) Phytochemical Screening of Garhwal Himalaya Wild Edible Fruit *Ficus palmata*. Int J Pharm Tech Res 4:1185–1191

Saleem R, Ahmad M, Hussain SA, Qazi AM, Ahmad SI, Qazi MH, Ali M, Faizi S, Akhtar S, Husnain SN (1999) Hypotensive, hypoglycaemic and toxicological studies on the flavonol C-glycoside shamimin from Bombax ceiba. Planta Med 65:331–334

Saleem S, Jafri L, ul Haq I, Chang LC, Calderwood D, Green BD, Mirza B (2014) Plants *Fagonia cretica* L. and *Hedera nepalensis* K. Koch contain natural compounds with potent dipeptidyl peptidase-4 (DPP-4) inhibitory activity. J Ethnopharmacol 156:26–32

Salem A, Olivares M, Lopez S, Gonzalez-Ronquillo M, Rojo R, Camacho L, Cerrillo S, Mejia H (2011a) Effect of natural extracts of *Salix babylonica* and *Leucaena leucocephala* on nutrient digestibility and growth performance of lambs. Anim Feed Sci Technol 170:27–34

Salem A, Salem M, Gonzalez-Ronquillo M, Camacho L, Cipriano M (2011b) Major chemical constituents of *Leucaena leucocephala* and *Salix babylonica* leaf extracts. J Trop Agric 49:95–98
Samarghandian S, Borji A, Hidar Tabasi S (2013) Effects of *Cichorium intybus* Linn on blood glucose, lipid constituents and selected oxidative stress parameters in streptozotocin-induced diabetic rats. Cardiovasc Haematol Disord-Drug Targets 13:231–236
Samejo M, Memon S, Bhanger M, Khan K (2012) Chemical composition of essential oils from *Alhagi maurorum*. Chem Nat Compd 48:1–3
Sánchez-González M, Lozano-Mena G, Juan ME, García-Granados A, Planas JM (2013) Liquid chromatography–mass spectrometry determination in plasma of maslinic acid, a bioactive compound from *Olea europaea* L. Food Chem 141:4375–4381
Saneja A, Sharma C, Aneja K, Pahwa R (2010) *Gymnema sylvestre* (Gurmar): a review. Pharm Lett 2:275–284
Sangi SMA, Sulaiman MI, El-wahab MFA, Ahmedani EI, Ali SS (2015) Antihyperglycemic effect of thymoquinone and oleuropein, on streptozotocin-induced diabetes mellitus in experimental animals. Phcog Mag 11:S251
Sanz-Biset J, Campos-de-la-Cruz J, Epiquién-Rivera MA, Cañigueral S (2009) A first survey on the medicinal plants of the Chazuta valley (Peruvian Amazon). J Ethnopharmacol 122:333–362
Saqib Z, Mahmood A, Malik RN, Mahmood A, Syed JH, Ahmad T (2014) Indigenous knowledge of medicinal plants in Kotli Sattian, Rawalpindi district, Pakistan. J Ethnopharmacol 151:820–828
Sarikaphuti A, Nararatwanchai T, Hashiguchi T, Ito T, Thaworanunta S, Kikuchi K, Oyama Y, Maruyama I, Tancharoen S (2013) Preventive effects of *Morus alba* L. anthocyanins on diabetes in Zucker diabetic fatty rats. Expe Ther Med 6:689–695
Sarkar S, Pranava M, MARITA AR (1996) Demonstration of the hypoglycemic action of *Momordica charantia* in a validated animal model of diabetes. Pharmacol Res 33:1–4
Sasidharan S, Sumathi V, Jegathambigai NR, Latha LY (2011) Antihyperglycaemic effects of ethanol extracts of *Carica papaya* and *Pandanus amaryfollius* leaf in streptozotocin-induced diabetic mice. Nat Prod Res 25:1982–1987
Savona G, Bruno M, Paternostro M, Marco JL, Rodríguez B (1982) Salviacoccin, a neo-clerodane diterpenoid from *Salvia coccinea*. Phytochemistry 21:2563–2566
Sawant L, Prabhakar B, Pandita N (2012) Quantitative HPLC analysis of ascorbic acid and gallic acid in *Phyllanthus emblica*. J Anal Bioanal Tech 1(3):2–4
Sedigheh A, Jamal MS, Mahbubeh S, Somayeh K, Mahmoud R, Azadeh A, Fatemeh S (2011) Hypoglycaemic and hypolipidemic effects of pumpkin (*Cucurbita pepo* L.) on alloxan-induced diabetic rats. Afr J Pharm and Pharmcol 5:2620–2626
Semwal D, Bamola A, Rawat U (2007) Chemical constituents from some antidiabetic plants. Univ J Phytochem Ayur Heig 2(3):40–48
Sendl A (1995) *Allium sativum* and *Allium ursinum*: Part 1 Chemistry, analysis, history, Botany. Phytomedicine 1:323–339
Sengottaiyan A, Praburaman L, Manoharan K, Rajinikanth R, Govarthanan M, Selvankumar T (2012) Hypoglycemic and hypolipidemic activity of *Solanum nigrum* in alloxan induced diabetic albino rats. Int J Pharm Sci Res 3:2841
Shabbir A, Shahzad M, Arfat Y, Ali L, Aziz RS, Murtaza G, Waqar SA (2012) *Berberis lycium* Royle: A review of its traditional uses, phytochemistry and pharmacology. Afr J Pharm Pharmacol 6:2346–2353
Shad AA, Ahmad S, Ullah R, AbdEl-Salam NM, Fouad H, Rehman NU, Hussain H, Saeed W (2014) Phytochemical and biological activities of four wild medicinal plants. Sci World J 2014:7
Shafaghat A (2010) Phytochemical investigation of quranic fruits and plants. J Med Plants 3:61–66. فصلنامه علمی پژوهشی گیاهان دارویی
Shah GM, Khan MA (2006) Checklist of Medicinal Plants of Siran Valley, Mansehra, Pakistan. Ethnobot Leaflets 10:63–71
Shah NA, Khan MR (2014) Antidiabetic effect of *Sida cordata* in alloxan induced diabetic rats. Biomed Res Int 2014:15

Shah A, Pandey V, Eckhardt G, Tschesche R (1985) A 13-membered cyclopeptide alkaloid from *Zizyphus sativa*. Phytochemistry 24:2765–2767

Shah BA, Chib R, Gupta P, Sethi VK, Koul S, Andotra SS, Nargotra A, Sharma S, Pandey A, Bani S (2009) Saponins as novel TNF-α inhibitors: isolation of saponins and a nor-pseudoguaianolide from *Parthenium hysterophorus*. Org Biomol Chem 7:3230–3235

Shah K, Patel M, Patel R, Parmar P (2010) *Mangifera indica* (mango). Pharmacogn Rev 4:42

Shah A, Marwat SK, Gohar F, Khan A, Bhatti KH, Amin M, Din NU, Ahmad M, Zafar M (2013a) Ethnobotanical study of medicinal plants of semi-tribal area of Makerwal & Gulla Khel (lying between Khyber Pakhtunkhwa and Punjab Provinces), Pakistan. Am J Plant Sci 2013(4):98–116

Shah NA, Khan MR, Ahmad B, Noureen F, Rashid U, Khan RA (2013b) Investigation on flavonoid composition and anti free radical potential of *Sida cordata*. BMC Complement Altern Med 13:1

Shahat AA, Cuyckens F, Wang W, Abdel-Shafeek KA, Husseiny HA, Apers S, Van Miert S, Pieters L, Vlietinck AJ, Claeys M (2005) Structural characterization of flavonol di-O-glycosides from *Farsetia aegyptia* by electrospray ionization and collision-induced dissociation mass spectrometry. Rapid Commun Mass Spectrom 19:2172–2178

Shahiladevi S, Jayanthi G, Jegadeesan M (2006) Preliminary phytochemical studies on *Solanum surattense* Burm. F. Seeds. Anc Sci Life 26:59

Shahreen S, Banik J, Hafiz A, Rahman S, Zaman AT, Shoyeb A, Chowdhury MH, Rahmatullah M (2012) Antihyperglycemic activities of leaves of three edible fruit plants (*Averrhoa carambola, Ficus hispida* and *Syzygium samarangense*) of Bangladesh. Afr J Tradit Complement Altern Med 9:287–291

Shahzadi T, Abbasi MA, Ur-Rehman A, Riaz T, Khan KM, Ashraf M, Afzal I, Akhtar MN, Ajaib M (2013) Antioxidant and lipoxygenase inhibiting new iridoid glucosides from *Caryopteris odorata*. Nat Prod Res 27:302–313

Shanmugasundaram E, Rajeswari G, Baskaran K, Kumar BR, Shanmugasundaram KR, Ahmath BK (1990) Use of *Gymnema sylvestre* leaf extract in the control of blood glucose in insulin-dependent diabetes mellitus. J Ethnopharmacol 30:281–294

Sharma U, Kumar A (2011) Anti-diabetic effect of *Rubus ellipticus* fruits extracts in alloxan-induced diabetic rats. J Diabetol 2(4):4

Sharma N, Patni V (2013) In vivo and in vitro qualitative phytochemical screening of Grewia species. Int J Biol Pharm Res 4:634–639

Sharma P, Singh R (2014) Effect of *Momordica dioica* fruit extract on antioxidant status in liver, kidney, pancreas, and serum of diabetic rats. Pharm Res 6:73

Sharma OP, Dawra RK, Ramesh D (1990) A triterpenoid acid, lantadene D from *Lantana camara* var. aculeata. Phytochemistry 29:3961–3962

Sharma B, Salunke R, Balomajumder C, Daniel S, Roy P (2010) Anti-diabetic potential of alkaloid rich fraction from *Capparis decidua* on diabetic mice. J Ethnopharmacol 127:457–462

Sharma M, Siddique M, Shamim AM, Gyanesh S, Pillai K (2011) Evaluation of antidiabetic and antioxidant effects of seabuckthorn (*Hippophae rhamnoides* L.) in streptozotocin-nicotinamide induced diabetic rats. The Open Conf Proc J 2:53–58

Sharmin R, Khan M, Akhtar MA, Alim A, Islam M, Anisuzzaman A, Ahmed M (2012) Hypoglycemic and hypolipidemic effects of cucumber, white pumpkin and ridge gourd in alloxan induced diabetic rats. J Sci Res 5:161–170

Shen W, Xu Y, Lu Y-H (2012) Inhibitory effects of citrus flavonoids on starch digestion and antihyperglycemic effects in HepG2 cells. J Agric Food Chem 60:9609–9619

Shetty K (1997) Biotechnology to harness the benefits of dietary phenolics; focus on Lamiaceae. Asia Pac J Clin Nutr 6:162–171

Sheweita S, Mashaly S, Newairy A, Abdou H, Eweda S (2016) Changes in oxidative stress and antioxidant enzyme activities in streptozotocin-induced Diabetes mellitus in rats: Role of *Alhagi Maurorum* extracts. Oxidative Med Cell Longev 2016:5264064

Shinwari ZK, Gilani SS (2003) Sustainable harvest of medicinal plants at Bulashbar Nullah, Astore (Northern Pakistan). J Ethnopharmacol 84:289–298

Shori AB, Baba AS (2014) Comparative antioxidant activity, proteolysis and in vitro α amylase and α-glucosidase inhibition of *Allium sativum*-yogurts made from cow and camel milk. J Saudi Chem Soc 18:456–463
Shrestha PM, Dhillion SS (2003) Medicinal plant diversity and use in the highlands of Dolakha district, Nepal. J Ethnopharmacol 86:81–96
Shukla R, Anand K, Prabhu K, Murthy PS (1994) Hypoglycaemic effect of the water extract of *Ficus benghalensis* in alloxan recovered, mildly diabetic and severely diabetic rabbits. Int J Diab Dev Ctries 14:78–81
Shukla S, Chatterji S, Mehta S, Rai PK, Singh RK, Yadav DK, Watal G (2011) Antidiabetic effect of *Raphanus sativus* root juice. Pharm Biol 49:32–37
Shunying Z, Yang Y, Huaidong Y, Yue Y, Guolin Z (2005) Chemical composition and antimicrobial activity of the essential oils of *Chrysanthemum indicum*. J Ethnopharmacol 96:151–158
Siddhuraju P, Vijayakumari K, Janardhanan K (1996a) Chemical composition and nutritional evaluation of an underexploited legume, *Acacia nilotica* (L.) Del. Food Chem 57:385–391
Siddhuraju P, Vijayakumari K, Janardhanan K (1996b) Chemical composition and protein quality of the little-known legume, velvet bean (*Mucuna pruriens* (L.) DC.). J Agric Food Chem 44:2636–2641
Siddiqi R, Naz S, Ahmad S, Sayeed SA (2011) Antimicrobial activity of the polyphenolic fractions derived from *Grewia asiatica*, *Eugenia jambolana* and *Carissa carandas*. Int J Food Sci Technol 46:250–256
Siddiqui S, Hafeez F, Begum S, Siddiqui BS (1988) Oleanderol, a new pentacyclic triterpene from the leaves of *Nerium oleander*. J Nat Prod 51:229–233
Siddiqui BS, Afshan F, Gulzar T, Sultana R, Naqvi SN-H, Tariq RM (2003) Tetracyclic triterpenoids from the leaves of *Azadirachta indica* and their insecticidal activities. Chem Pharm Bull 51:415–417
Siddiqui BS, Hasan M, Mairaj F, Mehmood I, Hafizur RM, Hameed A, Shinwari ZK (2014) Two new compounds from the aerial parts of *Bergenia himalaica* Boriss and their anti-hyperglycemic effect in streptozotocin-nicotinamide induced diabetic rats. J Ethnopharmacol 152:561–567
Simpson R, Morris GA (2014) The anti-diabetic potential of polysaccharides extracted from members of the cucurbit family: A review. Bioact Carbohydr Diet Fibre 3:106–114
Sindhu RK, Arora S (2013) Phytochemical and Pharmacognostical Investigations on Aerial roots of *Ficus lacor* Buch. Ham Int J Phytomed 5:267
Singh V, Kori ML (2014) Antidiabetic effect of hydroalcoholic combined plant extract of *Portulaca oleracea* and *Caralluma attenuata* in streptozotocin induced diabetic rats. Indo Am J Pharm Res 4:1391–1396
Singh K, Chandra V, Barthwal K (1975) Letter to the editor: Hypoglycaemic activity of *Acacia arabica*, *Acacia benthami* and *Acacia modesta* leguminous seed diets in normal young albino rats. Indian J Physiol Pharmacol 19:167
Singh S, Majumdar D, Yadav M (1996) Chemical and pharmacological studies on fixed oil of *Ocimum sanctum*. Indian J Exp Biol 34:1212–1215
Singh R, Pandey B, Tripathi M, Pandey V (2001) Anti-inflammatory effect of (+)-pinitol. Fitoterapia 72:168–170
Singh G, Maurya S, De Lampasona M, Catalan C (2006a) Chemical constituents, antifungal and antioxidative potential of *Foeniculum vulgare* volatile oil and its acetone extract. Food Control 17:745–752
Singh N, Mahmood U, Kaul V, Jirovetz L (2006b) A new phthalic acid ester from *Ajuga bracteosa*. Nat Prod Res 20:593–597
Singh R, Singh B, Singh S, Kumar N, Kumar S, Arora S (2008) Anti-free radical activities of kaempferol isolated from *Acacia nilotica* (L.) Willd. Ex. Del. Toxicol In Vitro 22:1965–1970
Singh RK, Mehta S, Jaiswal D, Rai PK, Watal G (2009) Antidiabetic effect of *Ficus benghalensis* aerial roots in experimental animals. J Ethnopharmacol 123:110–114
Singh D, Singh B, Goel RK (2011) Traditional uses, phytochemistry and pharmacology of *Ficus religiosa*: a review. J Ethnopharmacol 134:565–583

Singh R, Verma PK, Singh G (2012) Total phenolic, flavonoids and tannin contents in different extracts of *Artemisia absinthium*. J Intercul Ethnopharmacol 1:101–104

Singh D, Mukhija M, Singh S, Aggarwal A, Sundriyal A (2014) Anti-diabetic effect of hydroalcoholic extract of *Ficus palmata* Forsk leaves in streptozotocin-induced diabetic rats. Int J Green Pharm 8:276

Sinha SN, Paul D (2014) Antioxidant potentials of parthenium hysterophorus l leaf extracts. Sci Res J India 3(2):82–86

Sivashanmugam AT, Chatterjee TK (2013) In vitro and in vivo antidiabetic activity of *Polyalthia longifolia* (Sonner.) Thw. leaves. Orient Pharm Exp Med 13:289–300

Somani R, Kasture S, Singhai AK (2006) Antidiabetic potential of *Butea monosperma* in rats. Fitoterapia 77:86–90

Song Y, Wu T, Yang Q, Chen X, Wang M, Wang Y, Peng X, Ou S (2014) Ferulic acid alleviates the symptoms of diabetes in obese rats. J Funct Foods 9:141–147

Sonwa MM, König WA (2001) Chemical study of the essential oil of *Cyperus rotundus*. Phytochemistry 58:799–810

Sotohy S, Sayed A, Ahmed M (1997) Effect of tannin-rich plant (*Acacia nilotica*) on some nutritional and bacteriological parameters in goats. Deut Tierarzt Wochenschr 104:432–435

Souliman AM, Barakat HH, El-Mousallamy AM, Marzouk MS, Nawwar MA (1991) Phenolics from the bark of *Tamarix aphylla*. Phytochemistry 30:3763–3766

Sreelatha S, Inbavalli R (2012) Antioxidant, Antihyperglycemic, and Antihyperlipidemic Effects of *Coriandrum sativum* Leaf and Stem in Alloxan-Induced Diabetic Rats. J Food Sci 77:T119–T123

Sridevi M, Senthil S, Pugalendi K (2007) Antihyperglycemic effect of *Solanum surattense* leaf-extract in streptozotocin induced diabetic rats. J Pharmacol Toxicol 2:621–629

Sridhar T, Josthna P, Naidu C (2011) In vitro antibacterial activity and phytochemical analysis of *Solanum nigrum* (Linn.)-An important antiulcer medicinal plant. J Exp Sci 2(8):24–29

Srinivasan K, Sherief SH, Khule Shahu NK, Sivakumar T (2013) An evaluation of anti-diabetic and its biochemical properties of *Sida cordata* extract in experimental animal models. Indo Am J Pharm Res 3:6103–6111

Srivastav S, Singh P, Mishra G, Jha K, Khosa R (2011) *Achyranthes aspera*-An important medicinal plant: A review. J Nat Prod Plant Resour 1:1–14

Srivastava S, Srivastava M, Misra A, Pandey G, Rawat A (2015) A review on biological and chemical diversity in Berberis (Berberidaceae). EXCLI J 14:247

Stalin C, Dineshkumar P, Nithiyananthan K (2012) Evaluation of antidiabetic activity of methanolic leaf extract of *Ficus carica* in alloxan-induced diabetic rats. Asian J Pharm Clin Res 5:85–87

Street RA, Sidana J, Prinsloo G (2013) *Cichorium intybus*: Traditional uses, phytochemistry, pharmacology, and toxicology. Evid Based Complement Alternat Med 2013:1–13

Stuppner H, Moller E (1993) Cucurbitacins with unusual side chains from *Picrorhiza kurroa*. Phytochemistry 33:1139–1145

Suba V, Murugesan T, Arunachalam G, Mandal S, Saha B (2004) Anti-diabetic potential of *Barleria lupulina* extract in rats. Phytomedicine 11:202–205

Subramanian S, Priya N, Thamizhiniyan V (2011a) Biochemical evaluation of hypoglycemic, hypolipidemic and antioxidant properties of *Lippia nodiflora* leaves studied in alloxan-induced experimental diabetes in rats. Res J Pharmacol Pharmacod 3:299–304

Subramanian SP, Bhuvaneshwari S, Prasath GS (2011b) Antidiabetic and antioxidant potentials of *Euphorbia hirta* leaves extract studied in streptozotocin-induced experimental diabetes in rats. Gen Physiol Biophys 30:278–285

Sultan MT, Butt MS, Karim R, Iqbal SZ, Ahmad S, Zia-Ul-Haq M, Aliberti L, Ahmad AN, De Feo V (2014) Effect of *Nigella sativa* fixed and essential oils on antioxidant status, hepatic enzymes, and immunity in streptozotocin induced diabetes mellitus. BMC Complement Altern Med 14:1

Sultana N, Ata A (2008) Oleanolic acid and related derivatives as medicinally important compounds. J Enzyme Inhib Medl Chem 23:739–756

Sundaram R, Mitra S (2007) Antioxidant activity of ethyl acetate soluble fraction of *Acacia arabica* bark in rats. Indian J Pharm 39:33

Sunmonu TO, Afolayan AJ (2013) Evaluation of antidiabetic activity and associated toxicity of *Artemisia afra* aqueous extract in wistar rats. Evid Based Complement Alternat Med 2013:929074

Surendran S, Vijayalakshmi K (2011) GC-MS analysis of phytochemicals in *Cyamopsis tetragonoloba* fruit and Cyperus rotundus rhizome. IJPPR 2012:4

Survay NS, Ko E, Upadhyay CP, Mi J, Park SW, Lee D, Jung Y-S, Yoon D-Y, Hong S (2010) Hypoglycemic effects of fruits and vegetables in hyperglycemic rats for prevention of type-2 diabetes. Korean J Hortic Sci Technol 28:1–7

Swaroop A, Bagchi M, Kumar P, Preuss HG, Tiwari K, Marone PA, Bagchi D (2014) Safety, efficacy and toxicological evaluation of a novel, patented anti-diabetic extract of *Trigonella Foenum-graecum* seed extract (Fenfuro). Toxicol Mech Methods 24:495–503

Tabuti JR, Kukunda CB, Kaweesi D, Kasilo OM (2012) Herbal medicine use in the districts of Nakapiripirit, Pallisa, Kanungu, and Mukono in Uganda. J Ethnobiol Ethnomed 8:1

Takaya Y, Kondo Y, Furukawa T, Niwa M (2003) Antioxidant constituents of radish sprout (kaiware-daikon), *Raphanus sativus* L. J Agric Food Chem 51:8061–8066

Tanaka M, Misawa E, Ito Y, Habara N, Nomaguchi K, Yamada M, Toida T, Hayasawa H, Takase M, Inagaki M (2006) Identification of five phytosterols from *Aloe vera* gel as anti-diabetic compounds. Biol Pharm Bull 29:1418–1422

Tang G-Y, Li X-J, Zhang H-Y (2008) Antidiabetic components contained in vegetables and legumes. Molecules 13:1189–1194

Taniguchi H, Kobayashi-Hattori K, Tenmyo C, Kamei T, Uda Y, Sugita-Konishi Y, Oishi Y, Takita T (2006) Effect of Japanese radish (*Raphanus sativus*) sprout (Kaiware-daikon) on carbohydrate and lipid metabolisms in normal and streptozotocin-induced diabetic rats. Phytother Res 20:274–278

Tanira M, Ali B, Bashir A, Chandranath I (1996) Some pharmacologic and toxicologic studies on *Rhazya stricta* Decne in rats, mice and rabbits. Gen Pharmacol: Vas S 27:1261–1267

Tannin-Spitz T, Grossman S, Dovrat S, Gottlieb HE, Bergman M (2007) Growth inhibitory activity of cucurbitacin glucosides isolated from *Citrullus colocynthis* on human breast cancer cells. Biochem Pharmacol 73:56–67

Taştekin D, Atasever M, Adigüzel G, Keleş M, Taştekin A (2006) Hypoglycaemic effect of *Artemisia herba-alba* in experimental hyperglycaemic rats. Bull Vet Inst Pulawy 50:235–238

Tazin TQ, Rumi JF, Rahman S, Al-Nahain A, Jahan R, Rahmatullah M (2014) Oral glucose tolerance and antinociceptive activity evaluation of methanolic extract of *Vigna unguiculata* ssp. unguiculata beans. World J Pharm Pharm Sci 3:28–37

Teffo LS, Aderogba M, Eloff JN (2010) Antibacterial and antioxidant activities of four kaempferol methyl ethers isolated from *Dodonaea viscosa* Jacq. var. angustifolia leaf extracts. S Afr J Bot 76:25–29

Teklehaymanot T (2009) Ethnobotanical study of knowledge and medicinal plants use by the people in Dek Island in Ethiopia. J Ethnopharmacol 124:69–78

Thakur R, Misra L, Bhattacharyya S, Sen N, Sethi K (1990) Essential oils of Indian Artemisia. Proceedings of the 11th international congress of essential oils, fragrances and flavours. New Delhi, India, 12–16 November, 1989 Vol. 4 Chemistry-analysis and structure. Aspect Publishing, New York, pp 127–135

Thiruvenkatasubramaniam R, Jayakar B (2010) Anti-hyperglycemic and anti-hyperlipidaemic activities of *Bauhinia variegata* L on streptozotocin induced diabetic rats. Pharm Lett 2:330–334

Thomson M, Al-Amin ZM, Al-Qattan KK, Shaban LH, Ali M (2007) Anti-diabetic and hypolipidaemic properties of garlic (*Allium sativum*) in streptozotocin-induced diabetic rats. Int J Diabetes Metab 15:108–115

Tianaka S, Takido M (1988) Studies on the constituents of the seeds of *Cassia obtusifolia* L. The structures of two naphthopyrone glycosides. Chem Pharm Bull 36:3980–3984

Topçu G, Erenler R, Çakmak O, Johansson CB, Çelik C, Chai H-B, Pezzuto JM (1999) Diterpenes from the berries of *Juniperus excelsa*. Phytochemistry 50:1195–1199

Trakoon-osot W, Sotanaphun U, Phanachet P, Porasuphatana S, Udomsubpayakul U, Komindr S (2013) Pilot study: hypoglycemic and antiglycation activities of bitter melon (*Momordica charantia* L.) in type 2 diabetic patients. J Pharm Res 6:859–864

Tripathi P (2011) Diuretic activity of pistia stratiotes leaf extract in rats. Int Res J Pharm 2(3):249–251

Tripathi AK, Kohli S (2014) Pharmacognostical standardization and antidiabetic activity of *Syzygium cumini* (Linn.) barks (Myrtaceae) on streptozotocin-induced diabetic rats. J Complement Integra Med 11:71–81

Tripathi M, Jain L, Pandey VB, Ray AB, Rücker G (1996) Pindrolactone, a lanostane derivative from the leaves of *Abies pindrow*. Phytochemistry 43:853–855

Udayakumar R, Kasthurirengan S, Mariashibu TS, Rajesh M, Anbazhagan VR, Kim SC, Ganapathi A, Choi CW (2009) Hypoglycaemic and hypolipidaemic effects of *Withania somnifera* root and leaf extracts on alloxan-induced diabetic rats. Int J Mol Sci 10:2367–2382

Uddin G, Rauf A, Qaisar M, Latif A, Ali M (2011) Preliminary phytochemical screening and antimicrobial activity of *Hedera helix* L. Middle-East J Sci Res 8:198–202

Uddin G, Rauf A, Arfan M, Ali M, Qaisar M, Saadiq M, Atif M (2012) Preliminary phytochemical screening and antioxidant activity of *Bergenia caliata*. Middle-East J Sci Res 11:1140–1142

Ullah I (2012) A review of phytochemistry, bioactivities and ethno medicinal uses of *Rhazya stricta* Decsne (Apocynaceae). Afr J Microbiol Res 6:1629–1641

Ullah S, Khan MR, Shah NA, Shah SA, Majid M, Farooq MA (2014) Ethnomedicinal plant use value in the Lakki Marwat District of Pakistan. J Ethnopharmacol 158:412–422

Uma B, Prabhakar K, Rajendran S (2009) Invitro antimicrobial activity and phytochemical analysis of *Ficus religiosa* L. and *Ficus benghalensis* L. against Diarrhoeal Enterotoxigenic *E. coli*. Ethnobot Leaflets 2009:7

Uniyal SK, Singh K, Jamwal P, Lal B (2006) Traditional use of medicinal plants among the tribal communities of Chhota Bhangal, Western Himalaya. J Ethnobiol Ethnomed 2:1

Upadhyay D, Dash RP, Anandjiwala S, Nivsarkar M (2013) Comparative pharmacokinetic profiles of picrosides I and II from kutkin, *Picrorhiza kurroa* extract and its formulation in rats. Fitoterapia 85:76–83

Vadivel V, Kunyanga CN, Biesalski HK (2012) Antioxidant potential and type II diabetes-related enzyme inhibition of *Cassia obtusifolia* L.: effect of indigenous processing methods. Food Bioprocess Technol 5:2687–2696

Vats V, Grover J, Rathi S (2002) Evaluation of anti-hyperglycemic and hypoglycemic effect of *Trigonella foenum-graecum* Linn, *Ocimum sanctum* Linn and *Pterocarpus marsupium* Linn in normal and alloxanized diabetic rats. J Ethnopharmacol 79:95–100

Vats V, Yadav S, Grover J (2004) Ethanolic extract of *Ocimum sanctum* leaves partially attenuates streptozotocin-induced alterations in glycogen content and carbohydrate metabolism in rats. J Ethnopharmacol 90:155–160

Veerapur V, Prabhakar K, Parihar VK, Bansal P, Srinivasan K, Priyadarsini K, Unnikrishnan M (2010a) Antidiabetic, hypolipidaemic and antioxidant activity of *Dodonaea viscosa* aerial parts in streptozotocin-induced diabetic rats. Int J Phytomed 2(1):59–70

Veerapur V, Prabhakar K, Thippeswamy B, Bansal P, Srinivasan K, Unnikrishnan M (2010b) Antidiabetic effect of *Dodonaea viscosa* (L). Lacq. aerial parts in high fructose-fed insulin resistant rats: a mechanism based study. Indian J Exp Biol 48:800

Veerapur V, Prabhakar K, Thippeswamy B, Bansal P, Srinivasan K, Unnikrishnan M (2012) Antidiabetic effect of *Ficus racemosa* Linn. stem bark in high-fat diet and low-dose streptozotocin-induced type 2 diabetic rats: a mechanistic study. Food Chem 132:186–193

Velayutham R, Sankaradoss N, Ahamed KN (2012) Protective effect of tannins from *Ficus racemosa* in hypercholesterolemia and diabetes induced vascular tissue damage in rats. Asian Pac J Trop Med 5:367–373

Venâncio T, Oliveira A, Silva L, Machado O, Fernandes K, Xavier-Filho J (2003) A protein with amino acid sequence homology to bovine insulin is present in the legume *Vigna unguiculata* (cowpea). Braz J Med Biol Res 36:1167–1173

Venkatesh S, Reddy GD, Reddy BM, Ramesh M, Rao AA (2003) Antihyperglycemic activity of *Caralluma attenuata*. Fitoterapia 74:274–279

Verardo V, Bonoli M, Marconi E, Caboni MF (2008) Distribution of Bound Hydroxycinnamic Acids and Their Glycosyl Esters in Barley (*Hordeum vulgare* L.) Air-Classified Flour: Comparative Study between Reversed Phase-High Performance Chromatography– Mass Spectrometry (RP-HPLC/MS) and Spectrophotometric Analysis. J Agric Food Chem 56:11900–11905

Verma SK, Kumar A (2011) Therapeutic uses of *Withania somnifera* (Ashwagandha) with a note on withanolides and its pharmacological actions. f Asian J Pharm. Clin Res 4:1–4

Verma V, Mahmood U, Singh B (2002) Clerodane diterpenoids from *Ajuga bracteosa* Wall. Nat Prod Lett 16:255–259

Verma G, Gupta R, Prakash Gupta S (2014) Development of HPLC densitometric method for estimation of Quercetin in *Bombax ceiba* L. leaves. JASR 5(2):50–52

Vijayakumari K, Siddhuraju P, Janardhanan K (1997) Chemical composition, amino acid content and protein quality of the little-known legume *Bauhinia purpurea* L. J Sci Food Agric 73:279–286

Vogel P, Machado IK, Garavaglia J, Zani VT, de Souza D, Morelo SDB (2015) Polyphenols benefits of olive leaf (*Olea europaea* L) to human health. Nutr Hosp 31:1427–1433

Vujicic M, Nikolic I, Kontogianni VG, Saksida T, Charisiadis P, Orescanin-Dusic Z, Blagojevic D, Stosic-Grujicic S, Tzakos AG, Stojanovic I (2015) Methanolic extract of *Origanum vulgare* ameliorates type 1 diabetes through antioxidant, anti-inflammatory and anti-apoptotic activity. Br J Nutr 113:770–782

Wadood A, Wadood N, Shah S (1989) Effects of *Acacia arabica* and *Caralluma edulis* on blood glucose levels of normal and alloxan diabetic rabbits. J Pak Med Assoc 39:208–212

Waheed A, Miana G, Ahmad S, Khan MA (2006) Clinical investigation of hypoglycemic effect of *Coriandrum sativum* in type-2 (NIDDM) diabetic patients. Pak J Pharmacol 23:7–11

Wambugu SN, Sagini MN, Githinji CG, Kariuki HN, Ngeranwa JJ, Mwonjoria JK (2014) Ethno medicinal, phytochemical and pharmacological aspects of *Solanum incanum* (lin.). Int J Pharmacol Toxicol 2(2):17–20

Wang H, Yang J (1992) [Studies on the chemical constituents of *Arctium lappa* L]. Yao xue xue bao. Acta Pharm Sin 28:911–917

Wani H, Shah SA, Banday JA (2014) Chemical composition and antioxidant activity of the leaf essential oil of *Artemisia absinthium* growing wild in Kashmir, India. Aust J Pharm 3(2):90–94

Wasfi I, Bashir A, Amiri M, Abdalla A (1994) The effect of *Rhazya stricta* on glucose homeostasis in normal and streptozoticin diabetic rats. J Ethnopharmacol 43:141–147

Weththasinghe P, Liyanage R, Vidanarachchi J, Perera O, Jayawardana B (2014) Hypocholesterolemic and Hypoglycemic Effect of Cowpea (*Vigna unguiculata* L. Walp) Incorporated Experimental Diets in Wistar Rats (Rattus norvegicus). Agricul Agricultur Sci Proc 2:401–405

Whiting DR, Guariguata L, Weil C, Shaw J (2011) IDF diabetes atlas: global estimates of the prevalence of diabetes for 2011 and 2030. Diabetes Res Clin Pract 94:311–321

WHO. 2016. Diabets: Facts sheet. http://www.who.int/mediacentre/factsheets/fs312/en/index.html

Widharna R, Soemardji A, Wirasutisna K, Kardono L (2010) Anti diabetes mellitus activity in vivo of ethanolic extract and ethyl acetate fraction of *Euphorbia hirta* L. herb. IJP-Int J Pharma 6:231–240

Williams CA, Goldstone F, Greenham J (1996) Flavonoids, cinnamic acids and coumarins from the different tissues and medicinal preparations of *Taraxacum officinale*. Phytochemistry 42:121–127

Wu H-E, Li Y-H, Wei Z-Y, Liang H-Y (2008) A comparative study on the chemical constituents of volatile oil from *Artemisia Indica*, *Artemisia Feddei* and *Artemisia Argyi* in Guangxi Province. China Medical Herald 5:23–26

Wu L-Y, Gao H-Z, Wang X-L, Ye J-H, Lu J-L, Liang Y-R (2010) Analysis of chemical composition of *Chrysanthemum indicum* flowers by GC/MS and HPLC. J Med Plant Res 4:421–426

Xiang L, Xing D, Wang W, Wang R, Ding Y, Du L (2005) Alkaloids from *Portulaca oleracea* L. Phytochemistry 66:2595–2601

Xu M-L, Henan Province X, Wang L, Xu G-F, Wang M (2011) Antidiabetes and angiotensin converting enzyme inhibitory activity of *Sonchus asper* (L) hill extract. Korean J Pharmacogn 42(1):61–67

Xu X, Shan B, Liao C-H, Xie J-H, Wen P-W, Shi J-Y (2015) Anti-diabetic properties of *Momordica charantia* L. polysaccharide in alloxan-induced diabetic mice. Int J Biol Macromol 81:538–543

Yadav S, Mukundan U (2011) In vitro antioxidant properties of *Salvia coccinea* Buc'hoz ex etl. and *Salvia officinalis* L. Indian J Fundam Appl Life Sci 1:232–238

Yadav KS, Yadav NP, Shanker K, Thomas SC, Srivastav S, Srivastava S, Rai VK, Mishra N, Sinha P (2013) Assessment of antidiabetic potential of *Cissampelos pareira* leaf extract in streptozotocin–nicotinamide induced diabetic mice. J Pharm Res 6:874–878

Yang Y-C, Lim M-Y, Lee H-S (2003) Emodin isolated from *Cassia obtusifolia* (Leguminosae) seed shows larvicidal activity against three mosquito species. J Agric Food Chem 51:7629–7631

Yang T, Wang C, Liu H, Chou G, Cheng X, Wang Z (2010) A new antioxidant compound from *Capparis spinosa*. Pharm Biol 48:589–594

Yap VA, Loong B-J, Ting K-N, Hwei-San Loh S, Yong K-T, Low Y-Y, Kam T-S, Lim K-H (2015) Hispidacine, an unusual 8, 4′-oxyneolignan-alkaloid with vasorelaxant activity, and hispiloscine, an antiproliferative phenanthroindolizidine alkaloid, from *Ficus hispida* Linn. Phytochemistry 109:96–102

Yaseen G, Ahmad M, Zafar M, Sultana S, Kayani S, Cetto AA, Shaheen S (2015) Traditional management of diabetes in Pakistan: Ethnobotanical investigation from traditional health Practitioners. J Ethnopharmacol 174:91–117

Yasir M, Jain P, Debajyoti D, Kharya M (2010) Hypoglycemic and antihyperglycemic effect of different extracts of *Acacia arabica* lamk bark in normal and alloxan induced diabetic rats. Int J Phytomed 2

Yasir M, Shrivastava R, Jain P, Das D (2012) Hypoglycemic and antihyperglycemic effects of different extracts and combinations of *Withania coagulans* Dunal and *Acacia arabica* Lamk in normal and alloxan-induced diabetic rats. Pharmacogn Commun 2:61–66

Yassin MM, Mwafy SN (2007) Protective potential of glimepiride and *Nerium oleander* extract on lipid profile, body growth rate, and renal function in streptozotocin-induced diabetic rats. Turk J Biol 31:95–102

Yen H-F, Hsieh C-T, Hsieh T-J, Chang F-R, Wang C-K (2015) In vitro anti-diabetic effect and chemical component analysis of 29 essential oils products. J Food Drug Anal 23:124–129

Yi W, Wei Q, Di G, Liang J-Y, Yang-Li L (2012) Phenols and flavonoids from the aerial part of *Euphorbia hirta*. Chin J Nat Med 10:40–42

Youn UJ, Chai X, Park E-J, Kondratyuk TP, Simmons CJ, Borris RP, Mirza B, Pezzuto JM, Chang LC (2013) Biologically active withanolides from *Withania coagulans*. J Nat Prod 76:22–28

Youssef MKE, Youssef HM, Mousa RM (2013) Evaluation of antihyperglycaemic activity of citrus peels powders fortified biscuits in albino induced diabetic rats. Food Public Health 3:161–167

Yu Y-M, Chang W-C, Chang C-T, Hsieh C-L, Tsai C (2002) Effects of young barley leaf extract and antioxidative vitamins on LDL oxidation and free radical scavenging activities in type 2 diabetes. Diabete Metab 28(2):107–114

Yuwai KE, Rao KS, Kaluwin C, Jones GP, Rivett DE (1991) Chemical composition of *Momordica charantia* L. fruits. J Agric Food Chem 39:1762–1763

Zang Y, Sato H, Igarashi K (2011) Anti-Diabetic Effects of a kaempferol glycoside-rich fraction from unripe soybean (Edamame, *Glycine max* L. Merrill.'Jindai') Leaves on KK-Ay Mice. Biosci Biotechnol Biochem 75:1677–1684

Zeng L, Zhang R-Y, Wang X (1987) Studies on the constituents of *Ziziphus spinosus* Hu. Acta Pharm Sin 22:114–120

Zhang W, Guo Y-W (2006) Chemical studies on the constituents of the chinese medicinal herb *Euphorbia helioscopia* L. Chem Pharm Bull 54:1037–1039

Zhang H, Timmermann BN (2016) Withanolide Structural Revisions by 13C NMR Spectroscopic Analysis Inclusive of the γ-Gauche Effect. J Nat Prod 79(4):732–742

Zhang Y-J, Nagao T, Tanaka T, Yang C-R, Okabe H, Kouno I (2004) Antiproliferative activity of the main constituents from *Phyllanthus emblica*. Biol Pharm Bull 27:251–255

Zhang M, Chen M, Zhang H-Q, Sun S, Xia B, Wu F-H (2009) In vivo hypoglycemic effects of phenolics from the root bark of *Morus alba*. Fitoterapia 80:475–477

Zhang W, Zhao J, Wang J, Pang X, Zhuang X, Zhu X, Qu W (2010) Hypoglycemic effect of aqueous extract of seabuckthorn (*Hippophae rhamnoides* L.) seed residues in streptozotocin-induced diabetic rats. Phytother Res 24:228–232

Zhang Y, Sun Y, Xi W, Shen Y, Qiao L, Zhong L, Ye X, Zhou Z (2014) Phenolic compositions and antioxidant capacities of Chinese wild mandarin (*Citrus reticulata* Blanco) fruits. Food Chem 145:674–680

Zhang Y, Wu L, Ma Z, Cheng J, Liu J (2015) Anti-diabetic, anti-oxidant and anti-hyperlipidemic activities of flavonoids from corn silk on stz-induced diabetic mice. Molecules 21:7

Zhang L, Wei G, Liu Y, Zu Y, Gai Q, Yang L (2016) Antihyperglycemic and antioxidant activities of total alkaloids from *Catharanthus roseus* in streptozotocin-induced diabetic rats. J For Res 27:167–174

Zhao M, Xu L-J, Che C-T (2008) Alisolide, alisols O and P from the rhizome of *Alisma orientale*. Phytochemistry 69:527–532

Zhao L, Lan Q, Huang Z, Ouyang L, Zeng F (2011) Antidiabetic effect of a newly identified component of *Opuntia dillenii* polysaccharides. Phytomedicine 18:661–668

Zia T, Hasnain SN, Hasan S (2001) Evaluation of the oral hypoglycaemic effect of *Trigonella foenum-graecum* L.(methi) in normal mice. J Ethnopharmacol 75:191–195

Zia-Ul-Haq M, Ćavar S, Qayum M, Imran I, Vd F (2011) Compositional studies: antioxidant and antidiabetic activities of *Capparis decidua* (Forsk.) Edgew. Int J Mol Sci 12:8846–8861

Zia-Ul-Haq M, Raza Shah M, Qayum M, Ercisli S (2012a) Biological screening of selected flora of Pakistan. Biol Res 45:375–379

Zia-Ul-Haq M, Shahid SA, Muhammed S, Qayum M, Khan I, Ahmad S (2012b) Antimalarial, anti-emetic and antidiabetic potential of *Grewia asiatica* L. leaves. J Med Plant Res 6:3087–3092

Zia-Ul-Haq M, Ahmad S, Amarowicz R, De Feo V (2013a) Antioxidant activity of the extracts of some cowpea (*Vigna unguiculata* (L) Walp.) cultivars commonly consumed in Pakistan. Molecules 18:2005–2017

Zia-Ul-Haq M, Stanković MS, Rizwan K, Feo VD (2013b) *Grewia asiatica* L., a food plant with multiple uses. Molecules 18:2663–2682

Zohura Talukder F, Ahmed Khan K, Uddin R, Jahan N, Ashraful Alam M (2012) In vitro free radical scavenging and anti-hyperglycemic activities of *Achyranthes aspera* extract in alloxan-induced diabetic mice. Drug Discov Ther 6:298–305

Ethno-ecology, Human Health and Plants of the Thandiani Sub Forest Division, Abbottabad, KP, Pakistan

Waqas Khan, Shujaul Mulk Khan, and Habib Ahmad

Introduction

This chapter delivers an outline to the TsFD and its main physiographic plus floristic features with reference to present vegetation types and sampling locations. The Himalayas mountain ranges stretch over some 2500 km across five central Asian countries. It includes one of the earth's supreme diverse, complex, and remarkable biomasses which are characterized by moderately harsh climate, a solid degree of resource seasonality, and variety of both plant species and groups (Kala and Mathur 2002; Oommen and Shanker 2005). Northwestern region of Pakistan is one of the places having high phytogeographic and floristic importance prevailing in these world's largest mountain ranges. Due to their location, rocky lands, and critical geopolitical conditions, many of the distant mountainous valleys in this area have not yet experienced thorough floral studies. Moreover, most of the studies on vegetation based on qualitative data without proper quantification (Dickoré and Nüsser 2000; Ahmad et al. 2009; Signorini et al. 2009). Far less emphasis has been made to offer quantitative accounts of the plant communities and ethno-ecologies along geo-climatic and environmental gradients, to explain the main factors explaining local vegetation designs (Dasti et al. 2007; Malik and Husain 2008; Wazir et al. 2008; Saima et al. 2009).

W. Khan
Department of Botany, Post Graduate College Abbottabad, Abbottabad, Pakistan

S. M. Khan (✉)
Department of Plant Sciences, Quaid-i-Azam University, Islamabad, Pakistan

Islamia College University, Peshawar, Pakistan
https://www.qau.edu.pk/profile.php?id=804024

H. Ahmad
Islamia College University, Peshawar, Pakistan

M. Ozturk, K. R. Hakeem (eds.), *Plant and Human Health, Volume 1*,
https://doi.org/10.1007/978-3-319-93997-1_13

The TsFD is part of the moist temperate forests of Pakistan with a rich biodiversity. It is situated in the Galis forest division of Abbottabad, bounded by Siran forest division from the west, Muzaffarabad and Garhi Habebullah from the North, Abbottabad sub forest division from the south, and Berangali forest range from the east, located between 3329°–3421° north latitude and 7255°–7329° east longitude spreading over an area of 24987 hectares in which 2484 hectares are reserve forests and 947 hectares Guzara forests. The whole area is administratively under the reserve forest division of Khyber Pakhtunkhwa province, which is responsible to preserve the valuable plant and animal species of the area (Khan et al. 2016a). These forests are situated at 8600 ft (2600 m) above sea level. The highest point of the area is Thandiani top with an elevation of 2626 m from sea level. Most of this area is covered with pine forests and may be divided into three elevation ranges, namely, high elevation range (2200–2600 m), medium elevation range (1700–2200 m), and lower elevation range (1200–1700 m). Agriculture is the main source of economy, contributing about 30% to the GNP and employing 55% of the labor force. Main crops of the area include maize, wheat, potatoes, fruits and other off season vegetables. The high elevation range exhibit very less human population (Khan et al. 2011) while the density of population increases toward lower elevation range. Some of the note worthy villages include Gurlania, Tarheri, Riyala, Tarnawai, Balolia, Neelor, Kalapani, Mandroch, Bhoji, Larri, and Pahge. This is a difficult area in terms of communication and road services. The main tribes in the surrounding villages are Gujjars, Jadoons, Abbasis, Karrlals, Awans, Khokhars, and Rajputs with main local languages of Gujri, Hindko and Pothohari. Gujjars and Abbasis are mostly concerned in the uphill and cultivate the rained slopes of the forest ecosystem (Khan et al. 2017). They are usually more attentive concerning the traditional knowledge of plants and native environment (Khan et al. 2015). Floristically, the TsFD has been documented as a significant part of the western Himalayan province with some vegetation topographies that are Sino-Japanese in environment due to the effect of the rain-bearing monsoon winds (Ali and Qaiser 1986; Takhtadzhian and Cronquist 1986).

Exploitation of trees on commercial basis has not been recommended and is also banned through the directives of the federal as well as provincial governments. Only the removal of dead, dry, and diseased trees is carried out but on a limited scale by Forests Development Corporation (FDC). Plantation on roads/paths and building construction must be part of the main developmental works so that the forests may keep intake. Ensuring the effective protection of forests, the present status of the blocks and beats can be maintained details of range-wise distribution of which are as follows (Table 1).

Table 1 The distribution of TsFD in blocks and beats

S. No.	Name of range	Name of block	Name of beat
1	Thandiani	Kakul	Rayala
2			MairaNamal
3			BandiPahar
4			KakulGuzara
5		Sattu	Garliania
6			Larri
7			Kakul Reserve Forests
8			PogranGuzara
9		Terarri	Terarri
10			Maira Patti
11			Banda Guzara
12			Lagan ban

Source: Working plan for Galis reserved forest 1987–1997

Ethno-ecological Studies of TsFD

Food biodiversity and food capes establish actual foundations of the human practice and donate in a variety of ways to the holistic well-being of individuals. However, biodiversity is as such in a state of decline universally. The important factors contributing to this damage comprises industrialization trends, unsustainable farming, fishing, forest practices, and a marvelous hyper-consumption of land. In addition to, global climate change utilization of selected plant types by people is altering the natural environment and bringing evenness in different biota. Moreover the introduction of new range of products from industrial agriculture make the Human more vulnerable in the region. In this changing scenario, conservation of biodiversity, i.e., the improvement of the resilience of socio-ecological organizations (Berkes and Turner 2006), should have clear objectives globally. In the ethnobiological approaches to food and medicinal plants, the emphasis must be not only on natural environments but also on history, culture, and philosophy as well which are involves behind this diversity. In a nut shell, the *holistic* wildlife of food socio-ecological systems (Berkes et al. 2003) shapes what we today call Epicureanism, referring to the 1825 definition of the French scholar Jean Anthelme Brillat-Savarin (Brillat-Savarin 1960; Petrini 2007). Moreover, this "dynamic" conservation policy of food legacy allows the incessant process of evolution and coadaptation through suitable and sound organization practices. On the other hand, folk/traditional knowledge systems are recurring as a priority concern at the global level, as they are progressively being documented as establishing not only the palpable but especially the imperceptible heritage (UNESCO 2003) and what we may describe as unnoticeable imitations of local communities. Local communities are losing traditional knowledge/folk and their values, which go hand in hand with a decline in cultural diversity and the dilution of a true sense of community. This process of impoverishment of the social dimension of diversity and social unity is reproduced in the present global food system. Hinging on the idea that local small-scale agriculture must also serve the global market that alters food plants from a mere product to a multi purpose commodities and forces people to go for slow, organic and traditional food systems. In this situation, cultural, social, and environmental masses must be motivated to adopt research trails meant for slow food slogans based on folk/traditional knowledge and heritage, which can in turn provide locals with better health and understanding of their dynamic cultural wealth. Moreover, Barthel et al. (2013) introduced the concept of bio-cultural allopatry, i.e., places retaining precise, dense, social memories related to food security and stewardship of biodiversity. These hotspots of food bio-cultural heritage have shaped specific landscapes, which have been maintained through a mosaic of management practices that have coevolved in relation to local environmental variations. Ingenious knowledge related to breeding plants/animals, folk culinary processes and recipes, sociability's attached to traditional feasting frameworks, relics, written and oral accounts, as well as personified rituals and art is actually the Traditional Environmental knowledge, TEK that descends through a precise genetic and social reservoirs related to food, shelter, medicines and is shared between scientists, farmers, gatherers, shepherds, fishermen, environmentalists, consumers and the society at large via sound practices.

The vegetation type is the production of the habitat and environmental conditions. The data on flora help to analyse the effect of plant diversity in biological maintenance, and ecological organization that ultimately influence valuation and management performance of an ecosystem. It gives the foundation for prognosis of probable future deviations (Kent and Coker 1995). The measurable environmental trainings explore the current assembly, species diversity, and soil-plant association and ensure facts on spatial and time-based deviation in existing nutrients. There have continuously been needs to examine and elucidate the plant societies, on different coverage, and to collect firsthand evidences around the plant life of this floristically diverse region. The term conservation applies both in the vigorous and submissive senses. In a vigorous sense, the conservation is taking the action to confirm that things that are valued are in position to have a chance of continuing into the forthcoming generations. It can be closely related to rebuild the origional status of these natural resources and spreading efforts for their maintainance above from just trying to protect those aspects of the plant world that have significance and to encourage their conservation values. In an acquiescent sense, the term conservation refers to actions helpful for plant management carried out by people. There are various characters of the world of plants that can be subjected to conservation implications, especially at species level. In brief, conservation should be endorsed as a culture (Han et al. 2005). The plant conservation should not just be a cross field but also an ambition in terms of how people usually respond to it. The concept of structure used in all bio-research is a complementary concept to function. This is associated to physiological manners, anatomy, and morphology of the objects under study (Mueller-Dombois and Ellenberg 1974). The investigators are concerned in verifying the structure and floristic arrangement of forest environments. The combined influence of the plant height, density, basal area, and number of species on "complexity index" in the assessment of vegetation's physiognomy and the vegetation complex varies from season to season and year to year. The variation proposes a response by each species population to heat, moistures and light as adapted by the vegetation itself that is also well estimated by the traditional users. Consequently, the exploration actions give comprehensive information to the traditional users in relation to the biodiversity, vegetation composition, and other characteristics of plant communities (Singh and Singh 2010).

Climate

The tract generally forms the western outskirts of the northwestern lesser Himalayan series of mountains. Its climate represents both the subtropical and temperate zones. The part of the tract situated west of Abbottabad at lower elevations exhibit subtropical sort of climate, whereas rest of the region falls in the cold temperate zones. January and February remain the coldest months, while March and April are cooler; May and June are the hotter and drier months of the year (Khan et al. 2016b). The monsoon rains tend to start from the mid or early July and continue till the end of August or middle of September. The moisture, precipitation, and temperature conditions of the region are extremely favorable for rich undergrowth. The range of snowfall is from a few inches to 2–4 ft. during winter season in the months of December to March.

According to Pakistan Meteorological Centre, Abbottabad, the standard monthly rainfall statistics for the last 5 years of the Kakul Station are shown below (Figs. 1, 2, and 3).

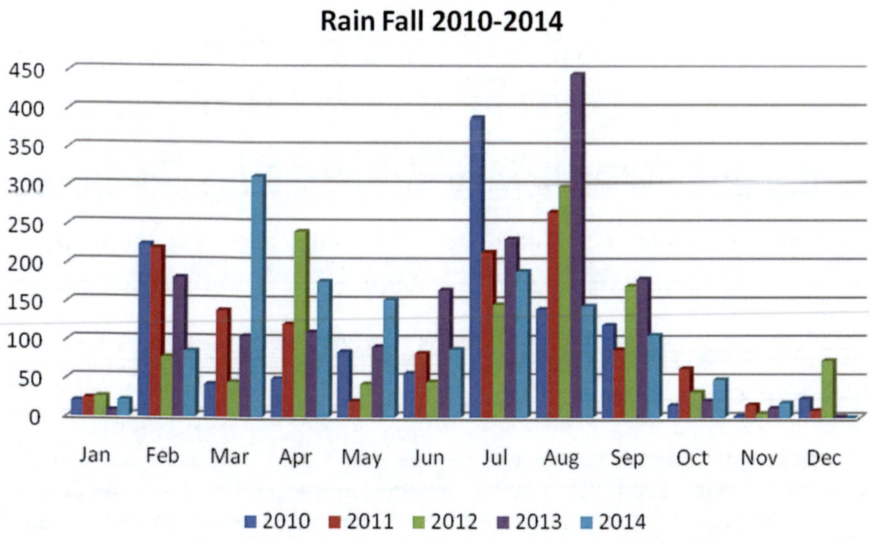

Fig. 1 Month-wise rainfall graph of 2010–2014

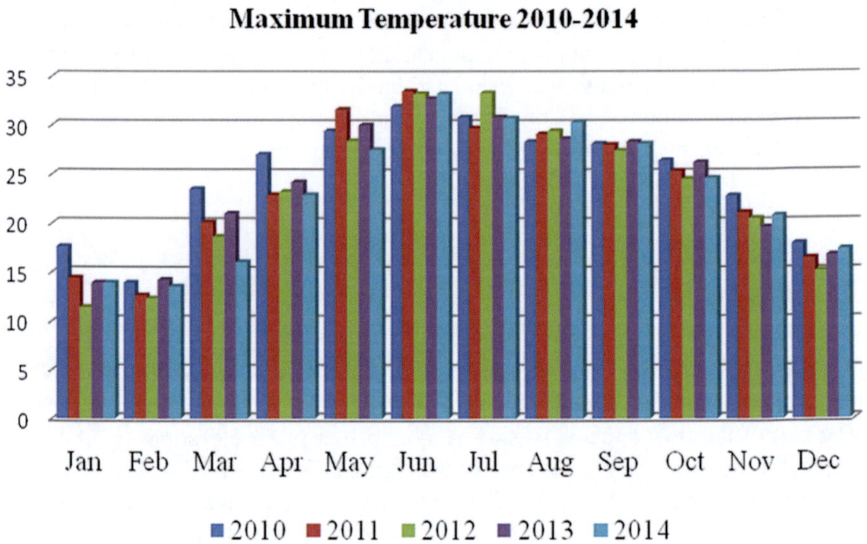

Fig. 2 Maximum monthly temperature recorded from years 2010 to 2014

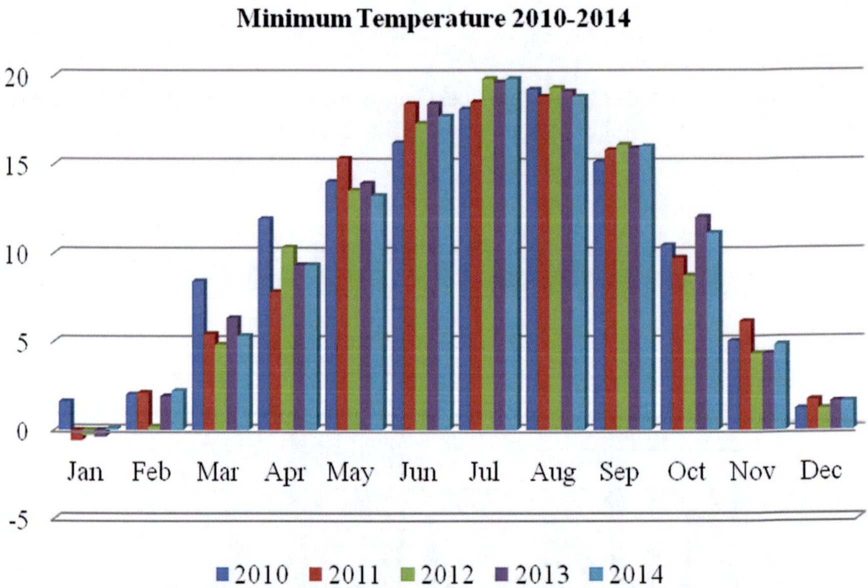

Fig. 3 Graph shows minimum monthly temperature of 5 years from 2010 to 2014

The above data shows that the tract is considerably cold and humid in winter and warm humid in monsoon season of the summer. Tendency toward lesser humidity is generally noticed during the months of March–June and October–December each year (Pakistan Meteorological Department Station Kakul, Abbottabad). The

Thandiani Sub Forest Division is situated at an altitude of 1800 m along a main ridge between Abbottabad and Kunhar basins. Sikher (2626 m) and Sattu (2610 m) peaks are the highest peaks in the region. In general, the area is rugged with steep slopes. The main ridge is interrupted by lateral spurs and nallahs, creating numerous side valleys. In Pakistan in general, and Himalayan forests in particular, ecosystem studies of natural environment in relation to cultural diversity have not been done widely. Champion and Khattak (1965) defined for the first time the forest types of Pakistan, using the following broad categories: swamps, dry subtropical forests, tropical thorn forests, subtropical pine forests, Himalayan moist temperate forests, Himalayan dry temperate forests, subalpine forests, and alpine scrub. Beg (1975) defined main habitat types as tropical swamps, tropical thorn forests, tropical dry deciduous forests, subtropical semi-evergreen forests, subtropical forests, moist temperate forests, dry temperate forests, subalpine forests, alpine vegetation, and cold desert. All of these vegetation types, excluding the swamps, can be found in northern part of the country (Champion and Khattak 1965; Beg 1975). The TsFD is located on the edge of the western Himalayas and thus forms a part of the globally established western Himalayan floristic province of the western Asiatic sub-region of Irano-Turania. Its geographical, geomorphological, geological, climatic, and vegetational setting give it a unique ecotonal position among the world's three largest mountain ranges i.e., the Himalayas, Hindu Kush, and Karakorum ranges. This lends specific phytogeographical diversity to the TsFD and its vegetation.

Vegetation

The vegetation of western Himalayan province is predominantly under the effect of monsoon winds and can be classified into different vegetational zones on the basis of temperature, humidity, and altitude (Champion and Khattak 1965; Takhtadzhian and Cronquist 1986). A brief description of the habitat types that occur within the TsFD and their associated vegetation are as follows.

According to Champion and Khattak (1965), the forests of Galis can mainly be classified into the following three types of forests:

1. Dry subtropical broad-leaved forest
2. Subtropical pine forests
3. Himalayan moist temperate forest

The Thandiani lies in the third category of Himalayan moist temperate forest which is further divided into the following subcategories:

1. Low-level blue pine forest
2. Western mixed conifer forest

The Thandiani and adjoining reserve and Guzara forests are typically representative of rich Himalayan mountain moist temperate ecosystem, supporting a large number of plants and animal life.

Dry Subtropical Broad-Leaved Forests

The dry subtropical broad-leaved forests occupy 6288 acres (32%) of the total area of forests included in this plan. The bulk of the area under this sort is situated in the lower Tanawal, and a small part has been found in the Berangali range on the right bank of the Jhelum River. This forest extends from about 300 to 1000 m elevation merging with the subtropical pine in the upper ridges. These forests attain a higher altitude on the relatively hotter and drier southern aspect slopes. The vegetation is mainly composed of *Dodonaea viscosa*, *Acacia modesta*, *Ziziphus jujuba*, *Berberis lycium*, and *Punica florida.* Significant species which are rapidly being eliminated are *Olea ferruginea*, *Pistacia integerrima*, and *Acacia arabica*. *Dodonaea viscose* (Sannatha) is the dominnant species established in these firewood forests. Its abundant seed-generating power and sprouting ability, together with the productive coppicing habit, permits the species to survive under the effects of frequent wounding and looping. Sannatha due to its combustibility and significant fattening values is highly important as firewood. It is also used for fencing, and is therefore, extensively cut and twisted as an result it is never permitted to grow to higher sizes but somewhat leftovers bushy forms. These forests consequently host stunted shrubs which are frequently intermittent by grassy rocks or bare rocks. More valued species like *Pistacia integerrima* (Kao Kanger), *Olea cuspidate*, and *Acacia modesta* (Phulai) have been carried to the edge of extinction. The old remnant appropriately conserved on account of spiritual affection of such species can be traced in the graveyard, and this is adequate proof of their indigenous existence (Fig. 4).

Subtropical Pine Forests

The subtropical pine forests can be located in between the monotone temperate and the subtropical broad leaved forests. The altitudinal range of existence of chir pine starts from 800 m and ends at 1900 m from sea level. However the species as happening in this area seems to have exceeded the upper altitudinal border and accesses into the blue pine zone, perhaps under the influence of global climate change and warming. Blue-pines forests have given way to the chir pine at shady slopes of the mountains. Chir pine in the Bagnotor Guzara forests on the western slopes are dominating above 1900 m altitude at sea level. The region below chir pine is 4624 acres which is about 24% of the total region. The development of chir pine is usual, and its restoration is properly up to the mark. In the lower limits of its occurrence, chir has given way to brushwood and has already been wiped off from a larger part of its habitat in Abbottabad range. The dominant associated species of chir are *Quercus incana* (white oak) and other broad-leaved species for example *Celtis australis*, *Zanthoxylum armatum*, and *Cornus macrophylla*. Common undergrowth consists of *Berberis lycium*, *Skimmia laureola*, *Daphne muricatus*, *Ziziphus jujuba*, *Ziziphus sativa*, and *Dodonaea viscosa.* Herbaceous or ground flora in chir

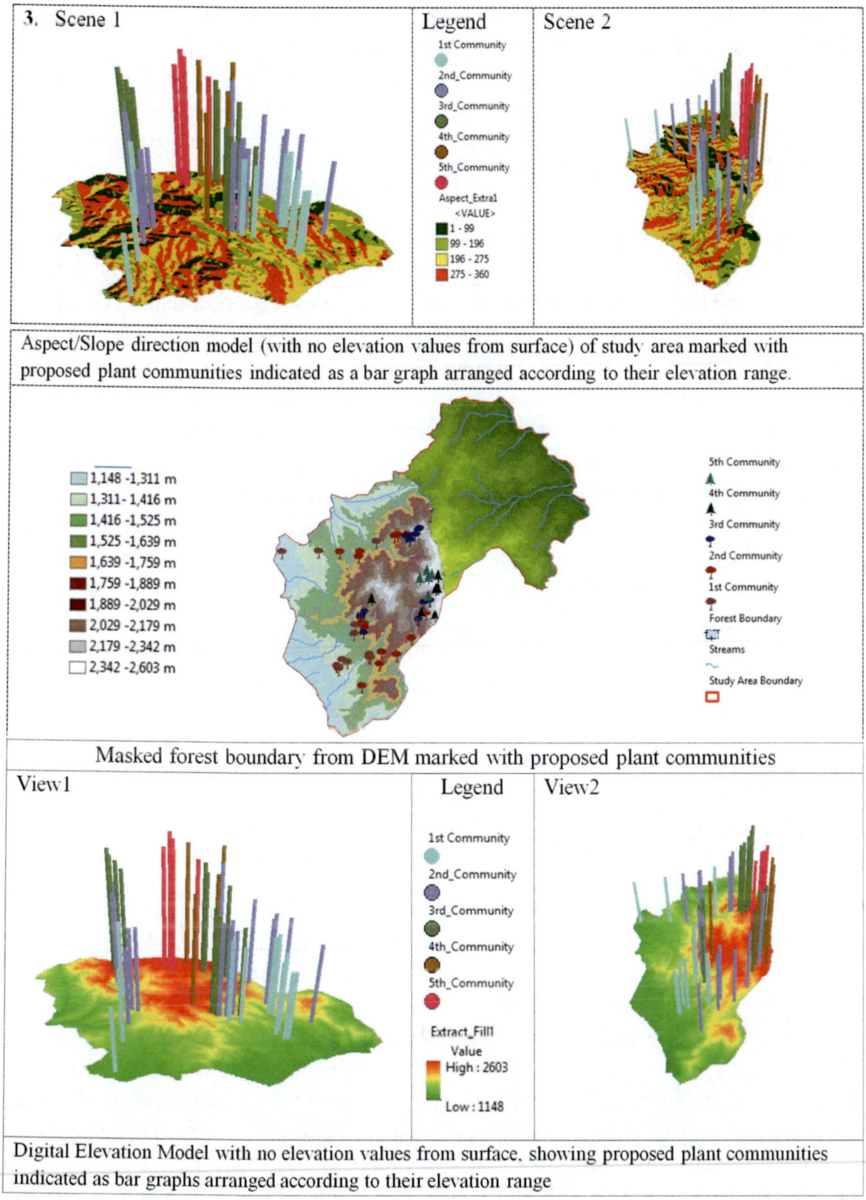

Fig. 4 Vegetation mapping according to different environmental attributes (aspect/elevation)

forests is comprised of *Adiantum incisum*, *Ajuga bracteosa*, *Potentilla fruticosa*, *Euphorbia wallichii* and *Bergenia ciliata* species.

Low-Level Blue Pine Vegetation

The range occupied by low-level blue pine woodland spreads over 8682 acres which make 44% of the studied region. A separate feature of Guzara blue pine is that they have improved stocking than any other coniferous forest types. Blue pine Guzara on the entire terrain exhabit adequate regeneration and seedling classes with an acute shortage of mature and sub mature trees. The Guzara forests of Tajwal, Darwaza, Inderseri, Sarbanna, and Dabban contain predominantly pole crops and show signs of fair stocking (Iqbal et al. 2018). The feature of the harvest is nearly alike the adjacent reserve forest vegetation (Khan et al. 2012). In pure blue pine there are number of frequently occurring and rare under wood species as well. Certain species which are noticeable only in the moist nallah beds or depression are *Prunus padus* (bird cherry), *Quercus incana* and *Quercus dilatata* (oak), *Diospyros lotus* (amlok) and *Cornus macrophylla* (Kandar). The common undergrowth plant species in the region consist of *Viburnum grandiflorum* (Guch), *Spirea vestita* (Amrer), *Berberis lyceum* (Sumbal) and *Rubus fruticosus* (garacha), respectively. It comprises mainly of *Viola canescens* (violet), *Fragaria vesca* (strawberry) and *Artemisia absinthium* (choo) besides many other grasses.

Mixed Coniferous Forest Vegetation

Most of the studied region occupied by western mixed coniferous forest vegetation is 3.44 hectares which establish about 2.1% of the whole region of the woodlands. This forest type occurs in Phalkot-4, Lassan-8, Birangali-9, and Inderseri-1. Composition of crops varies at different places, but the bulk of the growing stock consists of blue pine and silver fir; the latter is confined to cooler aspect slopes. Deodar occurs sporadically in Berangali, Inderseri, and Pichbhanna and Riala Guzaras, but on the whole it constitutes an insignificant proportion of the total growing stock. Fir crop is marked by an open canopy with scattered occurrence of mature and some overmature trees. Regeneration of silver fir and its younger age classes are either absent or deficient. The mature blue pine trees occur occasionally and seedlings classes occur frequently (Khan et al. 2011) (Figs. 5 and 6).

Fig. 5 Pictorial view of the dense coniferous forests

Fig. 6 Uprooting of trees due to extreme wind effects

Agroecology and Important Agricultural Crops

The punitive weather in TsFD is abridged to a very short summer, and due to which agricultural practices are restricted. The whole area is mono-cropic, i.e., only one crop (kharif) can be grown in a year during the season of summer only. Noteworthy crops are potatoes and peas. Other crops include maize, rye, fodder, wheat, beans, and cauliflower with less frequency and production. Fruit trees, counting pears, plums, walnuts, and apples are also grown in the area. Lack of financial support, poor means of communication and transport are the main difficulties to develop extensive land growth and agriculture. The river Kunhar and its branches provide water for irrigation in the summer, whereas natural springs and streams are the main sources of drinking water (Qureshi 2005).

Ethnology

Several tribes comprising Sardars, Gujjars, Jadoons, Qureshi and Dhoond inhabit the TsFD. The most significant between these are the Sardars who are renowned for their exclusive ethos, way of life, sacraments and courage. The Sardars are focused in the upper parts in most of the forests in TsFD where they improve rain-fed hills and are usually more conscious of traditional knowledge about plant use and local ecosystem (Ahmad et al. 2009). The Sardars were nominated by the British as a martial race that were thought to be logically strong in battle and owning qualities like courage, correctness, self-sufficiency, physical forte, discipline, and determination (Khan et al. 2013). Inhabitants of the region are very hard-working people and easily face any ruthless situation, particularly natural hazards and climatic constraints. The Sardars have their own exact language called Pahari which is among one of the most historic languages of the world. It is also measured to be the mother tongue of the present-day Urdu and Punjabi languages (Wayne 1996). Most of the people speak Pahari or Hindko languages in the region.

Livelihoods

Human life in the TsFD is a continuous challenging effort for survival. Usually, people have more than one type of occupation in order to maintain a sustainable livelihood. Generally, every household keeps cattle, the numbers and types of which vary from a few to hundreds. In the lower and middle forests, the second most common occupation is agriculture. Most of the people adapt these two professions for their livelihood. Having cultivable lands and a more reasonable climate, people grow crops in the lower forests which progressively decrease along the forests and increase in elevation and disappear at the forest upper margins. People of the

Table 2 Sources of information about the use of medicinal plants in the Thandiani forests

Source of information	Local elders	Farmers	Group discussions	Questionnaires	Interviews/ semi-structured interviews	Conventional healers
Numbers	13	31	04	34	18	2 women and 1 man

upper TsFD mainly trust on nurture of livestock and collection of native plants. Grazing livestock is dominant in the TsFD (Khan et al. 2015).

Ecosystem Services

Humans have long history of utilizing some of the plant species as food sources and the others as curing agents for diseases and injuries. A total of 252 plant species belonging to 97 families were recorded, out of which 47 plant species have high medicinal value with minimum importance value index (IVI) curing different diseases (Tables 2 and 3, Plates 1, 2, 3, 4, and 5). These 47 species have rare occurance in the region while having high use values in ethnomedicines and were therefore, primarily focused for documentation in this chapter. Disappearance of these species due to any reason may cause depletion of the traditional knowledge related to these species. Conservation priority must be given to these plants based upon their rarity in the region. The dominant families were Rosaceae with five species followed by Ranunculaceae with three species and Moraceae, Buxaceae, and Apiaceae having two species each, while the remaining families have one species each.

Majority of the plant species were found in mid-altitudes (Fig. 2). The altitudinal gradients were complex and involve many co-environmental variables such as topography, soil, and climate (Pausas and Austin 2001). In Himalayas the number of unique plant species are predictable from high altitude due to rapidly changing gradient of habitats (Shrestha and Joshi 1996). Various therapeutically important plant species flourish in the upper elevations of Thandiani forest region, reflecting great ranks of floral variety. Study in this region necessitate practices of modern approaches to the traditional systems that can lead to some novel information of taxa conservation as well as utilization (Palumbi 2001). Majority of the species in this region has significant uses in traditional healthcare system with a relatively inexpensive values, safety and sustainibility. Much care had been rewarded to records of the Himalayan region's medicinal plants, but no long term planned approach has been outlined so far. The plant resources are declining rapidly due to fragility of the natual ecosystem as well as unwise usage by the foothill societies for therapeutic purposes (Shinwari 2010). The major factors intimidating floral biodiversity in the region include changing climatic situations, overharvesting and overgrazing. The individuals existing in the buffer regions nearby the core park region mainly depend on normal resources and community facilities.

Table 3 Traditional uses of medicinal plants by local communities of Thandiani forests

S. No.	Botanical name	Local name	I.V.I	Family	Disorder treated
The medicinal importance of rarest species of 1st community (Khant et al., 2016b and 2017)					
1	*Hedera nepalensis*	Belrri	2.44	Araliaceae	Skin disorders
2	*Jacaranda mimosifolia*	Nelagul	4.33	Bignoniaceae	Syphilis and vulnerary
3	*Clematis amplexicaulis*	Churanhar	4.47	Ranunculaceae	Anti-inflammatory, cytotoxic, and antimicrobial effects
4	*Cuscuta reflexa*	Akashbail	4.5	Cuscutaceae	Eczema and scabies
5	*Lonicera bicolor*	Foota	5.4	Caprifoliaceae	Emiticocathartic, tonic, and diuretic
6	*Vitex negundo*	Marwand	5.8	Verbenaceae	Hair color
7	*Cyperus rotundus*	Deela	5.8	Cyperaceae	stomachic, emmenagogue, deobstruent, and emollient
8	*Celtis australis*	Batkarar	5.9	Celastraceae	Amenorrhea, lenitive, colic, diarrhea, dysentery, and peptic ulcers
9	*Polygonum amplexicaule*	Masloonrr	6.3	Polygonaceae	Infectious diseases, inflammation, gastrointestinal disorders, and cancer
10	*Buxus papillosa*	Angaroo	6.8	Buxaceae	Joints pain, skin disorder, and baldness
11	*Senecio chrysanthemoides*	Ragwort	8.05	Asteraceae	Antiseptic and rheumatic pain
12	*Foeniculum vulgare*	Sonf	9.41	Apiaceae	Constipation
The medicinal importance of rarest species of 2nd community (Khant et al., 2016b and 2017)					
1	*Aesculus indica*	Bankhorr	2.38	Hippocastanaceae	Rheumatism and colic pain
2	*Platanus orientalis*	Chinar	2.71	Platanaceae	Astringent, ophthalmic, and vulnerary
3	*Rubus* spp.	Chal	2.85	Rosaceae	Diarrhea and dysentery
4	*Pistacia integerrima*	Kangarr	3.34	Anacardiaceae	Antimicrobial, antioxidant and analgesic,
5	*Jasminum officinale*	Chambeli	3.75	Oleaceae	Aphrodisiac, sedative, antidepressant, antispasmodic, and analgesic
6	*Sarcococca saligna*	Ladan	3.87	Buxaceae	Laxative, blood purifier, and muscular pains
7	*Convolvulus prostrates*	Ilrra	3.87	Convolvulaceae	Purgative, diuretic, and laxative
8	*Solanum nigrum*	Kachmach	4.04	Solanaceae	Diuretic, diaphoretic, anodyne, and expectorant alternative

(continued)

Table 3 (continued)

S. No.	Botanical name	Local name	I.V.I	Family	Disorder treated
9	*Bupleurum* spp.	Beichaihu	4.06	Apiaceae	Common cold, bronchitis, and pneumonia
10	*Rhus punjabensis*	Sumac	4.25	Rosaceae	Diarrhea, hemorrhoids, leucorrhea, ophthalmia, conjunctivitis, and diuresis
11	*Buddleja asiatica*	Booi	4.6	Berberidaceae	Abortifacient
The medicinal importance of rarest species of 3rd community (Khant et al., 2016b and 2017)					
1	*Rubus fruticosus*	Chal	3.23	Rosaceae	Menstruation disorders
2	*Malva neglecta*	Sonchal	4.06	Malvaceae	Diarrhea and piles
3	*Ailanthus altissima*	Darawa	4.08	Simaroubaceae	Astringent, demulcent, aphrodisiac, and expectorant
4	*Morus nigra*	Kala Toot	4.16	Moraceae	Diuretic and expectorant
5	*Paeonia emodi*	Mamekh	4.23	Paeoniaceae	Joint pain
6	*Papaver somniferum*	Poppy	4.32	Papaveraceae	Sedative, analgesic, and antitussive
7	*Thalictrum cultratum*	Momyrun	4.49	Ranunculaceae	Ophthalmia and gastritis
8	*Hedera nepalensis*	Behri	4.51	Araliaceae	Cathartic, diaphoretic, skin and stimulant
9	*Rosa moschata*	Jungligulab	4.55	Rosaceae	Astringent, tonic and piles
10	*Punica granatum*	Darunna	4.74	Punicaceae	Cooling, refrigerant and breast development
11	*Morus alba*	Safeed toot	4.98	Moraceae	Antirheumatic, antispasmodic, diuretic, alterative, and diaphoretic
The medicinal importance of rarest species of 4th community (Khant et al., 2016b and 2017)					
1	*Geranium wallichianum*	Rattanjot	1.10	Geraniaceae	Vision problem, blood purification, jaundice, kidney and spleen problems
2	*Podophyllum emodi*	Bankhakhrri	1.81	Podophylaceae	Jaundice, liver ailment, fever, syphilis, hearing loss, and cancer
3	*Jasminum officinale*	Chambeli	1.82	Oleaceae	Aphrodisiac, sedative, antidepressant, antispasmodic, and analgesic
4	*Aesculus indica*	Bankhorr	3.29	Hippocastanaceae	Skin diseases, rheumatism, astringent, and narcotic and headaches
5	*Buddleja crispa*	Booi	3.33	Berberidaceae	Abortifacients
6	*Zanthoxylum armatum*	Timber	3.33	Rutaceae	Antiseptic, disinfectant, and deodorant properties

(continued)

Table 3 (continued)

S. No.	Botanical name	Local name	I.V.I	Family	Disorder treated
7	*Rhus punjabensis*	Sumac	3.46	Rosaceae	Diarrhea, ulcer, hemorrhoids, hemoptysis, conjunctivitis, and diuresis
8	*Clematis amplexicaulis*	Churanhar	3.63	Ranunculaceae	Anti-inflammatory, cytotoxic, and antimicrobial effects
9	*Berberis* spp.	Sumblo	3.73	Berberidaceae	Stomach ache
10	*Ailanthus altissima*	Darawa	3.84	Simaroubaceae	Antidiarrheal, antispasmodic, astringent, and diuretic
11	*Rosa moschata*	Chal	4.44	Rosaceae	Antispasmodic and antidiarrheal
12	*Robinia pseudoacacia*	Kekar	8.47	Papilionaceae	Diuretic, emetic, emollient laxative, purgative, and tonic
The medicinal importance of rarest species of 5th community (Khant et al., 2016b and 2017)					
1	*Acacia nilotica*	Kikar	1.91	Mimosoideae	Searing, sweltering, and torrid
2	*Cotoneaster minuta*	Bansathra	2.42	Rosaceae	Antipyretic and calmative
3	*Populus ciliata*	Safeeda	2.95	Salicaceae	Anti-inflammatory and febrifuge
4	*Sorbaria tomentosa*	Kaanhaji	3.39	Sonneratiaceae	Burns and wounds
5	*Verbescum thapsis*	Kutteykan	3.42	Scrophulariaceae	Emollient
6	*Thalictrum cultratum*	Momyrun	3.56	Ranunculaceae	Stomach pain and gastric trouble
7	*Bupleurum* spp.	Beichaihu	4.15	Apiaceae	Common cold, bronchitis, and pneumonia
8	*Cuscuta reflexa*	Akashbel	4.32	Cuscutaceae	Urine problems and constipation
9	*Capsella bursa pastoris*	Shufrt purse	4.46	Brassicaceae	Hemorrhages
10	*Arisaema flavum*	Adbis	4.53	Araceae	Expectorant, chronic tracheitis, bronchi ectasis, tetanus, and epilepsy
11	*Aquilegia pubiflora*	Koo-kuk	4.91	Ranunculaceae	Skin burns and wound healing

The anthropogenic factors and pressure on natural resources make it necessary to develop a comprehensive and maintainable forest management strategy (Hagler Bailly Pakistan 1999). The growing tourist traffic and activities in Abbottabad region at high-altitude pastures may increase further the threats to indigenous flora and would be another alarming factor in near future.

This study has numerous significant insinuations for the project and organization of reserves and Guzara forest plots. First, it proves that TsFD within upland forest

Aesculus indica (Comb) Hook

Aquilegia pubiflora Wall ex Royal

Arisaema flavum H.K.f

Berberis lycium Royal

Plate 1 Important rarely occuring medicinal plants of 1st community

Clematis montana L

Convolvulus arvinces Forssk

Cotoneaster minuta Klotz

Foeniculum vulgare Mill

Plate 2 Important rarely occuring medicinal plants of 2nd community

vegetation are certainly an appreciated reserve for the conservation of plant biodiversity, due to high levels of indigenous species, higher rate of productivity and a number of infrequent species of ethnoecological importances. Plant communities in these environments characterize distinctive mixtures of different forest species. The moist habitats in Thandiani forests permit recognition and safety to the fragile indicator vegetation. In fact, there is a considereable and perfect chance to preserve varied societies as well as natural diversity in a sustainable manner using community

Podophyllum emodi Wall

Sarbaria tomentosa L

Rubus fruticosus HK.f

Senecio chrysenthemoides DC

Plate 3 Important rarely occuring medicinal plants of 3rd community

Geranium wallichianum D.Don

Paeonia emodi wall

Punica granatum L

Morus nigra L

Plate 4 Important rarely occuring medicinal plants of 4th community

conservation procedures (Moral and Jones 2002). Conservation practitioners and ecologists have frequently ignored small lands in forest vegetation perhaps due to an absence of clear descriptions of habitat types and their respective indicators. It is therefore, hereby emphasized that indicators for a given region even at micro-habitat scale of forestlands play a vital role in ethnoecological understanding of that ecosystem. The forest vegetation explained by the present work here is merely not the documentation but also the cry of the day for maintenance of these extremely

Plate 5 Important rarely occuring medicinal plants of 5th community

important habitats. Our devised forestland types may also assist land managers to conserve the immense natural wealth of these kinds of habitats and ecosystems (Pyke et al. 2001). Conserving the different community types may lead to different conservation objectives and approaches. For example, Neelor and Barriback had lower local species richness (1.1), than other types, but the highest species richness value was found in upper Balolia (2.48), and the index of diversity values diverge from 11.1 to 39.44. These features make it significant to protect these micro-habitats, rather than larger regions, and to make them a focal point in landscapes conservation priorities (Flinn et al. 2008). The regional strategies, may include all of the varied types of TsFD vegetation, which may ultimate contribute considerably to the maintenance of plant biodiversity in particular and other types in general.

References

Ahmad H, Khan SM, Ghafoor S, Ali N (2009) Ethnobotanical study of upper Siran. J Herbs Spices Med Plants 15:86–97

Ali SI, Qaiser M (1986) A Phyto-geographical analysis of the Phenerogames of Pakistan and Kashmir. Proc Roy Soc Edinb 89:89–101

Barthel S, Crumley CL, Svedin U (2013) Bioculturalrefugia: combating the erosion of diversity in landscapes of food production. Ecol Soc 18:71

Beg AR (1975) Wildlife habitats of Pakistan. Bulletin 5

Berkes F, Turner NJ (2006) Knowledge, learning and the evolution of conservation practice for social-ecological system resilience. Hum Ecol 34:479–494

Berkes F, Colding J, Folke C (2003) Navigating social-ecological systems: building resilience for complexity and change. Cambridge University Press, Cambridge

Brillat-Savarin JA (1960) The physiology of taste. Dover, New York

Champion HG, & Khattak GM (1965) Forest Types of Pakistan
Dasti AA, Saima S, Athar M, Attiq-ur-Rahman, Malik SA (2007) Botanical composition and multivariate analysis of vegetation on the Pothowar Plateau, Pakistan. J Bot Res Inst Texas 1:557–568
Dickoré WB, Nüsser M (2000) Flora of Nanga Parbat (NW Himalaya, Pakistan). An annotated inventory of vascular plants with remarks on vegetation dynamics. Englera 19:1–253
Flinn KM, Lechowicz MJ, Waterway MJ (2008) Plant species diversity and composition of wetlands within an upland forest. Am J Bot 95(10):1216–1224
Hagler Bailly Pakistan (1999) Pakistan's National Communication to the UNFCC. Report prepared for the Ministry of Environment, Islamabad, Pakistan
Han L, Chunlin L, Sixiang Z, Zhilin L (2005) Polyploid induction of Cymbidium iridioides and its biological characteristics. Acta Horticulturae Sinica 32(5):853
Iqbal M, Khan SM, Khan MA, Ahmad Z, Ahmad H, (2018) A novel approach to phytosociological classification of weeds flora of an agroecological system through Cluster, Two Way Cluster and Indicator Species Analyses. Ecological Indicators, 84:590–606
Kala CP, Mathur VB (2002) Patterns of plant species distribution in the trans-Himalayan region of Ladakh, India. J Veg Sci 13:751–754
Kent M, Coker P (1995) Vegetation description and analysis: a practical approach. Wiley, Chichester
Khan W, Ahmad H, Shah GM (2011) Phytosociology and Geographical distribution of Thandiani Forests. LAP LAMBERT Academic Publishing GmbH & Co. KG, Saarbrücken. ISBN: 978-3-8465-1016-2
Khan SM, Page S, Ahmad H, Harper D (2012) Anthropogenic influences on the natural ecosystem of the Naran Valley in the western Himalayas. Pak. J. Bot, 44:231–238
Khan W, Majid A, Afzal M, Islam M, Bibi F (2013) Degree of aggregation and index of similarity of plant communities recorded at Thandiani Hills, District Abbottabad KPK, Pakistan. Eur J Agricul Sci 11:68–72
Khan W, Khan SM, Ahmad H (2015) Altitudinal variation in plant species richness and diversity at Thandiani sub forests division, Abbottabad, Pakistan. J Biol Environ Sci 7(1):46–53
Khan W, Khan SM, Ahmad H, Ahmad Z (2016a) Floral biodiversity and vegetation composition of the Western Himalayas, Thandiani sub forests division, Abbottabad. Con Bio Pak 1(1):1–9
Khan W, Khan SM, Ahmad H, Ahmad Z, Page S (2016b) Vegetation mapping and multivariate approach to indicator species of a forest ecosystem: a case study from the Thandiani sub forests division (TsFD) in the western Himalayas. Ecol Indic 71(2016):336–351
Khan W, Khan SM, Ahmad H, Shakeel A, Page S (2017) Ecological gradient analyses of plant associations in the Thandiani forests of the Western Himalayas, Pakistan. Turkish Journal of Botany, 41(3):253–264
Malik RN, Husain SZ (2008) Linking remote sensing and ecological vegetation communities: a multivariate approach. Pak J Bot 40:337–349
Moral RD, Jones C (2002) Vegetation development on pumice at Mount St. Helens, USA. Plant Ecol 162:9–22
Mueller-Dombois D, Ellenberg H (1974) Aims and methods of vegetation ecology. Wiley, New York, London
Oommen MA, Shanker K (2005) Elevational species richness patterns emerge from multiple local mechanisms in Himalayan woody plants. Ecology 86:3039–3047
Palumbi SR (2001) The ecology of marine protected areas. Marine Commun Ecol 9:509–530
Pausas JG, Austin MP (2001) Patterns of plant species richness in relation to different environments: an appraisal. J Veg Sci 12(2):153–166
Petrini C (2007) Slow Food Nation. Why our food should be good, clean, and fair. Rizzoli, New York
Pyke CR, Condit R, Aguilara S, Lao S (2001) Floristic composition across a climatic gradient in a neotropical lowland forest. J Veg Sci 12:553–566

Qureshi AS (2005) Climate change and water resources management in Pakistan. Climate change and water resources in South Asia. AA Balkema Publishers, Leiden, pp 197–230
Saima S, Dasti AA, Hussain F, Wazir SM, Malik SA (2009) Floristic compositions along an 18 km long transect in ayubia National Park district Abbottabad, Pakistan. Pak J Bot 41:2115–2127
Shinwari ZK (2010) Medicinal plants research in Pakistan. J Med Plant Res 4(3):161–176
Shrestha TB, Joshi RM (1996) Rare, endemic and endangered plants of Nepal. WWF. WWF Nepal Program, Kathmandu, Nepal
Signorini MA, Piredda M, Bruschi P (2009) Plants and traditional knowledge: an ethnobotanical investigation on Monte Ortobene (Nuoro, Sardinia). J Ethnobiol Ethnomed 5:6
Singh E, Singh MP (2010) Biodiversity and phytosociological analysis of plants around the municipal drains in Jaunpur. World Acad Sci Eng Technol 4:01–28
Takhtadzhian AL, Cronquist A (1986) Floristic regions of the world. University of California Press, London
UNESCO (2003) Convention for the safeguarding of the intangible cultural heritage. http://www.unesco.org/culture/ich/index.php?lg=EN&pg=00022. Accessed 13 Apr 2015
Wayne EL (1996) Writing Gojri: linguistic and sociolinguistic constraints on a standardized orthography for the Gujars of South Asia, Master Thesis. University of North Dakota, USA
Wazir SM, Dasti AA, Saima S, Shah J, Hussain F (2008) Multivariate analysis of vegetation of Chapursan valley: an alpine meadow in Pakistan. Pak J Bot 40:615–626

Ethnobotanical Uses of Some Plants of Families Apocynaceae and Asclepiadaceae from the Northwestern Region of Ahmednagar District, Maharashtra

Mohan B. Waman and Mahendra S. Khyade

Introduction

The importance of plants in different cultures is an unquestionable reality. They are used for the basic needs, viz., feeding, clothing, sheltering, hunting, and nursing (Garbi et al. 2015). However, the advent of modern technology has caused a great loss in the indigenous traditional knowledge which was used to be transmitted orally. However, the collection of information about natural flora, classification, management, and use of plants by the people hold importance among the ethnobotanists. Besides, the researchers and local people, too, face difficulties in documenting this knowledge (Ford 1978).

The northwest part of Ahmednagar district includes the entire Akole tehsil and a little hilly part of Sangamner tehsil situated in the Sahyadri ranges of northern Western Ghats. Earlier researchers including Billore and Hemadri (1969) and Pradhan and Singh (1999) have studied the phytodiversity of the region. Moreover, few ethnobotanical endeavor are carried out in forest villages by different workers (Petkar et al. 2002; Wabale and Petkar 2005; Khyade et al. 2010, 2011). Despite the vast work done by many authors on different aspects, very poor attention has still been paid on the families, Apocynaceae and Asclepiadaceae regarding their use in food and medicine. Keeping this view, the present study has been undertaken to survey and document the plants used in food and medicine.

M. B. Waman
Dr. D.Y. Patil Arts, Commerce and Science College, Pune, Maharashtra, India

M. S. Khyade (✉)
Department of Botany, S. N. Arts, D. J. M. Commerce and B. N. S. Science College, Sangamner, Maharashtra, India

M. Ozturk, K. R. Hakeem (eds.), *Plant and Human Health, Volume 1*,
https://doi.org/10.1007/978-3-319-93997-1_14

Materials and Methods

Study Area

Ahmednagar district lies between 18.2° to 19.9° northern latitude and 73.9° to 75.5° eastern longitude. The district comprises of 14 tehsils of which the entire Akole tehsil and a little part of Sangamner tehsil fall in the northwestern part of the Ahmednagar district (Fig. 1). This zone spreads along with crest of Sahyadri ranges of northern Western Ghats, one of the global biodiversity hotspots. The hill ranges of Adula, Baleshwar, and Harishchandragad lie in this region, and various high peaks are found in the same region. Kalsubai, one of the highest peaks in the Sahyadri of Maharashtra, lies in this region.

The study region is inhabited mostly by tribal population (Mahadev Koli and Thakar Tribe) with smallholdings and is economically backward. The vegetation of the study area is of semi-evergreen forest, moist deciduous forests, dry deciduous forests, and ravine vegetation type (Pradhan and Singh 1999). The average rainfall of the study area is about 4182 mm per annum. The temperature rises up to 40 °C in midsummer and falls up to 7 °C in winter (in the month of December).

Survey and Data Collection

Survey Methodology

The frequent field visits were conducted during 2014–2016 in the major villages and some small hamlets situated in and around the hilly areas of the study area. The places included Baleshwar, Bhandardara, Ghatghar, Akole, Kotul, Rajur, Murshet, Panjaare, Ratanwadi, Sambrad, and Shendi. The questionnaires were used in local language (in Marathi) to obtain information about the utility of plants for food and medicine with their local names, plant parts used, mode of preparation, and administration (Jain and Rao,1967). The Informants were chosen on the basis of their indigenous knowledge of plants used for different purposes. The informants were

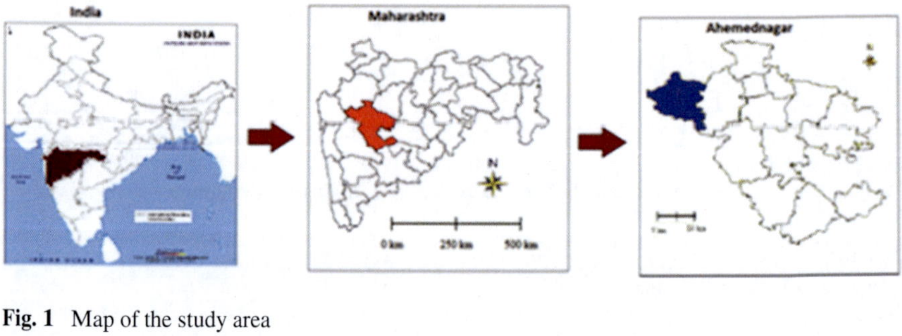

Fig. 1 Map of the study area

requested to indicate plants in the field for identification. The photographs were taken with the help of Canon SX50 HS camera. The collected specimens were identified by referring to various floras (Pradhan and Singh 1999; Cooke 1967; Sharma et al. 1996; Singh and Karthikeyan 2000; Singh et al. 2001). Moreover, the recent names of the plant species based on the International Plant Names Index (IPNI – www.ipni.org) have been given.

Tabulation and Data Analysis

The botanical names of the plants from both families are tabulated in alphabetical order along with vernacular names. Besides, the collected data is also tabulated in terms of habit, plant parts used and ailments treated, mode of preparations and roots of administration for each species (Table 2).

Use Value (UV)

The use value (UV), a quantitative method that demonstrates the relative importance of species known locally, was also calculated using the following formula:

$$UV = \frac{\sum U}{N}$$

where UV, the use value of a species; U, the number of citation per species; and N, the number of informants (Upadhyay et al. 2011).

Informants Consensus Factor (ICF)

For the data analysis, informant consensus factor (ICF) was employed to indicate how homogenous the information is. All the citations were placed into primary and secondary categories for which the plants were claimed to be used. ICF values will be lower (near 0) if plants are chosen randomly or if informants do not exchange information about their use. Values will be high (near or more than 1) if there is a well-defined selection criterion in the community and/or if information is exchanged between informants.

The ICF is calculated as in the following formula (Gazzaneo et al. 2005):

$$ICF = Nur - Nt / Nur - 1$$

where "Nur" is the number of used citations in each category and "Nt" is the number of species used.

Literature Review

Literature review was consulted for relevant ethnobotanical information. Field studies in the state of Maharashtra and India available in scientific journal and reference books were consulted. The literature was obtained from various international scientific databases such as Web of Science, MEDLINE/PubMed, Google Scholar, and NISCAIR Online Periodicals Repository.

Results and Discussion

Knowledge Distribution Among Informants

The demographic characteristics of the informants were recorded through face-to-face interviews. The number of practitioners between the age groups 50 and 59 was high when compared to the other groups. In this study, the sample size of females is low as their responses depended on the permission of their male partner. It has resulted into the eneven distribution of male-female ratio. Out of the 47 informants, 36 belonged to Mahadev Koli and 11 from Thakar tribe (Table 1). During the ethnobotanical surveys carried out in different villages of the study area, a total of 31 plant species belonging to both Asclepiadaceae and Apocynaceae have been documented. For each species botanical name, family, local name, parts used, methods of preparation, and use value were discussed in detail (Table 2).

Table 1 Sex and age character of people interviewed in the study area

Ethnic group	Age group	Gender		No. of person	Percentage
		Male	Female		
Mahadev Koli	30–39	1	2	03	8.33
	40–49	5	5	10	27.77
	50–59	7	4	11	30.55
	60–69	6	3	09	25.00
	70–79+	2	1	03	8.33
	Total	**21**	**15**	**36**	
Thakar	30–39	1	0	01	9.09
	40–49	2	1	03	27.27
	50–59	2	1	03	27.27
	60–69	2	1	03	27.27
	70–79+	1	0	01	9.09
	Total	**08**	**03**	**11**	

Table 2 Enumeration of plants used for food and medicine along with their UV reports in the study area

Botanical name with family	Local name	Habit	Parts used	Ethnobotanical uses	UV	Similar reported literature
Apocynaceae						
Carissa congesta Wight.	Karwand	Shrub	Fruit	Ripe fruits are eaten as raw	1.00	Waman (2005), Patil and Patil (2006), Datar and Upadhye (2016)
				Also unripe fruits used to make pickles	0.36	Watt (1889–1893), Anonymous (1948–1976), Datar and Upadhye (2016)
				Alcoholic beverage (wine) is prepared	0.36	Patil and Patil (2006)
			Leaves	Paste of leaves in water is applied in skin diseases	0.14	
Catharanthes pusilus (Murr.) G.Don	Ran sadaphuli	Herb	Leaves	2–3 leaves chewed two times to lowers fever	0.10	
Catharanthes roseus L.	Sadaphuli	Herb	Leaves	Paste is applied on skin diseases	0.44	Shende et al. (2014)
				Paste is applied on wounds	0.65	
Holarrhena pubescens (Buch-Ham) Wall.ex G.Don.	Kala kuda	Tree	Flower	Flowers are cocked as vegetables	1.00	Reddy (2012), Patale et al. (2015), Datar and Upadhye (2016)
			Bark	Dried powdered extract given internally in kidney stone	0.82	
Plumeria alba L.	Pandhara chapha	Tree	Latex	Latex directly applied on skin diseases	0.57	
Wrightia tinctoria R.Br.	Pandhara Kuda	Tree	Bark	Extract given internally in snakebite	0.27	Patil and Patil (2006)
Asclepiadaceae						
Calotropis gigantea (L.) Ait. Hort. Kewed.	Mothi Rui	Shrub	Latex	Latex applied in skin diseases	0.48	Kosalge and Fursule (2009), Desale et al. (2013)

(continued)

Table 2 (continued)

Botanical name with family	Local name	Habit	Parts used	Ethnobotanical uses	UV	Similar reported literature
				Latex applied in wounds	0.65	Waman (2005)
				Latex applied on swellings	0.57	
Calotropis procera (Ait.) R.Br.	Rui	Shrub	Latex	Latex applied on forehead in headache	0.14	Patil and Biradar (2011)
Caralluma adscendens (Roxb.) Haw	Shindal makad	Herb	Stem	Shoots eaten as raw	1.00	Vartak (1981), Waman (2005)
				Also cocked as vegetable	1.00	Waman (2005), Datar and Upadhye (2016)
Ceropegia attenuata Hook	Kharpudi	Herb	Tuber	Roasted tubers eaten	0.51	Yadav and Kamble (2008)
Ceropegia bulbosa Roxb.	Kharpudi	Climber	Tuber	Eaten as general tonic	0.34	Vartak (1981), Yadav and Kamble (2008), Jagtap et al. (2008a), Anonymous (1948–1976), Datar and Upadhye (2016)
Ceropegia hirsuta Wight & Arn	Haman	Climber	Tuber	Roasted tubers eaten	0.46	Waman (2005), Yadav and Kamble (2008), Datar and Upadhye (2016)
Ceropegia lawii Hook.	Kharpudi	Herb	Tuber	Roasted tubers eaten	1.00	Vartak (1981), Waman (2005), Yadav and Kamble (2008)
Ceropegia maccanii Ansari.	Tilori	Herb	Tuber	Roasted tubers eaten	0.27	Waman (2005), Yadav and Kamble (2008)
Ceropegia media (Huber) Ansari	Kharpudi	Climber	Tuber	Roasted tubers eaten	0.23	Yadav and Kamble (2008)
Ceropegia oculata Hook.	Haman	Climber	Tuber	Roasted tubers eaten	0.72	Waman (2005), Yadav and Kamble (2008), Datar and Upadhye (2016)
Ceropegia odorata Hook.	Kharpudi	Climber	Tuber	Roasted tubers eaten	0.51	Waman (2005), Jagtap et al. (2008b)

(continued)

Table 2 (continued)

Botanical name with family	Local name	Habit	Parts used	Ethnobotanical uses	UV	Similar reported literature
Ceropegia rollae Hamadri	Kharpudi	Herb	Tuber	Roasted tubers eaten	1.00	Waman (2005), Yadav and Kamble (2008)
Ceropegia sahyadrica Ansari & Kulkarni	Kharpudi	Herb	Tuber	Roasted tubers eaten	1.00	Waman (2005), Yadav and Kamble (2008)
Ceropegia mahabalei Hem & Ansari.	Kharpudi	Climber	Tuber	Roasted tubers eaten	0.23	Waman (2005), Yadav and Kamble (2008)
Cryptolepis buchanani Roem. & Schult.	Mothi Kavali	Climber	Root	Half glass of root extract is taken orally early in the morning till to cure the fever	1.00	Patil and Patil (2006), Jain (1991)
Cynanchum callialatum Ham, ex, Wight	Kavali	Climber	Leaves	Fresh 3–5 leave are eaten in high fever	0.14	
Frerea indica Dalz.	Makadshingi	Herb	Stem	Fresh shoots eaten as raw	0.57	Waman (2005)
Gymnema sylvestre (Retz) R.Br. ex Schultes.	Bedki	Climber	Leaves	2–3 leaves eaten two times a day in diabetes	1.00	Anonymous (1948–1976), Waman (2005), Bhosle et al. (2009), Kamble et al. (2010), Gayake et al. (2013), Shaikh et al. (2014)
Hemidesmus indicus (L.) Schult	Anantmul	Climber	Root	Ground root extract given internally against snakebite	0.70	Vartak (1981), Vaidya and Dhumal (2004), Jain (1991), Khyade et al. (2011)
Hoya alexicaca (Jacq) Moon		Climber	Leaves	Externally paste of leaves in water is applied on skin diseases	0.25	
Leptadenia reticulata (Retz.) Wt.et Arn	Hirandodi	Climber	Leaves	Externally paste of leaves in water is applied on skin diseases	0.53	Anonymous (1948–1976)
Sarcostema viminale (L.) R.Br.	Sabar	Climber	Latex	Externally applied for wound healing	0.76	Tayade and Patil (2005)

(continued)

Table 2 (continued)

Botanical name with family	Local name	Habit	Parts used	Ethnobotanical uses	UV	Similar reported literature
Tylophora dalzellii Hook f.	Kavali	Climber	Root	A glass of water extract given orally to fever	1.00	
Tylophora indica Buch-Ham.	Kavali	Climber	Root	A glass of water extract given orally against fever	1.00	
Wattakaka volubilis (L.f.) Stapf	Kavali	Climber	Latex	Latex directly applied on skin diseases	0.57	

Fig. 2 Plant species used

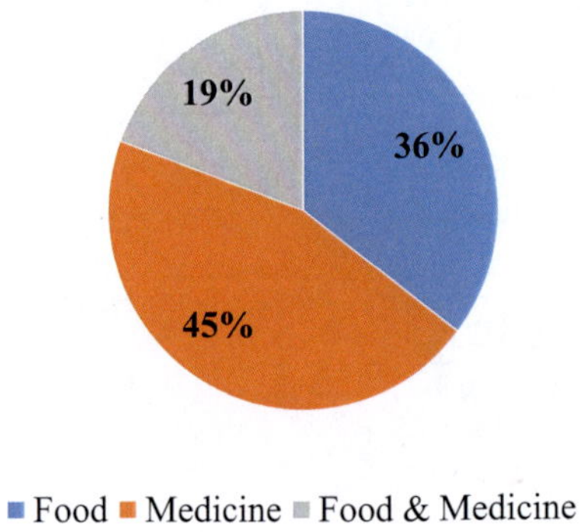

Analysis of the Data

The traditional knowledge of two ethnic communities of the study area has a tremendous importance. Among the 31 plants recorded, 14 species were used in medicine (45%), 11 species in food (36%), and the remaining 6 (19%) species were used in both food and medicine (Fig. 2). The highest number of species belonged to Asclepiadaceae (25 species each) followed by Apocynaceae (6 species each). Climbers are used more often (16 species), followed by herbs (9 species), shrubs (3 species), and tree (3 species) as shown in Fig. 3.

The collected plants were used for medicinal and food purposes; aerial parts and whole plants were also used in case of small herbaceous plants. Among plant parts tubers (35.48%) were the most frequently used part followed by the leaves (22.58), latex (16.12), root (12.9), bark and stem (6.45), and flower and fruits (3.22), respectively (Fig. 4).

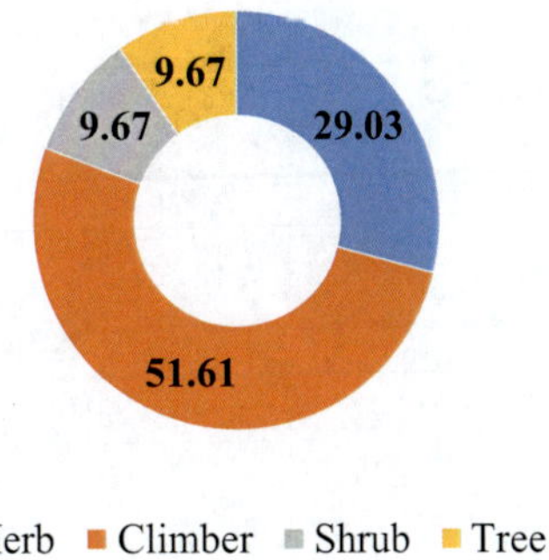

Fig. 3 Growth Forms of Plants

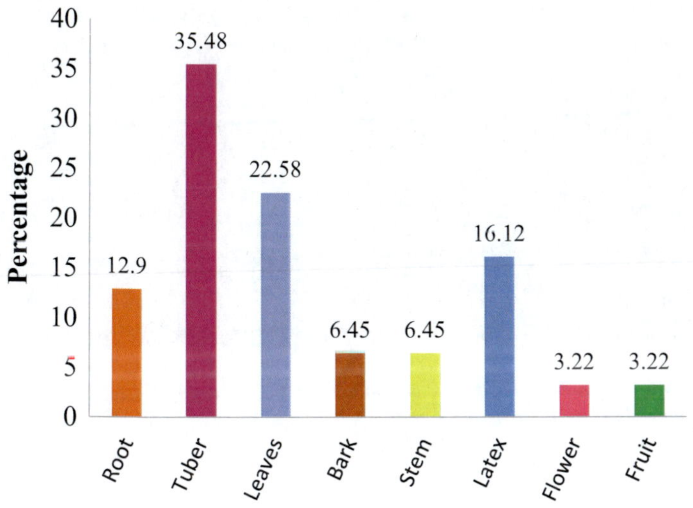

Fig. 4 Plant parts used

Use Value (UV)

UV is calculated for all the species and the single use by the single informant is not considered for analysis. The plant species, *Carissa congesta*, *Holarrhena pubescens*, *Caralluma adscendens*, *Ceropegia lawii*, *Ceropegia rollae* (Fig. 5), *Ceropegia sahyadrica*, *Cryptolepis buchanani* (Fig. 6), *Gymnema sylvestre*, *Tylophora dalzellii*, and *Tylophora indica* scored a high UV (1.00 each), followed by *Sarcostemma viminale* (0.82), *Ceropegia oculata* (0.72), *Hemidesmus indicus* (0.70), *Wattakaka volubilis*, *Frerea indica*, *Plumeria alba* (0.57 each), *Leptadenia reticulata* (0.53), and *Ceropegia odorata* and *Ceropegia attenuata* (0.51 each). A plant with high UV is the indication of its importance among the tribal communities under study as it is highly cited by many informants (Table 3). However, low citation by the informants for particular species does not indicate their less utility. Appearance of knowledge or unavailability of the plants in their vicinity could be the factors behind the low citation.

Fig. 5 (**a**) *Ceropegia oculata*; (**b**) *C. attenuata*; (**c**) *C. lawii*; (**d**) *Caralluma adscendens*; (**e**) *Holarrhena pubescens*; (**f**) *Carissa congesta*

Informants Consensus Factor (ICF)

In the ethnobotanical studies, informant consensus analysis provides a measure of reliability for the given claim of evidence (Malla and Chhetri 2012). The applications of plant in both food and medicine are classified into 14 categories, and the ICF values for each category are given (Table 3). ICF values obtained for the reported categories indicate the degree of shared knowledge for the treatment of each ailment category and utility category of food. The highest ICF scored in medicine is for diabetes, headache, kidney stone, swelling, and tonic (ICF = 1) followed by fever, wounds and snakebite (0.97), and skin disease (0.95). The highest ICF values for food were for pickle and beverage (ICF = 1) followed by vegetable (ICF = 98), raw (ICF = 98), and roasted food material (ICF = 96). ICF value is a proportion between the uses of different plants by various people for a particular use category. It suggests the level of sharing the knowledge of the use of plants among the people. High ICF value indicates the use of some plants by many tribal people, whereas, low value means use of different plants by many people.

Fig. 6 (**a**) *Hemidesmus indicus*; (**b**) *Cryptolepis buchanani*; (**c**) *Sarcostemma viminale*; (**d**) *Wrightia tinctoria*

New Reports

The present surveyed information of the plants for utility in food and medicine was compared with those gathered by earlier published reports on ethnobotanical surveys of state of Maharashtra and the important literature pertaining to Indian medicinal plants and raw materials (Watt 1889–1893; Anonymous 1948–1976; Jain 1991; Patil and Patil 2006). A review of literature indicated that 7 out of the 31 plant species are reported with new medicinal uses (Table 2). The ethnobotanical claims associated with these six species are *Catharanthus pusillus*, *Cynanchum callialatum*, *Tylophora indica*, and *Tylophora dalzellii* were reported for the first time for fever. Similarly, *Hoya alexicaca* and *Wattakaka volubilis* against *Plumeria alba* to cure skin diseases have not yet been reported and thus are new claims to the ethnobotanical knowledge (Table 2).

Table 3 Informant consensus factor (F_{IC}) by categories of utility in food and medicine

Primary category of use	Secondary category of use	Use citation (Nur)	Number of plant taxa used (Nt)	Informant consensus factor (ICF)
Medicine				
	Diabetes	47	1	1.00
	Headache	07	1	1.00
	Kidney stone	39	1	1.00
	Swelling	27	1	1.00
	Tonic	16	1	1.00
	Fever	155	5	0.97
	Wounds	98	3	0.97
	Snakebite	46	2	0.97
	Skin disease	142	7	0.95
Food				
	Alcoholic beverage	13	1	1.00
	Pickle	17	1	1.00
	Raw	121	3	0.98
	Vegetable	94	2	0.98
	Roasted food material	280	10	0.96

Rare, Endemic, and Threatened Species

Among the 31 plant species reported, some plant species come under rare, endemic, and threatened categories and are listed in various recent scientific literatures and red data list (Nayar and Sastry 1987–1989). *Ceropegia attenuata* (endemic to India and vulnerable), *Ceropegia bulbosa* (endemic to India), *Ceropegia lawii* (endemic to Maharashtra and endangered), *Ceropegia maccanii* (endemic to Maharashtra and endangered), *Ceropegia mahabalei* (endemic to Maharashtra and critically endangered), *Ceropegia media* (endemic to Maharashtra and endangered), *Ceropegia oculata* (endemic to Western Ghats and vulnerable), *Ceropegia odorata* (endemic and critically endangered), *Ceropegia rollae* (endemic to Maharashtra and critically endangered), *Ceropegia sahyadrica* (endemic to Maharashtra and critically endangered), *Frerea indica* (endemic to Maharashtra and critically endangered), etc. are reported from the study area.

Conclusion

The present ethnobotanical study revealed that the two ethnic communities residing in the hilly areas of the northwestern region of Ahmednagar district depend on forest resources to meet their requirements in the form of food and medicine. This study has explored the utility of 31 species that belong to both Asclepiadaceae and

Apocynaceae families. In medicine category, the plants with high ICF must be chemically investigated for the identification of bioactive compounds which can lead to designing new drugs of therapeutic importance. The plants reported for food category must be studied by nutraceutical point of view and could provide interesting opportunity for mountain agriculture.

Acknowledgments We are thankful to the authorities of the Forest Department and Wildlife Department, Nashik, for allowing us to carry out this study in the study area. We also would like to thank the local people of the study for sharing their valuable cultural knowledge with me. The author of MBW is grateful to the BCUD, Savitribai Phule Pune University, Pune (MS), India (Grant No: 15SCI001248), for providing financial assistance in this project.

References

Anonymous (1948–1976) The Wealth of India. A dictionary of Indian raw materials and industrial products. Vol. 1–11. Publication and Information Directorate, CSIR, New Delhi

Bhosle SV, Ghule VP, Aundhe DJ, Jagtap SD (2009) Ethnomedical knowledge of plants used by the tribal people of Purandhar in Maharashtra, India. Ethnobot Leaflets 13:1353–1361

Billore KV, Hemadri K (1969) Observation on the flora of Harishchandragarh Sahyadri range. Maharashtra Bull Botan Surv India 11:335–346

Cooke T (1967) (Reprint). Flora of Presidency of Bombay, Vol. 1–3. Botanical Survey of India (BSI), Calcutta

Datar MN, Upadhye AS (2016) Forests foods of Northern Region of Western Ghats. Maharashtra Association for the Cultivation of Science (MACS) Agharkar Research Institute, Pune

Desale MK, Bhamare PB, Sawant PS, Patil SR, Kamble SY (2013) Medicinal plants used by the rural people of Taluka Purandhar, district Pune, Maharashtra. Indian J Tradit Knowl 12:334–338

Ford RI (1978) The Nature and Status of Ethnobotany, Anthropological Papers No. 67 Museum of Anthropol. University of Michigan, Michigan

Garbi MI, Elbadri EO, Ahmed SK (2015) Anticancer activity of *Bauhinia rufescens* (Lam.) leaf extracts on MCF-7 Human Breast Cancer Cells. J Med Plants Stud 3:103–106

Gayake DN, Awasarkar UD, Sharma PP (2013) Indigenous traditional medicinal plant resources from Ahmednagar District, Maharastra, India. Asian J Biomed Pharmaceut Sci 3:1–5

Gazzaneo L, de Lucena R, de Albuquerque U (2005) Knowledge and use of medicinal plants by local specialists in a region of Atlantic forest in the state of Pernambuco (North-Eastern Brazil). J Ethnobiol Ethnomed 1:1–18

Jagtap SD, Deokule SS, Bhosle SS (2008a) Ethnomedicobotanical uses of endemic and RET plants utilised by the Korku tribe of Amravati district, Maharashtra. Indian J Tradit Knowl 7:284–287

Jagtap SD, Deokule SS, Bhosle SS (2008b) Ethnobotanical uses of endemic and RET plants by Pawra tribe of Nandurbar district, Maharashtra. Indian J Tradit Knowl 7:311–315

Jain SK (1991) Dictionary of Indian folk medicine and ethnobotany. Deep Publication, New Delhi

Jain SK, Rao RR (1967) A handbook of field and herbarium methods. Today and Tomorrow Printers and Publishers, New Delhi

Kamble SY, Patil SR, Sawant PS, Sawant S, Pawar SG, Singh EA (2010) Studies on plants used in traditional medicine by Bhilla tribe of Maharashtra. Indian J Tradit Knowl 9:591–598

Khyade MS, Awsarkar UD, Deshmukh RR, Petkar AS (2010) Ethnobotanical reports about few important diseases from Akole Tahasil of Ahmednagar District (MS), India. Asian J Exp Biol Sci 1:393–403

Khyade MS, Takate YA, Divekar MV (2011) Plants used as an antidote against snakebite in AkoleTaluka of Ahmednagar District (MS), India. J Nat Remed 11(2):182–192

Kosalge SB, Fursule RA (2009) Investigation of ethnomedicinal claims of some plants used by tribals of Satpuda Hills in India. J Ethnopharmacol 121:456–461
Malla B, Chhetri RB (2012) Indigenous knowledge on medicinal non-Timber forest products (NTFP) in Parbat district of Nepal. Indo Global J of Pharmaceutical Sci 2(2):213–225
Nayar MP, Sastry ARK (eds) (1987–1989) Red data book of Indian plants. Vol. I–III. Botanical Survey of India (BSI), Calcutta
Patale CK, Nasare PN, Narkhede SD (2015) Ethnobotanical studies on wild edible plants of Gond, Zhalba and Kawar tribes of Salekasa Taluka, Gondia district, Maharashtra state, India. Int Res J Pharm 6:512–518
Patil JU, Biradar SD (2011) Folkloric medicinal plants of Hingoli district, Maharashtra. Indian J Nat Prod Resour 2:97–101
Patil DA, Patil MV (2006) Ethnobotany of Nasik District, Maharashtra. Daya Publishing House, Delhi
Petkar AS, Wabale AS, Shinde MC (2002) Some Ethnomedicinal plants in the tribal areas of Akole and Sangamner talukas of Ahmednagar District (M.S.). J Ind Botan Soc 81:213–215
Pradhan SG, Singh NP (1999) Flora of Ahmednagar District (M.S.). Bishen Singh Mahendra Pal Singh, Dehradun
Reddy BM (2012) Wild edible plants of Chandrapur district, Maharashtra, India. Indian J Nat Prod Resour 3:110–117
Shaikh RU, Dukare DD, Sarwade KP, Sarwade PP (2014) Ethnobotanical study of folk medicinal plants used by villagers in Nanded district of Maharashtra (India). Int J Ayur Herb Med 4:1585–1595
Sharma BD, Karthikeyan S, Singh NP (1996) Flora of Maharashtra State–onocotyledons. Botanical Survey of India (BSI), Calcutta
Shende JJ, Rajurkar BM, Mhaiskar MN, Dalal LP (2014) Ethnobotanical studies of Samudrapur Tahsil of Wardha District. IOSR J Pharm Biol Sci 9:16–23
Singh NP, Karthikeyan S (2000) Flora of Maharashtra State–Dicotyledons, Vol. I. Botanical Survey of India (BSI), Calcutta
Singh NP, Lakshminarasimhan P, Karthikeyan S, Prasanna PV (2001) Flora of Maharashtra State–Dicotyledons, Vol. II. Botanical Survey of India (BSI), Calcutta
Tayade SK, Patil DA (2005) Hitherto untapped plantlore from Nandurbar district (Maharashtra). Nat Prod Radian 4:46–50
Upadhyay B, Singh KP, Kumar A (2011) Ethno-veterinary uses and informants consensus factor of medicinal plants of Sariska region, Rajasthan, India. J Ethnopharmacol 133:14–25
Vartak VD (1981) Observation on wild edible plants from hilly regions of Maharashtra and Goa: Resume and Future Prospects. In: Jain SK (ed) Glimpses of Indian Ethnobotany. IBH, Oxford, pp 261–271
Vaidya RR, Dhumal KN (2004) Ethnomedicinal Plant Antidotes Used by Koli Tribe. In Ghate, V.S., Sane, Hema & Ranade, S.S. (Eds.) Focus on Sacred groves and Ethnobotany 170–173
Wabale AS, Petkar AS (2005) Ethnomedicinal plants used against Jaundice by the Tribals of Akole Taluka (M.S.). J Phytol Res 2:259–261
Waman MB (2005) Ecological Studies of Kalsubai, Harishchandragarh, Ratangarh Forest. Thesis in Botany. Faculty of Science, University of Pune, Pune (MS), India
Watt George (1889–1893) A Dictionary of Economic Products of India Vol. 1–6 Periodical Expert, Shahadara, New Delhi, India.
Yadav SR, Kamble MY (2008) Threatened ceropegias of the Western Ghats and strategies for their conservation. In: Rawat GS (ed) Special habitats and threatened plants of India. ENVIS bulletin: wildlife and protected areas, Vol. 11(1). Wildlife Institute of India, Dehradun, pp 123–134

Plants, Food, and Health: Some Untold Truths

Ægir B. Kristiansson

Introduction

The relation between food and health is obvious, for an absence of food leads to certain death. The caritative association named CCFD estimates 795 million personnel suffering from food deprivation. On the other side, the WHO estimates that more than 1.9 billion adults, 18 years and older, were overweight; of these over 600 million were obese (WHO 2017b). Food quality also affects health; for example, according to the WHO, 125,000 children under 5 years of age die every year because of food-borne disease (WHO 2017a).

Another obvious point is that most foods rely directly or indirectly on plants. Of course fruits, vegetables, and cereals directly come from plants. Plants also feed animals which are used for producing meat, milk, or eggs. Fungi are cultivated on vegetal substrate. It is actually harder to find foods that do not rely on plants; salt and water are minerals; some edible algae, such as *Ascophyllum* sp., *Laminaria* sp., and *Fucus* sp., are actually chromists and not plants.

Once the relation between plants, food, and health has been presented, it should be precise that food is also an important academic field. The Journal Citation Reports includes 86 journals with the word "food" in their title (Thomson Reuters 2016). For a comparison, "health" is found in 236 titles, 67 for "plant," 386 for "medical/medicine," 37 for "informatics," or 251 for "engineering". Food is also a business, with some big companies, for example, Nestlé with a market capitalisation of 239 billion dollars (compared to Apple, 604 billion dollars) and an annual revenue of 89 billion dollars (compared to Walmart, 378 billion dollars).

Food research is therefore a field with important health responsibilities and a huge financial interest. A scientist might face a dilemma between the financial interest, considering that issues with companies can be prejudicial for funding, and the

Æ. B. Kristiansson (✉)
Institute of Life and Environmental Sciences, University of Iceland, Reykjavík, Iceland
e-mail: gbl1@hi.is

M. Ozturk, K. R. Hakeem (eds.), *Plant and Human Health, Volume 1*,
https://doi.org/10.1007/978-3-319-93997-1_15

transcendental quest of the truth, which the scientific rigour researchers should seek. The aim of this chapter is not to pretend that all food companies are making an obscure business detrimental for health, with complicity of scientists. It exists in fair-behaving companies in food sciences, for example, Andros who is reputed in Europe for its regulation of the practice, by suing every other company when it performs visible frauds, and then, it has to ensure no trouble exists with its own business, in order to avoid revenge from all the concurrences. Maybe the last sentence should not have been written, as well as most of the coming chapter that will present untold truths that no food scientist would publicly acknowledge.

Vitamins and Nutriment, an Ocean of Bias

Some parts of the population are consuming organic food for personal health reason, while some other parts find their motivation in the taste (Phuong 2013). Those two aspects seem different, but their essence is actually related. A produce is considered healthier because of its higher concentration in healthy chemicals, such as vitamins and antioxidant, which are supposed to improve health. On the other side, the "taste" is related to the concentration of organoleptic compounds. The untold truth is that both are related to the same bias.

Leaving aside the organoleptic side on the problem, for instance, in order to focus on the nutritional side, dosage of vitamins such as ascorbic acid is done as a routine procedure in most of the laboratories studying fruits and vegetables (Kampfenkel et al. 1995). Then results, largely shared in non-scientific network, show that organic food are much healthier than conventionally produced vegetables. Critical reviews of the scientific studies are not as optimistic (Williams 2002). The untold truth is that the organic vegetables have a lower water content that conventionally produced vegetables. Therefore, all vitamins, minerals, and organoleptic compounds are in higher concentration on a fresh matter base, without having an observable difference if computed on a dry matter base. This magnifying glass effect is one of the more frequent biases in postharvest agronomy.

Philosophical question remains, which one is relevant? Fresh matter or dry matter-based nutriment concentration? Obviously, when dealing with vegetable storage, a dry matter base is advisable, for a withered vegetable will not have any commercial value event with a higher vitamin content. On the other side, when dealing with freshly harvested vegetables, eaten according to their fresh weight, the question is open.

Gustatory quality might be seen as a related issue, if we consider only the "taste", but might be more complex if some physical properties, such as texture, are considered. This point should be considered, but before we should detail if chemical concentration should be based on fresh or dry matter, if any is relevant.

Fresh or Dry Bias?

The question of fresh or dry weight base is not new; indeed, the first published plant water content (WC) experiment goes back to the early twentieth century and provided both bases (Livingston and Brown 1912). The dry-based WC is an analogy with geology, with a mass of water smaller compared to dry weight. When dealing with plants, mostly composed of water, a dry matter-based WC is always over 100% and does not look legitimate. Nowadays, in review the most used is fresh wet water content, e.g. Lamikanra and Shamaila (2005), but some scientists still use a dry matter-based WC, e.g. Hitmi et al. (2000). The latter scientists would try to convince you that a dry-based WC is much more informative that a fresh-based one. Even if you told or demonstrate them both as just bijective and none is more precise than the other is, they would not listen to you. The demonstration can be found in Box 1.

Box 1 Demonstration of Bijectivity Between Water Contents Expressed on a Fresh or Dry Base

Noting :

$WC_d = Z;\ \ WC_f = Y;\ \ m_f - m_d = a$ et $m_d = b.$

Then:

$$Y = \frac{a}{a+b} \quad \text{and} \quad Z = \frac{a}{b}$$

$$a = Zb$$

$$Y = \frac{Zb}{Zb+b}$$

$$Y = \frac{Z}{Z+1}$$

$\forall Z_1 \in \mathbb{R}^{+}$ and $\forall Z_2 \in \mathbb{R}^{+}$ with $Z_2 > Z_1$:

$$Y_2 - Y_1$$

$$= \frac{Z_2}{Z_2+1} - \frac{Z_1}{Z_1+1}$$

$$= \frac{Z_2(Z_1+1)+Z_1(Z_2+1)}{(Z_1+1)(Z_2+2)}$$

$$= \frac{Z_2 - Z_1}{(Z_1+1)(Z_2+2)} > 0$$

We can then write the conversion formula:

$$Y = \frac{a}{a+b}$$

$$Y = \frac{Zb}{Zb+b}$$

$$Y = \frac{Z}{Z+1}$$

$$WC_f = \frac{WC_d}{1+WC_d}$$

and in a similar way:

$$WC_d = \frac{WC_f}{1-WC_f}$$

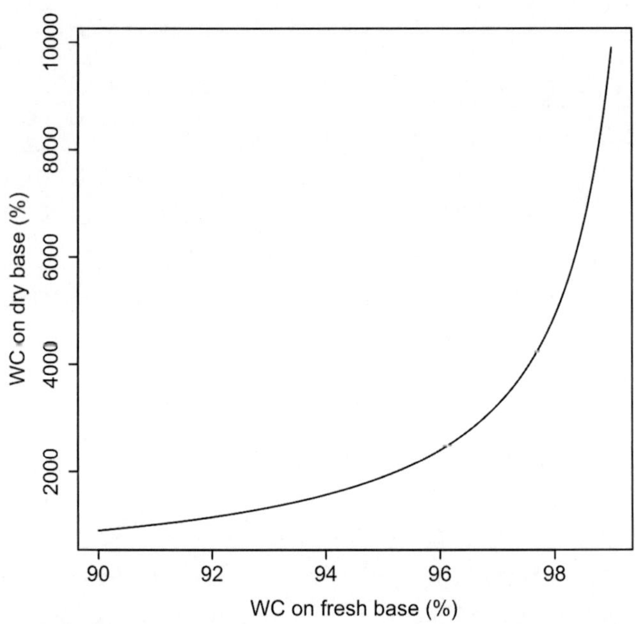

Fig. 1 Bijective relation between water content expressed on a fresh or dry matter base

For a simple conversion, Fig. 1 presents the relation between dry matter-based WC and fresh matter-based WC. Water content in itself can be considered as an old-fashioned measurement, as useless as cytogenetics in a world of molecular biology. Nevertheless, it can detect some untold issues about physiology, related to plants as

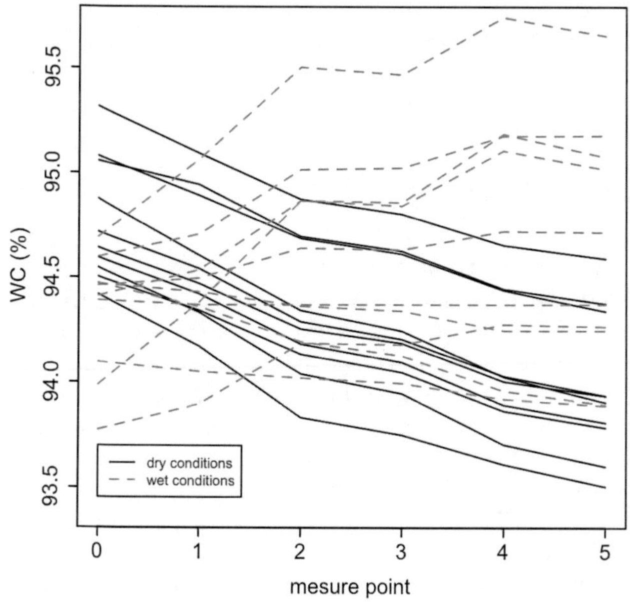

Fig. 2 Individually computed WC for single lettuces in dry condition (full black lines) or wet condition (dashed grey lines)

food, and health. A simple experiment, performed in a second-class team specialised in postharvest agronomy, outlines the profound bias in a normalisation on dry matter.

The objective was to make a kinetic of WC of lettuce (*Lactuca sativa* L.) in different humidity conditions over 3 days, with equal temperatures. Lettuces were weighted several times a day, and on the fourth day, they were put in a vacuum oven (104 °C) until stable weight. Then, considering dry matter loose through respiration insignificant, the WC at every point can be calculated, using the weight at that time. Results of the experiment are presented in the Fig. 2.

No one would be surprised that in dry conditions, WC decreased strongly, whereas in wet conditions, it was kept stable. What was not expected was that the water content at the initial point was statistically different. All the lettuces went out the same lot and should have similar WC on the starting point. The only possibility is that during the experiment, dry matter also changed and in a different way according to the treatment.

Out of being an epic fail, this experiment shown that (1) dry matter loss through respiration is not insignificant, (2) humidity influences respiration rates, and (3) biologist should not consider to a kinetic of plants' WC and ask their graduate student to make one. The dry matter of the plant can be globally divided in two groups of elements: (1) the structural elements and cell machinery and (2) the reserve of the cells (e.g. starch). Structural elements and cell machinery can be supposed as constant; on the other hand, reserves of the cells are variable. Therefore, concentration of any compound given on a dry matter base is not exempt of bias;

measures of biologic activity depend only on cell machinery and suffer the same bias. The untold truth is that actually there is not any reliable base to compare values of any concentration or activity in plants.

Assessing Gustative Quality

Gustative quality of a plant can be evaluated in several ways: physical parameters such as colour or texture can be measured in a reproducible way with adapted equipments more or less reliable. On the other hand, taste could be evaluated by some specific dosage of organoleptic compounds, depending on the species, usually expensive and rare. Few teams work in such a way, and this approach, out of the problem of variation of dry matter explained above, will not be detailed (considering potential confidentiality clauses author could face).

A more usual method is the sensorial approach. A jury composed of few trained judges evaluates the quality according to a scale foreseen in advance. Regular practice is a blind jury; evaluating alone the produces without any information on the sample, and rating of each judge, is used for giving the quality evaluation. Unfortunately, it is common that the "trained jury" is composed of the experimentalist, a lab technician, and the wife of the housekeeper, who received a 30-min formation on the quality of the produce by the experimentalist just before the rating. One who wants to blame this experimentalist should consider that his/her hierarchy could be blamed on him/her for taking some staff to help on such an experiment and that he/she could be forced to do it alone. The scale should have been created in a previous observation, before the true experiment, but in some case, the scale is created while doing the evaluation and is adapted during the course of experiment. Any potential reviewer would say such an approach would make the article rejected. Of course, during the redaction of any article, it would be presented as a trained blind jury evaluation.

Partial Results Presenting and Hidden Truth

Have you ever noticed in reading articles dealing with several vegetables that the presented results are not the same depending on the plant species authors are dealing with? An example can be found in some articles dealing with vegetable misting, e.g. Dieckmann et al. (1993). This research deals with five vegetables: field lettuce, head lettuce, broccoli, chicory, and carrots. They present ascorbic acid dosages only for two species, broccoli and head lettuce, and nothing about other species. For the two presented species, the results shown a better maintenance on ascorbic acid. Isn't this questionable?

Looking at another research about impact of misting on vegetables may bring some missing information. An experiment on radishes shows that misting delays senescence (Leduc 2011). Such a device maintains colour, chlorophyll concentration, and fresh aspect of the vegetables. Fresh aspect does not mean fresh product; indeed,

> **Box 2 Extract of EU Laws About Fresh Vegetables**
> Article 113(1)(b) and (c) of Regulation (EC) No 1234/2007 authorise the Commission to provide marketing standards for fruit and vegetables and processed fruit and vegetables, respectively. Pursuant to Article 113a(1) of that Regulation, fruit and vegetables, which are intended to be sold fresh to the consumer, may only be marketed if they are of sound, fair, and marketable quality and if the country of origin is indicated. To harmonise the implementation of that provision, it is appropriate to set out details of and provide for a general marketing standard for all fresh fruit and vegetables. Commission implementing regulation (eu) no 543/2011of 7 June 2011.

if chlorophyll is maintained, oxidative stress is maintained. If oxidative stress is maintained, ascorbic acid is oxidised and loses all its nutritional interest. The fresh-like aspect of a vegetable, associated with a lose of nutriments, is a fraud against the consumers, and is actually against laws in several countries, e.g. all countries of the European Union, as shown in Box 2.

The last visible element in any research paper presented such partial results in food science can be found in the "Acknowledgement" part of all these papers. They are funded, at least partially, by industrials, usually the one making benefit of the device or the process. From this observation, we can draw a few conclusions: (1) the food scientist working on such a subject has confidentiality clauses forbidding him to publish some results that could harm the company; (2) even in absence of such a clause, a food scientist publishing such a paper would never find any funding for future researches; and (3) defending the health of consumers is not a concern for (at least some) industrials who will prefer to hide detrimental information than having to drop a highly profitable device or process.

Those untold truths are only known by industrials and some scientists; therefore, such processes and devices can be used for a very long time, as long as no regulation agency will ever know the issues with them. Eventually, if the untold truth becomes known after few decades, when such processes and devices are widely used and consumers used to them, any action against them cannot be imagined. This is not the only perverse effect of private funding, out of promoting unfair processes and devices; it is detrimental for research that would improve global health, if they are detrimental for business.

Healthy Food Are on Horizon

Horizon is a virtual line where earth meets sky, by definition; it is impossible to reach it. Healthy foods are a prospect for optimistic food scientists, and many approaches are possible, with a very slow progress. Better than trying in vain to give an exhaustive list of all of them, detailed information on a case would be informative on the situation.

Table 1 Essential amino acid content of several cereals (g/100 g)

	Lymegrass	Maize	Rye	Barley	Soft red wheat
Source	(1)	(1)	(2)	(2)	(2)
Threonine	0.64	0.34	0.29	0.42	0.34
Valine	0.84	0.48	0.32	0.61	0.50
Methionine	0.25	0.27	0.15	0.24	0.17
Isoleucine	0.64	0.37	0.21	0.46	0.40
Leucine	1.27	1.35	0.56	0.85	0.76
Phenylalanine	1.00	0.60	0.43	0.7	0.50
Histidine	0.48	0.31	0.19	0.28	0.26
Lysine	0.62	0.20	0.29	0.46	0.31

Source: (1) Griffin and Rowlett (1981) (2) USDA (2011)

Perennial cereal is a perfect example of non-progressing field. As an illustration, a comparison of review articles over 20 years outlines no real new information (Wagoner 1990; Cox et al. 2002, 2006; Glover et al. 2010). Among perennial cereals, sea lymegrass (*Leymus arenarius*) is the one with the more available information. Lymegrass presents a nutritional interest; indeed, its grains show an interesting amount of essential amino acids (Table 1) and high concentration in oligoelements such as iron. It is impossible, for instance, to know if such high-protein content is specific to lymegrass or in general for all perennial grains nor to know if it is an artefact induced by lower yield.

An indirect health effect of perennial cereals comes from its environmental protection potential. Indeed, thanks to their extensive root system, irrigation and fertilisation needs are largely decreased. A single amendment using 500–600 Kg. ha^{-1} (20% phosphorus, 20% nitrogen, and 10% potassium) is enough for lymegrass cultivation (Wright 1994). This amendment is needed only during the first year, as well as the use of herbicide, for it is not needed when the plantation is settled. Considering a 5–10-year cycle, the use of pesticides is reduced accordingly. Out of the reduction of pollution, it favours biodiversity.

From a practical point of view, during cultivation of perennial cereals, during a 5–10-year period, only one ploughing is needed, only one amendment is needed, and only one herbicidal treatment is needed. From the second year, the only required step is the harvest. Then, carburant cost and carbon dioxide emission are lowered.

Why would such an appealing research field keep presenting so few progresses over time? The untold truth is that cultivation of perennial cereals only requests to seed once every 5–10 years; seed-producing companies would not accept to fund such a project, which will ultimately decrease their benefit. Phytosanitary companies would also suffer from profit drop if such crops were available. Private funding is therefore excluded. Public funding relies on project evaluation by reviewer of the field that are most of the time connected to private interest for their own projects. A reviewer in such a situation is more likely to oppose a funding, and without any funding the project is put aside and sometime never resumed.

Such a view of the system can seem dark, but it is actually the surface. A short view inside the system will show elements no sane scientist would believe.

What a Scientist Should Not Say

Food scientists are gathered in some highly specialised teams. Forgive me if I use the first person for describing untold elements I personally witnessed.

When collaborating with another research centre that had a related project with the same funding company, I was asked to gather previous data from that team; that is fine. I was then asked to use those data in order to validate our system; it is potentially acceptable as long as it is internal. Unfortunately, the different method of data acquisition made them incompatible. Then, I was asked to run the dataset through a model I designed, in order to obtain theoretical values that I was asked to consider as my own experimental values. That is not fine; it is both a stilling and a faking of data. I refused and was blamed for that. Untold truth is that some teams do not hesitate to publish imaginary results.

Another example of misbehaving scientist was a case when I sent a document to my superior for a proofreading before sending to the company requesting the experiments. Once the document was back, out of some typo and rephrasing, some experimental values had been edited. It is not certain if the aim was to fraud or just a mark of incompetence; anyway both options are not acceptable by scientists.

Facing such situation, I eventually signalled it to the scientific authorities of the university this team belonged to. The answer I received was I was harming the reputation of the university, and I should stop that. The untold truth is that a university might prefer to cover up misbehaving of a team in order to protect its reputation. In such cases, grad students are considered as fuses, and no one cares if in such a team, there is a grad burnt out per year. Who would have trusted that food science could have an effect on mental health?

Fresh Healthy Information for Fresh Healthy Diet

A good bibliography is like a wedding: it needs something new, something old, something given, and something blue (the blue element is open to interpretation). It is important to have something recent to prove that we are dealing with the state of the art of current research. Old papers are important in order to prove that we are looking at the bases and sources of information; furthermore when dealing with ethnobotanic, some very old observations on plant uses that might have been lost since can be precious. The given reference proves you have been discussing and obtaining an opening and the blue reference will avoid feeling such a colour. Another important point is that every information gets a source and the original one. Several researchers might have observed the same phenomenon; in such case several references should be provided including recent ones to show that it is still up to date and old ones to show that it is something known for a long time.

Food scientist does not observe such rules. Unquoted references are common, as an example after having written that high humidity favour wound healing of vegetables, information often found without any references (e.g. Gaffney et al. (1985)), the only relevant paper (I could find) that indeed observed it was Artschwager (1931). The reviewer has not appreciated such a quote, writing "Nothing newer on the subject?" Several philosophical questions can be asked: Should we drop an information accepted by many because there is no acceptable reference for it? In such a case, bibliography is incomplete. Should we give the information without any source? In such a case, nothing proves we are not making up information. Should we quote the paper where we read that information? Then, we attribute to someone an observation that is not his own. In addition, should we consider that the technical progress made in few decades has changed the way sweet potatoes react to humidity? Out of specific cases, such as Manns (1920), observing a microbial decay, related to confined air, warmed up by plant physiology, there is no reason to reject old observations as long as there is no evidence to reject them.

In such a field, all papers, out of the material and methods, should be less than 5 years old or up to 10 for very important papers. The young food scientist, who gave me this advice while I was looking at some 20-year-old articles, had a bad surprise a few weeks after, discovered an article on the same subject as that of his last published one, using a not so different method on the same vegetable, which had obtained the same results, and drew the same conclusion. The paper was a little bit more than 10 years old, so he had not read it before someone sent it to him (Do not forget the importance of the given reference.). Fortunately, for him, the reviewers were also ignoring few-years-old papers.

Attributing a given work to its original author seems also facultative in food science. In order to discover a new field, review articles and books are supposed to give reliable information with references to the original work and references to book and are acceptable for introductions of research articles but highly questionable for other books or review articles. While reading a book entitled *Water Activity in Foods* (Barbosa-Cánovas et al. 2007), looking for references, alert scientist will notice references to several chapters of another book entitled *Water Activity and Food* (Troller and Christian 1978), which was published 30 years earlier. Dedicated scientists would maybe consider reading this book; unfortunately, it would be regarded as a waste of time by his team, and availability of such a book might be an issue.

Conclusions

The examples given in this chapter are only provided in order to present some dysfunctions in the food science research and their consequences on health. Many others could be found if the barrier of confidentiality could be crossed. They just allow presenting a quick image of the field.

In a first instance, anyone should remind that a food company's first aim is to make money. The only case they will care about health is when they can use it in order to increase sales and benefit. Whenever a produce is developed, if it has a potential for high benefit, and some detrimental effects on health, which are known only by the producer, he will prefer to hide or deny them.

In a second instance, anyone reading an academic paper about food science should be aware that such an article would only present positive points.

A short reading of the Acknowledgement section of these articles will show that producing companies are funding the research; therefore the scientists have an interest conflict and cannot affirm anything that would be detrimental for the funding providers.

In a third instance, no food scientist would break the silence about such a practice. There are several points that many scientists are aware of a some simple fact, such as the fact that no process of storage at room temperature can maintain vitamin level of vegetable or that maintaining the green colour of leafy vegetables at room temperature is enhancing the degradation of antioxidant compounds. We can add that improving the yield is usually detrimental to the quality (in terms of taste or nutritional interest). If fraud control organisations knew these simple points, many companies would turn out of business, and scientist would lose research projects.

In a fourth instance, a project that could be a healthy solution for agriculture is most likely to keep unfunded. Any funding by private organisations is unthinkable, for it would attack their business. Funding by public research fund will pass by a reviewing process. The reviewers are likely to have interests with some concurrent companies or to have an antagonist interest if the given research goes against his own work.

Eventually, the advice to conclude this chapter would be a quote from *Candide* of Voltaire: “we must cultivate our garden”. However, this quote is not to be taken in the philosophical sense of the satire but really literally. Cultivating your own crops will favour biodiversity, will guaranty short time between harvest and consummation (and higher antioxidant content), and will avoid the vegetables to pass through questionable process.

References

Artschwager E (1931) Suberization and wound-periderm formation in sweet potato and gladiolus is affected by temperature and relative humidity. J Agr Res 43:353–364

Barbosa-Cánovas GV, Fontana AJ Jr, Schmidt SJ, Labuza TP (2007) Water activity in foods: fundamentals and applications, 1st edn. IFT Press, Blackwell Pub, Hoboken, NJ OCLC: ocm84838771

Cox T, Bender M, Picone C, Van Tassel D, Holland J, Brummer E, Zoeller B, Paterson A, Jackson W (2002) Breeding perennial grain crops. Crit Rev Plant Sci 21:59–92

Cox T, Glover J, Van Tassel D, Cox C, DeHaan L (2006) Prospects for developing perennial grain crops. Bioscience 56:649–659

Dieckmann A, List D, Zache U (1993) Cold water mist humidification to preserve the quality of fresh vegetables during retail sale. Lebensm Wiss U Technol 26:340–346

Gaffney JJ, Baird CD, Chau KV (1985) Influence of airflow rate, respiration, evaporative cooling, and other factors affecting weight loss calculations for fruits and vegetables. ASHRAE Trans 91:690–707

Glover J, Reganold J, Bell L, Borevitz J, Brummer E, Buckler E, Cox C, Cox T, Crews T, Culman S et al (2010) Increased food and ecosystem security via perennial grains. Science 328:1638–1639

Griffin L, Rowlett R (1981) A "lost" Viking cereal grain. J Ethnobiol 1:200–207
Hitmi A, Coudret A, Barthomeuf C, Sallanon H (2000) Role of intracellular water retention strength in freezing tolerance of *Chrysanthemum cinerariaefolium* vis cell cultures. J Plant Physiol 157:47–53
Kampfenkel K, Vanmontagu M, Inzé D (1995) Extraction and determination of ascorbate and dehydroascorbate from plant tissue. Anal Biochem 225:165–167
Lamikanra O, Shamaila M (2005) Water and its relation to fresh produce. In: Produce degradation: pathways and prevention. Taylor & Francis, Boca Raton, pp 267–291
Leduc B (2011) Ultrasonic misting influence on slowing senescence and improving quality retention in radish leaves during simulated retail storage (PDF download available). In: IFT Annual Meeting & Food Expo https://www.researchgate.net/publication/232185085_Ultrasonic_misting_influence_on_slowing_senescence_and_improving_quality_retention_in_radish_leaves_during_simulated_retail_storage Online version by editor were here: http://www.ift.org/Meetings-and-Events/Past-Meeting-Resources/Technical%20Abstract%20Search%20Details.aspx?id=52600
Livingston BE, Brown WH (1912) Relation of the daily march of transpiration to variations in the water content of foliage leaves. Bot Gaz 53:309–330
Manns TF (1920) Sweet potato storage in Delaware. Del Agr Expt Sta Bull 127:1–64
Phuong NT (2013) Consumers' perceptions of organic food in Australia and other countries: a review. J Agric Econ Dev 2:44–54
Thomson Reuters (2016) Journals in the 2016 release of Journal Citation Reports™. http://scientific.thomsonreuters.com/imgblast/JCRFullCovlist-2016.pdf
Troller JA, Christian JHB (1978) Water activity and food. Academic Press, New York OCLC: 683187598
USDA (2011) Cereal grains and pasta. In: Technical Report 20. USDA
Wagoner P (1990) Perennial grain development: past efforts and potential for the future. Crit Rev Plant Sci 9:381–408
WHO (2017a) WHO | food safety. http://www.who.int/mediacentre/factsheets/fs399/en/
WHO (2017b) WHO | obesity and overweight. http://www.who.int/mediacentre/factsheets/fs311/en/
Williams CM (2002) Nutritional quality of organic food: shades of grey or shades of green? P Nutr Soc 61:19–24
Wright S (1994) Beach wildrye: planting guide for Alaska. Alaska Dept. of Natural Resources. Division of Agriculture. Plant Materials Center, Palmer, AK

Potential Medicinal Plants Used in the Hypertension in Turkey, Pakistan, and Malaysia

Munir Ozturk, Volkan Altay, Abdul Latiff, Samreen Shareef, Farzana Shaheen, and M. Iqbal Choudhry

Introduction

In the industrialised nations, hypertension is recorded as the fourth contributor in the premature deaths, but in the industrialising nations, it is number 7 in the list (Deepa et al. 2003). The reports published by Reddy (1996) and Nissinen et al. (1988) have fully stressed the fact that the prevalence of hypertension is rapidly increasing in industrialising countries and is one of the leading causes of death and disability in industrialised countries. According to Kearney et al. (2005), Pradeepa and Mohan (2008), and Mittal and Singh (2010) in the year 2000, nearly 1 billion adults have been recorded as suffering from hypertension. The number is predicted to go up to 1.56 billion by 2025.

The effective richness of local traditional knowledge in the treatment of hypertension has been demonstrated in many countries (Heinrich 2000; Houghton et al. 2007; Joubert et al. 2008; Eddouks and Zeggwagh 2012). The studies carried out in these countries have provided an exhaustive list of herbals used as remedies in hypertension. Even in vivo and in vitro pharmacological studies have confirmed the effective hypotensive effect of herbals and the natural products derived from these

M. Ozturk (✉)
Vice President of the Islamic World, Academy of Sciences, Amann, Jordan

Department of Botany and Centre for Environmental Studies, Ege University, Izmir, Izmir, Turkey

V. Altay
Faculty of Science and Arts, Department of Biology, Hatay Mustafa Kemal University, Hatay, Turkey

A. Latiff
Faculty of Science and Technology, Universiti Kebangsaan Malaysia, Selangor, Malaysia

S. Shareef · F. Shaheen · M. Iqbal Choudhry
International Center for Chemical and Biological Sciences, University of Karachi, Karachi, Pakistan

M. Ozturk, K. R. Hakeem (eds.), *Plant and Human Health, Volume 1*,
https://doi.org/10.1007/978-3-319-93997-1_16

(Ziyyat et al. 1997; Kalus et al. 2000; Khayyal et al. 2002; Al-Qattan et al. 2003; El Bardai et al. 2003; Eddouks et al. 2002, 2005; Tahraoui et al. 2007; Eddouks and Zeggwagh 2012).

The traditional use of plant-derived preparations indicates a higher probability that the traditional practitioners found the remedy to be effective (Marles and Farnsworth 1995). In view of this, the studies on the traditional antihypertensive plants can prove very helpful. The discovery of locally available plant-derived medicine to treat hypertension in developing countries can prove highly fruitful. The commercial development of such botanical hypertensive agents and adjuncts to antihypertensive therapy will be highly beneficial (Marles and Farnsworth 1995). The objective in the preparation of this chapter has been to analyse the distribution of the plants used traditionally in the treatment of hypertension in Turkey, Pakistan, and Malaysia.

Study Areas

Turkey is located in the West Asia connecting Asia and Europe in the northern hemisphere, regarded as a meeting place of different phytogeographical regions and a meeting place for two centres of diversity and origin: the Near East and the Mediterranean (Fig. 1). It lies in the subtropical zone between 36–42° latitudes N and 26–45° longitude E, with an area of 783.562 km^2. The topography shows significant variations and ecological features change much over very short distances, exhibiting great variations in plant diversity with different types of ecosystems, occupying different habitats, and showing different vegetational characteristics (Ozturk et al. 2006, 2012a, b 2017a, b; Adak 2017). More than 11.000 flowering plant taxa are reported to show distribution in the country. This number is very near to that recorded from the European continent (Güner et al. 2012). The varying geomorphological, topographical, and climatic characteristics are responsible for the

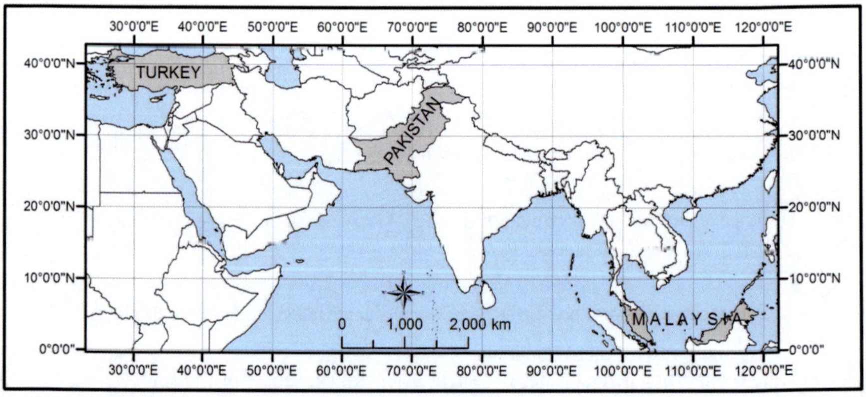

Fig. 1 The map showing the countries included in this study

varying habitats and plant diversity as well as endemism. The number of endemic taxa is reported to lie around 3035, comprising 31.12% of the total flora and confined to narrow and restricted ecological niches (Güner et al. 2012; Ozturk et al. 2016).

Pakistan has an area of 79.61 million ha. It is located at the western edge of South Asian subcontinent. The altitudinal variations are wide ranging from sea level to 8.000 m, and it extends about 1700 km northwards towards the great mountain chains of Hindukush, Himalayas, and Karakorum, starting from the Arabian Sea coast and the mouth of Indus near the Tropic of Cancer (Ahmad 2007a). The country in general experiences arid and semiarid climates with diverse plant cover ranging from subtropical to temperate and alpine forests. Average annual rainfall varies from 50 mm in arid and semiarid areas to 2.000 mm in moist forests, whereas the temperatures vary from below freezing in northern mountains during winter to 35–50 °C during summer in central and southern plains (Ahmad 2007a). The flora is characterised by Palaearctic and Indo-Malayan elements. The country possesses a unique biodiversity. There are nine major ecological zones within a peculiar geographical position (Fig. 1). The variations in the geographical position, altitude, soil, and climate are responsible for its rich floristic diversity. There are about 6000 species of higher plants, and out of these, 400 are endemic to the country (Nasir and Ali 1970–1995; Ahmad 2007a).

Malaysia lies just in the north of equator with a total area of 329.847 km^2 (Fig. 1). It experiences a warm equatorial climate, being hot and humid all around the year. There is no seasonality. The country is divided into two regions of similar size by the South China Sea. Part one is Peninsular Malaysia connected to mainland Asia on the western half. Part two covers the states of Sabah and Sarawak on the island of Borneo in the east (Saw and Chung 2015). The country is rich in biological resources, with a unique heritage of tropical rainforests. The latter has evolved over millions of years, possesses varying flora and fauna. There are over 14,500 species of flowering plants belonging to nearly 1500 genera. The number of tree species crosses the level of 2500. There are 800 species of orchids, 500 species of ferns, and 60 species of grasses and bamboos (www.fao.org).

Comparative Evaluation on Country Basis

Turkey

The studies undertaken on the use of plants in the traditional medicine have revealed that nearly 132 taxa grouped under 39 families and 92 genera have been evaluated in the treatment of hypertension. An alphabetical list of these taxa with their botanical name, part used, and information on the preparations used has been presented here in Appendix 1. The families represented by the largest number of taxa are Lamiaceae (30), Rosaceae (22), Asteraceae and Apiaceae (10 each), and Fabaceae

Table 1 The parts of herbs used in three countries and the number of taxa

Part used	Code	Turkey	Pakistan	Malaysia
Leaves	L	45	25	31
Aerial parts	AP	40	5	1
Fruit	FR	34	13	10
Flower	FL	17	3	3
Root	R	4	12	20
Seed	SE	11	7	2
Shoot	SH	4	2	2
Stem	ST	2	5	2
Cones	CO	2	–	–
Tuber	T	2	2	–
Whole plants	WP	7	23	11
Bulb	B	3	4	1
Branches	BR	1	–	–
Buds	BD	1	–	–
Rhizome	RH	–	2	–
Oil	OO	1	–	–
Resin	RS	1	–	–
Underground parts	UP	3	–	–
Pedicel	PE	2	–	–
Pith	PT	–	–	1
Corn silk	CS	–	–	1

and Polygonaceae (5 each), whereas the genera with highest number of taxa are *Thymus* (10 taxa), *Crataegus* (6 taxa), and *Rumex*, *Rubus*, and *Urtica* (4 taxa each) (Appendix 1). *Allium sativum*, *Viscum album* ssp. *album*, *Urtica dioica*, and *Olea europaea* have been recorded as the most commonly used taxa (Appendix 1). The leaves have been used from 45 taxa, aerial parts from 40 taxa, fruits from 34 taxa, and flowers from 17 taxa (Table 1). The preparations used include infusion (51 taxa), followed by decoction (46 taxa), fresh (15 taxa), and raw (12 taxa).

Pakistan

In all 86 taxa from 41 families and 76 genera have been evaluated in the herbal preparations for treating hypertension in the traditional medicine of the country. An alphabetical list is given here together with their botanical names, parts used, ailments treated, and information on the preparations used (Appendix 2). The largest number of taxa has been recorded from the families Asteraceae (9), Fabaceae (9), Amaranthaceae (4), and Amaryllidaceae, Apiaceae, Apocynaceae, Cucurbitaceae, Lamiaceae, Papaveraceae, Poaceae, Solanaceae, and Zygophyllaceae (3 each). *Allium* (3 taxa) and *Boerhavia*, *Caralluma*, *Erigeron*, *Fagonia*, *Fumaria*, *Sonchus*,

Table 2 The preparations used in three countries together with the number of taxa

Preparations	Code	Turkey	Pakistan	Malaysia
Decoction	DE	46	10	34
Infusion	IN	51	4	3
Eaten fresh	EF	15	–	6
Raw	R	12	–	1
Cooked	C	3	2	2
Boiled	BO	4	–	3
Crushed	CR	–	1	–
Dried	DR	3	–	–
Powdered	PW	2	1	–
Poultice	PU	1	–	–
Fried	FI	–	1	–
Pounded	PN	3	–	3
As tea	AT	1	–	6
Extract	EX	–	12	–
Juice	JU	4	4	4

Xanthium, and *Ziziphus* (2 taxa each) are the genera used with highest number of taxa (Appendix 2). *Allium sativum*, *Cannabis sativa*, *Ajuga integrifolia*, *Caralluma edulis*, *Taraxacum campylodes*, *Heliotropium strigosum*, and *Tribulus terrestris* are the most commonly used taxa (Appendix 2).

The numbers used on the basis of parts are as follows: leaves (25 taxa), whole plants (23 taxa), fruits (13 taxa), and root (12 taxa) (Table 1), the most common preparations being extract (12 taxa), followed by decoction (10 taxa), infusion (4 taxa), and juice (4 taxa) (Table 2).

Malaysia

A total of 61 taxa of medicinal plants belonging to 34 families and 53 genera are evaluated for treating hypertension in the traditional medicine. These are presented alphabetically in Appendix 3, which includes botanical name, part used, and information on the preparations used. Acanthaceae and Phyllanthaceae (4 each) and Apocynaceae, Asteraceae, Fabaceae, Lamiaceae, Malvaceae, and Oxalidaceae (3 each) are the families used at maximum level. *Phyllanthus* (3 taxa) and *Averrhoa*, *Blumea*, *Hibiscus*, *Morinda*, *Rauvolfia*, and *Tacca* (2 taxa each) are the genera with the highest number of taxa. *Andrographis paniculata*, *Centella asiatica*, *Eurycoma longifolia*, *Gynura procumbens*, *Solanum torvum, and Tetracera indica* are the commonly used taxa. The numbers of taxa on the basis of parts used are leaves (31 taxa), root (20 taxa), whole plants (11 taxa), and fruits (10 taxa) (Table 1). The decoction (34 taxa), followed by boiled (13 taxa), and fresh (6 taxa) are the most common preparations used.

Table 3 The medicinal plant taxa used in three countries for the treatment of hypertension

Taxa	Turkey	Pakistan	Malaysia
Allium cepa	x	x	
Allium sativum	x	x	x
Avena sativa	x	x	
Coriandrum sativum	x	x	
Foeniculum vulgare	x	x	
Fumaria officinalis	x	x	
Momordica charantia		x	x
Ocimum basilicum	x	x	
Phyllanthus acidus		x	x
Punica granatum	x		x
Tribulus terrestris	x	x	
Urtica dioica	x	x	
Ziziphus jujuba	x	x	

Table 4 Jaccard similarity index related to hypertension in three countries

	Turkey-Pakistan	Pakistan-Malaysia	Turkey-Malaysia
Hypertension disease group	9.17%	4.08%	2.07%

The distribution of taxa from three investigated countries shows that the number used in Turkey is 132 taxa; it is only 86 taxa in Pakistan and 61 taxa in Malaysia. Between Turkey-Pakistan ten taxa are used commonly, between Pakistan and Malaysia only three taxa, and between Turkey and Malaysia only two taxa (Table 3). A comparison reveals that in all these countries, *Allium sativum* is the only taxon commonly used for treatment of hypertension (Table 3).

A perusal of the data presented above reveals that Jaccard similarity index can be applied to calculate the similarity ratios of the herbal preparations used for hypertension (Table 4) in three countries (González-Tejero et al. 2008).

Conclusions

The herbal preparations have become a potential source of therapeutic aids and are playing a significant role in global healthcare systems for all living beings, not only during diseased conditions but also as potential material for maintaining proper health. A major factor acting as a barrier in the development of herb-based industries in less-developed countries seems to be the lack of information on both social and economic benefits that could be derived from the industrial utilisation of herbal drugs. Except for the use of herbs for local healthcare needs, not much information is available on their market potential and trading possibilities; therefore, governments as well as entrepreneurs have not exploited the real potential of the herbal preparations (Verma and Singh 2008).

Determination of the biological activity of herbs used in traditional medicine is of great help to the rural populations and in informal settlements (Verma and Singh 2008). Several studies are being carried out these days to isolate the active compounds by bioassay-guided fractionation from the herbs showing high biological activity during screening. In view of this, such scientific investigations can be evaluated to develop drugs for different diseases. There is a greater need for further research to isolate the compounds responsible for the observed biological activity (Verma and Singh 2008).

Instead of common ethnobotanical studies, a set of similar practices in the preparation, administration of the herbal preparations, or the ailments treated with plants will enlighten the common heritage in different regions (González-Tejero et al. 2008). Our investigations constitute the first comparative study performed with ethnobotanical data gathered in Turkey, Pakistan, and Malaysia.

González-Tejero et al. (2008) have prepared an exhaustive list for the species indicating the areas where each plant was mentioned. It underlines the ethnobotanical richness of the countries under question. However, there is a need for broadening such investigations to other areas in Asia as well as other parts of the world. This will constitute a base for future phytochemical/pharmacological investigations leading to new therapeutic products.

Acknowledgments First author of this chapter would like to express his deepest thanks to the Rectorate of Universiti Putra Malaysia, Selangor, Malaysia, and Director of ICCBS, University of Karachi, Pakistan, for their short-term appointments at their campuses.

Appendix 1: Herbs used for hypertension in the traditional medicine in Turkey

	Familia/Taxa	Parts used	Preparation	References[a]
	Amaranthaceae			
1	*Beta vulgaris* var. *altissima*	T		2
	Amaryllidaceae			
2	*Allium cepa*	B, L	R	2, 51
3	*Allium sativum*	B, ST	C, EF, DR, R	1–8, 27, 35, 40, 49, 52–54
4	*Allium scorodoprasum* ssp. *rotundum*	B	EF, DR	9
	Anacardiaceae			
5	*Rhus coriaria*	L, FR	DE, R, IN	2, 33, 54, 55
	Apiaceae			
6	*Anethum graveolens*	AP		2
7	*Coriandrum sativum*	FR, SE	R	2, 56
8	*Cuminum cyminum*			25

(continued)

	Familia/Taxa	Parts used	Preparation	References[a]
9	*Daucus carota*	R		2
10	*Diplotaenia cachrydifolia*	AP	DE	57
11	*Foeniculum vulgare*			14
12	*Oenanthe pimpinelloides*	AP	DE	18
13	*Petroselinum crispum*	AP	R	2, 58
14	*Pimpinella anisum*	FR	IN	21
15	*Prangos ferulacea*	ST	C	38
	Apocynaceae			
16	*Vinca herbacea*	AP	IN	59
	Arecaceae			
17	*Phoenix dactylifera*	FR	DR	2
	Asteraceae			
18	*Anthemis pseudocotula*	L, FL	DE	1
19	*Artemisia absinthium*	L	DE	2, 27
20	*Cichorium intybus*	AP	BO, DE, IN	11, 60, 61
21	*Cnicus benedictus*	L, FL	DE	12, 13
22	*Gundelia tournefortii*	SH		2
23	*Helichrysum compactum*	L	DE	28
24	*Lactuca sativa*	L	EF	2
25	*Matricaria* sp.	FL	DE	2
26	*Scorzonera semicana*	AP, L	R	12, 13, 41
27	*Tripleurospermum oreades* var. *oreades*	AP	DE	38
	Brassicaceae			
28	*Capsella bursa-pastoris*	L	DE	47
29	*Nasturtium officinale*	AP	C	32, 33
30	*Raphanus sativus*	SE		2
	Cannabaceae			
31	*Celtis australis*	L	DE	1
	Caprifoliaceae			
32	*Valeriana dioscoridis*	FL	IN	32
	Convolvulaceae			
33	*Convolvulus betonicifolius* ssp. *peduncularis*	AP	IN	55
	Cornaceae			
34	*Cornus mas*	FR	DE	35
	Cucurbitaceae			
35	*Bryonia multiflora*	FR	R	57
	Cupressaceae			
36	*Juniperus foetidissima*	CO, RS	DE, PW	40, 62
	Dioscoreaceae			
37	*Dioscorea communis*	AP	BO	60
	Elaeagnaceae			
38	*Elaeagnus angustifolia*	FL, FR		2

(continued)

	Familia/Taxa	Parts used	Preparation	References[a]
	Equisetaceae			
39	*Equisetum arvense*	AP	IN	22, 60
	Ericaceae			
40	*Arbutus unedo*	L	IN	1
41	*Erica manipuliflora*	L, FL	DE	1
	Fabaceae			
42	*Astragalus angustifolius* ssp. *angustifolius*	FL	IN	10
43	*Glycyrrhiza glabra* var. *glabra*	UP	PN	20, 26
44	*Glycyrrhiza glabra* var. *glandulifera*	UP	PN	16
45	*Phaseolus vulgaris*	SE		2
46	*Trifolium arvense*	AP		2
	Hypericaceae			
47	*Hypericum perforatum*	AP	DE, IN	3, 63
	Lamiaceae			
48	*Lavandula pedunculata* ssp. *cariensis*	L, FL	IN	1
49	*Lavandula stoechas* ssp. *stoechas*	L, FL	DE, IN	1, 2, 15, 19, 30, 35
50	*Mentha x piperita*	L	IN	1
51	*Mentha pulegium*	L		2
52	*Ocimum basilicum*	L	R	11
53	*Origanum onites*	AP, L, FL	DE, IN	1, 30, 60
54	*Origanum vulgare*	AP, L	DE, IN	9, 55
55	*Rosmarinus officinalis*	AP, L, FL	DE, IN	2, 19, 27
56	*Salvia fruticosa*	L	IN	2
57	*Satureja hortensis*	L	IN	55
58	*Satureja spicigera*	L	IN	54
59	*Sideritis lanata*	AP	IN	64
60	*Sideritis perfoliata*	AP	IN	42
61	*Stachys cretica* ssp. *mersinaea*	AP	IN	60
62	*Teucrium chamaedrys*	AP	IN	2
63	*Teucrium polium*	AP	IN	3, 65
64	*Thymbra capitata*			14
65	*Thymus* sp.	AP	IN	2, 3
66	*Thymus cilicicus*	L	IN	25
67	*Thymus fallax*	AP	DE, IN	9
68	*Thymus migricus*	AP	IN	9
69	*Thymus kotschyanus* var. *kotschyanus*	AP	DE, IN	9
70	*Thymus longicaulis*	AP	IN	22
71	*Thymus praecox* ssp. *grossheimii*	AP	IN	9
72	*Thymus sipyleus*	AP	IN	64

(continued)

	Familia/Taxa	Parts used	Preparation	References[a]
73	*Thymus transcaucasicus*	AP, WP	IN	9, 43
74	*Thymus zygioides*	AP	DE	44
75	*Vitex agnus-castus*	L		51
76	*Ziziphora capitata*	AP	DE	53
77	*Ziziphora taurica* ssp. *taurica*	AP, FL	IN	9, 36
	Lythraceae			
78	*Punica granatum*	FR	JU	2, 51
	Malvaceae			
79	*Althea officinalis*	FL		2
80	*Hibiscus* sp.	FL	IN	29
81	*Malva* sp.	L	DE	31
82	*Malva sylvestris*	AP	IN	25
	Myrtaceae			
83	*Myrtus communis* ssp. *communis*	L	DE	1
	Nitrariaceae			
84	*Peganum harmala*	AP, SE	DE	2, 60
	Oleaceae			
85	*Olea europaea* var. *europaea*	BD, FR, L, OO, SH	DE, IN	1, 2, 14, 15, 30, 34, 36, 40, 51, 53, 60, 63, 65
	Papaveraceae			
86	*Fumaria officinalis*	AP, WP	DE, IN	1–3, 21, 22, 27, 35
87	*Papaver bracteatum*	SE		57
	Pinaceae			
88	*Pinus nigra* ssp. *pallasiana*	CO	DE	66
	Plantaginaceae			
89	*Plantago major* ssp. *major*	L	DE, IN	11, 37, 50
	Poaceae			
90	*Avena sativa*	FR		2
91	*Oryza sativa*	FR		2
	Polygonaceae			
92	*Rheum ribes*	R, SH	DE, PN	2, 9
93	*Rumex acetosella*	AP, L	BO, R	54, 58
94	*Rumex crispus*	L	BO	58
95	*Rumex scutatus*	AP, L	DE	11, 38
96	*Rumex tuberosus*	L	IN, R	33, 58
	Rhamnaceae			
97	*Ziziphus jujuba*	FR	EF	15
	Rosaceae			
98	*Alchemilla sintenisii*	L	IN	1
99	*Cerasus avium*	PE	DE	2
100	*Cerasus vulgaris*	PE	DE	2
101	*Crataegus* spp.	L, FL, FR, SE	IN	1, 2, 15

(continued)

	Familia/Taxa	Parts used	Preparation	References[a]
102	*Crataegus azarolus*	FL, FR	DE, EF, IN	11, 16, 53
103	*Crataegus monogyna* var. *monogyna*	L, FL, FR, R	DE, EF, IN	12, 17–23, 40
104	*Crataegus orientalis* ssp. *orientalis*	FR	EF	7
105	*Crataegus orientalis* ssp. *szovitsii*	FR	EF	24
106	*Crataegus tanacetifolia*	FR	AT	58
107	*Cydonia oblonga*	L, SE	IN	19, 48
108	*Laurocerasus officinalis*	L	R	54
109	*Malus pumila*	FR	EF	58
110	*Mespilus germanica*	FR, L	DE, R	1, 54
111	*Prunus divaricata*	FR	EF	35, 65
112	*Prunus spinosa*	FR, SE	DE	15, 19
113	*Pyrus elaeagnifolia* ssp. *elaeagnifolia*	FR	IN	67
114	*Rosa canina*	FR, SE	EF	1, 2, 15
115	*Rubus caesius*	FR	JU	58
116	*Rubus canescens*	FR, R, SH, UP	DE, JU	9, 19, 25, 39, 58
117	*Rubus sanctus*	FR	EF	1, 68
118	*Rubus saxatilis*	FR	JU	58
119	*Sorbus aucuparia*	FR	EF	35
	Rutaceae			
120	*Citrus aurantium*	FR		2
121	*Citrus limon*	FR	EF, IN	2, 70
122	*Citrus sinensis*	FR		2
	Santalaceae			
123	*Viscum album* ssp. *abietis*	L	IN	59, 65
124	*Viscum album* ssp. *album*	BR, FR, L, WP	DE, IN	1, 2, 6, 7, 10, 15, 17, 18, 21, 22, 27, 29, 37, 46
125	*Viscum album* ssp. *austriacum*	L		44
	Solanaceae			
126	*Solanum tuberosum*	T		2
	Urticaceae			
127	*Urtica dioica*	AP, L, SE, WP	DE, EF, IN, PU, PW	1, 2, 4, 9, 11, 13, 22, 27, 31, 37, 38, 45, 51, 69
128	*Urtica membranacea*	WP	DE	1
129	*Urtica pilulifera*	WP	DE	1
130	*Urtica urens*	WP	DE	1
	Vitaceae			
131	*Vitis vinifera*	L		2
	Zygophyllaceae			
132	*Tribulus terrestris*	AP, FR, L, SE	DE, IN	1, 36, 53, 55

(continued)

[a]References: 1: (Tuzlacı 2006); 2: (Yücecan et al. 1988); 3: (Akaydın et al. 2013); 4: (Çakılcıoğlu et al. 2010); 5: (Çakılcıoğlu and Türkoğlu 2009); 6: (Ezer and Arısan 2006); 7: (Öztürk and Dinç 2005); 8: (Ezer and Avcı 2004); 9: (Altundag and Ozturk 2011); 10: (Ertuğ et al. 2004); 11: (Tetik et al. 2013); 12: (Çakılcıoğlu et al. 2007); 13: (Çakilcioglu and Turkoglu 2010); 14: (Bulut and Tuzlacı 2005); 15: (Sarı et al. 2010); 16: (Doğan and Bağcı 2011); 17: (Akan et al. 2005a); 18: (Genç and Özhatay 2006); 19: (Everest and Ozturk 2005); 20: (Akan et al. 2005b); 21: (Baytop 1984); 22: (Şanlı 2006); 23: (Savran et al. 2009); 24: (Sarper et al. 2009); 25: (Özçelik and Balabanlı 2005); 26: (Akan et al. 2013); 27: (Kultur 2007); 28: (Kargıoğlu et al. 2010); 29: (Çömlekçioğlu and Karaman 2008); 30: (Gürdal and Kültür 2013); 31: (Şimşek et al. 2004); 32: (Demirci and Özhatay 2012); 33: (Polat et al. 2012); 34: (Kıran 2006); 35: (Polat and Satıl 2012); 36: (Ertuğ 2002); 37: (Çakilcioglu et al. 2011); 38: (Özgen et al. 2012); 39: (Türkan et al. 2006); 40: (Bulut 2008); 41: (Khatun et al. 2012); 42: (Alpınar 1999); 43: (Güneş and Özhatay 2011); 44: (Deniz et al. 2010); 45: (Akan et al. 2008); 46: (Kahraman and Tatlı 2004); 47: (Ugulu et al. 2009); 48: (Koyuncu et al. 2009); 49: (Polat et al. 2013); 50: (Toksoy et al. 2010); 51: (Ozturk et al. 2013); 52: (Han and Bulut 2015); 53: (Sağiroğlu et al. 2013); 54: (Polat et al. 2015); 55: (Hayta et al. 2014); 56: (Yeşilyurt et al. 2017); 57: (Mükemre et al. 2015); 58: (Korkmaz and Karakurt 2015); 59: (Özdemir and Alpınar 2015); 60: (Sargin et al. 2015b); 61: (Kaval et al. 2014); 62: (Sargin et al. 2015a); 63: (Bulut and Tuzlaci 2013); 64: (Şenkardeş 2014); 65: (Sargin 2015); 66: (Günbatan et al. 2016); 67: (Arı et al. 2015); 68: (Yeşilyurt et al. 2017); 69: (Karcı et al. 2017); 70: (Akbulut and Bayramoglu 2014)

Appendix 2: Medicinal plants used in the treatment of hypertension in the traditional medicine in Pakistan

	Familia/Taxa	Parts used	Preparation	References[a]
	Aizoceae			
1	*Trianthema portulacastrum*	L, R	EX	1, 2
	Amaranthaceae			
2	*Achyranthes aspera*	WP		1, 11
3	*Aerva javanica*	WP		12
4	*Amaranthus viridis*	L, ST, WP	C, DE	12
5	*Chenopodium album*	WP	EX	13
	Amaryllidaceae			
6	*Allium cepa*	B		3
7	*Allium griffithianum*	B		4
8	*Allium sativum*	B, L, WP	DE	2, 3, 5, 6, 7, 8, 9, 10
	Apiaceae			
9	*Coriandrum sativum*			14
10	*Foeniculum vulgare*	L	EX	2
11	*Trachyspermum ammi*	SE		11
	Apocynaceae			
12	*Caralluma edulis*	L, ST	EX	2, 6, 15
13	*Caralluma tuberculata*	WP		13
14	*Carissa spinarum*	B, FR, L, R		16, 17

(continued)

	Familia/Taxa	Parts used	Preparation	References[a]
	Araceae			
15	*Colocasia esculenta*			14
	Asteraceae			
16	*Erigeron bonariensis*			10
17	*Erigeron canadensis*	WP		18
18	*Lactuca serriola*	WP		13, 18
19	*Sonchus arvensis*	WP		18
20	*Sonchus asper*	AP	DE, IN	10
21	*Taraxacum campylodes*	L, R, SH	DE	13, 18, 19
22	*Vernonia cinerea*	FR		20
23	*Xanthium spinosum*	FR, L, R		13
24	*Xanthium strumarium*	L	EX, JU	2, 16
	Balsaminaceae			
25	*Impatiens bicolor*	FR, SE		11
	Boraginaceae			
26	*Heliotropium strigosum*	WP	EX	1, 2, 17
27	*Trichodesma indicum*	L	EX	2, 4
	Brassicaceae			
28	*Brassica oleracea*			10
	Burseraceae			
29	*Boswellia serrata*	WP		19
	Cannabaceae			
30	*Cannabis sativa*	FL, L	JU	2, 13, 16, 17
	Caprifoliaceae			
31	*Valeriana jatamansi*	R	JU	21
	Crassulaceae			
32	*Bryophyllum pinnatum*	L		1
	Cucurbitaceae			
33	*Citrullus colocynthis*	AP	DE, IN	10
34	*Momordica charantia*	FR	C, EX	22
35	*Mukia maderaspatana*	AP	DE, IN	10
	Cyperaceae			
36	*Cyperus rotundus*	RH, T		1, 2, 23
	Datiscaceae			
37	*Datisca cannabina*	L, R	JU	24
	Dioscoreaceae			
38	*Dioscorea deltoidea*	T		24
	Euphorbiaceae			
39	*Acalypha wilkesiana*			25
40	*Ricinus communis*	L	EX	17
	Fabaceae			
41	*Albizia lebbeck*	L, SE		1
42	*Acacia nilotica*	L, WP	EX	17, 26
43	*Alhagi maurorum*	WP		19

(continued)

	Familia/Taxa	Parts used	Preparation	References[a]
44	*Astragalus propinquus*	R		13
45	*Crotalaria burhia*	WP	EX	2
46	*Medicago polymorpha*	L, SH	CR, FI	5
47	*Melilotus indicus*			14
48	*Sophora mollis*	R		13
49	*Taverniera nummularia*			14
	Geraniaceae			
50	*Geranium wallichianum*	RH	EX	13
	Grossulariaceae			
51	*Ribes himalense*			27
	Lamiaceae			
52	*Ajuga integrifolia*	AP, L, ST, WP	DE, PW	6, 7, 20, 24
53	*Mentha spicata*			9
54	*Ocimum basilicum*	WP		17
	Malvaceae			
55	*Hibiscus trionum*	L, FL		13
56	*Sida cordata*	R		18
	Meliaceae			
57	*Azadirachta indica*	FR, L		12
58	*Melia azedarach*	WP		20
	Myrtaceae			
59	*Syzygium cumini*	SE		12
	Nyctaginaceae			
60	*Boerhavia diffusa*	WP		16
61	*Boerhavia procumbens*	R		12
	Papaveraceae			
62	*Fumaria indica*	WP		12, 17
63	*Fumaria officinalis*			4
64	*Papaver somniferum*	FR, SE	DE	28
	Pedaliaceae			
65	*Pedalium murex*	FR		19
	Phyllanthaceae			
66	*Bridelia retusa*			25
67	*Phyllanthus acidus*			25
	Poaceae			
68	*Avena sativa*			
69	*Cynodon dactylon*	WP		1
70	*Desmostachya bipinnata*			14
	Pteridaceae			
71	*Adiantum incisum*	WP		16
	Ranunculaceae			
72	*Nigella sativa*			9
73	*Ranunculus muricatus*			10
	Rhamnaceae			

(continued)

	Familia/Taxa	Parts used	Preparation	References[a]
74	*Ziziphus jujuba*	AP	IN	10
75	*Ziziphus oxyphylla*	FR, R	DE	29
	Rosaceae			
76	*Rosa indica*	FL	DE	14, 28
	Salicaceae			
77	*Populus tremula*			14
	Solanaceae			
78	*Atropa acuminata*	L, R		11
79	*Solanum americanum*	FR, L, ST		16, 30
80	*Withania somnifera*	WP		31
	Urticaceae			
81	*Urtica dioica*	FR, L, ST		13
	Xanthorrhoeaceae			
82	*Asphodelus tenuifolius*	L, SE		1
	Zingiberaceae			
83	*Elettaria cardamomum*	FR		32
	Zygophyllaceae			
84	*Fagonia cretica*	WP		16
85	*Fagonia indica*			10, 14
86	*Tribulus terrestris*	FR, L, SE		1, 18, 19

[a]References: 1: (Mahmood et al. 2011a); 2: (Mahmood et al. 2013a); 3: (Ullah et al. 2014); 4: (Shah 2015); 5: (Badshah and Hussain 2011); 6: (Shinwari et al. 2003); 7: (Hamayun 2007); 8: (Sher et al. 2015); 9: (Ishtiaq et al. 2015); 10: (Ahmed et al. 2014); 11: (Gilani et al. 2001); 12: (Ahmad 2007b); 13: (Ullah et al. 2013); 14: (Ahmed et al. 2015); 15: (Zia-Ur-Rahman et al. 2004); 16: (Mahmood et al. 2011b); 17: (Mahmood and Mahmood 2012); 18: (Shinwari and Khan 2000); 19: (Shafi et al. 2001); 20: (Shah et al. 2015); 21: (Kayani et al. 2015); 22: (Hsu et al. 2013); 23: (Hussain 2013); 24: (Mahmood et al. 2012); 25: (Khalil et al. 2014); 26: (Mahmood et al. 2013b); 27: (Shinwari et al. 2006); 28: (Bibi et al. 2014); 29: (Zada Khan et al. 2015); 30: (Ullah et al. 2013); 31: (Jabeen et al. 2010); 32: (Gilani et al. 2008)

Appendix 3: Medicinal plants used for the treatment of hypertension in the traditional medicine in Malaysia

	Family/Taxa	Part used	Preparation	References[a]
	Acanthaceae			
1	*Acanthus ebracteatus*	L		1
2	*Andrographis paniculata*	L, R, WP	DE, IN, PN	2, 3, 4, 5, 6, 7
3	*Clinacanthus nutans*	L		1
4	*Strobilanthes crispus*	L	AT, BO	2
	Amaryllidaceae			
5	*Allium sativum*	B	BO, C	2

(continued)

	Family/Taxa	Part used	Preparation	References[a]
	Annonaceae			
6	*Annona muricata*	FR	JU	5, 8
7	*Polyalthia bullata*	FL, L, R	PN	3
	Apiaceae			
8	*Apium graveolens*	L, SE, WP	BO	2
9	*Centella asiatica*	L, WP	DE	2, 5, 9, 10
	Apocynaceae			
10	*Catharanthus roseus*	L, R, WP	AT, DE	2, 4
11	*Rauvolfia serpentina*			11
12	*Rauvolfia verticillata*	L, WP	AT	2
	Araceae			
13	*Homalomena rostrata*	R	DE	12
14	*Lasia* sp.	R	DE	7
	Araliaceae			
15	*Polyscias scutellaria*	L	DE	9
	Asteraceae			
16	*Blumea balsamifera*	L, WP	BO	2
17	*Blumea riparia*	L	AT, BO	13
18	*Gynura procumbens*	L, SE, WP	BO, EF	1, 2, 5
	Balsaminaceae			
19	*Impatiens balsamina*	WP	DE	3, 5
	Caricaceae			
20	*Carica papaya*	FL, L, SH	C, DE, EF, IN	4, 5
	Connaraceae			
21	*Cnestis* sp.	R	DE	7
	Cucurbitaceae			
22	*Momordica charantia*	FR	JU	9
	Dilleniaceae			
23	*Tetracera indica*	ST, R, L	DE	3, 4, 5
	Dioscoreaceae			
24	*Tacca* sp.	R	DE	7
25	*Tacca integrifolia*	L, R, WP	BO	2
	Euphorbiaceae			
26	*Macaranga pruinosa*	L		1
	Fabaceae			
27	*Archidendron jiringa*	R	DE	7
28	*Parkia speciosa*	R	DE	5, 7
29	*Sindora coriacea*	FR	DE	3
	Gesneriaceae			
30	*Cyrtandra pendula*	R	DE	12
	Lamiaceae			
31	*Orthosiphon aristatus*	L, R	AT, BO, DE	5, 13
32	*Plectranthus scutellarioides*	L	DE	3
33	*Vitex pinnata*	L, SH	EF	10

(continued)

	Family/Taxa	Part used	Preparation	References[a]
	Lythraceae			
34	*Punica granatum*	FR		2
	Malvaceae			
35	*Durio zibethinus*	R	DE	7
36	*Hibiscus rosa-sinensis*	FL	DE	3
37	*Hibiscus sabdariffa*	FR, L, WP	BO	2
	Menispermaceae			
38	*Fibraurea tinctoria*	ST	AT, BO	13
39	*Tinospora crispa*	L, WP	BO, PN	3, 10
	Moraceae			
40	*Artocarpus altilis*	L	BO	2
41	*Ficus deltoidea*	L	DE	5, 6
	Moringaceae			
42	*Moringa oleifera*	FR, L	BO	2
	Musaceae			
43	*Musa* sp.	PT	DE	7
	Ophioglossaceae			
44	*Helminthostachys zeylanica*	R	DE	12
	Oxalidaceae			
45	*Averrhoa bilimbi*	L, FR	DE, JU	4, 5
46	*Averrhoa carambola*	L	DE	7
47	*Oxalis barrelieri*	R	DE	6
	Phyllanthaceae			
48	*Phyllanthus acidus*	L	DE	5
49	*Phyllanthus niruri*	AP	DE	5, 10
50	*Phyllanthus pulcher*	R	DE	12
51	*Sauropus androgynus*	L	DE	5
	Piperaceae			
52	*Piper* sp.	L	R	7
	Poaceae			
53	*Lophatherum gracile*	R	DE	14
54	*Zea mays*	CS	DE	3
	Rubiaceae			
55	*Morinda citrifolia*	FR	EF, JU	4, 15
56	*Morinda corneri*	FR	EF	5
	Simaroubaceae			
57	*Eurycoma longifolia*	R	DE	7, 14, 16
	Solanaceae			
58	*Physalis minima*			10
59	*Solanum torvum*	FR, R	DE, EF, IN	3, 5, 6
	Urticaceae			
60	*Leucosyke capitellata*	L	DE	10
	Zingiberaceae			
61	*Zingiber officinale*			17

(continued)

[a]References: 1: (Kassim et al. 2016); 2: (Latiff 2016); 3: (Ong and Nordiana 1999); 4: (Ong et al. 2011a); 5: (Ong et al. 2011b); 6: (Ramli et al. 2015); 7: (Azliza et al. 2012); 8: (Samuel et al. 2010); 9: (Alsarhan et al. 2012); 10: (Ahmad and Holdsworth 2003); 11: (Jamal 2006); 12: (Ong et al. 2012); 13: (Kulip 2003); 14: (Mohammad et al. 2012); 15: (Lin 2005); 16: (Mitra et al. 2007); 17: (Alsarhan et al. 2014)

References

Adak MS (2017) The importance plant genetic resources of Turkey. J Plant Genet Breed 1:e101

Ahmad Z (2007a) Country report on plant genetic resources for food and agriculture. Pakistan Agricultural Research Council, Islamabad

Ahmad SS (2007b) Medicinal wild plants from Lahore-Islamabad motorway (M-2). Pak J Bot 39(2):355–375

Ahmad FB, Holdsworth DK (2003) Medicinal plants of Sabah, East Malaysia-part I. Pharm Biol 41(5):340–346

Ahmed N, Mahmood A, Tahir SS, Bano A, Malik RN, Hassan S, Ashraf A (2014) Ethnomedicinal knowledge and relative importance of indigenous medicinal plants of Cholistan desert, Punjab Province, Pakistan. J Ethnopharmacol 155(2):1263–1275

Ahmed N, Mahmood A, Mahmood A, Sadeghi Z, Farman M (2015) Ethnopharmacological importance of medicinal flora from the district of Vehari, Punjab province, Pakistan. J Ethnopharmacol 168:66–78

Akan H, Aslan M, Balos MM (2005a) GAP yöresindeki tıbbi ve aromatik bitkiler. TUBITAK Proje No: TBAG/Ç. SEK 22 (103-T009)

Akan H, Aslan M, Balos MM (2005b) Şanlıurfa kent merkezindeki semt pazarlarında satılan bazı bitkiler ve kullanım amaçları. Ot Sistematik Botanik Dergisi 12(2):43–58

Akan H, Korkut MM, Balos MM (2008) Arat Dağı ve çevresinde (Birecik, Şanlıurfa) etnobotanik bir araştırma. Fırat Üniversitesi Fen ve Mühendislik Bilimleri Dergisi 20(1):67–81

Akan H, Balos MM, Tel AZ (2013) The ethnobotany of some legume plants around Birecik (Şanlıurfa). Adyutayam 1(1):31–39

Akaydın G, Şimşek I, Arıtuluk ZC, Yeşilada E (2013) An ethnobotanical survey in selected towns of the Mediterranean subregion (Turkey). Turk J Biol 37:230–247

Akbulut S, Bayramoglu MM (2014) Reflections of socio-economic and demographic structure of urban and rural on the use of medicinal and aromatic plants: the sample of Trabzon province. Ethno Med 8(1):89–100

Alpınar K (1999) Ayvalık (Balıkesir) ve yakınındaki adaların floristik ve etnobotanik açıdan değerlendirilmesi. TUBITAK, TBAG-1407, Turkey

Al-Qattan KK, Khan I, Alnaqeeb MA, Ali M (2003) Mechanism of garlic (*Allium sativum*) induced reduction of hypertension in 2K-1C rats: a possible mediation of Na/H exchanger isoform-1. Prostaglandins Leukot Essent Fatty Acids 69(4):217–222

Alsarhan A, Sultana N, Kadir MRA, Aburjai T (2012) Ethnopharmacological survey of medicinal plants in Malaysia, the Kangkar Pulai region. Int J Pharmacol 8(8):679–686

Alsarhan A, Sultana N, Al-Khatib A, Kadir MRA (2014) Review on some Malaysian traditional medicinal plants with therapeutic properties. J Basic Appl Sci 10:149–159

Altundag E, Ozturk M (2011) Ethnomedicinal studies on the plant resources of east Anatolia, Turkey. Procedia Soc Behav Sci 19:756–777

Arı S, Temel M, Kargıoğlu M, Konuk M (2015) Ethnobotanical survey of plants used in Afyonkarahisar-Turkey. J Ethnobiol Ethnomed 11(1):84

Azliza MA, Ong HC, Vikineswary S, Noorlidah A, Haron NW (2012) Ethno-medicinal resources used by the Temuan in ulu Kuang Village. Ethno-Medicine 6(1):17–22

Badshah L, Hussain F (2011) People preferences and use of local medicinal flora in district tank, Pakistan. J Med Plants Res 5(1):22–29
Baytop T (1984) Türkiye'de Bitkiler ile Tedavi. İstanbul Üniversitesi Yayınları No: 3255, Eczacılık Fakültesi No: 40, Istanbul
Bibi S, Sultana J, Sultana H, Malik RN (2014) Ethnobotanical uses of medicinal plants in the highlands of Soan Valley, salt range, Pakistan. J Ethnopharmacol 155(1):352–361
Bulut GE (2008) Bayramiç (Çanakkale) yöresinde etnobotanik araştırmalar. Ph.D. thesis, Marmara Üniversity, İstanbul
Bulut EG, Tuzlacı E (2005) An ethnobotanical study in Bozcaada (Çanakkale-Turkey). IVth International Congress of Ethnobotany, ICEB 2005, pp 581–583
Bulut G, Tuzlaci E (2013) An ethnobotanical study of medicinal plants in Turgutlu (Manisa-Turkey). J Ethnopharmacol 149(3):633–647
Çakılcıoğlu U, Türkoğlu İ (2009) Çitli Ovası (Elazığ) ve çevresinin etnobotanik özellikleri. e-J New World Sci Acad Ecol Life Sci 4(2):81–85
Çakilcioglu U, Turkoglu I (2010) An ethnobotanical survey of medicinal plants in Sivrice (Elazıg-Turkey). J Ethnopharmacol 132:165–175
Çakılcıoğlu U, Türkoğlu İ, Kürşat M (2007) Harput (Elazığ) ve çevresinin etnobotanik özellikleri. Doğu Anadolu Bölgesi Araştırmaları:22–28
Çakılcıoğlu U, Şengün MT, Türkoğlu I (2010) An ethnobotanical survey of medicinal plants of Yazıkonak and Yurtbaşı districts of Elazığ province, Turkey. J Med Plants Res 4(7):567–572
Çakilcioglu U, Khatun S, Turkoglu I, Hayta S (2011) Ethnopharmacological survey of medicinal plants in Maden (Elazig-Turkey). J Ethnopharmacol 137:469–486
Çömlekçioğlu N, Karaman Ş (2008) Kahramanmaraş şehir merkezindeki aktar'larda bulunan tıbbi bitkiler. KSU J Sci Eng 11(1):23–32
Deepa R, Shanthirani CS, Pradeepa R, Mohan V (2003) Is the 'rule of halves' in hypertension still valid? Evidence from the Chennai urban population study. J Assoc Physicians India 51:153–157
Demirci S, Özhatay N (2012) An ethnobotanical study in Kahramanmaraş (Turkey), wild plants used for medicinal purpose in Andırın, Kahramanmaraş. Turk J Pharm Sci 9(1):75–92
Deniz L, Serteser A, Kargıoğlu M (2010) Uşak Üniversitesi ve yakın çevresindeki bazı bitkilerin mahalli adları ve etnobotanik özellikleri. AKÜ Fen Bilimleri Dergisi 1:57–72
Doğan G, Bağcı E (2011) The plants benefit to public in some residential area of Elazığ (Turkey) using traditional ecological knowledge and their ethnobotanical features. Fırat Univ Fen Bilimleri Dergisi 23(2):77–86
Eddouks M, Zeggwagh NA (2012) Phytotherapy of hypertension in Morocco. In: Phytotherapy in the management of diabetes and hypertension, vol 1. Bentham Science Publishers, Ann Arbor, p 193
Eddouks M, Maghrani M, Lemhadri A, Ouahidi ML, Jouad H (2002) Ethnopharmacological survey of medicinal plants used for the treatment of diabetes mellitus, hypertension and cardiac diseases in the south-east region of Morocco (Tafilalet). J Ethnopharmacol 82(2–3):97–103
Eddouks M, Maghrani M, Zeggwagh NA, Haloui M, Michel JB (2005) *Fraxinus excelsior* L. evokes a hypotensive action in normal and spontaneously hypertensive rats. J Ethnopharmacol 99(1):49–54
El Bardai S, Morel N, Wibo M, Fabre N, Llabres G, Lyoussi B, Quetin-Leclercq J (2003) The vasorelaxant activity of marrubenol and marrubiin from *Marrubium vulgare*. Planta Med 69(1):75–77
Ertuğ F (2002) Bodrum yöresinde halk tıbbında yararlanılan bitkiler. 14. Bitkisel ilaç hammaddeleri toplantısı. 29–31 Mayıs 2002, Eskişehir
Ertuğ F, Tümen G, Çelik A, Dirmenci T (2004) Buldan (Denizli) etnobotanik alan araştırması 2003. TÜBA Kültür Envanteri Dergisi 2
Everest A, Ozturk E (2005) Focusing on the ethnobotanical uses of plants in Mersin and Adana provinces (Turkey). J Ethnobiol Ethnomed:1–6
Ezer N, Arısan OM (2006) Folk medicines in Merzifon (Amasya, Turkey). Turk J Bot 30:223–230

Ezer N, Avcı K (2004) Çerkeş (Çankırı) yöresinde kullanılan halk ilaçları. Hacettepe Üniversitesi Eczacılık Fakültesi Dergisi 24(2):67–80
Genç GE, Özhatay N (2006) An ethnobotanıcal study in Çatalca (European part of Istanbul) II. Turk J Pharm Sci 3(2):73–89
Gilani SA, Qureshi RA, Farooq U (2001) Ethnobotanical studies of Ayubia national park district Abbottabad, Pakistan. J Biol Sci 1(4):284–286
Gilani AH, Jabeen Q, Khan AU, Shah AJ (2008) Gut modulatory, blood pressure lowering, diuretic and sedative activities of cardamom. J Ethnopharmacol 115(3):463–472
González-Tejero MR, Casares-Porcel M, Sánchez-Rojas CP, Ramiro-Gutiérrez JM, Molero-Mesa J et al (2008) Medicinal plants in the Mediterranean area: synthesis of the results of the project Rubia. J Ethnopharmacol 116(2):341–357
Günbatan T, Gürbüz İ, Özkan AMG (2016) The current status of ethnopharmacobotanical knowledge in Çamlıdere (Ankara, Turkey). Turk J Bot 40(3):241–249
Güner A, Aslan S, Ekim T, Vural M, Babaç MT (2012) Türkiye Bitkileri Listesi (Damarlı Bitkiler). Istanbul-Türkiye. Nezahat Gökyiğit Botanik Bahçesi ve Flora Araştırmaları Derneği Yayını
Güneş F, Özhatay N (2011) An ethnobotanical study from Kars (eastern) Turkey. Biol Diversity Conserv 4(1):30–41
Gürdal B, Kültür Ş (2013) An ethnobotanical study of medicinal plants in Marmaris (Mugla, Turkey). J Ethnopharmacol 146:113–126
Hamayun M (2007) Traditional uses of some medicinal plants of Swat Valley, Pakistan. Indian J Tradit Knowl 6(4):636–641
Han Mİ, Bulut G (2015) The folk-medicinal plants of Kadişehri (Yozgat-Turkey). Acta Soc Bot Pol 84(2):237–248
Hayta S, Polat R, Selvi S (2014) Traditional uses of medicinal plants in Elazığ (Turkey). J Ethnopharmacol 154(3):613–623
Heinrich M (2000) Ethnobotany and its role in drug development. Phytother Res 14(7):479–488
Houghton PJ, Howes MJ, Lee CC, Steventon G (2007) Uses and abuses of *in vitro* tests in ethnopharmacology: visualizing an elephant. J Ethnopharmacol 110(3):391–400
Hsu CL, Fang SC, Liu CW, Chen YF (2013) Inhibitory effects of new varieties of bitter melon on lipopolysaccharide-stimulated inflammatory response in RAW 264.7 cells. J Funct Foods 5(4):1829–1837
Hussain MI (2013) Studies on plant biodiversity conservation and sustainable utilization of medicinally important plant species of Johi & Gorakh Hill. Ph.D. thesis, University of Karachi, Karachi, Pakistan
Ishtiaq M, Mahmood A, Maqbool M (2015) Indigenous knowledge of medicinal plants from Sudhanoti district (AJK), Pakistan. J Ethnopharmacol 168:201–207
Jabeen S, Shah MT, Khan S, Hayat MQ (2010) Determination of major and trace elements in ten important folk therapeutic plants of Haripur basin, Pakistan. J Med Plants Res 4(7):559–566
Jamal JA (2006) Malay traditional medicine. Tech Monitor (Special Feature: traditional Medicine: S & T Advancement). pp 37–49
Joubert E, Gelderblom WCA, Louw A, de Beer D (2008) South African herbal teas: *Aspalathus linearis, Cyclopia* spp. and *Athrixia phylicoides* - a review. J Ethnopharmacol 119(39):376–412
Kahraman A, Tatlı A (2004) Umarbaba dağı (Eşme-Uşak) ve çevresindeki bazı bitkilerin mahalli adları ve etnobotanik özellikleri. Ot Sistematik Botanik Dergisi 11(2):147–154
Kalus U, Pindur G, Jung F, Mayer B, Radtke H, Bachmann K, Mrowietz C, Koscielny J, Kiesewetter H (2000) Influence of the onion as an essential ingredient of the Mediterranean diet on arterial blood pressure and blood fluidity. Arzneimittelforschung 50(9):795–801
Karcı E, Gürbüz İ, Akaydın G, Günbatan T (2017) Folk medicines of Bafra (Samsun-Turkey). Turk J Biochem 42(4):381–399
Kargıoğlu M, Cenkci S, Serteser A, Konuk M, Vural G (2010) Traditional uses of wild plants in the middle Aegean region of Turkey. Hum Ecol 38:429–450

Kassim DHA, Raduan SZ, Aziz MA, Chelum A, Morni AAM, Wahab RA (2016) Indigenous knowledge of medicinal plants used and its implication towards health-seeking behavior among the Melanau in Pulau bruit, Sarawak, Malaysia. J Adv Res Soc Behav Sci 4(2):136–145
Kaval I, Behçet L, Cakilcioglu U (2014) Ethnobotanical study on medicinal plants in Geçitli and its surrounding (Hakkari-Turkey). J Ethnopharmacol 155(1):171–184
Kayani S, Ahmad M, Sultana S, Shinwari ZK, Zafar M, Yaseen G et al (2015) Ethnobotany of medicinal plants among the communities of alpine and sub-alpine regions of Pakistan. J Ethnopharmacol 164:186–202
Kearney PM, Whelton M, Reynolds K, Muntner P, Whelton PK, He J (2005) Global burden of hypertension: analysis of worldwide data. Lancet 365:217–223
Khalil AT, Shinwari ZK, Qaiser M, Marwat KB (2014) Phyto-therapeutic claims about euphorbeaceous plants belonging to Pakistan; an ethnomedicinal review. Pak J Bot 46(3):1137–1144
Khatun S, Parlak KU, Polat R, Cakilcioglu U (2012) The endemic and rare plants of Maden (Elazig) and their uses in traditional medicine. J Herb Med 2:68–75
Khayyal MT, el-Ghazaly MA, Abdallah DM, Nassar NN, Okpanyi SN, Kreuter MH (2002) Blood pressure lowering effect of an olive leaf extract (*Olea europaea*) in L-NAME induced hypertension in rats. Arzneimittelforschung 52(11):797–802
Kıran Ö (2006) Kozan yöresi florasındaki tıbbi bitkiler ve bunların halk tıbbında kullanılışı. M.sc. thesis, Çukurova University, Adana
Korkmaz M, Karakurt E (2015) An ethnobotanical investigation to determine plants used as folk medicine in Kelkit (Gümüşhane/Turkey) district. Biol Diversity Conserv 8(3):290–303
Koyuncu O, Yaylacı ÖK, Tokur S (2009) Geyve (Sakarya) ve çevresinin etnobotanik açıdan incelenmesi. Ot Sistematik Botanik Dergisi 16(1):123–142
Kulip J (2003) An ethnobotanical survey of medicinal and other useful plants of Muruts in Sabah, Malaysia. Telopea 10(1):81–98
Kultur Ş (2007) Medicinal plants used in Kırklareli Province (Turkey). J Ethnopharmacol 111:341–364
Latiff A (2016) Antihypertensive plants of Malaysia. Unpublished personal report, Malaysia
Lin K (2005) Ethnobotanical study of medicinal plants used by the Jah hut peoples in Malaysia. Indian J Med Sci 59(4):156
Mahmood A, Mahmood A (2012) Indigenous wild medicinal plants used by local people of Dudial area, District Mirpur, Azad Jammu and Kashmir, Pakistan. Afr J Microbiol Res 6(17):3898–3903
Mahmood A, Mahmood A, Shaheen H, Qureshi RA, Sangi Y, Gilani SA (2011a) Ethno medicinal survey of plants from district Bhimber Azad Jammu and Kashmir, Pakistan. J Med Plants Res 5(11):2348–2360
Mahmood A, Qureshi RA, Mahmood A, Sangi Y, Shaheen H, Ahmad I, Nawaz Z (2011b) Ethnobotanical survey of common medicinal plants used by people of district Mirpur, AJK, Pakistan. J Med Plants Res 5(18):4493–4498
Mahmood A, Mahmood A, Malik RN (2012) Indigenous knowledge of medicinal plants from Leepa valley, Azad Jammu and Kashmir, Pakistan. J Ethnopharmacol 143(1):338–346
Mahmood A, Mahmood A, Malik RN, Shinwari ZK (2013a) Indigenous knowledge of medicinal plants from Gujranwala district, Pakistan. J Ethnopharmacol 148(2):714–723
Mahmood A, Rashid S, Malik RN (2013b) Determination of toxic heavy metals in indigenous medicinal plants used in Rawalpindi and Islamabad cities, Pakistan. J Ethnopharmacol 148(1):158–164
Marles RJ, Farnsworth NR (1995) Antidiabetic plants and their active constituents 1. Phytomedicine 2(2):137–189
Mitra R, Orbell J, Muralitharan MS (2007) Agriculture-medicinal plants of Malaysia. Asia-Pacific Biotech News 11(02):105–110
Mittal BV, Singh AK (2010) Hypertension in the developing world: challenges and opportunities. Am J Kidney Dis 55(3):590–598

Mohammad NS, Milow P, Ong HC (2012) Traditional medicinal plants used by the Kensiu tribe of Lubuk ulu Legong, Kedah, Malaysia. Ethno-Medicine 6(3):149–153

Mükemre M, Behçet L, Çakılcıoğlu U (2015) Ethnobotanical study on medicinal plants in villages of Çatak (van-Turkey). J Ethnopharmacol 166:361–374

Nasir E, Ali SI (1970–1995) Flora of Pakistan. National Herbarium/NARC and Department of Botany, University of Karachi, Islamabad and Karachi

Nissinen A, Bothig S, Granroth H, Lopez AD (1988) Hypertension in developing countries. World Health Stat Q 41:141–154

Ong HC, Nordiana M (1999) Malay ethno-medico botany in Machang, Kelantan, Malaysia. Fitoterapia 70(5):502–513

Ong HC, Ruzalila BN, Milow P (2011a) Traditional knowledge of medicinal plants among the Malay villagers in Kampung Tanjung Sabtu, Terengganu, Malaysia. Indian J Tradit Knowl 10(3):460–465

Ong HC, Zuki RM, Milow P (2011b) Traditional knowledge of medicinal plants among the Malay villagers in Kampung Mak Kemas, Terengganu, Malaysia. Ethno-Medicine 5(3):175–185

Ong HC, Faezah AW, Milow P (2012) Medicinal plants used by the Jah hut orang Asli at Kampung Pos Penderas, Pahang, Malaysia. Ethno-Medicine 6(1):11–15

Özçelik H, Balabanlı C (2005) Burdur İlinin Tıbbi ve Aromatik Bitkileri. I. Burdur Sempozyumu, Burdur/Turkiye. pp 1127–1136

Özdemir E, Alpınar K (2015) An ethnobotanical survey of medicinal plants in western part of central Taurus Mountains: Aladaglar (Nigde–Turkey). J Ethnopharmacol 166:53–65

Özgen U, Kaya Y, Houghton P (2012) Folk medicines in the villages of Ilıca District (Erzurum, Turkey). Turk J Biol 36:93–106

Öztürk M, Dinç M (2005) Nizip (Aksaray) bölgesinin etnobotanik özellikleri. Ot Sistematik Botanik Dergisi 12(1):93–102

Ozturk M, Guvensen A, Aksoy A, Beyazgul M (2006) An overview of the soils and sustainable land use in Turkiye. Proceedings of the Fifth International GAP Engineering Congress, Sanlıurfa, pp 1548–1555

Ozturk M, Altundağ E, Gücel S (2012a) Medicinal and aromatic plants (Turkey). Ethnopharmacology, Encyclopedia of Life Support Systems (EOLSS). http://www.eolss.net/sample-chapters/c03/e6-79-48.pdf

Ozturk M, Gucel S, Altundag E, Mert T, Gork C, Gork G, Akcicek E (2012b) Chapter 7: An overview of the medicinal plants of Turkey. In: Singh R (ed) Genetic resources, chromosome engineering and crop improvement: medicinal plants. CRC Press, LLC, Taylor & Francis, Boca Raton, FL, pp 181–206

Ozturk M, Uysal I, Gucel S, Altundag E, Dogan Y, Baslar S (2013) Medicinal uses of natural dye-yielding plants in Turkey. Res J Text Appar 17(2):69–80

Ozturk M, Altay V, Aksoy A (2016) Ecology of endangered endemic plant taxa of Turkiye in relation to climate change. In: Conservation and Sustainable use of genepool of plant World in Eurasia at the present stage International Scientific Conference within Day of Kazakhstan. September 3, EXPO-2016, Antalya, Turkiye

Ozturk M, Altay V, Gücel S, Altundağ E (2017a) Plant diversity of the drylands in southeast Anatolia-Turkey: role in human health and food security. In: Ansari AA, Gill SS (eds) Plant biodiversity: monitoring, assessment and conservation. CABI, UK, pp 83–124

Ozturk M, Altay V, Gonenç TM (2017b) Chapter 24: Herbal from High Mountains in the East Mediterranean. In: Bhojraj S et al (eds) Drug discovery from herbs – approaches and applications centre for scinece & technology of the non-aligned and other developing countries (NAM S & T Centre). DAYA Publishing House, New Delhi, pp 327–367

Polat R, Satıl F (2012) An ethnobotanical survey of medicinal plants in Edremit Gulf (Balıkesir-Turkey). J Ethnopharmacol 139:626–641

Polat R, Selvi S, Çakılcıoğlu U, Açar M (2012) Investigations of ethnobotanical aspect of wild plants sold in Bingöl (Turkey) local markets. Biol Diversity Conserv J 5(3):155–161

Polat R, Cakılcıoğlu U, Satıl F (2013) Traditional uses of medicinal plant in Solhan (Bingöl-Turkey). J Ethnopharmacol 148:951–963
Polat R, Cakilcioglu U, Kaltalioğlu K, Ulusan MD, Türkmen Z (2015) An ethnobotanical study on medicinal plants in Espiye and its surrounding (Giresun-Turkey). J Ethnopharmacol 163:1–11
Pradeepa R, Mohan V (2008) Hypertension & pre-hypertension in developing countries. Indian J Med Res 128(12):688–690
Ramli MR, Milow P, Chooi OH (2015) Traditional knowledge of a practitioner in medicinal plants of masjid Ijok village, Perak, Malaysia. Ethno-Medicine 9(1):59–66
Reddy KS (1996) Hypertension control in developing countries. Genetic issues. J Hum Hypertens 10:S33–S38
Sağiroğlu M, Topuz T, Ceylan K, Turna M (2013) An ethnobotanical survey from Yahyalı (Kayseri) and tarsus (Mersin). SAÜ Fen Edebiyat Dergisi 2:13–37
Samuel AJSJ, Kalusalingam A, Chellappan DK, Gopinath R, Radhamani S, Husain HA et al (2010) Ethnomedical survey of plants used by the orang Asli in Kampung Bawong, Perak, West Malaysia. J Ethnobiol Ethnomed 6(1):5
Şanlı BZ (2006) Bursa ve çevresinden toplanan ve ticareti yapılan bazı ekonomik bitkiler. M.sc. thesis. Uludağ University, Bursa
Sargin SA (2015) Ethnobotanical survey of medicinal plants in Bozyazi district of Mersin, Turkey. J Ethnopharmacol 173:105–126
Sargin SA, Selvi S, Büyükcengiz M (2015a) Ethnomedicinal plants of Aydincik District of Mersin, Turkey. J Ethnopharmacol 174:200–216
Sargin SA, Selvi S, López V (2015b) Ethnomedicinal plants of Sarigöl district (Manisa), Turkey. J Ethnopharmacol 171:64–84
Sarı AO, Oğuz B, Güvensen A, Bilgiç A, Şenol SG, Tort N (2010) Ege ve Güney Marmara bölgelerinde halk ilacı olarak kullanılan bitkiler. Anadolu J AARI 20(2):1–21
Sarper F, Akaydın G, Şimşek I, Yeşilada E (2009) An ethnobotanical field survey in the Haymana District of Ankara Province in Turkey. Turk J Biol 33:79–88
Savran A, Bağcı Y, Kargıoğlu M (2009) Gemerek (Sivas) ve çevresindeki bazı bitkilerin yerel adları ve etnobotanik özellikleri. Afyon Kocatepe Üniversitesi Fen Bilimleri Dergisi 8(1):313–321
Saw LG, Chung RCK (2015) The flora of Malaysia projects. Rodriguesia 66(4):947–960
Şenkardeş I (2014) Nevşehir'in güney ilçelerinde (Acıgöl, Derinkuyu, Gülşehir, Nevşehir-merkez, Ürgüp) etnobotanik araştırmalar. Ph.D. thesis, Marmara University, Istanbul
Shafi MS, Ashraf MY, Sarwar G (2001) Wild medicinal plants of Cholistan area of Pakistan. Pak J Biol Sci 4(1):112–116
Shah AH (2015) Cultural drivers of plant biodiversity of district tor Ghar. Ph.D. thesis, Hazara University, Pakistan
Shah AH, Khan SM, Shah AH, Mehmood A, Rahman IU, Ahmad H (2015) Cultural uses of plants among Basikhel tribe of district tor Ghar, Khyber Pakhtunkhwa, Pakistan. Pak J Bot 47:23–41
Sher H, Aldosari A, Ali A, de Boer HJ (2015) Indigenous knowledge of folk medicines among tribal minorities in Khyber Pakhtunkhwa, northwestern Pakistan. J Ethnopharmacol 166:157–167
Shinwari MI, Khan MA (2000) Folk use of medicinal herbs of Margalla hills national park, Islamabad. J Ethnopharmacol 69(1):45–56
Shinwari ZK, Khan AA, Nakaike T (2003) Medicinal and other useful plants of District Swat, Pakistan. WWF, Pakistan, p 68
Shinwari ZK, Watanabe T, Rehman M, Youshikawa T (2006) A pictorial guide to medicinal plants of Pakistan. KUST, Kohat
Şimşek I, Aytekin F, Yesilada E, Yıldırımlı Ş (2004) An ethnobotanical survey of the Beypazarı, Ayas ansd Güdül District towns of Ankara Province (Turkey). Econ Bot 58(4):705–720
Tahraoui A, El-Hilaly J, Israili ZH, Lyoussi B (2007) Ethnopharmacological survey of plants used in the traditional treatment of hypertension and diabetes in South-Eastern Morocco (Errachidia province). J Ethnopharmacol 110(1):105–117
Tetik F, Civelek S, Cakilcioglu U (2013) Traditional uses of some medicinal plants in Malatya (Turkey). J Ethnopharmacol 146:331–346

Toksoy D, Bayramoglu M, Hacisalihoglu S (2010) Usage and the economic potential of the medicinal plants in Eastern Black Sea Region of Turkey. J Environ Biol 31(5):623–628
Türkan Ş, Malyer H, Aydın SÖ, Tümen G (2006) Ordu ili ve çevresinde yetişen bazı bitkilerin etnobotanik özellikleri. Süleyman Demirel Üniversitesi Fen Bilimleri Enstitüsü Dergisi 10(2):162–166
Tuzlacı E (2006) Şifa Niyetine Türkiye'nin Bitkisel Halk İlaçları. Alfa Yayınları, Istanbul
Ugulu I, Baslar S, Yorek N, Dogan Y (2009) The investigation and quantitative ethnobotanical evaluation of medicinal plants used around Izmir province, Turkey. J Med Plants Res 3(5):345–367
Ullah M, Khan MU, Mahmood A, Malik RN, Hussain M, Wazir SM et al (2013) An ethnobotanical survey of indigenous medicinal plants in Wana district south Waziristan agency, Pakistan. J Ethnopharmacol 150(3):918–924
Ullah S, Khan MR, Shah NA, Shah SA, Majid M, Farooq MA (2014) Ethnomedicinal plant use value in the Lakki Marwat District of Pakistan. J Ethnopharmacol 158:412–422
Verma S, Singh SP (2008) Current and future status of herbal medicines. Vet World 1(11):347–350
www.fao.org
Yeşilyurt EB, Şimşek I, Akaydin G, Yeşilada E (2017) An ethnobotanical survey in selected districts of the Black Sea region (Turkey). Turk J Bot 41(1):47–62
Yücecan S, Akgün B, Çiftçi N, Aksoy C, Tayfur M, Taşçı N (1988) Plants used in the treatment of hypertension in the East Southeast Anatolia. Hacettepe Üniversitesi Eczacılık Fakültesi Dergisi 8(1):23–31
Zada Khan MP, Ahmad M, Zafar M, Sultana S, Ali MI, Sun H (2015) Ethnomedicinal uses of edible wild fruits (EWFs) in Swat Valley, Northern Pakistan. J Ethnopharmacol 173:191–203
Zia-Ur-Rahman LA, Begum S, Ishfaq M, Shinwari ZH (2004) Folk uses medicinal plants of Babuzai village in Sakra Mountain range (district Mardan). International symposium medicinal plants: linkages beyond national boundaries, September 7–9, 2004, pp 95–104
Ziyyat A, Legssyer A, Mekhfi H, Dassouli A, Serhrouchni M, Benjelloun W (1997) Phytotherapy of hypertension and diabetes in oriental Morocco. J Ethnopharmacol 58(1):45–54

Epimedium elatum (Morr & Decne): A Therapeutic Medicinal Plant from Northwestern Himalayas of India

Sajad Ahmad Lone, Ajai Prakash Gupta, Malik Muzafar Manzoor, Pooja Goyal, Qazi Pervaiz Hassan, and Suphla Gupta

Epimedium Species: An Introduction

The genus Epimedium of the family Berberidaceae comprises more than 65 accepted deciduous and hardy perennial species (The Plant List 2013). They are characterised by heart-shaped leaves commonly referred to as three branches–nine leaves grass. Majority of them have four-parted 'spider-like' flowers (bishop's hat) blooming in spring season (Ma et al. 2011). Epimedium is predominantly found in North temperate Zone of Northern Hemisphere. Most of the Epimedium species are endemic to China (48), Japan and Korea, while some are found in European and African countries (Ma et al. 2011; Zhang et al. 2016). They are known by common English names like rowdy lamb herb, barrenwort, bishop's hat, fairy wings and horny goat weed (Ma et al. 2011).

Several species of Epimedium plants are used as ground cover during spring season due to their tough and evergreen herbage (Ma et al. 2011). Extracts of Epimedium species have been widely used as a tonic, aphrodisiac and anti-rheumatic

S. A. Lone · S. Gupta (✉) · P. Goyal
Plant Biotechnology Division, CSIR-Indian Institute of Integrative Medicine, Jammu, India

Academy of Scientific and Innovative Research (AcSIR), New Delhi, India

A. P. Gupta
Academy of Scientific and Innovative Research (AcSIR), New Delhi, India

Quality Control and Quality Assurance Division, CSIR-Indian Institute of Integrative Medicine, Jammu, India

M. M. Manzoor
Plant Biotechnology Division, CSIR-Indian Institute of Integrative Medicine, Jammu, India

Q. P. Hassan
Academy of Scientific and Innovative Research (AcSIR), New Delhi, India

Plant Biotechnology Division, CSIR-Indian Institute of Integrative Medicine, Srinagar, India

M. Ozturk, K. R. Hakeem (eds.), *Plant and Human Health, Volume 1*,
https://doi.org/10.1007/978-3-319-93997-1_17

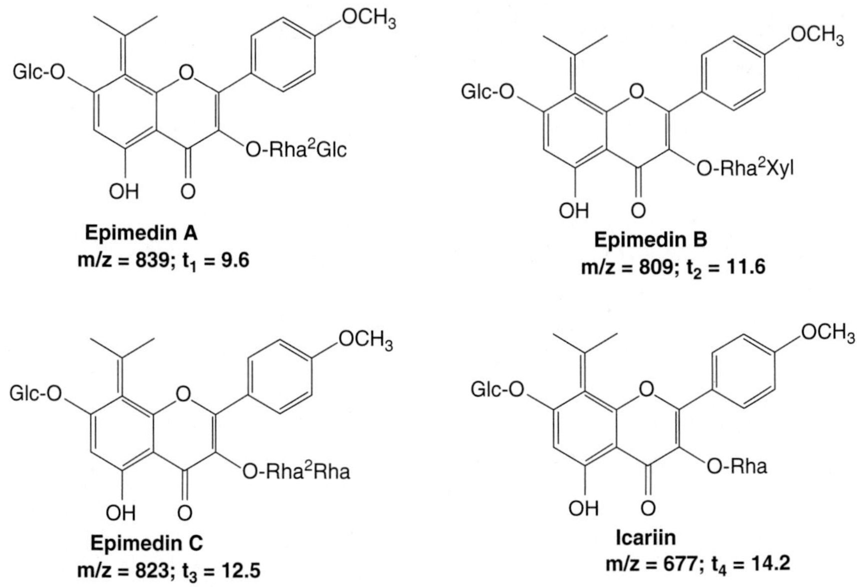

Fig. 1 Chemical structures of four major chemotaxonomic markers in Epimedium

preparations in China, Japan and Korea for more than 2000 years (Ma et al. 2011). Dried aerial parts of some Epimedium species are used for the preparation of *Herba Epimedii*, a well-known botanical supplement in Chinese pharmacology. Currently, more than 15 Epimedium species are used extensively in traditional Chinese medicine, but only 5 (*E. brevicornum*, *E. sagittatum*, *E. koreanum*, *E. pubescens* and *E. wushanense*) are recorded as the official source of *Herba Epimedii* (Pei et al. 2007; Chinese Pharmacopoeia Commission 2010).

Phytochemical characterisation of Epimedium species has led to the isolation of more than 260 chemical compounds (Ma et al. 2011); however, epimedin A, epimedin B, epimedin C and icariin (Fig. 1) are regarded as the major chemotaxonomic markers (Wu et al. 2003; Pei et al. 2007; Wang et al. 2007; Xu et al. 2013a, b; Chen et al. 2015a, b). Of these molecules, icariin is a well-known phosphodiesterase-5 (PDE-5) inhibitor, an enzyme known to cause erectile dysfunction in humans (Ma et al. 2011). Epimedium species have widely been used in the treatment of impotence, spermatorrhoea, women sterility, amenorrhoea, rheumatic arthritis, chronic bronchitis, etc. (Wu et al. 2003; Chen et al. 2015a, b). Besides, a wide range of pharmacological actions are attributed to phytochemical compounds in Epimedium species, including improving cardiovascular and cerebrovascular functions, modulating immunological function as well as having anti-osteoporosis, anti-oxidation, antitumour, antiviral and anti-ageing effects (Ma et al. 2011; Chen et al. 2015a, b). These activities have been compiled in under the section Pharmacological Importance of Epimedium species. Literature reveals that among the 65 accepted species, only 5 species (*E. brevicornum*, *E. sagittatum*, *E. koreanum*, *E. pubescens* and *E. wushanense*) have been studied extensively.

Epimedium elatum

Epimedium elatum (**Morren & Decne**) of Berberidaceae family has not been investigated much as evident in published literature. However, recent research efforts have tried to explore its medicinal potential. This medicinal plant species was first reported from Indian Himalayas in 1834 by European botanists (Morren and Decaisne 1834; Decne 1844; Stewart 1972). According to some reports, *E. elatum* is the only species in Epimedium genus, endemic to shady high-altitude coniferous forests of Northwestern Himalayas in India and Pakistan (www.gbif.org; Nasir and Ali 1980–2005; Perveen and Qaiser 2010; Tantry et al. 2012; Naseer et al. 2015; Arief et al. 2015, 2016). It is known by several local names like Saul sumbal and Chhal kambli. Commonly, it is also known as 'mosquito herb' as the extract of this species is sprayed as repellent for mosquitoes in some Himalayan communities (Arief et al. 2015). Phytochemically, *E. elatum* has been shown to contain high concentration of epimedin A, epimedin B, epimedin C and icariin (Sofi et al. 2014; Naseer et al. 2015; Arief et al. 2015, 2016). The chemical structure of four quality-determining bioactive glycoside flavonoids is given in Fig. 1.

The wild resources of medicinal Epimedium species have been dramatically reduced due to years of overharvesting and habitat destruction since the 1990s in China; as a consequence of it, some of the Epimedium species have even become endangered (Ward 2004; Xu et al. 2008). Several medicinal plants have also become threatened in Northwestern Himalayas due to anthropogenic threats (Tali et al. 2015). Like several other plants of Northwestern Himalayas, *E. elatum* too can become threatened in near future if timely conservation measures are not taken.

Botanical Description

Epimediums are usually low-growing, deciduous, perennial plants with leathery leaves and they spread by underground stems. The leaves are scale-like, alternate, long petiolated and alternately divided twice. The leaflets are ovate, acuminate, cordate, and up to 2.5–13.5 cm long and 1.5–7.5 cm wide, with setose margins (Ma et al. 2011). *E. elatum* is a sparsely branched, low-growing, deciduous perennial, glabrous herb (Fig. 2). The plant reaches a height of about 100 cm in shady habitats. The plant has characteristic 'three branches and nine leaves' pattern which is a characteristic morphological character of almost all Epimedium species (Ma et al. 2011). It has a thick spreading rhizome under the soil. The scale-like leaves are alternate and cordate with long petioles. Leaflets are usually ovate, acuminate and cordate with serrated margins. Flowers are hermaphrodite, pale yellowish or yellowish white, usually pollinated by bees. Panicle is irregularly branched, glabrous to glandular-hairy. The flowers of this plant resemble a bishop's hat (pendant shaped), having long spurs with varying colours in different habitats. They are 1–2 in. wide and linear to lanceolate in shape. The flower has eight pieces of sepals;

Fig. 2 *Epimedium elatum* growing in wild

outer four sepals are unequal and the inner four sepals are petaloid, reflexing at flowering time. Sepals are ovate to lanceolate with acute margins. Petals are lanceolate-elliptic with innermost being shorter, shoe shaped and honey bearing. The flower has four stamens, present opposite to petals, and one ovary with several ovules. Anthers are linear with long filaments, opening by two up-curved valves, usually crowded together. Capsules usually contain 2–3 biseriate, reniform, black seeds (indiabiodiversity.org; Stearn et al. 2002).

Distribution of Epimedium Species

Epimedium species are distributed mainly in China, which is also the centre of diversity and origin for this genus. About 80% species (48) of Epimedium are found in different regions of China (Zhang et al. 2016). They are also reported from other regions of Asia, Caucasian mountains, European Alps and North Africa (Zhang et al. 2007a; Zeng-li 2012; www.gbif.org). *E. alpinum* is found in the Alps and other mountain ranges of the Balkan Peninsula. *E. perralderianum* is found in Algeria whereas *E. pubigerum* is found along the north coast of Turkey. *E. pinnatum* is found in the Caucasus region (Zhang et al. 2007a). In China, about 15 Epimedium species are circulated in the crude drug markets for use as Ying Yang Huo (Ma et al. 2011). Among these 15, only 5 are regarded as the official source of *Herba Epimedii*. The distribution pattern of Epimedium species is summarised in Table 1.

Table 1 Distribution of Epimedium plants in the north temperate world

Distribution	Name of Epimedium species
China	*E. davidii, E. baojingense, E. dolichostemon,* ***E. koreanum****, E. fargesii, E. elongatum,* ***E. acuminatum****, E. simplicifolium, E. brachyrrhizum, E. multiflorum, E. enshiense, E. fangii, E. reticulatum, E. sagittatum var. glabratum, E. hunanense, E. latisepalum, E. ogisui, E. chlorandrum, E. platypetalum, E. franchetii, E. truncatum, E. borealiguizhouense, E. leptorrhizum, E. rhizomatosum,* ***E. pubescens****,* ***E. sagittatum****, E. sagittatum var. sagittatum, E. pauciflorum, E. lishihchenii, E. shuichengense, E. sutchuenense, E. myrianthum, E. flavum,* ***E. wushanense****, E. ecalcaratum, E. glandulosopilosum, E. parvifolium, E. stellulatum, E. brevicornum, E. ilicifolium, E. mikinorii, E. zhushanense, E. epsteinii, E. jinchengshanense, E. baiealiguizhouense, E. circinatocucullatum, E. dewuense, E. leptorrhizum*
Japan	*E. cremeum, E. diphyllum, E. grandiflorum, E. trifoliatobinatum, E. setosum, E. grandiflorum var. thunbergianum, E. kitamuranum, E. macranthum, E. grandiflorum var. higoense, E. grandiflorum var. coelestre, E. sempervirens, E. sempervirens var. multifoliolatum*
Europe	*E. alpinum, E. pubigerum, E. pinnatum, E. pinnatum subsp colchium, E. canrabrigensis, E. perralderianum*
North Africa	*E. perralderianum, E. pinnatum*
India	*E. elatum*
Korea	*E. koreanum*

Those highlighted in bold are the official source of *Herba Epimedii* in traditional Chinese medicine

The distribution of *E. elatum* is poorly documented. As per the local floras on Kashmir Himalaya, *E. elatum* has been reported from Pahalgam, Gulmarg and Harwan, forest ecozones of the Kashmir valley (Singh and Kachroo 1987; Sharma and Jamwal 1988). Recently, geographic distribution of *E. elatum* was pioneered and reported from 20 wild habitats of Kashmir Himalayas, by Lone et al. (2017) for the first time.

Among other countries of Asia, in Pakistan its distribution is scanty and has been observed in the Pallas valley in Kohistan district and Kaghan Valley (Manschra), North-West Frontier Province, Pakistan (Rafiq 1995; Jamal 2009; www.gbif.org). Recently it was reported from Utror and Gabral valleys, which are situated in the remote northwestern part of Pakistan (Jan et al. 2015; www.gbif.org).

Habitat Characteristics

Habitat of Epimedium species is generally cool shade of moist coniferous forests and also under cool shade of thickets. They are usually found near streams and wet land while some prefer to grow on cliffs near rock crevices. They prefer acidic soils and direct sunlight induces dwarf character in them and therefore, during dry and hot seasons, they must be watered to protect the major constituents within the plant (Ma et al. 2011). Lone et al. (2017) studied the habitat characteristics of *E. elatum* and it was shown to prefer moist shady habitats under coniferous forests in Northwestern Himalayas of India. According to this study, *E. elatum* accessions growing in cool and shady environment, under natural protection, were found to be comparatively taller under the cool shade of other plant species, ranging from tall pine trees and shrubs to grasses. The plants growing in subalpine regions were shown to display better growth characteristics than those growing near foothills and open sunny conditions. The populations growing under direct sunlight or subjected to anthropogenic disturbances like grazing were observed to be shorter in height. The study concluded that *E. elatum* has a very dwindling population status in Kashmir Himalayas. It is poorly distributed across most of the surveyed habitats due to habitat shrinkage. Kashmir Himalayas has recently witnessed depletion of medicinal flora due to excessive harvesting and anthropogenic pressures (Tali et al. 2015). In this regard, further studies are needed for the assessment of threat status as far as its conservation is concerned in Northwestern Himalayas in India (Lone et al. 2017).

Propagation of Epimedium Species

Epimediums are commonly propagated by rhizome to preserve cultivar identity and overcome low seed viability (Lubell and Brand 2005). Not much work has been published on the habitat characteristics and propagation of Epimedium species in general and *E. elatum* in particular. Propagation strategy of *E. elatum* at high-attitude medicinal gardens would be an important project for germplasm conservation especially in Himalayan mountain range.

Pharmacological Importance of Epimedium Species

Ethno-pharmacological Use

Epimedium species have a long history of ethno-pharmacological use in traditional Chinese medicine (TCM). The medicinal usage (Table 1) of *Herba Epimedii* was first recorded in Shen Nong Ben Cao Jing, the oldest book of Materia Medica in China, and it was considered as a 'middle-grade' herb in Ben Cao Gang Mu, another famous Chinese medicinal document (Ma et al. 2011). According to ancient TCM documents, the aphrodisiac property of Epimedium species was discovered by a Chinese goat herder who noticed more sexual activity in his goats on eating these plants (Ma et al. 2011). This is the reason that Epimedium plants are known as horny goat weeds. Dried leaves of more than 15 Epimedium species are being used as botanical supplement '*Herba Epimedii*', widely used as a tonic, aphrodisiac and anti-rheumatic preparations in China, Japan and Korea for more than 2000 years (Ma et al. 2011). *Herba Epimedii* has the effects of reinforcing the kidney yang;

strengthening the tendons and bones; relieving rheumatic conditions; weakening of the limbs, rheumatoid arthralgia with numbness and muscle contracture; and climacteric hypertension (Pharmacopoeia Commission of PRC 2010).

There are numerous commercial products containing *Herba Epimedii* which are sold as health supplements. Epimedium species like *E. sagittatum* and *E. grandiflorum* have been traditionally used to treat impotence, prospermia, hyperdiuresis, osteoporosis, menopause syndrome, rheumatic arthritis, hypertension and chronic tracheitis in China and Japan (Ma et al. 2011). In Korea, *E. koreanum* has been traditionally used for impotence, spermatorrhoea and forgetfulness (Liu and Xu 1984). The aerial parts of some Epimedium species such as *E. myrianthum*, *E. acuminatum* and *E. leptorrhizum* are also used by certain localities in China (Xie et al. 2010) while underground parts of Epimedium plants are widely used as anti-rheumatic medicine in ancient Chinese folk medicines (China Herb Compilation 1975). *E. pinnatum* and *E. alpinium* have also been used in folk medicines by some forest communities in some European and Asian countries (Ma et al. 2011) (Table 2).

E. elatum has been traditionally used to treat cold, cough, running nose, toothache, tooth decay and diseases of bones and joints in Northwestern Himalayan region of India (Arief et al. 2015). Literature cites Arief et al. (2015) confirming the ethnomedicinal usage of *E. elatum* for the first time from Northwestern Himalayan region. However, more ethnobotanical surveys are needed to explore ethnomedicinal uses prevalent among tribal forest communities living in Himalayas. The time-tested traditional medicinal uses in modern confirmatory experiments can boost its chances of becoming a potent medicinal plant in coming decades. However, for that, much research efforts are needed from all stakeholders.

Aphrodisiac Activity

Epimedium species have been used traditionally in various Asian medicines for a long time, to enhance erectile dysfunction and other impotence conditions in human beings. This has been demonstrated experimentally in various animal modelling studies. In several studies, Epimedium flavonoids increased sexual activity by enhancing sexual arousal, increasing vitality and improving sperm counts in vitro and in vivo (Ma et al. 2011; Chen et al. 2015a; Li et al. 2015; Shin et al. 2015). Total flavonoids from *E. brevicornum* were shown to promote male reproductive system and reproductive endocrine activities, thus possessing testosterone-like effects (Wang et al. 2001; She et al. 2003: Zhang and Yang 2006). The aqueous extracts of *E. brevicornum* have been shown to protect the structure and function of sperm membranes by improving their superoxide dismutase (SOD) vitality and intervening in lipid peroxidation (Yang et al. 2007). Epimedium flavonoids have also been shown to increase testicular weight, sperm counts and sperm motility (Yuan et al. 2014). Pharmacological studies have shown that aqueous extract decreases the corpus cavernosum smooth muscles by increasing the amount of cyclic guanosine monophosphate (cGMP), consequently inhibiting the phosphodiesterase type

Table 2 Reported biological action and bioactivity of Epimedium extract

S. No.	Biological action(s) or pharmacological activity	Extract or constituent used	References
1.	Aphrodisiac PDE-5 inhibitor	Extracts of *Epimedium* sp. Icariin	Chen (2009)
	Enhancer of testosterone secretion and kidney tonic	Decoction of *Epimedium* sp.	Wang et al. (2001)
2.	Effect on bone metabolism Anti-osteoporosis	Total flavonoids and icariin Water extract Total flavonoids	https://doi.org/10.1016/j.jep.2018.04.035
	Proliferation of primary osteoblasts	Total flavonoids, polysaccharides and main chemical constituents of Epimedium sp.	Zhang et al. (2008)
	Apoptotic inducer of osteoclast cells	Total flavonoids of Epimedium sp.	Zhang et al. (2008)
	Anti-rheumatoid arthritis	Ikarisoside A	Choi et al. (2010)
	Proliferation of bone marrow-derived stroma cells (BMSCs)	Icariin	Wu et al. (2009)
	Promoter of cartilage growth	Crude extract of *Epimedium brevicornum*	Li et al. (2012b)
3.	Effect on immune system Thymus activator	Methanolic extract(roots and rhizomes) of *E. alpinium*	
	Macrophage activator	Total flavonoids of Epimedium sp.	Zhang and Yu (1999)
	NK and LAK cell enhancer(s)	Total flavonoids and polysaccharides of *Epimedium* sp.	Meng et al. (1996)
	Enhancer of antibody response	Extract of aerial parts and epimedin C	Liang et al. (1997a, b)
4.	Effect on cardiovascular system inhibitor of cerebral ischaemia and hypoxia	Icariin	Song et al. (2011)
	Anti-myocardial ischaemia	Water extract of *Epimedium* sp.	
	Anti-hypersensitive activity	Aqueous extracts of *Epimedium* sp. Total flavonoids of *Epimedium* sp.	Mahboubi et al. (2013)
	Anti-arrhythmia	Crude water extract	
	Angiogenesis enhancer	Leaf and stem extract	Wang et al. (2007)
5.	Anticancer activity Inhibition of tumour cell proliferation	Icariin	Wang et al. (2007)
	Inhibition of tumour cell proliferation	Icariin	Cheng et al. (2007)
6.	Apoptosis inducer	Icariin	Ikeda et al. (2017)
8.	Anti-ageing	Aqueous extracts Total flavonoids Total flavonoids	Wu et al. (2003a) Ikeda et al. (2017) Meng et al. (1996)
9.	Anti-fatigue	Total flavonoids	Ma et al. (2009)

(continued)

Table 2 (continued)

S. No.	Biological action(s) or pharmacological activity	Extract or constituent used	References
10.	Anti-hypoxia	Total flavonoids	Zhang et al. (2009)
11.	Antibacterial	Icariin	Yan and Qiu (2005)
12.	Antiviral	Total flavonoids	Fang et al. (2003)
13.	Anti-inflammatory	Total flavonoids	Diaz et al. (2012)
14.	Anti-hepatotoxic	Icariside II Icariin	Cho et al. (1995)

5 enzyme (Chiu et al. 2006). According to the modern pharmacological studies, phosphodiesterase-5 (PDE-5) is the target protein for inhibition to treat erectile dysfunction. Studies have demonstrated that flavonoid glycosides, icariin and epimedin C, have been shown to possess an inhibiting effect on PDE-5 (Chen 2009). Modification of icariin by addition of two hydroxyethyl ethers moieties enhanced the PDE-5 inhibitory activity by 80-fold in one of the studies (Dell'Agli et al. 2008). Icariin and its synthetic analogues were shown to have a strong phosphodiesterase-5 (PDE-5) inhibitory activity and hence they were thought to cure erectile dysfunction in human beings (Dell'Agli et al. 2008). Epimedium herbs are popular testosterone adjuncts and could be used in future to treat hypogonadism in modern clinical circles (Jiang et al. 2016). Overall, Epimedium herbs have a strong potential in treating various problems related to sexual dysfunction disorders in human beings. *E. elatum* also contains the major flavonoid glycosides reported to have the above bioactivity (Zhengzheng et al. 2010). The species can be exploited for this activity and thus can become a potent aphrodisiac medicinal species from India in years to come.

Anti-osteoporosis Activity

Experimental studies have confirmed that Epimedium and its metabolites have a profound effect on bone metabolism, such as preventing calcium loss, stimulating the proliferation of osteoblasts, inhibiting bone resorption and promoting bone formation (Ma et al. 2011). Extracts from the dried aerial parts of Epimedium have long been used for bone health. Epimedium flavonoids are known to enhance bone growth by increasing the differentiation of bone mesenchymal stem cells (BMSCs) into osteoblasts and also by increasing the proliferation and maturation of osteoblasts (Liu et al. 2005). The water extract of Epimedium flavonoids has been shown to inhibit the proliferation, differentiation and bone resorption of osteoclasts by inducing G2/M arrest and apoptosis (Zhang et al. 2012). Icariin is known to induce the BMSCs into osteoblasts by up-regulating the expression of some osteoblast-related cytokines, such as transforming growth factor, bone fibroblast growth factor, insulin-like growth factor 1, osterix and Runx-2 (Liu et al. 2006a; Wu et al. 2010; Zhai et al. 2010). In addition, it has been shown to improve the secretion of alkaline phosphatase (ALP), colony-forming unit-fibroblast, osteocalcin and collagen I in BMSCs (Liu et al. 2006a; Zhai et al. 2010; Bian et al. 2011). Besides, it remarkably promoted the proliferation and maturation of osteoblasts (Hsieh et al. 2010; Ming et al. 2013) and modulated bone formation mainly via BMP-2/Smad4 signalling pathway (Hsieh et al. 2010; Liang et al. 2012a). Icariin was also shown to inhibit the formation and activity of osteoclasts by promoting osteoclastic apoptosis, inducing G2/M cell cycle arrest (Zhang et al. 2012; Ming et al. 2013), and decreasing the activity of osteoclast differentiation marker (Hsieh et al. 2011). It was shown to deter osteoclast differentiation and bone resorption by inhibiting interleukin 6 (IL-6), tumour necrosis factor-α (TNF-α) and cyclooxygenase type 2 (COX-2) expression (Hsieh et al. 2011).

Some of the recent studies have explained the oestrogen-like mechanism of Epimedium in preventing and treating osteoporosis (Xu et al. 2016; Chen et al. 2016a; Wang et al. 2016). Nine prenylflavonoid compounds from Epimedium species have been reported to target oestrogen signalling and other bone morphogenesis pathways in mesenchymal stem cell, osteoblast and osteoclast cell lineages (Indran et al. 2016). Recent clinical trials have reported positive effects of these medicinal plants on bone health, suggesting that compounds or extracts from them have the potential to be developed as agents, alone or in combination with other drugs, to prevent or delay the onset of osteoporosis and reduce the risk of hip fractures (Indran et al. 2016). Experimental evidence has proved that Epimedium species have a promising role to play in treating osteoporosis in coming decades. This raises expectation and interest in Epimedium plants for having valuable role in treating bone-related disorders in human beings.

Anticancer Activity

Modern pharmacology is slowly and surely discovering important bioactivities of epimedin flavonoids. According to recent pharmacological studies, icariin, icaritin, icariside II and its derivatives have strong anti-proliferative activities against various cancer cell lines (Tan et al. 2016) such as prostate carcinoma PC-3 cells (Huang et al. 2007a), hepatica carcinoma SMMC-7721 and HepG2 cells (Li et al. 2010a), gastric cancer BGC-823 cells (Wang et al. 2010), Leydig tumour MLTC-1 cells (Wang et al. 2011), gallbladder carcinoma GBC-SD and SGC-996 cells (Zhang et al. 2013a). Several studies have shown that the compound icaritin inhibits the proliferation of a variety of cancer cell lines, such as chronic myeloid leukaemia K562 cells (Feng Zhu et al. 2011); endometrial cancer Hec1A cells (Tong et al. 2011); breast cancer MCF7 cells (Guo et al. 2011); human acute myeloid leukaemia NB4, HL60 and U937 cells (Li et al. 2013); hepatica carcinoma HepG2 cells (He et al. 2010a; Sun et al. 2015); and osteosarcoma SaOS2 cells (Wang and Wang

2014). Baohuoside I, another key flavonoid glycoside in Epimedium species, has been shown to possess anti-proliferative properties against melanoma B16 (Wu et al. 2013) and A375 cells (Wu et al. 2012a), lung cancer A549 cells (Song et al. 2012), acute myeloid leukaemia U937 cells (Kang et al. 2012) and multiple myeloma U266 cells (Kim et al. 2011).

In recent years, there has been great interest in scientific investigation of the reported anticancer properties of icariin and its derivatives. Data reported from in vitro and in vivo studies demonstrated significant anticancer activity of epimedin compounds against a wide range of cancer cells which occurs through various mechanisms such as apoptosis, cell cycle modulation, anti-angiogenesis, anti-metastasis and immune modulation (Tan et al. 2016). According to pharmacological studies, epimedin C has been shown to enhance the immune system by improving the response of spleen antibody-forming cells to near-normal, up-regulating lymphocyte proliferation, and promoting the recovery of IL-2 production (Liang et al. 1997a). In addition, it shows anti-cancer effects by inducing G0/G1-phase cell cycle block via the increase in expression of p21Cip1, and p27Kip1, and decrease in c-Fos expression (Liu et al. 2006b). All the four major markers of Epimedium species, viz. epimedin A, epimedin B, epimedin C and icariin, have shown a strong anticancer activity in another study conducted on *E. koreanum* (Yasukawa et al. 2016). Thus, the Epimedium herbs may be useful in cancer prevention. All the parts of *E. elatum* are rich source of flavonoid glycoside compounds. This species needs to be bio-prospected for future use.

Neuroprotective Activity

Epimedium extracts and their purified compounds have demonstrated promising neuroprotective activities (Ma et al. 2011). Epimedium flavonoids have protective effect on 'neural synaptic structure and function' by promoting the expression of synapse-related proteins, such as synaptophysin and postsynaptic dense material 95 (Chu et al. 2008). They also promote proliferation and differentiation of hippocampal neural stem cells (Yao et al. 2010). Icariin, one of the main flavonoids of Epimediums, enhances the protein expression of brain-derived neurotrophic factor and tyrosine kinase tropomyosin receptor kinase B by increasing SOD activity and decreasing malondialdehyde (MDA) content in the hippocampus (Luo et al. 2007; Li et al. 2010b). The molecule showed inhibitory effects on age-related cognitive degeneration by increasing the expression of 5-bromo-2-deoxyuridine enabling polysialylated neural cell adhesion molecules to activate quiescent neural stem cells (Wu et al. 2012b). The apoptosis of neural stem cells is also inhibited by icariin by ameliorating mitochondrial membrane potential and blocking the activation of p38 (Liu et al. 2011). The molecule also improves the learning and memory abilities by increasing the levels of acetylcholine and choline acetyltransferase in the cerebral cortex (Gao et al. 2012; Zhang et al. 2014a). In addition, icariin has been shown to protect the induced neurotoxicity and cognitive degeneration by decreasing the

TNF-α and COX-2 expression (Guo et al. 2010; Zhang et al. 2015), increasing the monoamine levels, decreasing the acetylcholinesterase activity and inhibiting oxidative damage (He et al. 2010b).

Some of the studies have indicated the neuroprotective properties of icariin and icariside II in mice, and according to a study by Chen et al. (2016a, b) both might become a promising multi-target drug in the protection against Alzheimer's disease (Chen et al. 2016b). These recent activities of major flavonoid glycosides indicate that Epimedium plants have good potential in treating neurodegenerative diseases. Studies in this direction will help in understanding the underlying mechanism of neurodegenerative diseases and in discovering the ways for their treatment.

Preventing Cardiovascular Diseases

The extracts from the Epimedium species have shown to affect the blood pressure, blood rheology, arrhythmia, myocardial ischaemia, coronary blood flow and heart rate (Ma et al. 2011). Epimedium flavonoids have shown to prevent myocardial infarction by enhancing the level of serum creatine phosphokinase, lactate dehydrogenase and superoxide dismutase (SOD), and decreasing the content of malondialdehyde (MDA) (Huang et al. 2006). Icariin induces the cardiomyocyte differentiation of murine embryonic stem cells through various mechanisms involving up-regulated expression of p38, extracellular signal-regulated protein kinases (ERK), activator protein-1 and transcription factors 'c-jun' and 'c-fos' (Wo et al. 2008). Also, it has been known to enhance the calsequestrin and sodium–calcium exchanger (Ding et al. 2008; Sun et al. 2011). It stimulates angiogenesis by activating MEK-/ERK-dependent and phosphatidylinositol 3-kinase/Akt/endothelial nitric oxide synthase-dependent signalling pathways (Chung et al. 2008; Jin et al. 2010). Furthermore, icariin ameliorates left ventricular dysfunction and cardiac re-modelling by down regulating matrix metalloproteinase 2 (MMP-2), matrix metalloproteinase 9 (MMP-9) and myocardial apoptosis in rats with congestive heart failure (Song et al. 2011). Recent studies have indicated cardiac ischaemia and reperfusion injury can be alleviated by icariin treatment (Zhai et al. 2015). The modern lifestyle has triggered growth in lifestyle-related diseases including heart. Many clinical studies have indicated usefulness of Epimedium plant-based flavonoid glycosides in preventing these ailments. However, more studies are required in this direction but definitely Epimedium plant can be exploited in this course.

Anti-inflammatory Activity

Epimedium flavonoids, particularly icariin, have shown to possess potent anti-inflammatory activities (Ma et al. 2011). Icariin significantly alleviated the pathological changes of colitis by suppressing the phosphorylation of signal transducers and activating transcriptions 1 (p-STAT1) and 3 (p-STAT3) in CD4þT cells (Tao et al. 2013). It has been shown to attenuate acute lung inflammation by inhibiting the expressions of TNF-IL-6, and inducible nitric oxide synthase (Wu et al. 2009; Xu et al. 2010). Studies have shown that icaritin interferes with T-cell activation by inhibiting the proliferation of CD4+T cells, down-regulating CD25, IL-2 and interferon expression (Li et al. 2012a). In vivo studies have demonstrated that icariin attenuates the chronic airway inflammation through the mediation of Th17/Treg function (Wei et al. 2015).

Anti-ageing, Anti-fatigue and Anti-hypoxia Activities

Excellent reviews by Ma et al. (2011) and Chen et al. (2015a) on Epimedium plants have highlighted that epimedin flavonoids possess anti-ageing, anti-fatigue and anti-hypoxia activities. These flavonoids were shown to affect ageing by regulating the immune and endocrine systems, thus improving metabolism and organ function (Ma et al. 2011). They have been shown to delay natural senescence in animals in cell denucleation and cell fusion experiments (Meng et al. 1996). Epimedium was known to protect the mitochondrial DNA from oxidative damage in aged rats (Wang et al. 1996). One study has demonstrated that polysaccharides from *E. wushanense* have pronounced the effect on the production of superoxide dismutase (SOD) enzyme complex and glutathione peroxidase activity (Zeng et al. 1997), thereby playing a role in anti-ageing. The flavonoids from Epimedium possess anti-hypoxia activity by (a) prolonging the survival time of the normobaric hypoxic mice, (b) lessening encephaledema and pneumonedema and (c) raising the contents of haemoglobin and leukocytes (Zhang et al. 2009). Total flavonoids of Epimedium display anti-fatigue activity. This was demonstrated by prolonged swimming in mice, accompanied by increase in blood glucose level, haemoglobin and glycogen levels (Ma et al. 2009; Wang et al. 2014a).

Antimicrobial and Antiviral Activities

Epimedium flavonoids have significant antimicrobial and antiviral activities (Ma et al. 2011). They have strong inhibitory effects on a variety of microbes (Ma et al. 2011). Icariin has been shown to inhibit the activities of food-pollutant bacteria (Yan and Qiu. 2005). These flavonoids inhibit growth of several microorganisms including Micrococcus pyogenes var. albus, *Staphylococcus aureus*, *Diplococcus pharyngis communis, Micrococcus catarrhalis* and *Haemophilus influenzae* (Ma et al. 2011). Epimedium flavonoids showed positive effect on asthma in young children caused by viral infections in one of the studies conducted by Fang et al. (2003). The antimicrobial activity of chemical compounds isolated from *E. elatum* has been reported on six microbial strains, viz. *Escherichia coli*, *Salmonella typhi*, *Shigella dysenteriae*, *Klebsiella pneumoniae* and *Pseudomonas aeruginosa* (Tantry et al. 2012). Epimedin C possesses specific antimicrobial activity against *S. aureus* whereas quercetin was shown to have a broad antimicrobial activity against all the tested microbial strains with the exception of *S. aureus*. Elatoside A and Elatoside B [two unusual acylated flavonol glycosides found in *E. elatum*] showed antimicrobial activity against *S. aureus* and *P. aeruginosa* (Tantry et al. 2012). Epimedium polysaccharides possess synergistical antiviral action in chicken (Fan et al. 2011). Phytochemicals from *E. koreanum* have recently been known to display a strong antiviral activity against porcine epidemic diarrhoea virus (PEDV)-related diseases in pigs (Cho et al. 2012). In another study conducted by Cho et al. (2015), total aqueous extracts from *E. koreanum* showed a wide array of antiviral actions against a number of viruses both under in vitro and in vivo conditions. Thus, the use of *E. koreanum* as an orally antiviral agent has the potential to be an effective herbal remedy for prophylaxis and therapeutic applications in both humans and livestock (Cho et al. 2015). Further, more species of this genera may be assessed for their antimicrobial and antiviral activities.

PPAR-γ Ligand-Binding Activity

Wide ranges of natural products from medicinal plants have been screened for their peroxisome proliferator-activated receptor (PPAR) activating potential (Wang et al. 2014b). PPARs belong to a subfamily of the nuclear receptor superfamily of ligand-inducible transcription factors and to date three PPAR isotypes encoded by separate genes have been identified, viz. PPAR-α, PPAR-β/δ and PPAR-γ (Wang et al. 2014b). PPAR-γ ligand-binding activity of chemical compounds isolated from *E. elatum* was reported by Tantry et al. (2012). The most potent PPAR-γ ligand-binding activity was found with icariin, epimedin B, epimedin C, icariside II, icaritin and ikarisoside A. Studies on the activity of these PPAR-γ activators imply that the compounds having prenyl units are necessary for the appearance of the potent activity (Tantry et al. 2012). This was the first report wherein isolated chemical compounds from any Epimedium species have been evaluated for their peroxisome proliferator-activated receptor gamma ligand-binding activity. These findings encourage bioprospection and bio-evaluation of *E. elatum* that might provide interesting ligands in the future which might be useful in treating diabetes myelitis and related diseases.

Phenotypic Plasticity

Phenotypic plasticity in plants is defined as the ability of a single genotype to develop multiple phenotypes under different eco-edaphic environmental conditions (Palacio-López et al. 2015). These plastic responses to the environment can include changes not only in morphology but also in physiology, behaviour, life history, growth and demography of plant species (Miner et al. 2005). They can occur within the lifespan of an individual or across generations (Miner et al. 2005). High-altitude medicinal plants in Northwestern Himalayas are known to show a significant phenotypic plasticity in different habitats and under different environments (Badola and Aitken 2003). For example, *Podophyllum hexandrum* and *Sinopodophyllum hexandrum*, the highly prized medicinal plants in India and China, have been known to show a marked morphological variation in leaf polymorphism (Badola and Aitken 2003; Alam et al. 2008; Sultan et al. 2010; Liu et al. 2014). *Picrorhiza kurroa*, another high-altitude medicinal plant found in Indian Himalayas, has been known to produce variations in leaf morphology in different eco-edaphic environments of Northwestern Himalayas of India (Nautiyal et al. 2001). The narrow leaf variants in *P. hexandrum* and *P. kurroa* grow commonly in open pastures and along springs whereas the broad-leaf variants are found in shady habitats under shrub and scrub canopies (Badola and Aitken 2003; Sultan et al. 2010). Interestingly, only broad-leaf varieties in both medicinal species are considered pharmaceutically beneficial (Badola and Aitken 2003), which indicates the importance of studying morphological and genetic traits in identifying superior germplasm in the pursuit of either in situ or ex situ conservation (Badola 2002). Literature cites meagre information on phenotypic plasticity in Epimedium species but now there is renewed interest in documenting the habitat and morphology in both wild and cultivated Epimedium species. Studied phenotypic plasticity of 17 Epimedium species collected from Korea (9), Japan (7) and China (1) using six morphological characteristics. All the species were shown to exhibit a high level of morphological variation. Ming Feng (2008) was able to observe continuous morphological variation in 18 Epimedium species collected from different geographical regions of China, revealing several difficulties in their taxonomic identification. Yong et al. (2010) investigated the morphological variation in three Epimedium species, viz. *E. brevicornum*, *E. sagittatum* and *E. pubescens*, all showing significant phenotypic variability. The study showed that resource allocation to different plant parts under different environmental conditions was variable. The highest proportion of biomass was allocated to rhizomes and leaves and then to stems and roots. Among the three Epimedium species that he studied, *E. pubescens* was shown to produce highest phenotypic plasticity in different habitats and was thus better adapted for transplantation and cultivation trials (Yong et al. 2010). Quan et al. (2011) investigated the comparative phenotypic plasticity in two species of wild and cultivated Epimedium: *E. pubescens* and *E. wushanense*. Both the medicinal species were shown to be differing in their morphological characters. The leaf area and branch height per plant were remarkably higher in wild *E. pubescens* and *E. wushanense* than the cultivated species of the two, while the leaf number and branches per plant in these species were shown to be higher under cultivated conditions. The root, stem and leaf biomass per plant were observed to be higher in cultivated species than the wild plants of *E. wushanense*. The morphology of both the Epimedium species showed dwarfness under cultivated under high light intensities. They had more branches and leaves in cultivation than in wild habitats. The plants under shade were shown to produce larger leaves than the plants grown under open sunny conditions, which confirm the shade-loving habitat (Lone et al. 2017) of Epimedium species. Also, shady conditions have been known to enhance the content of bioactive components (epimedin ABC and icariin) in Epimedium species (Liang et al. 2012b).

Horie et al. (2012) investigated phenotypic plasticity in many populations of *E. diphyllum* and *E. sempervirens* species inhabited in different habitats of Japan. Both species were shown to be morphologically diverse in all habitats and showed frequent hybrid formation among them. All of the hybrids were shown to have a considerable morphological variation in their floral characteristics. The results of this study confirmed natural course of hybridization in Epimedium species in confirmation with earlier report of Sheng et al. (2011). Liang et al. (2013) reported significant morphological diversity in natural populations of *E. sagittatum* found in China. The morphological characters were shown to persist in the above species even in the common garden experiments, indicating the available diversity of different phenotypes in Epimedium species. Similarly, LanYing et al. (2014) investigated the morphological variation in three transplanted Epimedium species, viz. *E. wushanense*, *E. pubescens* and *E. acuminatum*, and all three species produced significant phenotypic plasticity in plant height and leaf morphology, under cultivation. They showed varying degrees of adaptability to the growing environment under various environmental conditions. Similar to this study, Zhou et al. (2014) studied morphological characteristics in the growth cycle of *E. acuminatum*. The study showed that proper habitat and good nutrition conditions were ideal for shortening the growth cycle and producing maximum herbage for medicinal purposes.

Xu et al. (2014) reported continuous morphological variations in different populations of *E. sagittatum* species complex. Xu et al. (2013b) had earlier reported variation in their non-glandular hairs. The study by Chen et al. (2015b) on the variations in morphological characteristics of *E. sagittatum* confirmed the variability in the plants growing in different geographical regions, particularly in the leaves and flowers. A difference in plant morphology of *E. pseudowushanense* has been documented recently by Pan and Guo (2016). In this study, seedlings of the species were shown to produce phenotypic plasticity under different light intensities. In shady areas, it had larger leaves than in open sunny conditions, thus confirming Epimedium species as shade-loving medicinal species as mentioned earlier. Xuemei et al. (2016) reported the impact of simulated warming on growth and floral characteristics of *E. wushanense* and *E. acuminatum*. Plant height, leaf growth and floral quantitative characters were shown to vary significantly under the influence of different warming temperatures in both the Epimedium species and they were shown to respond differently to temperature changes under cultivation, thereby displaying wide phenotypic plasticity. All of these investigations revealed that Epimedium species change their phenotypes under different environmental conditions, which also indicates the existence of different genotypes in Epimedium species. These studies on different species under different environmental conditions would throw light on the habitat characteristics and growth conditions of this less explored medicinal species. It would subsequently aid in the identification of its pristine habitats and their subsequent conservation from further exploitation.

Taxonomic Classification of Epimedium Species

The traditional methods of species identification in Epimedium systematics are inadequate due to complex taxonomical procedure and difficulties in identification of specimens based on their floral and leaf morphometry (Zhang et al. 2016). Earlier, leaves borne on the peduncle and corolla dimensions (floral character) were used for taxonomical identification of Epimedium species (Sun et al. 2005). Extensive morphological variation and frequent hybrid formation impeded their taxonomic identification (Sheng et al. 2011; Horie et al. 2012; Zhang et al. 2016). Biochemical markers have been used previously for understanding the molecular characterisation of Epimedium species (Xu et al. 2007; Zeng et al. 2010). Xu et al. (2007) investigated the genetic variability and population genetic structure of 11 natural populations of three Epimedium species, *E. pubescens*, *E. sagittatum* and *E. wushanense*, using allozymes as biochemical markers. Based on the study, they concluded *E. pubescens* to be more closely related to *E. sagittatum* than to *E. wushanense* in UPGMA-based cluster analysis, which was in agreement with the morphological characters and the recent phylogenetic analysis of these species. Further, the study proposed various factors responsible for the genetic variation in population structure of Epimedium species. Some of the significant factors were (1) mixed breeding system, (2) long-lived perennial life form, (3) ancient evolutionary history and (4) seed dispersal by ants (Xu et al. 2007). This was the only study where allozymes had been used for understanding population genetic structure of three medicinal Epimedium species from Hubei province in China. However, biochemical markers are not used frequently for genetic diversity analysis due to several disadvantages and hence are not preferred. The biochemistry of the plant is influenced by expression of genes which are dependent on environment and developmental stages and are under the influence of epistatic and pleiotropic interactions (Sarwat et al. 2012).

The most recent and updated system of classification was proposed by Stearn et al. (2002), in which Epimedium genus was classified into two subgenera, four sections and four series predominantly based on geographical distribution, leaf and flower morphology (Table 3).

Molecular markers and molecular taxonomical approaches have been barely used for understanding the phylogenetic relationship of different Epimedium species in general and *E. elatum* in particular. Literature cites very few studies where molecular markers have been employed to comprehend the phylogeny of *E. elatum*. Zhang et al. (2007a) reconstructed the phylogeny of 38 Epimedium species using ITS and atpB-rbcL spacer sequences as molecular markers. In this study, almost all sections recognised in the most recent classification of the genus Epimedium were found to be monophyletic, but subgenus Epimedium was found to be paraphyletic in relation to subgenus Rhizophyllum. *E. elatum* (western Himalaya) was shown to form a distinct lineage in all phylogenetic dendrograms, thereby indicating distinctness.

De Smet et al. (2012) recently gave additional evidence for origin and divergence of Chinese Epimedium species in section Diphyllon by using AFLP molecular markers in addition to nuclear and chloroplast genomic markers. According to their

Table 3 Stearn's (2002) classification of 54 Epimedium species, recognized up to 2002

I. Subgenus *Epimedium*	28. *E. zhushanense* K. F. Wu and S. X. Qian
i. Section *Diphyllon* (Kom.) Stearn	29. *E. baojingense* Q. L. Chen and B. M. Yang
A. Series *Campanulatae* Stearn	30. *E. gladulosopilosum* H. R. Liang
1. *E. campanulatum* M. Ogisu	**D. Series *Brachycerae* Stearn**
2. *E. platypetalum* K. Meyer	31. *E. pubescens* Maxim
3. *E. ecalcaratum* G. Y. Zhong	32. *E. brevicornu* Maxim
4. *E. shuichengense* S. Z. He	33. *E. reticulatum* C. Y. Wu
B. Series *Davidianae* Stearn	34. *E. sagittatum* (Sieb. and Zucc.) Maxim
5. *E. davidii* Franch	35. *E. myrianthum* Stearn
6. *E. fangii* Stearn	36. *E. stellulatum* Stearn
7. *E. hunanense* (Hand.-Mazz.) Hand.-Mazz.	37. *E. dolichostemon* Stearn
8. *E. flavum* Stearn	38. *E. fargesii* Franch
9. *E. ilicifolium* Stearn	39. *E. elachyphyllum*
10. *E. epsteinii* Stearn	40. *E. truncatum* H. R. Liang
11. *E. latisepalum* Stearn	41. *E. coactum* H. R. Liang and W. M. Yan
12. *E. ogisui* Stearn	42. *E. boreali-guizhouense* Yang
13. *E. pauciflorum* K. C. Yen	43. *E. lobophyllum* L. H. Liu and B. G. Li
14. *E. mikinorii* Stearn	**ii. Section *Macroceras* C. Morren & Decne**
C. Series *Dolichocerae* Stearn	44. *E. grandiflorum* C. Morren
15. *E. elongatum* Kom	45. *E. sempervirens* Nakai ex F. Maek
16. *E. membranaceum* K. Meyer	46. *E. koreanum* Nakai
17. *E. rhizomatosum* Stearn	47. *E. macrosepalum* Stearn
18. *E. lishihchenii* Stearn	48. *E. trifoliolatobinatum* (Koidz.) Koidz.
19. *E. acuminatum* Franch	49. *E. diphyllum* Lodd.
20. *E. franchetii* Stearn	**iii. Section *Polyphyllon* (Kom.) Stearn**
21. *E. enshiense* B. L. Guo and Hsiao	50. *E. elatum* Morr. & Decne.
22. *E. sutchuenense* Franch	**iv. Section *Epimedium***
23. *E. chlorandrum* Stearn	51. *E. alpinum* L.
24. *E. wushanense* T. S. Ying	52. *E. pubigerum* Morr. & Decne.
25. *E. leptorrhizum* Stearn	**II. Subgenus *Rhizophyllum* (Stearn)**
26. *E. brachyrrhizum* Stearn	53. *E. pinnatum* Fisch.
27. *E. simplicifolium* T. S. Ying	54. *E. perralderianum* Coss

study, all Epimedium species in section Diphyllon were shown to exhibit a high level of genetic diversity, and were proposed to be still in the process of differentiation in China. They also described four types of leaflet pubescence for the first time as a new taxonomic tool towards a modern interpretation of Epimedium taxonomy. They investigated the taxonomic position of *E. elatum* but it was shown to be inconsistent in phylogenetic trees (dendrograms) and so its taxonomy remained unresolved till date. In the Bayesian analysis, *E. elatum* was shown as a sister species to a clade congruent with Stearn's (2002) section Diphyllon, whereas in the maximum likelihood bootstrap (ML-BS) method it was shown as a sister species to the rest of Epimedium genus (De Smet et al. 2012). Different markers complicated its phylogenetic relationship, which was also depicted in the dendrograms. According to chloroplast data of De Smet et al. (2012), *E. elatum* was found as a sister to section Macroceras whereas nuclear data classified it in a clade in section Epimedium and subgenus Rhizophyllum, respectively. Currently, *E. elatum* is a single species in section Polyphyllon, but it needs further assessment to elucidate its origin and evolution in Northwestern Himalayas of India. Studies are therefore needed to determine its evolutionary relationship with the rest of the Epimedium genus.

Zhang et al. (2014b) utilised the AFLP fingerprinting technique for resolving taxonomic problems in Epimedium species. In the study, about 144 accessions from 58 Epimedium species and 1 accession of *Vancouveria hexandra* were selected for the study. The study proved highly significant in reconstructing the phylogenetic relationship of almost all Epimedium species. In the above report, the two subgenera and four sections of Epimedium were shown to be monophyletic based on dendrogram topology. The Chinese section Diphyllon was divided into five well-supported clades, which was shown to be related to flower characteristics (Zhang et al. 2014b). Section Diphyllon, containing all Chinese species, was first shown to cluster with *E. elatum* (section Polyphyllon), but later it was shown to form a trichotomy with section Macroceras and section Epimedium. The above research effort by Zhang et al. (2014b) was proposed to have a great implication in facilitating the utilisation of natural germplasm of Epimedium, especially for further development of new cultivars for ornamental and medicinal purposes. Epimedium genus thus needs more effective molecular markers to investigate the relationships between different species and their origin, evolution, migration and dispersal of the genus (Zhang et al. 2016).

DNA barcoding of Epimedium species is poorly known and there are only few reports where nuclear 5S rDNA intergenic spacer region was used for the authentication of Epimedium species (Sun et al. 2004, 2005). Later on, Yu et al. (2011) were able to successfully distinguish 37 specimens belonging to ten species of Epimedium using the psbA-trnH region. The authors suggested psbA-trnH sequence as the potential DNA barcode for whole Epimedium genus. Currently, more than 7390 nucleotide sequences of Epimedium species have been submitted in the Genbank (2017) and a rich identification database has been established.

E. elatum has usually been regarded as a single species in Northwestern Himalayas in India and DNA barcoding was needed to confirm this notion at molecular level. More efforts are required for authentication and conformity for this species from different parts of Himalayas where it has been reported to be growing.

Genetic Diversity

Genetic diversity is the consequence of genetic differences between individuals and is manifested as a change in the morphological, biochemical, physiological and DNA sequence characteristics. Genetic diversity can be thought of as reflecting the balance between appearance and disappearance of alleles (genetic variants). New alleles can appear at each generation by spontaneous mutation due to DNA replication errors or mutagen-induced DNA damage or recombination events (Ellegren and Galtier 2016). There are various factors such as selection (both artificial and natural), genetic drift and gene flow that act on alleles in different populations to cause variations. Ecogeographical factors such as climatic, edaphic and biotic factors including species-specific factors like ploidy level, breeding systems and population size have a definite role to play in determining the genetic diversity available within a species (Ellegren and Galtier 2016). With more variation, it is more likely that some individuals in a population will possess variations of alleles that are suited for the environment giving advantage to individuals as they are more likely to survive to produce offspring bearing that allele. The population will continue for more generations because of the success of these individuals (Frankham 2005).

Cataloguing of germplasm and information on genetic diversity are essential in formulating germplasm conservation programmes, whether in situ or ex situ. Thus assessment and characterisation of genetic diversity is an essential prerequisite for systematic documentation of plant species. In this context, molecular markers are vital in the assessment of genetic diversity, from gene and allele frequencies (genotype information) to nucleotide level like SNPs (Sarwat et al. 2012).

Non PCR and PCR-based molecular markers like RAPD, ISSR and EST-SSRs have been used previously for determining the genetic diversity in Epimedium species (Table 4). Nakai et al. (1996) analysed the genetic characterisation of 8 Epimedium species using a combination of RAPD and RFLP markers. In this study, all the investigated Epimedium species were shown be genetically different from each other, based on the unique RAPD and RFLP fingerprints. The study could identify *E. sagittatum* and *E. koreanum* easily based on their molecular banding patterns. Phylogenetic dendrogram analogy showed *E. sagittatum* to be a totally different species in comparison to other eight Epimedium species. RAPD and RFLP markers were proposed to aid in chemotaxonomical classification of Epimedium species. Utilised RAPD primers for investigating the genetic diversity of 17 Epimedium species collected from Korea (9), Japan (7) and China (1). Phylogenetic dendrograms and genetic distance results characterised all species at the genetic level. All species were shown to exhibit high level of genetic diversity. Ming Feng

Table 4 DNA fingerprinting status of Epimedium

Epimedium species	Fingerprinting markers	References
8 *Epimedium* sp.	RAPD and RFLP	Nakai et al. (1996)
17 *Epimedium* sp.	RAPD	
18 *Epimedium* sp.	RAPD and RFLP	Ming Feng (2008)
18 *Epimedium* sp.	RAPD	Li et al. (2011a, b)
53 *Epimedium* sp.	AFLP	De Smet et al. (2012)
E. acuminatum	ISSR	Mu-Dan et al. (2009)
7 *Epimedium* sp.	ISSR	Yan-Ying (2012)
E. brevicornum	SSR	Xu et al. (2008)
52 *Epimedium* sp.	EST-SSR	Zeng et al. (2010)
13 *Epimedium* sp.	EST-SSR	Yousaf et al. (2015)

(2008) investigated the genetic diversity of 18 Epimedium species employing RAPD and RFLP markers. The study established unique RAPD and RFLP fingerprints in all Epimedium species and showed them to have a good genetic diversity. Genetic diversity was correlated with their geographic distribution in China. Both the markers were proposed to aid in understanding the phylogenetic relationship in Epimedium species in combination with morphological markers.

In another study, RAPD based genetic diversity was studied by Li et al. (2011a, b) and Lim et al. (2000) in 18 Epimedium species. Genetic diversity was found to correlate with the geographical distribution of all Epimedium species and they were further classified into large-flowered and small-flowered species. Similarly, Ming-Jun (2011) investigated the genetic diversity of 17 Epimedium species using RAPD fingerprinting. Here, PCR technique could display a very high polymorphism (98.5%). They also reported amplification of some unique bands in some species and accessions, which were subsequently proposed to aid in the identification of Epimedium species.

Slowly with the advent of more reliable and reproducible techniques, ISSR was preferred over RAPD. ISSR technique offers several advantages over RAPD in being reproducible and easy to do and with no pre-sequence knowledge required. ISSR primers were used for the optimisation of PCR amplification in one of the studies conducted by Mu-Dan et al. (2009) in *E. acuminatum*. The authors, however, did not report its genetic diversity using the same marker system. Chen et al. (2012) investigated the genetic diversity in 12 populations of 7 Epimedium species using ISSR molecular markers. His study revealed a high genetic diversity among them (87.11%). The genetic diversity of 20 accessions of *E. elatum* has been reported by Lone et al. (2017), wherein 20 ISSR markers were utilised for generating unique DNA fingerprints. The study reported high genetic polymorphism (91.1%) in 20 populations of *E. elatum*. The unique bands amplified in some accessions were proposed to aid in its easy identification in future molecular fingerprinting studies (Lone et al. 2017). However, research is needed to utilise highly efficient markers like SSRs and EST-SSRs to correlate characters with genetic sequence for the complete genetic polymorphism in the different populations of *E. elatum*.

De Smet et al. (2012) analysed the genetic diversity and evolutionary relationship of 53 Epimedium species based on their AFLP fingerprinting revealing significant level of morphological and genetic variation in the collection. Liang et al. (2013) also utilised AFLP markers to demonstrate the genetic diversity in natural populations of *E. sagittatum* correlating morphological diversity with AFLP fingerprint data. Considering the drawbacks associated with AFLP fingerprinting, other molecular markers should be explored for their capacity to delimit *E. elatum* (Xu et al. 2008). These molecular insights can then be combined with clearly defined morphological characters in order to generate useful species descriptions in the above species.

Xu et al. (2008) investigated the genetic diversity of 38 accessions of *E. brevicornum*, utilising 17 pairs of SSR primers. The medicinal plant showed genetic polymorphism with number of alleles per locus. He could detect two to eight alleles with an average of 4.86, the expected heterozygosity (HE) ranging from 0.03 to 0.81, and the observed heterozygosity (HO) from 0.05 to 0.81. Moreover, SSR primers generated in the study were also investigated for their cross-species amplification in four Epimedium species, namely *E. koreanum*, *E. brevicornum*, *E. pubescens* and *E. wushanense*. Polymorphic microsatellite loci developed in this species using SSR technique were proposed to aid in the assessment of genetic diversity and genetic structure of *E. brevicornum* for its further evaluation and exploration in China.

Zeng et al. (2010) utilised EST-SSR marker system for analysis of genetic diversity of 52 Epimedium species in China displaying a high level of genetic diversity in these plants. They also studied cross-genus amplification of markers and used them for reconstructing Epimedium phylogeny for the first time. They proposed the use of EST-SSR markers in combination with other types of molecular markers for resolving the phylogeny of Epimedium species. Recently, Yousaf et al. (2015) also used EST-SSR markers for assessment of genetic diversity in 13 Epimedium species. They observed a high level of genetic variation in the investigated species and proposed the marker system in combination with morphological markers for effective resolution of the existing taxonomic problems in Epimedium species.

A retrotransposon-based marker system coined as internal primer binding site (iPBS) was used by Chen et al. (2015b) to assess the genetic diversity in *E. sagittatum* populations. Ten iPBS primers yielded 46 highly reproducible polymorphic markers. However, this marker system could not show significant genetic variation in *E. sagittatum*. The dendrogram showed no correspondence between chemotype and genotype of *E. sagittatum* populations. However, genetic clustering was shown to coincide with its geographical distributions in China.

Phytochemical Characterisation of Epimedium Species

Phytochemical characterisation of Epimedium species was first reported by Akai (1935) and since then more than 260 chemical moieties have been identified (Ma et al. 2011). Majority of the literature cites epimedin A, epimedin B, epimedin C and icariin as the dominant and characteristic markers in Epimedium species (Wu et al. 2003; Pei et al. 2007; Wang et al. 2007; Xu et al. 2013a; Chen et al. 2015a). Besides lignans, ionones, phenol glycosides, phenylethanoid glycosides, sesquiterpenes, essential oils, fatty acids and phytosterols are also being reported from different Epimedium species (Chen et al. 2015a). The summary of the phytochemical characterisation of Epimedium species done so far has been listed in Table 5.

Epimedium species are known to accumulate different concentrations of ABCI markers and only those Epimedium species which meet the quality standards are generally regarded as the potential medicinal plants for *Herba Epimedii* (Chen et al. 2015a). Several studies on phytochemical quantification of epimedin ABC and icariin have demonstrated that their concentration varies in different accessions of the same Epimedium species collected from different localities and also among the species (Liu et al. 2006c; Chen et al. 2007; Islam et al. 2008; Wu et al. 2008; Zhang

Table 5 Summary of systematic phytochemical characterisation of Epimedium species using HPLC and LCMS fingerprinting techniques

Name of the species	Marker quantified	References
E. elatum	Epimedin ABC and icariin	Sofi et al. (2014), Naseer et al. (2015)
E. elatum	Icariin, icariside I	Arief et al. (2015, 2016)
Herba Epimedii	Icariin	Sheng et al. (2008), Pei and Pei-Gen (2008)
Herba Epimedii	Epimedin C and icariin	Sun and Liang (2011), Peng et al. (2007)
Herba Epimedii	Epimedin C	Wang et al. (2003)
Herba Epimedii	Epimedin ABC and icariin	Guo et al. (1996), Shen et al. (2007), Pei et al. (2007), Zhang et al. (2007b), Huang et al. (2007b), Chen et al. (2008a), Islam et al. (2008), Nurul et al. (2009), Yu et al. (2010), Han and Lee (2017), Jin et al. (2009), Xie and Sun (2010), Xu et al. (2013a), Chen et al. (2007), Chen et al. (2008b), Wu et al. (2008), Bo et al. (2013)
Herba Epimedii	Breviflavone A and B	Hong et al. (2009)
E. koreanum	Flavonoids	Chen et al. (1996)
E. koreanum	Epimedin ABCI	Dong et al. (2010), Jia et al. (2010)
E. brevicornum	Epimedin B–C and icariin	Liang et al. (1997b), Yao et al. (2012), Xia et al. (2009)
E. brevicornum	Icariin	Quan et al. (2010)
E. acuminatum	Epimedin C and icariin	Lin et al. (2010)
E. wushanense	Epimedin C and icariin	Xie et al. (2007b)
E. wushanense	Epimedin ABC and icariin	Xie et al. (2011), Li et al. (2011b), Zhou et al. (2013)
E. sagittatum	Epimedin ABC and icariin	Liang et al. (2012b), Chen et al. (2015a)

et al. 2008; Sheng et al. 2008). Quan et al. (2011) reported the HPLC quantification of the flavonoid and icariin contents in aerial (leaf, stem) and underground parts of Epimedium species. The comparative study on contents of icariin and total flavonoid of *E. acuminatum* in different habitats and parts (leaf, root, stem and rhizome) was also reported by Zhou et al. (2012). According to this study, highest icariin content was found in leaf > root > stem > rhizome and the highest total flavonoid content was found in leaf (Fig. 3).

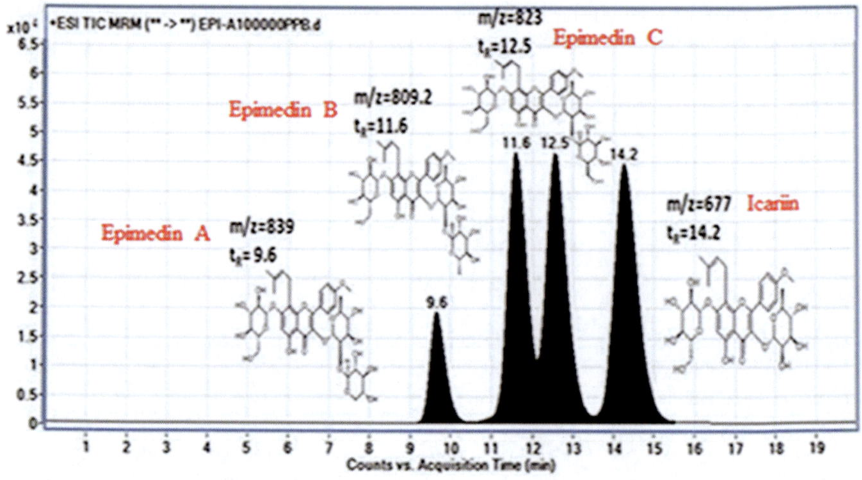

Fig. 3 MRM chromatogram of standard mixture of epimedin A, B and C and icariin

Large variation in ABCI contents has been reported among species. Chen et al. (2007) reported that ABCI multiglycoside concentration in four Epimedium species by HPLC and icariin content was shown to vary from 0.83 to 14.71 mg/g. Epimedin A varied from 0.35 to 3.31 mg/g, and epimedin B varied from 0.32 to 7.74 mg/g. Similarly, epimedin C was shown to vary from 1.22 to 27.63 mg/g whereas Baohuoside I content varied from 0.09 to 2.57 mg/g. In the above study, epimedin C was shown to be the dominant marker corroborating with earlier results of Pei et al. (2007), Guo et al. (1996) and Xie et al. (2007a).

Liang et al. (2012b) quantified icariin content in different populations of *E. sagittatum* and all were shown to produce a significant variation in ABCI multiglycosides under cultivation condition. The range of variation identified in different chemotypes of *E. sagittatum* spectrum of medicinal components could provide a valuable source material for selection and breeding of its novel varieties and also for its large-scale cultivation (Liang et al. 2012b). Xu et al. (2013a) reported an extensive variation in ABCI flavanoid glycosides in ten populations of *E. brevicornum* collected from different ecogeographical zones in China. In the study, epimedin B concentration ranged from 6.67 to 55.7 mg/g, while the concentration of epimedin C ranged from 5.39 to 23.0 mg/g in some of the chemotypes. The total flavonoids (ABCI) varied from 29.1 to 123 mg/g in ten populations. According to the Chinese standard for Epimedium species, almost all populations studied by Xu et al. (2013a) were medicinally rich in ABCI multiglycosides. Wang et al. (2013) studied the content of icariin and flavonoids in different accessions (plant parts) of *E. koreanum*. Icariin content in leaves and stem was reported to be lower than the Chinese standards of 0.5% (5 mg/g). However, the content of total flavonoids (ABCI) was reported to be higher in leaves (66.8–88.5 mg/g). Chen et al. (2015b) recently studied the phytochemical variation of 4 flavonoid glycosides (ABCI multiglycosides) in 11 wild populations of *E. sagittatum*. The study identified five elite chemotypes which showed differences in accumulation of these chemical markers under varying environments. This chemotype variation was shown to have a genetic basis in common garden experiment. The average concentrations of four markers among different populations were shown to vary from 7.80 to 26.90 mg/g for epimedin A, 8.73–20.89 mg/g for epimedin B, 11.18–55.71 mg/g for epimedin C and 3.45–20.68 mg/g for icariin. Epimedin C was found as the dominant flavonoid glycoside in *E. sagittatum* with concentration ranging from 49.77 to 56.54 mg/g in some elite chemotypes. The importance of these bioactive constituent quality and quantity reports in other Epimedium species is required to highlight the medicinal potential of *E. elatum*.

Chemoprofiling Studies of *Epimedium elatum* from Kashmir Himalayas

In the very first bio-prospection study of *E. elatum* from Northwestern Himalayas, ethanolic extract from the whole plant led to the isolation of 23 known natural products and 2 new acylated flavonol glycosides, i.e. elatoside and elatoside A (Tantry et al. 2012). Sofi et al. (2014) later collected this species from three ecogeographical regions in Kashmir Himalayas and reported the isolation, identification and quantification of ABCI multiglycosides in their aerial and underground parts. The concentration of epimedoside A ranged from 3 to 18 mg/g and it was reported to be abundant in underground parts. The concentration of icariin in aerial parts was 13 mg/g, much higher than underground parts. The concentration of epimedin A was shown to vary from 25.3 to 39.5 mg/g, higher than Zhang et al. (2008), who had reported ABCI variation in 20 Epimedium species. Both epimedin B and C were found as the major chemical markers in *E. elatum*. Epimedin C was found to be the most predominant flavonoid in *E. elatum* with a maximum value of 64.2 mg/g in aerial part (Sofi et al. 2014). The content of total 'ABCI multiglycosides' ranged between 16.9 and 105.5 mg/g as compared to the rest of Epimedium species. Based on their study, the optimal season for collecting Epimedium species was proposed to be from end of summer to beginning of the autumn season (Zhang et al. 1995). But Sofi et al. (2014) reported post-flowering season as the best harvesting time for *E. elatum* in Kashmir Himalayas. According to him, this season was found to be the ideal for production of 'ABCI multiglycosides' in *E. elatum* due to maximum herbage. This study confirmed *E. elatum* as a potential and promising medicinal plant as it matched standards set by Chinese Pharmacopeia Commission (2010) for Epimedium species.

In another study on chemo-biological standardisation of *E. elatum*, ABCI multiglycosides were isolated and simultaneously quantified by Naseer et al. (2015). The study found the concentration of different markers to be lower in root (icariin 5 mg/g, epimedin A 6.2 mg/g, epimedin B 5.2 mg/g and epimedin C 3.7 mg/g) than shoot where their concentration varied among 62.7 mg/g in epimedin C, 32.4 mg/g in epimedin A and 1.8 mg/g in epimedin B. Also, icariin was found in lower quantities in the aerial parts. Limitation in all of these studies was small sample size. This was taken care by another investigation where 20 accessions of *E. elatum* were studied for their phytochemical variation (Lone et al. 2017). The LCMS-based fingerprinting revealed significant variation among four investigated prenylated flavonoids (epimedin ABC and icariin) in different plant parts. The content of epimedin A was shown to vary between 0.89 and 10 mg/g, while epimedin B and C contents ranged between 0.95 and 20.5 mg/g, respectively. Icariin was shown to vary from 5.27 to 0.14 mg/g. In contrast to several earlier studies, rhizome was shown to accumulate higher content of icariin (2.69 mg/g) as compared to leaves (1.73 mg/g). Overall, aerial parts of *E. elatum* were reported to be rich in epimedin B and epimedin C, while underground parts were reported to be rich in icariin contents. Epimedin C was reported as major prenylated flavonoid glycoside in confirmation with other species like *E. sagittatum*, *E. pubescens*, *E. wushanense*, *E. acuminatum* and *E. myrianthum* (Pei et al. 2007; Guo et al. 1996; Xie et al. 2007a, b; Xu et al. 2013a, b). In this study, the average concentration of four prenylated flavonoid glycosides in leaves was shown to vary among 6.9 mg/g in epimedin C, 6.82 mg/g in epimedin B, 3.13 mg/g in epimedin A and 1.7 mg/g in icariin. However, further screening of several habitats to collect its elite chemotype from Northwestern Himalayas would give a comprehensive picture of *E. elatum* growing naturally in temperate and alpine regions of Himalayas.

Antioxidant and Biochemical (Total Flavanoid and Phenolic) Analysis

Reactive free radicals are the by-products in uncoupled electron flow in respiration and they rapidly attack molecules in nearby cells and damage lipids in cell membranes, proteins in tissues and DNA (Zhang et al. 2013b). Generation of reactive oxygen species (ROS) is related to the ageing process and degenerative diseases such as cancer and heart diseases (Pietta 2000; Chen et al. 2005). Since ancient times, the herbal medicines have been widely used as functional foods or pharmaceuticals all over the world. Interest in functional foods has been growing fast over the last few decades, leading to the discovery of new functional components or products that may help to retard ageing or prevent diseases (Zhang et al. 2014c). In nature, antioxidative enzymes (endogenous) such as superoxide dismutase have evolved to prevent the accumulation of free radicals by metabolising them. However, this protection is not sufficient under the prevalent stress conditions. Antioxidant herbal sources such as flavonoids and phenolics can provide additional power to enhance the ability in combating free radicals formed in the body. Multiple studies have focused on the antioxidant activities by in vitro assays and among the antioxidant compounds flavonoids and phenolics have been shown to possess antioxidant activities (Amico et al. 2008).

Flavonoids in Epimedium plants are well known to be the powerful antioxidants (Sze et al. 2010; Zhang et al. 2013b; Zhang et al. 2014c). The antioxidant activity is due to the scavenging or quenching of free radicals by chelating metal ions, or by inhibiting the enzymatic systems responsible for producing free radicals (Bláha et al. 2004; Dias et al. 2005). The aerial parts of Epimedium species especially leaves have high contents of total phenolic and total flavonoids which have been thought to be agents responsible for antioxidant activities (Zhang et al. 2013b; Zhang et al. 2014c). The study conducted by Lee et al. (2016) investigated the total phenol compounds (TPC) and total flavonoid compounds (TFC) in different plant parts of *E. koreanum* (root, stem, leaf) and among them root part was shown to exhibit highest TPC and TFC values. They also assessed the 2,2-diphenyl-1-picrylhydrazyl (DPPH), ferric reducing antioxidant power (FRAP) and 2,2'-azino-bis-3-ethylbenzthiazoline-6-sulphonic acid (ABTS) radical scavenging activities in the above medicinal plant and showed that leaf possessed the strongest radical scavenging activities.

Recently, Lone et al. (2017) in their study on *E elatum* demonstrated that TFC in leaves vary from minimum of 8.36 mg/g to maximum of 17.23 mg/g. In rhizomes, its value was found to vary from 7.20 to 25.73 mg/g. The flavonoid content of some elite accessions of *E. elatum* was found to be at par with the published reports of Zhang et al. (2014c) but was found to be lower than the contents reported by Mahboubi et al. (2013) in *E. pinnatum* and Zhang et al. (2013b) in *Herba Epimedii*. The total phenolic content in leaf, on the other hand, was shown to vary between a minimum of 1.19 mg/g and a maximum of 2.56 mg/g, while in rhizomes it varied from 1.56 to 3.67 mg/g. Since all accessions were collected from different ecogeographical habitats of Kashmir Himalayas, the considerable differences in flavonoid and phenolic contents in *E. elatum* could be due to the developmental stage, harvest season, drying processes and environmental factors in comparison to the published reports. The study presented the first report on the comparison of TFC and TPC in underground rhizome and aerial part (leaves) in *E. elatum* (Lone et al. 2017). Antioxidant activities (using DPPH and FRAP assay) of *E. elatum* in leaves and rhizomes of different accessions were also reported for the first time by the same authors. Almost all accessions were shown to have a significant antioxidant activity. DPPH scavenging activity was shown to vary from 5.93 to 90.30 in leaves while in rhizomes it ranged from 48.2 to 98.76, respectively. FRAP values in leaves were shown to vary from 17.18 to 76.55, while in the rhizome it was reported to be in the range from 98.16 to 116.12, respectively. This study reported the good antioxidant potential in rhizomes of *E. elatum*. The accumulation of higher antioxidants in underground parts of this species needs to be worked out for its future medicinal efficacy. Earlier reports have shown Epimedium leaves to be a good source of antioxidant molecules (Zhang et al. 2014c; Zhang et al. 2013b).

Herba Epimedii is used as a nutraceutical ingredient in many Asian countries particularly China, Japan and Korea and recent findings have also confirmed that their constituents can be used in treating neurodegenerative disorders such as Alzheimer's disease (Jan et al. 2015). Therefore, more studies are needed to assess the antioxidant characterisation of rare medicinal plant, which will not only bring the plant under captive cultivation but also help in its conservation and commercialisation.

References

Akai S (1935) Constituents of Epimedium macranthum Morr and Decne I. Chemical constitution of a new glucoside of Epimedium macranthum Morr and Decne I. J Pharm Soc Jpn 55:537–599

Alam A, Naik PK, Gulati P, Gulati AK, Mishra GP (2008) Characterization of genetic structure of Podophyllum hexandrum populations, an endangered medicinal herb of Northwestern Himalaya, using ISSR-PCR markers and its relatedness with podophyllotoxin content. Afr J Biotechnol 7:8

Amico V, Chillemi R, Mangiafico S, Spatafora C, Tringali C (2008) Polyphenol-enriched fractions from Sicilian grape pomace: HPLC–DAD analysis and antioxidant activity. Bioresour Technol 99(13):5960–5966

Arief ZM, Munshi AH, Shawl AS (2015) Evaluation of medicinal value of *Epimedium elatum* on the basis of pharmacologically active constituents, Icariin and Icariside-II. Pak J Pharm Sci 28:5

Arief ZM, Shawl AS, Munshi AH (2016) Altitudinal variation in pharmacologically active compounds of wild and cultivated populations of *Epimedium elatum*. J Appl Res Med Aromat Plants 3(2):48–51

Badola HK (2002) Endangered medicinal plant species-priorities and action. Theme Paper, International Workshop, Endangered Medicinal Plant Species in Himachal Pradesh, GBPIHED, Mohal-Kullu, India (18–19 March, 2002). p. 11

Badola HK, Aitken S (2003) The Himalayas of India: a treasury of medicinal plants under siege. Biodiversity 4(3):3–13

Bian Q, Huang JH, Liu SF, Ning Y, Yang Z, Zhao YJ, Shen ZY, Wang YJ (2011) Different molecular targets of Icariin on bMSCs in CORT and OVX-rats. Front Biosci 4:1224–1236

Bláha L, Kopp R, Šimková K, Mareš J (2004) Oxidative stress biomarkers are modulated in silver carp (Hypophthalmichthys molitrix Val.) exposed to microcystin-producing cyanobacterial water bloom. Acta Vet Brno 73(4):477–482

Bo LY, Kipletting TE, Jing J, Jun ZY (2013) Quantitative determination of multiple components in Herba Epimedii using a single reference standard: a comparison of two methods. Anal Methods 5(15):3741–3746

Chen CY (2009) Computational screening and design of traditional Chinese medicine (TCM) to block phosphodiesterase-5. J Molec Graph Model 28(3):261–269

Chen C, Sha M, Yang S (1996) Quantitative changes of flavonoids in Epimedium koreanum Nakai in different collecting periods. China J Chin Mater Med 21(2):86

Chen H, Zhang M, Xie B (2005) Components and antioxidant activity of polysaccharide conjugate from green tea. Food Chem 90(1):17–21

Chen XJ, Guo BL, Li SP, Zhang QW, Tu PF, Wang YT (2007) Simultaneous determination of 15 flavonoids in Epimedium using pressurized liquid extraction and high-performance liquid chromatography. J Chromatogr A 1163(1):96–104

Chen Y, Zhao Y, Jia X, Ding A (2008a) Simultaneous determination of five main flavonoids in Herba Epimedii from different species by RP-HPLC. China Pharm 19(6):431

Chen XJ, Ji H, Zhang QW, Tu PF, Wang YT, Guo BL, Li SP (2008b) A rapid method for simultaneous determination of 15 flavonoids in Epimedium using pressurized liquid extraction and ultra-performance liquid chromatography. J Pharm Biomed Anal 46(2):226–235

Chen J, Xu Y, Wei G, Liao S, Zhang Y, Huang W, Yuan L, Wang Y (2015a) Chemotypic and genetic diversity in Epimedium sagittatum from different geographical regions of China. Phytochemistry 116:180–187

Chen XJ, Tang ZH, Li XW, Xie CX, Lu JJ, Wang YT (2015b) Chemical constituents, quality control, and bioactivity of epimedii folium (Yinyanghuo). Am J Chin Med 43(05):783–834

Chen YJ, Zheng HY, Huang XX, Han SX, Zhang DS, Ni JZ, He XY (2016a) Neuroprotective effects of Icariin on brain metabolism, mitochondrial functions, and cognition in triple-transgenic Alzheimer's disease mice. CNS Neurosci Therap 22(1):63–73

Chen SH, Wang XL, Zheng LZ, Dai Y, Zhang JY, Guo BL, Yang ZJ, Yao XS, Qin L (2016b) Comparative study of two types of herbal capsules with different Epimedium species for the prevention of ovariectomised-induced osteoporosis in rats. J Orthop Transl 4:14–27
Chiu JH, Chen KK, Chien TM, Chiou WF, Chen CC, Wang JY, Lui WY, Wu CW (2006) Epimedium brevicornum Maxim extract relaxes rabbit corpus cavernosum through multitargets on nitric oxide/cyclic guanosine monophosphate signaling pathway. Int J Impot Res 18(4):335–342
Cho NJ, Sung SH, Lee HS et al (1995) Anti-hepatotoxic activity of icariside II, a constituent of Epimedium koreanum. Arch Pharm Res 18:289. https://doi.org/10.1007/BF02976415
Cho WK, Kim H, Choi YJ, Yim NH, Yang HJ, Ma JY (2012) Epimedium koreanum Nakai water extract exhibits antiviral activity against porcine epidermic diarrhea virus in vitro and in vivo. Evid Based Complement Alternat Med 29:2012
Cho WK, Weeratunga P, Lee BH, Park JS, Kim CJ, Ma JY, Lee JS (2015) Epimedium koreanum Nakai displays broad spectrum of antiviral activity in vitro and in vivo by inducing cellular antiviral state. Viruses 7(1):352–377
Choi HJ, Eun JS, Park YR, Kim DK, Li R, Moon WS, Park JM, Kim HS, Cho NP, Cho SD, Soh Y (2010) Ikarisoside A inhibits inducible nitric oxide synthase in lipopolysaccharide-stimulated RAW 264.7 cells via p38 kinase and nuclear factor-kappaB signaling pathways. Eur J Pharmacol 601(1-3):171–178. https://doi.org/10.1016/j.ejphar.2008.09.032
Chu J, Zhang L, Ye CF, Zhao L, Ya BL, Li L (2008) Effects of Epimedium flavonoids on synapse related protein in brains of dementia transgenic mice. Zhonghua Yi Xue Za Zhi 88(1):31–35
Chung BH, Kim JD, Kim CK, Kim JH, Won MH, Lee HS, Dong MS, Ha KS, Kwon YG, Kim YM (2008) Icariin stimulates angiogenesis by activating the MEK/ERK-and PI3K/Akt/eNOS-dependent signal pathways in human endothelial cells. Biochem Biophys Res Commun 376(2):404–408
Compilation CH. China herb compilation People's medical publishing house. 1975.
De Smet Y, Goetghebeur P, Wanke S, Asselman P, Samain MS (2012) Additional evidence for recent divergence of Chinese Epimedium (Berberidaceae) derived from AFLP, chloroplast and nuclear data supplemented with characterisation of leaflet pubescence. Plant Ecol Evolut 145(1):73–87
Decne (1844) In Jacquem. Voy Bot Himal 9:8
Dell'Agli M, Galli GV, Dal Cero E, Belluti F, Matera R, Zironi E, Pagliuca G, Bosisio E (2008) Potent inhibition of human phosphodiesterase-5 by icariin derivatives. J Nat Prod 71(9):1513–1517
Dias AS, Porawski M, Alonso M, Marroni N, Collado PS, González-Gallego J (2005) Quercetin decreases oxidative stress, NF-κB activation, and iNOS overexpression in liver of streptozotocin-induced diabetic rats. J Nutr 135(10):2299–2304
Diaz P, Jeong SC, Lee S, Khoo C, Koyyalamudi SR (2012) Antioxidant and anti-inflammatory activities of selected medicinal plants and fungi containing phenolic and flavonoid compounds. Chin Med 7(1):26
Ding L, Liang XG, Hu Y, Zhu DY, Lou YJ (2008) Involvement of p38MAPK and reactive oxygen species in icariin-induced cardiomyocyte differentiation of murine embryonic stem cells in vitro. Stem Cells Dev 17(4):751–760
Dong ZY, Peng Y, Huang Y, Li X, Zhao LY, Zhang W (2010) Determination of contents of four components in the extract of Epimedium koreanum Nakai by RP-HPLC. J Shenyang Pharmaceut Univ 8:015
Ellegren H, Galtier N (2016) Determinants of genetic diversity. Nat Rev Genet 17(7):422–433
Fan Y, Liu J, Wang D, Hu Y, Yang S, Wang J, Guo L, Zhao X, Wang H, Jiang Y (2011) Epimedium polysaccharide and propolis flavone can synergistically inhibit the cellular infectivity of NDV and improve the curative effect of ND in chicken. Int J Biol Macromol 48(3):439–444
Fang F, Xu MY, Jiang JJ, Xu YL, Wei HX (2003) Clinical and empirical research of "Chuan Ke Zhi" in treating childhood respiratory viral infection. Shanghai J Trad Chin Med 37:36–37
Feng Zhu J, Jian Li Z, Sen Zhang G, Meng K, Yong Kuang W, Li J, Fu Zhou X, Juan Li R, Ling Peng H, Wen Dai C, Shen JK (2011) Icaritin shows potent anti-leukemia activity on chronic

myeloid leukemia in vitro and in vivo by regulating MAPK/ERK/JNK and JAK2/STAT3/AKT signalings. PLoS One 6(8):e23720
Frankham R (2005) Genetics and extinction. Biol Conserv 126(2):131–140
Gao L, Tang Q, He X, Bi M (2012) Effect of icariin on learning and memory abilities and activity of cholinergic system of senescence-accelerated mice SAMP10. China J Chin Mater Med 37(14):2117–2121
Guo BL, Wang CL, Chen JM, Xiao PG (1996) Determination of 9 Flavonoidsin 5 species of Epimedium recorded in Chinese Pharmacopoeiaby HPLC [J]. Acta Pharm Sin:4
Guo J, Li F, Wu Q, Gong Q, Lu Y, Shi J (2010) Protective effects of icariin on brain dysfunction induced by lipopolysaccharide in rats. Phytomedicine 17(12):950–955
Guo Y, Zhang X, Meng J, Wang ZY (2011) An anticancer agent icaritin induces sustained activation of the extracellular signal-regulated kinase (ERK) pathway and inhibits growth of breast cancer cells. Eur J Pharmacol 658(2):114–122
Han F, Lee IS (2017) A new flavonol glycoside from the aerial parts of Epimedium koreanum Nakai. Nat Prod Res 31(3):320–325
Han S, Xie YY, Wang YM, Liang QL, Luo GA (2012) Comparative study on chemical quality of main species of Epimedium. Acta Pharm Sin 47(4):502–507
He J, Wang Y, Duan F, Jiang H, Chen MF, Tang SY (2010a) Icaritin induces apoptosis of HepG2 cells via the JNK1 signaling pathway independent of the estrogen receptor. Planta Med 76(16):1834–1839
He XL, Zhou WQ, Bi MG, Du GH (2010b) Neuroprotective effects of icariin on memory impairment and neurochemical deficits in senescence-accelerated mouse prone 8 (SAMP8) mice. Brain Res 1334:73–83
Hong X, Wang X, Yong EL, Gong Y (2009) Determination of breviflavone A and B in Epimedium herbs with liquid chromatography–tandem mass spectrometry. J Pharm Biomed Anal 49(3):853–857
Horie S, Suzuki K, Maki M (2012) Quantitative morphological analysis of populations in a hybrid zone of Epimedium diphyllum and E. sempervirens var. rugosum (Berberidaceae). Plant Ecol Evol 145(1):88–95
Hsieh TP, Sheu SY, Sun JS, Chen MH, Liu MH (2010) Icariin isolated from Epimedium pubescens regulates osteoblasts anabolism through BMP-2, SMAD4, and Cbfa1 expression. Phytomedicine 17(6):414–423
Hsieh TP, Sheu SY, Sun JS, Chen MH (2011) Icariin inhibits osteoclast differentiation and bone resorption by suppression of MAPKs/NF-κB regulated HIF-1α and PGE 2 synthesis. Phytomedicine 18(2):176–185
Huang XL, Wang W, Zhou YW (2006) Protective effect of Epimedium flavonoids injection on experimental myocardial infarction rats. Chin J Integr Tradition West Med 26(1):68–71
Huang X, Zhu D, Lou Y (2007a) A novel anticancer agent, icaritin, induced cell growth inhibition, G 1 arrest and mitochondrial transmembrane potential drop in human prostate carcinoma PC-3 cells. Eur J Pharmacol 564(1):26–36
Huang H, Liang M, Zhang X, Zhang C, Shen Z, Zhang W (2007b) Simultaneous determination of nine flavonoids and qualitative evaluation of Herba Epimedii by high performance liquid chromatography with ultraviolet detection. J Sep Sci 30(18):3207–3213
Ikeda M, Ishima Y, VTG C, Ikeda T, Kinoshita R, Watanabe H, Ishida T, Otagiri M, Maruyama T (2017) Apoptosis induction of poly-S-nitrosated human serum albumin in resistant solid tumor under hypoxia can be restored by phosphodiesterase 5 inhibition. Nitric Oxide 69:28–34
Indran IR, Liang RL, Min TE, Yong EL (2016) Preclinical studies and clinical evaluation of com pounds from the genus Epimedium for osteoporosis and bone health. Pharmacol Therapeut 162:188–205
Islam NM, Yoo HH, Lee MW, Dong MS, Park YI, Jeong HS, Kim DH (2008) Simultaneous quantitation of five flavonoid glycosides in herba Epimedii by high-performance liquid chromatography–tandem mass spectrometry. Phytochem Anal 19(1):71–77
Jamal Z (2009) Biodiversity, ethnobotany and conservation status of the flora of Kaghan Valley Mansehra, NWFP Pakistan. Doctoral dissertation, Quaid-i-Azam University, Islamabad

Jan S, Hamayun M, Khan SA, Ahmad N, Ahmad I, Wali S (2015) Plant diversity of Hindu Kush mountain region of Utror and Gabral, Northern Pakistan. Pak J Weed Sci Res 21(2):247–271

Jia XB, Jin XY, Wang JJ, Shao ZZ, Lan XL (2010) Comparison of the content of main component in Epimedium koreanum decoction pieces from different manufactories. Chin Pharm 21:1006–1008

Jiang J, Zhao BJ, Song J, Jia XB (2016) Pharmacology and clinical application of plants in Epimedium L. Chin Herb Med 8(1):12–23

Jin X, Jia X, Sun E, Wang J, Chen Y, Cai B (2009) Research on variation regularity of five main flavonoids contents in Epimedium and processed Epimedium. China J Chin Mater Med 34(21):2738–2742

Jin MS, Shi S, Zhang Y, Yan Y, Sun XD, Liu W, Liu HW (2010) Icariin-mediated differentiation of mouse adipose-derived stem cells into cardiomyocytes. Mol Cell Biochem 344(1-2):1–9

Kang SH, Jeong SJ, Kim SH, Kim JH, Jung JH, Koh W, Kim JH, Kim DK, Chen CY, Kim SH (2012) Icariside II induces apoptosis in U937 acute myeloid leukemia cells: role of inactivation of STAT3-related signaling. PLoS One 7(4):e28706

Kim SH, Ahn KS, Jeong SJ, Kwon TR, Jung JH, Yun SM, Han I, Lee SG, Kim DK, Kang M, Chen CY (2011) Janus activated kinase 2/signal transducer and activator of transcription 3 pathway mediates icariside II-induced apoptosis in U266 multiple myeloma cells. Eur J Pharmacol 654(1):10–16

LanYing C, YunXiang L, YiFan Q, QiuMei Q (2014) A comparative research of plant morphology, the relative content of chlorophy II (SPAD value) and stomatal conductance of three transplanted barrenwort species. J South China Agricul Univ 35(2):110–114

Lee S et al (2016) Antioxidative effects of extracts from different parts of Epimedium koreanum Nakai. J Korean Soc Food Sci Nutr 45(2):188–193

Li S, Dong P, Wang J, Zhang J, Gu J, Wu X, Quan Z (2010a) Icariin, a natural flavonol glycoside, induces apoptosis in human hepatoma SMMC-7721 cells via a ROS/JNK-dependent mitochondrial pathway. Cancer Lett 298(2):222–230

Li F, Gong QH, Wu Q, Lu YF, Shi JS (2010b) Icariin isolated from Epimedium brevicornum Maxim attenuates learning and memory deficits induced by d-galactose in rats. Pharmacol Biochem Behav 96(3):301–305

Li MJ, Du MF, Chen QF (2011a) Study on RAPD genetic diversity of Genus Epimedium L. Seed 6:014

Li HF, Guan XY, Ye M, Xiang C, Lin CH, Sun C, Guo DA (2011b) Qualitative and quantitative analyses of Epimedium wushanense by high-performance liquid chromatography coupled with diode array detection and electrospray ionization tandem mass spectrometry. J Sep Sci 34(12):1437–1446

Li X, Hu Y, He L, Wang S, Zhou H, Liu S (2012a) Icaritin inhibits T cell activation and prolongs skin allograft survival in mice. Int Immunopharmacol 13(1):1–7

Li D, Yuan T, Zhang X, Xiao Y, Wang R, Fan Y, Zhang X (2012b) Icariin: a potential promoting compound for cartilage tissue engineering. Osteoarthr Cartil 20(12):1647–1656

Li Q, Huai L, Zhang C, Wang C, Jia Y, Chen Y, Wang J (2013) Icaritin induces AML cell apoptosis via the MAPK/ERK and PI3K/AKT signal pathways. Int J Hematol 97(5):617–623

Li C, Li Q, Mei Q, Lu T (2015) Pharmacological effects and pharmacokinetic properties of icariin, the major bioactive component in Herba Epimedii. Life Sci 126:57–68

Liang HR, Vuorela P, Vuorela H, Hiltunen R (1997a) Isolation and immunomodulating effect of flavonol glycosides from Epimedium hunanense. Planta Med 63(04):316–319

Liang HR, Siren H, Reikkola ML, Vuorela P, Vuorela H, Hiltunen R (1997b) Characterization of flavonoids in extracts from four species of Epimedium by micellarelectrokineticcapillary chromatography with diode-array detection. J Chromatogr Sci 35(3):117–125

Liang W, Lin M, Li X, Li C, Gao B, Gan H, Yang Z, Lin X, Liao L, Yang M (2012a) Icariin promotes bone formation via the BMP-2/Smad4 signal transduction pathway in the hFOB 1.19 human osteoblastic cell line. Int J Mol Med 30(4):889–895

Liang Q, Wei G, Chen J, Wang Y, Huang H (2012b) Variation of medicinal components in a unique geographical accession of horny goat weed Epimedium sagittatum Maxim (Berberidaceae). Molecules 17(11):13345–13356

Liang Q, Zhang YJ, Xu YQ, Huang HW, Wang Y (2013) Morphological variations and genetic diversity of Epimedium sagittatum populations. Plant Sci J 31:422–427

Lim JD, Seong ES, Choi KJ, Kim SK, Chung IM, Heo K, Yu CY (2000) Morphological characteristics and RAPD analysis of Epimedium spp. Korean J Med Crop Sci 8(2):102–108

Lin Y, Zhang MS, Wu Q, Fang ZQ, Teng MD (2010) Content determination of icariin and Epimedin C in whole plant of Epimedium wushanense and Epimedium acuminatum Franch from Guizhou by RP-HPLC. Shenyang Pharmaceut Univ 27:453–456

Liu CR, Xu LX (1984) Analysis of active ingredients of traditional Chinese herbal drug. Assay of icariin in Epimedium. Chin J Pharm Anal 4:81–84

Liu Y, Zang H, Zhang H, Chen J (2005) Effect of Herba Epimedii Flavone on expression of OPG and RANKL in rat osteoblasts. J Chin Med Mater 28(12):1076–1078

Liu HJ, Wang XP, Lin J, Yu YC, Jiang XQ, Zhang XL, Zhou ZT, Zhang WD (2006a) The effect of icariin and astragalosid I on the proliferation and differentiation of bone marrow stromal cells. J Chin Med Mater 29(10):1062–1065

Liu TZ, Chen CY, Yiin SJ, Chen CH, Cheng JT, Shih MK, Wang YS, Chern CL (2006b) Molecular mechanism of cell cycle blockage of hepatoma SK-Hep-1 cells by Epimedin C through suppression of mitogen-activated protein kinase activation and increased expression of CDK inhibitors p21 Cip1 and p27 Kip1. Food Chem Toxicol 44(2):227–235

Liu JJ, Li SP, Wang YT (2006c) Optimization for quantitative determination of four flavonoids in Epimedium by capillary zone electrophoresis coupled with diode array detection using central composite design. J Chromatogr A 1103(2):344–349

Liu B, Zhang H, Xu C, Yang G, Tao J, Huang J, Wu J, Duan X, Cao Y, Dong J (2011) Neuroprotective effects of icariin on corticosterone-induced apoptosis in primary cultured rat hippocampal neurons. Brain Res 1375:59–67

Liu W, Yin D, Liu J, Li N (2014) Genetic diversity and structure of Sinopodophyllum hexandrum (Royle) Ying in the Qinling Mountains, China. PLoS One 9(10):e110500

Lone SA, Kushwaha M, Wani A, Kumar A, Gupta AP, Hassan QP, Chandra S, Gupta S (2017) Genetic diversity, LCMS based chemical fingerprinting and antioxidant activity of *Epimedium elatum* Morr & Decne. J Appl Res Med Aromat Plants 5:72–81

Lubell JD, Brand MH (2005) Division size and timing influence propagation of four species of Epimedium L. Hortic Sci 40(5):1444–1447

Luo Y, Nie J, Gong QH, Lu YF, Wu Q, Shi JS (2007) Protective effects of icariin against learning and memory deficits induced by aluminium in rats. Clin Exp Pharmacol Physiol 34(8):792–795

Ma JQ, Ma XQ, Zhang XY (2009) Mechanism of resisting exercise fatigue by Epimedium flavones in mice. Chin J New Drugs 18:553–555

Ma H, He X, Yang Y, Li M, Hao D, Jia Z (2011) The genus Epimedium: an ethnopharmacological and phytochemical review. J Ethnopharmacol 134(3):519–541

Mahboubi M, Kazempour N, Hosseini H, Mahboubi M (2013) Antimicrobial and antioxidant activity of Epimedium pinnatum. Herba Polonica 59(2):24–34. https://doi.org/10.2478/hepo-2013-0009

Meng X, Zeng N, Zhang Y, Lan X, Ren C, Cheng L (1996) Studies on effect of active constituents of Herba Epimedii on hypothalamic monoamines neurotransmitter and other brain functions in aging rats. China J Chin Mater Med 21(11):683–685

Miner BG, Sultan SE, Morgan SG, Padilla DK, Relyea RA (2005) Ecological consequences of phenotypic plasticity. Trends Ecol Evol 20(12):685–692

Ming Feng DU (2008) DNA genetic diversity of genus Epimedium L. and its phylogeny, Master Thesis. Guizhou Normal University, China.[In Chinese]

Ming LG, Chen KM, Xian CJ (2013) Functions and action mechanisms of flavonoids genistein and icariin in regulating bone remodeling. J Cell Physiol 228(3):513–521

Ming-Jun LI (2011) Establishment of DNA fingerprints on Epimedium L. J Anhui Agricul Sci 18:037

Morren C, Decaisne J (1834) Observations sur la Xore du Japon suivies de la monographie du genre *Epimedium*. Anal Sci Nature Botaniq 2:347–361

Mu-Dan LI, Xu SH, Zhang YJ, Ping GU (2009) Optimization of ISSR-PCR system in Epimedium acuminatum Franch. J Mount Agricul Biol 6:008

Nakai R, Shoyama Y, Shiraishi S (1996) Genetic characterization of Epimedium species using random amplified polymorphic DNA (RAPD) and PCR-restriction fragment length polymorphism (RFLP) diagnosis. Biol Pharm Bull 19(1):67–70

Naseer S, Lone SH, Lone JA, Khuroo MA, Bhat KA (2015) LC–MS guided isolation, quantification and antioxidant evaluation of bioactive principles from *Epimedium elatum*. J Chromatogr B 989:62–70

Nasir, E., Ali, S.I., Fl. Pakistan. Ther Ber 87, 1-31 1980–2005University of Karachi, Karachi

Nautiyal BP, Prakash V, Chauhan RS, Purohit H, Nautiyal MC (2001) Assessment of germinability, productivity and cost benefit analysis of Picrorhiza kurrooa cultivated at lower altitudes. Curr Sci 81(5):579–585

Nurul I, Lee SK, Jeong SY, Kim DH, Jin CB, Yoo HH (2009) Quantitative and pattern recognition analyses for the quality evaluation of Herba Epimedii by HPLC. Bull Kor Chem Soc 30(1):137–144

Palacio-López K, Beckage B, Scheiner S, Molofsky J (2015) The ubiquity of phenotypic plasticity in plants: a synthesis. Ecol Evol 5(16):3389–3400

Pan J, Guo B (2016) Effects of light intensity on the growth, photosynthetic characteristics, and flavonoid content of Epimedium pseudowushanense BL Guo. Molecules 21(11):1475

Pei BLGLK, Pei-Gen XIAO (2008) Further research on taxonomic significance of flavonoids in Epimedium (Berberidaceae). J Syst Evol 46(6):874–885

Pei LK, Huang WH, He TG, Guo BL (2007) Systematic studies on quality of main species of Herba epimedii. China J Chin Mater Med 32(21):2217–2222

Peng YD, Huang WH, Guo BL (2007) Research on quality of Epimedium extract in market. China J Chin Mater Med 32(18):1858–1861

Perveen A, Qaiser M (2010) Pollen flora of Pakistan–LXV. Berberidaceae. Pak J Bot 42(1):1–6

Pharmacopoeia Commission of P. R. China (2010) Pharmacopoeia of the People's Republic of China, vol I. China Medico-Pharmaceutical Science & Technology Publishing House, Beijing, p 306

Pietta PG (2000) Flavonoids as antioxidants. J Nat Prod 63(7):1035–1042

Quan QM, Wu W, Li YX, Cai QR (2010) Variation in icariin and flavonoid contents of Barrenwort species. J Med Plant Res 4(6):471–476

Quan QM, Fang ZL, Wu W, Li YX (2011) Comparative analysis of morphological characteristics and effective composition content of wild and cultivated Epimedium pubescens and Epimedium wushanense (Berberidaceae). J Med Plant Res 5(29):6523–6527

Rafiq RA (1992–1995) The flora of Palas valley and plant conservation priorities. Report on the botanical studies in Palas valley. Report prepared for Himalayan Jungle Project, Palas Valley, Kohistan

Sarwat M, Nabi G, Das S, Srivastava PS (2012) Molecular markers in medicinal plant biotechnology: past and present. Crit Rev Biotechnol 32(1):74–92

Sharma BM, Jamwal PS (1988) Flora of upper liddar valley of Kashmir Himalaya. J Econ Taxon Bot 1

She BR, Qin DN, Wang Z, She Y (2003) Effects of flavonoids from herba epimedii on the reproductive system in male rats. Chin J Androl 9:294–296

Shen P, Guo BL, Gong Y, Hong DY, Hong Y, Yong EL (2007) Taxonomic, genetic, chemical and estrogenic characteristics of Epimedium species. Phytochemistry 68(10):1448–1458

Sheng MY, Chen QF, Yang QX (2008) Variation in icariin and flavonoid contents of barrenwort accessions native to Guizhou, China. Biochem Syst Ecol 36(9):719–723

Sheng M, Chen Q, Wang L, Tian X (2011) Hybridization among Epimedium (Berberidaceae) species native to China. Sci Hortic 128(3):342–351

Shin YS, Zhao C, Zhang LT, Park JK (2015) Current status and clinical studies of oriental herbs in sexual medicine in Korea. World J Men's Health 33(2):62–72

Singh G, Kachroo P (1987) Forest flora of Srinagar and plants of neighbourhood. Delhi: Periodical Expert Book Agency x, 278p.-illus., map. En Plant records Geog, 6
Sofi SN, Shakeel-u-Rehman QPH, Lone SH, Bhat HM, Bhat KA (2014) Isolation, identification, and simultaneous quantification of five major flavonoids in *Epimedium elatum* by high performance liquid chromatography. J Liq Chromatogr Relat Technol 37(8):1104–1113
Song YH, Cai H, Gu N, Qian CF, Cao SP, Zhao ZM (2011) Icariin attenuates cardiac remodelling through down-regulating myocardial apoptosis and matrix metalloproteinase activity in rats with congestive heart failure. J Pharm Pharmacol 63(4):541–549. https://doi.org/10.1111/j.2042-7158.2010.01241.x
Song J, Shu L, Zhang Z, Tan X, Sun E, Jin X, Chen Y, Jia X (2012) Reactive oxygen species-mediated mitochondrial pathway is involved in Baohuoside I-induced apoptosis in human non-small cell lung cancer. Chemico-Biol Interact 199(1):9–17
Stearn WT, Shaw J, Green PS, Mathew B (2002) Genus Epimedium and other herbaceous Berberidaceae. Timber Press, Oregon
Stewart RR (1972) Ann. Cat. Vas. Pl. West Pakistan and Kashmir. p 281
Sultan P, Shawl AS, Rehman S, Ahmed SF, Ramteke PW (2010) Molecular characterization and marker based chemotaxonomic studies of Podophyllum hexandrum Royle. Fitoterapia 81(4):243–247
Sun C, Liang GY (2011) HPLC simultaneous determination of epimedin C and icariin in 21 species of Epimedium. Chin J Pharmaceut Anal 5:931–934
Sun Y, Fung KP, Leung PC, Shi D, Shaw PC (2004) Characterization of medicinal Epimedium species by 5S rRNA gene spacer sequencing. Planta Med 70(03):287–288
Sun Y, Fung KP, Leung PC, Shaw PC (2005) A phylogenetic analysis of Epimedium (Berberidaceae) based on nuclear ribosomal DNA sequences. Mol Phylogenet Evol 35(1):287–291
Sun X, Sun X, Jin X, Zhang X, Liu C, Lei L, Jin L, Liu H (2011) Icariin induces mouse embryonic stem cell differentiation into beating functional cardiomyocytes. Mol Cell Biochem 349(1–2):117–123
Sun L, Peng Q, Qu L, Gong L, Si J (2015) Anticancer agent icaritin induces apoptosis through caspase-dependent pathways in human hepatocellular carcinoma cells. Mol Med Rep 11(4):3094–3100
Sze SC, Tong Y, Ng TB, Cheng CL, Cheung HP (2010) Herba Epimedii: anti-oxidative properties and its medical implications. Molecules 15(11):7861–7870
Tali BA, Ganie AH, Nawchoo IA, Wani AA, Reshi ZA (2015) Assessment of threat status of selected endemic medicinal plants using IUCN regional guidelines: a case study from Kashmir Himalaya. J Nat Conserv 23:80–89
Tan HL, Chan KG, Pusparajah P, Saokaew S, Duangjai A, Lee LH, Goh BH (2016) Anti-cancer properties of the naturally occurring aphrodisiacs: icariin and its derivatives. Front Pharmacol 7
Tantry MA, Dar JA, Idris A, Akbar S, Shawl AS (2012) Acylated flavonol glycosides from *Epimedium elatum*, a plant endemic to the Western Himalayas. Fitoterapia 83(4):665–670
Tao F, Qian C, Guo W, Luo Q, Xu Q, Sun Y (2013) Inhibition of Th1/Th17 responses via suppression of STAT1 and STAT3 activation contributes to the amelioration of murine experimental colitis by a natural flavonoid glucoside icariin. Biochem Pharmacol 85(6):798–807
The Plant List (2013). Version 1.1. Published on the Internet; http://www.theplantlist.org/ Accessed 1 Jan 2013.
Tong JS, Zhang QH, Huang X, Fu XQ, Qi ST, Wang YP, Hou Y, Sheng J, Sun QY (2011) Icaritin causes sustained ERK1/2 activation and induces apoptosis in human endometrial cancer cells. PLoS One 6(3):e16781
Wang XF, Wang J (2014) Icaritin suppresses the proliferation of human osteosarcoma cells in vitro by increasing apoptosis and decreasing MMP expression. Acta Pharmacol Sin 35(4):531–539
Wang XM, Fu H, Liu GX (1996) Effect of Herba Epimedii and Fructus Lyciionmito- chondrial DNA deletion, activity of respiratory chain enzyme complexes and ATP synthesis in aged rats. J Peking Univ Health Sci 21:683–685

Wang F, Zheng Y, Xiao HB, Zhou M, Liu YX, Li GQ (2001) Effect of administration of herba Epimedium at the optimal time levels on sexual hormones. J Tradit Chin Med 42:619–621

Wang MQ, Bi ZM, Li P, Ji H, Cheng FL (2003) Determination of epimedin C and icariin in Herba Epimedii by HPLC. China J Chin Mater Med 28(11):1025–1027

Wang GJ, Tsai TH, Lin LC (2007) Prenylflavonol, acylated flavonol glycosides and related compounds from Epimedium sagittatum. Phytochemistry 68(19):2455–2464

Wang Y, Dong H, Zhu M, Ou Y, Zhang J, Luo H, Luo R, Wu J, Mao M, Liu X, Zhang J (2010) Icariin exterts negative effects on human gastric cancer cell invasion and migration by vasodilator-stimulated phosphoprotein via Rac1 pathway. Eur J Pharmacol 635(1):40–48

Wang Q, Hao J, Pu J, Zhao L, Lü Z, Hu J, Yu Q, Wang Y, Xie Y, Li G (2011) Icariin induces apoptosis in mouse MLTC-10 Leydig tumor cells through activation of the mitochondrial pathway and down-regulation of the expression of piwil4. Int J Oncol 39(4):973

Wang J, He L, Lu J (2013) Active component determination of Epimedium koreanum from different growing areas of Liaoning Province [J]. Liaoning Forestry. Sci Technol 2:004

Wang B, Yan F, Cai L (2014a) Anti-fatigue properties of icariin from Epimedium brevicornum. Biomed Res 25(3):297–302

Wang L, Waltenberger B, Pferschy-Wenzig EM, Blunder M, Liu X, Malainer C, Blazevic T, Schwaiger S, Rollinger JM, Heiss EH, Schuster D (2014b) Natural product agonists of peroxisome proliferator-activated receptor gamma (PPARγ): a review. Biochem Pharmacol 92(1):73–89

Wang L, Li Y, Guo Y, Ma R, Fu M, Niu J, Gao S, Zhang D (2016) Herba Epimedii: an ancient Chinese herbal medicine in the prevention and treatment of osteoporosis. Curr Pharmaceut Design 22(3):328–349

Ward BJ (2004) The plant Hunter's garden: the new explorers and their discoveries. Timber Press, Oregon, p 134

Wei Y, Liu B, Sun J, Lv Y, Luo Q, Liu F, Dong J (2015) Regulation of Th17/Treg function contributes to the attenuation of chronic airway inflammation by icariin in ovalbumin-induced murine asthma model. Immunobiology 220(6):789–797

Wo Y, Zhu D, Yu Y, Lou Y (2008) Involvement of NF-κB and AP-1 activation in icariin promoted cardiac differentiation of mouse embryonic stem cells. Eur J Pharmacol 586(1):59–66

Wu H, Lien EJ, Lien LL (2003) Chemical and pharmacological investigations of Epimedium species: a survey. In progress in drug research. Birkhäuser, Basel, pp 1–57

Wu CS, Guo BL, Sheng YX, Zhang JL (2008) Simultaneous determination of seven flavonoids in Epimedium by liquid chromatography–tandem mass spectrometry method. Chin Chem Lett 19(3):329–332

Wu JF, Dong JC, Xu CQ (2009) Effects of icariin on inflammation model stimulated by lipopolysaccharide in vitro and in vivo. Chin J Integr Trad West Med 29(4):330–334

Wu H, Zha ZG, Yao P (2010) Experimental study of icariin in inducing bone marrow mesenchymal stem cell differentiation. Chin J Integr Tradit West Med 30(4):410–415

Wu J, Guan M, Wong PF, Yu H, Dong J, Xu J (2012a) Icariside II potentiates paclitaxel-induced apoptosis in human melanoma A375 cells by inhibiting TLR4 signaling pathway. Food Chem Toxicol 50(9):3019–3024

Wu B, Chen Y, Huang J, Ning Y, Bian Q, Shan Y, Cai W, Zhang X, Shen Z (2012b) Icariin improves cognitive deficits and activates quiescent neural stem cells in aging rats. J Ethnopharmacol 142(3):746–753

Wu J, Xu J, Eksioglu EA, Chen X, Zhou J, Fortenbery N, Wei S, Dong J (2013) Icariside II induces apoptosis of melanoma cells through the downregulation of survival pathways. Nutr Cancer 65(1):110–117

Xia AJ, Huang HZ, Gao G, Zhang H, Zhu ZY, Chai YF (2009) Simultaneous determination of five flavone constituents in Epimedium Brevicornum by HPLC [J]. J Pharm Pract 4:014

Xie J, Sun W (2010) Separation and simultaneous quantification of seven prenyl flavones from organs of different epimedii using gradient reverse-phase HPLC. Nat Prod Res Dev 22:820–825

Xie JP, Wang ZD, Sun WJ (2007a) The investigation of epimedin C and icariin in the leaves of 9 species of Epimedium. Chin Tradit Herb Drug 38:613–614

Xie JP, Xu DB, Sun WJ (2007b) RP-HPLC determination four components in E. wushanense TS Ying. Chin J Pharmaceut Anal 27(3):437

Xie PS, Yan YZ, Guo BL, Lam CW, Chui SH, Yu QX (2010) Chemical pattern-aided classification to simplify the intricacy of morphological taxonomy of Epimedium species using chromatographic fingerprinting. J Pharm Biomed Anal 52(4):452–460

Xie JP, Xiang JM, Wang HD (2011) Effects of different processed products on the main ingredient of epimedin C and icariin of Epimedium wushanense [J]. Chin J Pharmaceut Anal 4:016

Xu Y, Li Z, Wang Y, Huang H (2007) Allozyme diversity and population genetic structure of three medicinal Epimedium species from Hubei. J Genet Genom 34(1):56–71

Xu Y, Huang H, Li Z, Wang Y (2008) Development of 12 novel microsatellite loci in a traditional Chinese medicinal plant, Epimedium brevicornum and cross-amplification in other related taxa. Conserv Genet 9(4):949–952

Xu CQ, Liu BJ, Wu JF, Xu YC, Duan XH, Cao YX, Dong JC (2010) Icariin attenuates LPS-induced acute inflammatory responses: involvement of PI3K/Akt and NF-κB signaling pathway. Eur J Pharmacol 642(1):146–153

Xu Y, Li Z, Yuan L, Zhang X, Lu D, Huang H, Wang Y (2013a) Variation of epimedins A–C and icariin in ten representative populations of Epimedium brevicornu Maxim., and implications for utilization. Chem Biodivers 10(4):711–721

Xu YQ, Cai WZ, Hu SF, Huang XH, Ge F, Wang Y (2013b) Morphological variation of non-glandular hairs in cultivated Epimedium sagittatum (Berberidaceae) populations and implications for taxonomy. Biodivers Sci 21:185–196

Xu YQ, Xu Y, Shi HJ, Hu SF, Ge F (2014) Taxonomic research on Epimedium sagittatum species complex and discussion. Chin Trad Herb Drug 22:029

Xu F, Ding Y, Guo Y, Liu B, Kou Z, Xiao W, Zhu J (2016) Anti-osteoporosis effect of Epimedium via an estrogen-like mechanism based on a system-level approach. J Ethnopharmacol 177:148–160

Xuemei Z, Zengli F, Qiumei Q (2016) Impact of simulated warming on growthand floral characteristics of two varieties of medicinal Epimedium. Braz Arch Biol Technol 59(SPE)

Yan ZK, Qiu H (2005) Antimicrobial tests of Icariin. China Food Additives 4:65–68

Yan-Ying C, Yun-Xiang L, Qiu-Mei Q et al (2012) Genetic diversity and relationship among 7 species of genus epimedium in Sichuan revealed by ISSR Analysis[J]. Bull Bot Res 32(2):208–212

Yang X, Zhang YH, Ding CF, Yan ZZ, Du J (2007) Extract from Epimedium brevi-cornum Maxim. against injury to function of human sperm membrane in vitro. Chin J Clin Pharmacol Therapeut 12.663–667

Yao R, Zhang L, Li X, Li L (2010) Effects of Epimedium flavonoids on proliferation and differentiation of neural stem cells in vitro. Neurol Res 32(7):736–742

Yao HM, Dong YH, Wu ZJ, Jiang RP (2012) An RP-HPLC determination of epimedin A, B, C and icraiin in self-emulsifying soft capsules of Epimedium brevicornum extracts [J]. Chin J New Drug 6:028

Yasukawa K, Ko SK, Whang WK (2016) Inhibitory effects of the aerial parts of Epimedium koreanum on TPA-induced inflammation and tumour promotion in two-stage carcinogenesis in mouse skin. J Pharm Nutr Sci 6(2):35–42

Yong NIE, Li Y-X, Ma Y-H, Quan Q-M (2010) Comparison of morphological characteristics and biomass allocation of module in medicinal plants of Epimedium from Sichuan [J]. Bull Botan Res 4

Yousaf Z, Hu WE, Zhang YA, Zeng S, Wang Y (2015) Systematic validation of medicinally important genus Epimedium species based on microsatellite markers. Pak J Bot 44(2):477–484

Yu J, Ding C-b, Zhang L, Yang R, Zhou Y, Tang L (2011) Identification of the genus Epimedium with DNA barcodes. J Med Plant Res 5(28):6413–6417

Yu X, Song J, Xiong Z, Li F (2010) Simultaneous assay of epimedin A epimedin B, epimedin C and icariin in herba epimedii by QAMS. China J Chin Mater Med 35(24):3310–3313

Yuan D, Wang H, He H, Jia L, He Y, Wang T, Zeng X, Li Y, Li S, Zhang C (2014) Protective effects of total flavonoids from Epimedium on the male mouse reproductive system against cyclophosphamide-induced oxidative injury by up-regulating the expressions of SOD3 and GPX1. Phytother Res 28(1):88–97

Zeng N, Meng X, Zhang Y, Lai X, Zheng J, Chen L, Ren C (1997) Antioxidative effect of constituents of herba Epimedii (ESPS). China J Chin Mater Med 22(1):46–48

Zeng S, Xiao G, Guo J, Fei Z, Xu Y, Roe BA, Wang Y (2010) Development of a EST dataset and characterization of EST-SSRs in a traditional Chinese medicinal plant, Epimedium sagittatum (Sieb. Et Zucc.) Maxim. BMC Genomics 11(1):1

Zeng-li FA (2012) Research advances in the pharmacological effects of HERBA EPIMEDII [J]. J Anhui Agricul Sci 17:029

Zhai Y, Ge B, Ma H, Ming L, Li Z, Cheng G, Zhou J, Chen K (2010) Icariin promotes osteogenic differentiation of rat bone marrow stromal cells in vitro. China J Chin Mater Med 35(23):3219–3222

Zhai M, He L, Ju X, Shao L, Li G, Zhang Y, Liu Y, Zhao H (2015) Icariin acts as a potential agent for preventing cardiac ischemia/reperfusion injury. Cell Biochem Biophys 72(2):589–597

Zhang ZB, Yang QT (2006) The testosterone mimetic properties of icariin. Asian J Androl 8(5):601–605

Zhang Y, Xiao CH, Meng XL (1995) Flavonoids contents determined and resources utilized of 8 species for sale native to Sichuan. Chin J Chin Mater Med 20:201–202

Zhang ML, Uhink CH, Kadereit JW (2007a) Phylogeny and biogeography of Epimedium/ Vancouveria (Berberidaceae): Western North American-East Asian disjunctions, the origin of European mountain plant taxa, and East Asian species diversity. Syst Bot 32(1):81–92

Zhang H, Gao X, Lu DY, Wang Y (2007b) Simultaneous determination of epimedin A, B, C and icariin in herba epimedii by high performance liquid chromatography. J Instrum Anal 26(2):198

Zhang HF, Yang TS, Li ZZ, Wang Y (2008) Simultaneous extraction of epimedin a, B, C and icariin from Herba Epimedii by ultrasonic technique. Ultrason Sonochem 15(4):376–385

Zhang RX, Jia ZP, Li FS, Li MX, Qiu JG (2009) Antihypoxic effect of total flavonoids of Herba Epimedii on hypoxia model mice. J Chin Med Mat 32:1736–1738

Zhang D, Zhang J, Fong C, Yao X, Yang M (2012) Herba epimedii flavonoids suppress osteoclastic differentiation and bone resorption by inducing G2/M arrest and apoptosis. Biochimie 94(12):2514–2522

Zhang DC, Liu JL, Ding YB, Xia JG, Chen GY (2013a) Icariin potentiates the antitumor activity of gemcitabine in gallbladder cancer by suppressing NF-κB. Acta Pharmacol Sin 34(2):301–308

Zhang W, Chen H, Wang Z, Lan G, Zhang L (2013b) Comparative studies on antioxidant activities of extracts and fractions from the leaves and stem of Epimedium koreanum Nakai. J Food Sci Technol 50(6):1122–1129

Zhang L, Shen C, Chu J, Zhang R, Li Y, Li L (2014a) Icariin decreases the expression of APP and BACE-1 and reduces the β-amyloid burden in an APP transgenic mouse model of Alzheimer's disease. Int J Biol Sci 10(2):181–191

Zhang Y, Yang L, Chen J, Sun W, Wang Y (2014b) Taxonomic and phylogenetic analysis of Epimedium L. based on amplified fragment length polymorphisms. Sci Hortic 170:284–292

Zhang L, Zhang Z, Luo Q, Lu H, Liang Y (2014c) Evaluation and prediction of the antioxidant activity of Epimedium from multi-wavelength chromatographic fingerprints and chemometrics. Analyt Method 6(4):1036–1043

Zhang D, Wang Z, Sheng C, Peng W, Hui S, Gong W, Chen S (2015) Icariin prevents amyloid Beta-induced apoptosis via the PI3K/Akt pathway in PC-12 cells. Evid Based Complement Alternat Med 2015:235265

Zhang Y, Du L, Liu A, Chen J, Wu L, Hu W, Zhang W, Kim K, Lee SC, Yang TJ, Wang Y (2016) The complete chloroplast genome sequences of five Epimedium species: lights into phylogenetic and taxonomic analyses. Front Plant Sci 7:306

ZhengZheng Z, JianGuang L, BaoLin G, LingYi K (2010) Analysis of bioactive flavonoids and resources research of main Herba Epimedii species from different origins. J China Pharmaceut Univ 41(2):146–150

Zhou T, Zhang X, Guo L, Lin G, Jiang W, Ai Q, Zhang C (2012) Variation of icariin and total flavonoid of Epimedium acuminatum in different parts and habitats. China J Chin Mater Med 37(13):1917–1921
Zhou YX, Zhang HR, Fan DD (2013) HPLC analysis of different parts in Epimedium wushanense from different regions. Chem Eng 2:6–8
Zhou T, Xiong HX, Lin G, Jiang WK, Guo LP, Lei M, Chen CY, Zhang XB (2014) Morphological character of growth cycle for Epimedium acuminatim and icariin content analysis. China J Chin Mater Med 39(1):48–51

Indian Herbal Drug Industry: Challenges and Future Prospects

Musadiq H. Bhat, Ashok K. Jain, and Mufida Fayaz

Introduction

Natural plant products are a source of new chemical diversity and are the choice of today's world. There have been increasing trends in the trade of natural plant products and their extraction at industrial level, particularly that of non-wood natural plant products. Increasing trade in natural plant products has supported economic growth and has helped in reducing poverty in a number of emerging countries. There is strong evidence that natural plant products play a significant role in the livelihoods of the world's rural poor. Natural plant products are the main source of income for the forest-dwelling population in many countries. A high trade status of natural plant products at global, national level as well as regional level can improve the economy of that particular country. Natural plant products consist of goods of biological origin other than wood, derived from plant. They have been used by forest dwellers for their welfare or livelihood commodity since long and are traditional sources of food, fibre, medicine, etc. The forest dwellers mainly belong to tribal communities. It is estimated that 25% of people of most of the countries depend on the plant's resources for their livelihood, and many of them living in or near dense forest depend highly on them for their livelihood (Killman 2003). Some 80% of the people living in developing countries depend on non-wood natural plant products, such as fruits and herbs, for their primary health and nutritional needs. The most used categories of non-wood natural plant products are medicines and edibles (Bouri and Mukharjee 2013). The therapeutic potential of plant products can be traced back to over 5000 years as there is evidence of its use in the treatment of diseases and for

M. H. Bhat (✉) · M. Fayaz
School of Studies in Botany, Jiwaji University Gwalior, Gwalior, MP, India

A. K. Jain
Institute of ethnobiology, Jiwaji University, Gwalior, MP, India

M. Ozturk, K. R. Hakeem (eds.), *Plant and Human Health, Volume 1*,
https://doi.org/10.1007/978-3-319-93997-1_18

revitalizing body systems in Indian and other civilizations (Mahesh and Satish 2008). India is one of the mega-diversity hot spots with rich heritage of traditional knowledge of medicines. India has about 4.5 million plant species, and among them only 250,000–500,000 plant species have been investigated phytochemically for biological or pharmacological activity. Still a large number of higher plants as a source for new therapeutics are to be explored. The potential for developing phytomedicine into various health-care products appears rewarding, both from the perspective of economy and safety. Many plant extracts are quite effective than the synthetic ones with no or least side effects, and very little scientific research on their biological activity has been worked out (Friedman et al. 2007). The non-timber forest products (NTFPs) play an important role in the livelihoods of millions of rural and urban people across the globe. It is well established that NTFPs fulfil multiple roles in supporting human well-being. The NTFPs provide the products for food, shelter, medicines, fibres, energy and cultural artefacts for many of the world's poorest population and a significant number of the less poor. The contribution of these resources to livelihoods typically ranges from 10 to 60% of total household income. The products also provide many households with a means of income generation, either as supplementary income to other livelihood activities or as the primary means of cash generation (Shackleton et al. 2015; Malhotra and Bhattacharya 2010; Pandey et al. 2011; Hegde and Enters 2000; Kaimowitz 2003; Saxena 2003a; Vantomme 2003; Saha and Sundriyal 2012; Asfaw et al. 2013; Babulo et al. 2009; Areki and Cunningham 2010; Mahapatra et al. 2005; Marshall et al. 2006; Shackleton et al. 2008). Despite attractive economic growth in some countries in South Asia, poverty remains a major development challenge, particularly in the Hindu Kush–Himalaya (HKH) region, for several biophysical and socioeconomic reasons (Hunzai et al. 2011). Earning a livelihood in mountain areas is considerably more complex and difficult because of inaccessibility, fragility and vulnerability arising from topographical complexity, altitudinal gradient and poor physical and socioeconomic facilities (Jodha 2005; Rasul and Kollmair 2010). Rural mountain people in developing countries seem to be prone to poverty (Wymann von Dach et al. 2006). Of the 210 million people in the HKH region, 66 million live in poverty and are highly vulnerable to climate and other global changes (Hunzai et al. 2011). An overwhelming majority of the poor in the HKH live in rural areas and depend heavily on nature-based goods and services, including forests and non-timber forest products (NTFPs). In India alone, 270 million people depend directly or indirectly on NTFPs, including medicinal and aromatic plants (MAPs), for their livelihoods (Mahapatra et al. 2005). Thus, the livelihood security of rural people living near forests depends greatly on the status and condition of the natural resources available in the forests and their optimum utilization through enterprise development (Rasul et al. 2008; Tiwari and Campbell 1997; Saxena 2003b).

Contribution of Non-timber Forest Products (NTFPs) to Rural Economy

Non-timber forest products have attracted great global attention in recent years due to increase in recognition of their contribution to domestic economies. NTFPs can provide important needs for rural livelihood and regional economies. Several million households all around the world depend a lot on NTFPs for both sustenance and usual income. Even though individually NTFPs make a small contribution, collectively they represent larger proportion of the rural economy and can add significantly to export revenues. India is an agriculture-driven country where 70% of the population lives in rural areas. It is a fact that most tribals live in forest regions and their livelihood is either partly or fully derived from collection from forests. In Southeast Asian countries, which are active in trade of NTFPs, India has a leading role with maximum number of persons and villages involved with sale of $199710.20 (Table 1).

Forest gatherers include tribals, forest dwellers, women and other marginalized groups. Most of the botanicals are sourced from nearby forests, shrubby areas, waste lands and fields. Forest-based small-scale activity provides an opportunity for employment for rural, tribal and marginalized groups which are based mainly upon the collection and processing of these non-timber forest products.

Importance of Medicinal Plants

In India, of the 17,000 species of higher plants, 7500 are known for medicinal uses. This proportion of medicinal plants is the highest proportion of plants known for their medical purposes in any country of the world for the existing flora of that respective country. Medicinal plants form a large group of economically important plants which provide basic raw materials for medicines. These plants and their products not only serve as valuable source of income for small holders and entrepreneurs but also help the country to earn valuable foreign exchange through export. Medicinal plants are rich in secondary metabolites and

Table 1 Southeast Asia: NTFPs trade

Country	Number of enterprises	Number of persons involved	Number of villages	Sales (USD)
India	27	2232	370	199710.20
Indonesia	29	1452	58	109796.84
Philippines	64	1946	68	97237.48
Cambodia	38	1649	81	97156.53
Total	**163**	**7400**	**600**	**448913.55**

(Source: NTFP's Annual Report 2011) (NTFP-Exchange Programme for South and Southeast Asia 2011)

act as possible source of drugs. These secondary metabolites include alkaloids, glycosides, flavonoids, steroids, etc. India is one of the richest countries where almost all the known medicinal plants can be found. Some plants are having great demand in the country and abroad as well like opium poppy, tropane alkaloid-bearing plants, sapogenin-bearing yams, senna, cinchona, etc. The ancient Indian system of medicine is mainly plant-based making use of most of our local plants. The Indian System of Medicine offers the most appropriate therapy against many diseases like jaundice, asthma, diabetes, gastric disorders, etc. Ayurveda, the oldest medicinal system in India, has reported about 2000 medicinal plant species, followed by Siddha and Unani. The *Charaka Samhita* (written document on herbal therapy) had reported the production of 340 herbal drugs along with their uses. Presently, about 25% of drugs are plant derived, and many others are synthetic analogues built on model compounds isolated from plant species in modern pharmacopoeia. India is already a major exporter of medicinal plants to various developed nations of the world. It is estimated that ₹860 billion worth of raw materials and drugs obtained from medicinal plants are exported from India. In developing countries like China, Nepal and India, medicinal plants are important revenue-generating resources and provide income to economically marginalized and indigenous people particularly people living in or around forests. Medicinal plants can meet the basic needs of the poor people living in rural areas. Most important Indian medicinal plants have been identified on the basis of their medicinal importance, commercial value and potential for research (Table 2).

Table 2 Priority species of medicinal plants based on commercial value

S. no.	Plant	Common name
1.	*Plantago ovata*	Isabgol
2.	*Bacopa manner*	Brahmi
3.	*Centella asiatica*	Mandukaparni
4.	*Withania somnifera*	Ashwagandha
5.	*Andrographis paniculata*	Kalmegh
6.	*Swertia chirata*	Chirata
7.	*Tinospora cordifolia*	Guduchi
8.	*Emblica officinalis*	Amla
9.	*Commiphora wightii*	Guggul
10.	*Phyllanthus amarus*	Bhumyamalaki
11.	*Podophyllum*	Papra
12.	*Asparagus racemosus*	Shatavari
13.	*Picrorhiza kurroa*	Kutki
14.	*Streblus asper*	Shakhotaka

Source: Exim Bank 1997 (Export Import Bank of India 1997)

Emerging Markets

Growing interest by multinational pharmaceutical companies and domestic manufacturers of herbal-based medicines is contributing significant economic growth of the global medicinal plant sector. In 1997, 3500 tons of the bark of *Prunus africana* ($220 million) used in the treatment of prostate disease was exported from Africa. The global market value of *Aloe vera* used to treat burns and added to skin creams and cosmetics has tremendously increased. In the three Himalayan countries of Nepal, Bhutan and Bangladesh, several thousand tons of medicinal plants are collected from forests providing earnings of millions of dollars each year. India, which is the hub of the regional trade, is at the national level of up to 40% of the state forest-based economy, and 70% of forest export revenues come from medicinal and aromatic plants (MAPs) and non-timber forest products (NTFPs), mostly in unprocessed and raw forms. Medicinal and aromatic plants worth US$ 18–20 million are exported with maximum quantity to India in raw form. India is expected to be among the top five countries in the world with respect to the absolute growth in terms of revenue. The Indian pharmaceutical markets show a huge growth of over 200% from a USD 6bn market in 2005 to a USD 20bn market in 2015 (Table 3).

Emerging markets, which mainly consists Russia, Brazil, China and Mexico as well as countries like South Africa, Turkey and Indonesia, are growing at a faster rate as compared to developed markets. According to IMS, a well-known industry research firm, these emerging markets will increase their global share from 16% in 2009 to 24% in 2014 (Fig. 1). This 50% rise in market share makes it an attractive market for many pharmaceutical companies operating the global arena. Thus,

Table 3 Ranking of countries according to percent growth in pharmaceutical markets

Rank	Country	Market in USD bn 2005	Market in USD bn 2015	Absolute growth	% Growth
1	USA	248	444	196	79
2	China	13	38	25	192
3	Japan	68	82	14	21
4	France	32	46	14	44
5	**India**	**6**	**20**	**14**	**233**
6	UK	19	32	13	68
7	Canada	13	25	12	92
8	Spain	14	25	11	79
9	Brazil	9	20	11	122
10	Mexico	10	19	9	90
11	Turkey	7	15	8	114
12	Germany	31	38	7	23
13	South Korea	8	15	7	88
14	Italy	20	25	5	25

Source: IMS World Review, Mckinsey Pharma Model (Indian Pharmaceuticals 2015; IMS Health Information and Consulting Services India Pvt. Ltd n.d.)

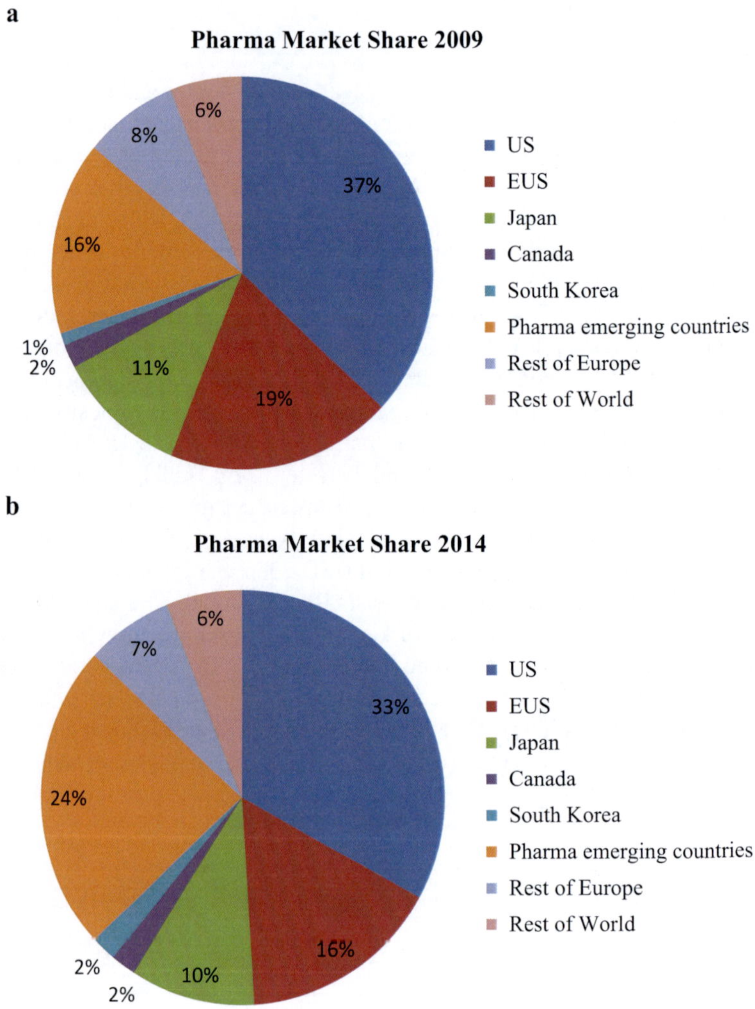

Fig. 1 Pharmaceutical market share of various nations (**a**) 2009, (**b**) 2014 (IMS Health Information and Consulting Services India Pvt. Ltd n.d.). Global market of medicines has doubled from 2000 to 2010, i.e. 30 USD bn to 60 USD bn, and it is expected to touch 500 USD bn in 2050 (Fig. 2)

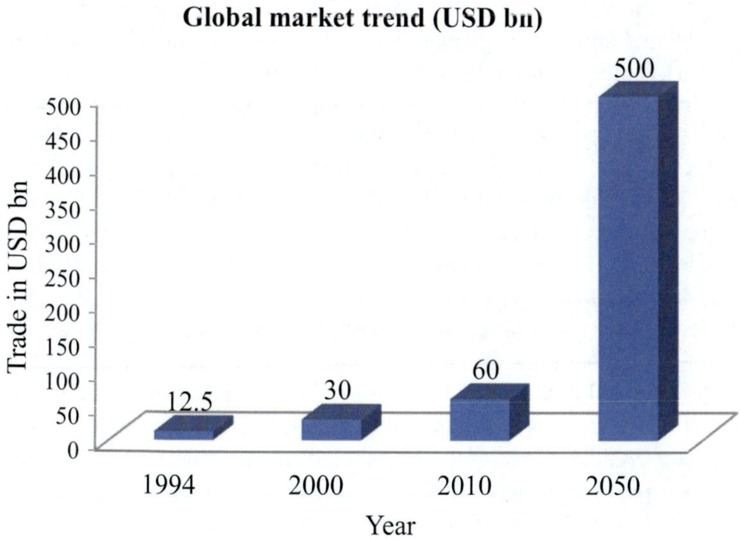

Fig. 2 Global market trend of medicinal plants

emerging markets are the next big thing for the drug fraternity. The sales in emerging markets have shown growth nearly at 100% from US $100bn in 2010 to a huge US $ 200 bn in 2015.

Trade of Medicinal Plants: Indian Scenario

India, as described earlier, is one of the richest regions as far as the diversity of plant species is concerned. India is the largest exporter, next to China, accounting for about 13% of the global exports. The USA is the principal market for Indian medicinal plants, accounting for 50% of exports. According to EXIM study, there are 880 medicinal plants species involved in national trade. Out of these, 48 species are exported and about 42 species are imported. Another survey conducted by the Ministry of Environment and Forests, Government of India, reveals that there are over 8000 species of medicinal plants grown in the country, 70% of which are found in the tropical forests in the Western and Eastern Ghats. The Export-Import Bank of India, in its report for the year 1997, puts medicinal plant-related trade in India at $.5.5 billion which is growing enormously.

The World Bank in its recent report on the potential of Indian forests to generate income has praised efforts of two states, viz. Madhya Pradesh and Assam, in marketing medicinal plants. However, the report also states that the country's natural resources are not being fully exploited. The market for medicinal plants in most states of the country is weak and focuses mainly on local trade. Over the past 10 years, there has been a considerable increase in interest in the use of herbal medicines in the world. Tuticorin region of Tamil Nadu state at present exports medicinal plants and herbal extracts worth Rs.40 crore annually. Psyllium husk (Isobgul husk) emerged as the largest item of exports registering a record growth of 162.80%. The export of jajoba seed, which happened to be the largest item of export in the year 2003–2004, declined drastically in 2004–2005 by reaching a low level of ₹8.53 crore as against ₹89.07 crore in 2003–2004, registering thereby a steep decline by 90.42%. In spite of considerable rate of growth in the medicinal plant and product market, India's share in world market of medicinal plant products is very low, i.e. 2.5% only. The export market is increasing faster than the domestic market. However, the export is being carried out in form of plants or their parts but not as value-added products. Countries that import plant and plant parts are poor in biodiversity but are technologically sophisticated which show a higher potential to develop value-added products from the imported materials. The exporters of Indian subcontinent face major problems while exporting medicinal plants and related products. The cyclical nature of medicinal plant market makes it difficult for them to maintain demand and supply in the global markets. Competition from synthetic substitutes, poor quality control, inadequate information on market trends and the use of poor technology along with other factors creates the problem of export of medicinal plants or their products from India (Table 4).

It is worrying that India has negligible patents on plant-based formulations, extracts and products. Regarding the export of medicinal plants, India's contribution to the international market is comparatively very low. Utilizing our plant diversity and proper planning, Indian products can enter the more and more overseas markets. This can be achieved only through proper development of medicinal plants, standardization of the extracts and keeping the quality. As per India Brand Equity Foundation (IBEF) reports, the domestic Indian pharmaceutical industry is estimated to be $26 billion in 2014 growing at nearly 20% and is expected to reach nearly $50 billion in 2020. There are about 200 companies which manufacture medicines for the largest population in the world. The scenario of generic drug market of the world shows how India is acting as a key player of the global pharmaceutical market. This drug market is expected to grow at about 60% by the year 2016–2017 and contribute nearly 36% of the total global market (Table 5). The contribution of the generic drug market to the world was expected to increase from 28.5% in 2012–2013 to 36% in 2016–2017.

Indian medicinal exports mainly comprise of bulk drugs, formulations and herbal products to the developed nations. Formulations contribute to 72% of the total Indian drug exports, which is nearly $11 bn as per statistics available from Pharmexcil for 2013–2014.

The growth rates of the Indian formulation exports in 2010–2011 over the period of 2009–2010 have been nearly 30%. This rate has managed to sustain this growth rate in the time period of 2011–2012 to 27% over the period of 2010–2011. The growth rate of Indian drug formulations has been nearly 20% in the financial year (FY) 2012–2013. Indian bulk drugs and herbal drugs have witnessed a decline in growth rate since the FY 2012–2013 on account of market factors and emerging competition from other Asian nations (Fig. 3).

Table 4 Important plants with reference to trade (Sharma et al. 2008)

S. No.	Plant name	Common name	Plant part	Estimated consumption (Tones)
1.	Aconitum heterophyllum	Atis	Root	20
2.	*Acorus calamus*	Vacha	Rhizome	150
3.	*Aloe vera*	Aloes	Leaf	200
4.	*Anacyclas pyrethrum*	Akkarkara	Fruit	50
5.	*Andrographis paniculata*	Kalmegh	Aerial part	250
6.	*Asparagus recemosus*	Satavatri	Root	500
7.	*Berberis Aristata*	Daru haldi	Root	500
8.	*Cedrus deodara*	Deodar	Heart Wood	203
9.	*Chlorophytum borivilianum*	Safed musli	Root	25
10.	*Cinnamomum zeylanicum*	Dalchini	Bark	200–300
11.	*Commipphora wrightii*	Guggul	Gum resin	500
12.	*Crocus sativus*	Keasr	Stigma	5
13.	*Cyprus rotundus*	Nagar motha	Rhizome	150
14.	*Eclipta alba*	Bhringraj	Aerial part	500
15.	*Elettaria cardamomum*	CAedamon	Seed	60
16.	*Embelia ribes burm*	Vidanga	Fruit	200
17.	*Glycyrrhiza glabra*	Milathi	Root	5000
18.	*Hedychum spicatum*	Kapurkachri	Rhizome	400
19.	*Hemidesmus indicus*	Anantmool	Root	200
20.	*Holarrhena pubescens*	Kurchi	Bark	150
21.	*Justicia adhatoda*	Vasaka	Leaf	500
22.	*Mucuna pruriens*	Kaunch beej	Seed	200
23.	*Myristica fragrans*	Jaiphal	Fruit	500
24.	*Nardostachy gradiflora*	Jatamansi	Root	200
25.	*Embelica offtcinalis*	Amla	Fruit	10,000
26.	*Picrorhiza kurroa*	Kutki	Root	200
27.	*Piper cubeba*	Cubeb	Fruit	150
28.	*Piper longum*	Pipramul	Fruit	200
29.	*Piper nigrum*	Black pepper	Fruit	150
30.	*Plumbago zeylanica*	Chitrak	Root	500
31.	*Pueraria tuberose*	Vidarikanda	Root	200
32.	*Saraca indica*	Ashoka	Bark	1200
33.	*Senna Alexandrian*	Senna	Leaf & pod	1000
34.	*Strychnos nux vomica*	Luchia	Seed	1000
35.	*Swertia chirayta*	Chirayita	Whole plant	500
36.	*Syzygium aromaticum*	Clove	Flower bud	150
37.	*Syzygium cumini*	Jaman beej	Seed	300
38.	*Trachyspermum ammi*	Ajwain	Fruit	200
39.	*Terminalia bellrica*	Bahera	Fruit	500

(continued)

Table 4 (continued)

S. No.	Plant name	Common name	Plant part	Estimated consumption (Tones)
40.	*Termmatia chebula*	Harar	Fruit	500
41.	*Tinospora cardifolia*	Guduchi	Stem	1000
42.	*Valeriana jatamansi*	Tagar	Root & Rhizome	150
43.	*Withania somnifera*	Ashwgandha	Root	500
44.	*Zingiber officinalis*	Ginger	Rhizome	500

Table 5 Scenario of generic drug market

	2012–2013	2016–2017	% Growth
Global pharma market (in $ bn)	962	1200	24.7
Global generic market (in $ bn)	274	432	57.7
Indian pharma generic drug exports (in $ bn)	15	25	66.7

Source: Report by Rashmi Pant January 6, 2016, Business standard

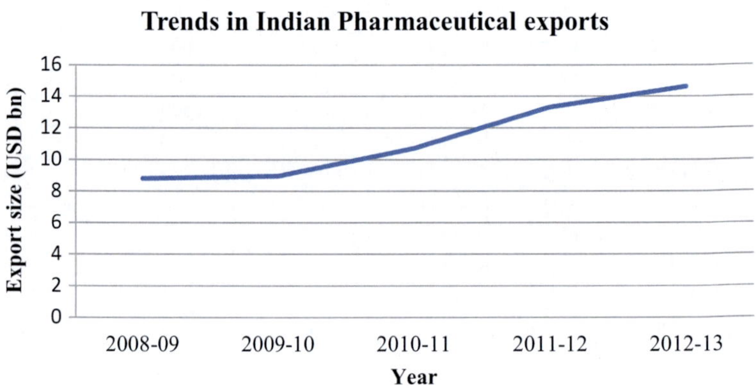

Fig. 3 Trends in Indian pharmaceutical exports (Source: Pharmexil)

Indian pharmaceutical market increased at a CAGR (compound annual growth rate) of 17.46% during 2015–2016 and is expected to increase at a CAGR of 15.92% to US$ 55 billion by 2020 (Fig. 4). About 20% of the global generic medicine market exports in terms of volume is supplied by India making it the largest supplier of generic medicines in the world (Fig. 5).

The current domestic trade of AYUSH industry is about Rs. 80–90 billion. The Indian medicinal plants and their products account for exports of about Rs. 10 billion. An increasing trend is observed in the export and import of AYUSH-related items in India. The export of AYUSH items has increased in 2011–2012 and 2012–2013 with an annual growth rate of 30%; however, it has decreased in 2013–2014 with a decline of 36%. Similarly, the import of AYUSH products increased by 23% annually in 2011–2012 and 2012–2013 and has further increased in 2013–2014 with an annual growth rate of 25% (Fig. 6) (Department of AYUSH n.d.).

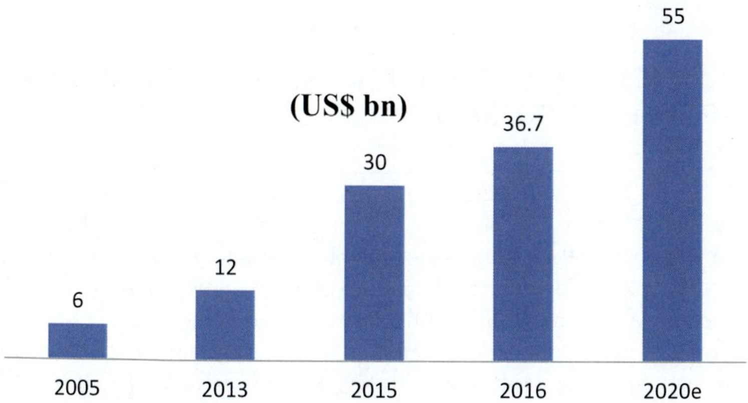

Fig. 4 Revenue of Indian pharmaceutical sector. Source: Department of Pharmaceuticals, PwC, McKinsey, TechSci Research

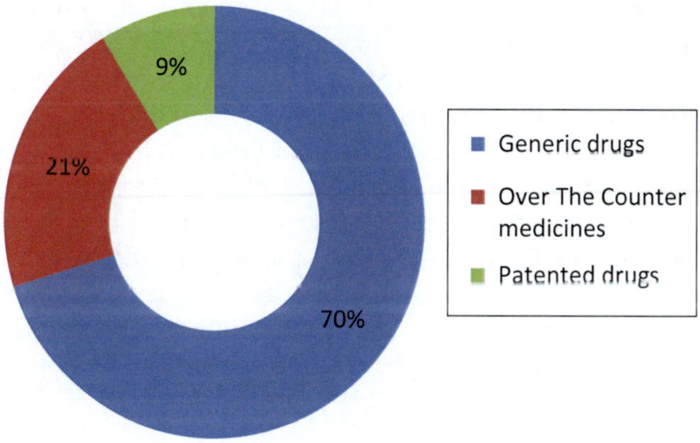

Fig. 5 Revenue share of Indian pharmaceutical subsegments in 2015. Source: Business Monitor International, FCCI Indian Pharma summit 2014–2015, TechSci Research

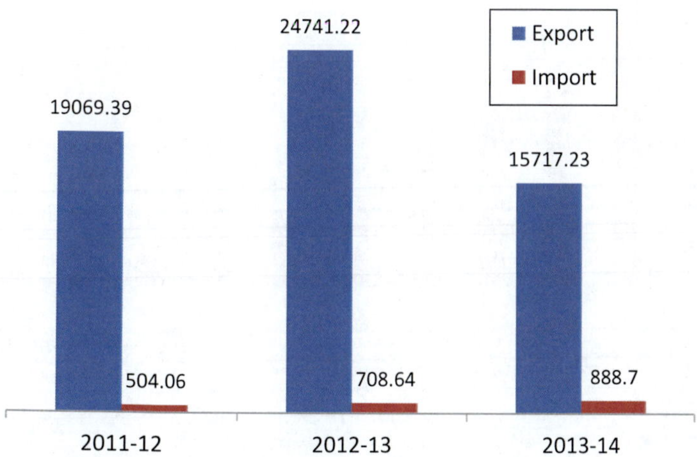

Fig. 6 Export and import of AYUSH items (in crores)

Medicinal Plants as a Source of Income in Madhya Pradesh: A Phytodiversity-Rich State of India

Madhya Pradesh, situated in the central region of the country, is one of the largest states of India. Due to vast diversity of topographical features, soil and climatic factors, diverse natural vegetation is available. The state is having varied types of forests ranging from dry thorny forests to tropical moist forests. A number of minor forest products including medicinal are found in the natural forests. Minor forest produces have tremendous potential and make significant contribution in terms of income and employment in rural areas of nearby forests. It is estimated that Baiga tribes of this state collect maximum minor forest products among forest dwellers in the world. A large proportion of population in MP is tribal who largely dwell in forests and manage to survive on forest products. Some important minor forest produces have been nationalized recognizing their importance in generation of employment and economy of rural population, as well as their economic role in industries. These plant products often provide income to local population such as tribes and marginalized sections with low income and limited employment opportunities (Fig. 7).

Since a large quantity of these forest-based products are sold in an unofficial way, little information is available about exact prices and demand and supply pattern. These have the potential to become sustainable sources of economy for tribes and weaker sections of the forests. These economic benefits derived from these NTFPs have been identified as a major opportunity for Community Forestry Projects (CFP) and Joint Forest Management (JFM). The primary tribal markets are generally informal, and it is difficult for the rural people to have access to information

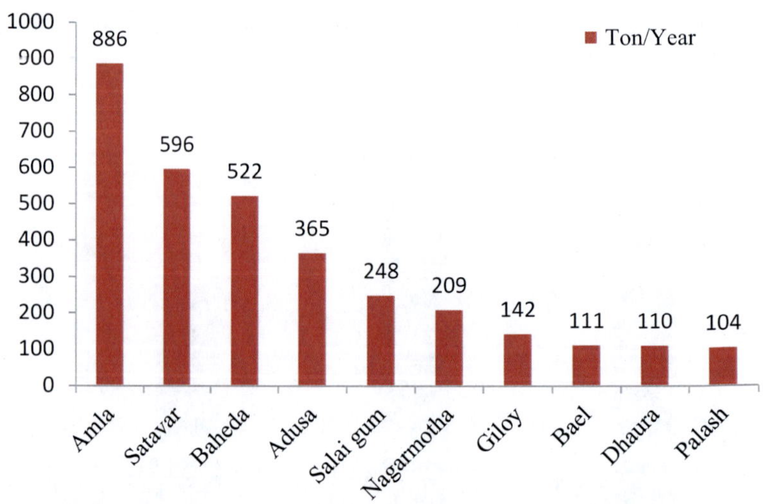

Fig. 7 Top traded herbs in MP state (Ghate and Dube 2014)

about potential markets and to have any control over the prices. NTFPs play an important role in both regional and national economies.

Opportunities in Developing the Medicinal Plants Sector

For developing the herbal industries, India possesses a rich diversity of medicinal plant species across the various forest types along an altitudinal gradient. Such a high diversity of medicinal plants would be helpful for further scientific research on exploring their medical efficacies. India has already established a reputation as a low-cost manufacturer of high-quality generic drugs in the global market. This fact can be used as an important tool for the marketing of herbal products. It is expected that India's aim to build a golden triangle between traditional medicine, modern medicine and modern science will be an advantage for developing the traditional herbal medicine and the medicinal plants sector.

Challenges in Medicinal Plants Sector

The continuous increase in human population in developing nations like India is one of the main causes for concern in meeting the daily needs of food and medicine as the economy and livelihood of societies primarily depend on forest products. This phenomenon leads to continuous erosion of forests and the forest products, thus making it tough to meet the requirements as well as to conserve useful natural resources. Although more and more species are being gradually added in the *Materia Medica*, the standard of their purity and authentic identification do not keep pace with the process of growth. The market prices for medicinal plant products provide only a limited insight into the exact status of the market, and not on the precise information of profits, supply and demand. Collectors and traders face problems in finding favourable markets due to price factors, quality and quantity considerations. Inadequate knowledge and incomplete information on products, markets and prices on the part of collectors and lack of standardization and consistency in quality for international marketing also create problems in the trade of medicinal plant products.

Demand for Medicinal Plants

The World Health Organization has estimated that the present demand for medicinal plants is approximately US $14 billion per year. The demand for medicinal plant-based raw materials is enormously growing at the rate of 15–25% annually, and according to an estimate of the WHO, the demand for medicinal plants is likely to increase to more than US $5 trillion in 2050. The demand for medicinal plants in India to meet both domestic and global markets was projected to increase at about 15–16% annually, between 2002 and 2005. If properly organized, the cultivation and management of medicinal plants becomes highly remunerative in economic perspectives for the small-scale growers. The estimated annual demand of some of the important herbal crude drugs used in preparation of Indian system of medicines is *Tinospora cordifolia* (10,000 T), *Pulchea racemosa* (3000 T), *Emblica officinalis* (2750 T), *Withania somnifera* (2500 T), *Asparagus racemosus* (2500 T), *Centella asiatica* (1500 T) and *Commiphora wightii*, *Cassia angustifolia*, *Terminalia chebula*, *Andrographis paniculata* and *Aloe vera* 1000 tons each.

The revenue from the major Indian systems of medicine, i.e. Ayurveda, Unani and Siddha, is estimated at more than half a billion dollars annually. In 2000, the gap between demand and supply of medicinal and aromatic plants was estimated to be about 40,000–200,000 tons; by 2005 the gap was expected to increase to 400,000 tons. Medicine-based industries create employment opportunities in poor, job-starved states, thereby increasing the cash earnings of local people. In India, the medicinal plant-related trade is estimated to be approximately US $1 billion per year. According to an estimate, the quantity of export of Ayurvedic products in India has increased thrice in recent years. The projected escalating demand of medicinal plants has led to the overexploitation of many plants from wild, which subsequently results in the loss of these plant species. For example, the large quantity of *Taxus baccata* has been gathered from the wild sources since its extract, Taxol, was established as a use in the treatment of ovarian cancer. *Aconitum heterophyllum*, *Nardostachys grandiflora*, *Dactylorhiza hatagirea*, *Polygonatum verticillatum*, *Gloriosa superba* and *Arnebia benthamii* are other examples of North Indian medicinal plants which have been overexploited for therapeutic uses. About 95% of the 400 plant species used in medicine preparation by various industries are harvested from wild populations in India. Harvesting medicinal plants for commercial use, coupled with the destructive harvest of underground parts of slow-reproducing, slow-growing and habitat-specific species, is the crucial factor to meet the goal of sustainability. Furthermore, rising demand with collapsing habitats may lead to the local extinction of many medicinal plant species (Table 6).

Existing Policies

In the National Five-Year Plans of India, the medicinal plants sector has been identified as an integral part of the Indian System of Medicine. In 1983, the National Health Policy recognized that the large stock of health manpower in Ayurveda, Siddha, Unani, homoeopathy and naturopathy had not been adequately utilized; therefore, steps need to be taken to move towards a meaningful integration of the indigenous and modern systems of medicine. The Planning Commission and the National Medicinal Plants Board (NMPB) of the Government of India have framed

Table 6 Annual demand for important Indian medicinal plants

S. No.	Plant	Demand (tonnes)		Annual growth rate (%)
		2001–2002	2003–2004	
1	Amla	22,730	41,783	22.5
2	Ashwagandha	7029	9128	3.1
3	Liquorice root	873	1360	15.9
4	Rauwolfia	424	589	11.6
5	Sandal wood	635	1073	19.1
6	Senna	6463	11,677	21.8
7	Andrographis	2005	2197	3.1
8	Bacopa	3823	6622	2101
9	Black nightshade	2078	2192	1.8
10	Guggul	1505	2549	19.2
11	Jatamansi	675	867	8.7
12	Picrorhiza	220	317	12.9
13	Indian aconite	322	3427	30
14	Long pepper	3993	6280	16.3

a policy on the commercial aspects of the medicinal plants sector. NMPB has prioritized 32, and the Planning Commission has enlisted 24 medicinal plant species for development in order to meet the desired aim of the medicinal plant sector. The Biodiversity Act 2002 has framed many rules for sustainable utilization of medicinal plants.

Suggestions

The export subsidy for the medicinal plants should be hiked to increase production and supply. The awareness campaign should be connected to the collectors living in rural areas every year. Funds should be allocated for conducting research and development to improve varieties of medicinal plants and enhance their availability. Set up export promotion zones exclusively for medicinal plants and herbal products in potential states. Harvesting, drying and storage of medicinal plants must ensure the safety against microbial contamination and quality deterioration. There should be a proper connectivity between growers and pharmaceutical companies to ensure marketability of raw drugs. The study emphasizes upon the need for assessment of demand and supply studies, so that resource management strategies and policy interventions could be properly customized. The Government should propagate and encourage the use of modern chemical fertilizers for the cultivation of medicinal plants. At the same time, there must be immediate efforts to ensure the conservation of diverse biological resources and the preservation and application of local and cultural knowledge on the use of these resources.

Conclusion

Medicinal plants constitute a vast, undocumented and overexploited economic resource, and they are the principal health-care resources for the majority of the people. Demand for medicinal plant produce is increasing in both developing and developed countries, and the bulk of the material trade is still from wild-harvested sources. The expanding trade in medicinal plants has serious implications on the survival of several plant species, with many under serious threats to become extinct. As India's share is less than 1% in the $62 billion market, which is growing 7–12% per annum, efforts should be made to adopt a package of best practices encompassing conservation, cultivation, quality control and standardization and research and development for medicinal and herbals for improving its marketing performance efficiently.

Acknowledgments The authors are highly thankful to Jiwaji University, Gwalior, for providing financial support in the form of Research Scholar Fellowship.

References

Areki F, Cunningham AB (2010) Fiji: commerce, carving and customary tenure. In: Laird SA, RJ ML, Wynberg RP (eds) Wild product governance: finding policies that work for non-timber forest products. Earthscan, London, pp 229–242

Asfaw A, Lemenih M, Kassa H, Ewnetu Z (2013) Importance, determinants and gender dimensions of forest income in Eastern highlands of Ethiopia: the case of communities around Jelo Afromontane forest. For Policy Econ 28:1–7

Babulo B, Muys B, Nega F, Tollens E, Nyssen J, Deckers J, Mathijs E (2009) The economic contribution of forest resource use to rural livelihoods in Tigray, Northern Ethiopia. For. Policy Econ. 11:109–117

Bouri T, Mukharjee A (2013) Documentation of traditional knowledge and Indigenous use of non-timber forest products in Durgapur forest range of Burdwan district, West Bengal. In: Paper presented in National Seminar on Ecology, Environment & Development 25–27 January, 2013, organized by Department of Environmental Sciences. Sambalpur University, Sambalpur

Department of AYUSH (n.d.) Ministry of Health and Family Welfare, Government of India. Available from http://www.indianmedicine.nic.in

Export Import Bank of India (1997) Indian medicinal plants: a sector study (Mumbai: Export Import Bank of India, Occasional Paper No 54)

Friedman M et al (2007) J Food Sci 72:M207–M213

Ghate U, Dube P (2014) Medicinal Plants Study in Madhya Pradesh (M.P.) State (Trade, Farming & Livelihoods) for biodiversity friendly development planning, Final Report of the supported by the M. P. State Biodiversity Board, Bhopal

Hegde R, Enters T (2000) Forest products and household economy: a case study from Mudumalai Wildlife Sanctuary, Southern India. Environ Conserv 27:250–259

Hunzai K, Gerlitz JY, Hoermann B (2011) Understanding mountain poverty in the Hindu Kush–Himalayas: regional report for Afghanistan, Bangladesh, Bhutan, China, India, Myanmar, Nepal, and Pakistan. International Centre for Integrated Mountain Development, Kathmandu, Nepal

IMS Health Information and Consulting Services India Pvt. Ltd (n.d.)

Indian Pharmaceuticals 2015, A Mckinsey & Company Report
Jodha NS (2005) Economic globalization and its repercussions for fragile mountains and communities. In: Huber UM, Reasoner MA, HKM B (eds) Global change and mountain regions. Springer, Dordrecht, the Netherlands, pp 583–591
Kaimowitz D (2003) Not by bread alone: forest and rural livelihood in Sub-Saharan Africa. In: Oksanen T, Pajar B, Tuomasjukka I (eds) Forest in poverty reduction strategies. European Forest Institute, Joensuu., ISBN: 952-9844-96-4, pp 45–64
Killman W (2003) "Non-wood News", March 2003, p. 1
Mahapatra AK, Albers HJ, Robinson EJZ (2005) The impact of NTFP sales on rural households' cash income in India's dry deciduous forest. Environ Manag 35:258–265
Mahesh B, Satish S (2008) Antimicrobial activity of some important medicinal plant against plant and human pathogens. World J Agric Sci 4(Suppl. 1):839–843
Malhotra KC, Bhattacharya P (2010) Forest and livelihood. CESS Publisher, Hyderabad, p 246
Marshall E, Schreckenberg K, Newton AC (2006) Commercialization of non-timber forest products: Factors influencing success: lessons learned from Mexico and Bolivia and policy implications for decision-makers. UNEP World Conservation Monitoring Centre, Cambridge http://www.odi.org/sites/odi.org.uk/files/odi-assets/publications-opinion-files/3769.pdf
NTFP-Exchange Programme for South and Southeast Asia (2011). Annual report. p 11
Pandey AK, Bhargava P, Negi MS (2011) Sustainable management of non-timber forest produce through joint forest management. Indian Forester 137:105–113
Rasul G, Karki M, Sah R (2008) The role of non-timber forest products in poverty reduction in India: prospects and problems. Dev Pract 18(6):779–788
Rasul G, Kollmair M (2010) Sustainable livelihood promotion through agricultural development in the hills of South Asia. In: Fonseka H (ed) Hill agriculture in SAARC countries, constraints and opportunities. SAARC Agriculture Centre, Dhaka, pp 167–182
Saha D, Sundriyal RC (2012) Utilization of non-timber forest products in humid tropics: Implications for management and livelihood. For. Policy Econ. 14:28–40
Saxena NC (2003a) Livelihood diversification and non-timber forest products in Orissa: wider lessons on the scope for policy change? ODI Working Paper No. 223, Overseas Development Institute, London
Saxena NC (2003b) Livelihood diversification and non-timber forest products in Orissa: wider lessons on the scope of policy change. Livelihood Strategy Paper. Department for International Development and Overseas Development Institute, London
Shackleton CM, Pandey AK, Ticktin T (2015) Ecological sustainability for non-timber forest products: dynamics and case studies of harvesting. Routledge, London., ISBN-13: 978–04157285 91, p 294
Shackleton S, Campbell B, Lotz Sisitka H, Shackleton C (2008) Links between the local trade in natural products, livelihoods and poverty alleviation in a semi-arid region of South Africa. World Dev 36:505–526
Sharma A, Shanker C, Tyagi LK, Singh M, Rao CV (2008) Herbal medicine for market potential in India: an overview. Acad J Plant Sci 1(2):26–36
Tiwari DD, Campbell JY (1997) Economics of non-timber forest products. In: Kerr JM, Marothia DK, Singh SK, Ramaswamy C, Bentley WB (eds) Natural resource economics: theory and application. India Book House, New Delhi
Vantomme P (2003) Forest products division. Forestry Department, Food and Agriculture Organization, Rome
Wymann von Dach S, Ott C, Klaey A, Stillhardt B (2006) Will international pursuit of the millennium development goals alleviate poverty in mountains? Mt Res Dev 26(1):4–8

Ethno-ecology of the Healing Forests of Sarban Hills, Abbottabad, Pakistan: An Economic and Medicinal Appraisal

Farhana Ijaz, Inayat Ur Rahman, Zafar Iqbal, Jane Alam, Niaz Ali, and Shujaul Mulk Khan

Introduction

Study Area

Abbottabad city is located in the Hazara division of the Khyber Pakhtunkhwa, Pakistan. The city can be located in the beautiful valley at 34°09′N latitude and 73°13′E longitude at an elevation of 4120 ft (1260 m) and 110 km north of the country's Capital Islamabad, and 150 km northeast of Peshawar the province's capital. Kashmir lies to east of the city. The total area of Abbottabad district is 179,653.5 (KfW). The city is well known throughout Pakistan for its pleasant weather, high-standards' educational institutions, and military establishments. It remains a popular hill station fascinating hundreds of tourists each year.

Geography of the Region

The city is bordered on four sides by the Sarban Hills, from which inhabitants and visitors can see amazing view of the area and city (Fig. 1). The scene of the city and the hills allows Abbottabad to pleasurable weather in the summer and cold winters. The Dor stream (a small river) South of Abbottabad passes through the town of Harnoi and reaching Tarbela Dam, positioned West of Abbottabad.

F. Ijaz · I. U. Rahman · Z. Iqbal · J. Alam · N. Ali
Department of Botany, Hazara University, Mansehra, Pakistan

S. M. Khan (✉)
Department of Plant Sciences, Quaid-i-Azam University, Islamabad, Pakistan
Email: https://www.qau.edu.pk/profile.php?id=804024

M. Ozturk, K. R. Hakeem (eds.), *Plant and Human Health, Volume 1*,
https://doi.org/10.1007/978-3-319-93997-1_19

Fig. 1 Map of the study area. (**a**) Abbottabad, (**b**) the Sarban Hills, (**c**) and (**d**). View of Abbottabad city from the Sarban Hills (Ijaz et al. 2016)

Population of the area

In 1901, the population of the town was 7764, while in 1911 it reached 11,506. The population has been raised between 1951 and 1998 tremendously, ranging from 319,000 to 881,666. Annual report show growth rate of 3.75% (MSU 2000:11).

Soil

Soil of Abbottabad district is rocky type and rich in minerals, comprising deposits of granite, limestone, phyllite, schist, slate, soapstone and quartz.

Climate

Abbottabad has a humid subtropical sort of climate with mild to warm temperature during the spring and autumn months, hot in the months of June and July, and cool to mild during the winter season. Temperature can rise as high as 38 °C (100 °F) during the mid-summer months and drops below −5 °C (23 °F) during December to February.

Snowfall occurs occasionally in the months of December and January, while heavy rainfall occurs during the monsoon season from July to September.

Languages

Hindko, Gujri Pahari and Pashto are the predominant local languages of the district. According to a survey conducted in 1998, major portion of the population comprises of Hindko speakers.

Economy and Tourism

Its economy is mostly based on tourism. The area is renowned for its shady gardens, churches, and wide streets in the old cantonment area, which was established during the British rule. Abbottabad has been attracting tourists to the city since the colonial era, as it is a major transit point to all major tourist regions of Pakistan like Nathia Gali, Ayubia, Naran, etc.

Like the northern areas of the country, tourism is one of the important sources of livelihood for the natives. In the summer when temperature rises to around 45 °C in Punjab and Khyber Pakhtunkhwa, a large number of tourists travel to Abbottabad. The Karakoram, Himalayas, and Hindu Kush ranges can be approached from Abbottabad, and it continues to be a passage city for tourists, serving as a base for visiting numerous nearby places, such as Hunza, Gilgit, Skardu, and Indus Kohistan of the Karakoram Range. Some popular tourist destinations are Dor River Valley (Harnol/Harnoi), Nathia Gali, Bara Gali, Dunga Gali, Thandiani, Ilyasi Mosque, Khaira Gali, and Shimla Hill.

Education

Abbottabad has one of the country's highest literacy rates with 67% in 2007, ranking first in Khyber Pakhtunkhwa and ninth overall in Pakistan. The city also has a young demographic (ages 15–30) due to a large number of students who have come across the country for studies in different institutes like PIPS, Army Burn Hall College, Army Public College, and Abbottabad Public School. The city has a wide variety of postsecondary institutions, such as Ayub Medical College and COMSATS University of Engineering and Technology.

Agriculture

About 4.82% of the land is used for agricultural purposes. The cool, temperate climate is suited for fruit production. Abbottabad district produces 5537 tons of fruit annually. Forest of Abbottabad covers an area of 36,394.6 ha, which is 21.4% of the total area, while official figures show that the forest area is higher than 36,441 ha. Three main forests of the Abbottabad are Himalayan moist temperate forest, subtropical pine forest, and subtropical broad-leaved scrub forest (Agriculture Statistic 1999–2000). Biodiversity of Abbottabad is very unique and is rich due to difference in vegetation zones and various environments having varied sort of habitat types. Further, some 1300 different plant species are reported in Abbottabad. Nasir and Ali (1982) reported 80 endemic species from Hazara division; of these, 50 species are also found in Abbottabad.

Livestock

Animal rearing is the oldest practice in Abbottabad. Most of the people have cow, buffaloes, and goats used for source of milk, butter, and "ghee". Besides that mules are also reared for the purpose of loading materials on the upper hilly areas where transport system is not available.

Socioeconomic Status

Most of the people have government jobs, some in the education sector and the others in field of construction but majority of the people are abroad for job and other purposes. A small proportion of the populations are related to the field of agriculture, growing vegetables and fruits to fulfill their needs (IUCN 2004).

Biodiversity

In 1985, Walter G. Rosen defined the biodiversity as life on earth. He includes all ecosystems and life forms that are parts of biodiversity (Rahman et al. 2016a, b; Khan et al. 2012a, b). Many raw materials such as fertilizers, food, shelter, medicines, and fibers for clothing materials as well as source energy in the form of animal traction all are assertion of biodiversity. For the survival of human and maintaining balanced environment plant biodiversity play an important role.

Ethnobotany in its historical perspective

Ethnobotany is the study of plants and people that demonstrates how people use plants and for what purpose. Significant uses of plants either medicinally or else and their occurrence in the confined environment depend upon the notion of the people. Records found in Babylon (1770 BC) and in ancient Egypt (1550 BC) indicate that the usage of medicinal plants started from very ancient times. The prehistoric people of Egypt believed that medicinal plants are very valuable for their Pharaohs (Egyptian King) and believed that medicinal plants were even effective in the afterlife of their Pharaohs, which is evident from the plants found in Giza pyramids.

A question arises in everyone's mind: how people use plants and what are the reasons they use them? To solve this problem, ethnobotanists collect information from the existing peoples. In this way an appreciative result found is not only of the current uses of plants but also of the significance of plants for construction, food, and medicine in their ancient actuality. It also provides an idea of people's old-style natural knowledge specially related to plants and the effect of this knowledge on the research and methods used in ethnobotany. Ethnobotany is the field of study, which is based on knowledge coming through generations from primitive and aboriginal people. It gives original ideas and informations of plant in relation to the day today lives, civilizations, religion, beliefs and everyday applied concerns of the people. It also explains how they use the plants around them for traditional purposes that is food, treatments and other cultural uses.

Jones described the future rebirth of ethnobotany by pointing out that ethnobotany is entirely revolves around the interrelation of human and plants. Ethnobiology has been further explained by ethnobiologists whom also suggested some strategies for investigators and interviewers. These recommendations contain plant documentation, comparative richness and accessibility of the plants, local names of the plants, purposes of use, season of collection, whether or not the plant is inherent to the area, the commercial importance, species not used in the region, and importance of the plant in the economy or the culture.

Later on, Jones' definition was reformed by Ford to accommodate the development of the field and established that "Ethnobotany is the study of direct interrelations between humans and plants." Accumulation of the term "direct" actually encompases persistent interaction with plants, allowing them to organize in their way and to produce traditional rules for influencing the plants and their confined environments. Omission of the word "primitive" certified the extension of the ground study. Ethnobotany at this point was concerned with the traditional facts of predominantly non-western people. Ethnobotany sustains a multidisciplinary character: botanically, focus on the plant and plant uses, although ecological patterns, plant distributions, resources consumption, and agricultural and cultivated designs have become popular paths of the study among ethnobotanists.

Anthropologically, ethnobotanists focus on realizing human collaborations with plants through original symbols, folklore, ceremonies, and plant categorizations among non-western peoples. Archeologically, ethnobotanists focus on agronomic origins and ancient plant uses and relied on plant and coprolite remains, fossilized pollen, and old food stores to regulate primitive plant uses and associations. Recently ethnobotanists developed a holistic methodology habitually relying on several concepts in order to conduct their investigation, comprising data from botany, musicology, architecture, linguistics, pharmacology, anthropology, conservation biology, and many others, depending on the questions being asked in the study.

Ethnobotanists through close interaction with plants of an area are capable to relay local and specific plant nomenclatures and to study all the physical assets of the plants. They pay courtesy to traditionally related conceptual and representative properties of the plants in an area. Ecological associations within the plant community are essential to these studies as well as the more plant-human relationship in relation to community economics. Here, anthropological economic theory plays an important role in that it helps the ethnobotanists measure and calculate human necessities and their influence on a confined environment.

Traditional Medicines

Traditional medicine is defined as "the sum total of all the awareness and practices, whether rational or not, used in analysis, anticipation and rejection of corporeal, psychological or collective inequity and have faith completely on applied practice and opinion tendered down from generation to generation, whether orally or in text".

Traditional medicine or more properly old system of health precaution has been resumed in the past 20 years. Every area had a form of traditional medicine or a traditional system of health care. Arabic, African, Chinese, and other traditional medicines are extremely surrounded in a specific sociocultural background which fluctuates between societies. This difference between societies gives traditional medicine its assorted and pluralist nature. The devotion formerly given by governments to extensive health-care submission has given a new motion to research, asset, and project of programs in this field in some rising countries in Africa and elsewhere.

Medicinal Plants

For thousands of years, to cure different diseases, plants are used by humans. According to the World Health Organization, most populations still depend on traditional medicines for their mental and corporeal health desires, since they cannot afford the products of Western medicinal industries, together with their side effects and lack of health care services. Rural areas of many emerging countries still depend on old herbal system of medicines and have found a place in daily life. These drugs are comparatively harmless and inexpensive than artificial or new medicine (Ahmad et al., 2009). People living in rural areas from their personal skill know that these traditional medicines are valuable source of natural products to sustain human health, but they may not realize the science behind these drugs, but knew that some medicinal plants are highly effective only when used at beneficial quantities (Pieroni and Quave, 2014).

There is a huge demand of herbal medicines in both developed and developing countries as a foundation of most important health care owing to their aspects having wide natural and medicinal activities, high safety limitations, and minor expenditures. Herbal fragments are safe and would overcome the conflict created by the pathogens, as they exist in a collective form or in a pooled form of more than one fragment in the protoplasm of the plant cell. Even with the arrival of new or allopathic medicine, it has been noted that a number of important modern medicines were derived from plants used by ethnic people. Traditional use of medicine is familiar as a way to learn about prospective future remedies. Scientists have identified a number of combinations used in typical medicine, which were derived from "ethnomedical" plant sources. Plants are used therapeutically in different countries and are a source of many strong and influential drugs.

Ethnoveterinary uses of medicinal plants

Ethnoveterinary medicines lead to the modern approaches of animal health maintenance that is primarily based on the knowledge, abilities, techniques, practices, and beliefs about animal health care found among indigenous communities. Such knowledge vary not only in section to section but also within different societies. It has been established through experiments and error and thoughtful investigations. Therefore, it is less organized, less formalized and not comprehensively documented as an effective method of disease control in animals. There are the records of original knowledge in many countries; it has been transferred across generations by a verbal tradition and therefore is in danger of destruction. Local healers have less insight to deal with treatment and control of prevalent and rare infectious diseases like foot and mouth disease, anthrax, septicemia, acute life-threatening bacterial diseases, and render pest, they can manage with a reasonable range of mutual diseases such as wounds, colds, diarrhea, worms, coccidiosis, and generative syndromes.

Ethnobotanical knowledge is more commonly known in livestock owners, which has formed the foundation for transmission of plant materials as possible sources of medical remedies. About 60 diseases of animals are identified by the herders of the Turkana and Samburu communities in Kenya, and they identify them as curable and non-curable by using local medicines. Some how 35 of these diseases were curable, including cough, streptothricosis, and diarrhea. Scientific explorations and research on farmers in Trinidad and Tobago found that adding preparations from plants such as *Momordica charantia* to drinking water develops the yield and viability of broilers. Paw-paw latex (*Carica papaya*) has been used effectively as an anthelmintic remedy in goats. Despite such achievements, very little of this old knowledge has been recognized in emerging countries, and ethnoveterinary awareness has had no place in ordinary veterinary medicine. In current years, increasing attention has been paid to ethnoveterinary knowledge and local veterinary practices. There is a growing approval that some of these practices have therapeutic value and that they should be documented before this knowledge is misplaced.

35,000 to 70,000 plant species are used in traditional medicines to treat different diseases (Mehmood et al. 2015). About 70–80% of the world inhabitants still use traditional medicine. Ali (2008) and Ali and Qaiser (1986) reported that 1572 genera and 5521 species have been found in Pakistan, generally restricted to the hilly regions.

In Pakistan, the field of ethnobotany is nascent as compared to other countries, but in recent years, a lot of work has been done in this field by many researchers in different areas of Pakistan (Haq and Hussain 1993; Abbasi et al. 2013; Abbas et al. 2016; Bibi et al. 2016), but no information and documentation are available especially on Sarban Hills, Abbottabad.

So this research is an effort to document and preserve this folk asset, to explore the cultural knowledge of the study area, to enlist the indigenous medicinal plants used by local people for common day ailments, to create the awareness among the local community about the protection of native medicinal flora, and to collect medicinal plants of the area for proper identification and future references.

Materials and Methods

Present study was carried out February to September 2013. During this time the research area was visited twice a month for collection of data relevant to ethnoecologically important plant species, and conservation of the area. Every study trip was well planned and performed effectively.

The research was completed in three stages:

- Field trips for data collection
- Literature collection
- Documentation of data obtained from research study area

Field Trips and Field Work

Fieldwork was carried out in order to investigate the traditional plants and biodiversity of various parts of Sarban Hills in different seasons. Timings for the fieldwork were selected according to the growth and collection season of the plants. The fieldwork includes interviews, observations, and collection of various plant species and their photography present in the investigated area. All the plants during flowering and fruiting stage were collected and photographed. Collection numbers were given to the specimens and field data including scientific names, vernacular name, family names, habit, habitat, locality, parts used, economic importance, and other relevant information which was noted in the field notebook. The plant specimens were serially tagged and properly pressed for drying placed between blotting papers (Khan et al. 2012a, b, 2013a, b, c, Ijaz et al. 2015, 2016; Rahman et al. 2016a, b; Ahmad et al. 2016).

Collection of Plant Materials

The plant materials will be collected aseptically with gloves into sterile polyethylene pouches. The taxonomic features and medicinal use of botanicals will be noted. The data regarding collection and storage of botanicals on local shops will also be obtained.

Questionnaires

During fieldwork interviews were conducted with the local inhabitants, selected informers, the herbalists “Hakims” (local physicians of the eastern system of medicine), and pansaries (medicinal plant sellers in the local markets). Questions concerning the utility of different plants, quantity of plants used, plant part used, economic or market value and fuel wood, fodders, and head loads had been asked.

Herbarium Work

Plants in the presser are kept for about 2 weeks at room temperature. After each 48 hours, the blotting papers were changed. The specimens were poisoned using mercuric chloride and absolute alcohol (2 g mercuric chloride dissolved in 100 mL of absolute alcohol) and mounted on the standard-sized herbarium sheets. The data noted in the field notebook during field survey was transferred to the labeling slip pasted on the herbarium sheets on the right side at the bottom of herbarium sheets.

Plants Identification

The plants were identified with the help of taxonomic literature (Stewart 1972: Ali and Qaiser 1986, Parker 1918, Frazer-Jenkins 1991: Khan et al. 2015). Previous published literature was also searched upon to evaluate the conservation status of various species in the field (Hamayun 2005: Ahmad et al. 2015). Important plant species were also photographed. Voucher specimens were deposited in the herbarium of Hazara University, Mansehra. For the authenticity of identification, herbarium of Hazara University, Mansehra, was consulted.

Market Assessments

Economic, commercial, and medicinal value of indigenous plants utilized in the study area have acquired the due place in local markets of Abbottabad and Mansehra. In this regard a questionnaire has been adopted to interview the local plant collectors and medicinal plant sellers in the local markets. Marketing chain for the economic plant collectors and people involved in medicinal plant trade was investigated. A list of economic plants was prepared with emphasis on plant market availability status, collection methods, and local prices of these plants. The available literature on the market survey was also collected and reviewed, comparing it with present findings.

Documentation of Research Data

In the last phase, data collected during fieldwork was documented. The economic plant collected data during field observation was checked and also compared with the available literature. The economical inventory includes botanical names, vernacular names, families, habit, habitat, plant diversity, medicinal plants, and other economic importance.

Results

Plant-diversity of the Sarban Hills

Sarban Hills are unique and rich in biodiversity. Villages like Khota Qabar, Barra morre, Salhad, Nogakhi, Khrella, Khayal and Khunj are situated in the Sarban hill slopes. About 147 plant species were collected from the area which are being used for different purposes, viz., medicinal, fodder, and fuel.

These 147 plant species belong to angiosperms, gymnosperms, pteridophytes. Pteridophyte contains (2) species of family Adiantaceae, and angiosperms contain (142) species belonging to 56 families in which 6 are monocot and 50 are dicot families. Three species are of gymnosperm (*Pinus roxburghii*, *Cedrus deodara*, and *Cupressus sempervirens*). *Pinus roxburghii* and *Cupressus sempervirens* were collected, while single spp. *Cedrus deodara* is present on top of the hill which shows the presence of *Cedrus deodara* vegetation in the history. On the basis of plant habit, 147 plant species are divided into 4 growth forms: herbs 58.2%, shrubs 30.8%, trees 19.8%, and climbers 5.4% (Fig. 2).

Angiosperm consists of 56 families in which 6 families are monocot Alliaceae, Araliaceae, Commelinaceae, Liliaceae and Poaceae. Dicot families are Acanthaceae, Amaranthaceae, Apiaceae, Apocynaceae, Berberidaceae, Bignoniaceae, Brassicaceae, Caesalpiniaceae, Cannabaceae, Cappridaceae, Caprifoliaceae, Caryophyllaceae, Celastraceae, Chenopodiaceae, Convolvulaceae, Commelinaceae, Cucurbitaceae,

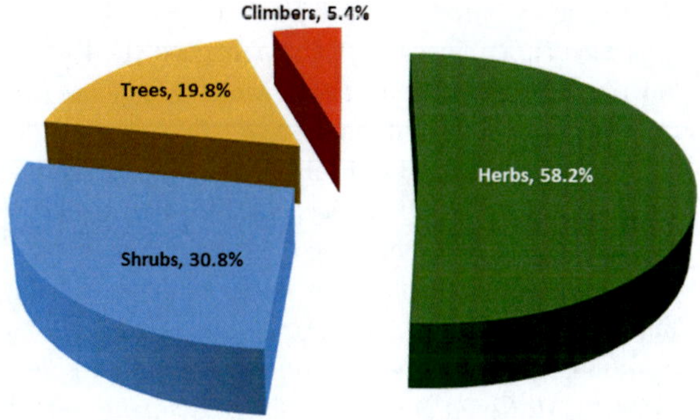

Fig. 2 Plants on the basis of habit

Ebenaceae, Eurphorbiaceae, Fumariaceae, Geraniaceae, Lamiaceae, Lythraceae, Malvaceae, Meliaceae, Menispermaceae, Mimosaceae, Moraceae, Myrsinaceae, Myrtaceae, Oleaceae, Onagraceae, Oxalidaceae, Papilionaceae, Plantaginaceae, Polygonaceae, Portulacaceae, Primulaceae, Punicaceae, Ranunculaceae, Rhamnaceae, Rosaceae, Rubiaceae, Rutaceae, Scrophulariaceae, Spindaceae, Solanaceae, Urticaceae, Verbenaceae, and Violaceae. Gymnosperm families are Pinaceae and Cupressaceae.

Asteraceae family leads by having the maximum number of plant species which are 19, Lamiaceae having 15, and Papilionaceae family having 13 species. Other important families are Brassicaceae, Amaranthaceae, Euphorbiaceae, Poaceae, and Acanthaceae. The flora of Sarban is in danger due to decreases of plant species day by day due to cutting of shrubs and trees for the purpose of fuel wood, construction, and fodder and also due to overgrazing. Many of these families are near to be extinct from the area.

In present plant collection from Sarban Hill area, 147 plant species were collected which were then pressed, dried, poisoned, and mount on the standard-sized herbarium sheet. After that plants are identified by flora of Pakistan and also with the help of taxonomist of the Botany Department of Hazara University. There were 2 or more than 2 specimens of the same species collected in some cases, e.g., 3 specimens of *Oenothera rosea*, 2 specimens of *Rydingia limbata*, and 2 specimens of *Berberis lycium* were collected, so after identification of all collected plant specimens, 147 different species related to different families were found. These plants were further divided on the basis of their life form, viz., herbs, shrubs, and trees.

Ethno-ecology of the Sarban Hills

Medicinal Plants

Local people of the Sarban Hills use 107 plants as medicinal source (73.2% of plants are used medicinally in total plant collection). These plants belong to families Mimosaceae (1), Chenopodiaceae (3), Amaranthaceae (6), Papilionaceae (13), Spindaceae (1), Asteraceae (19), Urticaceae (2), Lamiaceae (15), Acanthaceae (4), Adiantaceae (1), Ebenaceae (1), Rosaceae (3), Berberidaceae (3), Myrtaceae (3),

Solanaceae (4), Moraceae (3), Brassicaceae (8), Fumariaceae (1), Apocynaceae (2), Apiaceae (1), Rubiaceae (2), Cannabaceae (1), Ranunculaceae (3), Geraniaceae (2), Convolvulaceae (3), Tiliaceae (1), Menispermaceae (1), Tiliaceae (1), Commelinaceae (1), Araliaceae (1), Bignoniaceae (10, Oleaceae (4), Verbenaceae (2), Caprifoliaceae (1), Nyctaginaceae (1), Malvaceae (5), Celastraceae (2), Euphorbiaceae (4), Myrsinaceae (1), Oxalidaceae (1), Portulacaceae (1), Plantaginaceae (2), Anacardiaceae (1), Lythraceae (1), Scrophulariaceae (3), and Rhamnaceae.

Plants used medicinally on large scale in the study area are *Acacia modesta*, *Zanthoxylum armatum*, *Berberis lycium*, *Dalbergia sissoo*, *Mentha arvensis*, *Punica granatum*, *Ajuga bracteosa*, *Bauhinia variegata*, *Solanum xanthocarpum*, *Olea ferruginea*, *Mallotus philippensis*, *Artemisia absinthium*, *Amaranthus viridis*, *Foeniculum vulgare*, *Rumex hastatus*, etc. Plants are considered as the main source of medicines throughout the history. By the earliest information, 80% of the people in the world still depend on medicinal remedies. Similarly the local society of study area still depends on plants as the source of medicines. They use plants by different procedures to cure diseases and prefer the herbal medicines on artificial pharmaceutical medicines. Peoples of the area use plants to treat headache pain to kidney infections. They use roots of *Berberis lycium* as blood purifier agents. *Rydingia limbata* whole plant is use to cure throat problems. *Ziziphus nummularia* is use for hair growth. *Adiantum capillus-veneris* (fronds) is used in cough syrups. Extracts of *Fumaria indica* are used by diabetic patients. Similarly other plants from the area are used to cure different diseases, viz., liver, fever, cough, mouth problems, gum problems, eye diseases, constipation, hair growth, diuretic, stomach diseases, and abdominal diseases, many plants were used in condition of diarrhea/dysentery, and some are used for ulcers (Fig. 3). Many plants perform dual functions, viz., *Fumaria indica* is used by diabetic patients and it also helps to purify the blood. Similarly, *Acacia modesta* gum is used for back pain, and it is also used to treat dysentery (Table 1). Twelve plant parts were used to cure various diseases (Fig. 4).

Collection of Medicinal Plants

Medicinal plants are usually collected by the poor society of the area, which earns their living by collecting medicinal herbs and selling them to the local healers. The collectors are mostly man and women. They collect the herbal plants and then clean, wash, dry, and pack it. In the process of collection, they also destroy a large number of medicinal plants due to unawareness. Like medicinal plants this area is also rich with fodder for livestocks, wood for fuel, and function. Plant collection is further divided on the basis of their uses:

- Medicinal plants
- Fodder plants
- Fuel plants
- Ornamental plants
- Poisonous plants
- Insect repellant plants

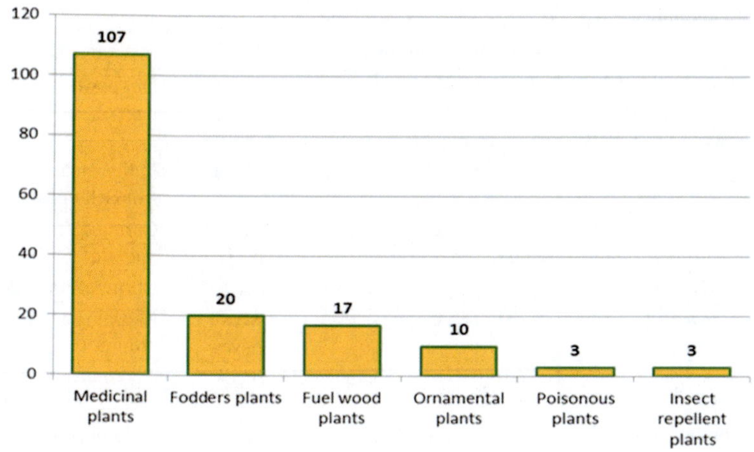

Fig. 3 Plants used for curing various diseases

Table 1 Medicinal plants collected from the Sarban Hills

Botanical name	Local name	Family	Part used	Medicinal uses
Acacia modesta wall.	Phulahi	Mimosaceae	Gum	Tonic use for back pain, dysentery
Achyranthes aspera L.	Kutri	Amaranthaceae	Whole plant	Cough and asthma, kidney problem
Adiantum capillus-veneris L.	Fern	Adiantaceae	Fronds	Bronchial disorder
Ajuga bracteosa Wallich ex. Benth	Koori buti	Labiate	Whole plant	Eye pain, ear pain and throat pain
Albizia lebbeck (L.) Benth.	Siris	Fabaceae	Wood bark	Abdominal caner
Allium griffithianum Boiss.	Jangle piaz	Alliaceae	Whole plant	Colic, vomiting
Amaranthus viridis L.	Ganihar	Amaranthaceae	Leaf	Eye sight problems, diuretic
Anagallis arvensis L.	Bili buti	Primulaceae	Whole plant	Anti-inflammatory
Artemisia absinthium L.	Chaw	Asteraceae	Leaves	Carminative, cold, fever
Aesculus indica (Wall ex. Camb)	Ban khor	Hippocastanaceae	Leaves	For cleaning teeth
Baccharoides anthelmintica (L.) Moench	Kalijeeri	Asteraceae	Seeds	Headache, ear and teeth pain, falling hairs
Bauhinia variegata L.	Sumbal	Berberidaceae	Root bark	Roots and fruits blood purifier
Berberis lycium Royle	Kachnar, Kalyar	Fabaceae	Root	Roots use to prevent fatness

(continued)

Table 1 (continued)

Botanical name	Local name	Family	Part used	Medicinal uses
Brugmansia suaveolens (Humb. & Bonpl. ex Willd.) Bercht. & J. Presl	Bel buti	Solanaceae	Leaves	Dry powder of leaves with oil, reduce pain
Cannabis sativa L.	Bhang	Cannabaceae	Leave	Indigestion, liver, stomach inflammation
Capsella bursa pastoris (L.) Medic	Gule pancha	Brassicaceae	Leaves, seed	Curing diarrhea, diuretic
Carissa spinarum L.	Granda	Apocynaceae	Leaves	Hormonal disorder
Cedrus deodara	Deodara	Cupressaceae	Wood, leaves	Neurological, fever, asthma
Chenopodium ambrosioides L.	Baljawain	Chenopodiaceae	Seeds	Abdominal diseases, headache
Cichorium intybus L.	Kashni	Asteraceae	Leaves, roots	Typhoid, constipation
Cissampelos pareira L.	Ghore sumi	Menispermaceae	Leaves	Antidiabetic
Clematis grata Wall.	Granda	Ranunculaceae	Roots	Boils
Commelina benghalensis L.	Angalara	Commelinaceae	Paste of leaves	Diuretic, laxative, swelling of skin
Convolvulus arvensis L.	Liali	Convolvulaceae	Whole plant	Constipation
Conyza Canadensis less.	Mirche buti	Asteraceae	Flower	Rheumatism
Cotoneaster microphyllus Wall. ex Lindl.	Lounri	Rosaceae	Fruit	Stomachache
Cupressus sempervirens	Pencil pine	Pinaceae	Stem	Heal wounds, toxin removal
Cyperus rotundus L.	Muther	Cyperaceae	Leaves, seeds	Respiratory infection
Dalbergia sissoo DC.	Tahli	Papilionaceae	Leaves, wood	Dandruff, expectorant, timber
Daphne mucronata Royle	Rutti lal	Amaranthaceae	Leaves, fruit	Fruit purgative, roots gastro intestinal, wood gun powder
Debregeasia saeneb (Forssk.) Hepper & J.R.I. Wood	Chengal	Urticaceae	Fruit	Jaundice, antifungal, diarrhea
Dicliptera bupleuroides Nees		Acanthaceae	Arial part	Used as tonic
Diospyros kaki L.f.	Bara amlook	Ebenaceae	Ripe fruit	Laxative
Dodonaea viscosa L.	Sanatha	Spindaceae	Bark	Joint swelling
Duchesnea indica (jacks.) Focke	Jangle strawberry	Rosaceae	Fruit	Nerve tonic, laxative, diarrhea
Eucalyptus globulus Labill	Gond	Myrtaceae	Leaves, trunks	Kill germs in wounds, fuel
Ficus carica L.	Anjeer	Moraceae	Fruit	Foot-ache, laxative

(continued)

Table 1 (continued)

Botanical name	Local name	Family	Part used	Medicinal uses
Ficus sarmentosa Buch.-Ham. ex Sm.	Doda bail	Moraceae	Stem	Tonic for blood
Foeniculum vulgare mill	Saunf	Apiaceae	Seeds, fruit	Abdominal diseases, cough
Fumaria indica (Hausskn.) Pugsley	Papra	Fumariaceae	Whole plant	Antiemetic, blood purifier, constipation
Galinsoga parviflora Cav.	Chota phool	Asteraceae	Extract of leaves	Leaves use as salad in some cases, leave extract given in fever
Galium aparine L.		Rubiaceae	Whole plant	Diuretic, use in complains of kidney and bladder
Geranium mascatense Boiss.	Bhanda	Geraniaceae	Whole plant	Diuretic
Glandularia canadensis (L.) small	Neeli buti	Asteraceae	Flower	Eye diseases
Grewia optiva J.R.Drumm. ex Burret	Tambar	Tiliaceae	Fruit	Increase blood level
Hedera nepalensis K. Koch	Arbam bal	Araliaceae	Leaves	Diabetes, rheumatism, fever
Himalaiella heteromalla (D.Don) Raab-Straube	Gajar mula	Asteraceae	Root	Powder use for ulcer
Incarvillea emodi (Royle ex. Lindl.) Chatterjee	Kaur	Bignoniaceae	Whole plant	Plant is febrifuge
Indigofera heterantha Brandis	Kainthi	Papilionaceae	Root, leaves, branches	Scabies, stomach problems
Ipomoea hederaceae Jacq.	Kala dana	Convolvulaceae	Seeds	Purgative, use as tonic, diuretic and expel intestinal worms
Jasminum humile L.	Peli chambeli	Oleaceae	Flower, root juice	Ornamental, flowers tonic for heart, root juice to remove ring worms
Jasminum sambac (L.) Aiton	White chambeli	Oleaceae	Whole plant	Ornamental, diuretic, flowers in headache, skin diseases
Justicia adhatoda L.	Bahker	Acanthaceae	Whole plant	Leaves and roots in cough, asthma indigestion
Lamium amplexicaule L.	Bushka	Lamiaceae	Leaves	Fever reducing, laxative
Lantana camara L.	Panch phul	Verbenaceae	Whole plant	Diaphoretic, carminative, antiseptic
Lathyrus aphaca L.	Kukar bhang	Papilionaceae	Seed	Wound healing
Lepidium virginicum L.	Halun	Brassicaceae	Seeds	Abdominal pain

(continued)

Table 1 (continued)

Botanical name	Local name	Family	Part used	Medicinal uses
Lonicera caprifolium L.	Phut	Caprifoliaceae	Seeds	Diuretic
Mallotus philippensis (Lam.) Muell	Kimila	Euphorbiaceae	Fruit	Powder for ulcer, constipation, kill intestinal worms.
Malva parviflora L.	Sonchal	Malvaceae	Leaves	Constipation, bronchial disorder
Malva sylvestris L.	Saunchal	Malvaceae	Whole plant	Plant is cooling, emollient
Malvastrum coromandelianum (L.) Garcke	Peli buti	Malvaceae	Leaves, roots	Jaundice, anti-inflammatory
Maytenus royleanus (Wall. ex Lawson) cufodontis	Garanda	Celastraceae	Bark	Bark decoction use for wounds
Medicago sativa L.	Singi	Papilionaceae	Leaves	Women problems, digestion
Melia azedarach L.	Dreik	Meliaceae	Whole plant	Anti-lice, blood purifier, gas trouble in cattle's
Mentha longifolia (L.) Huds	Jangli podina	Lamiaceae	Whole plant	Carminative, colic
Micromeria biflora (Buch.-Ham. ex D.Don) Benth.	Shamokai	Lamiaceae	Whole plant	Carminative, stimulant, headache
Mirabilis jalapa L.	Gul-e-abbas	Nyctaginaceae	Roots, leaves	Roots purgative, poultice for inflammation, leaves juice is useful in urine discharge
Myrsine africana L.	Khukan	Myrsinaceae	Leaves, fruit	Intestinal worms, purify blood, colic
Nerium oleander L.	Kanair	Apocynaceae	Whole plant	Poisonous, antidiabetic
Olea ferruginea Royle	Kahu	Oleaceae	Leaves	Throat pain, mouth, toothache, cough
Origanum vulgare L.	Ban ajwain	Lamiaceae	Whole plant	Used as tonic, carminative, colic, and antispasmodic and for stomachache
Oxalis corniculata L.	Kati buti	Oxalidaceae	Whole plant	Mouth taste, indigestion, bladder inflammation
Persicaria capitata (Buch.-Ham. ex D.Don) H.Gross		Polygonaceae	Arial part	Fever, diarrhea, urinary tract infection
Phaseolus lunatus L.	Bean	Papilionaceae	Seeds and pods	Boil pods are diuretic, seeds control blood control

(continued)

Table 1 (continued)

Botanical name	Local name	Family	Part used	Medicinal uses
Phaseolus vulgaris L.	Phali	Papilionaceae	Seeds	Edible
Pimpinella stewartii Nasir	Patli saunf	Apiaceae	Seeds	Carminative
Pinus roxburghii	Chir pine	Pinaceae	Stem, needle	Wound healing, diuretic
Pinus roxburghii Sarg.	Chirr	Pinaceae	Bark, extract	Anti-inflammatory, analgesic
Pistacia khinjuk stocks	Kangar	Anacardiaceae	Dry leaves	Religious ceremony
Plantago lanceolata L.	Aspagol	Plantaginaceae	Powder	Laxative
Polygonum plebeium R.Br.	Rani phal	Polygonaceae	Whole plant	Use for cough
Portulaca oleracea L.	Lunak	Portulacaceae	Leaves	Kidney treatment
Pterospermum acerifolium (L.) Willd	Kana Champa wild	Malvaceae	Leaves, flowers	Stop bleeding, insect repelling
Punica granatum L.	Daruna	Lythraceae	Fruit, tonic	Heart palpitation
Pyrus pashia Buch.-Ham. ex D.Don	Batang	Rosaceae	Fruit	Laxative
Randia tetrasperma (Roxb.) Benth. and Hook.f	Khukhri	Rubiaceae	Fruit	Anthelmintic
Ricinus communis L.	Arand	Euphorbiaceae	Seed	Constipation, dandruff, purgative
Rumex dentatus L.	Holla	Polygonaceae	Leaves	Edible
Rumex hastatus D.Don	Khatimar	Polygonaceae	Roots	Bone fracture
Rydingia limbata (Benth.) Scheen & V.A. Albert	Koi booi	Labiatae	Whole plant	Mouth gums and throat problems
Salvia moorcroftiana Wall. ex Benth	Kaljari	Labiatae	Leaves	Cough and diarrhea
Silybum marianum (L.) Gaertn.	Kandyara	Asteraceae	Seeds	Liver diseases
Sisymbrium irio L.	Khub kalan	Brassicaceae	Whole plant	Expectorant
Solanum nigrum L.	Kachmach	Solanaceae	Black fruit	Frackles, skin diseases, liver disorder
Solanum xanthocarpum L.	Morian	Solanaceae	Leaves, fruits extract	Typhoid
Sonchus asper (L.) hill	Dodak	Asteraceae	Shoot	Wound healing
Tagetes minuta L.	Satbarga	Asteraceae	Leaves	Kill germs in wounds, cough, stomach problems
Taraxacum officinale Weber.	Dohdal	Asteraceae	Leaves, roots	Jaundice

(continued)

Table 1 (continued)

Botanical name	Local name	Family	Part used	Medicinal uses
Tagetes erecta L.	Gutta	Asteraceae	Seeds	Women problems
Trifolium repens L.	Shaftal	Papilionaceae	Whole plant	Infusion of plant is used to treat fever
Urtica pilulifera L.	Bicho booti	Urticaceae	Roots	Antidiabetic
Verbascum thapsus L.	Gider Tabaco	Scrophulariaceae	Leaves	Diarrhea, antiseptic
Vitex negundo L.	Marwani	Verbenaceae	Leaves, seeds	Gas trouble and cholera, rheumatism
Xanthium strumarium L.	Katula	Asteraceae	Leaves	Malarial fever, carminative
Zanthoxylum armatum DC.	Timber	Rutaceae	Stem, fruit	Toothache, gums problems, cholera, indigestion
Ziziphus jujuba Mill.	Shingle	Rhamnaceae	Fruit	Laxative
Ziziphus nummularia (Burm. f.) Wight & Arn.	Bair	Rhamnaceae	Leaves	Scabies, carminative, sedative

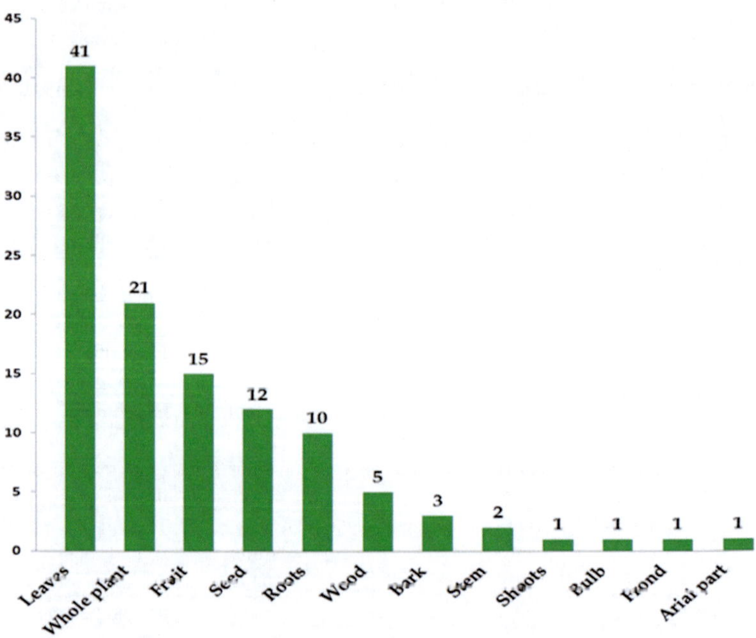

Fig. 4 Parts of plants used

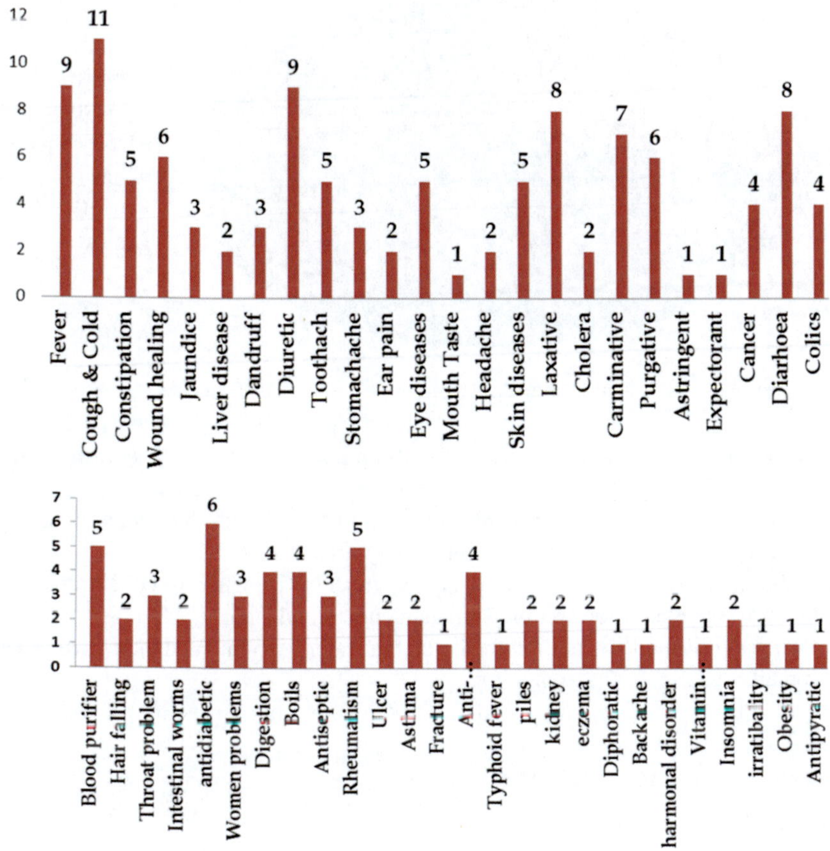

Fig. 5 Division of plant species recorded from Sarban Hills

Medicinal plants collected from Sarban area were 107 in number, fodder 20, fuel 17, ornamental 10, poisonous 3, and insect repellant 3 (number of plant species), respectively (Fig. 5).

Traditional Medicines

Our results revealed that 107 plant species are being utilized to cure different diseases. Its collection by the local people has also been described. It is evident that they waste plants during collection and/or preservation due to lack of awareness and proper training. Plants are used in various ways including making their extractions, infusions, decoctions, poultices, syrups, and direct utilization (Fig. 6).

- *Infusion:* Infusion is a liquid preparation made by dipping one or more herbs in hot water for a small period of time, and then this liquid is stained.

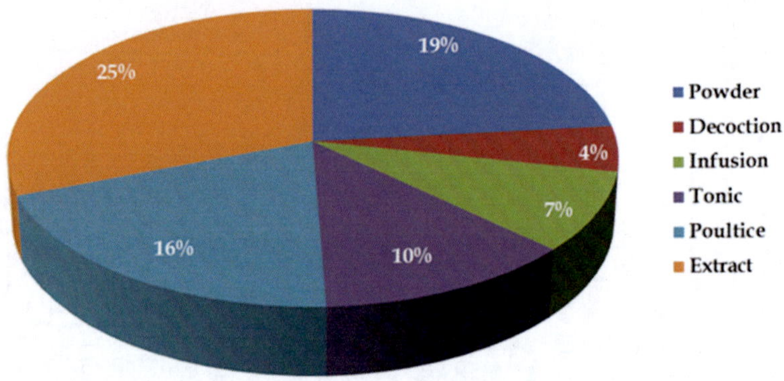

Fig. 6 Route of administration or medicinal plants

- Plant parts are boiled in water about half an hour and then strained. Infusion is made fresh every day.
- *Extraction:* Pure plant juice is filtered from the plant parts without adding anything and with all chemical constituents present.
- *Decoction:* Decoction is made by adding fresh or dry herbs in cold water; then this water is boiled for 15 min and again cooled and stained.
- *Poultice:* It is made by boiling fresh plant in water for 5 min and then creased and spread over the wound area to treat.
- *Syrup:* It is made by adding sugar in the infusion. This is used to treat cough.

Fodder Plants

The residents of Sarban Hill areas are mostly poor and depend on keeping livestock for milk and other requirements. They also get a considerable amount of money from the sale of different products they attained from these livestock. To nourish their livestock, they are dependent on fodder that is obtained from grasses growing on hills. They also collect green fodder from the surrounding forests for their livestock. Animal rearing is the old tradition of inhabitants. Fodders play an important role for animals from which they get energy and strength for them to fight with with cold seasons. In winter season, people from the area rush to the forest trees for fodder purposes to increase the dried fodder. Fodder obtained from the area is used directly in the fresh form. The people prefer fresh grasses and herbs as compared to dried fodder because they are more nutritious and also increase milk capacity of the cattle. The people prefer shrubs and grasses than fodder (Table 2). Some important fodder plants used to feed livestock in the research area are *Amaranthus caudatus*, *Bromus diandrus*, *Bromus sterilis*, *Cardamine pratensis*, *Carex spp.*, *Cyperus niveus*, *Lathyrus sativa*, *Lespedeza juncea*, *Melilotus indicus*, *Oenothera rosea*, *Parthenium hysterophorus*, *Ranunculus muricatus*, *Ranunculus arvensis*, *Vicia angustifolia*, and *Veronica arvensis*.

Table 2 Fodder plants collected from the Sarban Hills

Botanical names	Vernacular names	Family
Amaranths caudatus L.	Ghaner	Amaranthaceae
Arundo donax L.	Naar	Poaceae
Bromus diandrus Roth.	Grass	Poaceae
Bromus sterilis L.	Grass	Poaceae
Broussonetia papyrifera (L.) vent	Jangle toot	Moraceae
Cardamine pretense L.		Brassicaceae
Carex spp.	Grass	Cyperaceae
Coronopus didymus (L.) sm	Jangle haloon	Brassicaceae
Cyperus niveus Retz.	Grass	Cyperaceae
Lathyrus sativa L.	Jangle matar	Papilionaceae
Lespedeza juncea		Papilionaceae
Maytenus royleanus	Garanda	Celastraceae
Melilotus indica	Sinji	Papilionaceae
Oenothera rosea	Buti	Onagraceae
Parthenium hysterophorus	Chandni	Asteraceae
Ranunculus arvensis L.	Chechampa	Ranunculaceae
Ranunculus muricatus L.	Barea	Ranunculaceae
Veronica arvensis L.		Plantaginaceae
Vicia angustifolia		Papilionaceae

Timber and fuelwood plants

Many plants of Sarban are used as a source of timber and fuel. On the upper areas of Sarban Hills, people still depend on plants as a source of fuel wood, but the investigation shows that 20 plant species of the area were used as a source of fuel. The residents of the area are living in the forests and have an easy access to wood; that's why they prefer wood as source of energy. They extensively cut the trees increasing pressure on natural resources. Best quality furniture is also made from the wood of *Dalbergia sissoo*; it is also used as a source of fuel by the local community. *Pinus roxburghii* is also used as timber wood for construction and also as a source of fuel (Table 3). *Cupressus sempervirens* is used in making door and windows of houses. In summer season people of the area stock woods in their houses for cold/winter season to use as fuel source (Figs. 7 and 8).

Discussion

Plant-diversity

Biodiversity is a gift from nature for us and our future generation. Humans continuously depend on plants to fulfill their needs, e.g., food, fiber, shelter, and herbal medicines. Flora of Pakistan is very rich due to the different climatic conditions and

Table 3 List of fuel wood of Sarban Hills

Botanical names	Vernacular names	Family
Acacia modesta	Phulai	Papilionaceae
Arundo donax	Naar	Poaceae
Aesculus indica	Ban khor	Hippocastanaceae
Berberis lycium	Sumbal	Berberidaceae
Broussonetia papyrifera	Jangle toot	Moraceae
Callistemon citrinus	Bottle brush	Myrtaceae
Dalbergia sissoo	Tahli	Papilionaceae
Dodonaea viscosa	Sanatha	Spindaceae
Eucalyptus globulus	Gond	Myrtaceae
Ficus carica	Phugwara	Moraceae
Grewia optiva	Tamber	Tiliaceae
Melia azedarach	Daeik	Meliaceae
Morus alba	Safaid toot	Moraceae
Morus nigra	Kala toot	Moraceae
Pinus roxburghii	Chirr	Pinaceae
Punica granatum	Daruna	Lythraceae
Rydingia limbata	Kooibui	Lamiaceae

different soil textures. Roundabout 6000 flowering plant species are present in Pakistan. The number of dicot species is 4492, monocot 1140, gymnosperm 23, and pteridophyte 128. Further, 214 families are published in the flora of Pakistan which is near in completing process (Ali and Qaiser 1986). Asteraceae and Rosaceae families are still to be published in Flora of Pakistan. There are about 6 phytogeographic regions present in the flora of Pakistan like Mediterranean, Saharo-Sindian, Euro-Siberian, Irano-Turanian, Sino-Japanese, and Indian region. Asteraceae 649 species, Papilionaceae 439, Poaceae 597, Cyperaceae in the lower plants 202, and Brassicaceae 250 are the families containing the larger number of species. 189 pteridophytes are included in the lower plants.

Sarban Hills are one of the important hills due to its biodiversity and are still unexplored. It is clearly shown in the study of the flora of Pakistan. In the studies 147 plant species were collected from the area and submitted in the herbarium of Hazara University. From the moist temperate forest of Pakistan, the larger number of collections has been done which is preserved in the biggest herbariums of Pakistan, e.g., Karachi University Herbarium, Quaid-i-Azam University Herbarium, and National Herbarium NARC (S.I Ali, Yasin Nasir, Mir Ajab Khan, A.R. Beg and Saddique) have done huge number of plant collections from moist temperate region of Pakistan.

In Pakistan, larger numbers of medicinal plants are present which is about 400–600 in number. Larger numbers of papers are published in the Giliyat and Muree. Other papers on such topics are by Parker (1918), Stewart (1972), Hedge and Wendelbo (1978), Dhar and Kachroo (1984), Biswas (1987), and Blatter (1928–1929). On the pteridophyte of Pakistan, work is done by Frazer-Jenkins (1991).

Fig. 7 Grazing view of the Sarban Hills

15 genera and 43 species along with their English names, local names, collection time, and place of collection are published in the Flora of Rawalpindi District and West Pakistan.

In the research, 147 plant species were collected, identified, pressed, and poisoned mounted along with their vernacular names, botanical names, habit, habitat, flower color, traditional uses, and recipes which are mentioned and submitted in the herbarium of Hazara University. A similar study was conducted by Kazim and

Fig. 8 View of cutting trees for fuel and timber

Siddiqui (1953) in the upper Guraiz and Astore valley. They identified 87 medicinal plants and also enlisted their local names, distribution, elevation, and traditional uses. Zaman and Khan (1970) reported 100 medicinal plants of West Pakistan with their botanical name, family name, distribution, description, constituents, and traditional uses. A field survey in Mansehra was conducted in which 53 wild and 17 cultivated medicinal plants were collected by Haq (1993), and these plants were enlisted with botanical names, local or vernacular names, family, parts used, habit, habitat, distribution, constituents, and medicinal and traditional uses.

In this study, 2 species of pteridophyte, 3 species of gymnosperm, and 142 species of angiosperm were related to 56 families in which 50 families of dicot and 6 families of monocot were collected from the research area. A larger number of plants are related to Asteraceae family which has 19 species, Labiatae family which has 15 species, Papilionaceae family which has 13 species, and Brassicaceae family which has 8 species. In a similar research study conducted in the Palas valley, 600 plant species were collected in which there are 114 monocots, 456 dicots, 11 gymnosperms, and 19 pteridophytes. Later on study trip is arranged in the area to meet the local people of the area and know about the use of plants and their traditional recipes which they use in their daily life.

The species diversity present in the Sarban area is very important. Plants like *Justicia adhatoda*, *Vernonia anthelmintica*, *Taraxacum officinale*, *Sonchus asper*, *Rydingia limbata*, *Adiantum capillus-veneris*, *Verbena Canadensis*, *Verbascum thapsus*, and *Eucalyptus camaldulensis* are present in large amount in the study area. Ahmad (2003) discussed 31 (aromatic and medicinal) plant species were cultivated in 3 different locations. The groundwork results show that *Aconitum heterophyllum*, *Aconitum violaceum*, *Bistorta amplexicaulis*, *Colchicum luteum*, *Ginkgo biloba*, *Crocus sativus*, *Matricaria chamomile*, *Viola odorata*, *Podophyllum hexandrum*, and *Valeriana jatamansi* can be cultivated effectively.

The knowledge obtained from the field trip is collected from the indigenous people of the area, who know the uses of these plants and traditional recipes used in their everyday life and also know the ethnobotanical importance of these plants of the Sarban Hill areas. They use 107 plants as medicinal out of 147 plants related to 59 families. Similarly a field study was conducted in northern Chitral by Khan and Fevre (1996). The purpose of this study was indigenous knowledge of plant species for medicinal uses in Chitral. They revealed that 85 plant species from 39 families were used by local inhabitants for medicinal requirements.

In the study area before this, there is no work done with respect to biodiversity. Lot of work is needed to be done because this is a diversity-rich area. Three types of forests, alpine forest, subtropical thorn forest, and subtropical broad-leave forest, are present in the area. From the area 147 plants are collected in which 107 are used as medicinal and are related to 57 families. Similarly Ahmad and Sirajuddin (1996) divided the seven types of forests like alpine forest, subtropical chir forest, moist temperate forest, subtropical dry deciduous forest, and subtropical broad-leave forest in the swat district; from the area of research they found 135 families in which 1541 species of plants and 48 species of medicinal plants are found.

Many historic people create the methods to conserve the animals, plants, ecosystem resources, local knowledge, and cultural traditions. To know about the traditional knowledge of local people in relation to biodiversity resources, first of all it is necessary to develop the mountainous region. It plays an important role to increase the value of natural plant resources.

Ethno-ecology and Traditional Knowledge

Ethnobotany of the area is very important since large amounts of medicinal plants are present in the area. So to know the traditional knowledge of the plants and ethno-ecology of the area a field survey should be conducted in such potential areas. Questionnaires must be precised that could cover all the traditional knowledge of the local community. Such knowledge can also be documented by interviewing the local pansaries and hakims and saved in the written form. Also they provided the traditional recipes of some medicinal plants. Similarly, Singh and Pandey (1980) reported 125 medicinal plant species lore of the tribes of eastern Rajasthan belonging to 104 genera and 54 families. They gathered the information about these plants from local peoples of the tribal areas including hakims, vaids, herbalists, and forest officials. Some medicinal folk recipes regarding these plants were also documented. Also a field survey was conducted in Khairpur District, Sindh Pakistan. They provided information on 35 plant species belonging to 23 families possessing high medicinal properties. All these plant species are widely spread in the whole district. All the provided information about plants (medicinal values and traditional uses) was collected during survey from local people, Hakims, of the area and concerned literature.

Ethnobotany is directly related to the common man with their society, traditions, and religion. Ethnobotanist creates and publishes a lot of books in which they show the importance of plants in different aspects of life. Plants used as medicine for different diseases, used for healing wounds, used in religious ceremonies, used as fodder and fuel, used for making instruments, used in construction, and used as a source of energy, all these aspects are highlighted in ethnobotany. In earlier years too much work are done on ethnobotany in western countries. More knowledge is explored. Much work is done on ethno-narcotic (Lewin 1964), investigation of traditional veterinary phytomedicines, ethnoveterinary medicines, and ethnopediatrics, all work accompanied.

Medicinal Plants

In this research project, 147 plant species were collected in which the people of Sarban Hill used 107 plants as medicine to cure different diseases, 20 plant species were used as fodder, and 17 plants were used as a source of fuel. Plants are the major source of medicine all over the world throughout the history. Similarly, an ethnobotanical survey was conducted in Kaghan Valley and Mansehra District, KP, Pakistan. They explored the traditional knowledge regarding usage of medicinal plants in particular area. They found 48 plant species in the valley in which only 26 plant species were used by local people for the treatment of various diseases. The local people also use 21 indigenous plant species as shelter and food and more than 3 species as fodder. Different types of effective chemicals are present in the plants which are used in the medicine industry and insecticides and food industry and as a source of medicines as old as life on the earth. Hamilton (2004) pointed out that plants are used as medicines on a large scale in the future.

The aboriginal people of the Sarban Hills know different plant species used as medicine. The method and techniques about the use of plants for medicinal remedies are still alive. They know the uses of hundreds of species in medicines. They also know about the use, parts used either in the form of extraction, decoction, tonic, infusion or poultice. No one knows the exact number of plants which are used daily in the medicines but estimatedly 20% of the global vascular plants that is about 50,000 species are used as traditional medicines (Baillie et al. 2004). Large no of medicinal plants are used in traditional recipes. In the Sarban Hill areas, the medicinal use of plant is different as compared to the traditional recipes of other areas. Some plants are taken directly and some are used in different recipes, as use by the historic culture of the area. WHO (2002) reported that majority of the world population currently depends on traditional medicine for their primary health and needs. Medicinal plants are widely used as household remedies and by practitioners of traditional systems of medicines, particularly in the developing world where public health-care services may be limited. At the same time, interest in traditional and modern and alternative medicine in industrialized countries has grown rapidly.

Marketing of Medicinal Plants

By investigating the study area and market, it was found that many of the plant species which was used locally also have value in the market. Local people collect plants and sell these in the market to get some cash money. Different parts of plants are sold depending on their usage. Like the fruit and stem of *Zanthoxylum armatum* (timber) is sold. Fruit and roots of *Berberis lycium*, dry leave powder of *Cannabis sativa*. Similarly (Shinwari and Khan 1998) in the present studies, it is found that 26 species of medicinal plants collected from the Margalla Hills National Park Islamabad are sold in the market. Besides that, it is stated 5000 tons of medicinal plants related to 500 families are reported from the swat district (Chaudhry et al. 2000). Local people of Utror and Gabral valley collect 42 medicinal plants species in which they sell 15 species to the Madyan, Mingora, and Kalam and exported 27 species to the international markets (Hamayun 2005).

From the study area of Sarban Hill, it is found that most of the collection of medicinal plants is done by the women of local area (75%), but from the upper areas, medicinal plants are collected by man (25%), and then plants are sold in the local markets from where it is exported to the other parts of the country. Some of the plants or plant parts are stored by people in dried, preserved, stored and packed form and utilize it while the prices in local market higher. Schippmann (2001) studied that gathering of medicinal plants can provide source of income for local people.

In the study it is found that medicinal plants are widely grown in the Sarban Hill areas, in which this has great value in the national and international markets. If the government takes steps to develop the medicinal plant trade in other countries, the people of the Sarban areas can get more benefit from herbs. Important example of such extracts are Forskolin remedy and artemisins etc which commonly used in Indian herbal products and clinical herb for numbers of different purposes (Lange 1998).

Medicinal plants are used all over the world to cure diseases and via local knowledge and traditional recipes which have sometimes very high prices. In the world market the worth of medicinal plants and its products is about 60 billion US dollor (WHO 2002). Locally, 71 wild medicinal plants are imported from swat by which 24 medicinal plants species are transported to the big herbal markets of Pakistan and 40 in the local market of swat (Sher 2002).

Fodder Availability for Livestock

Due to its rich biodiversity, Sarban Hills play an important role for the availability of fodder for livestock. Literature survey of the region shows that many of the plants are used as fodder through which animals get mineral and nutrients and people of the area in return earn money by selling good quality milk, butter and fats. It was found that on the hill slopes wild grasses are widely grown naturally and cut for their livestock's fodder by the local farmers. About 20 plant species from the Sarban Hill area were identified which are used for livestock's feed. Animal rearing is the old tradition of these areas. A huge number of animals are reared like goat, sheep, and cattle from which they get milk, butter, and animal fats. Shrubs and tree are good source of fodder for the livestock. Similarly, investigated that local people of Margalla Hills National Park, Islamabad, depends on the indigenous plant resources for shelter, food, fodder, and medicines.

Fuel and Timber Woods

About 17 plants in the collections are used as fuel wood. The local community of the upper areas of Sarban still depends on wood as a source of fuel. For this purpose they regularly cut down trees, and this has decreased the number of trees and shrubs. Mostly for the winter season, people store huge amounts of wood in the summer by cutting down trees and shrubs and destroying the diversity of plants. Martin (1995) proposed that main threats to the shrubs and trees are basically the fuel deficiency in Sulaiman Range Mountains. In severe and long winter season, a massive quantity of wood for fuel purpose is used, and about 13,000 people span their life on selling *Pinus gerardiana* plant seeds in a good yielding year. Ethnobiologists investigated that in Hindu-Himalayas biodiversity was badly affected by the deficiencies of fuel. Different solutions were recommended for the harmful impact of fuel insufficiency by applying certain strategies at state, regional, and local level. Exploration of different other fuel sources, plantation of rapid-growing trees, and endangered plant species conservation were also recommended.

About 20 plant species collected from the Sarban Hill are used as source of timber wood which is used in making furniture, instruments, doors, and windows and used in constructions as beam in the houses. Many plants perform two functions at a time, for example, *Acacia modesta* (wood used as fuel and gum used for back pain and weak bones) and *Zanthoxylum armatum* (fruits used for stomachache, stem used as miswak, and dry wood used as a source of fuel). Similarly (Hamayun and Khan 2003) an ethnobotanical survey was conducted to study shrubs and trees of Buner District as the area has diverse flora and high ethnobotanical potential. It was found that 94 different plant species are used for medicinal, timber, fuel wood, fodder, ornamental, agricultural tool, thatching, fencing, naming (folk lore), and fruit-yielding purposes. Bulk of plant species show multiple uses like *Juglans regia*: wood is used for making furniture, gun woody parts, and carving and as fuel; root bark (Dandasa) is used for cleaning and sparkling teeth; leaves are used in women folk for coloring lips (make up); and nuts are edible and are traded to other parts of the country. Shinwari and Khan (1998) studied 24 species of shrubs and 27 species of trees used by the locals of Margalla Hills National Park, Islamabad, as food, shelter, fodder, cultural purposes, and local medication.

Ethnoveterinary Importance of the Region

In the earliest study, there are many plant species found in the area that are used for ethnoveterinary purposes. Many plants of this area were used from the ancient times to cure the diseases of animals. A significant part of Sarban Hill population still depends on the livestock to fulfill their basic needs. So for this purpose, the health of livestock is also very important. Many of the diseases, which are common in animals, are colic, fever, and foot and mouth diseases which are very common. To cure these diseases, people of the area use different plants like *Zanthoxylum armatum*, *Salvia moorcroftiana*, *Vitex negundo*, *Aesculus indica*, and *Cissampelos pareira* which are important medicinal ethnoveterinary plants. The seeds of *Zanthoxylum armatum* are used to cure the foot and mouth diseases of animals; *Salvia moorcroftiana* is used in case of fever in animals. The seeds of *Aesculus indica* are used to cure the colic in donkeys. Medicinal plants are used from historical times as ethnoveterinary product. But due to lack of knowledge, people couldn't rescue and/or develope the ethnoveterinary plants in written form. Similarly, from the Samahni valley district Shimber, Azad Kashmir, found treatments for veterinary by using traditional plant remedies. Sudarsanum et al. (1995) stated that 106 plants were used to cure animal diseases in Andhra Pradesh. Cousins (1995) checked the antiviral, antibacterial, and antifungal activities of plants which are used in human and veterinary medicines and also in crop protection. Inhabitants of coastal region use the fresh and dried seaweeds in large amounts.

Conservation of the Sarban Plant Resources

Pakistan is rich in natural resources due to its diverse environment and soil conditions; about 6000 flowering plant species are reported in Pakistan in which majority of species are found in the northern areas. Sarban Hills are rich in floral diversity. People of the area are relatively low educated and don't have awareness about the uses of plants. Due to this, majority of diversity is destroyed. Although three types of forests are present in the area and there are a lot of plant species, which are important as medicinal resources, due to their unawareness, these resources are at the verge of destruction (Ijaz 2014). Present attempt was made to introduce the area nationally as well as internationally and to help the native people in terms of education and awareness about the uses of plant and selling in national and international markets.

References

Abbas Z, Khan SM, Abbassi AM, Pieroni A, Ullah Z, Iqbal M, Ahmad Z (2016) Ethnobotany of the Balti community, Tormik valley, Karakorum range, Baltistan, Pakistan. J Ethnob Ethnomed 12:38. https://doi.org/10.1186/s13002-016-0114-y

Abbasi AM, Khan SM, Ahmad M, Khan MA, Quave CL, Pieroni A (2013) Botanical ethnoveterinary therapies in three districts of the Lesser Himalayas of Pakistan. J Ethnob Ethnomed 9:84. https://doi.org/10.1186/1746-4269-9-84

Ahmad H (2003) Cultivation and sustainable harvesting of medicinal and aromatic plants through community involvement. Intern, workshop on conservation and sustainable use of medicinal and aromatic plants in Pakistan, WWF, MINFAL and Qarshi industries Pvt. Ltd. p 1

Ahmad H, Sirajuddin (1996) Ethnobotanical profile of Swat. In: Shinwari ZK, Khan BA, Khan AA (eds) Proceedings of the first training workshop on ethnobotany and its application to conservation, National Herbarium. PARC, Islamabad, pp 113–118

Ahmad H, Khan SM, Ghafoor S, Ali N (2009) Ethnobotanical study of upper Siran. J Herbs Spices Med Plants 15:86–97

Ahmad H, Turk MO, Ahmad W, Khan SM (2015) Status of natural resources in the Uplands of the Swat Valley, Pakistan. In: Ozturk M, Hakeem KR, Faridah-Hanum I, Efe R (eds) Climate change impacts on high-altitude ecosystems. Springer, New york, p 647 http://www.springer.com/us/book/9783319128580

Ahmad Z, Khan SM, Ali S, Rahman IU, Ara H, Noreen I, Khan A (2016) Indicator species analyses of weed communities of maize crop in district Mardan, Pakistan. Pak J Weed Sci Res 22(2):227–238

Ali SI (2008) Significance of Flora with special reference to Pakistan. Pak J Bot 40(3):967–971

Ali SI, Qaiser M (1986) A photographic analysis of the Phanerogams of Pakistan and Kashmir. Proc R Soc Edinburgh 89B:89–101

Baillie JEM, Hilton-Taylor C, Stuart SN (2004) 2004 IUCN red list of threatened species: global species assessment. IUCN, Switzerland and Cambridge

Bibi F, Ahmad H, Qureshi RA, Shaheen N, Khan SM, Shaheen S, Shaheen G (2016) Ethno medicinal attributes and antioxidant screening of some selected plant species of Tanawal area. Pak Int J Bios 9(1):237–254

Biswas AK (1987) Environmental concerns in Pakistan with special reference to water and forests. Environ Conserv 14:4

Blatter E (1928–1929) Beautiful flowers of Kashmir, vol 2. Jhon Bale and Danielsson, London

Chaudhry M, Ahmad S, Ali A, Sher H, Malik S (2000) Technical report. Market study of medicinal herbs in Malakand, Peshawar, Lahore and Karachi. SDC-Intercorporation, Peshawar, p 7
Cousins DJ (1995) Plants with antimicrobial properties (antiviral, antifungal and antimicrobial). International. pp 116–134
Dhar U, Kachroo P (1984) Alpine flora of Kashmir. In: Its photographic assessment. Printing Press, Jodhpur
Frazer-Jenkins CR (1991) The ferns and allies of far west Himalaya. Pakistan Syst 5(1–2):85–120
Hamayun M (2005) Studies on Ethnobotany, Conservation and plant diversity of Utror and Gabral Valleys District Swat, Pakistan. Ph.D. thesis Department of Plant Sciences, Quaid-I-Azam University, Islamabad, Pakistan
Hamayun M, Khan MA (2003) Common medicinal folk recipes of District Buner, NWFP, Pakistan. J Ethnobot Leaflets 17(7):2003–2007
Hamilton AC (2004) Medicinal plants, conservation and livelihoods. Biodivers Conserv 13:1477–1517
Haq I (1993) Medicinal plants of Mansehra District, NWFP, Pakistan. Hamdard Med 34(3):51–86
Haq I, Hussain M (1993) Medicinal plants of Mansehra. Hamdard Med XXXVI(3):63–100
Hedge IC, Wendelbo P (1978) Patterns of distribution and endemism in Iran. Notes from Royal Botanic Garden, Edinburgh 36:441–464
IUCN (2004) IUCN Red List Categories and Criteria, Version 3.1. IUCN Species Survival Commission. IUCN, Gland
Ijaz F (2014) Biodiversity and traditional uses of plants of Sarban Hills, Abbottabad. M. Phil. Thesis, Hazara University Mansehra, KP, Pakistan
Ijaz F, Iqbal Z, Alam J, Khan SM, Afzal A, Rahman IU, Afzal M, Islam M, Sohail M (2015) Ethno medicinal study upon folk recipes against various human diseases in Sarban Hills, Abbottabad, Pakistan. World J Zool 10(1):41–46
Ijaz F, Iqbal Z, Rahman IU, Khan SM, Shah GM, Khan K, Afzal A (2016) Investigation of traditional medicinal floral knowledge of Sarban Hills, Abbottabad, KP, Pakistan. J Ethnopharmacol 179:208–233
Kazim MA, Siddiqui IA (1953) The drug plants of Kalat state. Pak J Forest: 217–237
Khan SM, Ahmad H (2014) Role of indigenous Arqiyat distillery in conservation of Rosa species. Int J Phytoremediation 6(2):162–164
Khan SM, Ahmad H (2015) Species diversity and use patterns of the alpine flora with special reference to climate change in the Naran, Pakistan. In: Ozturk M, Hakeem KR, Faridah-Hanum I, Efe R (eds) Climate change impacts on high-altitude ecosystems, Springer, New York, p 647. ISBN:978-3-319-12858-0, http://www.springer.com/us/book/9783319128580
Khan AA, Fevre JL (1996) Indigenous knowledge of plants: a case study in Chitral. Proc. First Train. Workshop Ethnob. Appl. Conser., PARC, Islamabad, pp 136–151
Khan SM, Harper DM, Page S, Ahmad H (2011) Residual Value Analyses of the medicinal flora of the western Himalaya; The Naran Valley Pakistan. Pak J Bot 43(SI):97–104
Khan SM, Page S, Ahmad H, Harper DM (2012a) Anthropogenic influences on the natural ecosystem of the Naran valley in the Western Himalayas. Pak J Bot 44(SI):231–238
Khan SM, Page S, Ahmad H, Shaheen H, Harper DM (2012b) Vegetation dynamics in the Western Himalayas, diversity indices and climate change. Sci Tech Dev 31(3):232–243
Khan SM, Page S, Ahmad H, Shaheen H, Zahidullah MA, Harper DM (2013a) Medicinal flora and ethnoecological knowledge in the Naran Valley, Western Himalaya, Pakistan. J Ethnob Ethnomed 9:4. https://doi.org/10.1186/1746-4269-9-4
Khan SM, Page S, Ahmad H, Harper DM (2013b) Identifying plant species and communities across environmental gradients in the western Himalayas: method development and conservation use. J Ecol Inf 14:99–103
Khan SM, Page S, Ahmad H, Harper DM (2013c) Sustainable utilization and conservation of plant biodiversity in montane ecosystems; using the western Himalayas as a case study. Ann Bot 112(3):479–501

Khan SM, Page S, Ahmad H, Harper DM (2014) Ethno-ecological importance of plant biodiversity in mountain ecosystems with special emphasis on indicator species; a case study of the Naran Valley in the northern Pakistan. J Ecol Ind 37(Part A):175–185
Khan KU, Shah M, Ahmad H, Ashraf M, Rahman IU, Iqbal Z, Khan SM, Majid A (2015) Investigation of traditional veterinary phytomedicines used in Deosai plateau, Pakistan. Glob Vet 15(4):381–388
Lange D (1998) Europe's medicinal and aromatic plants: their use, trade and conservation. TRAFFIC International, Cambridge, p 77., 2 appendices
Lewin L (1964) Phantastica: narcotics and stimulating drugs. their uses and abuses. Dutton, New York
Martin GJ (1995) Ethnobotany, a people and plants conservation manual. Chapman and Hall, London
Mehmood A, Khan SM, Shah AH, Shah AH, Ahmad H (2015) First floristic exploration of District Torghar, Khyber Pakhtunkhwa, Pakistan. Pak J Bot 47(SI):57–70
Multi-Donor Support Unit (MSU) (2000) District population profile: Operationalising and interpreting population census data for planning (NWFP) Islamabad GoP
Nasir E, Ali SI (1982) Flora of Pakistan. Pan Graphics Ltd, Islamabad
Parker RN (1918. (Reprinted 1973) A Forest Flora of Punjab with Hazara and Delhi. Govt. Printing Press, Lahore
Pieroni A, Quave CL (eds) (2014) Ethnobotany and biocultural diversities in the Balkans: perspectives on sustainable rural development and reconciliation. Springer, Berlin
Rahman IU, Ijaz F, Afzal A, Iqbal Z, Ali N, Khan SM (2016a) Contributions to the phytotherapies of digestive disorders; traditional knowledge and cultural drivers of Manoor Valley, northern Pakistan. J Ethnopharmacol 192:30–52
Rahman A, Khan SM, Hussain A, Rahman IU, Iqbal Z, Ijaz F (2016b) Ecological assessment of plant communities and associated edaphic and topographic variables of the Peochar Valley district swat of the Hindu Kush Mountains. Mount Res Devel 36(3):332–341
Schippmann U (2001) Medicinal plants significant trade. CITES Project S-109, Plants committee document PC9 9.1.3. (rev.). BFN Skripten-39, Bundesamt fur Naturschutz, Bonn
Sher H (2002) Some medicinal and economic plants of Mahodand, Utror, Gabral valleys (district Sawat), Gabur, Brgusht valleys (district Chitral), Feb 2002. Report for Pak, mount, Areas conserv, Proj. IUCN, NWFP-Chitral
Shinwari MI, Khan MA (1998) Ethnobotany of Margalla Hills, National Park of Islamabad. Department of Biological Sciences Quaid-e-Azam University, Islamabad
Singh V, Pandey RP (1980) Medicinal plants-lore of the tribals of eastern Rajasthan. India J Econ Bot 1:137–148
Stewart RR (1972) An annotated catalogue to the vascular plants of West Pakistan and Kashmir. Fakhri Printing Press, Karachi
Sudarsanum G, Reddy MB, Nagaru N (1995) Veterinary crude drugs in Rayalaseema, Andhra Pradesh, India. Int J Pharm 33(1):52–60
WHO (2002) World Health Organization traditional medicine strategy 2002–2005, Geneva
Zaman MB, Khan MS (1970) Hundred drug plants of West Pakistan. Medicinal plants branch. PFI, Peshawar, pp 5–8

Exploring the Therapeutic Characteristics of Plant Species in the Chichawatni Irrigated Plantation Pakistan

Muhammad Arif, Waseem Razzaq Khan, Muhammad Khurram Shahzad, Amna Hussain, and Cao Yukun

Introduction

There is an estimated 2.85–2.50 (million) plant species across the globe (Paton et al. 2008; Joppa et al. 2011), and many of them are widely used in pharmaceutical materials, trade, and traditional remedies (World Health 1997; Belcher and Schreckenberg 2007; Shahidullah and Haque 2010; Sher et al. 2017). History of medicinal plants can be traced to ancient successive civilizations of Mesopotamia (Sumerian, Babylonian), China, and India. Therapeutic knowledge regarding plants has been investigated and practiced for the treatment of broad range of illnesses in the Mesopotamia. During 3000 BC, the study revealed that Babylonian and Sumerian pharmaceutical prescriptions were using curative plants for therapy of wide range of

M. Arif
Department of Forestry Economics and Management, Northeast Forestry University, Harbin, People's Republic of China

Punjab Forest Department, Government of Punjab, Lahore, Pakistan

W. R. Khan
Department of Forest Management, Faculty of Forestry, Universiti Putra Malaysia, Serdang, Selangor, Malaysia

M. K. Shahzad
Punjab Forest Department, Government of Punjab, Lahore, Pakistan

Department of Forest Management, Northeast Forestry University, Harbin, People's Republic of China

A. Hussain
Department of Forest Management, Northeast Forestry University, Harbin, People's Republic of China

C. Yukun (✉)
Department of Forestry Economics and Management, Northeast Forestry University, Harbin, People's Republic of China

M. Ozturk, K. R. Hakeem (eds.), *Plant and Human Health, Volume 1*,
https://doi.org/10.1007/978-3-319-93997-1_20

diseases. As per aged record of 2500 BC, Chinese and Indian traditional healers had also been utilizing these plants, for which they got benefited from ancient herbal proscriptions of Ebers Papyrus, written in circa 1550 BC (Nikbakht and Kafi, 2004; Saad et al. 2005; Azaizeh et al. 2010).

Cognizance of medicinal plants contains a history of over four millennia. At that time, people were restricted to reply only on natural substances. At first, this knowledge was transmitted orally, followed by the written material in the parchments, baked clay tablets, papyri, manuscripts, pharmacopeias, etc. (Petrovska 2012; Teall 2013). The Convention on Biological Diversity has accepted the dependence of international community on natural resources that mostly include medicinal plants. The conservation and the sustainable uses of biological diversity have also been recognizing the contribution of traditional knowledge (Maina 2012).

Importance and novelties in drug discovery have reemphasized the interest in therapeutic plants during the last decades. Pharmacological advancement and activities are screening several medicinal plant species every day. FAO has estimated an astonishing figure of over 50,000 curative plant species around the world, which comes to 18.9% of the total available flora (Schippmann et al. 2002). Medicinal plants serve as a primary health source for about 70–80% population of the world. Higher plants, despite the ancient nature of the tradition, are providing active ingredients for 25% of prescription drugs (World Health 1997; Shahidullah and Haque 2010). During the last three decades, herbal product demand has been increased in the worldwide market of traditional plants. Better results from the products of the medicinal plants have attracted the pharmacological industry, which has intensified the demand of herbal products. During 2002, it was estimated that an amount of US$ 60,000 million had been invested in the trade of herbal products in the international market (Sher et al. 2010).

Owing an extensive variation in soil, latitude, and climatic conditions, Pakistan has a rich and diversified flora. An extensive multiplying of vegetation occurs in this region, right from the mangroves of Arabian Sea to the alpine pastures of the northern high mountains. The forest ecosystems, therefore, possess profuse biodiversity of flora and fauna. Out of 6000 flora species recorded in Pakistan, 300 plants representing 149 genera are endemic. Ninety percent of the endemic species occur in the northern and western mountains at over 1200 m elevation (Rahim 2010). It has been forecasted that 10% of the indigenous flora is under threat of extinction (Shinwari 2010). Pakistan is a developing country where data collection is much painstaking as compared to the developed states. There are few controversies over the conservation status of indigenous plant species. Various researchers have mentioned diverse figures about endangered native flowering species in dissimilar time spans (Khan 2005). Some studies have been conducted for the indigenous uses of native species of different districts and reviewed by Shinwari (2010). Among the leading irrigated plantations of Punjab, Changa Manga plantation is the only compact plantation in Pakistan where its ethnomedicinal survey was carried out in 2014. It was identified that native residents use the local plants for the treatment of malaria, hepatitis, dysentery, cholera, burns, eye problems, ulcers, insomnia, insect stings, cough, and gastric problems (Ahmad et al. 2014). There is a dire need to understand a scientific study encompassing the medicinal uses of present flora with its current status.

The aim of the current study is to explore the availability of ethnobotanical plants in the compact forest of Chichawatni, Pakistan. Objectives of the study are to document the inventory of therapeutic plants within the study area; to appraise the present status of commercially important curative and lucrative plants; and to explore the possibilities for sustainable development of medicinal species for the future toxicological, photochemical, and pharmacological studies of the plantation.

The Study Area

Chichawatni irrigated plantation also known as Chichawatni Reserved Forest is situated in district Sahiwal, Punjab, Pakistan, between latitudes 30°-29′-32.91″N and 30°-33′-45.84″N and longitudes 72°-36′-00.25″E and 72°-46′-48.65″E. Plantation elevation level is 153.6–163.7 m above the sea level. Total stocked area is 3823.20 hectares out of gross 4666.8 hectares (Khaggah 2015). The climate of the plantation is dry with a mean temperature of 65–75 °F. Maximum rainfall is received in July and August as a summer monsoon. The indigenous plant species included *Acacia nilotica* (Kikar), *Acacia modesta* (Phulai), *Albizia procera* and *Albizia lebbeck* (Siris), *Azadirachta indica* (Neem), *Melia azedarach* (Bakain), *Salvadora oleoides* (Jal), *Tamarix articulata* (Frash), *Terminalia arjuna* (Arjan), *Prosopis cineraria* (Jand), and *Ziziphus mauritiana* (Ber). Prevailing plant communities comprises of *Bombax ceiba* (Simal), *Dalbergia sissoo* (Shisham), *Eucalyptus camaldulensis*, *Morus alba* and *Morus nigra* (Mulberry), *Prosopis glandulosa* and *Prosopis juliflora* (Mesquite), etc. (Arif et al. 2017; Arif 2018) (Figs. 1 and 2).

Field Survey

Current study was conducted during spring and summer of 2015. Forest's resources were explored with the help of working plans and compartment history files. General information of the investigated area was obtained from Punjab Forest Department, Pakistan. Frequent visits were made to different sites within the study area. The information was collected with the help of a questionnaire, which contains personal information and utilizations of forest resource. Facts about conventional uses of forest resources were gathered from the native herbalists, midwives, traditional healers, and local adult villagers. Ages of the respondents varied from 18 to 65 years with a high representation of elderly persons. Generally, one informant per five houses was contacted and interviewed. Their interests as collectors, traders of forest resources, and native uses were documented. Economic importance, distribution, frequency, and demand of plant species were noted. The plants were collected, dried, and preserved properly from different sites of the study area. Fully dried specimens were mounted on herbarium sheets with scotch tape; plants were identified. The following research tools/instruments were used during the research work:

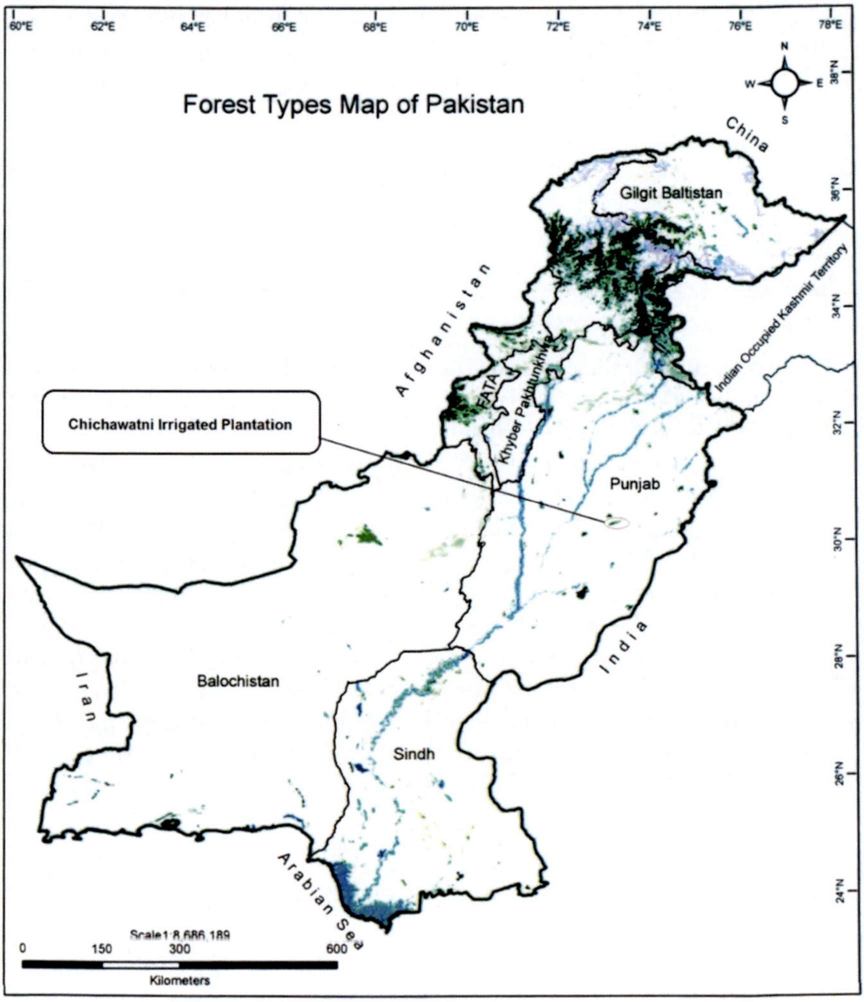

Fig. 1 Location of study area on the forest cover map of Pakistan

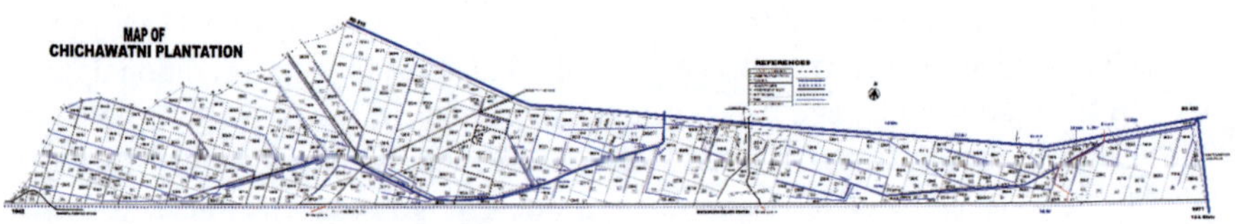

Fig. 2 Map of Chichawatni irrigated Plantation

quadrat, nylon threads, meter rod, pH meter, altimeter, hydrometer, soil air thermometers, digital camera, questionnaire, and computer with other accessories.

Present study revealed that the investigated area was traditionally rich in customary usage of plant legends and forest products, since time immemorial. This compact forest contains a total of 122 plant species. Shrubs and undergrowth are found in a maximum figure of 52 species (42.62%), followed by the presence of trees (47 species) (38.52%), grasses (14 species) (11.48%), and climbers (9 species) (7.38%, Fig. 3). 53 vascular plants belong to 31 families which exist within the study area. A study of 3 species (5.66%) was recorded as critically vulnerable, 14 species (26.40%) as endangered, 23 species (43.40%) as rarely distributed, and 13 species (24.52%) as secure or least concerned (Fig. 4). Sustainable harvesting of plants

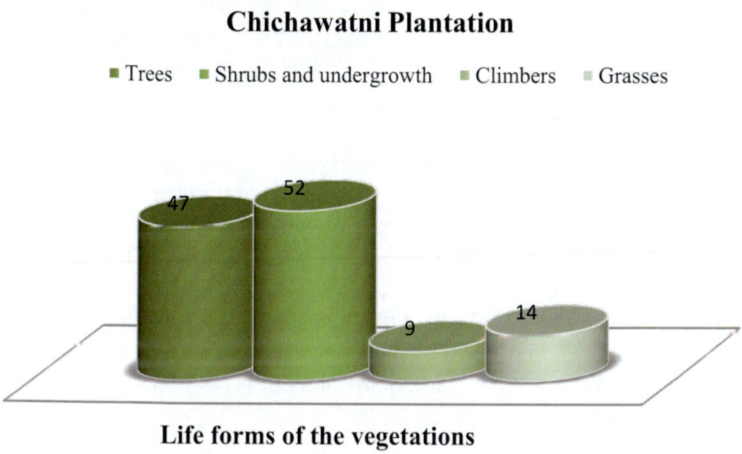

Fig. 3 Life forms of the plants in Chichawatni plantation

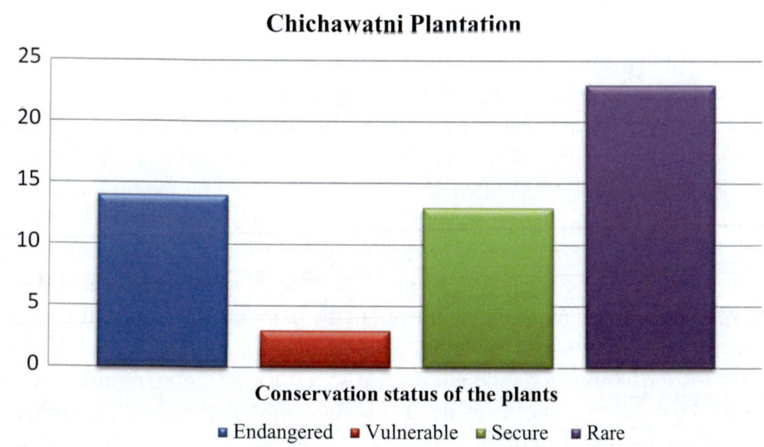

Fig. 4 Conservation status of the plants in Chichawatni plantation

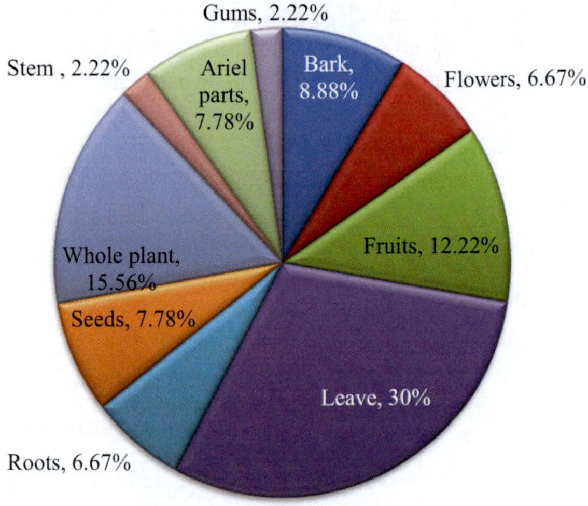

Fig. 5 Parts usage % age of the plants in Chichawatni plantation

depends on the time and method of harvesting and consumption of their specific parts. Therefore, information regarding the usage of particular medicinal and aromatic plant species and the time of their harvesting was also collected. Leaves are the most important part of 27 species (30%) that is utilized. The entire use of a plant was found at a second category in 14 species (15.56%), followed by the usage of consuming 11 species (12.22%) for fruits, 8 species (8.89%) for bark, 7 species (each) (7.78%) for seed and stem, 6 species (each) (6.67%) for flowers and roots, and 2 species (each) (2.22%) for aerial parts and gums (Fig. 5).

The study showed that the reported plants play a vital role in the conventional livelihood of local folk. The plants available within the study area serve as an antidote for abdominal pains, aching, cramps, cough, constipation, diarrhea, eyes infections, injury, insect stings, impotency, jaundice, kidney stones, nervous disorders, respiratory infections, renal insufficiency, skin diseases, sore throat chest pain, swellings, vomiting, and wounds (Table 1).

It is ascertained that species present in the study area belong to 28 families. Asteraceae is the leading family with 10 plant species, and it was followed by Fabaceae (6 spp.), Amaranthaceae (4 spp.), Brassicaceae (2 spp.), Malvaceae (2 spp.), Poaceae (2 spp.), etc. (Table 1). Prominent ethnomedical plants among these families are *Acacia nilotica*, *Albizia lebbeck* L., *Salvadora oleoides* Dene., *Tamarix articulata* L., *Terminalia arjuna* (Rox. Ex D.C), *Ageratum conyzoides* L., *Cannabis sativa* L., *Datura stramonium* L., *Fumaria officinalis* L., *Oxalis corniculata* L., *Prosopis cineraria* L., *Sonchus oleraceus* L., and *Ziziphus nummularia* (Burm. f.).

It is noted that the level of knowledge about the plant resources is realistic in local people, especially about aromatic and medicinal plants. Different people having the distinctive level of knowledge, depending on social group, gender, and occupation, i.e., the hakims and tenants, were having higher knowledge about medicinal and aromatic plants than other common people. Similarly, the men have extra knowledge than women, and among the men, especially the elder, they were more aware

Table 1 Important plants with medicinal uses in the Chichawatni compact plantation Pakistan

Plant family	Plant binomial species	Local name	Part (s) used	Conservation status	Medicinal uses
Amaranthaceae	*Achyranthes aspera* L.	Poth kant, ludhri	WP	Endangered	Tonic, ophthalmic, colic, and emollient
	Avera javanica (Burm.f.) Juss. ex Schult.	Javi	SD, L	Endangered	Emollient, diuretic, and antibacterial
	Chenopodium album L.	Bathu	WP	Rare	Tonic, carminative, ophthalmic, cathartic, aphrodisiac, and diuretic
	Suaeda fruticosa (L.) Forsskal.	Boi booti	L	Endangered	Antibacterial, anodyne, and colic
Apocynaceae	*Calotropis procera* (Aiton).	Akk	WP	Secure	Colic, alternative, acrid, anathematic, antimicrobial, and sedative
Asteraceae	*Ageratum conyzoides* L.	Boh	FL	Endangered	Astringent, laxative, antipyretic and antibiotic, and expectorant
	Ayapana triplinervis (M.Vahl)	Kamrakh	L	Vulnerable	Tonic, antipyretic, ophthalmic, and colic
	Cirsium arvense L.	Kandhari	FL	Rare	Emetic
	Conyza bonariensis L.	Loosan booti, namkeen booti	WP	Secure	Diuretic and styptic
	Conyza Canadensis L.	Karari	S	Endangered	Astringent, diuretic, and antirheumatic
	Galinsoga ciliata (Rafin.) Blake.	Kakoh	FL, L	Rare	Clotting agent, antibiotic, and tonic
	Parthenium hysterophorus L.	Booti	WP	Rare	Antipyretic and emollient
	Sonchus arvensis L.	Dodh bhatal	L, FR	Rare	Tonic
	Sonchus oleraceus L.		S, AP	Endangered	Laxative, antipyretic, and tonic
	Taraxacum officinale F.H. Wigg.	Kanfhul	WP	Vulnerable	Antibacterial, aperients, and analgesic

(continued)

Table 1 (continued)

Plant family	Plant binomial species	Local name	Part (s) used	Conservation status	Medicinal uses
Brassicaceae	*Coronopus didymus* L.	Thandi booti	L	Rare	Cooling
	Sisymbrium irio L.	Saag booti	SD	Secure	Tonic, alternative, and colic
Cannabaceae	*Cannabis sativa* L.	Bhang	L, FR	Endangered	Sedative, blood purifier, cooling, and antiseptic
Capparaceae	*Capparis decidua* L.	Kaluari	FR, AP	Endangered	Carminative, analgesic, bitter, and anti-inflammatory
Caryophyllaceae	*Stellaria media* L.	Washtah	L	Rare	Laxative, cathartic, expectorant, antibiotic, and cytostatic
Cleomaceae	*Cleome viscosa* L.	Chaskoo	L, SD	Endangered	Tonic, ophthalmic, anodyne, and carminative
Convolvulaceae	*Convolvulus arvensis* L.	Lali	WP	Rare	Aperients
Euphorbiaceae	*Euphorbia helioscopia* L.	Chhatri dodak, gandi booti	SD, R	Rare	Laxative and colic
Fabaceae	*Acacia modesta* (Linn.) Wall.	Phulai	G, S	Rare	Stimulant
	Acacia nilotica Willd.	Kikar,babul	WP	Secure	Astringent, stimulant, and tonic
	Albizia lebbeck L.	Siris	L, F	Rare	Acrid, stimulant, and tonic
	Albizia procera (Roxb.) Benth.	Siris	L	Rare	Stimulant and tonic
	Dalbergia sissoo Roxb.	Shisham	L, B, SD	Secure	Astringent and tonic
	Prosopis cineraria (L.) Druce.	Pirasoo	SD, S, B	Secure	Demulcent, pectoral, and tonic
Lamiaceae	*Mentha spicata* L.	Jangli pudina	L	Rare	Tonic, sedative, cordial, and aperients

(continued)

Table 1 (continued)

Plant family	Plant binomial species	Local name	Part (s) used	Conservation status	Medicinal uses
Malvaceae	*Abutilon theophrasti* L.	Peeli booti	WP	Secure	Diuretic, styptic, ophthalmic, and stimulant
	Malvastrum coromandelianum (L.)	Patakha	FR	Secure	Tonic, antibacterial, anti-inflammatory, cooling, and expectorant
Meliaceae	*Melia azedarach* L.	Bakain	FR	Secure	Stimulant
	Azadirachta indica L.	Neem	B, L, SD, S	Rare	Tonic, stimulant, expectorant, febrifuge, and alternative
Moraceae	*Morus alba* L.	Safaid toot	R, L, FR, B	Secure	Antibacterial, astringent, and ophthalmic
Myrtaceae	*Eucalyptus camaldulensis* Hook. f.	Sufaida	L, B	Secure	Sedative, anesthetic, antiseptic, and expectorant
Oxalidaceae	*Oxalis corniculata* L.	Khati booti	L, FR	Endangered	Sedative, anodyne, tonic, carminative, stimulant, and anathematic
Papaveraceae	*Ficus carica* L.	Injeer	FR, L	Rare	Stimulant and anti-inflammatory
Plantaginaceae	*Plantago ovata* Forssk.	Isphogol	L.FL	Rare	Demulcent, laxative, astringent, refrigerant, and emollient
Poaceae	*Cynodon dactylon* (L.) Pers.	Humrik booti	WP	Secure	Anti-inflammatory, stimulant, and colic
	Desmostachya bipinnata (L.) Stapf	Drabh	R, L	Secure	Tonic, febrifuge, and laxative
Polygonaceae	*Rumex Crispus* L.	Palak booti	L, FR	Rare	Laxative, tonic, and expectorant
Primulaceae	*Anagallis arvensis* L.	Sowa	WP	Endangered	Emollient and analgesic

(continued)

Table 1 (continued)

Plant family	Plant binomial species	Local name	Part (s) used	Conservation status	Medicinal uses
Ranunculaceae	*Ranunculus muricatus* L.	Rara	L	Endangered	Acrid, stimulant, and tonic
Rhamnaceae	*Ziziphus mauritiana* L.	Ber	FR	Rare	Cooling, astringent, and stimulant
Salicaceae	*Salix tetrasperma* Roxb.	Beesan	B, L, S	Rare	Anodyne and febrifuge
Simaroubaceae	*Ailanthus excelsa Roxb.* ex Willd.	Darawa	B, R, L	Endangered	Astringent
Solanaceae	*Datura stramonium* L.	Datura	WP	Endangered	helminths, soporific, anti-incendiary, anesthetic, antibiotic, and emollient
Verbenaceae	*Lippia nodiflora* L.	Rye	L, FL, R	Rare	Blood purifier, joint ache, tonic, stimulant, and anodyne

Keys: *AP* aerial parts, *B* bark, *FL* flowers, *FR* fruits, *G* gums, *L* leaves, *R* roots, *S* stem, *SD* seeds, *WP* whole plan

than the younger members from the community. It is further noticed that the distribution, pattern, and availability of medicinal plant species were depending on grazing, harvesting intensity, habitat loss, and forest stand.

The people living around the Chichawatni plantation are mostly agriculturists, and their socioeconomic life relies on this forest regarding timber, firewood, medicinal plants, and animal fodder. However, the people's faith in the curative benefits of herbal medicine is declining with the change in time. The introduction of allopathic medicines in the remote areas has faded people's trust, skill, and knowledge (Sher et al. 2015, 2016). Existing plant species at the same time, as recorded in the current study, do serve as multipurpose species as well. Amazingly, in spite of this dependence, important renewable natural resource has not been managed scientifically for its sustainable production. Resultantly, various species are going to be vanished in specific regions around the world (FAO 2016). Overexploitation at some places has led to a state where certain plants have acquired the status of endangered or even near extinction. It is reported that ruthless utilization of medicinal plants, non-systematic grazing, and unchecked harvesting can degrade the environment (Sher et al. 2014). Further it has raised the number of endangered species in the fragile ecosystem. Advance research in modern science has discovered multiple medicinal effects of the existing plants. This development, in turn, has further raised the frequency of cutting such tree species. It has become essential to take immediate remedial steps for conservation and stabilization of this resource. Being a common property resource, active participation of the local communities to regenerate the ecosystem under stress is of prime significance.

The current study further revealed that such multiple tree species had significant nutritional and economic value for local villagers. Adjacent living communities draw benefits of major and minor values from different species, i.e., *Plantago ovata* Forssk is used to control constipation problem in the study areas, while *Salix tetrasperma* was used for the treatment of fever. Traditional use and medicinal properties of some plant species have been previously reported by the national researchers, i.e., *Datura stramonium* as emollient and spasmolytic, used for relieving spasm of smooth muscle (Ahmad et al. 2012). An extensive survey should be carried out regarding distribution phonology, bearing, and ecological characteristics to formulate and undertake a conservation strategy for the species under threat of vulnerability. Species like *Taraxacum officinale*, *Ayapana triplinervis*, and *Fumaria officinalis* should be considered on priority.

Conclusion

The study documents the ethnopharmacological knowledge, which can be used as an important tool for the future toxicological, pharmacological, and photochemical studies of the area. This study will create awareness and inculcate knowledge regarding potential medicinal and aromatic plant resources, in particular, and forest resources in general. There is a strong need for the documentation of local knowledge of medicinal plants, and this transforming information should be prioritized from older generations to their descendants.

References

Ahmad KS, Qureshi R, Hameed M, Ahmad F, Nawaz T (2012) Conservation assessment and medicinal importance of some plants resources from Sharda, Neelum Valley, Azad Jammu and Kashmir, Pakistan. Int J Agric Biol 14:997–1000

Ahmad SS, Erum S, Khan SM, Nawaz M, Wahid A (2014) Exploring the medicinal plants wealth: a traditional medico-botanical knowledge of local communities in Changa Manga Forest, Pakistan. Middle-East J Scientific Res 20:1772–1779

Arif M, Shahzad MK, Elzaki EEA, Hussain A, Zhang B, Yukun C (2017) Biomass and carbon stocks estimation in Chichawatni irrigated plantation in Pakistan. Int J Agric Biol 19: 1339–1349

Arif M (2018) Resources assessment and sustainable management of Chichawatni irrigated plantation in Pakistan. Doctoral thesis, Northeast Forestry University, Harbin, P. R. China

Azaizeh H, Saad B, Cooper E, Said O (2010) Traditional Arabic and Islamic medicine, a re-emerging health aid. Evid Based Complement Alternat Med 7:419–424

Belcher B, Schreckenberg K (2007) Commercialisation of non-timber forest products: A reality check. Dev Policy Rev 25:355–377

FAO (2016) Global forest resources assessment 2015. How are the world's forests changing? , Rome, Food and Agriculture Organization
Joppa LN, Roberts DL, Pimm SL (2011) How many species of flowering plants are there? Proc Royal Soc Lond B: Biol Sci 278:554–559
Khaggah NH (2015) Working plan of Chichawatni Forest Division. Forestry, Wildlife and Fisheries Department, Government of Punjab, Lahore, Pakistan
Khan A (2005) Joint Forest Management Plan Madaklasht Valley Chitral (Qashqar Conservancy). In: Technical Report Prepared for Pakistan Mountain Areas Conservation Project
Maina CK (2012) Traditional knowledge management and preservation: intersections with library and information science. Int Inform Library Rev 44:13–27
Nikbakht A, Kafi M, (2004) The history of traditional medicine and herbal plants in Iran. In: VIII International people-plant symposium on exploring therapeutic powers of flowers, greenery and nature, vol 790, pp 255–258
Paton AJ, Brummitt N, Govaerts Rl, Harman K, Hinchcliffe S, Allkin B, Lughadha EN (2008) Towards Target 1 of the Global Strategy for Plant Conservation: a working list of all known plant species progress and prospects. Taxon 57:602–611
Petrovska BB (2012) Historical review of medicinal plants' usage. Pharmacognosy Rev 6:1
Rahim SMA (2010) Working plan of changa manga forest division. Forestry, Wildlife and Fisheries Department, Government of Punjab, Lahore, Pakistan
Saad B, Azaizeh H, Said O (2005) Tradition and perspectives of Arab herbal medicine: a review. Evid Based Complement Alternat Med 2:475–479
Schippmann U, Leaman DJ, Cunningham AB (2002) Impact of cultivation and gathering of medicinal plants on biodiversity: global trends and issues. In: Biodiversity and the ecosystem approach in agriculture, forestry and fisheries. FAO, Rome
Shahidullah AKM, Haque CE (2010) Linking Medicinal Plant Production with Livelihood Enhancement in Bangladesh: Implications of. J Transdisciplinary Environ Stud 9(2)
Sher H, Aldosari A, Ali A, de Boer HJ (2014) Economic benefits of high value medicinal plants to Pakistani communities: an analysis of current practice and potential. J Ethnobiol Ethnomed 10
Sher H, Aldosari A, Ali A, de Boer HJ (2015) Indigenous knowledge of folk medicines among tribal minorities in Khyber Pakhtunkhwa, northwestern Pakistan. J Ethnopharmacol 166:157–167
Sher H, Al-Yemeni M, Sher H (2010) Forest resource utilization assessment for economic development of rural community in northern parts of Pakistan. J Med Plant Res 4:1786–1789
Sher H, Barkworth ME, de Boer HJ (2017) Medicinal and aromatic plant cultivation in the Swat valley, north-western Pakistan, for economic development and biodiversity conservation. Genet Resources Crop Evol 64:237–245
Sher H, Bussmann RW, Hart R, de Boer HJ (2016) Traditional use of medicinal plants among Kalasha, Ismaeli and Sunni groups in Chitral District, Khyber Pakhtunkhwa province, Pakistan. J Ethnopharmacol 188:57–69
Shinwari ZK (2010) Medicinal plants research in Pakistan. J Med Plant Res 4:161–176
Teall EK (2013) Medicine and doctoring in ancient mesopotamia. Grand Valley J Hist 3:2
World Health O (1997) Essential medicines and health products information portal: A World Health Organization resource. World Health Organization

Therapcutic Characteristics of Murree Plants: An Emerging Feature

Waseem Razzaq Khan, Muhammad Arif, Sadaf Shakoor, M. Nazre, and M. Muslim

Introduction

Globally 350,000–400,000 species have been estimated as medicinal plants (Joppa Roberts and Pimm 2010; Paton et al. 2008), and many of them are valuable as future of safe drugs (Abbasi et al. 2010; Bussmann et al. 2008). Sustainability of this renewable natural resource is a vital issue, which needs to be discussed worldwide. As it not only provides raw materials for herbal remedies, for around 80% of the world's population, but also helpful for primary health care in rural and remote hilly areas (Ntie-Kang et al. 2013). Pakistan has a unique position in the developing world due to potential of a variety of aromatic and therapeutic plants and varied ecosystem and climate, which reflect valuable plant heritage and biodiversity. There is calculated a value of billion US dollars for the trade of fragrant and medicinal plants in 2013 (Hamilton 2013). Due to increasing demand for herbal medicines (Khan et al. 2011), this trade is likely to rise largely by the year 2050 (Lange 1997). In South Asia, millions of people, especially low-income groups and women, are involved in the harvesting, cultivation, and trade of medicinal plants (de Boer et al. 2012; Larsen and Olsen 2007). Due to the cultural acceptability of plant-based

W. R. Khan (✉) · M. Nazre
Department of Forest Management, Faculty of Forestry, Universiti Putra Malaysia, Serdang 43300, Selangor, Malaysia

M. Arif
Department of Forestry Economics and Management, Northeast Forestry University, Harbin, People's Republic of China

S. Shakoor
Department of Nutrition and Dietetics, Faculty of Medicine and Health Sciences, Universiti Putra Malaysia, Serdang 43300, Selangor, Malaysia

M. Muslim
Pakistan Forest Institute, Peshawar, Pakistan

M. Ozturk, K. R. Hakeem (eds.), *Plant and Human Health, Volume 1*,
https://doi.org/10.1007/978-3-319-93997-1_21

products and their economic potential, interest has been growing in medicinal plants in Pakistan during the past few decades (Sher et al. 2014, 2015).

It has been reported that more than 6000 species of flowering plants are found in Pakistan. Majority of ethnobotanical plants are located in northwestern and northern parts of the country. Out of all, phytochemical properties have been recognized in 1012 plant species; whereas 350–400 plant species are traded in different drug markets of the country, which are used by 28 leading manufacturing units of Homeopathic, Ayurvedic, and Greco-Arabic medicines. More than 200 drug plants are being used by unregistered practitioners, which are scattered in the remote hilly and rural areas. These plants are used for primary health care and as household remedies against several diseases. The uncontrolled and haphazard exploitation of important medicinal and aromatic plants, such as *Saussurea lappa*, *Podophyllum emodi*, *Rheum emodi*, *Dioscorea deltoidea*, *Atropa acuminata*, *Angelica glauca*, *Digitalis* sp., *Aconitum* sp., *Valeriana jatamansi*, etc., presents imminent danger of annihilation and lead to these species being scheduled in CITES (Convention of International Trade of Endangered Species). There is a need for the protection of natural habitats of medicinal and aromatic plant species to conserve genetic diversity for the continuation of the evolutionary process. These genetic reservoirs can be utilized for improving yield, resistance to diseases of field crops, genetic manipulation, and biomedical research to fight fatal diseases like AIDS, hepatitis, cardiovascular, cancer, etc. Similarly various plants have high-quality potential to manufacture lifesaving and painkilling drugs from their derivatives like digitoxin and digoxin derived from *Digitalis purpurea* used in the treatment of cardiovascular disease. Azizol and Chang (2000) have worked over the manufacturing quality of herbal medicines, and they reported appropriate cultivation and harvesting of plant material being chief factors in good quality of medicines.

Murree is located in the district of Rawalpindi in Pakistan (Fig. 1). The area is dominated by hills and slope. Lying in Himalayan foot hills at 33°54′30" north latitude and 73°26′ east longitude, it has an altitude of 2286 m (Ahmed et al. 2013). Having a cool and humid climate throughout the year supports the growth of variety of trees and shrubs. Subtropical and moist temperate forests such as blue pine and chir pine are abundant in the area. Angiosperms, ferns, and mosses are mainly found as a vast spread on the ground.

In Murree, Pakistan, medicinal plants have been the main source of the life for the people living within and around the study area. These people have deep knowledge of the abundance, distribution, use, harvest, and market value of medicinal plants. The aim of the current chapter is to reveal the availability of ethnobotanical plant in the forests of Murree, Pakistan. The objectives of the chapter are to make an inventory of medicinal, economic, and aromatic plants for the socioeconomic development of the study area, to assess the present status of commercially important medicinal and economic plants, and to explore the possibilities for sustainable development of a medicinal plant by the involvement of rural communities (Table 1).

As per Table 2, 70 species of plants have been reported for high medicinal values, which were further categorized into 12 subgroups based on their specific disease treatment. Ahmed et al. (2013) reported similar results; most of the plants were

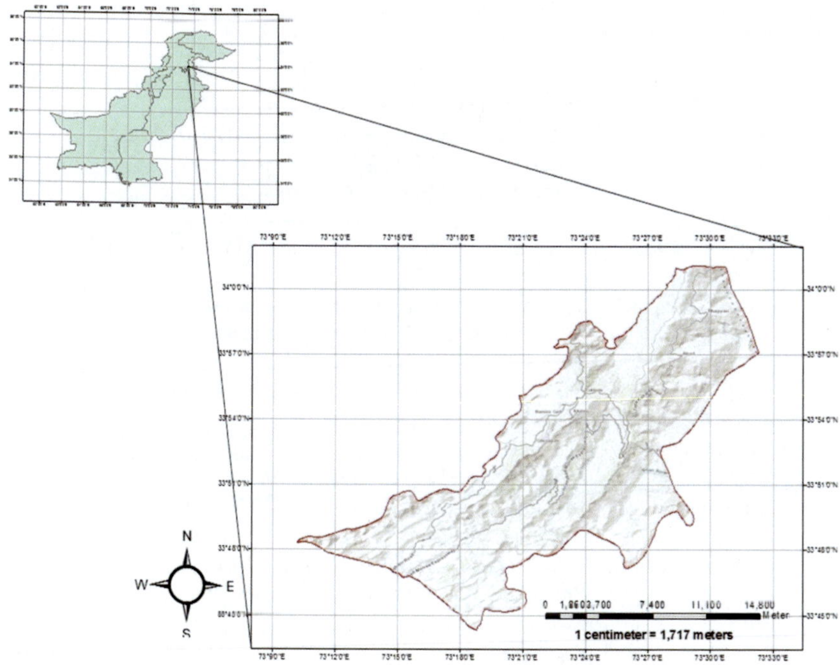

Fig. 1 Map of study area (outlined in red boundaries): Murree, Pakistan

Table 1 Classification of plants in Murree based on ethnobotany

Sr. No.	Ethnobotanical uses	Number of species	Percentage
1	Medicinal species	70	51.3
2	Multipurpose plants	33	20
3	Fodder/forage species	27	16.6
4	Fuel wood species	21	14
5	Vegetables	8	6.6
6	Timber wood species	07	6
7	Roof thatching species	07	4.6
8	Wild fruits	06	4
9	Toolmaking species	06	4
10	Poisonous species	06	4
11	Veterinary important plants	05	2.6
12	Religious plants of Muslim	04	2.6
13	Fences/hedges species	05	2.6
14	Mud supporter	04	2.6
15	Blooming species	03	2
16	Evil repellent species	02	1.3
17	Honey bee species	02	1.3

Table 2 Ethnobotanical use of plants in Murree, Pakistan

Sr. No.	Diseases	Botanical name	Total species
1	Stomach and gastrointestinal problems	*Mentha longifolia, Mentha viridis, Polygonatum multiflorum, Trachyspermum ammi, Bunium persicum, Acorus calamus, Plantago lanceolata, P. ovata, Bergenia ciliata, Rosa webbianum, R. moschata*	15
2	Skin diseases, itching, scabies, and eczema	*Verbascum thapsus, Aesculus indica, Urtica dioica, Bergenia ciliata, Chenopodium album, Pistachio, Melia azedarach*	10
3	General body tonic; backache, arthritis	*Mows nigra, Paeonia emodi, Solanum nigrum, Asparagus* spp., *Viola serpens, Bistorta amplexicaule, Juglans regia*	10
4	Chest pain, fever, and sore throat	*Berberis vulgaris, Allium sativum, Dioscorea deltoidea, Juniperus excelsa*	07
5	Kidney stone and kidney problem	*Aconitum leave, A. heterophyllum, Bergenia ciliata, Dioscorea deltoidea, Gentiana kurroo*	06
6	Anthelmintic	*Valeriana jatamansi, Artemisia brevifolia, Bistorta amplexicaule, Amaranthus viridis*	05
7	Nervous disorder	*Cichorium intybus, Podophyllum hexandrum, Aconitum leave, Atropa acuminata, Hypericum perforatum*	04
8	Liver diseases and jaundice	*Cichorium intybus, Morus alba/nigra, Onosma hispida, Rheum webbianum*	04
9	Diabetes	*Hedera nepalensis, Rheum webbianum*	03
10	Swelling	*Verbascum thapsus, Salvia*	02
11	Cut and wound	*Berberis vulgaris, B. lycium*	02
12	Laxative	*Corydalis govaniana, Digitalis lanata*	02

used for therapeutic purpose, followed by fodder, while spiritual significance was practiced by a few elders (Orr et al. 2017). The mentioned plants are categorized on the basis of their medicinal effects against the diseases such as stomach and gastrointestinal problems (15 spp.); skin disease, scabies, eczema, and aching (10 spp.); general body tonic and arthritis (10 spp.); sore throat, chest pain, and fever (7 spp.); kidney stone (6 spp.); anthelmintic (5 spp.); liver diseases (4 spp.); swelling (2 spp.); cut and wound healing (2 spp.); nervous disorder (4 spp.); laxative (2 spp.); and for the control of diabetes (3 spp.). Sustainable harvest of plants depends on time, method of harvesting, and plant parts used. Therefore, information regarding the use of specific parts of medicinal and aromatic plant species and their harvest are also collected. Out of 70 medicinal plant species, 20 species have been categorized on their parts used as roots and rhizomes, 18 as leaves, 16 as whole parts, 8 as flowers, 6 as fruit and seeds, and 2 as barks.

The local medicinal uses of few high-value MAP species in terms of its traditional medicinal uses as the local community has strong faith on these few species only the remaining are used in rare cases. Therefore, the current use of each species is discussed as follows: the mixture of ghee and dried powdered rhizomes of

Aconitum leaf is helpful for curing lumbago problem and also for anorexia. Its dried powdered rhizomes are cooked in goat's ghee and taken orally for the treatment of myalgia. Some of the local hakims (especially women) use their roots for abortion. *Aconitum heterophyllum* has been similarly used to aconitum leaf but of inferior quality. The fresh rhizomes of *Bistorta amplexicaule* are taken orally with a glass of water for curing of joint pains in old age people. The fruits of *Bunium persicum* are mainly collected for commercial purposes. It is used as a spice and condiment agent at local and national levels. The rhizomes of *C. govaniano* are used to treat all kinds of eye diseases and to improve eye sight. *Ferula narthex* is mixed with chicken soup in the form of dried ground roots to be used orally for the treatment arthritis. *Ephedra gerardiana* is used for fuel and snuff. *Hippophae rhamnoides* fruit is mainly collected for commercial purposes, while in some sites its dried fruits are taken orally with water for curing hypertension. *Hyoscyamns niger*, although locally considered as poisonous, some local hakims use its fruits for curing epilepsy. The leaves and young stem of *Mentha longifolia* are locally administered for curing of diarrhea and dysentery and for aiding digestion. Young shoots of *Nepeta cataria* are mainly collected for commercial purposes. Indigenous people also use the powdered stem for external wound healing. The rhizomes of *Rheum webbianum* are used for removal of kidney stone. The fruits of *Trachyspermum ammi* are used as a stimulant carminative and antispasmodic agent. The flowers of *Viola canescens* are helpful to improve eyesight.

According to Ahmed et al. (2013), ethnobotanical statistic shows that most plants are used for medicinal and fodder purposes followed by fuel, fruit, vegetable, and ethnoveterinary. There is also an established association of medicinal use of plants to the fruits used. Non-woody plants have a high tendency toward the medicinal use of the plants as compared to woody plants. Annual plants are less likely to be directly associated with the medicinal use of plants in the surveyed vegetation. Underground plant parts are also to be used for medicinal purposes as revealed from the local discussion. In short, hilly areas, plants are very important by nutritional and medicinal aspect (Orr et al. 2017). Further plants ethnography is better described by (Casagrande 2002, Voeks 2009).

Local Knowledge About Plant Resources

It has been noted in Murree that local people have longed realized about aromatic and medicinal plants. Out of 216 plant species, 160 species had been known locally for various purposes, while the remaining 56 species were unknown to the inhabitants of the area. Different people have the different level of knowledge, depending on the social group, gender, and occupation, i.e., the Hakims having higher knowledge about medicinal and aromatic plants than other people. Similarly, men usually have extra knowledge than women, especially the elderly.

Availability and Distribution

MAP species and other forest resources like fodder, fuel, wood, and timber wood spread in almost whole of the area under investigation. It is further noted that pattern and distribution of individual species were different according to differences in habitats, aspect, and altitude. Some species are quite abundant, and some are in the restricted habitat of the study area. i.e., *Bistorta amplexicaule*, *Mentha viridis/longifolia*, *Paeonia emodi*, *Polygonatum multiflorum/verticillatum* exist abundantly. *Aconitum* leave. *Valeriana jatamansi*, *Podophyllum hexandrum*, *Dioscorea deltoid*, and *Berberis vulgaris* have been recorded in few sites. Similarly, *Corydalis govaniana*, *Primula denticulate*, *Mentha longifolia*, *Viola serpents*, and *Valerian jatamansi* are restricted to moist and shady locations.

The distribution, pattern, and availability of valuable MAP species and other economically important forest resources are depending on grazing, harvesting intensity, habitat loss, and coniferous forest stand. For example, the population size of some aromatic and medicinal plants (*Dioscorea deltoidea*, *Podophyllum hexandrum*, *Viola serpens*, *Paeonia emodi*, *Valeriana jatamansi*, and *Berberis vulgaris*) are under high threat near residential area due to overgrazing and anthropogenic activities. Higher rates of grazing and unsustainable harvesting were found in residential areas due to the rapid destruction of habitat for expansion of settlement and agriculture purposes. Population size of these species was increasing with the rise in altitude and in areas with dense coniferous forest. Locals have revealed the availability of these forest resources being very highly 30 years ago.

Berberis vulgaris

Bergenia ciliata

Digitalis purpurea

Dioscorea Deltoidea

The Relation Between Exploitation of MAPs and Socioeconomic Factors

The correlation analysis was conducted between exploitation and socioeconomic factors of the study area. The results' parameters showed significant negative correlation with the family education level, which is defined in terms of family education index (FEI). FEI simply indicates that harvesting of MAPs decreases as the education level of family increases (Table 3). Farmers were found more involved in the harvesting of MAPs. It was also found that the size of the landholding has a significant negative correlation with the harvesting of MAP species. People having enough food products and sources of income and do not prefer to practice the harvest and use of MAPs. Resultantly, the period of food shortage has a significant positive correlation with the harvesting of MAPs. In addition, people with large livestock herds have a significant positive correlation as they were harvesting more MAP species more than others. The movements of livestock herders were found different at different altitudes in different seasons. Household income from business has revealed significant negative correlation. Harvestings of MAPs are found in those people who are also having an alternate source of cash income. Because harvesting of MAPs is not productive like other businesses as results are shown in table.

Deterioration of Ethnobotanical Species

Ethnobotanical information discusses the use of plants in a definite culture which has been experienced throughout generations. Inhabitants of the area largely depend on this plant association. Existing plant species do serve as multipurpose species: a cure for ailments, fuel for cooking, wood for construction food and fodder for humans and livestock, respectively. In spite of such dependence, these important renewable natural resource has not been managed scientifically for its sustainable production. Consequently various species are under risk of extinction in specific

Table 3 Correlation coefficients between exploitation of MAPs and socioeconomic factors of medicinal and aromatic plants

Socioeconomic variables	Correlation coefficients	Significance level
Education index (family)	−0.399	0.020*
Income generated from labor work	−0.325	0.010*
Land ownership	−0.412	0.050*
Income (total)	−0267	Not significant
Food shortage period	−0.423	0.020*
Livestock unit	−0.456	0.010*

D.F = 33, NS = Non significant, *= Significant at 95% confidence interval regions of the world (Sher et al. 2017; MacDicken 2015). Overexploitation at some places has led to a state where certain plants have acquired the status of endangered or even near extinction.

National flora has drastically deteriorated due to unchecked and unplanned exploitation in the form of grazing beyond carrying capacity, illegal cutting, frequent collection of medicinal plants, urbanization, and exploitation of agriculture lands at the expense of forest ecosystem. Sher et al. (2015) reported that prevailing activities like ruthless utilization of medicinal plants, nonsystematic grazing, and harvesting had degraded the environment. Further, it has raised the endangered species in the fragile ecosystem. Advance research in modern science has discovered multiple medicinal effects of the existing plants. This development, in turn, has further raised the frequency of cutting such tree species. It became essential to take immediate steps for conservation and stabilization of trees as trees could play a multipurpose role as well as medicinal and fodder. Being a common property resource, active participation of the local communities to regenerate the ecosystem under stress is very important.

Nutritional and Economic Value

It was revealed that multiple tree species have significant nutritional and economic value for local villagers. The communities residing in the area draw benefits of major and minor values from different species. Al-Yemeni and Sher (2010) reported that in Sri Lanka, food scarcity had made bread fruit (*Artocarpus otitis*) as a supplement food and even a substitute for a staple like rice. Numerous local fruit trees had been used for the horticultural improvement of common varieties. Worth mentioning here is the use of root stock from such an indigenous flora. An extensive survey should be carried out regarding nutritious species and their role in the daily life of the people. Species like *Juglans regia*, *Ficus carica*, *Diospyros lotus*, *Pyrus communis*, *Morus nigra*, and *Morus alba* are considered as nutritious species, and steps should be taken for the conservation of these species.

There is a dire need to protect such deteriorating ecosystem to maintain the balance of a wide range of biodiversity and for the ultimate safe survival of our own generations to come. Sher et al. (2014, 2015) also reported that northern areas of Pakistan, including Swat, are major production sites for medicinal plants of economic value. Trade business of these plants, as noticed in this study, is not simple as it involves many stockholders. It includes a chain of beneficiaries, right from collectors to middlemen to retailers and wholesalers. A large share of profits is fetched by the wholesalers. Price of the production will considerably rise, provided it gets an opportunity to access the international market.

Conclusion

Rosa damascene and *Rheum webbianum* are abundantly available in the Murree area. Chinese scientists have been treating cancer with the oil extracted from the *Rheum webbianum* species. Juice of fruits has a nutritional value and is taken as a tonic. Furthermore, by developing small-scale enterprise like preparation of jams and pickles, China has been carrying an annual profit of five million US dollars from this single species. *Rheum webbianum* harness similar scope in Pakistan, particularly in Murree and allied areas. A multinational and integrated endeavor is required to reverse the process of environmental degradation. It should keep in front sociocultural attitudes and economic uplift of the local people while devising rehabilitation strategies. From a preservation point of view, in the initial phase, it would be urgent and appropriate to establish conservation plots of adequate dimensions to protect and nurse the medicinal and aromatic plants. Establishing community-based small-scale enterprise is essential to local conservation. Profit sharing within the community will raise the feeling of ownership regarding common property resource. This part leads to creating awareness and inculcates knowledge regarding potential medicinal and aromatic plant resources, in particular, and forest resources. Documentation of local knowledge about medicinal plants transferring from older generations to their descendants must also be prioritized.

References

Abbasi AM, Khan MA, Ahmed M, Zafar M (2010) Herbal medicines used to cure various ailments by the inhabitants of Abbottabad district, North West Frontier Province, Pakistan. Indian Journal Traditional Knowledge 9:175–183

Ahmed E, Arshad M, Saboor A, Qureshi R, Mustafa G, Sadiq S, Chaudhari SK (2013) Ethnobotanical appraisal and medicinal use of plants in Patriata, New Murree, evidence from Pakistan. J Ethnobiol Ethnomed 9:13

Azizol AK, Chang YS (2000) Production of quality herbal medicine towards the 21st century

Bussmann R, Sharon D, Ly J (2008) From garden to market? The cultivation of native and introduced medicinal plant species in Cajamarca, Peru and implications for habitat conservation. Ethnobot Res Appl 6:351–361

Casagrande, D. G. (2002) Ecology, cognition, and cultural transmission of Tzeltal Maya medicinal plant knowledge. Uga

de Boer HJ, Lamxay V, Björk L (2012) Comparing medicinal plant knowledge using similarity indices: a case of the Brou, Saek and Kry in Lao PDR. J Ethnopharmacol 141:481–500

Hamilton A (2013) Plant conservation: an ecosystem approach. Routledge

Joppa LN, Roberts DL, Pimm SL (2010) How many species of flowering plants are there? Proc Royal Soc Lond B Biol Sci rspb20101004

Khan B, Abdukadir A, Qureshi R, Mustafa G (2011) Medicinal uses of plants by the inhabitants of Khunjerab National Park, Gilgit, Pakistan. Pak J Bot 43:2301–2310

Lange D (1997) Trade in plant material for medicinal and other purposes. Traffic Bull 17:21–32

Larsen HO, Olsen CS (2007) Unsustainable collection and unfair trade? Uncovering and assessing assumptions regarding Central Himalayan medicinal plant conservation. Biodivers Conserv 16:1679–1697

MacDicken KG (2015) Global forest resources assessment 2015: what, why and how? Forest Ecol Manage 352:3–8

Ntie-Kang F, Lifongo LL, Mbaze LMA, Ekwelle N, Owono LCO, Megnassan E, Judson PN, Sippl W, Efange SMN (2013) Cameroonian medicinal plants: a bioactivity versus ethnobotanical survey and chemotaxonomic classification. BMC Complement Altern Med 13:147

Orr DJ, Jernigan KA, Belichenko OS, Kolosova VB (2017) Naukan ethnobotany in post-Soviet times: lost edibles and new medicinals. J Ethnobiol Ethnomed 13:61

Paton AJ, Brummitt N, Govaerts R l, Harman K, Hinchcliffe S, Allkin B, Lughadha EN (2008) Towards target 1 of the global strategy for plant conservation: a working list of all known plant species—progress and prospects. Taxon 57:602–611

Sher H, Aldosari A, Ali A, de Boer HJ (2014) Economic benefits of high value medicinal plants to Pakistani communities: an analysis of current practice and potential. J Ethnobiol Ethnomed 10:71

Sher H, Aldosari A, Ali A, de Boer HJ (2015) Indigenous knowledge of folk medicines among tribal minorities in Khyber Pakhtunkhwa, northwestern Pakistan. J Ethnopharmacol 166:157–167

Sher H, Al-Yemeni M, Sher H (2010) Forest resource utilization assessment for economic development of rural community in northern parts of Pakistan. J Medicinal Plants Res 4:1786–1789

Sher H, Barkworth ME, de Boer HJ (2017) Medicinal and aromatic plant cultivation in the swat valley, North-Western Pakistan, for economic development and biodiversity conservation. Genet Resour Crop Evol 64:237–245

Voeks RA (2009) Traditions in transition: African diaspora ethnobotany in lowland South America. In: Alexiades M (ed) Mobility and migration in indigenous Amazonia: contemporary ethnoecological perspectives, vol 11. Berghahn, London, pp 275–294

Plant Resources and Human Ecology of Tarnawai area, District Abbottabad, Pakistan

Raheela Taj, Inayat Ur Rahman, Abbas Hussain Shah, Shujaul Mulk Khan, Aftab Afzal, Niaz Ali, Zafar Iqbal, and Farhana Ijaz

Introduction to Human Ecology and Natural Resources

Tarnawai is one of the major part of the Union Council Banda Pir Khan of Abbottabad District. It is located at an elevation of 1524 m. It is at distance of about 27 km from Abbottabad city. It is situated between 34°16′15’ North and 73°17′47’ East. Banda Pir Khan is one of the 51 union councils of Abbottabad District in KPK, Pakistan. It is located in the north west of Abbottabad city near the border with Mansehra District. According to the 1998 Census, the total population of Banda Pir Khan is 25,385. This union council consists of the following areas: Banda Pir Khan, Mandroch, Samesar, Bandi Dhundan, Tarnawai and Garlaniyan. Qalandarabad is the main junction of the union council. Abbottabad District takes its name from the city of Abbottabad named after Major James Abbott, the first deputy commissioner of Hazara who served from 1849 to 1853 under the British colonial administration of India. The area is scenic and its location at the base of the Himalayas. It has a temperate climate almost throughout the year. It is situated between 33°50′ and 34°23’ north and 73°35′ and 73°31′ east. Abbottabad is bordered by Mansehra District in the north, Muzaffarabad and Rawalpindi districts in the east, Haripur and Rawalpindi districts in the south, and Haripur district in the west. It is spread over an area of 1967 km^2. The average elevation of peaks ranges from 2500 to 2700 m in the district. These mountains form a part of lesser Himalayas and dominate the

R. Taj · I. U. Rahman · A. Afzal · N. Ali · Z. Iqbal · F. Ijaz
Department of Botany, Hazara Universit, Mansehra, Pakistan
e-mail: hajibotanist@hu.edu.pk

A. H. Shah
Department of Botany, GPGC, Mansehra, Pakistan

S. M. Khan (✉)
Department of Plant Sciences, Quaid-i-Azam University, Islamabad, Pakistan
e-mail: smkhanqau@qau.edu.pk, https://www.qau.edu.pk/profile.php?id=804024

M. Ozturk, K. R. Hakeem (eds.), *Plant and Human Health, Volume 1*,
https://doi.org/10.1007/978-3-319-93997-1_22

landscape (Pastakia 2004). Miranjani and Mushkpuri are the highest peaks of the district. The soil of Abbottabad is rich in minerals, containing biotite, granite, limestone, slate, soapstone, quartz, etc. Phosphate stones are extracted from Kakool and Tarnawai. Most of the soil is gray in color (under moist forest) and coarse in texture. Abbottabad has a humid subtropical climate, with mild to warm temperatures during spring and autumn, hot temperatures during June and July, and cool to mild temperatures during winter. The temperature rises as high as 38 °C (100 °F) during mild summer months and drops below -5 °C (23 °F) during extreme cold waves. Snowfall occurs occasionally in December and January. Abbottabad District lies within active monsoon zone and due to which receives heavy rainfall in the monsoon. The population of the Abbottabad District was 881,000 according to the 1998 Census. The most frequently spoken language in Abbottabad is Hindko and Gujri. Besides this, Panjabi, Pashto, and Urdu are also spoken here. Most of the district's tribes belong to Abbasi, Dhund, Sardar, Jadoon, Karlal, Syed, and Tanoli tribes.

The area of Abbottabad District is 1967 km^2 and is situated in the Orash Valley at an elevation of 4120 feet (1260 m). It is located in predominantly mountainous terrain. The average peak elevation in the district ranges from 2500 to 2700 m. The elevation of the Miranjani peak is 3313 m, and it is the highest peak (Pastakia 2004). On the north of Abbottabad side, Kaghan valley is located.

The soil of Abbottabad is rich in minerals. It contains deposits of limestone, granite, phosphate, and soapstone. These minerals are deposited in the hills as well as deposited on the valley floors. The soil is coarse in texture and gray in color. The big mines of phosphate are present in Abbottabad. The phosphate stones are extracted from Kakool and village Tarnawai.

Abbottabad lies in the humid subtropical zone. Abbottabad's summer starts a little late and is relatively mild. Temperature begins to rise in the month of May and begins to drop in the months of September and October. In severe winter, heavy snowfall occurs in the higher elevations causing the snowline to drop to around 1650 m. The average precipitation received by land is 60% during July to August and 40% received during September and June. The population of Abbottabad District was 881,000 in the 1998 Census, and according to 2017, the urban population of Abbottabad is 157,699, and the population of rural area is 723,301.

The Abbottabad District has high standard educational institutions, and people are attracted toward its educational systems all over Pakistan. Sometimes Abbottabad is also known as "The city of Schools." It is a home to a number of training institutes, schools, and colleges. The literacy ratio is 56.6% in district, which includes 74.5% male and 39.1% female. The local language in Abbottabad is Hindko. It is spoken by 94% of the rural population and 75% by urban population. Pashto and Urdu also have been spoken by urban population. According to the 1998 Census, overall employment in the district is as high as 31.13%. 31.84% of the district's men and only 1.05% of women are unemployed. 30.4% of the rural people are unemployed while 33.5% of the city dwellers are jobless. About 1072 health institutes have been established in Abbottabad District. Poverty is prevalent in the district and its proportion is assuming with the passage of time. About 51% of the district's population lies below the poverty line, and 30% of the population belongs to middle group.

Abbottabad District accounts for 5.4 of the province forest resources. Forests of Abbottabad District cover 36,394.6 acres. Parks and protected areas are also present here, e.g., Ayubia National Park and Qalandarabad game reserve area. Tarnawai game reserve forest is also present in the study area. The economy of Abbottabad depends upon natural resources and agriculture. Small part of the land is under cultivation. Fruits and nut crops are also cultivated in Abbottabad, e.g., pear, plum, apricot, walnut, etc. Abbottabad is rich with highly medicinal plants with 43 varieties. Abbottabad has been attracting the tourist to its beautiful scenery since the colonial era, and it is a major transit point to all major tourist regions in Pakistan such as Ayubia, Nathiagali, Dunga Gali, Harnoi, etc. On October 8, 2005, Abbottabad District was badly affected by an earthquake. Although most of the citizens survived, many older buildings were destroyed or severely damaged. Flora and fauna of the area was also adversely affected.

Ethnobotany

Indigenous knowledge is as old as human civilization. The word "ethnobotany" literally means the study of botany of the primitive human race. Ethnobotany is the association between cultural anthropology and botany, investigating the uses of plants as medicine, nourishment, and natural resources. In 1986, the term ethnobotany was published by Hershberger and suggested that it is a field which "indicates the cultural positions of tribes who used the plants for food, shelter, or clothing." In this way this term began to be used, and a new field was opened.

Ethnobotany is the study about tribal and rural people for recording their unique information about plant wealth and for search of sources of herbal medicines, edible, and other aspects of plants (Jain 1991). In many cases the purpose of ethnobotanical studies by botanists has followed the tradition of early naturalist who reported the various uses of plants by primitive cultures for possible inclusion in home country economy. Schulte has drawn attention to his tradition by quoting Hershberger's advice that by studying primitive cultures, we can find the valuable characteristics they have used in their wildlife, which may fill some niche in our own.

Ethnobotany is an interdisciplinary science, which includes aspects of both the science and humanities. Ethnobotany serves as a gateway to various disciplines. The listing of plants of ethnobotanical value is important for evaluating human-plant relationship (Bye 1979) and for understanding the regional human ecology relations to their environment (Alcorn 1981). Thus ethnobotany has emerged as a multidisciplinary science involving sociology, ecology, botany, anthropology, forestry, agriculture, mythology, pharmacology, economics, literature, phytochemistry, and veterinary medicine.

Pakistan has nine different ecological zones, which have over 6000 unique plant species. Due to its climate, Pakistan is quite rich in medicinal plants mostly herbs which are spread over a large area. These herbs are mostly growing in the wild, and methodical selection also has been made to collect and cultivate them.

About 5691 known species of flowering plants are present in Pakistan, and 1010 of them can be classified as medicinal (Stewart 1972). In Pakistan mostly medicinal herbs are primarily used by Tibbi Dawakhanas. Allopathic herbal medicines are prepared from these medicinal plants and have high cost. Medicinal plants are also used as a traditional medicine in rural areas as a more appropriate method of treatment.

Local people are fundamentally involved in all stages of research, so it is a better chance of "buy in" and more robust solutions. Among the parts of plant species, seeds, fruits, flowers, leaves, roots, bark, frond, tubers, rhizome, and bulbs are commonly used. A Hazara region is the most important market for crude drugs. These drugs are easily available on cheap prices. Among the commercially exploited drugs, *Viola canescens*, *Berberis lycium*, *Papaver somniferum*, *Bergenia ciliata*, *Acacia modesta*, and *Pistacia chinensis* fetch a high price. Rates of *Viola canescens*, *Berberis lycium*, *Acacia modesta*, and *Bergenia ciliata* are increasing day by day due to decline in availability. Dafni et al. (2005) conducted a survey on ethnobotany of *Ziziphus spina-christi* in the Middle East for various aspects. Historical, religious, linguistic, as well as pharmacological evidences from Muslims, Jews, and Christians customs suggested that this is the only tree considered "Holly" by Muslims in addition to its status as "sacred tree" in the Middle East. It has also a special status as a "blessed tree" among Druze.

The study area being the part of Hazara division shows rich floristic diversity, and this study will explore the ethnobotanical resources of the area and prove very much helpful in future research as well.

The study area was selected for ethnobotanical study because this area was unexplored due to difficult paths and shortage of a lot of facilities, e.g., roads, vans, hotels, restaurants, etc. So, the current study was designed to explore the floral checklist and document the indigenous knowledge of ethnoecologically important species of the study area.

Field Survey and Field Work

A field survey was conducted during 2016–2017 to access the different plants in Tarnawai, Abbottabad District. A total of eight trips were made to the study area for the collection of data regarding plant species. A questionnaire method was adopted to collect information about plants from local communities. Face-to-face interviews were also taken from different peoples and data collected from different places in the study area. Data were based on common name of the plant, botanical name, family, habit, habitat, locality, part used, and pattern of use. Interviews were taken randomly from 89 people including males and females belonging to different tribes from different localities. The areas under survey were Biyang, Chikiar, Gidar banna, Kalu Bandi, Rarkandi, and Jabbri in Tarnawai, Abbottabad. Field notebook, pen or pencil, polythene bags, tags, trowel, scissors, camera, newspaper, presser, cutter, umbrella, and GPS were used during the survey (Ijaz 2014).

The plants were collected in spring season (March to April) and summer season (June to July). The plants were properly dried and pressed by using a newspaper for about 2–4 weeks at normal temperature. Then the plants were treated or poisoned with chemical solution for preservation and mounted on standard herbarium sheets. Then data were shifted from a field notebook on herbarium label of standard herbarium sheets. This herbarium label is always pasted on the right side of the herbarium sheets (Ijaz 2014; Rahman et al. 2016a, b). The size of the herbarium sheet is standardized (11.5 × 16.5) (Ijaz 2014). The specimens were identified by using relevant materials and Flora of Pakistan (Stewart 1967–72). Plants species were also photographed. The specimens were also verified. The specimens were deposited in the Herbarium, Department of Botany, Govt. Post Graduate College, Mansehra (Pakistan).

Survey on Cultural Knowledge

The culturally significant plants were collected from Tarnawai area. The plants were classified on the basis of their utility in the area. For ethnobotanical information of the area, local people were interviewed. A total of 89 people were interviewed including 51% female, followed by 45% male and 4% hakims. A questionnaire was constructed, and a survey was conducted at different localities in the study area. The data noted on questionnaires were according to field observations (Ijaz 2014).

Documentation and Ethnobotanical Exploration

Field visits were undertaken together with local communities to identify the plant specimens of household importance, medicinal importance, and economic value. Plant species used for different purposes were recorded along with the data on their local names, parts used, and modes of utilization. A comprehensive list was prepared. In order to verify the information, queries were made for the same plant and same uses, from different individuals at different areas.

Statistical Analysis

The recorded data were analyzed via using statistical indices, Use Value Index (UVI), and Relative Frequency Citations (RFCs) (Ijaz et al. 2016; Rahman et al. 2016a, b).

Use Value Index (UVi)

To know about the importance of ethno species, the use value index (UVi) was applied to give a quantitative form of its relative importance to the plants neutrally (Philips et al. 1994; Rahman et al. 2016a, b). This study evaluates the relative importance of each ethno species, based on relative use between informants. The use value was calculated by applying the following formula.

$$\mathrm{UVi} = \sum \mathrm{Ui} / \mathrm{Ni}$$

Ui is the number of use for a given plant species reports cited by each informant.
N is the total number of informants interviewed for a given plant species.

Relative Frequency Citations (RFCs)

Relative frequency of citation was used to determine the local importance of each plant species by using the following formula (Rahman et al. 2016a, b).

$$\mathrm{RFCs} = \mathrm{FCs} / N$$

FCs is the number of informants who mentioned the use of plant species.
N is the total number of informants.

Observations

Ethno-ecological data was recorded from 89 informants including 40 (45%) male, 45 (51%) female, and 4 (4%) hakims. These informants are belonging to different tribes such as Sardar, Karlal, Awan, Syed, Abbasi, and Dhund tribes. The most frequently informants belong to the Sardar Gujjars families. Old age people have more information about traditional plants than the youngsters and non-educated.

Floristic Diversity

In the present study, 71 plant species were documented from the study area. These are being potentially used ethno-ecologically by local communities. Based on habit form these species can be categorized into 18 trees (25% share), 8 shrubs (11% share), and 45 herbs (64% share) (Fig. 1). The recorded plant species are

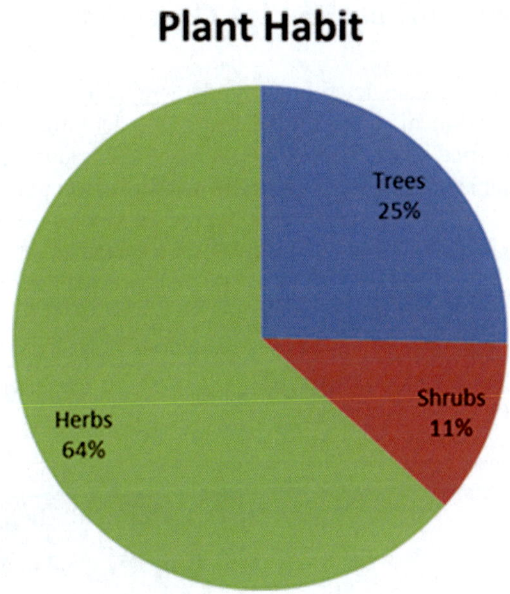

Fig. 1 Plant habit of the recorded plant species of the study area

belonging to 37 families and 63 genera. In study area, a total of 37 families are recorded. Rosaceae is the largest represented plant family with 8 plant species followed by Poaceae (5 spp.), Asteraceae (4 spp.), Brassicaceae (3 spp.), Lamiaceae (3 spp.), Solanaceae (3 spp.), Moraceae (3 spp.), Fabaceae (3 spp.), Amaranthaceae (2 spp.), Euphorbiaceae (2 spp.), Polygonaceae (2 spp.), Umbelliferae (2 spp.), Pinaceae (2 spp.), Rutaceae (2 spp.), Caryophyllaceae (2 spp.), Liliaceae (2 spp.), Rhamnaceae (2 spp.), Pteridaceae (2 spp.), Berberidaceae (1 sp.), Saxifragaceae (1 sp.), Punicaceae (1 sp.), Rubiaceae (1 sp.), Meliaceae (1 sp.), Valerianaceae (1 sp.), Violaceae (1 sp.), Vitaceae (1 sp.), Fagaceae (1 sp.), Oleaceae (1 sp.), Juglandaceae (1 sp.), Ebenaceae (1 sp.), Oxalidaceae (1 sp.), Malvaceae (1 sp.), Plantaginaceae (1 sp.), Cannabaceae (1 sp.), Convolvulaceae (1 sp.), Fumariaceae (1 sp.), and Chenopodiaceae (1 sp.) in terms of species utilization by the local communities. A total of 63 genera have been identified from the study area, including family 4 from family Rosaceae, followed by families like Poaceae (5 genera), Asteraceae (4), Brassicaceae (3), Solanaceae (3), Fabaceae (3), Lamiaceae (3), Moraceae (2), Pteridaceae (2), Liliaceae (2), Amaranthaceae (2), Rutaceae (2), Caryophyllaceae (2) Euphorbiaceae (2), Umbelliferae (2), Punicaceae (1), Rhamnaceae (1), Valerianaceae (1), Rubiaceae (1), Meliaceae (1), Violaceae (1), Polygonaceae (1), Vitaceae (1), Fagaceae (1), Oleaceae (1), Juglandaceae (1), Ebenaceae (1), Berberidaceae (1), Saxifragaceae (1), Punicaceae (1), Oxalidaceae (1), Malvaceae (1), Plantaginaceae (1), Cannabaceae (1), Convolvulaceae (1), Fumariaceae (1), and Chenopodiaceae (1).

Ethno-Ecological Use Categories

Survey conducted in the study area shows the highest number of plant species are used medicinally by local people belonging to diverse families. This survey was carried out in the field to document the traditional knowledge pertinent to ethnobotanical and ethnoecological uses prevailing in the area. The documented data revealed that a maximum number of species were reported to be used for medicinal purposes, 43 (43%) followed by vegetables, 20 (19%), fodder, 13 (12%), fuel, 15 (14%), timber, 12 (11%), and ornamental purposes, 3 (3%) as mentioned by the informants (Figs. 2, 3, 4, and 5).

Fig. 2 Ethnoecological categories of the recorded plant species of the study area

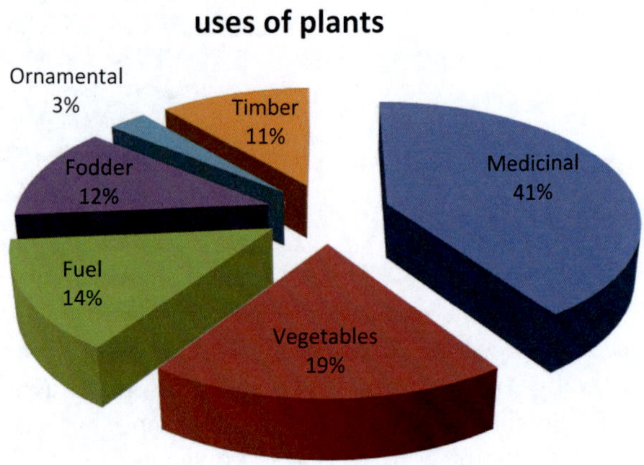

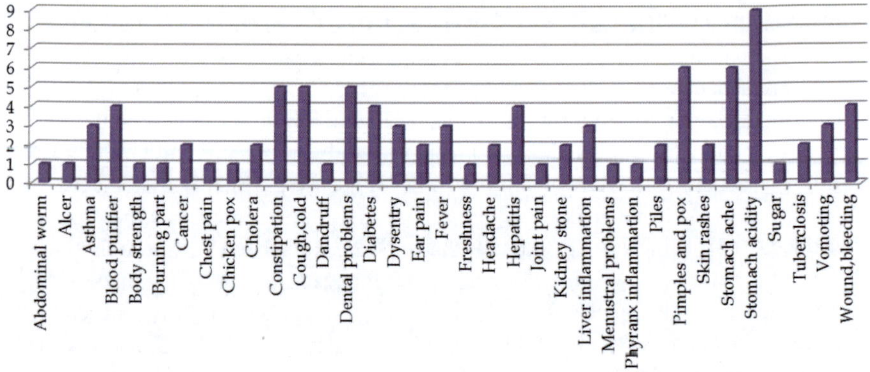

Fig. 3 Range of ethnobotanical categories

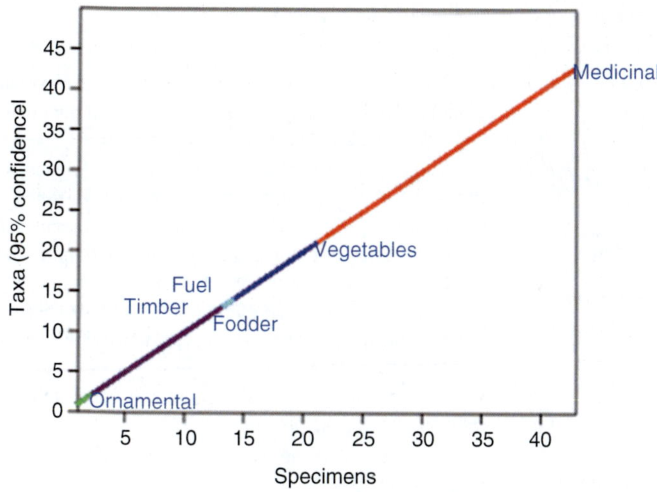

Fig. 4 Distribution of species in association with ethnobotanical categories

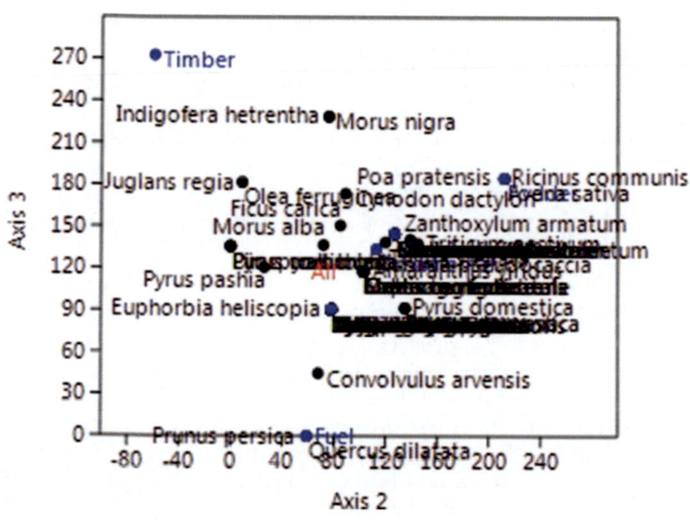

Fig. 5 Correlation among ethnobotanical categories

Classification of Diseases

The reported plant species in the area shows that 43 plants species are used differently for 35 various diseases by inhabitants (Tables 1 and 2). Local people of the study area use plants for various purposes i.e., stomach acidity (9 species), stomachache (6 species), pimples and skin pox (6 species), dental problems (5 species), cough and cold (5 species), constipation (5 species), diabetes (4 species), hepatitis (4 species), wound and bleeding (4 species), blood purifier (4 species), dysentery (3

Table 1 Folk recipes of ethnomedicinal plant species of the study area

S. No	Botanical name	Folk recipe
1	*Achyranthes aspera* L.	Fresh juice of leaves is applied in piles and other wounds. Ash of plant is added to honey and used in treatment of cough and asthma. Roots are also used for cleaning teeth
2	*Adiantum raddianum* C. Presl.	Stem and leaves are crushed into powder, and mixed with water to use for hepatitis, stomach burn, heartburn, and headache
3	*Ajuga integrifolia* Buch.-Ham.	Leaves are crushed and mixed with water and then used as a juice before breakfast for stomach acidity and blood purification, and juice is also put in the ear for ear pain
4	*Aloe vera* (L.) Burm.f.	Juice of stem is used as a blood purifier and also used for stomach acidity and stomach burn. Juice is also taken for diabetes, cancer, and pimples on face
5	*Amaranthus viridis* L.	Aerial part of plant is cooked and used as a vegetable to cure the inflammation of the pharynx
6	*Artemisia absinthium* L.	Leaves are crushed and made into paste and placed under the teeth for cavities
7	*Berberis lycium* Royle.	Roots are dried, crushed, and made into powder and applied on wounds and also used for bone pain and joint pain. Bark of roots are put off and chewed for cough
8	*Bergenia ciliata* (Haw.) Sternb.	Roots are crushed into powder, and mixed with water to use for stomach burn and heartburn problems
9	*Cannabis sativa* L.	Leaves are crushed and mixed in water to make juice, added with milk and use to feel high and relax
10	*Capsella bursa-pastoris* L.	Whole plant is crushed, and one spoon of powder leaves are taken every month for menstrual cycle when too early (8, 10, 15 days early)
11	*Capsicum annum* L.	Seeds of fruits are placed under the teeth for tooth pain and cavities
12	*Centaurea iberica* Trevir. ex Spreng.	Plant fresh leaves are ground and mixed with water for use in case of gastric problems
13	*Citrus medica* L.	The lemon juice is mixed in curee (chicken soup) and taken as blood purifier. The fruit is rubbed on the face to remove pimples and on the face skin
14	*Convolvulus arvensis* L.	Roots are crushed to make a shake for use in case of diarrhea and stomachache. Hairs are washed with juice of *Convolvulus arvensis* to remove dandruff, and the roots can also be used to treat liver inflammation
15	*Cichorium intybus* L.	Roots are boiled in water to make tea for vomiting and fever. Roots are soaked in water during the night and then the juice is taken before breakfast for diabetes
16	*Cynodon dactylon* L.	Whole plant is crushed and taken with water for vomiting
17	*Dryopteris patens* (SW.)D. Ktze., C. Chr.	Leaves of *Dryopteris* are crushed and applied on burning part of the body
18	*Euphorbia helioscopia* L.	Plant milk is applied on skin and used to cure skin rashes. It is also used for cholera

(continued)

Table 1 (continued)

S. No	Botanical name	Folk recipe
19	*Foeniculum vulgare* Mill.	Edible fruit helps in digestion
20	*Fumaria indica* (Hauskn) Pugsley.	Whole plant is crushed to make powder and a spoon of that powder is taken with honey to cure vomiting
21	*Galium aparine* L.	Fresh plant juice is used to cure wounds, ulcers, and other skin problems
22	*Juglans regia* L.	Leaves and twigs of *Juglans regia* are used as a toothbrush for cleaning the teeth
23	*Malva neglecta* L.	The older leaves are boiled in water and then made into paste and used for headache
24	*Medicago truncatula* Gaerth.	Leaves are used as vegetables to cure the chest pain, cough, and cold
25	*Melia azedarach* L.	Fruits are crushed to make powder to be taken with water for constipation and piles. Fruits are also used in making shampoo
26	*Mentha longifolia* (L.) L.	Whole plant is dried and crushed to make powder that is used then in case of stomach acidity, stomach pain, cholera and dysentery. Leaves are crushed and mixed with yogurt for use as a food supplement
27	*Morus alba* L.	Edible fruits are used to cure the stomach acidity
28	*Nasturtium officinale* R. Br.	Leaves are cooked and eaten as vegetables for the death of cancerous cells
29	*Olea ferruginea* Royle.	Leaves are boiled and made into tea and taken for cough and cold
30	*Oxalis corniculata* L.	Leaves are used as vegetables eaten for stomach acidity and dysentery
31	*Plantago lanceolata* L.	Leaves are crushed and mixed into water to make juice to cure cholera, nausea, and dysentery. Juice is also used for ear pain
32	*Punica granatum* L.	Carp of fruit is crushed, made into powder, and mixed with water to cure dysentery and cholera. Seeds of fruits and green chillies are crushed and used as "chatni" in meal
33	*Pyrus pashia* Buch.-ham. ex D. Don.	Fruits are eaten to cure kidney stone and for dysentery
34	*Rubus ellipticus* Sm.	Roots of *Rubus ellipticus* and *Rumex hastatus* are boiled and used as a tea for hepatitis B and C
35	*Rumex hastatus* L.	Roots are boiled and made into tea and used for hepatitis, tuberculosis, and kidney stone. Stem and leaves are crushed into powder and applied on wounds and also used to control bleeding
36	*Solanum nigrum* L.	Leaves are crushed and mixed in water and used as juice for abdominal insects in children
37	*Triticum aestivum* L.	Grains of wheat are soaked in water, crushed, and then made into powder called "nishasta" used for body strengthening
38	*Valeriana wallichii* DC.	Leaves are crushed and made into powder and used with water for skin rashes and skin pox. Powder of roots are mixed with flour and given to cattle for tuberculosis

(continued)

Table 1 (continued)

S. No	Botanical name	Folk recipe
39	*Viola canescens* wall.	Flowers are boiled and used as tea for cough, cold, flu, and asthma
40	*Vitis vinifera* L.	Dried fruits of *Vitis vinifera* are eaten to cure chickenpox
41	*Zanthoxylum armatum* DC.	Twigs are used as a brush for cleaning teeth. Fruits are ground with *Punica granatum's* seeds and leaves of mint to make "chatni," to reduce stomach acidity and stomachache
42	*Ziziphus nummularia* (Burm.f.) Wight Qs Arn.	Leaves are crushed and mixed in water to make juice and is then used to reduce high sugar level
43	*Ziziphus oxyphylla* Edgew.	Roots are crushed into powder, and mixed with water to cure hepatitis. Leaves are crushed and applied on skin pox and skin tumor as well

species), asthma (3 species), liver inflammation (3 species), vomiting (3 species), fever (3 species), kidney stone (2 plant species), cholera (2 species), tuberculosis (2 species), skin rashes (2 species), headache (2 species), ear pain (2 species), cancer (2 species), piles (2 species), sugar (1 species), joint pain (1 species), burning part (1 species), abdominal worm (1 species), chickenpox (1 species), chest pain (1 species), menstrual problem (1 species), freshness (1 species), dandruff (1 species), pharynx inflammation (1 species), body strengthening (1 species), and ulcer (1 species) (Fig. 6).

Among vegetables, *Solanum nigrum*, *Malva neglecta*, *Lamium amplexicaule*, *Oxalis corniculata*, *Traxicum officinale*, *Medicago truncatula*, *Nasturtium officinale*, *Ficus carica*, *Capsicum frutescens*, *Amaranthus viridis*, and *Zanthoxylum armatum* have medicinal importances as well and hence used for by local communities for double purposes, e.g., to treat asthma, cough, fever, dental problems, ear pain, abdominal worms, death of cancers cells, pharynx inflammation, headache, and fever (Table 2). *Mentha longifolia* is used for digestive problems and also used in food. All the above plant species are wild but not cultivated specially for diseases except *Capsicum frutescens*, cultivated especially for vegetables. When all the above mentioned wild plant species would be cultivated on large scale by the inhabitants for vegetable and nmedicine purposes, this could be a profitable source as for their socioeconomies.

Prunus armeniaca, *Prunus domestica*, *Malus domestica*, *Rubus ellipticus*, and other many plant species were used as fodder for cattle in the study area (Table 3). Grazing is one of the common and usual practices for cows, goats, sheep, and buffaloes, and these domestic animals fulfill their dairy requirements from the local plant species as fodder and improve the local micro economy. As we know, grasses are the major source of fodder for cattles of the local people. Among grasses *Cynodon dactylon* is mostly grazed by the animals and is sometimes stored in dry form for winter use as well. *Triticum aestivum* (wheat) is the most important and cultivated crop in Pakistan. The fruits of various plants species are used by local inhabitants including *Ficus carica*, *Morus nigra*, *Morus alba*, *Malus domestica*,

Table 2 Ethnomedicinal plants of Tarnawai, Abbottabad District

S. No.	Botanical name	Vernacular name	Family	Habit	Part used	Medicinal use	UVi	RFCs
1	*Achyranthes aspera* L.	Puthkanda	Amaranthaceae	Herb	Whole plant	Respiratory problems, piles, wound, dental problems	0.71	0.27
2	*Adiantum raddianum* C. Presl	Kukwai	Pteridaceae	Herb	Stem, leaves	Hepatitis, stomach acidity, fever, and headache	0.69	0.22
3	*Ajuga integrifolia* Buch.-ham.	Kori boti, manji boti	Lamiaceae	Herb	Leaves	Stomach acidity, blood purifier, ear pain	0.89	0.35
4	*Aloe vera* (l.)Burm.f.	Koar gandal	Liliaceae	Herb	Stem	Blood purifier, digestive problems, anti-cancerous, diabetes	0.77	0.27
5	*Amaranthus viridis* L.	Chalaira	Amaranthaceae	Herb	Whole plant	For pharynx inflammation	0.71	0.22
6	*Artemisia absinthium* L.	Chauu	Asteraceae	Herb	Leaves	Dental problems	0.59	0.22
7	*Berberis lycium* Royle.	Sumbal	Berberidaceae	Shrub	Root	Joint pain, cough, and healing wound	0.95	0.44
8	*Bergenia ciliata*(haw.) Sternb	Bhatpy	Saxifragaceae	Herb	Root	Stomach acidity	0.92	0.39
9	*Cannabis sativa* L.	Phang	Cannabaceae	Herb	Whole plant	Freshness, excitement	0.83	0.34
10	*Capsella bursa pastoris* (L.) Medik	Chambraka/ chapatra	Brassicaceae	Herb	Leaves	Menstrual problems	0.66	0.24
11	*Capsicum annuum* L.	Neli mirch	Solanaceae	Herb	Fruit	Dental problems	0.84	0.29
12	*Centaurea iberica* Trevir. ex Spreng	Kanda Boti	Asteraceae	Herb	Leaves	Gastric problems	0.30	0.12
13	*Citrus medica* L.	Nimboo	Rutaceae	Shrub	Fruit	Blood purifier, skin problems	0.70	0.24
14	*Convolvulus arvensis* L.	Erli	Convolvulaceae	Herb	Root	Anti-inflammation for liver, dandruff	0.76	0.29
15	*Cichorium intybus* L.	Kasni	Asteraceae	Herb	Root	Vomiting, fever, diabetes	0.85	0.33
16	*Cynodon dactylon* L.	Khabal	Poaceae	Herb	Whole plant	Vomiting, nausea	0.37	0.15
17	*Dryopteris patens* (SW.) O.Ktze.,C.Chr	Alfi	Pteridaceae	Herb	Leaves	Burn part of the body	0.33	0.10
18	*Euphorbia helioscopia* L.	Dodhal	Euphorbiaceae	Herb	Stem	Skin rashes, cholera	0.49	0.20
19	*Foeniculum vulgare* Mill.	Sonf	Umbelliferae	Herb	Fruit	Digestive problems	0.90	0.40

(continued)

Table 2 (continued)

S. No.	Botanical name	Vernacular name	Family	Habit	Part used	Medicinal use	UVi	RFCs
20	*Fumaria indica* (Hausskn) Pugsley.	Pappra	Fumariaceae	Herb	Whole plant	Blood purifier, vomiting	0.47	0.20
21	*Galium aparine* L.	Unknown	Rubiaceae	Herb	Leaves, stem	Ulcer, wound, skin problems	0.61	0.21
22	*Juglans regia* L.	Khorri	Juglandaceae	Tree	Leaves, stem	Dental cavities	0.98	0.46
23	*Malva neglecta* L.	Sonchal	Malvaceae	Herb	Leaves	Headache, fever	0.80	0.31
24	*Medicago truncatula* Gaerth.	Sinjii	Fabaceae	Herb	Leaves, stem	Respiratory diseases	0.81	0.34
25	*Melia azedarach* L.	Daraik	Meliaceae	Tree	Fruit	Gastrointestinal problems	0.73	0.33
26	*Mentha longifolia* (L.) L.	Poodna	Lamiaceae	Herb	Leaves	Digestive problems	0.90	0.40
27	*Morus alba* L.	Chitta tut	Moraceae	Tree	Fruit	Stomach acidity	0.71	0.33
28	*Nasturtium officinale* R.Br.	Tarameera	Brassicaceae	Herb	Leaves	Anti-cancerous	0.76	0.28
29	*Olea ferruginea* Royle.	Kauu	Oleaceae	Tree	Leaves	Respiratory problems	0.77	0.34
30	*Oxalis corniculata* L.	Khatkurla	Oxalidaceae	Herb	Leaves, stem	Digestive problems	0.42	0.16
31	*Plantago lanceolata* L.	Chamchipatra	Plantaginaceae	Herb	Leaves	Ear pain, gastrointestinal problems	0.84	0.36
32	*Punica granatum* L.	Drunna	Punicaceae	Shrub	Fruit	Antidiarrheal, dysentery, cholera	0.91	0.44
33	*Pyrus pashia* Buch.-ham.Ex D. Don.	Batangi	Rosaceae	Tree	Fruit	Kidney stone and dysentery	0.84	0.29
34	*Rubus ellipticus* Sm.	Garacha	Rosaceae	Shrub	Root	Hepatitis B and C	0.38	0.15
35	*Rumex hastatus* L.	Khatimal	Polygonaceae	Herb	Root, stem, leaves	Hepatitis, tuberculosis, kidney stone, wound, and bleeding	0.89	0.28

S. No.	Botanical name	Vernacular name	Family	Habit	Part used	Medicinal use	UVi	RFCs
36	*Solanum nigrum* L.	Kachmach	Solanaceae	Herb	Leaves	Diabetes, abdominal worm	0.76	0.33
37	*Triticum aestivum* L.	Karrank	Poaceae	Herb	Seed	Body strengthened	0.72	0.24
38	*Valeriana wallichii* DC.	Mushk bala	Valerianaceae	Herb	Leaves, root	Skin rashes and skin pox	0.54	0.15
39	*Viola canescens* wall.	Banafsha	Violaceae	Herb	Flower	Cough, cold, and asthma	0.76	0.31
40	*Vitis vinifera* L.	Daakh	Vitaceae	Herb	Fruit	Chickenpox	0.61	0.25
41	*Zanthoxylum armatum* DC.	Timber	Rutaceae	Shrub	Fruit, stem	Stomachache, stomach acidity, dental problems	0.81	0.33
42	*Ziziphus nummularia* (Burm.f.) Wight Qs Arn.	Sinjli	Rhamnaceae	Tree	Fruits, leaves	Sugar	0.69	0.25
43	*Ziziphus oxyphylla* Edgew.	Phitni	Rhamnaceae	Shrub	Root, leaves	Hepatitis, skin pox	0.42	0.11

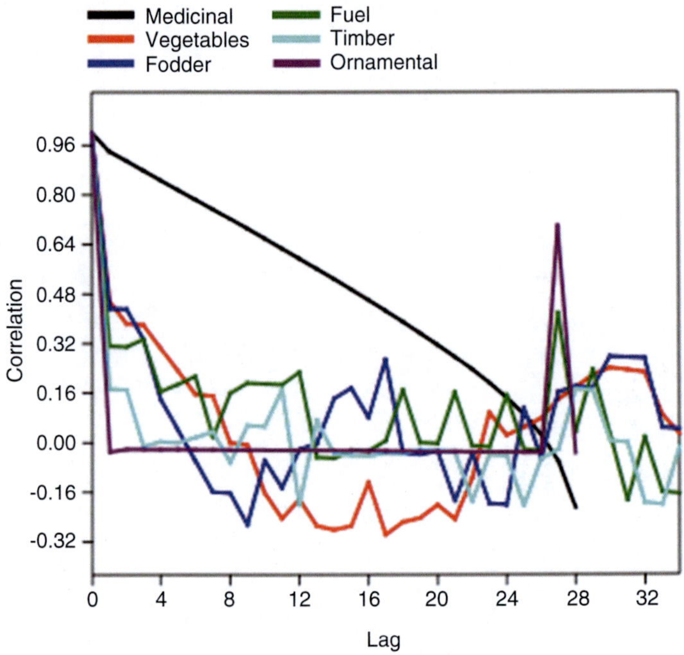

Fig. 6 Various ailments treated by plant species of the study area

Prunus persica, *P. domestica*, *P. armeniaca*, *Pyrus communis*, and *Vitis vinifera* in both dried and fresh form. Presently these species are under pressure due to deforestation. *Juglans regia*, *Pinus wallichiana*, *Pinus roxburghii*, *Quercus* spp., *Diospyros lotus*, and many other plants are used for timber as well as for fuel purpose by the inhabitants of village Tarnawai. Among ornamental plants, *Traxicum officinale*, *Tulipa stellata*, and *Robinia pseudoacacia* occur frequently in the study area.

Statistical Analysis

Use Value Index (UVi)

Use value is the quantitative method that indicates the relative importance of a plant species for a population (Vendruscolo and Mentz 2006). In this study the use value of ethnomedicinal species ranges from 0.30 to 0.98. The plant species with use values were reported: *Juglans regia* (0.98 UVi), *Berberis lycium* (0.95 UVi), *Ziziphus oxyphylla*, and *Bergenia ciliata* (0.92 UVi). Maximum used values of medicinal plants mentioned might be due to their common distribution and local practitioners' awareness which make the plant species a first choice for ailment (Rahman et al. 2016a, b).

Table 3 Ethno-ecological use categories of the recorded plant species use in the study area

S. No	Botanical name	Medicinal	Vegetables	Fodder	Fuel	Timber	Ornamental
1	*Achyranthes aspera* L.	+	–	–	–	–	–
2	*Adiantum raddianum* C. Presl	+	–	–	–	–	–
3	*Ajuga integrifolia* Buch.-ham.	+	–	–	–	–	–
4	*Aloe vera* (L.) Burm.f.	+	–	–	–	–	–
5	*Amaranthus viridis* L.	+	+	–	–	–	–
6	*Artemisia absinthium* L.	+	–	–	–	–	–
7	*Arundo donax* L.	–	–	–	+	+	–
8	*Avena sativa* L.	–	–	+	–	–	–
9	*Berberis lycium* Royle.	+	–	–	–	–	–
10	*Bergenia ciliate* (haw.) Sternb	+	–	–	–	–	–
11	*Brassica campestris* L.	–	+	–	–	–	–
12	*Cannabis sativa* L.	+	–	–	–	–	–
13	*Capsella bursa pastoris* (L.) Medik	+	–	–	–	–	–
14	*Capsicum Annuum* L.	+	+	–	–	–	–
15	*Centaurea iberica* Trevir. ex Spreng	+	–	–	–	–	–
16	*Chenopodium album* L.	–	+	–	–	–	–
17	*Citrus medica* L.	+	–	–	–	–	–
18	*Convolvulus arvensis* L.	+	–	–	+	–	–
19	*Coriandrum sativum* L.	–	+	–	–	–	–
20	*Cichorium intybus* L.	+	–	–	–	–	–
21	*Cynodon dactylon* L.	+	+	+	–	+	–
22	*Diospyros lotus* L.	–	–	–	+	+	–
23	*Dryopteris patens* (Sw.) O.Ktze., C.Chr.	+	–	–	–	–	–
24	*Duchesnea indica* (Jacks.) Focke	–	+	–	–	–	–
25	*Euphorbia helioscopia* L.	+	–	–	–	–	–
26	*Ficus carica* L.	–	+	+	+	+	–
27	*Foeniculum vulgare* Mill.	+	–	–	–	–	–
28	*Fumaria indica* (Hausskn) Pugsley	+	–	–	–	–	–
29	*Gallium aparine* L.	+	–	–	–	–	–
30	*Indigofera heterantha* Brandis	–	–	+	–	+	–
31	*Juglans regia* L.	+	–	–	–	+	–
32	*Lamium amplexicaule* L.	–	+	–	–	–	–
33	*Lycopersicum esculentum* Mill.	–	+	–	–	–	–

(continued)

Table 3 (continued)

S. No	Botanical name	Medicinal	Vegetables	Fodder	Fuel	Timber	Ornamental
34	*Malva neglecta* L.	+	+	–	–	–	–
35	*Medicago truncatula* Gaerth	+	+	–	–	–	–
36	*Melia azedarach* L.	+	–	–	–	–	–
37	*Menthe longifolia* (L.) L.	+	–	–	–	–	–
38	*Morus alba* L.	+	–	+	+	+	–
39	*Morus nigra* L.	–	–	+	–	+	–
40	*Nasturtium officinale* R.Br.	+	+	–	–	–	–
41	*Olea ferruginea* Royle.	+	–	–	–	+	–
42	*Oxalis corniculata* L.	+	+	–	–	–	–
43	*Pinus roxburghii* Sarg.	–	–	–	+	+	–
44	*Pinus wallichiana* Jackson	–	–	–	+	+	–
45	*Plantago lanceolata* L.	+	–	–	–	–	–
46	*Poa pratensis* L.	–	–	+	–	–	–
47	*Prunus armeniaca* L.	–	–	+	+	–	–
48	*Prunus domestica* L.	–	–	+	+	–	–
49	*Prunus persica* (L.) Batsch	–	–	–	+	–	–
50	*Punica granatum* L.	+	–	–	–	–	–
51	*Pyrus communis* L.	–	–	–	+	+	–
52	*Pyrus domestica* L.	–	–	+	+	–	–
53	*Pyrus pashia* Buch.-ham. Ex D.Don	+	–	–	+	+	–
54	*Quercus dilatata* Royle	–	–	–	+	–	–
55	*Ricinus communis* L.	–	–	+	–	–	–
56	*Robinia pseudoacacia* L.	–	–	–	–	–	+
57	*Rubus ellipticus* Sm.	+	–	–	–	–	–
58	*Rumex hastatus* D.Don.	–	+	–	–	–	–
59	*Rumex hastatus* L.	+	–	–	–	–	–
60	*Silene conoidea* L.	–	+	–	–	–	–
61	*Solanum nigrum* L.	+	+	–	–	–	–
62	*Stellaria media* (L.) Vill.	–	+	–	–	–	–
63	*Traxicum officinale* G.H.Weber ex Wiggers.	–	+	–	–	–	–
64	*Triticum aestivum* L.	+	–	+	–	–	–
65	*Tulipa stellate* hook.	–	+	–	–	–	+
66	*Valeriana wallichii* DC.	+	–	–	–	–	–
67	*Viola canescens* wall.	+	–	–	–	–	–
68	*Vitis vinifera* L.	+	–	–	–	–	–
69	*Zanthoxylum armatum* DC.	+	+	+	–	–	–
70	*Ziziphus nummularia* (Burm.f.) Wight Qs Arn.	+	–	–	–	–	–
71	*Ziziphus oxyphylla* Edgew	+	–	–	–	–	–

Relative Frequency Citations (RFCs)

A relative frequency citation (RFC) shows the local importance of each plant species with reference to informants who cited the plant species. In this study, the value of relative frequency citation of medicinal plants species ranges from 0.10 to 0.46. The reason for the maximum RFCs may be due to the wide range of distribution, easy availability, and cultural uses of plants for curing various diseases by using these medicinal plant species. In the present study, the relative frequency citation of medicinal plants was *Juglans regia* (0.46 RFC and 41 URs), *Berberis lycium* and *Punica granatum* (0.44 RFCs and 39 URs), *Pyrus pashia*, *Capsicum annuum*, and *Convolvulus arvensis* (0.29 RFC and 26 URs) each. Singnorini et al. (2009) reported that *Asparagus acutifolius was* the most used species (0.71 RFCs). Medicinal plant species with maximum RFCs should be progressed to biological, phytochemistry and pharmacological studies to access and prove their vitality.

Prunus domestica

Prunus armeniaca

Citrus medica

Juglans regia

Dryopteris patens

Adiantum radianum

Bergenia ciliata

Rumex dentatus

Stellaria media

Silene conoidea

Lamium amplexicaule

Zyzyphus numularia

Conclusions

The flora of Pakistan is diverse due to different ecological zones and diverse climatic and soil conditions. Pakistan is blessed with diverse flora, and a total 6000 flowering plant species were reported. Plants are the priceless gift of nature. The human beings take their food directly or indirectly from plants and totally depend upon plants. The human world mostly fulfills their needs from plants, i.e., medicine, fodder, food, fuel, timber, etc. The floral diversity play a key role as carbon sink around the globe (Health et al. 2005). The inhabitants of Tarnawai utilize 71 plants species, belonging to 64% herbs, 25% trees, and 11% shrubs. Due to moist subtropical climate, herbaceous flora is very rich in this area. In the ethnobotanical survey, herbaceous flora was frequently used by inhabitants as traditional medicines, and the reason is that the herbaceous form is dominant due to their easy availability; due to the presence of wide range of bioactive compound, medicinal and healing potential is high (Khattak et al. 2015). Herbaceous growth as the dominant growth habit was also reported by many other researchers, from Torghar District (Shah et al. 2015), from Manoor Valley (Rahman et al. 2016a, b), and from Sarban Hills in Abbottabad (Ijaz et al. 2015).

The reported plant species were belonging to 37 different families, and the leading family reported from the study area was Rosaceae with 8 species, followed by Poaceae with 5 spp. and Asteraceae with 4 spp. Abbottabad being part of humid subtropical and temperate zone of the lesser Himalayas and hence Rosaceae is the largest family as it is primarily present in the Northern Hemisphere and adopts varieties of habitats except deserts and tropical rain forest. Due to its multidimensional uses, plant species are mostly utilized as medicine, food, fuel, timber, and fodder by the local communities. Rosaceae family has 2500 plant species followed by 98 genera, including herbs, shrubs, and trees. Family Rosaceae was the leading family also reported by other researchers, while dissimilar results reported by Ijaz et al. (2016); they reported Asteraceae as the leading family used to cure various ailments. Mostly members of family Poaceae are used as fodder and as fuel in the study area. Members of Asteraceae were mostly used as ethnomedicine and Rosaceae for fuel and food purposes.

The indigenous knowledge pertinent to ethnobotanical uses of plants in the area was recorded. The data revealed that 41% medicinal plants, 19% vegetables plants, 3% ornamental, 12% fodder, 14% fuel, and 11% timber plants were used ethnobotanically by inhabitants of Tarnawai. Haider and Qaiser (2009) documented that 83 plants were used as traditional medicines in Chitral Valley. In Naran Valley 183 plant species were used as timber, fuel, food, fodder, and medicines by local communities reported by Khan et al. (2012). During research study it was found out that that 41% of the medicinal plants have variety of uses to cures disorders, and the most curable disease was the gastrointestinal disease which is cured by 15 spp. Nine ethnomedicinal spp. are used for stomach acidity, and six spp. are used to cure the stomachache. Five plant spp. are used for cough, cold, and asthma. As it is a rural area, people used the local flora as homemade drugs for the treatment of different diseases because there are no such health and other basic facilities to fulfill their basic needs; that's why their lives totally depend on plants. Similar results were reported by Abbasi et al. (2013) and Rokaya et al. (2010) that constipation, stomachache, diarrhea, and anthelminthic disease were the most curable diseases.

In this area, the old ones are more informative than the youngsters. Mostly females are more informative about traditional knowledge of medicinal plants than men. It was also observed that most of the informants were uneducated, some are included in primary and secondary education category, and very less number of informants are included in the bachelor and master's degree. Most of the respondents were belonging to Sardar tribes as it is the dominant tribe. Besides this, Dhund, Karlal, Abbasi, Awan, and Syed tribes were also found in the study area.

The reported data were quantitatively analyzed by using statistical method and by using software. Two formulas were applied on recorded data, i.e., use value index (UVi) and relative frequency citation (RFCs). The maximum use values were found in *Juglans regia* (0.98UVi) and in *Berberis lycium* (0.95 UVi), and the maximum relative frequency citations were found in *Juglans regia* (0.46 RFCs and 41 URs) and in *Berberis lycium* and *Punica granatum* (0.44 RFCs and 29 URs), while Ijaz et al. (2016) reported maximum use value in *Berberis lycium* (0.91) and *Cannabis sativa* (0.81), and the maximum relative frequency citations were reported in *Berberis lycium* (0.313) and in *Ziziphus nummularia* (0.276). The maximum use value and relative frequency citations were found in these plants species due to the wide range of their local distribution and easy availability of the plants for cultural uses as medicine or for other so many purposes.

Fumaria indica is used for vomiting, for constipation, and for blood purification. The same results are shown by Polio et al. (2008). *Lycopersicum esculentum* is commonly used for lycopene by the inhabitants especially in cooked tomatoes, used as vegetables. The fruit contains lycopene, and it is one of the most powerful antioxidants (Sesso et al. 2003). So the results are similar with (Sesso et al. 2003). *Bergenia ciliata* is commonly used for ulcers and stomach problems. The rhizome is crushed and used in stomach ulcers, internal infection, and muscular disorders and also as tonic in a specific quantity and recipes by the local Hakeems these results are quite similar with the findings of others in the adjacent regions. Generally, *Ricinus communis* is used as fodder locally, but oil is also extracted from it to use in skin problems, that was also reported by Sairam (2002), i.e., oil is used on skin for prevention of dryness and to protect oneself from damage and pricking of nails, lethargy, and infection.

Rumex hastatus is used to cure hepatitis, tuberculosis, and kidney stone and heal wounds, while dissimilar results were reported by Ijaz et al. (2016), i.e., *Rumex hastatus* is used in arthritis and for bone fracture. These results were reported from Sarban Hills in Abbottabad. *Juglans regia* is used in dental problems. The wood of *Juglans regia* is best in making furniture, and the bark is good for the gums and available as "Dandasa" in local market which is used for cleaning the teeth. These results are similar with (Afzal et al. 2009). *Amaranthus viridis* is used as vegetables and for the treatment of inflammation of the pharynx, while opposite results are reported by Shah et al. (2015), i.e., he reported it to be used as blood purifier in Basikhel Tribe of Torghar District, KP, Pakistan. These opposing results are due to the different geographical conditions and diversity in cultural practices.

Cynodon dactylon is reported here to be used for vomiting and cholera apart from good source of fodder while opposing results are reported by Mehmood et al. (2011), where the plant of the *Cynodon dactylon* has been reported as carminative that expels poison from body, and is used to treat small pox and headache in District Mirpur, AJK, Pakistan. These differences in results are due to opposing climatic and soil condition of both study areas. *Solanum nigrum* whole plant juice is used for ear pain and for abdominal worms, while dissimilar results are reported by Shah et al. (2015), i.e., whole plant juice is used for fever and eye disease and suitable for digestion and dysentery. Ahmed et al. (2014) reported that the leaves of *Ajuga bracteosa* are washed, boiled in water, and left overnight in dew and then this decoction is taken before breakfast for blood purification, while the same results were reported from Tarnawai, i.e., *Ajuga bracteosa* leaves' juice is taken by people before breakfast for blood purification and for ear pain. *Zanthoxylum armatum* is used for gastrointestinal problems. The fruits of *Zanthoxylum armatum* are used in diarrhea, in constipation, and in cholera. The same results are shown by Rahman et al. (2016a, b) and Khan et al. (2015).

The present study indicated that the study area has abundant ethnobotanical plants used as a wide spectrum of human resources. As a result, these valuable plants are increasingly entered into the threatened list by loss of habitats, overexploitation, and deforestation. The ratio of the medicinal plant is high in the area, and these medicinal plants are valuable global resources but unfortunately, due to overexploitation, are increasingly destroyed and included in the red list category. The major cause of overexploitation includes poverty, illiteracy, unsustainable use of plants, breakdown of traditional controls, and lack of all other basic facilities, i.e., hospitals, markets, Sui gas, roads, schools, and colleges. There must be guidelines for work to improve the local management system for all the plant resources. There must be resource ownership and decision-making processes within the local communities in Tarnawai. The people should be aware about the plant resource use, land use, and resource ownership category. Those who actually harvest the locally used wild plant resources must be within the management system. Mostly females are the most harvesters of wild medicinal plants and are the most unprivileged members in society.

Recommendations

Awareness projects should be initiated in Tarnawai, to educate the local communities about the importance of all natural resources including plants and how to utilize them in a sustainable maner. Overgrazing should be controlled in the study area. There must be a control on collection beyond a permisable limits of that plant species which is used in a larger quantity for a medicine or other purpose. Conservation of multi-purpose plant species must be declared necessary according to the IUCN recommendations which are being potentially used abundantly. There should be a control on deforestation due to which floras are being endangered in the study area.

References

Abbasi AM, Khan SM, Ahmad M, Khan MA, Quave CL, Pieroni A (2013) Botanical ethnoveterinary therapies in three districts of the lesser Himalayas of Pakistan. J Ethnobiol Ethnomed 9:84

Afzal S, Mir AK, Tayyaba S (2009) Ethnobotanical studied northern Pakistan. J Ayub Med Coll Abbottabad 21:52–57

Ahmed M, Sultana S, Hadi SF, Hadda TB, Rashid S, Zafar M, Khan MA, Khan MPZ, Yaseen G (2014) An ethnobotanical study of medicinal plants in high mountains region of Chail Valley (district swat Pakistan). J Ethnobiol 10:36

Alcorn JB (1981) Huastec non-crop resource management. Hum Ecol:395–417

Bye RA (1979) Incipient domestication of mustard in north-West Mexico. Kiva 44:237–256

Haidar A, Qaiser M (2009) The ethnobotany of Chitral Valley, Pakistan with particular references to medicinal plants. Pak J Bot 41(4):2009–2041

Health J, Ayres E, Possell M, Bardgatt RD, Black HIJ, Grant H, Ineson P, Kerstiens G (2005) Rising atmosphere CO_2 reducessequestration of root-devied soil carbon. Science 309:1711–1713

Ijaz F (2014). Biodiversity and traditional uses of plants of Sarban Hills, Abbottabad. M. Phil. Thesis Hazara University Mansehra, KP, Pakistan

Ijaz F, Iqbal Z, Alam J, Khan SM, Afzal A, Rahman IU, Islam M, Sohail (2015) Ethnomedicinal study upon folk recipes against various human diseases in Sarban Hills, Abbottabad, Pakistan. World J Zoology 10(1):41–46

Ijaz F, Iqbal Z, Rahman IU, Khan SM, Shah GM, Khan K, Afzal A (2016) Investigation of traditional medicinal floral knowledge of Sarban Hills, Abbottabad, KP, Pakistan. J Ethnopharmacol 179:208–233

Jain SK (1991) Dictionary of Indian folk medicine and ethnobotany. Deep Publ Delhi, pp 1–31

Khan SM (2012) Plant communities and Vegetation Ecosystem Services in Naran Valley. Western Himalaya. *PhD, University of Leicester,* University of Leicester 249

Khan SM, Ud Din N, Ilyas M, Sohail IU, Ijaz Rahman F, Iqbal Z, Ali Z (2015) Ethnobotanical study of some medicinal plants of tehsil Kabal, District Swat, KP, Pakistan. Med Aromat Plants 4:189

Khattak NS, Nouroz F, Rahman IU, Noreen S (2015) Ethnoveterinary uses of medicinal plants of district Karak, Pakistan. J Ethnopharmacol 171:273–279

Mehmood A, Riffat NM, Zabta KS, Aqeel M (2011) Ethnobotanical survey of plants from Neelum valley, Azad Jammu and Kashmir, Pakistan. Pak J Bot 43:105–110

Pastakia F (2004) Abbottabad state of environment and development. International union for the conservation of natural resources, Pakistan and the government of the north-west Frontier Province (NWFP). *Rosette Printers,* Karachi. ISBN: 969–8141-72.3- www. Asc. iucnp

Phillips OI, Hall P, Gentry AH, Sawyer SA, Va'squez R (1994) Dynamics and species richness of tropical forests. Proc Natle Acad Sci USA 91:2805–2809

Polio A, Natale AD, Appetiti E (2008) Continuity and change in the Mediterranean medical tradition: *Ruta* spp. (Rutaceae) in Hippocratic medicine and present practices. J Ethnopharmacol 116(3):469–482

Rahman IU, Ijaz F, Afzal A, Iqbal Z, Ali N, Khan SM (2016a) Contributions to the phytotherapies of digestive disorders; traditional knowledge and cultural drivers of Manoor Valley, Northern Pakistan. J Ethnopharmacol 192:30–52

Rahman IU, Ijaz F, Iqbal Z, Afzal A, Khan MA, Muhammad S, Qadir G, Asif M (2016b) A novel survey of the ethnomedicinal knowledge of dental problems in Manoor Valley (Northern Himalaya), Pakistan. J Ethnopharmacol 194C:877–894

Rokaya MB, Munzbergova Z, Timsina B (2010) Ethnobotanical study of medicinal plants from the Humla district of western Nepal. J Ethnopharmacol 130:485–504

Sairam K, Rao CV, Babu MD, Kumar KV, Agrawal VK, Goel RK (2002) Antiulcerogenic effect of methanolic extract of Emblica officinalis: an experimental study. J Ethnopharmacol 82:1–9

Sesso HD, Liu S, Gaziano JM, Buring JE (2003) Dietary lycopene tomato- based food products and vascular diseases in women. J Nutr 133(7):2336–2341

Shah AH, Khan SM, Shah AH, Mehmood A, Rahman IU, Ahmed H (2015) Cultural uses of plants among Basikhel Tribe of District tor Ghar, Khyber Pakhtunkhwa, Pakistan. Pak J Bot 47(SI):23–41

Singnorini MA, Piredda M, Bruschi P (2009) Plants and traditional knowledge: an ethnobotanical investigation on Monte Ortobene (Nuoro, Sardinia). J Ethnobiol Ethnomed 5:6

Vendruscolo GS, Mentz LA (2006) Estudo da concordancia das citasoes de uso e importancia das especiese familias utiliza das como medicinais pela communidade dobairro Ponta Grossa. Porto Alegre, RS, Brasil. Acta Bot Bras (2):367–382

Licorice in Middle Asia

F. O. Khassanov

Introduction

The last revision of the genus *Glycyrrhiza* was made by V. Grankina (2009) where 36 species were divided into two subgenera and 9 sections (excluding *Meristotropis* species). Afterwards, two more new species were described recently from Siberia (*G. krasnoborovii* Grankina, *G. orientalis* Grankina and Letjaeva). Genus *Meristotropis* was separated by A. Fischer and C. Meyer (1843) with the only species *M. triphylla*. There are several papers on the relationship between *G. glabra* and some other allied species showing allied position between genera *Glycyrrhiza* and *Meristotropis* (Barghi and Siljak-Yakovlev 1990; Yamazaki et al. 1994). On the other hand, embryological studies made by O. Ashurmetov (1987) confirmed separated position of these genera. Controversial views after studying of palynological characters were published by Meng L. and X.-Y. Zhu (2010). Accordingly genera *Glycyrrhiza* and *Glycyrrhizopsis* (Boiss.) Boiss. showed morphological differences but *Meristotropis* suggested to be merged in *Glycyrrhiza.*

Taxonomy (Figs. 1, 2, 3, 4, 5, 6, and 7)

Genus *Glycyrrhiza* L., Sp. Pl. 2: 741 (1753).
Section ***Glabrae*** Grankina, Novosti Sist. Vyssh. Rast. 40: 95 (2009). Type: *G. glabra* L.

1. ***G. glabra*** L. Sp. Pl.: 742 (1753) = *G. violacea* Boiss. and Noe, Diagn. Pl. Orient. ser. 2, 2: 23 (1856).

F. O. Khassanov (✉)
Institute of Botany, Academy of Science, Tashkent, Uzbekistan

M. Ozturk, K. R. Hakeem (eds.), *Plant and Human Health, Volume 1*,
https://doi.org/10.1007/978-3-319-93997-1_23

Fig. 1 *G. michajloviana* in the mountains of Tien-Shan

Fig. 2 *G. shiheziensis* in the mountains of Tien-Shan

Lectotype: Herb Linn 916.3 (LINN). Described from Europe.

Distribution: Europe, Mediterranean, Siberia, Iran, Afghanistan, Caucasus, Middle Asia, Mongolia, China.

2. ***G. hirsuta*** L., Sp. Pl.: 742 (1753) = *G. pallida Boiss.*, Diagn. Pl. Orient., ser. 2, 3(2): 22 (1856).

Type: Herb. Linn. NN 914.9–917.7 (LINN). Described from Orient.

Fig. 3 *G. soongoroca* in the mountains of Tien-Shan

Distribution: Southern Russia, Caucasus, Middle Asia (Kazakhstan).

Section *Arthrocarpaea* (Fisch. et C. A. Mey.) Grankina, Novosti Sist. Vyssh. Rast. 40: 102 (2009). Type: *G. asperrima* L. f.

3. ***G. brachycarpa*** Boiss., Diagn. Pl. Orient., ser. 1, 1(2): 28–32 (1843).

Type: "In Syria circa Damascum, 1830, Aucher. Pl., exs. N 996" (G, isolectotype LE, designated by Grankina in Novosti Sist. Vyssh. Rast. 40: 97 (2009)).

Distribution: Syria, Middle Asia (Kazakhstan).

4. ***G. michajloviana*** Grankina and E.V. Kuzmin, Izv. Natsional. Akad. Nauk Respubl. Kazakhstan, Ser. Biol. Med. 4: 18–20 (Fig. 1, map) (2006) et Novosti Sist. Vyssh. Rast. 40: 97 (2009) = *G. korshinskyi* Grig. in Bull. Jard. Bot. Princip URSS, 23, 1–2:94 (1930).

Type: Kazakhstan, Almaata prov., ad ripam sinistram humilior affientia Kurty, praedium saltuarum, 16.07.1959, V.P. Michailova (NS, isotypes LE, AA).

Distribution: Kazakhstan.

In the protologue two different gatherings designated as type in Latin and Russian parts of the article but later in Grankina (2009) senior author designated one type specimen making this taxon validated.

5. ***G. nadezhinae*** Grankina, Novosti Sist. Vyssh. Rast. 39: 215–217 (2007).

Fig. 4 *G. nadezhinae* in Uzbekistan

Fig. 5 *G. eglandulosa* in Altay area

Type: Kazakhstan, prov. Dzhambul, Kuragajskyi distr., ad ripam dextram fluvi Chu, pag. Uspenskoe, 13.07.1970, T.P. Nadezhina (NS, isotypes LE, AA).

Distribution: Kazakhstan, Uzbekistan, Tadzhikistan, Turkmenistan, China (Xinjiang).

Section ***Parvifloroides*** Grankina, Novosti Sist. Vyssh. Rast. 40: 98 (2009). Type: *G. uralensis* Fisch.

6. ***G. uralensis*** Fisch. in de Candolle, Prodr. 2:248 (1825).

Fig. 6 *G. glabra* in the mountains of Pamir-Alay

Fig. 7 *M. triphylla* in the mountains of Tien-Shan

Type: *Glycyrrhiza uralensis* m. Ural. Helm. Fructis non vidi, *G. asperrima* pr. G. uralensis Fisch. (G, isotype LE).

Distribution: Russia (Ural), Middle Asia (Kazakhstan).

Section ***Flexuosocarpae*** Grankina. Novosti Sist. Vyssh. Rast. 40: 102 (2009). Type: *G. soongorica* Grankina

7. ***G. shiheziensis*** X. Y. Li, Bull. Bot. Res. North-East. Forest. Inst. 9, 1: 29 (1989).

Holotype: "Xinjiang: Shihezi, 16.07.1981, N 810339, X. Y. Li" (SAC).
Distribution: China, Russia (Siberia), Middle Asia (Kazakhstan).

8. ***G. eglandulosa*** X. Y. Li, Bull. Bot. Res. North-East. Forest. Univ. 13, 1: 29 (1993).

Holotype: "Xinjiang: Shihezi, 4.07.1982, N 820171, X. Y. Li" (SAC).
Distribution: China, Russia (Altay), Middle Asia (Kazakhstan).

9. ***G. soongorica*** Grankina, Novosti Sist. Vyssh. Rast. 33: 148 (2001).

Holotype: "Mongolia Kobdo ajmak, Uentsch somon, 25 km ad meridiem a somone, ripa sinistra fluminis Uentsch, 6.08.1972, Nadezhina" (LE).
Distribution: Mongolia, Middle Asia (Kazakhstan, Kyrgyzstan), China (Xinjiang).

10. ***G. sergievskiana*** Grankina and Aralbaev, Izv. Natsional. Akad. Nauk Respubl. Kazakhstan, Ser. Biol. Med. 3: 29–31 (Fig. 1, map) (2006).

Type: "Kazakhstan orientalis, Zaissanica depression adjacentia pagi Kokpektinsk, locus salsus, 29.08.1988, Grankina and Pisareva" (NS, isotypes LE, TK, AA).
Distribution: Middle Asia (Kazakhstan).
Section ***Arthrocarpaea*** (Fisch. et C. A. Mey.) Grankina, Novosti Sist. Vyssh. Rast. 40: 102 (2009). Type: *G. asperrima* L. f.

11. ***G. aspera*** Pall. Reise Russ. Reich. 1: 449 (1771) = *G. laxiflora* X.Y. Li and D.C. Feng, Bull. Bot. Res., Harbin 13(1): 40 (1993) = *G. macrophylla* X.Y. Li, Bull. Bot. Res., Harbin 9(1): 30 (1989) = *G. nutantiflora* X.Y. Li, Bull. Bot. Res., Harbin 13(1): 36 (1993) = *G. prostrata* X.Y. Li and D.C. Feng, Bull. Bot. Res., Harbin 13(1): 38 (1993)

Lectotype: In aridissimo ausralioris deserti limo frequens planta, Pallas, 1770 (Herb. Linn. N 916.5 (LINN), designated by Grankina in Novosti Sist. Vyssh. Rast. 40: 103 (2009)).
Distribution: Southern Russia, Caucasus, Iran, Middle Asia (Kazakhstan), China.

12. ***G. hispida*** Pall., Reise 3: 72 (1776).

Lectotype: "Deserto inter Volgam et Iaikum abundat. Pallas, 177!" (Herb. Linn. N 916.5 (LINN), designated by Grankina in Novosti Sist. Vyssh. Rast. 40: 103 (2009)).
Distribution: Southern Russia, Caucasus, Iran, Middle Asia (Kazakhstan), China.

13. ***G. laxissima*** Vassilcz., Bot. Mater. Gerb. Bot. Inst. Komarova Akad. Nauk S.S.S.R. 11: 120 (1949).

Holotype: "Western Tien-Shan, Angren river valley (in Russian), 17.05.1880, Regel (LE).
Distribution: Middle Asia (Uzbekistan, Kazakhstan).
Section ***Monilîcarpae*** Grankina, Novosti Sist. Vyssh. Rast. 40: 103 (2009). Type: *G. zaissanica* Serg.

14. ***G. zaissanica Serg.*** in Animadvcrs. Syst. Herb. Univ. Tomsk. Nos. 1–2, 11 (1933).

Type: "Kazakhstania, Zaissanica depressio, circa opp. Kokpektinsk, in herbosis, 7.07.1929, Krylov et Sergievskaja" (TK).
Distribution: Middle Asia (Kazakhstan), China.

15. ***G. purpureiflora*** X.Y. Li, Bull. Bot. Res., Harbin 13(1): 34 (1993).

Holotype: "Xinjiang: Shihezi, 24.05.1990, N 90237, X. Y. Li" (SAC).
Distribution: Middle Asia (Kazakhstan), China.
Section ***Glycyrrhiza***.

16. ***G. echinata*** L. Sp. Pl.: 741 (1753) = *G. dioscoridis* Medik. in Vorles, Churpf. Phys. Ges. 2: 367 (1787).

Lectotype: In Gargano Apuliae, in deserto Nagico Tatariae (Herb. Linn. N 916.1. (LINN, designated by Chamberlain in Jarvis et al.: 50 (1993)).
Distribution: Europe, Caucasus, Middle Asia, Turkey, Iran, Afghanistan.

17. ***G. foetidissima Tausch***, Flora (Regensb.) 1, 13: 210 (1831).

Holotype: "Colitur in horto bot. Pragensi" (PRG).
Distribution: Europe, Southern Russia, Caucasus, Iran, Morocco, Middle Asia (Turkmenistan).
Genus ***Meristotropis*** Fisch. et C. A. Mey. in Index Seminum (St. Petersburg-Petropolitanus) 9. 95 (1843).

18. ***M. bucharica*** (Regel) Kruganova in Acta Inst. Bot. Acad. Sci. URSS, 1, 11:194 (1955); *G. gontscharovii* Maslenn. Trudy Tadzhikistanskoy Bazy 8:617 (1938); *G. kulabensis* M.M.Ivanova, Trudy Tadzhikistanskoy Bazy, 8:620 (1938).

Holotype: "Tadzhikistania, 1881, Regel" (LE).
Distribution: Middle Asia (Tadzhikistan, Uzbekistan, Turkmenistan).

19. ***M. triphylla*** (Fisch. et C. A. Mey.) Fisch. et C. A. Mey. in Index Seminum (St. Petersburg (Petropolitanus) 9:95 (1842) = *M. erythrocarpa* Vassilzc. Not. Syst. Herb. Inst. Bot. Acad. Sci. URSS 11: 121 (1948) = *M. xanthioides* Vassilcz., Bot. Mater. Gerb. Bot. Inst. Komarova Akad. Nauk S.S.S.R. 11: 120 (1949).

Holotype: "East bank of the Caspian Sea, Mangyshlak, Fisch (in Russian)" (LE).
Distribution: Middle Asia (Kazakhstan, Uzbekistan, Turkmenistan).

There are about 40 species of *Glycyrrhiza* and 3 species of *Meristotropis* naturally growing in the Old and New World and one of the main centers of diversity is located on the large area of deserts and mountains in Middle Asia, Siberia, and China. One of the most used medicinal plants—*G. glabra*—is mainly distributed in Middle Asia within the floodlands and valleys of large rivers and cannels (Ozturk et al. 2017). It also has fair quantities in the water meadows even in the deserts and steppes. *G. glabra* is a very common plant in the low-mountainous vegetation among other tall grasses (*Ferula* species, *Hordeum bulbosum* L., *Elytrigia trichophora* (Link) Nevski, *Origanum tyttanthum* Gontsch., *Crambe kotschyana* Boiss., *Haplophyllum acutifolium* (DC.) G. Don etc.) and shrubs (*Pistacia vera* L., *Amygdalus spinosissima* Bunge., *Rosa* species, *Atraphaxis sarawschanica* Pavlov, *Cotoneaster nummularioides* Pojark., etc.). Usually it grows on the awashed southern slopes up to 2000 m above sea level, but it can also be found even in juniper, apple, or maple forests as a subdominant. In the mountainous zone of Western Pamir-Alay it also grows together with *G. uralensis* having neighboring populations nearby the springs and rivers. In the Western Tien-Shan it is growing together with *Meristotropis triphylla* having isolated populations one of each other. *G. glabra* is one of the best pioneer plants on the landslips growing together with *Datisca cannabina* L. and *Tussilago farfara* L. It also could be named as a ruderal plant not only in natural plant associations, but also in the fruit or vegetable gardens and various crops.

Uses

Information about the use of glycyrrhizin-comprising species of genus *Glycyrrhiza* L. in the region of Middle Asia comes from the ancient times (Varganov et al. 1985; Ozturk et al. 2017). Licorice is well known from many recipients of Eastern medicine to have higher position than ginseng. Avicenna used *G. glabra* as one of the numerous compounds as a treatment against tertian ague, rash and scab, agnail, stomach ulcer, biliary dyspepsia, leprosy, toxin substances, paralysis, chill, cough, bronchitis, etc. (Sadykov et al. 1980). Glycyrrhizin acid is a well-known antagonist of poisons with antiphlogogenic action. Roots and rhizomes of licorice contain up to 23% of glycyrrhizin (potassium and calcium acid salts). Glycyrrhizin itself consists of triterpene saponin where aglycon is glycyrrhetic acid with sugary residuum of two molecules of glucuronic acid). Besides 27 different flavonoids (liquiritin, isoliquiritin), ascorbic acid (up to 30 mg) has been found as well as a small amount of essential oil, gum, resin, and asparagines (Abubakirov and Yatsin 1959). Chemicals made from licorice have a beneficial effect against viruses and natrium glycyrrhetic acid salt is very active against protozoa (Muravjev and Ponamarev 1962). Licorice is used not only in medicinal purposes but also in steel factories as a foam-forming admixture (Varganov et al. 1985; Ozturk et al. 2017).

In 1964 a special laboratory for studies of glycyrrhizin-comprising taxa was founded by Prof. K. Zakirov. He worked out original system of actions for melioration and mapping of wild populations of *G. glabra* in Middle Asia. All these wide researches have been supported by a special Department of Cabinet of Ministry of USSR—Sojuzlakritsa. This organization was one of the leading ones in the export of licorice root in the world. Researchers of this laboratory have been collecting all species of genus *Glycyrrhiza* growing in Soviet Union as well as numerous varieties of *G. glabra* and *G. uralensis*. They were studying anatomy, embryology, and physiology of highly productive forms and intraspecific hybrids in order to create new plantations (about 20,000 hectares) of licorice in salty steppes and deserts especially. Studying water balance of licorice in salty deserts it was revealed that planted licorice was promoting lowering of soil salinization (Nigmatov 1977). Licorice is a well-known plant in uzbek folk medicine as shirin-miya. People are collecting the roots of *G. glabra* for medicine purposes sailing on the markets. Usually one can buy licorice roots even in the markets in small towns and villages. In the mountainous zone local people are collecting *G. uralensis* together with *G. glabra* without any difference. Local healers are using licorice in the same way as Avicenna (Ozturk et al. 2017). The list of diseases is very wide (from stomach and liver problems up to cancer).

References

Abubakirov A, Yatsin V (1959) Issledovanie sredneaziatskih vidov Glycyrrgiza L. na soderzhanie glicyrizinovoy kisloty. Uzb Khim Zhurn 5:81–86

Ashurmetov O (1987) Embryology of some Glycyrrhiza species in the middle Asia. Avtoreferat doktorskoj dissert. Leningrad 48

Barghi N, Siljak-Yakovlev S (1990) Karyological study in three species of *Glycyrrhiza* genus (*G. glabra, G. lepidota* and *G. echinata*). Caryologia 43:223–234

Fischer A, Meyer C (1843) Addenda plantarum novarum. Index Seminum (St Petersburg, Petropolitanus) 9:95

Grankina V (2009) The system of the genus *Glycyrrhiza* L. (*Fabaceae*). Novosti systematiki, St Petersbourgh 40:89–109

Meng L, Zhu X-Y (2010) Palynological characters of *Glycyrrhiza, Glycyrrhizopsis*,and *Meristotropis (Leguminosae)*,with special reference to their taxonomic significance. J Syst Evol 48:6

Muravjev I, Ponamarev V (1962) Glycyrrizinovaja kislota I ee preparaty v kachestve novyh lekarstvennyh sredstv medicinskoy promyshlennosti SSSR. Med Prom 8:16

Nigmatov S (1977) Rastenija-melioranty. Tashkent, Izdateljstvo FAN 17

Ozturk M, Altay V, Hakeem KR, Akçiçek E (2017) Licorice: from botany to phytochemistry. SpringerBriefs in Plant Science. https://doi.org/10.1007/978-3-319-74240-3.

Sadykov and Others (1980) Avicanna—Kanon vrachebnoj nauki. Izdanie 2, Tom 4 153(235):322

Varganov L, Pauzner L, Chevrinidi S (1985) Perspectivy vvedenija v kulturu glycyrinsoderzhashih I dubiljnyh rastenij. In: Zakirov P, Khassanov O (eds) Puti povyshenija produktivnosti lekarstennyh I kormovyh rastenij v culture. FAN UzSSR, Izd, pp 3–9

Yamazaki M, Sato A, Shimomura K, Saito K, Murakoshi I (1994) Genetic relationships among *Glycyrrhiza* plants determined by RAPD and RFLP analyses. Biol Pharm Bull 17:1529–1531

Ethnobotany of Medicinal Plants for Livelihood and Community Health in Deserts of Sindh-Pakistan

Ghulam Yaseen, Mushtaq Ahmad, Daniel Potter, Muhammad Zafar, Shazia Sultana, and Sehrosh Mir

Introduction to Deserts of Sindh

Biogeography

A desert is moisture-deficit area under climatic conditions and has precipitation less than 20% of moisture loss through evaporation (Smith et al. 1995). Deserts include various regions along the tropics in southern as well as northern hemispheres (Murray 1999). In Pakistan, deserts occupy about 11 million hectares including 14% land area of the whole country. The major deserts are found in Sindh including Thar Desert. The Thar Desert is situated in Sindh and subdivided into *Nara*, *Achro*, and Thar located in southern Sindh. Nara occupies Khairpur and some areas of Sanghar districts, while *Achro* Thar is bordered to Nara and Indian Thar Desert. Nara and Achro Thar are topographically marked with plain areas, some sandy hills, and some lakes (Qureshi and Bhatti 2005a). The Thar Desert occupies four districts of Sindh, namely, *Tharparkar*, *Umerkot*, *Mirpur Khas*, and few areas of Badin. The Thar Desert shares its border with Barmer and Jaisalmer of India in the southeast and *Rann of Kutch* of India in the south, while the Mirpur Khas region of the Thar Desert shares its border with Rajasthan (Yaseen et al. 2015). The Nara Desert is linked to the Thar via Sanghar and Mirpur Khas. The vegetation is represented by primitive cryptogams to higher vascular plants, especially in the Nagarparkar. Kharan and Chagi deserts are sandy and mountainous deserts situated in Balochistan province in southwestern Pakistan. Most of the deserts in Pakistan cover sand dunes, scrub vegetation, weathered rocks, salt lakes, and small natural water dams.

G. Yaseen · M. Ahmad (✉) · M. Zafar · S. Sultana · S. Mir
Department of Plant Sciences, Quaid-i-Azam University Islamabad, Islamabad, Pakistan

D. Potter
Department of Plant Sciences, University of California Davis, Davis, CA, USA

M. Ozturk, K. R. Hakeem (eds.), *Plant and Human Health, Volume 1*,
https://doi.org/10.1007/978-3-319-93997-1_24

The deserts of Sindh include Thar desert, Nara desert and *Achro* Thar desert. Thar desert is in Badin, Tharparkar, Umerkot and Mirpurkhas. Nara desert is in Khairpur and Achro Thar desert is located in Sanghar and Khairpur. It is bordered to Thar Desert of India (Great Indian Desert) in the east, while in the west it shares borders with Kirthar Range. In the north it continues into the Cholistan Desert, in the east it extends into the Rajasthan Desert, and in the southern extremes, the Thar submerges into the *Rann of Kutch* toward the Arabian Sea. Overall, it occupies 18th position in all deserts while 9th position in subtropical deserts. Based on population density, it is most densely populated with 83 persons per square among all deserts of the world. According to an estimate, its population is nearly about 1.2 million (Khan et al. 2003). It covers about 30,000 km^2 area, while the rest of Thar Desert is situated in India. The topography of the area includes sand dunes, hill outcrops, and the alluvial flood plains. Like Kalahari Desert, its aridity is not very high where most of the areas have more than 100 mm annual rainfall rate. However, unlike Kalahari, it is characterized with diversity of arid lands, foot hills in Nagarparkar, valleys and delta of River Indus, and salt land of *Rann of Kutch* in the south (Khan et al. 2003).

Pictorial Representation of Deserts of Sindh-Pakistan

Climate

The climatic conditions include high temperature, low annual rainfall, and high rate of evaporation than precipitation. The annual temperature ranges from 20 °C to 51 °C in different seasons. In May to July, temperature may reach to peak, while lowest temperature may be recorded in January. Annual rainfall ranges from 88 to 135, while July to August are most favorable months for monsoon rain. However, winter rainfall with low intensity may shower during December, January, and February. In records, normal monsoon is 127.5 mm, while it reached to maximum 443.9 mm in 2011 due to abrupt climatic changes (Cheema et al. 2012). Water scarcity is observed in most of the areas that has resulted to limited agriculture, while groundwater may be found at 50–300 ft of depth with variation in salt concentration. The most favorable and easily available sources of water are ponds, dugouts, and tanks filled during monsoon rainfall (Qureshi and Bhatti 2005a). Climate change and global warming during the last half century contributed to the severity and exacerbation of floods in 2010 and 2011. The Intergovernmental Panel on Climate Change (2007) reported that climate projection reveals more frequent and more intense weather events due to global warming.

Plate 1 (**a**, **b**) Presents the panoramic views of deserts of Sindh

Plate 2 (**a**, **b**) Green view after moon soon rain fall

Plate 3 (**a**) Country side life in deserts of Sindh. (**b**) Cultural houses in villages

Plate 4 (**a**, **b**) Religious places in deserts of Sindh

Population

The deserts of Sindh including Thar Desert, Achro, and Nara sustain high level of livestock and human population. Most of the population comprised of small nomads commonly distributed according to need and rainfall pattern. The commonly speaking language is Sindhi, while mother language varies among tribes. Commonly, people of deserts of Sindh are known as *Thari*, and livestock keeping is the only livelihood of the inhabitants of the study area (Qureshi and Bhatti 2008). The main occupations of its inhabitants are either agriculture which doesn't go well because of lack of water and irrigated land or farming which causes overgrazing of land. The area is often in drought, despite that people do not have anything else to do. They depend on these professions because there is a serious lack of education in the region.

Cultural Heritage and Ethnography

The deserts of Sindh possess rich cultural heritage that is inherited from the generations and present a thrill for tourists and visitors. Demographically, the deserts of Sindh including Thar Desert, *Nara*, and *Achro* Thar are occupied by diversity of ethnic groups. Major ethnic groups include *Bhagora*, *Bheel*, *Kohli*, *Meena*, *Garasia*, *Damor*, and *Garasia* and possess strong cultural association with medicinal plants. The most dominant spoken languages are *Dhatki*, Gujarati, and Sindhi, while rare spoken language is Urdu. In culture, like other parts of Sindh, joint family system is most commonly established with few exceptions. Among family systems, early-age marriages, apothecary, and weddings of daughters without their consent are most common. Early-age marriage is considered as religious fact and families and caste are preferred over the others. In most of families, all three generations including grandparents, parents, and child live under one roof with joint family system. At family level, males are given duties related to plowing, weaving, animal grazing, and house construction, while the major duties are bringing water from wells and ponds, cooking, washing, and helping their family members in the field (Qureshi and Bhatti 2005b).

The deserts of Sindh exhibit various cultures. The most popular are *Rajasthani*, *Sindhi*, and *Gujarati*. In a variety of music, *Thari* is considered as most popular, impressive, and peculiar. In many carnivals and traditional programs, expert *Thari* singers commonly called *faqirs* are invited to show off their expertise. Among various folk dances, most common are *chakar* rand, *mitco*, *dandan* rand, and *rasooro*. In each custom, both Muslims and Hindus are directly interlinked without any discrimination of religious belief. Hindus and Muslims form the two ethnic groups in almost all regions of the deserts of Sindh. The Hindus form 62% of all population, and Muslims are represented by 38% (Iqbal et al. 2000), but their coexistence represents strong cultural relation without any differentiation of religion. The sociocultural interrelationships are observed during cultural celebrations. The major cultural celebrations are *Holi*, *Diwali* or *Deepawali*, *Krishna Janmashtami*, and *Maha Shivaratri*.

Health and Education

The deserts of Sindh have been facing a lot of issues, but lack of information has been a major obstacle in the empowerment of indigenous people of Thar. According to recent rankings, literacy rate is less than 17% in most of the regions especially in Tharparkar. According to *Alif Aiilaan* (www.alifailaan.pk/district_rankings), female literacy rate is only 7% among *Thari* populations. Various factors are found to be responsible, but most common includes poverty, lack of basic educational facilities, and cultural constraints. Young generation is bound to work with their parents in fields for financial assistance. Parents think that without the help of siblings in daily routines, they cannot afford their financial burden. People are not aware even of their basic rights. According to *Alif Ailaan* Pakistan District Education Ranking 2017, it has been ranked 113th in terms of education in Pakistan. Status of basic facilities at school level is very poor.

The deserts of Sindh lack basic health facilities in most of the area. Health centers with enough facilities are only available in district headquarters, while approach to such facilities is not possible for people living in flung areas due to lack of proper roads and transport facilities. In villages, doctors and nurses are available, while lower staff such as health technicians perform the duties of health specialists which is another major cause of deaths in children. Most of the women lack good physical health due to continuous work at field resulting in mental and health problems. In various regions, concept of medical checkup during pregnancy is very rare due to ignorance and unavailability of basic health facilities. At local level, various labor cases are dealt by untrained and unskilled technicians that result into prenatal deaths.

The Plant Diversity and Its Cultural Significance

The plant diversity provides complete to partial substitute for many products including NTFPs, fertilizers, allopathic drugs, pesticides, and many other important products used in daily life. The biodiversity of the deserts in Sindh is directly associated with the habitat diversity. Vegetation in most of the regions is sparse except Nagarparkar. In most of the areas, perennial herbs and thorny shrubs with drought resistance are commonly dominant. The most important species of plants include *Commiphora wightii*, *Euphorbia tirucalli*, *Tecomella undulata*, *Pedalium murex*, *Moringa concanensis*, *Cistanche tubulosa*, *Acacia jacquemontii*, *Senna oiccidentalis*, etc. In Nagarparkar, the vegetation is dense due to its topography and commonly found springs and waterfall. It seems lush green during monsoon. Overall, tree population is meager except few species such as *Acacia senegal*, *Prosopis cineraria*, and *Tecomella undulata*, while ephemerals grow up during monsoon and complete their life cycle before winter. Each habitat and landform in the desert supports distinctive plants and animals with their respective niches, which are endemic to the thar. The plant species of economic value have become either threatened or endangered (e.g., *Commiphora wightii* and *Tecomella undulata*). Other endemic species have ecological requirements and need to multiply and be conserved.

The deserts of Pakistan present a unique plant biodiversity, with estimated 400–600 medicinal plants out of a total well-known 7000 plant species in the country (Khatri et al. 2011; Yaseen et al. 2015). The biodiversity of the deserts is intimately associated with the habitat diversity. The vegetation is represented by primitive cryptogams to higher vascular plants, especially in the Nagarparkar of Thar Desert. Land degradation, overgrazing and overexploitation of natural resources, transition of human culture, and increased anthropogenic activities are serious issues in deserts of Pakistan resulting in loss of biodiversity and cultural integrity. The plant resources along with other biodiversity are degrading day by day due to various environmental factors such as low rainfall, scarcity of water, natural disasters, and mismanagement of land sources. These factors have resulted in habitat destruction and serious threats to plant diversity especially to those which are overexploited for trade, food, and shelter. Besides, these factors are threatening the plant species in the wild by changing their habitat.

Ethnobotany in Deserts of Sindh

The deserts of various regions represent great diversity in terms of both their traditional knowledge about medicinal plants and their geomorphological processes (Goudie 2002). In deserts such as Thar Desert, the dependency of local people on traditional herbal medicine is deeply rooted and has a very old history. Among rural populations, local people has rich knowledge about traditional use of medicinal plants in treating various diseases which are not easily treated using other medicines. In Thar Desert, plant-based medicines were practiced since the arrival of the Christian era perhaps during Mohenjo Daro and Harappan civilizations. Among rural communities of the deserts, various civilizations were dependent on herbal medicines (Katewa 2009). However, very little attention has been paid to the ethnobotanical studies among cultural communities of the deserts.

The deserts of Sindh present a unique plant biodiversity, with estimated 400–600 medicinal plants out of a total of 7000 plant species in the country (Khatri et al. 2011; Yaseen et al. 2015). In the desert communities of Sindh, traditional medicines obtained from plants are still more popular than available allopathic and biomedical

Plate 5 (**a**, **b**) Tourist visits in deserts of Sindh

medicine. Specifically, people with low income and socioeconomic status rely on traditional medicine because of the low or lack of availability of medicines and inaccessibility to modern health-care systems. The Thar Desert occupies a major transitional zone between the arid, Palearctic landscapes of Southwest Asia and the Sahara and the monsoonal, oriental biomes of South and East Asia, making it a critical location to investigate cross-cultural trends about medicinal plant across Southern Asia (Boivin 2012).

The medicinal plants-rich region, Thar Desert, is strongly associated with indigenous knowledge, and this is seriously depleted due to abrupt climatic changes, long-term drought due to short rainfall, famines, and natural disasters. Due to this, indigenous knowledge related to plant-based ethnomedicines in this desert of Pakistan especially in deserts of Sindh is under threat. In deserts of Sindh, local people use plants in various herbal recipes for curing various diseases due to lack of health facilities and less hygienic conditions and most probably due to transfer of indigenous knowledge from generation to generation. Table 1 presents the detail of some of the important plant species, their local names, plant part used, modes of utilization, and treated diseases and modes of administration.

Diversity of Medicinal Plants in Deserts of Sindh

In total, 45 important plant species belonging to 24 plant families are commonly used in the treatment of various diseases among rural communities in deserts of Sindh-Pakistan. The most commonly used families are Fabaceae, with six species, followed by Amaranthaceae with four species, Asteraceae with three species, Acanthaceae with three species, and Malavaceae with three species, while the rest of the families have one or two species. In growth forms, the most dominant are herbs (31 species) followed by trees (7 species). The most frequently cited species are *Blepharis scindica*, *Oxalis corniculata*, and *Portulaca oleracea*, while least documented species are *Achyranthes bidentata* followed by *Euphorbia caducifolia* (Table 1).

There are many factors regarding the diversity of plant families in any region. In this study, Fabaceae are highly distributed and possess a large number of species resulting in high number of reported species in the study, while Amaranthaceae represent well ecological adaptations and are surviving under stress conditions. The local people use the species belonging to these families for curing number of diseases. The most dominantly cited species of these families may have a wide variety of phytochemical compounds responsible for curative effect. In most of the studies, herbs are frequently reported for treating various diseases. According to Shrestha and Dhillion (2003), the common use of herbs is due to their easy manipulation in herbal recipes, while Ayyanar and Ignacimuthu (2005) and Uniyal et al. (2006) suggest that the more frequent use of herbs may be due to their diverse geographic range along roadsides, home gardens, and farmlands. Based on our findings and

Table 1 Ethnobotanically important medicinal plant species reported from deserts of Sindh and their documented uses

Taxonomic name	Local name	Plant part used	Mode of utilization	Diseases treated (body organ system treated/mode of administration)[1]
Abelmoschus esculentus (L.) Moench	Bhindi	Fruit Seeds	Powder Raw Decoction	Cough (RT/I), asthma (RT/I), diarrhea (DS/I), nerve tonic (NS/I), sexual weakness in male (RS/I), general debility (NS/I), spermatorrhea (RS/I), urinary disorders (RS/I)
Abutilon indicum (L.) Sweet	Peeli nair	Whole plant Leaves Seeds Flower Root	Decoction Powder Juice	Diarrhea (DS/I), gonorrhea (RS/I), urinary bladder burning (US/I), urethral inflammation (US/I), constipation (DS/I), jaundice (GS/I), piles (DS/E), body pain (NS/E), cough (RT/I), bronchitis (RT/I), leprosy (SO/E), sexual disorders (RS/I)
Acacia jacquemontii Benth	Bhaaori	Leaves Gum	Juice	Brain tonic (NS/I), sexual weakness in male (RS/I), menstrual disorders (RS/I), fever (CS/I), stomach pain (DS/I), kidney stone (US/I), diabetes (GS/I), jaundice (GS/I)
Acacia senegal (L.) Willd.	Khoonbhat	Gum Leaves Bark	Paste	Arthritis (MS/I), brain tonic (NS/I), jaundice (GS/I), toothache (DS/I), sexual weakness in male (RS/I), liver tonic (GS/I)
Achyranthes bidentata Blume	Ubat kandri	Leaves Flowers	Powder Paste	Pneumonia (RT/I), skin diseases (SO/E), nausea (DS/I), scorpion bite (CS/I), burning in urine (US/I), dysentery (DS/I), joints pain (MS/E), pimples (SO/I), boils (SO/E), contraceptive (RS/I)
Aerva lanata (L.) Juss.	Bhoo	Whole plant Root	Decoction	Toothache (DS/I), joint pain (MS/E)
Albizia lebbeck (L.) Benth.	Siraainh	Leaves Seeds Flower Bark	Powder Paste	Tuberculosis (RT/I), reddishness in eyes (SO/E), boils (SO/E), diabetes (GS/I), sexual impotency (RS/I), blood purifier (CS/I), asthma (RT/I), stomach ulcer (DS/I), piles (DS/I), leprosy (SO/E), gonorrhea (RS/I)
Alhagi maurorum Medik.	Kandero	Flower Leaves Root	Decoction	Piles (DS/E), eye diseases (SO/E), constipation (DS/I), abdominal pain (DS/I), diarrhea (DS/I), blood purifier (CS/I), skin allergy (SO/E), diabetes (GS/I), kidney stones (US/I)

(continued)

Table 1 (continued)

Taxonomic name	Local name	Plant part used	Mode of utilization	Diseases treated (body organ system treated/mode of administration)[1]
Amaranthus viridis L.	Lulhar	Leaves Seeds Whole plant Root	Paste Powder Decoction	Flu (RT/I), fever (CS/I), scorpion sting (CS/E), joint pain (MS/I), skin allergy (SO/E), abscess and boils (SO/I), amenorrhea (RS/I), eye sight (SO/I), piles (DS/I), constipation (DS/I), gallbladder stones (GS/I), kidney stones (US/I), diarrhea (DS/I), mouth ulcer (DS/I), contraceptive (RS/I)
Barleria acanthoides Vahl	Nar utangan	Whole plant Leaves	Juice Powder	Liver disorders (GS/I), diabetes (GS/I), cold (RT/I), malaria (CS/I), pharyngitis (RT/I), sexual weakness in male (RS/I), menstrual problems (RS/I)
Barleria cristata L.	Accholi	Roots Flowers	Juice	Cough (RT/I), rheumatism (MS/I), pneumonia (RT/I), snakebite (CS/I), ear and eye infections (SO/O), flu (RT/I)
Blepharis scindica Stocks ex T. Anderson	Utangan	Seeds	Powder	Earache (SO/E), spermatorrhea (RS/I), sexual impotency (RS/I), increased milk production in female (GS/I), dysmenorrhea (RS/I), brain tonic (RS/I)
Borago officinalis L.	Deniii	Leaves Seeds	Paste Powder Decoction	Skin infections (SO/I), wound healing (SO/I), body pain (NS/I), leucorrhea (RS/I), painful urine (US/I)
Caesalpinia bonduc (L.) Roxb.	Kharpat	Seeds Leaves Fruit	Decoction Powder	Asthma (RT/I), anemia (CS/I), wounds (SO/E), blood purifier (CS/I), typhoid fever (CS/I), pimples (SO/E), jaundice (GS/I)
Capparis decidua (Forssk.) Edgew.	Kirar	Flower Stem Root	Paste Juice Powder Decoction	Sexual weakness (RS/I), stomach ulcer (DS/I), cough (RT/I), asthma (RT/I), blood purifier and producer (CS/I), constipation (DS/I), toothache (DS/E), bronchial problems (RT/I), joint pain (MS/E), heart tonic (CS/I), bone fracture (MS/I)
Caralluma edulis (Edgew.) Benth. ex Hook.f.	Maddi thoor	Whole plant Leaves	Juice Powder	Jaundice (GS/I), dysentery (DS/I), stomachache (DS/I), constipation (DS/I), hepatitis B and C (GS/I), diabetes (GS/I), high blood pressure (CS/I), pimples (SO/I), blood purification (CS/I), rheumatism (MS/I), paralysis (NS/I)
Cassia fistula L.	Chilkani Kaathi	Seeds Fruit	Juice Decoction	Gastric problems (DS/I), constipation (DS/I), hepatitis A (GS/I), heart tonic (CS/I), leprosy (SO/I), chest infections (RT/I), dysentery (DS/I), diarrhea (DS/I), scabies (SO/E)

(continued)

Table 1 (continued)

Taxonomic name	Local name	Plant part used	Mode of utilization	Diseases treated (body organ system treated/mode of administration)[1]
Chenopodium album L.	Chhilh	Leaves Root Seeds Shoot	Juice Oil Powder Decoction	Jaundice (GS/I), urinary infections (US/I), snakebite (CS/I), stomach ulcer (DS/I), dyspepsia (DS/I), seminal weakness (RS/I), hemorrhoids (DS/I), cardiac disorders (CS/I), sexual weakness in male (RS/I), vitamin C source (NS/I), kidney stones (US/I), spleen enlargement (GS/I), constipation (DS/I)
Cleome viscosa L.	Chhiprii	Leaves Seeds Root	Juice	Earache (SO/E), stomach ulcer (DS/I), indigestion (DS/I), ear infections (SO/E), deafness (SO/E)
Cocculus hirsutus (L.) W. Theob.	Sar paarano	Fruit Leaves	Decoction Juice Paste	Constipation (DS/I), fever (CS/I), eye infections (SO/E), dysentery (DS/I)
Cucumis melo subsp. *agrestis* var. *agrestis* L.	Chhibhar	Fruit Seeds Leaves	Paste Powder Decoction	Skin infections (SO/E), stomach ulcer (DS/I), kidney pain (US/I), constipation (DS/I), urethra inflammation (US/I), spleen diseases (GS/I), jaundice (GS/I), painful urination (US/I), leucorrhea (RS/I), eczema (SO/E)
Cuscuta reflexa Roxb.	Be Paari	Whole plant	Juice Decoction	Jaundice (GS/I), cough (RT/I), blood purification (CS/I), bronchitis (RT/I), fever (CS/I), sex stimulation (RS/I)
Cymbopogon citratus (DC.) Stapf	Mitho Gaah	Leaves	Decoction Juice	Chronic fever (CS/I), gastric irritations (DS/I), insect bites (CS/I), bronchitis (RT/I), epilepsy (NS/I), skin diseases (SO/I), cholera (DS/I), nerve tonic (NS/I)
Echinops echinatus Roxb.	Kandy wari Booti	Seeds Leaves	Powder Juice	Sexual debility (RS/I), urinary infections (US/I), cough (RT/I), kill lice (MS/E), liver tonic (GS/I), bronchial disorders (RT/I), eye infections (SO/I), burning sensation in urine (US/I)
Euphorbia caducifolia Haines	Danda thoor	Latex Shoot	Powder Paste	Ear pain (SO/E), ringworm (SO/E), skin ulcers (SO/E), wounds (SO/E), may cause abortion (RS/I)
Euphorbia hirta L.	Kaazi dustar	Leaves Seed Flower	Juice Powder Decoction	Redness of eyes (SO/E), eye diseases (SO/E), hair tonic (MS/I), asthma (RT/I), respiratory infections (RT/I), cough (RT/I), measles (RT/I), athlete's foot (SO/E), premature ejaculation (RS/I), diarrhea (DS/I)

(continued)

Table 1 (continued)

Taxonomic name	Local name	Plant part used	Mode of utilization	Diseases treated (body organ system treated/mode of administration)[1]
Launaea procumbens (Roxb.) Ramayya and Rajagopal	Bhitar	Whole Plant Roots	Leaves Juice	Painful urination (US/I), kidney disorders (US/I), urethral inflammation (US/I), milk production in females (GS/I), skin allergy (SO/E), ringworm (SO/E)
Lawsonia inermis L.	Mehndi	Seeds Bark Leaves Flower Root	Paste Powder	Cooling agent (GS/I), athlete's foot (SO/E), jaundice (GS/I), stomach ulcer (DS/I), cough (RT/I), bronchitis (RT/I), dysentery (DS/I), headache (NS/E), burning sensation (SO/E), leprosy (SO/E), may cause sterility in male (RS/I)
Leucas aspera (Willd.) Link	Asairo	Leaves	Decoction	Fever (CS/I), malaria (CS/I), typhoid (CS/I), skin ulcer (SO/E), measles (SO/E)
Luffa acutangula (L.) Roxb.	Toori	Fruit	Paste	Stomach pain (DS/I), stomach ulcer (DS/I), constipation (DS/I), Flu (RT/I), nose allergy (RT/I)
Moringa oleifera Lam.	Swanjno	Bark Fruit Gum Root	Paste Decoction	Sexual weakness in male (RS/I), liver disorders (GS/I), spleen diseases (GS/I), tetanus (NS/I), Paralysis (NS/I), toothache (DS/E), earache (SO/E), rheumatism (MS/I), enhance urination (US/I), may cause abortion (RS/I)
Ocimum tenuiflorum L.	Tulsi	Leaves Seeds Flower	Oil Decoction	Mouth sores (DS/I), urinary diseases (US/I), cough (RT/I), indigestion (DS/I), depression (NS/I), induce sweating (GS/ I), gonorrhea (RS/I), low blood pressure (CS/I), bronchitis (RT/I), cold (RT/I), fever (CS/I), diarrhea (DS/I), ringworm (SO/I), stomach ulcer (DS/I), malaria (CS/I)
Opuntia dillenii (Ker Gawl.) Haw.	Naang Phan	Latex Fruit	Paste	Asthma (RT/I), cough (RT/), hepatitis (GS/I), gonorrhea (RS/I), paralysis (NS/I)
Oxalis corniculata L.	Dil wari	Leaves Root Fruit Seeds Whole plant	Juice Powder Decoction	Fever (CS/I); dysentery (DS/I); GIT worms (DS/I); stomach ulcer (DS/I); used as blood clotting (CS/E); skin diseases (SO/E); snakebite (CS/I); spermatorrhea (RS/I); premature ejaculation (RS/I); jaundice (GS/I); hepatitis A, B, and C (GS/I); eyesight (SO/I)

(continued)

Table 1 (continued)

Taxonomic name	Local name	Plant part used	Mode of utilization	Diseases treated (body organ system treated/mode of administration)[1]
Peganum harmala L.	Harmal	Root Seeds Leaves Stem Fruit	Powder Juice Smoke Decoction	Body spasm (NS/I), nervous disorders (NS/I), rheumatism (MS/I), cough (RT/I), stomachache (DS/I), measles (SO/E), asthma (RT/I), jaundice (GS/I), malaria (CS/I), laryngitis (RT/E), may cause abortion (RS/I)
Portulaca oleracea L.	Loonak	Shoot Seeds Leaves	Juice Decoction Powder	Kidney and urinary disorders (US/I), lung infections (RT/I), urinary bladder inflammation (US/I), jaundice (GS/I), typhoid (CS/I), skin allergy (SO/I)
Solanum nigrum L.	Pat peeron	Leaves Shoot Seed Fruit Stem	Powder Paste Juice Decoction	Enhance urination (US/I), hepatitis A (GS/I), sore throat (RT/I), skin diseases (SO/I), skin itching (SO/E), stomach ulcer (DS/I), flu (RT/I), cough (RT/I), fever (CS/I), ear infections (SO/E), piles (DS/E), constipation (DS/I), obesity (GS/I)
Tamarindus indica L.	Imli	Fruit Root	Juice	Hepatitis B (GS/I), jaundice (GS/I), blood purifier (CS/I)
Tecomella undulata (Sm.) Seem.	Rohiro	Shoot Flower	Decoction	GIT worms (DS/I), constipation (DS/I), menstrual problems (RS/I), wounds (SO/I), tetanus (NS/I)
Thespesia populnea (L.) Sol. ex Corrêa	Denaa	Root Flower Leaves	Juice Paste	Insect bites (CS/E), gonorrhea (RS/I), ringworm (SO/E), migraine (NS/I), headache (NS/I), scabies (SO/E)
Trichodesma indicum (L.) Lehm.	Gaaozaban	Whole plant Leaves Root	Decoction Juice Powder Paste	Snakebite (CS/I), enhance urination (US/I), joint pain (MS/E), dysentery (DS/I), blood purifier (CS/I), influenza (RT/I), cough (RT/I), urinary infections (US/I), flu (RT/I)
Typha angustifolia L.	Pann	Flowers	Powder Paste	Enhance urination (US/I), burning sensation (GS/I), sedative (NS/I), used as blood clotting agent (CS/E), kidney stones (US/I), painful menstruation (RS/I)
Withania coagulans (Stocks) Dunal	Asgand	Root Leaves	Powder	Nerve tonic (NS/I), stomach ulcer (DS/I), sexual weakness (RS/I), arthritis (MS/I), rheumatism (MS/I), skin ulcer (SO/E), asthma (RT/I), uterine infections (RS/I), sexual power (RS/I)
Xanthium strumarium L.	Jangli Dhatoori	Leaves Fruit Root Seeds Bark	Powder Paste	Fever (CS/I), induce sweating (GS/I), enhance urination (US/I), nerve relaxation (NS/I), earache (SO/E), smallpox (CS/I)

(continued)

Table 1 (continued)

Taxonomic name	Local name	Plant part used	Mode of utilization	Diseases treated (body organ system treated/mode of administration)[1]
Ziziphus nummularia (Burm.f.) Wight and Arn.	Bair	Fruit Leaves	Juice Ash Powder Decoction	Hair tonic (MS/E), constipation (DS/I), pimples (SO/E), snakebite (CS/I), measles (SO/I), bronchial disorders (RT/I), indigestion (DS/I), blood purifier (CS/I), skin infections (SO/I)

[1](“I” stands for internal and “E” stands for external mode of administration); *DS* digestive system, *RT* respiratory tract, *NS* nervous system, *CS* circulatory system, *MS* musculoskeletal system, *SO* sensory organs, *US* urinary system, *RS* reproductive system, *GS* glandular system

previous studies (Megersa et al. 2013), herbs are frequently used due to their easy accessibility in wild habitats and their common availability.

Diversity of Plant Part Used and Modes of Utilization

The leaves with 30 reports are commonly used in the herbal recipes, while least used plant part is latex followed by gum (three reports) and shoots (four reports). In modes of utilization, powders (26 reports) and decoctions (25 reports) are frequently used, while ash, smoke, raw forms, and oil are less frequently used modes of utilization. The mode of administration includes internal and external use as mentioned by study participants. The internal mode comprised of oral intake of recipe, while external application is directly used as topical. The internal mode of application is frequently cited in present study.

In plant part used, leaves are the most commonly used in herbal recipes for treating various diseases. This common use of leaves in herbal recipes may be due to their easily extractable variety of phytochemicals and secondary metabolites which may be used as valuable source for drug discovery (Verpoorte and Memelink 2002). Besides, leaves are the main photosynthetic organ and act as reservoir of photosynthetic products which are basis of active bioactive compounds (Balick and Cox 1996; Bhattarai et al. 2009). In herbal clinics and markets, seeds and roots are also commonly used for making herbal recipes, while in some cases, fruits in the form of powder and juice are also reportedly used for treating number disorders. There may be a wide variety of active phytochemical compounds stored in leaves, in roots, as well as in seeds which might play an important role in the treatment of various diseases. Meanwhile roots and seeds are easily traded in herbal markets and clinics, so preference is given to roots in various ethnic communities. When we talk about the folk use of plants among indigenous communities, people always use leaves because of their easily availability and ease in preparation of recipes, whereas in trade, herbal sellers and herbalists prefer roots or undergrounds parts due to their massive nature.

The data on modes of preparation reported by informants in the study shows that most of rural population of the study area used decoctions prepared at homes by boiling plant material in water, while THPs frequently provided herbal formulas in the form of powders. The prepared decoction contains a diversity of active compounds that plays important role in therapeutics. THPs prepare recipes and store them for a long time as they must sell them in the herbal market. So, they prefer powder instead of decoction. Many studies such as Costa Pinto and Maduro (2001), Medeiros et al. (2004), and Oliveira et al. (2010) have concluded that boiling plant material in water or any other solvent activates the phytochemical compounds or may harmonize synergetic effect of various compounds stored in that plant part.

Pictorial Representation of Some Important Plant Species

Cultural Significance of Medicinal Plants in Deserts of Sindh

The deserts of Sindh especially Thar Desert occupy a distinct geographical location in the deserts of Asia. The Thar Desert is particularly important for its endemic and medicinal plant species. Each habitat and landform in the desert supports distinctive plants with their respective niches (Khan et al. 2003). Different indigenous communities of deserts in Sindh are represented by nomads that are directly linked with wild plants. For most of the nomads, livestock is one of the major sources of income, and status of each nomad is determined on the basis of size of herd owned. The nomads use plants as forage, food, shelter, and folk medicine and for various other purposes.

The Thar Desert especially Nagarparkar region has diversity of medicinal plants ranging from lower nonvascular to angiosperms. The local people of deserts of Sindh are completely dependent on plants in daily life and possess close cultural associations with local plants. The indigenous possess rich traditional knowledge about medicinal plants acquired from their parents who had experience about use of plants in herbal medicine (Katewa 2009). In deserts of Sindh, most of the rural people depend on traditional medicinal systems for primary health care. The local old-aged people are experts in making herbal medicine, but documentation of this information from them is very difficult due to their belief that transfer of information may lose their expertise. The traditional health practitioners (THPs) of Thar Desert have a diversity of traditional knowledge on herbal recipes prepared from local plant species.

The local people throughout the deserts of Sindh use plants for their life activities including as herbal medicines, rituals, and cultural aspects. The knowledge of people and their perceptions on environment are thought to be important element of cultural integrity. In desert communities, these cultural associations have a long history which can be traced perhaps from the dawn of early civilizations. Early civilizations include Mohenjo Daro, Harappan, Aryan, and Dravidians. The civiliza-

Plate 6 (**a**, **b**) Interviews and group discussion during documentation of ethnobotanical knowledge in deserts of Sindh

Plate 7 (**a**, **b**) Some vendors selling local plants in local markets in Thar Deserts

Plate 8 (**a**) *Capparis decidua.* (**b**) *Withania coagulans*

Plate 9 (**a**) *Caesalpinia bonduc*. (**b**) *Cuscuta reflexa*

Plate 10 (**a**) *Euphorbia caducifolia*. (**b**) *Opuntia dillenii*

Plate 11 (**a**) *Peganum harmala*. (**b**) *Luffa acutangula*

tions had rich traditional knowledge about herbal medicine especially in Indian subcontinent (Albuquerque et al. 2017). In deserts of this subcontinent, indigenous medicines were commonly utilized during early civilizations, and early human populations were closely associated with plants used as food and medicine (Ramawat et al. 2009). Being closely associated to nature, local people have acquired traditional knowledge from their grandparents. The expertise about medicinal properties and herbal recipes and uses is enriched among rural populations, and many aspects are unknown outside these populations. The local plants among these indigenous populations are thought to be a cultural and integral part, and knowledge about such plants is passed on from generation to generation through oral folklore (Katewa 2009).

Current Threats to Traditional Knowledge in Deserts of Sindh

The rural indigenous communities are facing several problems that lead to direct or indirect loss of traditional knowledge (TK) possessed by local people. The major causes of threats include loss of pathways of transmission of traditional knowledge, change in livelihood practices, change in traditional religious and ritual beliefs, fast climatic changes resulting in loss of biodiversity linked with traditional knowledge, loss of knowledge about intellectual property rights among local communities, disappearance or nonfunctional institutions linked to preservation of TK, lack of legislative governmental policies and their implementation, impact of modern culture, fast shift in urbanization, and economic development pressure. A comprehensive understanding of the drivers of TK change and conservation options is lacking in most locations. All these factors single or in combination are threatening traditional knowledge linked with plant diversity found in deserts of Sindh. Many conservation and preservation actions are needed to save the precious traditional knowledge. To reduce the loss and threats to TK, important actions that may be applied are community-based TK conservation activities, traditional lifeway programs, indigenous capacity building, institutional development, self- government, self-organization, community alliance, indigenous financing, ethno-tourism or eco-cultural tourism, education and awareness building, customary education, research and documentation of TK, development of database, etc. These proposed actions should be applied at various levels to compensate for the threats and loss of TK in deserts of Sindh.

References

Albuquerque UP, Ramos MA, Júnior WSF, de Medeiros PM (2017) Ethnobotany for beginners. Springer, New York

Ayyanar M, Ignacimuthu S (2005) Traditional knowledge of Kani tribals in Kouthalai of Tirunelveli hills, Tamil Nadu, India. J Ethnopharmacol 102(2):246–255

Balick MJ, Cox PA (1996) Plants, people, and culture: the science of ethnobotany. Scientific American Library, New York, NY

Bhattarai S, Chaudhary RP, Taylor RS (2009) Wild edible plants used by the people of Manang district, central Nepal. Ecol Food Nutr 48(1):1–20

Boivin N (2012) The Palaeolithic occupation of the Thar Desert. University of Oxford, Oxford

Cheema S, Zaman Q, Rasul G (2012) Persistent heavy downpour in desert areas of Pakistan in South Asian Monsoon 2011. Pak J Meteorol 9(17):71–84

Costa Pinto AA, Maduro CB (2001) Produtos e subprodutos da medicina popular comercializados na cidade de Boa Vista. Roraima Acta Amazônica 33(2):281–290

Goudie A (2002) Great warm deserts of the world: landscapes and evolution, vol 1. Oxford University Press on Demand

Iqbal M, Farooq U, Bashir A, Khan N, Malik S (2000) A Baseline survey for the development of livestock sector in Cholistan. Joint Publication of AERU, AARI, Faisalabad, SSI, NARC, Islamabad and GTZ, Lahore

Katewa S (2009) Indigenous people and forests: perspectives of an ethnobotanical study from Rajasthan (India). In: Ramawat K (ed) Herbal drugs: ethnomedicine to modern medicine. Springer, Berlin, Heidelberg, pp 33–56

Khan T, Dular AK, Solomon DM (2003) Biodiversity conservation in the Thar Desert; with emphasis on endemic and medicinal plants. Environmentalist 23(2):137–144

Khatri A, Rathore A, Patil U (2011) *Prosopis cineraria* (L.) Druce: a boon plant of desert-an overview. Int J Biomed Adv Res 1(5):141–149

Medeiros MFT, Fonseca VD, Andreata RHP (2004) Plantas medicinais e seus usos pelos sitiantes da Reserva Rio das Pedras, Mangaratiba, RJ, Brasil. Acta Bot Bras 18(2):391–399

Megersa M, Asfaw Z, Kelbessa E, Beyene A, Woldeab B (2013) An ethnobotanical study of medicinal plants in Wayu Tuka district, east Welega zone of oromia regional state, West Ethiopia. J Ethnobiol Ethnomed 9(68):2–18

Murray MD (1999) Encyclopedia of television news. Greenwood Publishing Group

Oliveira GLD, Oliveira AFMD, Andrade LDHC (2010) Plantas medicinais utilizadas na comunidade urbana de Muribeca, Nordeste do Brasil. Acta Bot Bras 24(2):571–577

Qureshi R, Bhatti GR (2005a) Nara Desert, Pakistan. Part I: soils, climate, and vegetation. Rangelands 27(5):27–31

Qureshi R, Bhatti GR (2005b) Nara Desert, Pakistan. Part II: Human Life. Rangelands 27(5):32–35

Qureshi R, Bhatti GR (2008) Ethnobotany of plants used by the Thari people of Nara Desert, Pakistan. Fitoterapia 79(6):468–473

Ramawat KG, Dass S, Mathur M (2009) Herbal drugs: ethnomedicine to modern medicine. Springer, Berlin, Heidelberg

Shrestha PM, Dhillion SS (2003) Medicinal plant diversity and use in the highlands of Dolakha district, Nepal. J Ethnopharmacol 86(1):81–96

Smith M, Vellen L, Pask J (1995) Vegetation history from archaeological charcoals in central Australia: The late Quaternary record from Puritjarra rock shelter. Vegetation History and Archaeobotany 4(3):171–177

Uniyal SK, Singh K, Jamwal P, Lal B (2006) Traditional use of medicinal plants among the tribal communities of Chhota Bhangal, Western Himalaya. J Ethnobiol Ethnomed 2(1):1

Verpoorte R, Memelink J (2002) Engineering secondary metabolite production in plants. Curr Opin Biotechnol 13(2):181–187

Yaseen G, Ahmad M, Sultana S, Alharrasi AS, Hussain J, Zafar M, Ur-Rehman S (2015) Ethnobotany of medicinal plants in the Thar desert (Sindh) of Pakistan. J Ethnopharmacol 163(2):43–59

Index

M. Ozturk, K. R. Hakeem (eds.), *Plant and Human Health, Volume 1*,
https://doi.org/10.1007/978-3-319-93997-1

FSC
www.fsc.org
MIX
Papier | Fördert gute Waldnutzung
FSC® C083411

Kolibri 360 GmbH
Ferdinand-Jühlke-Straße 7
99095 Erfurt, Deutschland
produktsicherheit@kolibri360.de